Hands of Time

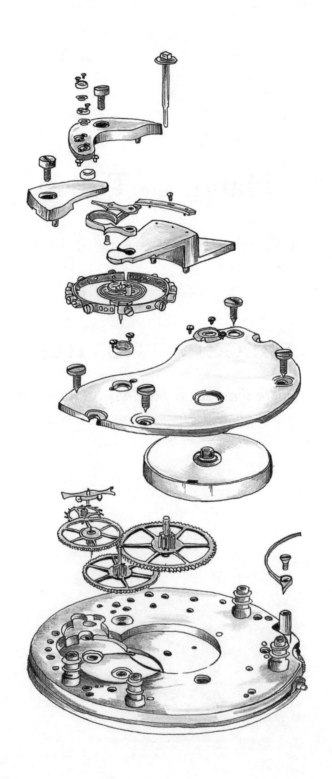

Hands of Time

A Watchmaker's History of Time

REBECCA STRUTHERS

*with illustrations by Craig Struthers
and photographs by Andy Pilsbury*

HODDER &
STOUGHTON

First published in Great Britain in 2023 by Hodder & Stoughton
An Hachette UK company

4

A CIP catalogue record for this title is available from the British Library

Hardback ISBN 9781529339031
Trade Paperback ISBN 9781529339000
eBook ISBN 9781529339017

Typeset in Bembo MT by Hewer Text UK Ltd, Edinburgh
Printed and bound in Great Britain by Clays Ltd, Elcograf S.p.A.

Hodder & Stoughton policy is to use papers that are natural, renewable
and recyclable products and made from wood grown in sustainable forests.
The logging and manufacturing processes are expected to conform
to the environmental regulations of the country of origin.

Hodder & Stoughton Ltd
Carmelite House
50 Victoria Embankment
London EC4Y 0DZ

www.hodder.co.uk

In memory of Adam Phillips and Indy Struthers

Contents

A Backward-facing Foreword*

Jeff 10/3/71

When I first started training as a watchmaker at nineteen, I was taught never to leave a trace of my presence within the watches I restored. And yet, these traces can tell us all kinds of stories about otherwise inanimate objects. For example, the vintage Omega Seamaster wristwatch on my workbench was repaired by a man called Jeff on 10 March 1971. I know this because Jeff scratched his name and the service date into the back of the dial, so that if the watch ever came back to him, he would know that he had previously worked on it and when.

Working as we do on objects only a few centimetres in diameter, a watchmaker's world is often not much bigger than a thumbnail. It is all-consuming. Sometimes a whole morning passes and I have barely shifted my gaze beyond the postage-stamp-sized mechanism I am working on. I suddenly realise the coffee next to me is cold and my eyes are dry from concentrating so hard I've forgotten to blink. My husband, Craig, is also a watchmaker and, although we work on benches that face each other, we can spend whole days in near silence, exchanging little conversation beyond orders for the kettle. When we make a new watch, whether from salvaged parts or from scratch, it can take us anywhere from six months to six years. We can measure sections of our lives by these watches, and sometimes find ourselves noticeably older when we are finished with them.

* With thanks to Alexander Marshack's *The Roots of Civilization*.

Our workshop is in an eighteenth-century goldsmiths' factory in Birmingham's historic Jewellery Quarter. Craftspeople have been producing work here for seven generations and the walls feel steeped in appropriate history. The rooms beneath us contain centuries-old presses, dies and design charts, as well as artisans still using this equipment to produce jewellery. Our small room on the top floor is bright and airy, with skylights and arched windows. When we first moved in and were preparing the space, we were told that during the Blitz a bomb had crashed through the roof and failed to detonate. I pulled down an old insulation screen covering the skylight and found a beam still charred from the blast, which I scrubbed and left on show. It now presides over our lathe bench and wheel-cutting machine, which, ironically, is German. We call her Helga. She sits on a long bench that extends down the whole side of our workshop and is covered with a variety of old machines. Underneath, there are copious drawers full of the dull gleam of old watch mechanisms and parts that we have sourced or rescued over the years – often from bullion traders who had taken them out of their cases, which were to be sold as scrap gold or silver, or from the workshops of watch-makers past being cleared out by their families. Our 'clean' watch-making benches are on the opposite side of the workshop, as far as possible from the swarf (metal shavings) and oil that occasionally spit from the machines.

We keep our workshop very clean, to avoid dust or dirt finding its way into the delicate watch mechanisms. In the state-of-the-art watch manufactories of Switzerland or East Asia, workshops will have double air-lock doors and sticky mats to remove dirt from shoes, and watchmakers wear compulsory lab coats and shower caps over their feet. We are a little more relaxed. Our dog, Archie, snoozes in a corner. By the end of a day making new watch parts the room smells of lathe oil, a distinctive aroma almost like that of a tomato vine, with notes of metallic copper and iron. There will be small mounds of brass or steel swarf scattered around our lathes, mills and drills, as well as oil- and coffee-ring-stained sketches of parts strewn liberally across the workbenches. We sweep the floor regularly, ready for the occasional team hunt for accidentally flicked parts – you could probably make up a complete watch from the parts stuck between floorboards or rolling under sets of drawers in most watchmaking workshops. Our floor is pale grey vinyl, the perfect contrast for the yellow of brass or bright red of a ruby jewel.* No one tells you that one of the key skills for any watchmaker is the ability to find tiny shiny things on the floor.†

The last residents of our workshop were enamellers, and traditional makers have worked in this space for over two centuries. In this room at least, not much has changed. Although we have modern computers, most of the tools and machines we work with are between 50 and 150 years old. Our skills too, are from a bygone era. In the 'golden age' of watchmaking that ran through the seventeenth and eighteenth centuries, Britain was the centre of the watchmaking world. Now watchmakers like Craig and me are a rare breed. In 2012 we set up on our own, becoming just one of a handful of firms in the UK with the skills to make mechanical watches from scratch and to restore antique watches from the last five centuries. But the course we trained on no longer exists. The Heritage Crafts Red List of Endangered Crafts (much like the Red List of Threatened Species but for craft) currently lists artisanal watchmaking as a critically endangered skill in the UK.

In part our skillset is disappearing because, in our technologically advanced age, computer numerical control (CNC) can virtually make a whole watch for you. And you might well ask why we bother with this old equipment when we could feed a computer design into a software-operated machine to do most of the manufacturing for us instead. But where's the fun in that? We love getting our hands dirty

---

* Many mechanical watches use bearings made from synthetic ruby, or corundum, as it makes for an incredibly hard surface. The small steel pivots that support each toothed wheel can rotate against the ruby without it wearing away from the friction.

† Our hunts are usually conducted under the watchful, if a little confused, eye of Archie, who has never quite come to terms with the notion of looking for things that aren't edible.

making things and fiddling with little parts to get them to work together. You build a closer relationship with what you're making when you work by hand. You can hear when the cutting speed of a lathe or drill is perfect, and you can feel from the resistance whether the pressure of your tool is correct. We like this sense of connection to the objects, and to the generations of artisans who came before us.

I have always been fascinated by time, but I never set out to be a watchmaker. At school I wanted to be a pathologist (this was long before TV crime dramas made forensics cool). I was an oddball who was fascinated by how things worked, particularly bodies. I wanted to help people, but I wasn't always very good at actually talking to them; working with the dead, I reasoned, would save me a lot of difficult conversations with patients. I liked the idea of figuring out why a body had stopped functioning. In the process I hoped I might help other people, perhaps by helping to bring about justice, or a deeper understanding of a deadly disease.

My career as a pathologist was never to be, but there is something forensic about working on old watches. Watch mechanisms contain tens if not hundreds, and occasionally thousands, of components, each of which has a specific task to perform. The most basic simply tell the time. The most complicated (the added functions in a watch that go beyond time-telling are in fact called 'complications') can chime the hours and minutes on gongs made from finely tuned wire, accurately maintain the date for over a century, or chart the stars. When any of these parts become faulty or need cleaning, or re-oiling, they stop the mechanism from functioning. As restorers we dissect to determine the cause of death, only with the bonus that, once repaired and reassembled, our subject has another chance at life. The final stage of reassembling a mechanical watch is to replace the balance, which makes the watch start to tick again. There is nothing quite like hearing life restored to a piece that has not worked for years, or even centuries, knowing that its tick sounds the same to me now as it did to the watchmaker who first assembled it. The pulse of the balance is referred to as its 'beat', and the coiled spring used to regulate its action 'breathes'.

As time went on, it felt very natural to me to begin switching between working on watches and thinking and writing about them and their history. I became the first practising watchmaker in the UK to pursue a doctorate in antiquarian horology (the study of the history of timekeeping). After all, restorers are, in part, historians. It's a practical kind of history: you have to know how something was made and how it once operated to return it to the way its maker intended. Now I found it worked the other way too: when Craig and I first started making our own watches from scratch, my historical research and writing influenced the watches we made, in a sort of horological cross-fertilisation. My research enlarged my tiny watchmaker's world. The focus of the watchmaker is often smaller than a grain of rice, but the inspiration for horology is the universe – I love this contrast of micro and macro. And poring over the construction of an eighteenth-century watch to discern what it could tell me about its provenance and owners made me keenly aware not only of how history had shaped the watch, but also of how the watch has shaped us.

It would not be a stretch to say that the invention of mechanical timekeepers has been as significant for human culture as the printing press. Imagine trying to catch a train by relying on the position of the sun. Or organising a Zoom conference of 200 people located all around the world, each trying to decipher the start time by hanging out of their window to be within earshot of the bells of the nearest public clock. Or, at the more life-and-death end of the scale, think of surgeons performing an organ transplant or removing a tumour with no accurate reference point to measure their patient's heart rate. Our ability to do business, structure our day, and access life-saving developments in science and medicine all rely on – are, in fact, made possible by – access to accurate time.

From its very beginning the watch both reflected and developed our relationship with time. Watches don't create time, they measure our cultural perception of time. All time-measuring devices, whether they are ancient carved bones or the watches I restore on my workbench, are a way of counting, measuring, analysing the world around us. The earliest timekeepers began by tracking naturally occurring phenomena in the world and solar system. Even now, the most up-to-date modelling devices we possess – smartwatches such as the Apple Watch – can still track a celestial routine, keeping pace with our planet as it hurtles round the sun over the course of a day. The systems we've developed to understand these processes, and our place within them, are our way of getting to grips with our universe, of applying a cosmic rational order that we can use to better live our lives.

What we call a watch – a tiny, wearable clock – is a miracle of engineering. Mechanical watches are among the most efficient machines ever created. I have worked on watches that haven't been serviced since the 1980s, and yet have only just stopped running. I struggle to name another mechanism that would work day and night for nearly forty years before requiring any maintenance from a mechanic. As of 2020, the most complicated watch in the world contains nearly 3,000 parts and is capable of measuring the Gregorian, Hebrew, astronomical and lunar calendars, chiming the hours and minutes together with fifty other complications, all in a device that fits in the palm of your hand. The smallest watch movement ever created was first made in the 1920s and fits ninety-eight parts into a volume of only 0.2 cubic centimetres. The first chronometer, a watch so accurate it could be used by sailors to calculate longitude at sea, was made over sixty years before the invention of the electric motor and over a hundred years before the first electric lighting. Watches have since accompanied humans to the summit of Everest, the depths of the Mariana Trench, both the North and the South Poles, and even to the moon.

Our concept of time is inseparable from our culture. In fact, the word *time* is the most commonly used noun in the English language. In Western, capitalist cultures time is something we have, or don't have, save or lose, it marches on, it drags, seems to stand still and flies. Time thrums constantly underneath everything we do. It is the backdrop and the context for our existence and our place in what is now a supremely mechanised world.

Slowly, over the course of tens of thousands of years, the power balance between humans and time has slowly shifted. What began with us living our lives around the natural phenomena the world threw at us developed to become an entity we have sought to control. Now it often feels as if time controls us. Time, we've discovered, is not as 'fixed' as we first believed. It might not be universal, constantly moving on, waiting for no man. It might be relative, personal and even, one day, reversible – medically speaking at least.

I knew early on that I wanted to be a watchmaker rather than a clockmaker. Watches have followed us through our daily lives for centuries, worn on or near our bodies. I've always been fascinated by the intimacy of that relationship. The connection between person and watch, its tick reflecting the beat of our very own pulse, the rhythm of our own body, was for a long time the closest relationship we had with a machine – until, of course, mobile phones came along. In many ways watches are an extension of us, a projection of our identity, our personality, our aspirations, as well as our social and economic status. A watch is an individual's timekeeper, but it is also a kind of diary: it holds in its restless hands our memories of the hours, days and years we have spent wearing it. It is an inanimate but uniquely human repository of life itself.

This book, too, is a history of timekeeping, and of time itself, but from my irregular perspective as a twenty-first-century watchmaker. We begin with the very earliest human-made timekeepers, crafted from bones, measuring shadows, channelling water, fire or sand. I'll then explore how inventors later found ways to combine natural power sources with artificial engineering. The first clocks as we understand them were a product of an extraordinary combination of curiosity, experimentation and highly sophisticated science. Their mechanisms, which were once so huge that they could only be accommodated in massive church towers, were the forerunners of the miniature timekeepers I handle every day on my workbench.

From there we'll turn to the wonder of watches. Each chapter explores a pivotal moment in the history of the watch, from its advent 500 years ago to the present day. I'll unpick the mind-boggling technical advances that enabled these machines to become portable and accurate enough to conquer the world. I'll show how these little instruments coordinated work, worship and wars, and how they helped European nations navigate and map the world, supporting global trade and enabling colonial expansion. We'll show how explorers and battle-weary soldiers depended on them for their very survival, and how decisive historical events were dictated by them. We'll also track their evolution from elite status symbol to popular tool to status symbol all over again. The watch is the metronome of Western civilisation itself, establishing a rhythm that has driven our history and continues to govern our time- and productivity-obsessed age.

It is also a personal history. My particular interests glimmer behind the dial. Many of the watches in this book I have handled, or even repaired, and the stories they tell are central to this narrative. I once worked on a watch, destined to be a wedding gift from parent to child, that had been in the same family since the eighteenth century. Holding it, working on it and considering its past and future, I felt like I was bridging time itself. Handling hundreds of years of horology creates an eerie sense of self-awareness. Working on an antique watch in minute detail, I feel an almost tangible connection to the people who made and wore it. Tiny traces of humanity stand out like signatures – initials or names like Jeff's, concealed within the mechanics, or a 250-year-old enameller's fingerprint, accidentally baked into the blue-green glass and hidden underneath the dial of a pocket watch. Conscious that I am another chapter in the story of an object created before I was born and which, if cared for, will live on for centuries after I'm gone, I collect these signs of life.

A watchmaker is a custodian, protecting these objects, absorbing their history, as well as preparing them for new connections yet to be created. Occasionally I see a watch again, for servicing or repair, years after I've first worked on it, and it's like reconnecting with an old friend. My memories of the watch's marks and idiosyncrasies come flooding back, and sometimes new ones are created. I've had a recently repaired watch returned with water damage; not long after it had been given as an eighteenth birthday present it was submerged in a swimming pool along with its mildly inebriated new owner on holiday in Mallorca. The mechanism inside has now been restored to its original state, but the slight stain around the three o'clock position on its dial is a permanent reminder of the perils of mixing tequila shots, chlorinated water and vintage precision mechanics.

Every morning, when I sit at my bench and start work, the watch in front of me is a new beginning. Each one has its own history. Engineering perfection aside, each knock and scrape, each hidden mark left by a past repairer, even the way they were designed, and the techniques used to make them, are clues to a story that extends far beyond the tiny object before me.

I

Facing the Sun

Kia whakatōmuri te haere whakamua
I walk backwards into the future with my eyes fixed on my past
Māori proverb

I've always been fascinated by nature. As a child I liked nothing more than gathering slugs in the garden, and getting covered in mud and slime. Above all, I loved to learn how things worked. One of my earliest memories is my dad first showing me his microscope. I was dazzled to discover another world secretly existing within my own but invisible to my naked eye. I loved it so much my parents bought me a children's version for Christmas, which was portable, so I could use it around the garden. We spent hours studying pond-water samples. I then drew the weird and wonderful creatures I'd see darting and crawling around on the slide.

I grew up in a suburb of Birmingham called Perry Barr. It was a densely populated sprawl of brick, concrete and tarmac, sliced in two by the A34 and its undulating flyovers and underpasses. The closest thing to a local landscape was a fly-tipped rubbish patch of wasteland where my sister and I snuck off to play. We called it the 'background' (it was quite literally the ground at the back of our house).

I don't remember much about the seasons in Perry Barr. Aside from the odd smattering of snow in winter, the spiky grass of the 'background' was reddish-brown all year round. In autumn, leaves gathered in slimy clumps on the pavement, while my parents debated whether it was too early to switch on the heating. At night the streetlights suffused the sky with a milky orange glow that made the stars almost invisible.

I will always be a Birmingham girl. But in my early thirties Craig and I had no choice but to leave. We were economic migrants forced *away* from the city, seeking lower house prices within the range of our freelance incomes. We bought an old weavers' cottage in a small town in the northernmost part of Staffordshire, bordering the Peak District. It was the cheapest place we could find within a 50-mile radius of our workshop.

Neither of us had ever lived so close to the countryside. We spent our first months exploring the fields, woods and moors around our new home while walking our dog. Archie's favourite route took us through a valley I later learned was called Little Switzerland – a fitting choice for two watchmakers and their watchdog. Ever drawn to the relics of an industrial past, we liked to stroll along the repurposed railway line that once connected Cheshire to Uttoxeter, through the woodlands of Dimmingsdale alongside the River Churnet. Archie's nose would be aquiver with the unfamiliar natural smells: badger, deer, weasel, owl and vole.

As the seasons changed, so did the walks. In winter, the low rays of sunlight pierced the skeletons of old oaks and frostbitten hedges. In spring, the woodland shadows filled with bluebells. Autumn brought mists so thick that sometimes we struggled to see more than a few metres ahead. I started to notice how animals were rotated in their fields – which times of year the cows would be out, and when the sheep would be lambing. I learned the hard way that Archie had to be kept away from certain areas during muck-spreading seasons: late winter and spring.

My first autumn in the cottage was spent working on an important watchmaking project, with a deadline set for Christmas. It was a particularly complicated and ambitious project and, as the days rolled on and my progress didn't, I kept telling myself, 'The year isn't coming to a close yet, I still have time.' But increasingly I wished

I'd invested my energy in inventing a time machine rather than machines to tell the time.

One late-autumn afternoon, I looked up and saw a flock of Canada geese flying in a rowdy 'V' formation across the sky. As the weeks went on, these flocks got larger and larger until one day, as I was walking through the woods, the whole sky was filled with beating wings and honking beaks. Archie tilted his head from side to side with a curious expression that I presume meant either 'what's that?' or 'that looks tasty: should I chase it?' I suddenly remembered standing in the 'background' as a child, looking up at a similar flock of geese. For a brief, bitter-sweet moment, past and present collided.

In the northern hemisphere flocking geese are a reliable sign that the year is coming to an end.* With my deadline looming, my strongest feeling was that I wanted them to stop; it was almost as if they were telling me that *my* time was running out. We were both, in a way, keeping time.

Archie watching a flock of Canada geese.

* I had always understood that geese were migrating. In fact, Canada geese are generally resident in the UK but still flock in the autumn.

The natural world surrounding us is riddled with temporal cues if you know where to look for them. It was our first clock, and it continues to tick around us for those that take notice. It was living with and saturated by nature that caused humankind to develop the first timepieces. If watches are personal time, then our first watch was our internal one. You could say that the watch emerged from our first efforts to align our inner sense of time with what we observed in the world around us.

The object that archaeologists currently consider to be the strongest contender for the earliest known timepiece is 44,000 years old. It was discovered in 1940 when a man collecting bat guano in modern-day South Africa discovered a cave in the Lebombo Mountains, nestled among the bush and scrub. The cave was filled with ancient human bones, some of which were 90,000 years old. The site, now called Border Cave, is one of the most important in the history of humankind. Border Cave, which was continuously inhabited by humans for 120,000 years, protected its inhabitants both in life and in death. Located high in the mountains overlooking the plains of what is now called Eswatini, it was an easy site to defend from predatory animals and other humans and offered a good vantage point from which to scout for prey. Archaeologists found more than 69,000 artefacts there, many showing a rich understanding of the natural world and how to engage with it: there were sticks used to dig for carbohydrate-rich tubers, sharpened bones for leatherwork-ing, pieces of jewellery made from ostrich eggs and marine shells, and straw bedding that, repeatedly layered on top of ash and camphor bush, was probably used to repel biting insects and parasites like ticks.

But for me, the most extraordinary discovery was a small piece of carved baboon fibula, about the length of an index finger, inscribed with twenty-nine clear notches, and polished by the hands of its owners through many years of use. It is the first clear evidence of calculation in human history. The Lebombo Bone dates back to long before the advent of agriculture or any evidence of planning for the seasons, and even longer before we conceived of anything close to a regular working day. It is a measuring device from a time when, so far as we know, there was very little to measure.

So what were our ancestors trying to calculate? We can't know for sure, but some scholars have a theory. After the passage of day and night, another likely division of time for our ancestors is thought to have been the phases of the moon. The marks on the bone consist of thirty spaces alternating with twenty-nine notches. The average lunar month is about 29.5 days. If our ancestors rotated measurement between the notches and the spaces that divided them, they would have reached the average of 29.5 days and therefore correctly calculated the lunar month. Some scholars have even suggested that its makers used the bone to track their reproductive cycles, or the length of a pregnancy. I like to think it was handled by our great-great – hundreds of greats – grandmother as she counted down the days.

Many ancient cultures believed that the two cycles – lunar and menstrual – were linked; indeed the belief persists to this day. A recent study found no definitive link, but hypothesised that modern lifestyles, particularly our exposure to artificial light, might have weakened the synchronicity. If so, we would be far from the only creatures whose internal clocks are aligned to the rhythms of the natural world.

My friend Jim,* a farmer and master whisky blender in western Scotland, and his wife Janet, a fourth-generation shepherdess, told me how the sharply decreasing daylight hours in November trigger their ewes to ovulate. With incredible predictability and to within a few days, all the ewes in the flock will follow virtually the same

---

* I got to know Jim through our mutual love of whisky and craft, one of the many unusual and fascinating connections you form over the years when you're in an equally unusual and eclectic line of work.

cycle. Within two cycles, most if not all of the ewes will be pregnant. Twenty-one weeks later, in around the first half of April, the lambs are born, aligning perfectly with the end of the bitter winter weather and the beginning of spring growth. Jim describes this work of coordination as 'getting the mouths ready to eat the grass'.

Just as the lambs arrive, one or two days either side of 17 April, swallows arrive from their 6,000-mile migration, escaping the heat of the South African summer. Between the new livestock and the nesting swallows, spring breaks on the farm in what Jim lovingly refers to as a 'huge burst of life'. In September the swallows queue up on telegraph cables and tree-branches, knowing, somehow, that it is time to leave.

Every living creature has an internal clock. Those of us who live with canine companions and work a regular schedule will have noticed their eerie ability to predict when we're about to come home after a day at work. This is thought to be because the moment we step out through the front door we leave a human smell timer for our dog, who learns that once our scent has diminished to a certain level, enough time has passed for us to be due home. The rooster crowing at daybreak, a timekeeper across the globe, is operating according to an inner circadian clock that has been found to run for an average of 23.8 hours, hence their crowing just before the dawn. Even organisms as small as plankton are known to move up and down in the water, from the depths to the surface, every dusk and dawn. We can be fairly certain that they sense changes in UV levels (in very strong sunlight they sink a little lower into the water to avoid damage) and so can tell day from night by the sun's light. They have even been observed, in controlled experiments in a dark aquarium, to continue to make their vertical

'Oh my, is that the time already?' A plankton heads to the surface.

migration for several days in the complete absence of light. In other words, they too have a biological clock that functions around a 24-hour system.

If anything, our ability to read our internal clock is weaker than most animals' because it's interfered with by our perception of time, which can be warped by emotions: happiness, novelty and absorption seem to speed it up, while boredom and fear appear to slow it down. This body clock undeniably exists, almost like a sixth sense, but it is not universally understood (my internal hour may not be your internal hour). Timekeepers are a symbol of our drive to share, quantify and externalise our intuitive awareness of time. The Lebombo Bone suggests that we were doing this as early as 40,000 years ago.

Ancient measuring devices have been found on almost every continent. Many of the oldest examples sprang up independently of each other, and their patterns suggest different purposes. The first humans to populate Europe, the Aurignacians, left behind what appear to be early calendars. In Baden-Württemberg, a small plaque made from an eagle's wing bone is believed to be the world's oldest known star chart.* In the Democratic Republic of Congo, a 25,000-year-old bone tool handle, known as the Ishango Bone, bears a series of carved notches suggesting mathematical calculations like addition, subtraction, doubling and prime numbers.

These hand-held devices seem to mark an important conceptual turning point for our species. As the philosopher William Irwin Thompson puts it, 'The human being was no longer simply walking in nature; it was miniaturizing the universe and carrying a model of it in its hand in the form of a lunar calendrical tally stick.' But I think they do more than this. By capturing cosmic events in a device that we can put on our wrist or hold in our hand, we are reassuring ourselves – perhaps misguidedly – that we can control the

---

* The eighty-six notches could represent the number of days that one of Orion's two prominent stars, Betelgeuse, is visible.

uncontrollable. They make us feel that we are no longer just existing in time but *using* it, to our advantage.

What did time feel like to ancient humans? Did they simply live 'in the moment', the fantasy of many a self-help devotee? It may well be that they lived in 'survival mode'. Anyone who has experienced extreme circumstances where food, warmth and safety are limited and under threat will tell you that their focus is purely on the here and now. But it is something of a 'progress' myth to assume that simply because we do not have evidence of early humankind externalising an under-standing of existing in a moment in time, that he did not have it. The development of cave art, from as early as 45,000 years ago but increas-ingly common from 35,000 years ago, perhaps demonstrates a concept of a more distant past and future. If you frequented a cave with pre-existing rock paintings, your thoughts might naturally turn to the ancestors who made them before you; and if you added your own marks to the walls, you might contemplate the generations that would see them after you were gone. But there is no way to know where the genesis of our shared time lies. The appearance of grave goods later, about 13,000–15,000 years ago, gives us more conclusive evidence of a belief in a time beyond our own. Burying loved ones with their treas-ured possessions – a favourite knife, jewellery, an infant's toy – implies that these objects might be required in a future afterlife.

A few years ago, archaeologists discovered a 23,000-year-old human camp on the fertile shores of the Sea of Galilee in Israel. They found 140 different species of plant there, including emmer, barley, oats and, crucially, the remnants of lots and lots of weeds (weeds flourish in disturbed soil and cultivated land, which is why they are the bane of every gardener's life).* This site is the earliest known evidence of basic agriculture, some 11,000 years earlier than previously believed.

---

* They also found a stone grinding slab and sickle blades, indicating that the cere-als were being grown, harvested and processed in an organised manner.

These pioneering farmers were probably also observing the position of the sun, the phases of the moon and the migration of animals. Above all, they clearly had a conception of the future: they understood that if they planted something in the present, they could reap their rewards months later.

This is still some way from the modern experience of time as defined by the hours on a clock dial. For our ancestors, time was divided not by abstract numbers but by natural *events*, such as seasons and their related weather conditions. This is how the Kenyan-born philosopher Rev. Dr John S. Mbiti described event-based time in relation to traditional African hunter-gatherer communities: 'There is the "hot" month, the month of the first rains, the weeding month, the beans month, the hunting month, etc. It doesn't matter whether the "hunting month" lasts twenty-five or thirty-five days: the event of hunting is what matters much more than the mathematical length of the month.' Cycles of a longer duration, like a year, would be measured by the repetition of the agricultural cycle, such as the passage of two wet and two dry seasons, with these four seasons making one year. The exact number of days in a year was not important, 'since a year is not reckoned in terms of mathematical days but in terms of events. Therefore one year might have 350 days while another has 390 days. The years may, and often do, differ in their length according to days, but not in their seasons and other regular events.' In many ways this system makes better sense than our attempts to bend nature's unpredictable patterns to our will. To pin one's hopes on a natural event occurring on a numerical day or hour of a human-constructed calendrical system is to be doomed to disappointment.

Storytelling also played an important role in event-based systems of recording time. Without a numerical calendar to reference, tales about ancestors – and their experiences of good and bad harvests, floods, droughts, eclipses – were invaluable ways by which history acquired its shape, how the past informed the present and predicted the future. For some coastal Aboriginal communities living in modern-day Australia, these stories go as far back as the rising of the ocean at the end of the last Ice Age 10,000 years ago. Māori culture likewise places supreme value on genealogy and ancestry – on all that went before – and uses the wonderful word 'whakapapa' (pronounced 'fakapapa') to describe it. For them, a meaningful future is unimaginable without knowledge of the past.

Nature still influences our relationship with time – even in our increasingly digital age. British Summer Time, the act of adjusting our clocks backwards or forwards by an hour every six months to increase the amount of daylight we have in the winter mornings, shows that the light, rather than the hour, remains the decisive factor in getting up in the morning. We still calculate the length of a pregnancy according to lunar months (ten lunar months is forty weeks) or end our day on the beach by judging the movement of the tide. The changing colours of the leaves or a sudden chill in the air tell us summer is over far more viscerally than any date on a calendar.

What's more, our timekeeping is *still* event- and story-based. We say, 'That was just before you were born'; 'It was the summer after my GCSEs'; 'It was the month after our wedding', locating things around milestone moments in our own lives. Current generations will for many years think of things as 'Before' or 'After Covid' – an almost universal mass event, even though for those stuck at home during the pandemic lockdowns time lost all distinction. As the milestones that might have marked the year – weddings and holidays, parties and exams, even Christmas – were all cancelled, the days felt oddly 'out of time'.

Close your eyes and think of a watch.

I suspect you're picturing an analogue clock face, with the dial divided into twelve.* Two hands are rotating 'clockwise'. And the whole thing is mounted on a wearable strap.

---

* Although this perhaps depends on your age! As mobile phones and computers have taken over as our go-to devices for telling the time, analogue dials are becoming less common in public spaces. Many train stations now use digital displays. The lack of common use of analogue timekeeping has meant many school classrooms now have digital clocks.

All of these elements were established in the ancient world. And all of them were arrived at through dialogue with nature. The Sumerians, the first known Mesopotamian civilisation (located in modern-day Iraq and Syria), are often credited with inventing the first numerical system for measuring time. They developed the first written number system based around the number sixty, which still dictates how we quantify minutes, hours, angles and geographical coordinates. This number was easily divisible without complicated fractions or decimals. It was also divisible by three – helpful because most humans have an inbuilt body calculator for the three times table. Each of our fingers has three joints or knuckles, so one hand (not counting thumbs!) accounts for twelve finger knuckles; together, both hands total twenty-four. This counting system may well be the origin of the twenty-four-hour day.

A thousand miles west of Sumer, in ancient Egypt, scholars started to use the sun and stars to divide time even further. The name of the ancient Egyptian god of the sky, Horus, whose right eye was believed to be the sun, is the origin of the modern word 'hour'. About 5,000 years ago, Egyptians discovered that the Earth's solar year – the time it takes our planet to revolve round the sun – influenced the rising waters of the Nile, and coincided with the summer solstice and the prominence of Sirius, the dog star, in the night sky.

Our very idea of 'clockwise' is also a function of the sun, as well as an accident of location. The civilisations which shaped contemporary timekeeping systems were generally located in the northern hemisphere. And if you want to follow the path of the sun across the heavens in the northern hemisphere, you have to face south. From that position, the sun moves from left to right over the course of the day, with the countering shadow it casts creeping around from right to left: in other words, clockwise. This simple observation surely led our ancestors to gauge the time from the lengths and angles of the shadows cast by the people, buildings or trees around them. The sundial, the first clock 'dial' as we know it, was an attempt to harness this very phenomenon, replacing random shadow-casting objects with a designed vertical rod or shape called a gnomon.

No one knows who first invented the sundial or shadow clock. They appear all over the world, from the Stone Circle at Stonehenge (c. 3000 BC) in England, positioned to align with the sun on the summer and winter solstices, to the painted sticks used to make calculations from shadows at the ancient astronomical site of Taosi in China (c. 2300 BC). In the ancient Egyptian burial site the Valley of the Kings, the divisions of a very early sundial were found etched into a flat sheet of limestone on the floor of a worker's hut dating to the middle of the second millennium BC. The gnomon, a separate vertical stick pushed into a hole at the centre of the dial, has been lost but would once have cast its shadow on a semicircle drawn in black and divided into twelve sections approximately 15 degrees apart. The rough divisions were enough to allow its owner to pinpoint the start of the working day, lunch, and time to pack up and head home before it got dark. It is this pairing of gnomon and dial that creates a 'true' sundial.

Sundials served another important purpose: they were a community's focal point. Often planted in the heart of towns and cities, they provided the local population with a shared sense of time – one that everyone could access and work to. This collective understanding of time proved crucial to the development of civilisation. By charting the skies and measuring the movements of the sun, we were able to divide the lives and routines of large groups of people into ever smaller and more accurate parcels of time. These divisions made it increasingly easy to work together and schedule our interactions with others – whether it was farming, trade, education or governance – and in turn help us make plans for the future.

In the hours of darkness, the ancient Egyptians looked to the stars, using them like a vast celestial clock face (we still use zodiac and star groups to measure the passage of time).* Astronomers identified at least forty-three different patterns that included sȝḥ (or Sah in English,

---

* In the night sky the constellation of the Great Bear, Ursa Major, marks the seasons clockwise. The bear's tail points eastward in the spring, to the south in the summer and northward in winter.

which includes parts of Orion), 'ryt (the transliteration of 'the jaws', modern-day Cassiopeia), knmt (possibly meaning cow and signifying Canis Major) and nwt (the Milky Way, and sky symbol of the goddess Nut). They also knew of the planets Mercury, Venus, Mars, Saturn and Jupiter; and they could calculate and predict lunar eclipses. Celestial calendars played an important role in planning the annual lunar festival, where swine were offered to the moon god and Osiris, the god of agriculture, at the season of the new moon.

When I imagine a watch, I always hear it ticking; a constant, nagging reminder of how fast our time on this planet is passing. Many earlier timekeepers likewise recorded the *passage* of time. Water clocks did this by harnessing the regular pace at which water flows through a hole. The earliest examples were surprisingly simple: they were essentially an earthenware vessel filled with a certain measure of water that then flowed into a second earthenware vessel. They relied on an accurate understanding of volume and flow rates. For these clocks, time would literally run out and need to then be refilled by a dutiful attendant. Alabaster and black basalt water clocks were used by the ancient Egyptians for this purpose, while Bronze Age clay items have been found on the coast of the Black Sea in modern-day Ukraine. Variations on this basic system developed all over the world, from ancient Babylon and Persia to India, China and Native North America and ancient Rome. In ancient Greece, *clepsydra* (meaning 'water thief') were used in the Athenian law courts to mark the time given to each speaker. Some of these clocks could even sound an alarm. In the design of a water clock dating from 427 BC, invented by Plato, a series of four vertically stacked ceramic urns is used that allows water in the top vessel to flow at a slow and controlled rate into the next vessel directly below it. When that urn is full, at a precise timed moment calibrated by its size and the speed of the flowing water that fills it, water floods through a siphon into the third urn below all at once. The sudden rush of water forces the air in the third vessel through a pipe near its top that whistles to sound the alarm. The fourth urn at the bottom of the stack collects the water ready for it to be reused.

In ninth-century England, King Alfred the Great of the West Saxons used candle clocks like a modern-day productivity guru, keeping a strict daily schedule consisting of eight hours for work, eight hours for study and eight hours for sleep.* His 'clock' consisted of six candles of uniform width and height. Each one took four hours to burn and was marked with twelve equally spaced divisions, each representing twenty minutes. So two candles marked the duration of Alfred's daily reading and writing (he was a passionate scholar who translated a number of Latin religious texts into Old English), and another pair stood guard as he planned battle tactics to defend his lands against invading Viking armies or mediated disputes between his subjects. The final two watched over the king as he slept.

As travel increased and people needed to tell the time on the go, many of these traditional timekeepers proved impractical: sundials were too static, clepsydras sloshed all over the place, and candle clocks were extinguished by the wind. In the second half of the Middle Ages, the hour- or sandglass were increasingly used alongside them. By the end of the thirteenth century, 'sand-clocks' were being used on ships. In his *Documenti d'Amore*, written between 1306 and 1313, Francesco da Barberino insisted that 'in addition to a lodestone, skilled helmsmen, good lookout, and chart, the sailor must have his sand-clock.' In the late fifteenth century, Christopher Columbus is said to have used a half-hour *ampoletta* (meaning ampoule, a kind of glass container), which was maintained by the helmsman and corrected using the midday sun as a reference. Sand-clocks helped sailors locate themselves not only in time, but also in space: by knowing *how long* it had been since you set sail, and the speed you were travelling (measured literally in knots, by extending a line with equally spaced knots overboard and timing how quickly the knots dragged out to sea), you could calculate roughly where

---

* Candle clocks, made from whale fat, probably originated in China around 200 BC. Their relatively stable and consistent burning rate made them useful indoors and at night.

you were and when you would reach land. This process is called dead reckoning and for centuries the sand-clock was the best device available. It would take another 500 years and a revolution in science and engineering before the mechanical clock could match the accuracy of the sandglass for the measurement of longitude at sea.

In the sixteenth century sundials became small and portable. Ring dials* (the smallest were about the size of a man's wedding band) were engraved metal rings that could be held up to the sun and read as the light shone through a small hole in the main band and cast a bright dot on a scale inside. The sides of the ring, formed from separate pieces of metal, could be rotated to adjust for the correct month and latitude to give an accurate reading. Their invention is credited to the sixteenth-century Dutch mathematician and philosopher Gemma Frisius (1508–1555), who in 1534 took his idea for an 'astronomer's ring' to the engraver and goldsmith Gaspard van der Heyden – a collaboration of science and craftsmanship that foreshadowed the watchmaker's art.

The wrist-strap we imagined earlier is a defining feature of the watch, because it makes time wearable and therefore personal. The ring dial was significant for the same reason. This was the first timekeeper that could be tucked in a pocket or suspended on a cord or chain and carried throughout the day. Petite, lightweight, and completely unaffected by the motion of their wearer, ring dials proved so practical that they were used for several centuries after the invention of the watch. They even get a cameo in Shakespeare's *As You Like It*, when Jacques describes meeting a fool in the forest who makes a big show of pulling 'a dial' out of his pocket, telling him 'very wisely "It is ten o'clock."'

That passage reminds me of *my* foolish moment in the forest, as I looked up at the geese and obsessed over my own prized timepiece.

---

* Properly termed *universal equinoctial ring dials*.

In the end, the deadline I was working to during that first winter in the countryside passed me by, just like the birds: I didn't finish the watch for another three years.

I always find it comforting to remember that however mechanised and digitised our experience of time seems today, it will always be underpinned by natural forces that remain completely out of our control. And that, in the end, some things still take as long as they take.

2

Ingenious Devices

'Measure what is measurable, and make measurable what is not so'
Attributed to Galileo (1564–1642)

When I was seventeen I dropped out of school and wound up on a silversmithing and jewellery course. As far as my friends and teachers were concerned, I might as well have run away to join the circus. I had always loved art, but I had taken all science A levels in pursuit of my dream of becoming a pathologist. School wasn't encouraging. My careers adviser implied that medicine was not really for kids from my working-class background. And though I loved science, the way it was taught felt so rigid, dry and cold. So little of it was hands-on. I'd spent all year looking forward to dissecting a sheep's heart in biology only to discover the exercise was being cancelled after another student fainted. Halfway through the course, I had my moment of rebellion: I decided if science wasn't going to have me, I'd run away to art school.

The tutor on the silversmithing course was a master goldsmith from Austria who had started his apprenticeship at thirteen and was due to retire the year I finished the course. Peter had such a wealth of knowledge that he was humbling to be around. I still adhere to many of his creative philosophies. He is the reason I only make watch cases out of precious metal. One day he saw me begin to make up a design with gilding metal (a copper-based alloy with similar working properties to 9ct gold but without the price tag) and invited me over to the workshop safe. I explained I didn't have the money to buy precious metals, so he sorted through trays of gold sheet and wire and gave me what I needed. 'Rebecca,' he said, 'you have taken a long time working on that design, and it is a beautiful design. You must only work in materials worthy of your efforts. Don't worry about the metal costs, we'll sort it out another time.' And so I made the piece in gold, a brooch in the form of a phoenix with black diamond eyes, blood-red rubies across its breast and a diamond in its tail. I pierced the individual feathers in its wings to create tiny windows, or cells, through the metal that I filled with glass enamel. He was right. It's daunting, not to mention expensive, to work with precious materials, but you could spend a whole year making something from gilding metal that would hold little value on completion. If you have the confidence, and capital, to make something in precious metal, its value will only increase. Even so, I ended up scrapping the brooch a few years later, during one of many desperate moments teetering on the brink of financial ruin. I still remember crying as I twisted it apart with pliers to break the stones from their settings.

Peter also taught me how to make mistakes. I was making my first solitaire ring with a rex claw setting – one of the most traditional single-stone engagement ring designs. I'd got a bit carried away filing the claws that would hold the stone in place and accidentally made them too short. Deflated, I asked Peter if there was a way the setting could be saved. He sat at my bench and proceeded to re-dress the claws, making them perfectly usable again. I remember calling him a genius and thanking him profusely for saving me a week of work. His response? 'Rebecca [he always started with "Rebecca . . ."], do you know how I know how to fix this? I know because I have made this mistake myself and learned from it. It's okay to make mistakes – as long as you learn from them.' I still think of those words most days.

Peter's jewellery course was where I learned metalworking skills like soldering and saw piercing – the art of using a fine saw blade just a millimetre or so in depth to cut intricate shapes. I was still interested in science and engineering and, as I became more confident, I started to incorporate hinges, pivots and other simple devices to introduce movement into my jewellery. As a visual thinker, being able to see how things work out in the real world made sense to me. In retrospect, that's how I've always learned. By experimenting, watching and testing the outcomes of my physical interactions with things.

As time went on, I started experimenting with basic automata – moving mechanics that imitate living things. I've always loved orreries – clockwork models of the solar system. To me they're some of the finest examples of nature as represented in mechanics: our very human way of containing the universe in a device small enough to be examined on your desk. So, for my final project in jewellery and silversmithing I designed an orrery where each of the planets was a removable and wearable piece of jewellery. Saturn was a pendant whose rings could turn independently of each other and in different directions. The sun was a ring made up of spiky 'flames' that spun and appeared to flicker. It would also have made a good knuckleduster in a fight.

At that stage I lacked sufficient understanding of the mechanics I needed to create a properly functional orrery that could chart the movements of the planets. I was also up against a tight deadline. I took enormous liberties with the planetary positions, and the rings that held them had to be rotated independently by hand, rather than an interconnected system powered by a motor or turning a handle. The real thing didn't live up to my design; it was nowhere near as advanced as I wanted it to be. I needed more time and knowledge to fully realise it, but I was nonetheless captivated by the workings of the interconnected moving parts.

During the end-of-year show, my orrery caught the attention of the horology students. A small group (one of them was Craig) made a beeline for me, seeing that I shared their interest in making small fiddly things that moved. Up until then, when I thought of watchmakers, if I thought of them at all, I'd imagined people changing batteries or straps in shopping centres. But in their workshop, surrounded for the first time by the whirring lathes and mills, and the smell of metal and swarf, I realised watchmaking could allow me to be an artist, designer, engineer and physicist all at once. By the end of the course I had joined them.

I wish I still had my orrery. But, as with the phoenix brooch, I ended up scrapping it to pay the rent.

My own early, fumbling attempts to make things move always play on my mind when I think about the earliest clockmakers. Without constant human maintenance, water, sand and candle clocks were little more than early egg-timers. 'True' clocks and watches require a self-renewing or mechanised power source. Although mechanical clocks and watches didn't appear until the fourteenth and sixteenth centuries respectively, the developments needed to make this leap were already under way over a millennium earlier. In ancient Mesopotamia, all kinds of engineering advances – from hydraulic machinery, crop irrigation systems and textile mills to mass production of bricks and pottery, wheeled chariots and even the plough – supplied the groundwork for mechanised timekeepers.

I like to imagine the astonishment that must have greeted the first hydromechanical timekeeper that appeared in Europe. In 802 AD one of King Charlemagne's favourite diplomats returned from a mission to Baghdad with a shower of gifts designed to dazzle and impress. They had been sent by Caliph Harun al-Rashid and epitomised the Islamic Golden Age that had flourished under his rule. While the star of the show was undoubtedly Abul-Abbas, the adult Asian elephant, who must have caused mayhem as he plodded through the streets of Aix-la-Chapelle,* Harun also gave Charlemagne a brass water clock with a mechanism capable of striking the hours. Witnesses described how at the end of each hour, a cymbal would clang while a model horseman emerged from one of twelve doors. We don't know exactly how the mechanism worked, but it would have been a system of weights and ropes both powered and controlled by the changing water level as

---

* Now Aachen, in western Germany.

it seeped out of a hole. For the lucky Europeans who beheld it, this clock would have seemed like magic.

In the eleventh century, the Spanish Muslim astronomer and inventor of scientific instruments, al-Zarqali, designed and created a water clock capable not only of telling the hours of the day, but also of displaying celestial information. Located in the ancient city of Toledo in central Spain, the clock was renowned for its ability to illustrate the current phase of the moon, using two basins that would gradually fill with water and then empty over the course of twenty-nine days, to echo the waxing and waning of the moon. Al-Zarqali managed the ever-changing water levels with a subterranean pipe system that compensated for water removed or added to the basins, should anyone attempt to interfere with them. I can just imagine a curious child (it would have been me) sneaking off to al-Zarqali's clock while their parents bargained with a nearby market stallholder. After removing some of its water to see what happened, they would then have watched in wonder as the dish magically refilled itself to exactly the right point.

My home city of Birmingham has a water feature at its heart, too. Designed by Indian sculptor Dhruva Mistry in 1992, a fountain called *The River* takes pride of place outside the town hall in Victoria Square. At the top, a giant reclining female nude, cast in bronze, holds a pitcher of water that flows into a palatial upper pool where it is retained until it overflows its lower edge and runs down a series of steps that empty into a second large pool at the bottom. Affectionately referred to by locals as the 'Floozie in the Jacuzzi', she provides a convenient meeting point that everyone knows. Some people offer her coins and make wishes, and, on a hot summer's day, a few even venture into the water (albeit illicitly) to splash around and cool down. Installations situated in public spaces invite people to interact with them.* These are shared

---

* Sometimes the interaction is a little too much. After a few years out of action, during which time the water was replaced with a flowerbed, the 'Floozie' was restored to its watery glory in 2022. But within a month it had been damaged by someone adding bubble bath – not a problem al-Zarqali ever faced.

spaces, over which we all have some sense of communal ownership.

Al-Zarqali's clock remained in Toledo until a later inventor, intrigued by its construction, was permitted to dismantle it for examination but apparently proved unable to put it back together again. I've always been moved by that story because I've worked on many timepieces that have had the same fate. The initial conversation with the owner usually starts with them saying, 'Well, this is a bit awkward, but I . . . and then I . . . so you see . . . cutting a long story short . . .' before admitting that their inherited grandparent's/ bought at auction/gift for a significant occasion watch is now a bag of bits that they have no idea how to put back together. Curiosity has been killing clocks and watches for centuries.

Nearly 4,000 miles east of Toledo in the Henan province of China, in 1088, the astronomer Su Song was commissioned by the emperor to create the world's finest water clock, intended to showcase a showpiece for the intellectual prowess of the Song dynasty. The brief for Su Song's clepsydra involved a number of complex celestial displays: at the time, dynastic houses were ruled according to a heavenly mandate, or *tianming*, which required an ability to track and predict astronomical events for interpretation, as a guide to inform bureaucratic decisions. With its bronze astronomical and armillary spheres, and automata mannequins beating gongs at important hours of the day, Su Song's clock didn't only confirm China's technological pre-eminence, but functioned as 'a hotline to divinity, a conduit through which heavenly wisdom flowed into the Imperial court'.

Starting with a scaled-down model prototype he'd made from wood designed in the style of a pagoda tower, Su Song and his team of craftsmen and engineers spent eight years building his hydromechanical celestial clepsydra. The finished clock was 40 feet high – around the height of a four-storey building – and was powered by a giant water wheel, 11 feet in diameter, with thirty-six buckets around its rim. As each bucket filled with water and became heavy enough to trip the mechanism, it would fall forward, rotating the wheel and placing the next empty bucket in the path of the flowing water supplied from a separate tank to keep the volume consistent.

What fascinates me, as a watchmaker, about Su Song's pioneering piece of engineering is that it represents the first ever escapement – the mechanism that alternately checks and releases the power of the gear wheels, giving the clock the potential for an indefinite duration. (Infinite, that is, so long as there were human monitors constantly available to maintain the water level in the supply tank and perform any necessary adjustments and servicing.) This group of components capable of locking and releasing motive energy from a source like water, gravity or a spring would become instrumental in the invention of the first fully mechanical clocks. It was also, I should add, the first moment in history when the 'tick, tock' sound of a clock was heard.

All of these early mechanical devices were suffused with the sheer joy of experimentation and discovery, the result of trial, error and endless possibility. At the turn of the thirteenth century, Ismail al-Jazari, a Muslim polymath, scholar and inventor from Upper Mesopotamia, took the development of mechanics to an entirely new level. The chief engineer of the Artuklu Palace in Turkey, al-Jazari is sometimes referred to as one of the 'fathers of robotics'; he was a master of automata. His *Book of Knowledge of Ingenious Mechanical Devices,* rumoured to have inspired Leonardo da Vinci some 250 years later, details around 100 mechanical inventions accompanied by delightfully vivid hand-painted illustrations. They include automata peacocks, a humanoid waitress powered by water who could serve drinks at parties, a band of musicians who 'played' music to party guests, and several complex candle and water clocks.

Where Harun sent a clock accompanied by an elephant, al-Jazari went one better and designed a water clock that took the form of an elephant. On the creature's back, perched on a Persian carpet, an Arabian scribe with a pen in his hand sits in a golden howdah. A mahout, or elephant trainer, rides at the front, and red Chinese dragons and an Egyptian phoenix adorn the top. Concealed in the elephant's belly is a water-filled basin. Floating on the water is a bowl, punctured with a small hole, attached by strings to a pulley mechanism. The bowl slowly fills with water and sinks down, causing the scribe to rotate once every minute. After half an hour the bowl fills completely and sinks to the bottom of the basin, triggering a see-saw mechanism that releases a ball into the mouth of one of the dragons. The weight of the ball causes the dragon to tip forwards, pulling the sunken bowl back up to the surface of the water below. This process also triggers a human statue on top of the carriage to raise his hand. A cymbal sounds every half-hour, the phoenix spins and the mahout moves his beaters. Once the cycle is complete, and the performance ends, the characters return to their original positions and wait for the bowl to fill again.

The elements of the clock were deliberately designed to embrace the combined global knowledge of engineering at this point. According to al-Jazari, 'the elephant represents the Indian and African cultures, the two dragons represent ancient Chinese culture, the phoenix represents Persian culture, the water work represents ancient Greek culture, and the turban represents Islamic culture'. The elephant clock continues to amaze today. In 2005 a monumental replica was built as the centrepiece of a shopping mall in Dubai. Sitting in the middle of a vaulted marble hall, surrounded by eager shoppers taking photographs, the Dubai elephant clock is once again a focal point of shared time.

When it came to timekeeping, medieval Europeans lagged some way behind their Chinese and Islamic counterparts. But as Europe entered the Renaissance, a growing number of astronomers emerged, within the Catholic Church and more generally from elite society, to push clockmaking into a thrilling new era. Deeming water an unreliable power source – in European summers it tended to evaporate, and in the winters it froze – they made major advances in the development of genuinely mechanical clocks.

The key to making a fully mechanical and comparatively low-maintenance timekeeper was finding a reliable power source. This problem was eventually solved with a little help from gravity. Some time in the fourteenth century – we don't know exactly when, or who invented it and where – a remarkable clock appeared. It was powered by a rope attached to a heavy weight at one end fixed to a horizontal arbor (a bit like the spool you wind thread into) and the mechanism at the other.* The oldest surviving clock with such a mechanism was constructed in 1386 and now lives in Salisbury Cathedral in England. In this case, power was supplied by two stone weights; as the weights descend, ropes unwind from large wooden spools. One spool drives the going train (responsible for keeping the time) while the other drives the striking train (which chimed the bell to give the time signal). The arbor could be wound manually with a crank, slowly coiling the rope around it and raising the weight upwards. Ratchet work allowed the arbor to turn one way but locked it, a bit like reeling in a fish on a line, preventing it from spinning back the other way. Once fully coiled, or wound, the force of gravity would then pull the weight down, causing the arbor to want to turn back on itself at an uncontrolled speed. So a new device to control the speed of the rotation was required: we call it a *verge escapement*.

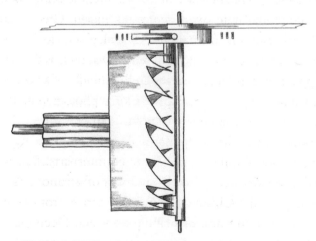

The verge escapement was the first to be used in fully
mechanical timekeepers and remained in use
until the early nineteenth century.

* There is a detailed account of another turret clock in Chioggia, near Venice, which also dates back to at least 1386 and survives – albeit in a currently decommissioned state.

The verge in these early clocks is made from a long, thin rod of steel (the staff) topped with a horizontal bar, making it look like a capital T. At the top and bottom of the staff are two rectangular 'flags' attached at about 90 degrees to each other. These flags are spaced so that when one of them swings with the oscillating staff, it catches a tooth of the crown wheel (so called on account of its jagged teeth, which echo the shape of a crown). As the wheel turns forward, one flag swings to allow a tooth through, while the other flag catches another tooth, thus controlling the release of power. Once the flag has made contact with the crown wheel, the momentum impels the verge to swing back the other way, allowing the other flag to catch a tooth while this time the first flag lets a tooth through, and so on, back and forth. This cycle repeats over and over, thousands of times every hour, controlling the release of power – and creating an audible *clunk, clunk, clunk* each time a flag catches a tooth.

The verge escapement made possible the first church clocks – devices whose workings lived in turrets and towers, and which loomed over towns and cities to be visible for miles around. The earliest clocks had no dials; rather, they struck the hours on a bell to indicate the time, in order to guide public routines such as worship. The Elizabethan dramatist Thomas Dekker described how the clock bell could be 'heard a farre off, whither we lye in our bed in the night, or in the day time we be farre from a Dial'. The word 'clock' itself derives from the Medieval Latin *clocca*, and French *cloche*, which both mean 'bell'. In late medieval and early modern Europe, right up until the seventeenth century, time was still a public rather than private matter and was delivered, literally, from on high. In an age when 90 per cent of Europe's population was peasantry, the idea of a personal clock with which you might self-determine your time was still a long way off.

Grand public turret clocks are about as far removed as you can get from the microscopic world of the watchmaker. A few years ago I had the pleasure of visiting the workshops of Smith of Derby, a turret clock restoration and manufacturing company that was founded in 1856 and is now managed by the fifth generation of the Smith family line. I imagine the experience would be quite surreal for most people but, coming from my own workshop, I felt like a Borrower who had just emerged from behind a skirting board into a realm of giants. While the tools we work with are fairly similar, Smith's are five, ten or twenty times the size of mine. It felt incredibly familiar, and yet completely different. When I fit the hands to a watch dial (one of the last things I do when I put a watch back together) they are often no longer than the tip of my little finger. In the workshops of turret clockmakers, giant hands, some taller than you or me, have to be winched back into place on dials the size of a double-decker bus.*

Aside from differences in scale, turret clockmakers face struggles that are almost inconceivable to watchmakers. They battle gales and freezing cold weather up church towers, swing in harnesses suspended off roofs, scrape away piles of acidic pigeon poo that have clogged up the mechanism, and occasionally face altercations with angry nesting seagulls. All of this makes me feel grateful to be working in a warm, safe workshop. But there is nevertheless something magical about turret clock movements. Their aesthetic is pure H.G. Wells. They're curious science-fictiony machines that clunk and clang and whir. The action of the movement is heavy and methodical, almost symbolic of the way 'clock time' would go on to dominate the modern world.

The verge escapement was not, in truth, very accurate – some clocks lost or gained several hours over the course of a week – but it

---

* To give you an idea of scale, the largest clock dial in existence can be found at the top of the Abraj Al Bait Towers in Saudi Arabia. Constructed in 2012, it measures 43 metres across – just short of the length of two male blue whales lined up nose to tail.

provided a foundation on which later inventors and engineers could build, making timekeepers that were more complicated, more impressive and more eye-catching than people only a century earlier could have imagined. Medieval church clocks were soon displaying a highly sophisticated range of information, including planetary positions, predicted eclipses, lunar phases and high and low tide times – often brought to life with elaborate automata. Financial records from Norwich Cathedral between 1321 and 1325 describe the commission and installation of a mechanical clock whose displays included a model of the sun and moon and fifty-nine moving sculptures carved from wood, including a choir and procession of monks.*
In the fourteenth and fifteenth centuries, the Church saw these complicated astronomical clocks as grand public representations of Christian cosmology, designed to illustrate, through ornament and automata, the meaning of the passage of time to followers of the faith.

Astronomy was central to the worldview of medieval Christians, who saw God as the divine architect of the universe. Pictorial depictions from the period show God as a geometer, dividing compasses in hand, mapping out his plan for the cosmos. Celestial events were also believed to have a direct effect on human life. Marriages, diplomatic decisions, even surgeries were performed with reference to the alignment of the moon and stars. Each of the twelve zodiac signs had an associated part of the anatomy and it was held to be dangerous to operate on a part of the body if the moon was in its associated zodiac sign. Devices like lunar volvelles could be consulted to calculate the position of the moon in relation to the sun, which when checked against a zodiac man – an illustration of a man's body on

---

* The inventories describe in detail the use of brightly coloured paints and gilding. The whole project required the talents of a wide range of skilled craftspeople, who were employed on the build for three years – from blacksmiths and carpenters to masons, plasterers and bell founders. The total cost of the clock was listed as £52, which at that time, had it been a one-man job, would have covered the wage of a single skilled master craftsperson for over fourteen years of solid work.

which the anatomical parts associated with each zodiac sign were marked – would tell you if the signs were auspicious. If you had an ingrown toenail and the volvelle said the moon was in Pisces (the sign associated with feet) then bad luck; you'd have to wait a month before you could have the nail removed.

Just as in ancient Egypt, where the study of horology and astronomy was largely conducted by priests, so in medieval Europe monks were among the fortunate few who, in the absence of worldly distractions and without having to worry about keeping a roof over their heads, were able to dedicate significant time to the furtherment of knowledge. A number of the earliest clockmakers were men of the church, including the monk and natural philosopher Richard of Wallingford (c. 1292–1336) who designed an astronomical clock in the 1320s, and the clergyman and astronomer Jean Fusoris (c. 1365–1436), the designer of a monumental astrological clock for Bourges Cathedral in France.

Astronomers drove clocks towards ever greater accuracy. In order to make measurements of an observed phenomenon, like a lunar eclipse or a passing comet, they needed a timekeeper that was capable of measuring perfectly equal time increments. Ancient methods of division often used the separation of day and night to mark the passing of each day, creating hours of varying length depending on the time of year and the distance between dawn and dusk. Mechanical clocks, however, relied on gearing to control the motion of the hand, meaning that (providing the clock was running correctly and didn't stop) there was no variance in how long it took each hand to make a full rotation. The regimented nature of a mechanical clock made a totally controlled uniformity possible.

Galileo, who is rumoured to have said 'measure what is measurable, and make measurable what is not so', is credited with discovering one of the most significant improvements in accuracy: pendulum isochronism (how a pendulum, in the absence of variables such as wind, will swing at a constant rate). The story goes that, while attending mass in Pisa Cathedral, the nineteen-year-old Galileo looked up and noticed the regular, repeated swinging of an altar lamp suspended from the ceiling. In that moment, it struck him that this swing could be used to trigger the regular release of power from a mechanism. The idea rumbled away at the back of his mind for years, until in 1637 he designed the first ever pendulum-regulated clock, which used a swinging weight to trigger the release of the escapement. Galileo died five years later, and never saw his ingenious concept become a reality. It would be another fifteen years before the Dutch physicist and mathematician Christiaan Huygens converted it into a working clock mechanism.

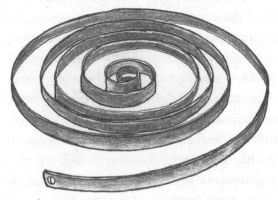

A typical traditional mainspring. This form was used from its advent until the twentieth century when its shape was refined to generate a more uniform torque as it unwound.

But if astronomical observations were to be made in a range of locations, clocks needed to be portable. The most significant advance to aid this process was the introduction of something called a mainspring, in which a tightly coiled steel spring replaced weights as the power source. We can't be sure who invented the mainspring, but it's highly likely the technology also came out of locksmithing and emerged from northern Italy. The earliest surviving spring-driven clock dates to 1430. These springs, which were long but wafer thin – a bit like a ribbon you would tie round a gift – were coiled into a drum or barrel, and wound around a central arbor using a crank or key. The spring coils like a tight spiral round the arbor, then, when released, pulls back as its elasticity forces it to unravel, dragging its outer hooking with it to create the all-important rotating motion. To prevent the mainspring from suddenly releasing all of its power through the train wheels at once, the speed of this rotation is controlled in exactly the same way as the earlier weight drive was, via an escapement.

Mainsprings allowed clocks for the first time to become independent of gravity and thus small enough to move around with us on our travels. If religion had supported the emergence of architectural mechanical clocks across Europe and scientists had made timekeepers more accurate and practical, the wealthy now rebranded these ingenious devices as status symbols. Over the course of the fifteenth century, clocks became a common sight in the homes of aristocrats and wealthy merchants – especially those with an interest in astronomy. They used these objects to show off their knowledge of cutting-edge technologies, just as we do today when we queue up for days outside an Apple store to buy the latest iPhone. Their cost, and therefore exclusivity, made them objects of desire, and soon materials like gilded brass and copper replaced the iron used in the earliest mechanisms. As engravers and gilders became involved, clocks grew ever more ornate.

The British Museum owns a remarkable clock, made in 1585 by Augsburg clockmaker Hans Schlottheim, probably for the Holy Roman Emperor Rudolf II. The clock, which is mounted on a gilded brass galleon, was designed to 'sail' down the middle of a busy banqueting table 'firing' miniature smoking automata cannons while humanoid figures moved around on deck. The hours were chimed on bells suspended below the crow's nest while a mechanism in the hull played music to the beat of a drum. In the midst of this spectacular performance, the tiny clock dial on the ship's bridge is almost entirely missed.

In the sixteenth century, these extraordinary clocks weren't, in a way, all that extraordinary: many master craftsmen around Europe were making such mechanical marvels to satisfy an insatiable demand among the elite. There was just one problem: these elaborate machines could only be admired at home. They relied on their owner's friends, colleagues and clients to come over for dinner. By this stage Europe's ruling classes were ready for objects that, while equally intricate and curious, were portable enough to take around the world. For that to become a reality, the clock needed to be small enough to wear.

Small objects are good at disappearing. They can be stolen, destroyed or misplaced. They can vanish into the bowels of private collections. They can hide in the backs of drawers, in shoeboxes under beds, and even under floorboards. But then, sometimes, they are discovered again by chance.

So it was with the world's oldest known watch, which was found in a box of old clock parts at a London flea market in 1987 for £10. To the untrained eye it didn't look like a watch. It was essentially a ball, roughly the size and weight of a hen's egg, formed from two hemispheres of sheet copper, which had been hammered, or 'raised', to form a near-perfect sphere. On top of the ball is a hoop, through which would have run a chain, so it could be worn round the neck. Underneath it, three little feet allow it to sit on a table without roll-ing over. The case is decorated with a crude engraving of figures, a village scene and leaves. The upper segment is pierced with a number of comma-shaped holes through which one can just see the dial inside. To tell the time you unclip the top of the ball, which hinges back to reveal a single hour hand (the earliest watches weren't accurate enough to warrant a minute hand, and smaller increments of time weren't as important to their owners as they are now) circling a track of Roman numerals on an engraved dial. The watch is marked with the initials 'MDVPHN'. This offers us the first clue as to its origins: MDV is the date – 1505; PH is for Peter Henlein, a watchmaker known for making small portable mechanical clocks at the time; and N is for Nuremberg, where this one was made.

The buyer of this unusual object initially doubted its authenticity and sold it a few years later. Its next owner took the device to an expert, and they too were told that it was fake. The watch was sold again, for an undisclosed but presumably low sum. The third buyer then subjected the watch to detailed scientific analysis that proved, beyond all reasonable doubt, that the piece was indeed made in or around 1505 and was likely to be genuine, and therefore the earliest surviving watch in history. The watch is now estimated to be worth between £45 and £70 million.

Of all the remarkable facts about this little device, what's perhaps most remarkable is that we know something of its maker. We know that Peter Henlein was born in Nuremberg in 1485, the son of a brass worker, and was apprenticed as a locksmith, as were many of the first watchmakers. We also know that the decisive turning point in his early life happened not in a workshop, but in a tavern. In 1504, when he was nineteen, Henlein was involved in a brawl in which a fellow locksmith, Georg Glaser, was killed. As one of those accused of the murder, he begged for asylum at the Franciscan Monastery of Nuremberg and sheltered there from 1504 to 1508.

In the fifteenth and sixteenth centuries, the city of Nuremberg in southern Germany was one of Europe's creative and intellectual powerhouses. It was a centre of the German Renaissance, home to celebrated goldsmith Wenzel Jamnitzer (1507/08–1585), and Albrecht Dürer, who had set up his workshop there in 1495. Its monastery was also a magnet for academics and craftsmen, and doubtless provided Henlein with access to new tools and techniques as well as the work of visiting mathematicians and astronomers. Henlein had, serendipitously, found himself in an environment conducive to exploiting his talent.

It was perhaps in the monastery that Henlein learned to make the miniature fusee that he added to his watch. The fusee (pronounced fyu-zi), which connected to the mainspring, might have first been developed as a crossbow mechanism, and appeared in a design by Leonardo in 1490. As the watch is wound, the spring coils more tightly, storing energy, which then turns the barrel containing it as it uncoils – like the rotating dancer in a musical box or a mechanical egg-timer. And just as a dancer's pirouettes become slower and slower over time, the rotational power in the watch spring starts out strong but then diminishes as it unwinds. The helter-skelter-shaped tapered fusee helped regulate this, using a principle similar to a bicycle's gears.

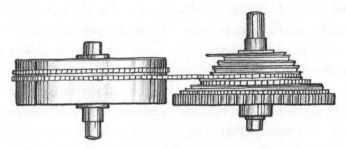

The fusee attached to a mainspring barrel by a fine bicycle-like chain. The earlier gut-line version worked in exactly the same way.

Henlein quickly developed a reputation for making clocks of extraordinary ingenuity and impeccable craftsmanship. He received commissions from the Nuremberg council for an astronomical device and made a public clock for the tower of Castle Lichtenau. But his speciality, it seems, was the creation of small ornate spherical watches, just like ours, that hung from chains as jewellery or as chatelaine brooches attached to clothing. In 1511, the scholar Johannes Cockläus described how Peter Henlein, 'out of a little iron constructs clocks with numerous wheels which can be wound up at will, having no pendulum, go for forty-eight hours, and strike, and can be carried in the purse as well as the pocket'.

It is harder to decipher the man behind these wonderful devices. As one of the few named watchmakers from that time, his character has repeatedly been mythologised. His modern fame is the result of Walter Harlan's play *The Nuremberg Egg* (1913), which was adapted into a film and released in 1939 as part of a Nazi propaganda drive to promote stories of German supremacy. The final cut was approved by Joseph Goebbels. The play and film make out Henlein to be a loving husband and dedicated artisan who dies from a weak heart. Later research, however, reveals a darker side. Analysis of this first watch movement has shown what appear to be the micro-engraved initials 'PH' marked over and over again on the metal of the mechanism, unreadable to the naked eye, which my psychologist friend suggests could indicate a sociopathic and narcissistic personality. We know that Henlein was capable of violence (the murder case against him was dropped after he paid blood money to the victim's family, rather than his innocence being proved), and that he offered fulsome support to his brother, Herman, who was beheaded for the murder of an eight-year-old beggar girl, in what is believed to have been a sexually motivated crime. Peter lacked any empathy for the murdered child and her family, and repeatedly attempted to have his brother pardoned. Put simply: Peter Henlein is one of the only celebrated watchmakers from history with whom I'd never want to go for a pint.

Henlein's watch is in some senses quite simplistic. The metals are raw – there is none of the advanced polishing or fine finishes associated with luxury modern watchmaking; the movements are made of iron – not an ideal material for a watch as, when struck, iron atoms tend to align themselves to the Earth's magnetic field, turning the movement into a magnet and disrupting the mechanism; and the engraving that adorns the surface is rustic and naive. And yet, as a modern-day watchmaker I am in awe of the skill that lies behind this device. Henlein's watch was made before high-quality magnification, digital measuring devices, motorised lathes or drills. Every element of his watch – every tooth of each wheel and pinion, and every last tiny screw – was made and assembled by hand. And, remarkably, five centuries later, watches like this one are still running, albeit with the care and attention of later restorers like me.

We have arrived at our first remarkable, mechanical, wearable timepiece. I see this watch as both a culmination and a beginning: the culmination of tens of thousands of years of a human journey towards personal, portable and mechanised time; and the beginning of a much faster-moving story – lasting little more than five centuries – about humans and machines.

3

Tempus Fugit

Time is more use to us than wealth or fate
Because it changes when appropriate.

Old age is an ill none can cure
And youth a good that nobody can store.
As soon as man is born his death is sure
And those who seem happy merely struggle more.

Mary, Queen of Scots, c.1580

At first sight the hollowed black sockets, gaping nasal cavity and fleshless rictus grin look ghoulish. But I try to set aside these initial impressions, to perform a kind of Etch-a-Sketch clearing of my mind, in order to imagine what this watch meant to its first owner. I'm looking at a sixteenth-century death skull watch, in the collection of the Worshipful Company of Clockmakers, which was once believed to have belonged to Mary, Queen of Scots.

The watch is silver, about the size of a small satsuma, and covered in delicate engravings. On the forehead of the skull, a sandglass- and scythe-brandishing skeleton stands with one foot at the door of a palace and the other at the door of a cottage – a reminder that Death haunts prince and pauper alike. On the back of the skull, the figure of Time – also holding a scythe – is flanked by a serpent eating his tail, and Ovid's classic phrase '*Tempus edax rerum*' ('Time devours all things'). The sides of the cranium are pierced by an intricate network of holes. Inside the skull lies a bell that the mechanism struck to sound the time. To read the time, you snap back the jaw to reveal a dial hidden in the roof of the mouth. The workings of the watch lie where, in a living skull, the brain would be.

This watch is a devotional object from a time when life was short and death an inescapable reality of everyday life. Merely to glance at it was to be reminded that at any moment you might be called to meet your maker. It's a sobering thought for any of us, but especially anyone who has ever brushed against the edges of their mortality.

I was raised a strict, practising atheist. We lived in a multicultural neighbourhood alongside Sikhs, Hindus, Muslims, Irish Catholics and Polish Jews, but my dad wanted nothing to do with his own Catholic upbringing. Dad – who had been a social security officer in nearby Handsworth before leaving work to bring up my sister and me – always told us that we should respect the religions of others while avoiding having one of our own. Our West Indian next-door neighbours were good family friends and, though they never discussed their Christian beliefs with us, they knew our leanings. Every so often they would gingerly

An example of an early-seventeenth-century silver skull watch.

post a flyer from their church through our letter box, inviting us to find the light of Jesus. I remember asking my dad why they did this when they knew we didn't believe in God. He explained that it was because they liked us, cared about us, and were worried that we were going to hell for not believing in God. It was a kind gesture, he said, and we should be grateful – if unconverted.

And yet – I've always been drawn to churches and cathedrals. Is it the peace that seems to penetrate the walls of worship, or the intense colours cast through a stained-glass window? Whatever it is, it's hard not to feel moved when you experience a full choir or organ recital in a sacred space. We can lose ourselves here, in time and in space, reminded of our own cosmic insignificance. Religion makes you feel small but at the same time part of something bigger. This was definitely true in the sixteenth century. A sense of belonging to God's universe shaped every aspect of a person's life back then, including attitudes to time – and therefore watches.

At the end of the eighteenth century a flurry of excitement spread among London's antique dealers when a watch believed to have belonged to Robert the Bruce, Scotland's Warrior King, was identified. This would have been a particularly impressive find given that Robert died in 1329, more than 150 years before the invention of the watch.

This story goes to illustrate that discovering the exact provenance of a watch is often tricky and distorted by hope. And Queen Mary's skull watch is no exception. Legend has it that she kept it with her at all times and only gave it to Mary Seton, her favourite lady-in-waiting, shortly before her execution.* But in the early 1980s, Cedric Jagger, former keeper of the Clockmakers' Company collection, discovered several watches that over the centuries had been claimed to be Mary's, in addition to a dozen copies. In all likelihood we'll never know whether any of the skull watches that have survived to this day were the one possessed by Mary – or if, indeed, her watch has survived at all.

But say our watch *was* Mary's, what would it have meant to her? It would have had monetary value, of course. For a queen, jewellery was a valuable commodity – a currency that could be used to fund

---

* Mary Seton, one of the famous 'Four Marys' who accompanied Mary, Queen of Scots from Scotland to France as a child, wasn't with her during her final imprisonment, so it is assumed that the watch was left to her servants along with other personal items such as jewellery, letters and small portraits, with instructions to distribute them when they themselves were eventually released.

wars, buy alliances or broker deals. It also probably had sentimental value. Mary was apparently given the watch by her first husband, the king of France, Francis II. Mary and Francis married in 1558, when they were just fifteen and fourteen years old respectively. Their marriage appears to have been exceptionally happy, albeit childlike, as the two had been raised together since childhood. But Francis, who had always suffered ill-health, died less than three years after their wedding and seventeen months into his reign, of an ear infection that led to an abscess in his brain. Mary lost her mother, whom she loved dearly, the same year. Grief-stricken, she had little reason to stay in France. In 1561 Mary returned to Scotland to reclaim her place on the throne, as a devout French Catholic in a now-Protestant country. Her watch might have been a reminder of all she had lost.

The watch would also have had religious significance. In the sixteenth and seventeenth centuries, skulls regularly appeared in still lives and portraits, often accompanied by timepieces – sandglasses, clocks or early watches. The juxtaposition is fitting. Death and time of course go hand in hand: time is the unstoppable beat that counts down the hours of our life. Baudelaire once described it as 'the watchful deadly foe, the enemy who gnaws at our hearts'. Many skull watches were likewise inscribed with Latin quotations such as *tempus fugit* (time flies), *memento mori* (remember you are mortal), *carpe diem* (seize the day) or *incerta hora* (the hour of death is uncertain) to remind their owners that the afterlife, for which mortal existence was just a preparation, was waiting.

For Mary, the skull might also have represented the one domain over which she truly retained full control: her mind. Women at the time had very little freedom. Mary, who was just six days old when she succeeded to the Scottish throne, was a political pawn from birth. At six months old she was betrothed to Henry VIII's only son, the Protestant Prince Edward (future Edward VI), but the match was opposed by Scottish Catholics. At five years old, she was matched to Francis, then the Dauphin of France, and sent to be raised at the French court. When she returned home to Scotland as queen after Francis's death, she was to rule for just five years before her forced abdication in 1567. Even while under house arrest she was implicated in multiple plots by Catholic conspirators to usurp the throne of Queen Elizabeth I, until finally Elizabeth reluctantly signed her cousin's death warrant.

Imprisoned as she was for the last nineteen years of her life, Mary's skull watch might have been a reminder that, for all of that lost time, another, better, *eternal* life would begin after death. The once beautiful and vibrant queen's health suffered terribly in those two decades. She was plagued by bouts of sickness, including an agonising pain in her right side that stopped her from sleeping and intermittent discomfort in her right arm that made even writing hard. Her legs became so painful that, by the time of her execution, she was permanently lame. Later medical scholars have suggested that she suffered from a form of hereditary liver disease, or porphyria (known to have also afflicted one of her later descendants, George III), which might also account for her regular emotional breakdowns, attributed at the time to hysteria or madness. Her letters and the quotes she embroidered during her imprisonment make clear that she was ready to move on from her earthly life, in which the threat of execution hung over her like the sword of Damocles, and be reborn in heaven.

Mary was beheaded for treason at Fotheringhay Castle, Northamptonshire, on a cold February morning in 1587. She knew that in death she would herself become a symbol, and used the moment to secure her legacy for centuries to come. As she was led towards the block, her ladies-in-waiting assisted her in removing her jet-black satin dress, embroidered with black velvet, and embellished with jet buttons detailed with pearls. When they stripped back her outer garments, Mary was revealed to be wearing a deep-crimson petticoat, a red satin bodice and a pair of red sleeves: the colour of blood, and the Catholic colour of martyrdom.

I often picture Mary in her freezing cold chambers the night before her death, kneeling before her Bible in candlelight and holding her watch. The slowness of her timepiece (like all sixteenth-century watches) would have created a 'tick' closer in pace to the beating of her heart. I like to think that, as she prepared for her execution, her watch provided her with some comfort.

What was a watch in the sixteenth century? The truth is, it wasn't much of a timekeeper. Verge watches are temperamental. They had no devices to compensate for changes in temperature and didn't respond well to shocks or sudden movements. This made them prone to stopping and losing minutes, even hours. A watch was more of an exotic rarity, available only to the wealthiest. Very few watches were then in circulation and for much of the three centuries that followed their advent, they were exceptionally expensive. In a portrait of King Henry VIII by Hans Holbein the Younger from 1536, a curious locket hangs from a gold chain round the monarch's neck, looking an awful lot like a sixteenth-century watch. Further watches are listed in an inventory of the royal wardrobes of Queen Elizabeth I, including another skull watch and what might have been one of the first wristwatches ever made. The list, drawn up in 1572, describes an 'armlet or skakell of golde, all over fairely garnished with rubys and dyamondes, having in the closing therof a clocke'. Unfortunately, this spectacular-sounding artefact hasn't survived.

The earliest watches typically fell into one of two styles that were fashionable to wear on a chain. One was more spherical, like Henlein's, and called a pomander after the cage-like balls containing fragrant incense that were swung in church. The other was housed in a flattened cylindrical case, known as a tambour from the French for 'drum', with hinged covers to the front and back. In the middle of the sixteenth century, technical advances allowed watchmakers to start exploring different case shapes, containing smaller mechanisms, as well as new techniques to decorate them. This was the beginning of form watches, so called because they mimicked the forms of other objects such as flower buds and animals. They also often doubled as devotional objects, taking the shape of crucifixes, Bibles and, of course, skulls.

Whenever I see form watches in museums or auction houses, I can't help but smile. They're such exquisite little objects. The perfect combination of intricate craftsmanship, design and engineering. If I had the freedom to create whatever I wanted and didn't need to worry about paying the bills, these are the things I would be making. They are as much *objets d'art* as watches, created collaboratively by masters of various crafts. They are often the products of enamellers, engravers, chasers, goldsmiths and lapidarists (cutters of precious stones) working together alongside conventional watchmakers, and are adorned with gold, silver, rubies, emeralds and diamonds. In these watches, function is still very much secondary to dazzling, unashamedly luxurious form. It's not hard to imagine how precious and emotive they would have been for their owners, and how they might have elicited a near-spiritual response to their cold stone or metal.

In 1912, workmen were demolishing dilapidated buildings on the corner of Cheapside in London when they saw something glittering under their feet. After carefully removing floorboards and soil, they unearthed 500 incredible pieces of Elizabethan and Jacobean jewellery – a discovery that was subsequently dubbed the 'Cheapside Hoard'. That story gives me goosebumps. One of the many magical properties of gold is that it doesn't tarnish with age or poor storage conditions. It's one of those materials that's instantly identifiable – too bright, too rich and warm in colour, and too heavy, to be anything else. Ornate vitreous enamel also keeps its colour like new, as do most gemstones. So despite spending nearly 300 years in the ground, this tangled heap of treasures would have gleamed almost as they did on the day they were buried.

At first the workmen decided not to declare their find to the property's owners, who happened to be the Worshipful Company of Goldsmiths.* Instead, they took the collection to jeweller George Fabian Lawrence, better known to London's navvies as 'Stony Jack'. They arrived with pockets, hats and handkerchiefs bulging with

* The Goldsmiths' Company is one of several wealthy London guilds that still own property all over the city.

jewels. Newspaper articles from the time describe them depositing 'many great lumps of caked earth' on the jeweller's floor, with one of the navvies exclaiming, 'We've struck a toy shop, I thinks guvnor!' Immediately spotting the significance of the collection, Lawrence set about securing the treasure for the London Museum.*

One of the most extraordinary items in the hoard was a form watch, its gilded movement set into a Colombian emerald the size of a cherry tomato. It is the only known surviving watch made with a solid emerald case. The precious green mineral has been cut in the shape of a hexagonal box to follow the prismatic shape of the raw crystal, so as to enhance its refraction of light and brilliance of colour. The dial, visible through its translucent stone lid, is of matching emerald-green enamel on a textured golden background embellished with a ring of Roman numerals. Unfortunately, the inner workings of the watch have not survived their subterranean adventure as well as their case has. Centuries of rust have fused the mechanism, tomb-like, inside its emerald crust.

What little we know about this watch has been deciphered from stereographic X-rays. The movement is typical of a finely made watch of its day, but unsigned, so we don't know who its maker was or where he was based. We know the emerald itself came from Muzo in Colombia – evidence of the international trade in luxury goods at that time – but we don't know who cut it. There were collectives of cutters capable of making a case like this in Seville, Lisbon, Geneva, and possibly London. Whoever made the case possessed astonishing skill. I trained briefly as a diamond grader, and know that to carve into a gemstone like this takes an incredible level of experience and ability. Lapidarists need an encyclopaedic knowledge of the properties of different stones. Every stone likes to be cut and finished in a different way – dictated by their molecular structure – and each one is like a fingerprint, with patterns of inclusions as unique as a snowflake. Lapidarists read the stones, studying their

---

* In 1976 the London Museum merged with the Guildhall Museum to become the Museum of London, which is where the Cheapside Hoard is now held.

structure in intimate detail, to figure out the shape of the final cut. Emerald is particularly tricky. It's part of the beryl family, and the shape of its molecules means it grows in hexagonal sticks, like a thick HB pencil. It's not as hard as ruby and sapphire, but chips easily. The larger and more precious the raw gemstone is, the greater the pressure on the lapidarist to get it right. One wrong move could chip a huge chunk out of the case or, worse, crack the whole thing in two – shattering its value at the same time.

We can be almost certain that this watch – in my view one of the finest jewels in existence – wasn't the work of one hand. Few form watches were: watchmakers in this era drew on the skills of a vast range of craftspeople to support and augment their work on the mechanism. To an extent Craig and I still do this today. Between us we have spent nearly four decades specialising in one craft, so we're painfully aware of what it takes to achieve mastery. We'll never be able to engrave, enamel or cut gemstones as well as a fellow artisan who has dedicated their life to the pursuit of their craft. If you want your work to be the very finest it can be, you have to collaborate.

Raw emerald naturally forms in hexagonal crystals.

Each time we design a watch, one of the first things we do is assemble an artisanal A-Team. Some of the people we work with are local (sometimes as local as the floor below us). Some are a few streets away. Some are from across Europe – true also of the watchmakers of the seventeenth century, although, unlike our horological predecessors, we met some of our contacts over Instagram. I have made a watch with a rock crystal case, sourcing the right stone from a dealer in London with the advice of a lapidarist in Birmingham who then cut it for me. I've collaborated with a dial enameller in Glasgow (the only commercial one left in the UK), sharing years of our research in order to help us achieve the perfect smooth finish.

I've commissioned a gun engraver in Germany to decorate our movements with ornate and minuscule acanthus-leaf scrollwork and text as small as the width of a grain of rice. Sharing your ideas with fellow specialists is a catalyst for innovation and creativity. For a job that can be so solitary, it's invigorating to be around others as dedicated to their crafts as you are to yours.

Acanthus leaf scroll engraving on a watch plate. The style has remained popular for hundreds of years – this is part of a design for a watch we made recently.

In the sixteenth century many of the best enamellers, engravers and goldsmiths worked in the small French city of Blois.* The Château de Blois was an official royal residence, and the local craftspeople catered to the royalty and nobility, who treated the city as a second home. This is where our skull watch was created. While these artisans undoubtedly earned their international reputation through their work, their renown was also assisted by their enforced movement around Europe during an era of violent religious persecution. It is powerfully ironic that Mary's watch, a potent religious talisman for a Catholic woman in an inhospitably Protestant country, was very likely made by a Protestant artisan in an equally hostile Catholic France.

---

* Blois had strong ties with the Medici family. Catherine de' Medici lived at the Château de Blois, dying there in 1589.

A great many exceptionally talented French craftspeople at that time, from goldsmiths and enamellers like those in Blois to watchmakers in Paris, were Huguenot – the popular name given to the French followers of John Calvin, founder of the protestant Calvinist Church. Catholic France under Catherine de' Medici and her sons was a brutal enemy of Protestantism. The nation's ruling Catholic aristocracy (many of whom were linked to Mary, Queen of Scots through blood or marriage, although Mary herself always practised and preached toleration between the Christian faiths) was responsible for supporting atrocities that today would be regarded as ethnic cleansing and genocide.

Trouble began for the Huguenots in 1547, when Mary's father-in-law, King Henry II of France, decided to take direct action against the Protestant threat, condemning over 500 Calvinist followers to death for heresy. In 1562, the year after Mary arrived in Scotland, her uncle, the Catholic factional leader Francis, Duke of Guise, sent his soldiers to disperse a group of Huguenots holding a forbidden religious service within the town of Vassy. In the bloody resistance that followed, many of the 1,200 Huguenots in the congregation, including women and children, were slaughtered with swords and muskets. The massacre (news of which cannot have endeared Mary to her new Protestant Scottish subjects) tipped persecution of the Huguenots into an all-out war of religion. In August 1572, another Guise-led attack saw the loss of several thousand Protestant lives at the St Bartholomew's Day Massacre in Paris. As many as 10,000 were killed in similar riots in Bordeaux, Lyon and other French cities. Over the next two decades of persecution, thousands of Huguenots fled across Europe. They were refugees, arriving with little more than the clothes on their backs and the skills in their heads and hands.

The British Museum possesses a watch made by the French watchmaker David Bouguet, dating back to around 1650, that bears evidence of such foreign talent. It is a small rounded case, the shape of a classic pocket watch, measuring just over 4.5 centimetres across, and covered with black enamel and a dense bed of brightly coloured enamelled flowers: blood-red roses, blue-and-yellow violas, variegated tulips and chequered red-and-white fritillaries connected by yellow gold and green scrolling vines. The front of the case, which covers the dial, is further embellished with ninety-two diamonds set in bands around the flowers. These old diamonds, each slightly irregular in shape, are cut in a style known as the Dutch rose. Crafted with far fewer facets than modern diamonds, whose many facets work together to reflect as much light back out of the stone as possible, they are duller but sparkle with a subtle grey fire, like droplets of water on the bonnet of a gloss black car.

More delights await inside. As you open the case, the cover above the dial reveals a painted country scene of a rambler walking with a cane, picked out in fine black linework on sky-blue enamel. The Roman numerals of the dial are in black on a white chapter ring that surrounds a full-colour miniature of two toga-clad figures in conversation amid an Elysian lakeside scene, while a flock of birds flies overhead. The decoration of the actual watch movement is, in contrast, quite restrained, though beautifully finished. It is very likely that the case of Bouguet's watch would have been created entirely by the artisans of the French city of Blois before being exported to London, where Bouguet, a newly arrived Huguenot watchmaker from France, created a movement to fit.

The Huguenot refugee community was close-knit. Its members shared knowledge, skills and crafts with their friends and passed them on to the next generation. David Bouguet's family is a perfect example of this process. Bouguet had arrived in England by 1622, and in 1628 he was admitted to the London Blacksmiths' Company. Two of his sons, David and Solomon, also became watchmakers. Another of his sons, Hector, apprenticed a Huguenot diamond cutter, Isaac Mebert/Maubert, who ended up marrying Bouguet's daughter Marie. Isaac's brother Nicholas married Bouguet's other daughter Suzanne, and another of his daughters, Marthe, married a jeweller (Isaac Romieu). I wouldn't be surprised if the diamonds on Bouguet's black watch were cut in his son-in-law's workshop.

Londoners often referred to Huguenots as 'liberties' or 'strangers', and didn't always make them feel welcome. In 1622, English watchmakers were so worried about the new arrivals that they petitioned James I to prevent them trading in the city and to establish a specialist livery company for watch- and clockmakers. The Worshipful Company of Clockmakers was duly founded in 1631, though Huguenots weren't actually excluded from it. In 1678 the Goldsmiths' Company, complaining that Huguenots were undercutting English workers and damaging their trade, attempted to prevent foreign Protestant craftspeople from working in certain places and accessing the seven-year apprenticeship that allowed them to become a member of the Company. As it happened, many of the anti-French campaigners employed descendants of the Huguenots and it became common for British craftsmen to send their sons to study with a Huguenot master or take on a Huguenot apprentice themselves. Bouguet's life must have been one of highs and lows: one minute he was working for wealthy patrons who admired, respected and valued his work; the next, he was being called a 'French dog' (or worse) in the street.

The rise of Reformed Protestantism fundamentally changed the nature of the watch. The first things to go were the signs of ostentatious wealth. Where Catholic watchmakers glorified God with elaborate decoration, the Reformed worldview saw decoration as a distraction from His true glory. John Calvin even banned his followers from wearing jewellery. This, ironically, prompted many jewellers to turn to watchmaking, leading to the rise in fine goldsmithing, enamelling and stone setting in Swiss watch cases.*

In Protestant England, the Civil Wars of the 1640s and Oliver

---

* Genevan watchmakers who wanted to make more elaborate pieces found markets for their work abroad. The small silver watch in the form of a pocket-sized lion made by Jean-Baptiste Duboule in around 1635 was very likely destined for Constantinople in the Ottoman Empire.

Cromwell's reign in the 1650s saw Puritans try to 'purify' the Church of England of every last vestige of Roman Catholicism, of which they felt too much remained under the rule of Charles I. Flamboyant dress was criticised for reeking of 'papery and develyshnes' and was seen as a symbol of pride and an incitement to lust. Everything from beard curling, imported fragrances and ruffs to cosmetics, tight doublets and presumptuously large codpieces came under attack. Even the powdered wigs that dominated fashions under Charles I were given the chop. Puritans dressed modestly, wearing sober colours, plain linen cuffs and collars (sometimes from home-woven cloth without any trimmings or buttons at all) and straight, simple hair.

A typical silver Puritan watch, plain and lacking in ornate decoration. They were a total contrast to the earlier form watches.

Watches were already such invaluable tools that even strict Puritans didn't want to give them up (Cromwell himself seems to have owned one). But they too were simplified significantly. Puritan watches were relatively small, measuring around 3 centimetres across and up to 5 centimetres in length, and often oval in shape. They were devoid of decoration or embellishment, without gemstones or fritillaries. Their cases were usually crafted from silver – gold would have been too ostentatious* – and completely plain, their smooth surfaces not unlike the gentle sheen of a weather-beaten pebble on a beach. Their dials, hidden inside the front cover, were also plain, with the exception of their purely functional chapter ring, indicating the hours of the day by a single hand.

In this new pared-back form, the watch proclaimed a very different understanding of time from the skull watch that started this chapter. The Protestants thought of time as a gift from God; to waste it was a sin. They believed that to prosper in the next life, you needed to use your time well in this one. Puritan values emphasised responsibility, self-control, hard work and efficiency. In the Puritan day there was no such thing as spare time, only time that should in fact be spent in God's service. It was even argued that 'spending time in leisure activities' was 'a form of theft, a defrauding of the master'.

In 1673, the influential Puritan church leader Richard Baxter published *A Christian Directory* in order to guide the faithful as to how a good Christian should manage his or her time:

> Time being man's opportunity for all those works for which he liveth, and which his Creator doth expect from him, and on which his endless life dependeth, the Redeeming or well improving of it, must needs be of most high importance to him: And therefore it is well made by holy *Paul* the great mark to distinguish the *Wise* from fools.

Time management, it would seem, was godliness. 'One of the greatest Time-wasting sins is idleness, or sloth,' he writes, condemning:

> He [who] spendeth his Time in fruitless wishes: He lyeth in bed, or sitteth idly, and *wisheth,* Would this were labouring: He feasteth his flesh, and *wisheth* that this were fasting: He followeth his sports and pleasures, and *wisheth* that this were prayer, and a mortified life. He lets his heart run after lust, or pride, or Covetousness, and wisheth that this were heavenly mindedness.

---

* Although, remarkably, exceptionally rare examples in gold do exist.

. . . See that ye walk circumspectly, says the Apostle . . . redeeming the time; saving all the time you can for the best purposes; buying up every fleeting moment out of the hands of sin and Satan, out of the hands of sloth, ease, pleasure, worldly business.

I've spent a large portion of my life feeling guilty: for not working hard enough, for oversleeping. Even on holiday I struggle to relax, because of the guilt of not working. I doubt that my ancient ancestors, living in caves carving star charts, felt the same sharp pang of shame when they took a moment to relax. Time guilt is rooted in social conditioning. Having a lie-in on a Sunday morning isn't going to make much of a difference to anything. Stealing a precious extra few moments in a cuddle with your child, or pausing in the garden to enjoy the sunshine on your face, might not get the work finished, or the washing-up done – but the guilt many of us feel about such activities is often completely disproportionate to their impact. Something in our cultural history has taught us to feel bad for not working. Reading Baxter on the mortal perils of enjoying a moment's pleasure, I can't help but suspect that just as the rituals of Catholicism continue to stir me, so sixteenth-century Puritanism runs through my non-believing bloodstream. Though this extreme interpretation of Christianity has been marginalised for more than 300 years, its teachings continue to infiltrate our experience of time. Puritanism spelled the beginning of the end of the 'work–life balance'.

Cromwell's Puritanical Commonwealth was short-lived. After his death in 1658, and a brief rule under his son Richard that lasted less than a year, a king returned to the English throne. Decadent timepieces predictably followed. Charles II was a great admirer of the artistry of horologists. He kept at least seven clocks in his bedroom – their poorly synchronised chimes drove his assistants to distraction – and another in the antechamber that also recorded wind direction. As his reign progressed and watchmaking flourished, he often insisted on being the first witness to the latest horological inventions.

In France, the Edict of Nantes in 1598, signed on the ascension of Henry IV, gave the Huguenots some years of relative peace. By the 1680s, however, although Louis XIV had pledged to support the Edict, a renewed campaign to purge France of Protestantism was under way. Through forced conversions, propaganda, the separation of Huguenot children from their families and the demolition of Protestant temples, life for Huguenots became increasingly difficult once again. Finally, in 1685, the 'perpetual and irrevocable' Edict collapsed. What followed was another Huguenot exodus. In the years that followed the Revocation, between 200,000 and 250,000 Huguenots escaped, while as many as 700,000 renounced their faith and converted to Catholicism. While the majority of refugees headed to the Dutch Republic, the second-most popular destination was Britain, where between 50,000 and 60,000 refugees are estimated to have fled. Switzerland also provided shelter for a great number of Huguenot settlers. These Huguenot migrants played a central role in the development of the watch industry both in the UK and in Switzerland, and their impact still resonates to this day. Ruled over by a clock-loving sovereign, and now enriched, by an influx of new talent, London was set for a golden age of watchmaking.

4

The Golden Age

Out of the right fob hung a great silver chain, with a wonderful kind of engine at the bottom . . . He called it his oracle, and said it pointed out the time for every action of his life.

Gulliver's Travels, Jonathan Swift, 1726

As a trainee watchmaker, the first thing you must do is make your tools. It makes sense that, before you're allowed anywhere near the incredibly fragile mechanics of a watch movement, you start on something robust that you'll need in the long run. My first project, on my three-year course run by the British Horological Institute, was making a 'centre cutter/scriber'. This was a pencil-like rod of steel with two different functions. I had to finish one end as a flat-headed screwdriver, only razor sharp (the scriber), and at the other I had to create three facets that come together at their tip (the cutter). This we would use to 'key' a metal surface, cutting a tiny recess to create a location point for a drill bit to bite into. Watchmakers need to drill a lot of holes.

We used our first tool to help make our next – a movement holder. Like a tiny vice, it is the bit of kit you use to hold a watch movement while you work on it. We had to file the holder by hand according to a precise technical drawing and our finished piece had to be accurate to those dimensions within three-tenths of a millimetre. This tiny amount of leeway for accuracy is called tolerance. It felt incredibly precise at the time but this was just a beginner's-level introduction to the microscopic world of watch-making. These days, we sometimes work within a tolerance of microns – thousandths of a millimetre.

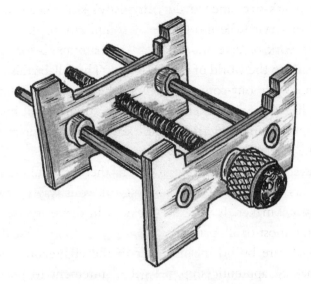

A brass movement holder designed to hold a
watch mechanism while I work on it.

In this way we inched through our training, only touching an actual pocket watch movement (which I worked on in my new movement holder) for the first time towards the very end of our first year. Several of my classmates had dropped out by this stage, fed up with only working with files and bits of metal for the best part of a year. For me, making the movement holder was like a very controlled form of jewellery and silversmithing, so I loved it. But even then I couldn't resist adding a decorative flourish to this most basic of tools. I decorated it with a scratch-brush grained finish I'd learned as a jeweller (rather than straight-grained, which is the 'proper' horo-logical way). I gilded it, and commissioned a lapis lazuli to be cut by our local lapidarist that I then set in the winding button you turn to open and close the jaws. Project number two and I was already doing things I hadn't been told to.

By the second year we were starting to make watch parts – over-size ones at first, using drawings from the syllabus as reference, as we refined our skills to work on components for the real watches on our bench that were ever smaller and more intricate. We tackled pocket watches at first, before slowly working our way down to smaller gent's-size and finally tiny lady's-size mechanisms. We learned to service the most basic of complications, like automatic winding work or date and calendar mechanisms, before gradually progressing to the world of chronographs. The syllabus also required we demonstrate our competence with the verge, the cylinder, and the English and Swiss lever escapements. Although all but the last are redundant in modern watchmaking, for a restorer they will never die out. After the foundation course, graduates could get an entry-level job in a workshop and from there it would take several more years of hard graft and luck to work your way up to being a master watchmaker. It was an education in dexterity, attention to detail and, most of all, patience.*

Things were by no means easier in the eighteenth century. A watchmaking apprenticeship, a legal requirement to practise as a watchmaker in the City of London, was seven years of almost monastic intensity (apprentices were not allowed to marry while training). This would then be followed by perhaps two or three years as a journeyman watchmaker honing your skills, until the completion of a 'master-piece' – a complete watch from scratch – earned you the title of watchmaker.

Highly skilled and inventive, the best watchmakers were in fierce demand and had begun to enjoy high status and renown. This was the golden age of English watchmaking, when Europe's most brilliant watchmakers bounced ideas off one another and vied to advance the accuracy and complexity of the watch. Many of the famous watch- and clockmakers of the day would have been known to each other through the Worshipful Company of Clockmakers, a livery company established in 1631. In archival documents from this

---

* I didn't know then that I would be one of the last to receive this training. A few years later the course was axed in favour of a more theory-heavy academic BA in horology.

era, I've found the names of the greats all listed together as signatories of the same letters from the Company, like a roll-call of the celebrated names in horology: Thomas Tompion, regarded as the 'Father of English Clockmaking', who worked with Robert Hooke to create some of the very first watches with balance springs; George Graham, Tompion's pupil and successor (he married Tompion's niece, Elizabeth Tompion), who, when he wasn't busy making scientific instruments for Edmond Halley, found time to invent the orrery and made considerable improvements to pendulum design; Daniel Quare, master of the repeating watch, and Thomas Mudge, once George Graham's apprentice, George III's Royal Watchmaker, whose lever escapement was quietly revolutionary. These are the celebrities of the watchmaking world – a workshop employing them all would be the equivalent of a Premier League fantasy football team. In a century that saw the invention of the piano, the steam engine, the hot-air balloon, the spinning jenny and the steamship, the watch kept pace, proving vital to solving some of the most pressing scientific questions of the age.

By the start of the eighteenth century timekeepers were physically, if not always financially, accessible, and familiar to all. The vast majority of English parishes now possessed a public clock in their church tower and clocks had begun to appear in inns, schools, post offices and almshouses. By the end of the century there would be clocks hung in every pub and tavern throughout the British Isles. As you walked the streets of cities such as London or Bristol a clock was never far from sight or hearing. Clocks were increasingly found in domestic homes – even servants, who still could not afford to own a clock themselves, would have been used to seeing them. For those who owned domestic clocks, the most popular location for them was the kitchen – one of the few rooms that every home has, regardless of your wealth and status.

Time didn't only enter public consciousness but became the subject of intense philosophical debate. While Isaac Newton believed that time was '[a]bsolute, true, and mathematical', others such as David Hume and John Locke argued that it was relative – that time essentially depended on the people perceiving it (these ideas were famously taken further in the twentieth century by Einstein's theory of relativity). In the same period, the writer Laurence Sterne gleefully played with time in his masterpiece, *The Life and Opinions of Tristram Shandy, Gentleman*, crafting a narrative in which time contracts, expands, and goes backwards as well as forwards.

Watches, from the second half of the seventeenth century onwards, had occupied an increasingly important (if not especially useful) role in the personal lives of the wealthy. Samuel Pepys, picking up his new watch from Briggs the scrivener ('and a very fine watch it is') in May 1665, was no less enamoured, distracted and ruled by his new watch than any of us might be with a new smartphone. 'So home and late at my office . . .' he wrote, as he so often did,

> But, Lord! to see how much of my old folly and childishnesse hangs upon me still that I cannot forbear carrying my watch in my hand in the coach all this afternoon, and seeing what o'clock it is one hundred times; and am apt to think with myself, how could I be so long without one; though I remember since, I had one, and found it a trouble, and resolved to carry one no more about me while I lived.

Two months later his watch was already back at the menders. Accuracy, for the watch, was still a work in progress.

Luckily for Pepys, the late 1600s saw two remarkable leaps forward in horological progress that kick-started a century obsessed with time. In 1657 the Dutch mathematician Christiaan Huygens invented the pendulum clock, successfully applying Galileo's theory of isochronism of 1637. The theory is that a pendulum will take the same duration to swing regardless of how big that swing is. This consistent swing triggered the equally consistent hold and release of the escapement, forging the way for the most accurate kind of clock that had ever been invented. The long pendulum swinging underneath lent itself to the longcase design, or 'grandfather' clocks, which became increasingly popular in the decades that followed. In 1675 polymath and scientist Robert Hooke invented the metal hairspring, or balance spring as it is also known, as revolutionary for the watch as the pendulum had been for the clock. A flat steel spiral of very fine wire, the metal hairspring* was designed according to the principle of Hooke's eponymous law of elasticity – '*ut tensio, sic vis*', meaning 'as the extension, so the force'. Force exerted on a spring causes an equal return of force by the spring. If I tighten a coiled spring beyond its resting position and then release it, the spring will fling itself outwards. But as it flings out, the force will cause it to fling out a bit too far and then want to coil back again in order to return to the nice comfy spiral in which it was first formed. This action causes the spring to 'breathe' in and out as it over- and under-coils rhythmically with the oscillation of the balance wheel, triggering a consistent release of the escape wheel teeth. The effect on accuracy was dramatic. It was this invention that, for the first time, made it worthwhile adding a minute hand to a watch – a milestone in the history of mechanical watchmaking. Metal hairsprings were quickly welcomed by an increasingly watch-loving public and it was not unusual for people to have metal hairsprings retrofitted to their older, pre-hairspring watches, which had little or no regulation. Pepys was soon timing his walks between Woolwich and Greenwich to the minute.

The City of London was the ticking heart of the watchmaking world at this time. In 1665, the Great Plague had devasted the population of London (in eighteen months it killed almost 100,000 people, nearly a quarter of the city's inhabitants), but our Huguenot artisans, fleeing the Revocation of the Edict of Nantes in 1685, had swelled its numbers. By the eighteenth century, watchmaking workshops in England were usually small collectives of journeymen and apprentices, all led by a master and supplemented by the work of

---

* Metal hairsprings replaced the literal hog's-hair bristle that preceded them.

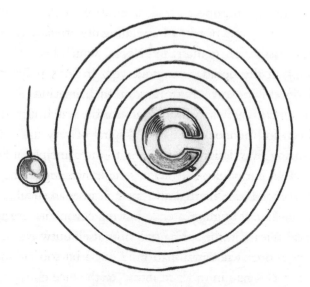

Early hairsprings were formed as a flat spiral. The collet at
the centre is used to secure it to the balance staff.

several other local craft workshops such as goldsmiths, engravers, chainmakers and springmakers. When I take apart a watch from this era, I can often count as many as four or five, or even more, different makers' marks and other signatures, concealed throughout the entirety of the movement, inside and out. In this way I can mentally piece together the creative colonies that existed in watchmaking centres like Clerkenwell in London, with watches and watch parts being shifted around multiple workshops all within a few streets' radius of each other. When I study old maps, looking up the addresses of registered makers, I can't help but notice there's often a tavern or inn in the middle of it all, and I like to imagine these craftspeople huddled together over jugs of ale in a smoky tavern discussing business and sharing ideas. There are still echoes of this in Birmingham's Jewellery Quarter, where I work today. Although local boys no longer run down the road to the assay office with wheelbarrows piled with gold jewellery to be hallmarked, we makers still look out for each other. We all know each other's business and sometimes like to meet for a pint of real ale.

Watch tools, components and later even whole movements (though not yet cased and ready for retail) were often made in workshops in Lancashire, particularly in Prescot, 8 miles east of Liverpool, where ample coal supplies, a long-standing tradition of metalworking and a good transport link to London fostered a cottage industry of suppliers, but the most advanced branches of the trade were concentrated in London.

Outside London, watch- and toolmakers rarely completed an apprenticeship, but within the city, the tightly controlled apprentice system created valuable opportunities for those trainee watchmakers who were lucky enough to find a position. Master watchmakers and their formal apprentices would move around different cities, widening their network of potential patrons. Their watches were charged at prices based as much on their social status as the quality of their work, so watchmakers (and their apprentices) from wealthier backgrounds were set up to be more successful from the start. It was a bit like the watchmaking version of public school – it was not just the education but who you met that conferred the advantages.

Fifteen-year-old Thomas Mudge, the son of a headmaster in Devon, arrived to take up his apprenticeship with the renowned clock- and watchmaker George Graham in the spring of 1730. On gaining his freedom of the Clockmakers' Company in 1738, he took lodgings and spent the first part of his career in the shadows, making exceptionally complicated watches on commission to other watchmakers, signing his works with their names – a standard practice at the time. He might have carried on in this way indefinitely were it not for a watch he had made for another celebrated maker of the day, John Ellicott. The watch, which could display the equation of time (the difference between the true solar day, which changes according to the position of the sun, and the mean or average solar day) with a range of additional calendar indications, was sold to King Ferdinand of Spain. But, so the story goes, someone in the royal court dropped the watch on the floor, damaging it so badly that it had to be sent back to Ellicott, who was unable to fix it. He was forced to send it back to its actual maker, Mudge, who undertook the necessary repairs. When King Ferdinand found out, he insisted on commissioning Mudge directly in the future. The king's patronage lifted Mudge from obscurity, encouraging him to make the leap to setting up his own workshop at 'the Dial and one crown' on Fleet Street in 1748.

Mudge became renowned for the mechanical innovation of his watches. He made King Ferdinand of Spain a grande sonnerie watch set into the top of a walking cane. Grande sonneries are considered to be one of the finest complications as their mechanism chimes both the hour and the quarter-hour, but can also chime out the time on demand if the owner wants to hear it in between. Mudge was also famous as the first watchmaker to integrate a perpetual calendar – so called because it compensates for variations in the duration of months and years to 'perpetually' show the correct day and date – into a watch. Thomas Tompion and George Graham had applied the perpetual calendar to a clock as early as 1695, but the process of shrinking the mechanism down into something small enough to fit into a pocket watch is credited to Mudge.* Just as powerful patronage had supported the development of the first mechanical clocks in the days of Su Song, so the technical advances of these ever-more intricate watches were bankrolled by the wealthy who desired them. It was the

---

* Perpetual calendars are possibly one of the most discreet, yet complex, horological indications invented. They take something we might take for granted on a modern watch, a date display, and turn it into a date that is displayed with accuracy in (almost) perpetuity. To do this they employ a series of gears that can count not only the days and months, but years and leap years. They memorise the number of days in each month for us in the teeth of their wheels – the slowest of which turns once every four years. Looking at one of Mudge's surviving pocket perpetual calendars, dating from around 1762, the care he took to ensure that the dial was functional and easy to read is evident. The date is marked by a gold marker set above the twelve-o'clock position, which sits against a rotating disc of dates on the very outer rim of the dial. In the centre of the dial the moon phase is shown above two crescent apertures that look onto dials showing the day and month. February has its own dial within a dial indicating whether it is a leap year. And yet, despite the level of detail the watch is carrying, nothing is fussy.

equivalent of investing in a tech start-up. Your investment helps the business innovate, and that innovation boosts the value of your shares. Only instead of shares you receive a very lovely watch you can wear as it (ideally) increases in value along with the renown of its maker.

Mudge's correspondence with another of his patrons, Count von Brühl, a Polish-Saxon statesman reputed to have owned the largest watch collection in Europe, gives us an insight into the mutual rewards of the patron–craftsman relationship. The two were in regular contact throughout the making process. In his letters to von Brühl, Mudge goes into a surprising level of technical detail, discussing engineering and material principles, temperature coefficients and other technical challenges he faced. For many horological patrons, commissioning a watch was more than just the purchase of a beautiful thing; it was a dynamic relationship in which the patron invested as much in the process as he or she did in the finished product. The patron often had a genuine desire to fully understand their watch's workings and feel part of its creation.

In our workshop, the involvement of collectors and patrons is still a key aspect of the making process. Working bespoke not only gives us the opportunity to meet a specific brief but also to iron out any requests and alterations while a watch is being made. We've remade and adjusted the size of winding crowns to suit the ergonomics of our clients' wrists, adapted them if they have arthritis and struggle with winding a watch, and adjusted them depending on whether they wear their watch on the right or left. We've altered colourways and proportions to refine legibility, we've repeatedly modelled and remodelled cases for clients who give us feedback on how they feel compared to their other watches. The client's involvement in the many mini-decisions that are part of the creative process is immensely useful for us, but it also makes space for them to become part of our workshop. They become as much a part of the finished product as our own hands have been.

The patron or collector who commissions a watch also brings a wider perspective to us as makers. One of the challenges of being a watchmaker is that we rarely earn enough to be able to afford the objects we make. The result is that we don't tend to have significant collections ourselves, certainly not at the level our patrons do. It's why I find our clients' input so valuable: they're the ones out there in the real world looking for pieces; they know what they choose and why, what pieces are actually like to own and use on a day-to-day basis.

In Mudge's era, a patron was also keen to be associated with the very latest scientific developments. Mudge's association with von Brühl led to a purchase by King George III, who in 1770 commissioned Mudge to make a watch for his wife, Queen Charlotte, featuring the earliest known example of Mudge's most ground-breaking invention, the detached lever escapement. The king, like von Brühl, took an active interest in the watches and clocks he commissioned, and was something of an amateur horologist himself. There are manuscripts in the Royal Collection written in the king's hand in which he describes the process for assembling and dissembling watches. Queen Charlotte too was an avid watch collector and something of a magpie. Her friend, the diarist Caroline Lybbe Powys, recalled seeing 'twenty-five watches, all highly adorn'd with jewels' in a case beside the queen's bed at Buckingham House in 1767. I can't help but read into the fact that she kept her collection in such a close and personal proximity, falling asleep by them each night and waking up to them every morning. Queen Charlotte clearly loved her watches.

The 'Queen's watch', as Mudge came to call his watch for Queen Charlotte, would certainly have won King George brownie points. Mudge later described it as 'the most perfect watch that can be worn in the pocket, that was ever made'. From a watchmaker's perspective, it is Mudge's detached lever escapement that is the true star.*

One of the great enemies of accuracy in the watch is friction, because it upsets the precision of the escapement. In the verge

---

* Mudge first developed the lever escapement in 1754, but it appeared for the first time in his Queen's watch.

escapement, the oscillating balance wheel was in near-constant engagement with the train wheels, creating varying friction. Mudge's great innovation was to invent an escapement that was 'detached' from the balance wheel. His lever worked by being knocked back and forth by a pin fixed to the underside of the oscillating balance wheel. As the wheel, and pin under it, swung back and forth they'd flick one end of the pivoted lever back and forth at the same time. At the other end of the lever, a pair of pallets would catch and release a tooth of the escape wheel with each 'tick'. This meant that the oscillating balance was exposed to friction for only the very briefest moment as the lever was flicked across.

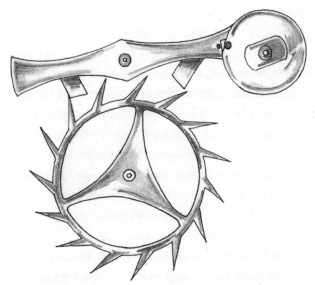

The English lever escapement – the commercial adaptation of Thomas Mudge's detached lever. The design was refined in the nineteenth century to form the Swiss lever escapement (spot the national rivalry), which is still used in almost all mechanical watches to this day.

Mudge himself was modest about the potential of this invention, if only because of its extreme technical intricacy. In a letter to von Brühl, he declared the level escapement:

requires a delicacy in the execution that you will find very few artists equal to, and fewer still that will give themselves the trouble to arrive at; which takes much from its merits. And as to the honour of the invention, I must confess I am not at all solicitous about it; whoever would rob me of it does me honour.

I'm not surprised that they gave Mudge a headache. Making lever escapements requires a high level of accuracy, and is challenging to pull off even today, when we have the benefit of modern engineering equipment. Yet we still do honour him by 'robbing' it. To this day, virtually every mechanical watch made in the world uses his invention.

This kind of high-level technical innovation meant that watchmakers began to play a crucial role in the development of other industries. Horologist Samuel Watson helped physician Sir John Floyer (1649–1734) design, make and sell the first pulse watches to aid physicians in counting their patients' pulses. Thomas Mudge created a watch for John Smeaton (1724-1792) with temperature compensation, to even out the expansion and contraction of metals as they move between hot and cold temperatures. Smeaton, the world's first self-proclaimed 'civil engineer', is famous for his work developing new forms of cement, which he used to improve the way Britain constructed its lighthouses. He used his Mudge watch to assist him in his survey work. (A device to compensate for changes in temperature also appears in Mudge's watch for Queen Charlotte). In 1777, watchmaker John Wyke supplied all the wheels, pinions and framework to make the first pedometer, designed by Matthew Boulton, a kind of antiquarian Fitbit capable of counting the wearer's steps. Watch- and clockmakers were recruited to help refine and maintain factory equipment. According to a 1798 account from a factory in Carlisle, '. . . the cotton and woollen manufactories are entirely indebted for the state of perfection to which the machinery used therein is now brought to the clock and watch makers, great numbers of whom have, for several years past . . . been employed in inventing and constructing as well as superintending such machinery . . .' But the most famous horological challenge of the eighteenth century, an age in which the British navy expanded exponentially, was that of the 'quest for longitude' – the essential navigational coordinate that until now had relied on the stars, dead reckoning, a sand-timer and guesswork.

The starter gun that announced the beginning of the British quest for longitude was fired early on in the century, triggered by one of the worst maritime disasters in British naval history. On a foggy night in 1707 four Royal Navy warships under the command of Sir Cloudesley Shovell were wrecked on rocks off the coast of the Isles of Scilly. The warships, on their return from Gibraltar after besieging the port of Toulon in France, sank, and most of their crew drowned, their bodies washing up onto the surrounding coast for days afterwards. As many as 2,000 men were lost. The cause of the tragedy was deemed to be a fatal combination of poor visibility and a mistake in their course caused by plotting their longitude incorrectly, meaning that they were completely unaware of the approaching hazard.

The ships had struggled to find their exact location or 'navigational position'. On land, the process of establishing a navigational position is relatively straightforward, using landmarks as reference. At sea navigators were more, well, *at sea*. Latitude – how far north or south of the equator you are – could be determined from the position of the sun in the sky. But to calculate the east–west position, known as longitude – how far you've travelled through imaginary lines that run from pole to pole – a navigator needed to be able to calculate speed and course from a given position at a given time (usually the home port and the time of departure). Wind, currents and tides all affected calculations, while motion and temperature could affect the accuracy of any timekeeper, with potentially fatal results.

The origins of pinpointing longitude date back thousands of years. Some of the stars we still use as navigational markers today are mentioned in Homer's *Odyssey*, in which the goddess Calypso tells the hero, Odysseus, how to steer his ship on a steady course to Phaeacia by keeping the stars of the Great Bear on his left. The credit for devising longitude at sea most likely originates with the Polynesians, who for thousands of years had been masters of oceanic navigation. Tupaia, a Tahitian Polynesian navigator recruited by Captain Cook in 1769 during his expedition onboard HMS *Endeavour* to map Terra Australis Incognita, astonished the crew with his almost instinctive ability to know precisely where they were using the stars and dead reckoning. He was also able to draw a now-famous map of vast tracts of the Pacific from memory – roughly the area of Europe including European Russia – including the names of seventy-four islands, and give detailed accounts of the complex Pacific wind system. What I find most striking about this is that the means by which Polynesian navigators like Tupaia found their way around the Pacific Ocean were natural, just like those we first used to discover and measure time. By the eighteenth century, however, Europeans had become so disconnected from the natural world surrounding them that they needed a machine to help them.

The Scilly naval tragedy proved a catalyst. In 1714 the Longitude Act announced a prize of £20,000, about £1.5 million in modern money, to the person who could solve the longitude problem. It laid down a challenge for Britain's greatest minds across the fields of science, engineering and mathematics. The committee sought advice from Isaac Newton, by then the mature age of seventy-two, and his friend Edmond Halley, whose travels mapping the stars made him an obvious choice. When Newton presented his remarks on the task ahead to the committee, he listed the methods that were currently in existence, albeit 'difficult to execute'. One method, he said, 'is by a Watch to keep time exactly. But, by reason of the motion of the Ship, the Variation of Heat and Cold, Wet and Dry, and the Difference of Gravity in different Latitudes, such a watch hath not yet been made.' And nor, in his opinion, was it likely to ever be.

Newton and his contemporaries felt sure the solution lay in astronomy – perhaps in studying the eclipses in Jupiter's satellites, or in predicting the disappearances of stars behind our moon, or by observations of lunar and solar eclipses. There was also the possibility of a lunar distance method, where longitude could be calculated by measuring the distance between the moon and the sun by day, or the moon and navigational stars by night. This feedback was used to dictate the terms of the prize, which awarded a first, second and third place to would-be applicants from any scientific or artistic discipline for inventions judged purely on their degree of accuracy when tested 'over the ocean, from Great Britain to any such Port in the West Indies as those Commissioners Choose . . .' (In other words, the Caribbean–UK leg of the transatlantic slave trade triangle . . .).

With the most renowned minds of the day called to action, no one expected the answer to come from an unapprenticed clockmaker from Yorkshire, or to arrive in the shape of a watch.

At first glance, H4 looks like a typical pocket watch of the era, although with an overall diameter of 16.5 centimetres, you would struggle to find a pocket large enough to fit it. Aesthetically too, it looks similar, with a plain polished silver case and a dial made from white enamel and marked with black numerals bordered by decorative scrolls of acanthus leaves detailed in black linework. But this was no ordinary watch. Weighing nearly 1.5 kilograms (close to four tins of baked beans), it contains an extraordinary movement.

Completed in 1759, H4 is the fourth of John Harrison's five experimental marine timekeepers, made to satisfy the demands of the Board of Longitude. Its predecessors, H1, 2 and 3, had been large, unwieldy clocks, though of sufficient promise and technical brilliance to warrant the invaluable interest and support of George Graham, Mudge's old master. Even with Graham's backing,

Harrison's labour was a dogged, solitary effort. He was his own harshest critic and, discerning a fault in H2, would not allow it to be tested. Twenty years elapsed between H2 and H3, as Harrison struggled on, beset by technical difficulties. With H3, lighter than its predecessors and comparatively compact at 2 feet high and 1 foot wide, it seemed as though he had pushed the size of a sea clock to its limits.

Thus the arrival of the comparatively diminutive H4 just a year after H3, and in the form of a watch, came as something of a surprise. Harrison's H4 mechanism involved unprecedented engineering. While he used mechanisms that were already in watches at the time, such as the verge escapement, he refined them to a level rarely seen even now. To reduce friction and improve durability, the steel flags that make the entrance and exit pallets of the verge were made from diamonds. The round balance wheel is huge compared to a normal eighteenth-century watch, but this technical adaptation makes it less susceptible to variations as the watch moves with the motions of the ship. A longer mainspring gave the H4 a running time of thirty hours from full wind.

H4 was first put to the test in 1761 when HMS *Deptford* set sail from Portsmouth, headed to Kingston, Jamaica. Harrison sent his timekeeper to travel on board the ship, along with his son William as its minder. H4 was quick to make an impression, helping to correctly calculate the time of their arrival at the port of Madeira on the outward journey, earlier than the crew's own predictions. The captain was so impressed, it's said he offered to buy Harrison's next timekeeper on the spot. Over the journey, which lasted eighty-one days and five hours, H4 lost a mere 3 minutes and 36.5 seconds. It was considered to have met the Board's exacting requirements – subject to a second test. While Harrison wrangled with the Board to meet conditions that seemed to become ever stricter, H4 became a template for its successor, H5.

Today, Harrison's first marine chronometers lie in state at the Royal Observatory in Greenwich. In their decoration, H4 and H5 are far plainer than personal pocket watches of their day. Harrison's

H4 still bears some of the characteristic acanthus-leaf scroll engraving and pierced embellishment that was popular at the time, but by the start of the nineteenth century this ornamentation had been lost from the production of chronometers.

It seems that the more accurate and functional the watch, the less it needed ornate decoration to justify its existence. Nevertheless, these first scientific instrument watches are exquisite, just in a different way. Their beauty is in their functionality. I have to remind myself that they were made without any of the modern technology I have access to today, and yet they are still more accurate than some run-of-the-mill mechanical watches currently on the market. And, as I handle them, I am awed by the care that has been taken in their finish. I often notice a technique once known as chamfering, and now commonly referred to in the Swiss-French as *anglage*, where sharp corners are carefully filed to a 45-degree angle around the plates, bridges and even tiny spokes of every wheel in the train to make them appear lighter and more refined. These chamfers are brightly polished, in contrast to the grained or frosted flat surface next to them, so they glisten as you move the mechanism in the light. Although they can now be done perfectly by machine, you can tell when they've been done by hand as they catch the light differently. I'm also excited to see what we watchmakers call *black polishing* on some of these early chronometers.

Black polishing is typically only done on very hard metals like steel. When the surface has been polished to a mirror-like state of perfection, without a single scratch or mark, it appears as black as onyx when it catches a shadow. Black polishing was, and still is, done by hand and is immensely time-consuming. Polishing parts is useful to reduce friction, but polishing to this level is a demonstration of sheer skill by the maker. Even as precision and functionality became the singular purpose of these timepieces, it delights me to see how their makers found ways to include their own personality and identity in the finishing of their work.

For me, chronometers are one of the purest examples of how the watch is so much more than a piece of scientific equipment. It is a piece of scientific equipment crafted by human hands, sometimes over a period of years. Their finishing is as idiosyncratic as a signature, a mark of pride that reveals the maker's personal investment in the work, one that goes beyond pure function and makes it a work of art.

As a student watchmaker, I was raised with the belief that John Harrison was a horological hero whose invention had saved incalculable lives at sea. In many ways that was true; Harrison *was* a brilliant inventor and his timepieces had a dramatic impact on navigation. But the H4 was not altogether the solution it is fabled to be.

In 1831 HMS *Beagle* carried twenty-two watches and chronometers, as well as a young and enthusiastic university graduate named Charles Darwin, onboard her second voyage to survey the coasts of the southernmost parts of South America. By the time she returned home in October 1836, only half of the timekeepers were still in good working order.

One of the problems early chronometers faced was accumulated inaccuracy over very long distances. If the inaccuracy was consistent — say, if the navigator knew the chronometer was gaining exactly five seconds a day — this would be easy enough to calculate and compensate for, but it was rarely that straightforward. In 1840, Henry Raper, a British navy lieutenant and authority on navigation who normally extolled the virtues of chronometers, observed how 'Chronometers are generally found to perform best at the beginning of their voyage; many subsequently become useless from irregularity, and some fail altogether. They are liable, also, to change their rates suddenly, and then to resume the former rate in a few days.'

The rigours of the open ocean posed further challenges. Most early-modern navigational instruments, chronometers included, suffered varying degrees of inconsistency caused by the dramatic changes in temperature, salty air, humidity and even magnetism from the many iron goods on board a ship.*

To keep them protected, chronometers were kept in wooden boxes, which were susceptible to warping. This meant that when the time came to wind the chronometer it was sometimes comically impossible to open the box. Their winding and monitoring was restricted to experienced ranking officers, and the boxes were locked to prevent curious hands from tinkering with anything. The lock introduced the potential for that most inbuilt and fundamental of human errors – losing keys. Astronomer William Bayly, who served Captain Cook onboard the *Adventure* on his second voyage through the Antarctic and Pacific, reported several occasions when he had to rescue the ship's chronometer from incarceration. The first incident occurred when an officer accidentally bent one of the catches in the lock, so the catch needed to be sawn through and repaired. Shortly after, a key snapped off inside the lock. Then, a month later, Bayly had to break in a third time after someone left the ship with the key.

And no amount of temperature regulation or spare keys could guard against the grave threat of the ship's cat. One of the most famous cats in exploration history, Trim belonged to Captain Matthew Flinders (although, from my experience of living with cats, it is perhaps more accurate to say Captain Flinders belonged to Trim) and accompanied him on the first circumnavigation of the continent now known as Australia. Flinders writes how:

> Trim took a fancy to nautical astronomy. When an officer took lunar or other observations, he would place himself by the time-keeper, and consider the motion of the hands, and apparently the uses of the instrument, with much earnest attention; he would try to touch the second hand, listen to the ticking, and walk all around

* Although, I should say, Harrison worked hard to accommodate temperature change and among his considerable contributions to the history of horology are huge advances in our understanding of, and compensation for, temperature variation.

the piece to assure himself whether or no it might not be a living animal.*

Cats aside, the chief challenge to the viability of early chronometers was the cost.† Although prices reduced over the years, at the end of the eighteenth century chronometers still cost between £63 and £105, which when compared to the maximum annual wage of a Royal Navy lieutenant – £48 – was still completely out of most sailors' reach. And even when they could afford them, navigators continued to use the old methods to calculate their whereabouts – a chronometer alone was just not reliable enough. They were often used in conjunction with sextants, hand-held instruments that use mirrors to measure the angular distance between two objects. Sextants helped sailors to make celestial observations, which remained an essential means of measurement, especially when sailing into the unknown. They would also use them to check the chronometers' timekeeping against astronomical references whenever a crew could make land (and find a level stationary surface from which to make very precise measurements). This became an indispensable back-up for the chronometer, almost like a factory reset.

For me, Harrison's greatest achievement wasn't the creation of H4, or whether he won the Longitude Prize. (The Board of

---

* Trim, it appears, was a little more restrained than the ship's cat aboard HMS *Discovery*, which under the captaincy of George Vancouver set about mapping the west coast of North America between 1791 and 1795. Vancouver's exploration was assisted by a number of chronometers, an astronomical regulator clock, sextants and a pocket watch with a seconds hand. It was this pocket watch that unfortunately fell foul of a curious kitten at the beginning of the voyage. Excusing the mischievous cat for breaking one of the ship's state-of-the-art timekeepers, the exploration's astronomer William Gooch wrote that 'she is a very young cat & perhaps its beating attracted her notice'.

† These early chronometers were not cheap. In 1769, watchmaker Larcum Kendall was paid £450 to make the first replica of H4. Called K1, it took him two years to complete, and he received a further bonus of £50 when it was finished. That total of £500 was just under a fifth of the purchase price of the entire HMS *Endeavour*, which had been bought by the navy for £2,800.

Longitude, equivocating to the end about whether he met the terms of their prize, eventually gave him a further final award of £8,750.) It was demonstrating, for the first time, that a mechanical time-keeper – a watch, no less – had the capacity to solve one of the greatest problems of the day.

Despite its limitations, the chronometer was a valuable tool that, in combination with other methods, made charting the world possible for all. Not only did the chronometer help sailors to find their way through the vast open oceans, but those navigations enabled geographers to map the world more accurately. Imagine how dangerous it would have been to travel from, say, Portsmouth to New York without knowing exactly what shape the United States' coastline is, or where precisely the port you're heading to is located. The world as we know it today was made through the adventures and exploration of the eighteenth and nineteenth century – exploration that, in many ways, was made possible through advances in timekeeping.

But this expansion of our horizons had a dark side. Antiquarian horologists have not always acknowledged the role some of these inventions played in more damaging developments. When you read about the history of the Board of Longitude, the focus is usually on the preservation of the lives of sailors making perilous long-distance voyages, and on the benefits the chronometer brought to cartography. Little mention is made of the other interest Western nations had in perfecting transatlantic trade in the eighteenth century, and how improvements in navigation aided and refined the systematic enslavement of millions of Africans and their subsequent transportation to the Americas. There is rarely an acknowledgement of the impact being 'discovered' had on indigenous Australians, or how naval advantages assisted in the colonisation of India or South America. And although the Board of Longitude existed to support the Royal Navy, because chronometers were so expensive they were far more likely to be used by highly profitable commercial trading bodies like the East India Company, which relied on slave labour and trafficked slaves from East and West Africa. In 1802, thirty years after the invention of the chronometer, just 7 per cent of Royal Navy vessels had actually been supplied with chronometers.

The leaps in accuracy achieved for the chronometer made a different kind of watch – and time – possible. And for this, Harrison should share his crown with Mudge. While Harrison's legend over-shadowed pretty much all of his golden age contemporaries, in fact no maker has left as lasting a mark on the advancement of the watch movement as Thomas Mudge. Mudge's lever escapement watches turned out to be more accurate and more reliable even than Harrison's. Had Mudge succeeded a few years earlier, he likely would have beaten Harrison and claimed the prize, and lasting celebrity, for himself. Nevertheless, his legacy lives on in the watch. It is the descendant of Mudge's lever escapement that is still ticking on your wrist.

5

Forging Time

After all we are a world of imitations; all the Arts that is to say imitate as far
as they can the one great truth that all can see. Such is the eternal instinct in
the human beast, to try & reproduce something of that majesty.

Virginia Woolf, in a letter to a friend, 1899

In 2008, I was a cataloguer at an auction house in Birmingham. After
graduating from horological college I had worked my way up from
the bottom to become their chief watch specialist. The job was always
varied and surprising. Some days I'd be dealing with infinitely precious
privately owned pieces making a rare outing from their home in a
bank vault, other days I'd be rifling through boxes of jumbled junk,
trying to identify anything with potential value. One morning, in a
box of antique silver tableware, I came across a silver pair-cased watch
with a hallmark dating it to 1783. The inner case housed the mechanism and had a hole in the back for winding. On the front, the dial was
visible under a domed bullseye crystal, named after its large eyeball-like
profile. This all sat snugly inside an outer, more durable case that
protected the delicate mechanism from the elements. I held the watch
up to my Anglepoise light and looked through my eyeglass. The dial
was an ornately engraved sheet of silver, with inlaid black wax numerals – a technique known as *champlevé*. Its centre had been pierced with
a delicate acanthus-leaf scroll detail to reveal the bright flash of blued
steel shim underneath. Just above the centre of the dial, framed in a
decorative scroll, was the watchmaker's name: John Wilter.

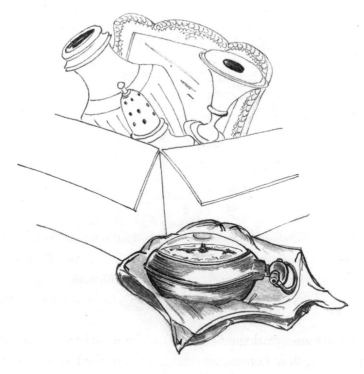

A cardboard box of silverware brought to the auction
house for valuation and cataloguing.

So far, I hadn't spotted anything hugely noteworthy. I removed the
outer case and opened the inner case to reveal the movement, which
was gilded brass and also engraved and decorated with areas of pierced
detail. It too was signed 'John Wilter, London'. My curiosity was now
piqued. The design was highly unusual for an eighteenth-century
English watch. The dial was arcaded (meaning that the minute track,
which is normally a perfect circle, had scalloped arches surrounding
each of the numerals) – a motif popular in Dutch clocks of the time,
but almost unheard of in England. The movement's components were
likewise crafted in a continental style and were of a lower quality than
you would expect to see in genuine London work. I pulled my bible
– Loomes' *Watchmakers and Clockmakers of the World* – off the shelf and
leafed through its pages until, eventually, I found the relevant entry:
'Wilter, John – perhaps a fictitious name'.

77

An arcaded *champlevé* watch dial signed 'Wilter, London'.
The acanthus scroll work in the middle was skeletonised
to reveal blued steel underneath.

It turned out that this was a type of watch I'd never heard of before: a so-called 'Dutch forgery'. These watches were low-quality fakes that most horologists either ignored or condemned. They typically purported to be English watches but were Dutch in style. Why, I wondered, would someone be forging an English watch in the Dutch style? And who was John Wilter? I found no evidence that anyone, let alone a watchmaker, with that name even existed at the time. I didn't know it then, but I would spend a whole decade of my life searching for John Wilter.* My quest would teach me a great deal about how watches became accessible to all.

I began my research at the Horological Study Room of the British Museum. I've spent countless hours there volunteering and researching; in many respects it's my spiritual home. To get there I had to weave my way through throngs of tourists browsing the galleries,

---

* John Wilter and Dutch forgery watches became the subject of my PhD.

and a muffled hum of voices in other languages. I confess, I always loved being watched by these curious visitors as I sidestepped the security barrier and unlocked the imposing oak double doors that led out of the gallery. It's strangely empowering to open a door that's many times taller than you.

The marble floor of the British Museum's Great Court is inscribed, in jet-black stone, with a quotation from Tennyson: 'And let thy feet millenniums hence be set in midst of knowledge.' For a tactile, object-obsessed thinker like me, the British Museum really is a temple of knowledge. No matter how many times I opened those doors it felt like I was opening the gates to a secret world – a vast network of hidden treasure.

Great museums are like icebergs, with only a tiny fraction of their vast collection visible to the public. As I closed the grand doors behind me with a reassuring thud, I left the tip of the iceberg behind me. The acoustics immediately changed. I walked down long, quiet corridors, flanked by floor-to-ceiling cases filled with antique books, behind rippled antique glass. As I descended into the basement, the temperature dropped. The walls were now covered with bright white tiles, an aesthetic somewhere between a London Underground station and a Victorian hospital. Here, between cabinets of Bronze Age pottery, was my destination: an inconspicuous door with a buzzer. I was met by the curator and escorted into a room lined with grandfather (or longcase, to use the correct name) clocks where, at the far end, were two long rows of mahogany cabinets positioned back to back. Those cabinets housed the museum's 4,500 watches, arranged in hundreds of specimen drawers. The collection spans the whole history of the watch, from its invention in the sixteenth century to the present day. It contains the work of almost every well-known maker as well as pieces made by artisans whose names have been lost to history. It also contains dozens of Dutch forgeries.

The horology collection is the only collection in the museum where conservation work is carried out by its curators, as horology is one of those rare subjects where theorists have to be practitioners in order to care for the objects of their study. I've found several kindred spirits among these curator-conservators. When I started volunteering in 2008 the then head curator David Thompson, who would go on to supervise my PhD, became my informal mentor. David had studied at the old Hackney College Horology School, which closed in the late 1990s, and spent thirty-three years working with the British Museum's collection. I owe David my taste for the horological hunt. His office was a desk hidden at the end of a warren of crowded bookcases filled with centuries of horological literature. He had a librarian's memory. If confronted with a question to which he didn't have an immediate answer (they were rare), he could instantly locate the relevant information from the shelves around him. I now model my own study on his office.

When he retired, Paul Buck took over. Paul is, to me, one of the most fascinating people on the planet. Every single time I speak to him, he tells me something that makes me gawp in amazement. That's no mean feat in an industry populated by what my husband calls 'woolly-hatted, cream-horn-eating, walking-sandal-and-sock-wearing garden-shed-tinkerers'. We watchmakers are not generally 'cool' people. We're the engineers of a bygone era who spend most of our day indoors working on eye-strainingly small objects with little to no human contact. But Paul (AKA Pablo Labritain, drummer in the punk rock band 999) is an exception. His specialist area is old cuckoo clocks (properly termed Black Forest clocks), but for many years he spent his lunch break practising the drums in the museum's Radium Room, where any objects containing radioactive substances are safely contained in steel chemical storage cabinets with extraction. I'm not sure anything could be more 'punk' than playing drums in a radioactive room.

Aside from his skills as a drummer, Paul is an exceptional restorer and teacher. It was Paul who taught me the painstaking process of repairing minuscule fusee chains. He showed me how to fashion the end of a round needle file into a three-sided cutter to cut through the rivet holding the tiny pins that secure each link in place. He taught me how to remove the pins and the damaged links either side of the break before threading the severed ends back together. He told me to use the steel of a sewing needle, as opposed to modern carbon steel, as his many years at the bench had taught him this better replicated the metal original watchmakers would have used centuries ago.*

Paul and David gave me permission to dismantle the Dutch forgeries. Like a forensic scientist, I used a combination of microscopic, X-ray and X-ray fluorescence scanning to build up a picture of their lives. All of them, like Wilter's, were verge watches.

When I first restored a verge watch as a student, I asked my tutor what sort of timekeeping I should aim for. 'Getting them running at all is a triumph,' he replied unencouragingly. Many modern repairers refuse to work on them. For one thing, they have frequently suffered from centuries of bodged repairs, which need to be carefully undone before the watch can be properly restored. I once handled a Dutch forgery whose outer case was adorned with a heavily worn depiction of the abduction of Helen of Troy in *repoussé*.† A previous restorer, however, had haphazardly re-engraved Helen's missing features, making her look less like the face that launched a thousand ships and more like Munch's *Scream*. The movements, meanwhile, are often clogged up with rust, dust, or bits of cardigan, all of which need to be painstakingly removed. The bearings are typically made of brass, as opposed to the harder-wearing materials like ruby used in higher-quality and more modern watches, so the mechanism wears itself out as it runs. As the movements weren't standardised, no spare parts are

---

* When Paul retired, although he's still touring with 999, the department was handed down to Oliver Cooke and Laura Turner, to whose continuing patience and support I am indebted.

† A relief design on a thin sheet of metal hammered through from the underside.

available; you can't even pinch a component from another watch of the same age without laborious customisation. Anything broken needs to be remade by hand. But I can't help loving these watches. Every single one of them is a character. Each one has its own clumsy idiosyncrasies, like an old car or a favourite pair of jeans that are falling apart but you can't bear to part with.

It was this handmade element, however, that initially made watches so expensive. Watches were complicated and time-consuming to produce: thirty or more individuals, each with different, interconnecting skills, would have been involved in the process of making a single watch. As a result, even the largest eighteenth-century British workshops only had the capacity to make a few thousand watches a year. And yet as the century progressed, a new, more affordable kind of watch began to appear in pawnbrokers' windows and on market stalls. At first, they appeared only in ones or twos, but by the end of the eighteenth century they far outstripped the numbers that Britain's established watch industry was capable of producing. Someone, somewhere, was making them more quickly and cheaply.

London's watchmaking world was, as we have seen, something of an elite boys' club. The long and expensive apprenticeships required by the Clockmakers' Company meant that training in the capital was only an option for the privileged few, which in turn placed a stranglehold on the City's watch production. This encouraged them to turn to the trade further afield to assist with their production. To keep up with the growing demand for watches, their makers became increasingly reliant on buying in what we now call *ébauches* (blank movements supplied ready to be finished and branded by another watchmaker), made in regional workshops in Lancashire and later Coventry, by artisans who had never served a formal apprenticeship. One worker in the north said it was 'only those who intended to become masters' who served their time formally while 'the rank and file of the workpeople never became formally indentured'.

Yet accounts from Prescot in Lancashire, where some of the finest horological tools together with parts and full *ébauches* were made, describe an exquisite level of craftsmanship achieved by non-apprenticed makers. One observer noted how these craftsmen formed minuscule 'bay leaf'-shaped gear teeth by eye despite the fact that they 'would stare at you for a simpleton to hear you talk about the epicycloidal curve'. This form of watchmaking was often a side hustle. Farmers who owned enough land to feed themselves and their families but had little left to trade for profit often engaged in other enterprises like spinning, weaving and, indeed, watchmaking to augment their income. Some local manufacturers set up workshops on small farms, running them in conjunction with their main business. These often beautifully executed watch parts would then be packed up and sent to watchmakers in London and across Britain, where they were finished and made up into watches.

British watchmaking capacity and ability to compete on prices was also limited by the exclusion of women from the workforce. Artisan culture was almost exclusively male. Some women were listed as masters and apprentices in the Clockmakers' Company, but most of them were actually milliners (milliners didn't have their own company at the time so were listed under others). The number of women serving formal apprenticeships was incredibly low across all trades, at just 1–2 per cent. One ongoing study has so far found only 1,396 women associated with watchmaking in the United Kingdom between the seventeenth and twentieth centuries: that might sound a lot, but when you consider that in 1817 alone there were over 20,000 people involved in watchmaking just in London, you realise it isn't.

Elsewhere in the world, for example in Switzerland, and later in the US, women were being welcomed into the workshop with open arms. As much as I would like to say this was an act of workplace equality, in reality women could be paid less, meaning the watches they made could also be sold for less.

There still aren't many female watchmakers. I was – and am – a rarity in the field. This has been a mixed blessing for someone who suffers from anxiety and a chronic sense of imposter syndrome. Being different makes it easier to be noticed. And being noticed comes with benefits and drawbacks. I have been lucky to have support from amazing friends and mentors, without whom it's unlikely I would have even finished my training. But I've also taken a lot of flak. In my very first workshop there were some who made it clear they thought I'd only got on the course because of tokenism. I once had a tutor petition an employer who'd offered me a summer placement to withdraw it and offer it to one of his male students. I've heard others say there's no point training women to be watchmakers as we have children and give up the profession. I've been told on several occasions, 'You're not special, you know.' This always makes me reflect: how could anyone think that I believe myself to be special? I suspect I'll never stop feeling like an outsider.

One morning in the car on my way into the workshop, I was listening to an interview on *Woman's Hour*. Cambridge scholar Morgan Seag was discussing the reasons why the British Antarctic Survey banned women from visiting the Antarctic until 1983. There were lots of predictable excuses: it was another era; they felt women wouldn't be interested due to the lack of toilets, shops and hairdressers; they were concerned about the impact of releasing females into a male colony. But what really struck me was her description of ice as a kind of stage of masculinity, which women, men feared, would undermine. Pioneering explorers like Robert Scott and Ernest Shackleton had created an illusion of the hero, of brave men tackling terrifying crevasses, extreme weather and starvation, to boldly strike out into the unknown. Professor Liz Morris, the first woman to work in the interior of the continent in 1987–88, said there were men who were resistant to her participation and quoted US Antarctic Leader George J. Dufek's view that if 'a middle-aged woman with no particular physical skills could hack it, then how could they [men] be heroes?'

It suddenly dawned on me that the same thing was happening in my field. Young watchmakers are also raised on the stories of heroes

– Huygens, Tompion, Graham, Mudge, Harrison – the geniuses of the golden age who designed extraordinary works of engineering to solve some of the greatest scientific problems of their day. Men – and they *were* all men – who socialised with aristocracy and wowed audiences with their mechanical ingenuity, created objects out of tiny pieces of metal that moved independently, as if by sorcery. The mechanism of a watch has been perceived by society as something so perfect and complex that it was used in an argument to counter the existence of God (usually assumed to be another man). When the Clockmakers' Company was founded in 1631, its charter declared that it would oversee the 'Fellowship of the Art and Mysterie of Clockmaking'. To this day, watchmaking is still viewed as a dark art that only a special few can understand. And yet, here I am: a socially awkward, tattooed woman, raised in a working-class household, with absolutely nothing 'special' about me – and I'm a watchmaker. If someone like me can become a master watchmaker, anyone can.

No wonder I found the dodgy John Wilter so appealing: I was an underdog drawn to another.

In the British Museum, I found the same *ébauche* makers' marks (called platemakers' marks and hidden under the dial) on movements signed with a range of fictitious names. It seemed that a relatively small number of workshops were making huge numbers of watches on a scale unseen in England at the time. Not even the most active workshops in the north of the country could match this pace. At first sight, the watches looked Dutch: the scalloping of the minute track and the shape of the balance bridge, the component that secured the upper pivot of the balance wheel, were stylistically Dutch rather than English in design. But the quantities didn't make sense. Although there were a number of very talented watchmakers in the Dutch Republic at this time, the Dutch industry was tiny in comparison to London's – and nowhere near big enough to produce the vast the number of 'Dutch forgeries' entering the market.

A two-footed balance bridge. This design is different to the
single foot normally found in eighteenth-century English
watches but was popular among watchmakers on the Continent,
including the Dutch Republic and Switzerland.

By studying the marks hidden in these watches, and comparing
them to archival records, I was able to work out what was happening.
Dutch merchants were indeed commissioning them in their national
style but, knowing that consumers preferred London-made watches
in the same way we might prefer German cars, Japanese cameras or
Belgian chocolate today, they signed them with English names, in the
hope of fetching higher prices. But, curiously, these fakes weren't
being made in the Dutch Republic or England. The assay marks,
hidden signatures, contemporary witness statements, newspaper
reports and even accounts from the places where they originated
made it absolutely clear: these watches were coming from Switzerland
where, since the start of the eighteenth century, a new approach to
watch production was being perfected. It was known as *établissage*.

If traditional watchmaking relied on small collectives of skilled
artisans passing items between different workshops, and Britain's
unofficial watch trade relied on cottage industry workers, *établissage*
brought a larger number of workers under a single roof, known as a
manufactory. In the manufactory, labour was organised as a production line with specific chainmakers, springmakers, wheel-cutters and
pinionmakers working alongside each other. Although the techniques
and equipment were much the same as those used in traditional
manufacturing methods, the *établissage* system dramatically

streamlined production under the management of a single firm. It was highly efficient and, as a result, manufactories could create huge numbers of watches. Where Britain's largest workshop could produce a few thousand watches a year, a Swiss manufactory could produce 40,000. This completely revolutionised the industry. As a result of *établissage*, European watch production rose dramatically over the course of the eighteenth century, reaching an estimated 400,000 per year in the last quarter of the century, possibly even more.

Switzerland's geography made it perfectly placed to supply the trade. The country was located on a major trans-European trade route, with Dutch, French and English merchants constantly passing through as they made their way from the River Rhine to the River Rhône, which acted as a natural transport link between the Baltic Sea in the north and the Mediterranean in the south.* The frequency with which watch manufactories appear in the linking land route between the two rivers is a strong indicator of a merchant-directed industry. Merchants, who were constantly travelling across Europe as

---

* I've read a number of stories about how these watches made their way out of Switzerland and around the world. The market for these Swiss-produced fakes was shadowed by a criminal underbelly willing to smuggle them to their final destination. Watches are small, making them easy to transport in large quantities, tucked away in trunks under linen, in empty wine barrels – or strapped to hungry dogs. In 1842 the director of French customs reported that his officers at the border were being set upon by packs of ferocious dogs, in a state of 'madness'. The dogs had apparently been taken across the mountainous Swiss–French border, deprived of food and beaten before being sent out into the night with watches strapped to their bodies. Laden with up to 12 kilograms of watches each, the dogs would dash back across the border straight to their master's house, where food and good treatment would await. I have to say I'm somewhat sceptical that 'wild' dogs could be trusted to traverse miles of mountainous terrain between abusive masters like masochistic carrier pigeons. My own friendly and moderately trained Staffordshire Bull Terrier cross, Archie, cannot be trusted to make it from one end of our workshop to the other without finding a distraction. The bulk of the hundreds of thousands of watches leaving Switzerland each year would have been shifted via more reliable methods – in carts on little-trodden mountain passes and with the ships and merchants trading across Switzerland between the Rhône and the Rhine.

well as further afield, were much more aware of changing fashions, and much more in tune with market demands, than overworked craftsmen stuck in their proverbial cottages. This led to a paradigm shift in watchmaking: one in which merchants told craftsmen what to do, rather than simply retailing on their behalf.

As I rifled through the British Museum's collection, I was constantly on the lookout for evidence of the Dutch forgeries' fictitious origins. I found several more John Wilter watches. Curiously, a small number of these were very high quality and English in style, but the majority were lower-quality Dutch-style forgeries. I also found all kinds of spelling mistakes on the movements, like the suspicious typos you often see in junk emails. Renowned father–son watchmakers Joseph and Thomas Windmills had been credited as 'Wintmills, London' on one watch and 'Jos Windemiels, London' on another. I also found 'Vindmill', 'Wintmill', 'Windemill' and 'Vindemill'. A Jonh Wilter and a John Vilter also featured, although, unlike the Windmills, the identity of the John Wilter they were apparently imitating remained a mystery to me. Nevertheless, as Dutch-speaking merchants commissioned English-named watches from French-speaking manufactories in Switzerland, something was clearly getting lost in translation.

But Dutch forgeries were immensely important. These watches, which often undercut their genuine London competitors by more than 50 per cent, represent the first mass-produced timekeepers. This was the moment when portable time ceased to be the preserve of the ultra-rich. These watches made no contribution towards the accuracy or reliability of the watch, nor were they technically or aesthetically innovative, but they were cheap – and that's what makes them interesting. For the first time since their invention, a way had been found to make watches affordable. By the end of the eighteenth century they were becoming an increasingly common accessory among ever wider social groups.

My Wilter watch is evidence of one of the most significant socioeconomic developments of the eighteenth century: imitation. From the

1760s onwards the Industrial Revolution gave rise to an emerging middle class with aspirations beyond their financial means. Theatres, parks and the emergence of free museums and art galleries created more opportunities for the wealthy and the aspiring to collide. And as literacy rates improved and print production increased, newspapers provided these people with a window into the lives and possessions of the upper classes. It drove the desire for luxury possessions. And as these material luxuries remained out of financial reach for most of the population, the solution was to fake them.

From painted blue Oriental-inspired ceramics to plated metal and steel cut and polished to look like diamonds, an entire industry rose up, producing pseudo-luxuries for a new and rapidly expanding group of people. Sheffield Plate,* invented around 1742, became a highly popular commodity for any committed social climber. Under the dim candlelight of a Georgian dinner party, and after a few too many glasses of wine, guests could easily be tricked into thinking their supper was being served on an exceptionally expensive solid silver service. Similarly, ormolu – the name given to gilded bronze or brass objects – might convince visitors that their host's home was filled with solid gold *objets*. Georgian industrialists made keeping up with the Joneses possible on almost every level – provided you didn't look too closely.

Dutch forgery watches were just one part of this larger process. In the eighteenth century, watches – which were worn prominently on chatelaines, an ornate style of chain that hung from a waistband – were conspicuous signifiers of wealth and status. So much so, in fact, that in 1797 the prime minister and chancellor of the exchequer William Pitt introduced a tax on watch ownership. He justified it by declaring that owning a watch was a mark of luxury and so proof that the owner could afford to bear the additional tax.

---

* Formed by rolling and pressing a thin sheet of silver onto a much cheaper copper base metal, plate could be used to create everything from candlesticks and other tableware to complete dinner services. To add to the web of illusion, makers' logos and marks were commonly designed with striking similarity to genuine hallmarks.

(Needless to say, the tax was exceedingly unpopular. The middle class rebelled, with some going as far as scrapping the gold cases on their watches and having new ones made in cheaper metals to avoid their watches being classed as eligible for taxation.)

If a century earlier most people's lives were lived by a communal clock, personal watches were now everywhere. We see them in visual art. Clocks and watches make regular appearances in the work of William Hogarth, who used them to track his characters' progress through his stories. His famous series *A Harlot's Progress* charts the corruption of an innocent country girl, Moll Hackabout. The closer Moll gets to her eventual demise (incarceration and then death from venereal disease), the closer the clocks in his engravings edge towards the eleventh hour. Interestingly Moll, a 'harlot', apparently owns a fancy 'repeating' pocket watch.

The proliferation of watches was matched by a surge in pickpockets and street robbers. Watches are the pickpocket's loot of choice in John Gay's *The Beggar's Opera* (1728). Timepieces were highly prized by thieves, commonly traded for services like prostitution or to pay off gambling debts, as happens in Plate VI of Hogarth's *A Rake's Progress* (published 1735). Records from the Old Bailey show that inns, taverns and gin booths were the preferred hunting grounds for pickpockets, with thefts peaking between 8 and 11 p.m. (when nearly half of all reports were made), dropping off after midnight, then rising again in the morning, around 7 a.m., when people woke up with a sore head to discover their watch was gone. Stolen watches were quickly and easily disseminated into London's criminal underground through pawn-brokers and second-hand shops. Unscrupulous jewellers readily took them in; there were even horologists willing to modify them by chang-ing names or hallmarks to avoid them being identified by their true owners. Daniel Defoe's irrepressible Moll Flanders in his 1722 novel, one of the most famous pickpockets in literary history, regularly uses a watch we can assume was pinched from an unsuspecting victim.

The Old Bailey's papers don't simply reveal increased levels of watch ownership and theft, but also an increased level of time awareness. Over the course of the eighteenth century, the specific time of an event became a more and more common part of witnesses' testimonies in crime reports. Thomas Hillier, who appeared in court in 1775 after his silver pair-cased watch was stolen by highwaymen between Hampstead and London, stated that the event occurred at 'about a quarter after nine at night' and the whole ordeal lasted around 'a minute or a minute and a half'. Although Hillier's account was unusually specific, it was one of thousands from Londoners in this period that had begun to give times, dates and durations of occurrences. Time awareness was slowly but surely growing.

The story of eighteenth-century watch-time has two very different sides. On the one hand, you have the glamorous golden age of horological advances, of marine chronometers, of watches crafted by some of the most highly educated scientists in society. But you also have a murky underbelly of fakes and forgeries that, in my view, is no less interesting and important. Dutch forgeries disrupted the connection between artisan watchmakers and wealthy patrons. They made a giant leap towards making watches affordable, paving the way for other later companies to make them truly accessible to all. For that reason, to me they are as significant in the story of horology as Harrison's chronometers. In the end, how can we argue an innovation is truly world-changing if only a small elite can access it? By widening access to time, cheap watches helped to close the divide between rich and poor, the aristocracy and the masses. They democratised time.

But what about John Wilter? Was he too an invention, a fake? Several years ago, I happened to be leafing through the minutes of an 1817 hearing at the House of Commons when I finally found a contemporary reference to him, from a person who claimed to have known the man behind the myth. I remember when I saw that name on the page – a name that had haunted me for years – I froze for a moment. I looked away, took a deep breath, and started reading. The witness – another watchmaker called Henry Clarke – spoke admiringly of a man who he claimed had, by then, passed away. He'd been making watches to commission for a merchant who had ordered that:

> [He] introduced the making of watches with the feigned name of 'Wilters, London,' on them; those watches were well made, and would have done credit to the maker, who should have put his name upon them; other persons speedily imitated the external appearance of the watches . . . [but] those had sham day of the month, dials and hands without and wheels to move them, and also the sham appearance of being jewelled in the pivot holes . . . The last I saw of those spurious watches were offered to me for sale at 34s. each, but really were good for nothing; whereas the first introducer of watches, with that feigned name, was not overpaid at eight guineas each.

Wilter, I discovered, was both real *and* fake. The name itself was probably the invention of a Dutch merchant who wanted something that sounded English but also couldn't be traced. But the man he originally commissioned *was* a genuine English watchmaker of considerable skill. 'John Wilter' became something of a brand. The merchant had then realised he could boost his profits by having Wilter watches made cheaply on the Continent. This tied up perfectly with the evidence in the British Museum, where I'd found a few high-quality Wilter watches as well as several more typical Dutch forgeries. This short passage filled in a gap I hadn't previously been able to explain. And these watches are a prime example of why fakes shouldn't be scorned, as the value they can add to our understanding of the world, of industries and even economies, is immense. As a scholar I studied them, and now I've even managed to collect a few of my own, including my most cherished ones, by my infamous John Wilter.

6

Revolution Time

At times the heart plays tricks and lets us down. The vigilant are right. For God – the mighty Breguet – gave us faith, and seeing it was good, improved it with a watchful eye.

Victor Hugo, *Les Chansons*, 1865

In August 2006, Rachel Hasson, the artistic director of the LA Mayer Museum of Islamic Art in Jerusalem, received a call from Zion Yakubov, a watchmaker in Tel Aviv: a cache of stolen antique timepieces had been found and he thought she should come and take a look. She'd had calls like this before; they were always a hoax. But this one was different.

Twenty-five years earlier, the museum's unparalleled watch collection had been stolen in a heist that had confounded the police, the Israeli Intelligence Community and even Samuel Nahmias, Israel's most accomplished detective. On the night of 15 April 1983, 106 irreplaceable watches and clocks, with one particular piece valued at more than $30 million, vanished. A huge search began but every line of enquiry yielded nothing. As the years passed, it seemed as though the objects had disappeared from existence.

Detectives had followed leads as far afield as Moscow and Switzerland, but it turned out the watches were in a storage facility in Tel Aviv, just an hour away from where they were taken. Hasson and Eli Kahan, a member of the museum's board, travelled to the office of Hila Efron-Gabai, a lawyer who had been asked to return the watches anonymously for her client. They identified the watches from their serial numbers; some were intact, others were damaged. But one watch brought tears to Hasson's eyes. There, wrapped in yellowing newspaper, lay 'the Mona Lisa of watches', made by Abraham-Louis Breguet for Marie Antoinette, the most complicated, beautiful and valuable watch ever made.

Police finally tracked the robbery to notorious Israeli thief Na'aman Diller, who had confessed the crime to his wife on his deathbed. They were disappointed not to have a chance to interview Diller, who had 'a unique style'. (In 1967 – with a short break to fight in the Six-Day War – he dug a 300-foot trench at the back of a Tel Aviv bank in order to blow open a vault.) Diller always acted alone. This time he had used a hydraulic jack to force the museum railings apart and a rope ladder and hooks to climb 10 feet to a small window just 18 inches high. Diller, who was 'whippet-thin', then slithered through the narrow opening before making off with most of the priceless horology collection.

This is one of the more recent chapters in the story of the 'Queen's watch', which began some 200 years ago.* In 1783 – the same year my humble John Wilter forgery was assembled – an anonymous admirer of Marie Antoinette sought out the services of the most famous watchmaker in Europe in order to commission a very special gift. No doubt the admirer hoped to gain favour with the queen, who was already an enthusiastic patron of Breguet's work. The commission was for a watch more complex than any that had gone before it. It was to include all the most advanced and complicated mechanisms of the day and no expense should be spared: gold should replace other metals wherever possible, including in the mechanism itself. It was to be a watch for its times, exemplifying the best and worst of the *ancien régime*.

The best: the lavish tastes of Louis XVI's court meant there were no boundaries on creativity. To make this watch nonpareil, Breguet

---

* Marie Antoinette's watch, like Queen Charlotte's Thomas Mudge watch, is commonly referred to as the 'Queen's watch.'

was given an unlimited budget, and as long as he needed. It is hard to express how exciting it would be, for a maker, to be presented with an open chequebook *and* an open calendar – pure scope in time and budget to test your skills and ingenuity to their limits. Even for the successful watchmaker – and by that time Breguet had been appointed *Horloger du Roy* to Louis XVI – cashflow is one of the hardest things to manage. An exquisite, handcrafted timepiece costs an obscene amount of money to buy and to make. For us today we can sometimes expect a six-figure sum – but by the time you've deducted overheads, taxes, material costs, outworkers' bills and divided it by the x number of years it's taken to make, it can sometimes barely cover a living wage for its maker.

For Breguet, the unlimited time allowance and budget enabled him to pour everything he'd spent his lifelong career mastering into a single piece. The watch had a total of twenty-three complications, those functions which are surplus to telling the time. It was self-winding and could strike the time out loud, sounding the hours, quarters and minutes on finely tuned gongs made from wire, and it displayed the equation of time. It had power reserve indication (it could run for forty-eight hours from full wind), a chronograph, a thermometer and a perpetual calendar à la Mudge. In total, the watch required 823 parts squeezed into a 6-centimetre diameter pocket watch and is still considered one of the five most complicated watches in the world. Its engineering was so spectacular (and exquisite) that the whole movement was left visible through the glass dial and casing to show the mechanism busily working away inside.

The worst: while Breguet worked on this infinitely luxurious, money-no-object watch, people were starving. France, still locked into a feudal system, taxed the peasants (96 per cent of the population, who lacked any political or economic power), while the clergy and aristocracy reaped the benefits. From 1787 to 1789 France weathered terrible harvests, droughts, cattle disease and skyrocketing bread prices. Meanwhile, the French government was bankrupt – their costly involvement in the American War of Independence, and Louis XVI's extravagant court, had taken a heavy toll on the national purse.

Tax increases lit the touch paper, the outrage of the citizens of Paris was unleashed, culminating in the French Revolution in 1789.

Marie Antoinette never got to see her watch. Breguet's workshop was disrupted by world events and it was not completed until 1827, thirty-four years after its beautiful, if politically insensitive, intended recipient had met her fate at the guillotine.

True master watchmakers were, and still are, rare creatures. In his entry on horology in Diderot's *Encyclopédie*, the celebrated watch-maker Ferdinand Berthoud (1727–1807) describes the demands of his craft: 'thorough mastery of *horology* requires *the theory of science, the skill of handwork, and the talent for design*, three qualities that are not easily fostered in the same individual'. I like his admission that it is hard to unite these qualities in the one human. Craig and I have always said that individually we're good watchmakers, but together we make one incredibly good watchmaker. We have complementary strengths: Craig is the illustrator, designer of the aesthetic, whereas I'm more mathematically minded and can turn his beautiful hand illustrations into precise technical drawings that we can work to. Craig has an incredible level of dexterity working with fine parts, like repairing damage to hairsprings, while I like finishing – putting the polished angles on the spokes of gear wheels, springs and plates by hand with tiny files. I love wheel cutting; positioning and cutting the hundreds of teeth in a watch by hand. I like the way you get into a repetitive, almost hypnotic motion with the cutter as you wind it forth and back, again and again, over and over. Craig finds it mono-tonous, but that's okay, because he has far more patience than I do for turning tiny balance staffs, just a few millimetres in length, by hand and polishing them to a hard, bright, mirror finish. In Breguet, however, all of those skills and more were uniquely combined.

In horology, it is something of a cliché to confess that your hero is Abraham-Louis Breguet. He is one of the greatest and most cele-brated minds in the history of our industry. He started his apprenticeship in Paris in 1762 at the age of fifteen and, by his death in 1823, he had revolutionised his profession. The barrister and horology buff (and one-time owner of the 'Queen's watch' together with many others that were stolen from the LA Mayer Museum) Sir David Lionel Salomons famously declared that 'to carry a fine Breguet watch is to feel that you have the brains of a genius in your pocket.'

Breguet's extraordinary technical skill and inventiveness is the stuff of horological legend. No other watchmaker created as many inventions that have remained in use to the present day. He was the originator of the *perpétuelle*, the first automatic winding mechanism, which powers the watch through the movement of the wearer using the swinging action of a weight inside the watch to power up the mainspring. He invented gongs made from wire in his repeating watches that chimed the hours and minutes, allowing for these watches to be much slimmer than the traditional bells that existed beforehand. He developed what is still referred to as the Breguet overcoil hairspring, which was a technical improvement to the flat spiral design that preceded it. Breguet found that raising the last outer turn of the flat spring above the rest significantly improved isochronism, or the equal duration of each breath, and consequently timekeeping and accuracy. As before, all these developments are still featured in watches to this day; some have been advanced further, but others, like the overcoil, have remained virtually unchanged for over 200 years. Thanks to his innovations, watch movements became slimmer. Gone were the pebble-like pair-cased pocket watches of old. Breguet's watches were slim enough to fit in the tailored pockets that were becoming fashionable among gentlemen.

Ironically, for a man who lived in an age synonymous with excess, Breguet stripped his designs right back. His work was restrained and embodied purity of purpose. The pared-back yet elegant watch hands he used are still named 'Breguet hands'. Often in electric blue steel, or in gold, they are exquisitely long and slim with a small hollow circle near their tip, and a little arrow-like point to precisely indicate the hours and minutes on the dial. He favoured a style of engraving known as engine-turning. Instead of hand-engraving the precious metal of his watch dials, he used manually operated rose and straight-line 'engines' to create intricate geometric patterns. These machines are, in my opinion, among the most beautiful ever invented. Rose engines slowly rotate a series of different-shaped discs, translating this movement into a pattern that gets cut into the surface of the watch dial.* You could compare the process to a bicycle with, say, an octagonal wheel. As you ride along, the seat will bump up and down according to the shape of the wheel. In a rose engine, the bicycle seat is your watch dial, which moves around against a stationary cutter. By moving the dial past the cutter many times, and altering the cutter's position slightly each time, an orderly pattern is inscribed into the metal. They are remarkable, and hypnotic to watch in action, and can produce a huge range of patterns, from the basket weave, which looks a little like a tiny chequered chessboard, to rosettes, which radiate from the centre in different shapes, like ripples across the surface of a pool escaping a dropped stone. They can engrave milled circles, which are used as bands to border the numerals on the dial or embellish the main band of the case with a pattern a bit like the edge of an old pound coin. The effect, at once detailed and discreet, gives the dial a subtle lustre, the height of low-key luxury. The machines are now as rare as hen's teeth. I've occasionally commissioned this kind of work from specialists, but even if I *could* get my hands on my own machine, I would rarely use this style. It is so associated with Breguet, it feels like trespassing.

Breguet's watches also revealed their own magical mechanical workings. Instead of the classic full plates – two round discs sandwiching (and concealing) most of the movement – he dissected these plates into a skeleton-like series of 'bars', which allowed you to see into the movement and watch the train wheels (which gear down the power from the mainspring) turn and the escapement ticking

---

* Straight-line engines use a similar principle but move up and down rather than round in a circle.

through.* Breguet's movements were still made from brass gilded with yellow gold but, instead of the delicate floral and foliate engraving that would have typically embellished them, he chose to *frost* his movements, a satin finish that is both iridescent and yet matte, depending on how it catches the light. It can be achieved by various means, from the application of acid to the use of a stipple brush or, as watchmakers do today, silicon sand blasting (our silicon blasting machine was retired from a dentist, who was using it to clean the residue off tooth moulds.)† As counterfeiters started to mimic Breguet's designs, he developed a minuscule secret signature that was etched into his dials and was near impossible to replicate. You needed a magnifying glass to see it properly.

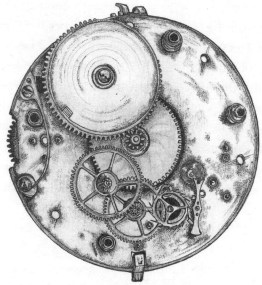

Train wheels running between the mainspring barrel at
the top, and the escapement on the lower right.

* It's worth noting that Breguet wasn't the inventor of this movement design, which had been popularised in the years preceding his rise to celebrity – and he never professed to be. But his innovations and notoriety have led to it becoming iconically 'his'.
† There is an overlap between dentistry and watchmaking tools. We both have a lot of pliers and we're used to working in small, fiddly places. In fact, between the seventeenth and nineteenth centuries some watchmakers used to offer dentistry – but not, for all his talents, Breguet.

Breguet's watches were precision scientific instruments, but many of these design quirks serve no functional purpose. They existed purely for delight, and joyously elevate engineering to become a work of mechanical art.

It was Breguet's stepfather, Joseph Tattet, who came from a family of watchmakers in Paris, who first introduced him to the profession. Breguet, born into a Huguenot family in Neuchâtel, Switzerland, in 1747, was the eldest of five and the only son. He lost his father at just eleven years old, and left school two years later. His mother remarried, and in 1762 young Breguet, like so many artisans before him, made his way across the 'permeable frontier' of the French–Swiss border to begin his training in his stepfather's family workshop in Paris.

Breguet attended evening classes in mathematics at the Collège Mazarin and his tutor, Abbé Joseph-François Marie, praised his talents in court circles so highly that word of his emerging talent made its way to the king and queen of France. Like George III across the Channel, King Louis XVI of France was fascinated by the mechanics of watchmaking. At the age of fifteen Breguet began an apprenticeship to a master watchmaker at the palace of Versailles.

When he was twenty-eight, Breguet founded his workshop on the ground floor of number 39 Quai de l'Horloge, the watch- and clockmaking quarter of Paris's bustling Île de la Cité. This was, to a great degree, made possible through the dowry he received from the wealthy and respected Parisian family of his wife-to-be, the 22-year-old Cécile Marie-Louise Lhuillier. Watchmakers often married later in life – in part because liaisons of any kind with the opposite sex were strictly forbidden within apprenticeship indentures – and it was not uncommon at that time for talented male artisans to establish their first solo ventures through dowries. They married that same year and set up home together, living and working in the same building.

Tragedy struck Breguet's life frequently. In the years that followed Abbé Joseph-François Marie, his former tutor and by now one of his closest friends, died in suspicious circumstances. He lost his mother, Suzanne-Marguerite, and stepfather, Joseph, in quick succession, leaving him responsible for his four younger sisters. He and Cécile welcomed a son, Antoine, in 1776, who was the only one of his children to survive into adulthood. Their time together was to be short-lived. Cécile died in 1780. She was twenty-eight. He never remarried. We know little, if anything at all, of Breguet's mental state throughout this time. It was an era when death was closer – it was not uncommon for children and young adults to be orphaned, or for wives to be lost to childbirth – but it is hard to believe that this onslaught of loss had no impact on him or his work. Perhaps watchmaking provided him with an escape and pouring himself into his practice was a coping mechanism. I certainly find that losing myself in the microscopic world of watchmaking is an effective way of disconnecting from reality and shutting off from the outside world.

Yet Breguet's workshop went from strength to strength. Marie Antoinette has been credited as the individual who 'caused Breguet's rapid success and sudden vogue'. She possessed his no. 2 watch,* the second watch he ever made as an independent watchmaker, and recommended him widely both at home and abroad. In 1785 he became Purveyor of Watches to King Louis XVI. Marie Antoinette and Mudge's patron, Queen Charlotte, were friends and exchanged many letters, although they never met face to face. I like to imagine that perhaps they discussed their love of watches, and perhaps Marie Antoinette's recommendation led to the watches Breguet later made for Queen Charlotte and King George III. And of course, with characteristic largesse, Marie Antoinette gave Breguet's watches to those closest to her. She gifted her no. 14 Breguet to Hans Axel von Fersen, a close friend and Swedish count – she had 'AF' in blue enamel added to the case – and commissioned for herself Breguet's no. 46, which, judging from the description, appears to have been a matching watch with the same initials – a little cosy. It's possible

---

* It's described as being self-winding, with day-of-the-month indication, and a repeater.

they were lovers, the watch's *pérpetuelle* symbolising a perpetual love. And some suggest that the anonymous admirer who commissioned the Marie Antoinette watch was von Fersen.

A portrait of Breguet shows a fresh-faced man with an early receding hairline and a look of kindly intelligence. He was clearly never a man for wigs, despite his dealings at court. Accounts of his nature emphasise his modesty and generosity. I particularly appreciate the fact that he was kind to his employees. He is said to have reassured his apprentices, 'Do not be discouraged, or allow failure to dishearten you,' a sentiment that should be framed and mounted above every horological workshop door. He regularly tipped his workers, adding tails to the zeros on their invoices to him to turn them into nines. While tipping is now an act of common courtesy, during the *ancien régime* the idea of volunteering cash to an outworker as a mark of respect would have been virtually unheard of.

Breguet's *pare-chute* shock setting – he would have been aware of Jean-Pierre Blanchard's ongoing experiments jumping out of a hot-air balloon and using a parachute to soften his fall. Similar to its namesake, the crook-shaped spring on the left of the balance cock allows the staff to bounce very slightly when the watch is knocked or dropped, softening the blow and protecting the pivots from damage.

He sometimes revealed himself to be a showman. In around 1790 he invented the first ever shock setting in a watch. It was a sprung setting that worked to protect the delicate balance pivots from bumps and knocks, which would frequently cause them to break and resulted in a very time-consuming repair. He called it the *pare-chute* (or parachute). To demonstrate how effective his new setting was, legend has it Breguet tested it in front of a crowd of esteemed guests at a party at the home of Charles Maurice de Talleyrand-Périgord, the Bishop of Autun and future First Prince of Benevento. An observer described how Breguet threw 'his watch to the ground, without seeming to harm it in the slightest. "This devil, Breguet," exclaimed the Prince, "is always trying to go one better."' When Breguet retrieved the watch from the floor, guests were astonished to see it was still in perfect working order.

Breguet 'had the power to render even kings the slaves of fashion'. As the century drew to a close, however, that association came to be problematic. In 1792, the French First Republic was born in blood. In the months that followed, the Terror gripped France. Members of the aristocracy, the clergy, and anyone perceived to embody or have an association with the ruling elite of old France were rounded up and imprisoned. By 1794, a series of massacres and mass executions had claimed the lives of thousands of men, women and children. While official figures place the death toll at around 17,000 formal executions, later historians have estimated that the total cost, including those who died in prison or on the run without having been tried, might be as high as 50,000 lives.

Although the most famous method of execution was the guillotine, a grim mascot of the Terror, most of the victims were killed by sword, rifle, pistol or bayonet. Many others died from starvation or disease in France's overcrowded and filthy prisons. Marie Antoinette was detained initially at the Temple prison, where she asked for, and was granted, 'a simple Breguet watch'. Later she was transferred to the Conciergerie in anticipation of her trial. She begged that she would not 'suffer long', but in fact she spent more than two months living in a dank and isolated cell before her trial and eventual execution for treason on 16 October 1793.

The accounts of the survivors who bore witness to the Terror are horrifying. One described how they witnessed a young woman who was forced to drink the blood of a recently executed victim in exchange for her father's liberation. Others tell of fresh blood streaming through pebbled courtyards, or how the condemned would be forced to walk over the dismembered bodies of fellow prisoners to meet their own fate. Executions were a daily public spectacle, attracting huge audiences in street venues like the Place de la Revolution (now the Place de la Concorde). Grisly testimonies made their way, via fleeing survivors, to be recounted in the courts and stately homes of the neighbouring countries who offered them refuge. Europe's ruling classes looked on in horror. The fear that the French Revolution would spread beyond her borders was felt across the continent.

Anyone associated with the king was at risk of losing their life – and Breguet was no exception. A fascinating aspect of Breguet's character is that he had friends in all sorts of places. Despite his cosy relationship with Versailles, he was also very close to political theorist, scientist and French Revolutionary leader Jean-Paul Marat. As the unrest and instability in France intensified even Marat, who had been one of the *ancien régime*'s fiercest critics, fell foul of popular opinion after publishing an attack on the king's finance minister. In April 1793, when an angry lynch mob gathered outside Marat's house to drag him to his fate, it was Breguet who devised a plan to aid his friend's escape. Marat was not aesthetically blessed (he suffered from debilitating dermatitis) and the two made a split-second decision to use his wizened face to their advantage. Using shawls and a dress, Breguet dressed him as an elderly woman and snuck him out of the house on his arm, weaving him through the baying crowds to make their flight to safety. Two months later, when Marat discovered that Breguet himself had been earmarked for the guillotine, he returned the favour by sending him warning and aiding his flight to Switzerland by pulling strings with the powers that be to afford him safe conduct along with what was left of his family: his surviving son Antoine – saving him from conscription into the Revolutionary army – and the sister of his late wife, on the pretext of making his annual business trip to Switzerland. Breguet would never see Marat again. On 13 July 1793 Marat was stabbed in his bathtub by Charlotte Corday, a young Girondin.*

* The Girondins were a more moderate faction of the revolutionaries, who also ended up facing mass executions during the Reign of Terror.

Breguet was one of many thousands of people who fled France for nearby countries in this period. And, even for Breguet, life in exile was not easy. Food and materials were all in short supply and refugees were often unwelcome. He would have had to leave most of his tools behind, something that would be immensely painful for any watchmaker, whose tool collections can number in the tens of thousands and feel as personal as fingerprints. Breguet set up a modest workshop, with just a few employees, in Le Locle, not far from his birthplace in Neuchâtel. We can only imagine what the local trade thought of having the most famous watchmaker in the world setting up a competing workshop on their doorstep. He later travelled to London, where for a brief spell he worked for George III. Interestingly, it was during exile that Breguet accomplished some of his greatest technical achievements. Short on tools but still obsessed with his subject, he focused his attention on the invention of new mechanical solutions, including a device known as the *tourbillon*. Breguet conceived the *tourbillon* to overcome something we refer to as positional error. This is the error caused by the effects of gravity on the mechanism as it shifts position with the watch-wearer's movements. This error is most keenly felt by the sensitive oscillating balance, which is also the most vital component for accuracy. Breguet's solution was to house the entire balance and escapement in a constantly rotating carriage, which then evened out the impact of gravitational pull and improved timekeeping accuracy. So important was this new invention that it is labelled on the dial of one his earliest examples, sold to King George IV of Great Britain. The word *tourbillon* was translated into English for the king as the rather delightful 'Whirling-about regulator'.

Nothing was spared in the deconstruction of the *ancien régime*, not even time itself. Time is a loaded construct, filled with associations, whether they be social, political, religious or cultural. Now the Republic rebelled against the authoritarian associations of the Gregorian calendar (the one we still use today), which was seen as a symbolic extension of the establishment. Time under the *ancien régime* had become a symbol of power and control – possessed by the leisured and wealthy at the expense of those less fortunate – so now they embraced a new era by redefining the measurement of time itself. The new Republic was to have a new calendar, with new months, weeks, days, hours and even minutes.

It's something that might seem wildly pointless today – after all, whether there are ten hours in a day or twenty-four, it will make no difference to how many of them we'll live for. But for the Revolutionaries it was an act of starting afresh. This attempt to manipulate time to signify a complete political rebirth has been attempted elsewhere – in Cambodia, the Khmer Rouge declared the year of their takeover 'Year Zero', even though for the rest of the world it was 1975. As Jamaican philosopher Charles Wade Mills phrased it, the resetting of the historical chronometer in the French Republican calendar reflected 'the triumph of reason, light, and equality over the irrationalities and injustice of the *ancien régime*'.

Decimal time, as it was known, literally restarted the clock with the Revolution, redefining September 1792 as 'Year I of the Republic'. This new calendar was still divided into twelve months, but each month was of an equal thirty-day length, with the remaining five days reserved for a series of festivals to mark the end of the year. Each of these thirty-day months was, in turn, divided into a series of three ten-day weeks. The days of the week were renamed, as were the months, to better reflect the seasons. In autumn, *Brumaire*, from the French *brume* meaning 'mist' or 'fog', begins in October and is followed by *Frimaire*, from *frimas* or 'frost', in November. The winter months were *Nivôse*, *Pluviôse* and *Ventôse*, which derived from 'snowy', 'rainy' and 'windy' respectively. Spring began in March, or *Germinal*, the month of germination, then *Floréal* (flowering) and *Prairial* (meadow). As the seasons draw to summer, *Messidor*, from the Latin 'harvest', starts at the end of June, followed by *Thermidor* (hot) and *Fructidor* (fruit) before descending back into autumn with *Vendémiaire*, or 'vintage', in what we would think of as late September.

By tying the seasons to something tangible, the Revolutionaries were liberating time from what they perceived as the oppression of religion and superstition. With many calendrical names linked to Roman gods (such as March to the god of war, Mars, and June to the goddess Juno, wife of Jupiter), the new system eradicated any connection with the gods of old. It was a calendar based on reason, giving time back to the people, to the natural world, and particularly to agriculture. If this is all sounding vaguely familiar, it's because it has striking similarities to event-based timekeeping, albeit with greater emphasis on the structured numerical division of days.

Changing time also changed the face of watches, literally. The new decimal system had further divided the day into just ten hours of an equal 100-minute duration, with each minute being made of 100 seconds, making the decimal second faster than the standard time we operate by today by 0.86 seconds to every decimal. New watches and clocks had to adhere to this new time and examples of the decimal watches made during this brief time survive to this day, including some made by Breguet. They are curious, surreal-looking things. It can take a few moments to realise the reason their dials look odd is because they display ten divisions rather than twelve.

One of Breguet's decimal timekeepers, now in the Frick Collection in New York, not only rises to the challenge of decimal time but also reflects Republican France's eagerness to dispense with myth. The hands of our twelve-hour watches turn in a clockwise direction, replicating our earliest observations of the sun's movement round the Earth as experienced in the northern hemisphere – a movement that since Copernicus (1473–1543) has been known to be based on a misapprehension. For Breguet, a fully metric timekeeper had to embrace rationality and fact, so he designed a dial with two rings – one showing the ten decimal hours, the other a hundred decimal minutes. While the watch hand moves clockwise, every ten decimal minutes the hour ring clicks counterclockwise to reflect the Earth's counterclockwise heliocentric rotation in mechanical form.

In the end, habit proved stronger than ideology, and the life of the decimal clock was short: it was abandoned in less than a year. From its introduction in 1792, the decimal calendar lasted fourteen years until it was ended in 1806 and the Gregorian calendar adopted once more.* Breguet too was only briefly in exile. His cover story about leaving Paris for business reasons only held for so long, especially when the army was looking for his son for conscription. When Breguet's business travel passport expired and he had not returned, he was declared a traitor and a Royalist, and his workshop in Quai de l'Horloge was confiscated and put up for sale. Breguet returned to Paris in April 1795, once the situation had stabilised. Timepieces were in high demand to equip the French army and navy, as well as for scientists, but the city's trade had ground to a near-standstill. The cards were stacked in Breguet's favour. He was a modest man, but he knew his worth, and so he set about not only negotiating the return of his workshop and home at the Quai de l'Horloge, but also demanding that the damage to his business be repaid by the state. Remarkably, with a little help from well-connected friends, the new government returned his house and refitted his workshops at the national expense. An almost unheard-of achievement. The only condition was that he would be up and running within three months. Breguet agreed – on the promise that his workers would be exempt from military service. The deal was done.

Now he extended his business across Europe. One of Breguet's ingenious ideas was that of creating *souscription* (or subscription) watches. For a 25 per cent up-front fee, clients could commission a reliable, everyday, no-frills watch from Breguet; it was still a luxury few could afford, but cheaper than bespoke. The down payment meant Breguet gathered capital up front and could create several similar watches together in a less expensive, serially produced way, which made his watches affordable to a wider range of clients. The scale of production was nowhere near the one we see with our Dutch

---

* Although the litre and metre units of measurement have endured, as did the French franc until recently, which survived until its replacement by the euro on 1 January 2002.

forgeries, but it was a significant move. There are few examples of master watchmakers at any time in history who, once achieving such a high level of celebrity, turned their hand to making their pieces less, not more, expensive. It was an immensely successful business model, with Breguet selling and making around 700 *souscription* watches over the turn of the century.

His chief clientele remained the elite – but a new elite. While in exile he had worked for George III, now he made watches for the bankers and officers of the new Republic; but also for royalty across Europe, including Alexander I, tsar of Russia. Breguet had a bit of a following among the Russian aristocracy and even gets a mention in Alexander Pushkin's *Eugène Onegin*:

A dandy on the boulevards—
Strolling at leisure,
Until his Breguet, ever vigilant
Reminds him it's midday.

Breguet was as brilliant a diplomat as he was a watchmaker, creating pieces for friends, lovers and sworn enemies alike. Just as he had managed to juggle being *Horloger du Roy* and a friend of Marat, now he made watches for both Napoleon and the Duke of Wellington. Napoleon was so taken with his work he visited Breguet's factory in disguise on several occasions. Arthur Wellesley, 1st Duke of Wellington, owned multiple Breguet watches, including at least one *montre à tact*, which would have enabled him to tell the time by the feel of the case in his pocket. This means it is quite possible that Abraham-Louis Breguet was the unofficial timekeeper of the Battle of Waterloo.

Despite mixing with the very highest members of society, Breguet lived a humble and quiet life. He was described as having a 'young spirit' even in old age. In his last years he was profoundly deaf, but remained cheerful. Breguet's ultimate project, his watch for Marie Antoinette, remained at the core of his focus. There is a note written in August 1832 that confirmed that the queen's watch was still on his bench a month before he died at the age of seventy-six. He had continued to work on it to the end. The piece was ultimately completed by his son and heir, Antoine-Louis.

Just as it's impossible to think of the eighteenth century without considering the French Revolution – a political uprising that sent shockwaves across Europe – so one cannot think about the history of the watch without considering Breguet. His raw talent alongside his political dexterity saved him in an era of continuous flux. In the centuries since, his reputation and his unparalleled watches have endured, with his name appearing in literature as a comment on characters' taste, style and affluence. He is mentioned twice in Alexander Dumas's *The Count of Monte Cristo* and namechecked by Jules Verne and by Thackeray in *Vanity Fair*. Stendhal declared Breguet's watches to be a finer piece of work than the human body: 'Breguet makes a watch that never goes wrong for 20 years, and yet this wretched machine, the body we live with, goes wrong and brings aches and pains at least once a week,' while Victor Hugo went one further, referring, in his 1865 poetry book *Les Chansons des rues et des bois*, to 'God – the mighty Breguet'. To make him a deity might be a stretch too far, but Breguet's watches were certainly a byword for accuracy, utility and divine beauty.

7

Working to the Clock

. . . a stern room, with a deadly statistical clock in it, which measured every second with a beat like a rap upon a coffin-lid.

Charles Dickens, *Hard Times*, 1854

When Craig and I first took our workshop in the Jewellery Quarter, there was a factory next door. It has since been knocked down by property developers, who are turning the area into residential and retail units, but it used to be a huge place: sprawling old red-brick Victorian buildings that had been clumsily connected over the years via 1970s office blocks and more modern aeroplane-hangar-like corrugated-iron warehouses. The factory did everything from making bus seats to large-scale metal pressing and forming and precision machining. It was loud – deafeningly loud – and the constant hum of the motors made our workshop buzz with white noise. It became such a part of the atmosphere that we stopped noticing it until the end of the day, when all the machines were simultaneously switched off and we'd be momentarily stunned by the silence.

A horn was used to announce the beginning of the day, the start and end of lunch, and the end of working hours. First thing in the morning, we'd listen out for the sound of the horn reverberating around the buildings like the whistle of a steam engine, followed by the deep, slow whir of motors and heavy machinery firing up. Using horns to mark time in factories was common through the Industrial

Revolution and beyond, as not much else could cut through the noise of the machines. Today that idea of clock-watching, waiting for the horn, a shared experience with everyone downing tools and walking out the gates together as a place of work grinds to a halt, feels almost quaint. Now that we have mobile phones, remote email access, social media, shift working, virtual meetings and flexitime, there aren't many businesses that just switch off like that. But the advent of clock time in our working lives was as seismic a shift for workers in the industrial era as the late-capitalist, post-pandemic 'wfh' departure from the communal workplace has been in ours. It changed our understanding of how time should be spent.

Factories dominated the skylines of industrialised cities.

Before industrialisation, Britain relied largely on the rhythms of the natural world to dictate the working day and its activities. Working with the land, or the tides, was task-oriented and seasonal. Longer hours could be worked during the extended daylight of the summer months, to make up for those lost to the late dawns and early dusks of winter. For farmers, the harvest was hard work late into the summer evening, which gave way to the cold months when the shorter days could be spent focusing on animal husbandry until things warmed up ready for planting the land again. Crofters – small-scale farmers working on rented plots ('crofts') – turned their hand to building and thatching and, when storms forced them indoors, they made cots or even coffins (quite literally caring for their communities from cradle to grave). Fishermen could mend nets and repair boats when the weather was too poor to catch fish. Life was harsh and unforgiving; but by attending to what was necessary in that moment, this way of working naturally created more labour-intensive seasons as well as those that allowed more time for relaxation and pleasure, albeit with less delineation of the working day. This continues to a greater or lesser extent in agricultural communities today.

The dawn of the predictable, mechanised world of the factory presented a stark contrast, and, from around 1760 through to the Victorian age, put England at the forefront of the Industrial Revolution. This change was less about a switch to wage-labour – even in feudal times, 'dayworkers' had been employed for a fee – than a new stringency about *timed* labour. The working day was no longer defined by sunrise and sunset. As factory processes developed and jobs became more specialised, precise timing became critical for synchronisation. Employees were another cog in the wheel, hired for a duration and budgeted in terms of productivity. Punctuality became profit.

Try to visualise a person of authority any time from the nineteenth to as late as the mid-twentieth century. Perhaps he (it would almost certainly be a 'he') is an industrialist, a factory owner; maybe he's a workshop manager; he could be a politician or trade union leader. Smartly dressed in his dark suit, white shirt buttoned up to the collar. He might be wearing a top hat, a bowler, or even a flat cap. He might have a beard, a moustache, or be clean-shaven. He's almost certainly wearing a waistcoat. If he's wealthier maybe it's silk and patterned, or perhaps it's heavy and woollen, more practical and less ostentatious.* Maybe you're picturing a man like Winston Churchill or Keir Hardie. Perhaps Prince Albert, Abraham Lincoln or even a fictional character like Arthur Conan Doyle's Dr John

---

* The ubiquity of waistcoats, even for men doing heavy manual work in boiling hot jewellery workshops, amazes me.

Watson or Harper Lee's Atticus Finch. Regardless of whether they were new money or old, their social background, or whether they sat on the political right or left, there is something that generally unites the style of these men. Next time you see their photographs or read a description, look out for the chain of a pocket watch pinned to the buttonhole of their waistcoat. (Prince Albert was such a fan of wearing his watch in this way that the chain is now known as an Albert chain.) In this era of industrial expansion, the watch became a symbol not only of its owner's affluence and education, but also of his structured attitude towards work.

A silver pocket watch on an Albert chain, allowing
it to be worn with a waistcoat.

Although Puritanism had disappeared from the mainstream in Europe by the time of the Industrial Revolution, industrialists, too, preached redemption through hard work – lest the Devil find work for idle hands to do. Now, though, the goal was productivity as much as redemption, although the two were often conveniently conflated. To those used to working by the clock, the provincial workers' way of time appeared lazy and disorganised and became increasingly associated with unchristian, slovenly ways. Instead 'time thrift' was promoted as a virtue, and even as a source of health. In 1757, the Irish statesman Edmund Burke argued that it was 'excessive rest and relaxation [that] can be fatal producing melancholy, dejection, despair, and often self-murder' while hard work was 'necessary to health of body and mind'.

Historian E.P. Thompson, in his famous essay 'Time, Work-Discipline and Industrial Capitalism', poetically described the role of the watch in eighteenth-century Britain as 'the small instrument which now regulated the rhythms of industrial life'. It's a description that, as a watchmaker, I particularly enjoy, as I'm often 'regulating' the watches I work on – adjusting the active hairspring length to get the watch running at the right rate – so they can regulate us in our daily lives. For the managerial classes, however, their watches dictated not just their own lives but also those of their employees.

In 1850 James Myles, a factory worker from Dundee, wrote a detailed account of his life working in a spinning mill. James had lived in the countryside before relocating to Dundee with his mother and siblings after his father was sentenced to seven years' transportation to the colonies for murder. James was just seven years old when he managed to get a factory job, a great relief to his mother as the family were already starving. He describes stepping into 'the dust, the din, the work, the hissing and roaring of one person to another'. At a nearby mill the working day ran for seventeen to nineteen hours and mealtimes were almost dispensed with in order to eke the very most out of their workers' productivity, 'Women were employed to boil potatoes and carry them in baskets to the different flats; and the children had to swallow a potato hastily . . . On dinners cooked and eaten as I have described, they had to subsist till half past nine, and frequently ten at night.' In order to get workers to the factory on time, foremen sent men round to wake them up. Myles describes how 'balmy sleep had scarcely closed their urchin eyelids, and steeped their infant souls in blessed forgetfulness, when the thumping of the watchmen's staff on the door would rouse them from repose, and the words "Get up; it's four o'clock," reminded them they were factory children, the unprotected victims of monotonous slavery.'

Human alarm clocks, or 'knocker-uppers', became a common sight in industrial cities.* If you weren't in possession of a clock with an alarm (an expensive complication at the time), you could pay your neighbourhood knocker-upper a small fee to tap on your bedroom windows with a long stick, or even a pea shooter, at the agreed time. Knocker-uppers tried to concentrate as many clients within a short walking distance as they could, but were also careful not to knock too hard in case they woke up their customer's neighbours for free. Their services became more in demand as factories increasingly relied on shift work, expecting people to work irregular hours.

Once in the workplace, access to time was often deliberately restricted and could be manipulated by the employer. By removing all visible clocks other than those controlled by the factory, the only person who knew what time the workers had started and how long they'd been going was the factory master. Shaving time off lunch and designated breaks and extending the working day for a few minutes here and there was easily done. As watches started to become more affordable, those who were able to buy them posed an unwelcome challenge to the factory master's authority.

An account from a mill worker in the mid-nineteenth century describes how: 'We worked as long as we could see in the summer time, and I could not say what hour it was when we stopped. There was nobody but the master and the master's son who had a watch, and we did not know the time. There was one man who had a watch . . . It was taken from him and given into the master's custody because he had told the men the time of day . . .'

James Myles tells a similar story: 'In reality there were no regular hours: masters and managers did with us as they liked. The clocks at factories were often put forward in the morning and back at night, and instead of being instruments for the measurement of time, they were used as cloaks for cheatery and oppression. Though it is known among the hands, all were afraid to speak, and a workman then was afraid to carry a watch, as it was no uncommon event to dismiss

* Knocker-uppers were still going in some towns in the north as late as the 1970s.

anyone who presumed to know too much about the science of horology.'

Time was a form of social control. Making people start work at the crack of dawn, or even earlier, was seen as an effective way to prevent working-class misbehaviour and help them to become productive members of society. As one industrialist explained, 'The necessity of early rising would reduce the poor to a necessity of going to Bed betime; and thereby prevent the Danger of Midnight revels.' And getting the poor used to temporal control couldn't start soon enough. Even children's anarchic sense of the present should be tamed and fitted to schedule. In 1770 English cleric William Temple had advocated that all poor children should be sent from the age of four to workhouses, where they would also receive two hours of schooling a day. He believed that there was:

> considerable use in their being, somehow or other, constantly employed for at least twelve hours a day, whether [these four-year-olds] earn their living or not; for by these means, we hope that the rising generation will be so habituated to constant employment that it would at length prove agreeable and entertaining to them . . .

Because we all know how entertaining most four-year-olds would find ten hours of hard labour followed by another two of schooling. In 1772, in an essay distributed as a pamphlet entitled *A View of Real Grievances*, an anonymous author added that this training in the 'habit of industry' would ensure that, by the time a child was just six or seven, they would be 'habituated, not to say naturalized to Labour and Fatigue.' For those readers with young children looking for further tips, the author offered examples of the work most suited to children of 'their age and strength', chief being agriculture or service at sea. Appropriate tasks to occupy them include digging, ploughing, hedging, chopping wood and carrying heavy things. What could go wrong with giving a six-year-old an axe or sending them off to join the navy?

The watch industry had its own branch of exploitative child labour which Sue Newman termed in her eponymous book, *The Christchurch*

Fusee Chain Gang. When the Napoleonic Wars caused problems with the supply of fusee chains, most of which came from Switzerland, an entrepreneurial clockmaker from the south coast of England, called Robert Harvey Cox, saw an opportunity. Making fusee chains isn't complicated, but it is exceedingly fiddly. The chains, similar in design to a bicycle chain, are not much thicker than a horse's hair, and are made up of links that are each stamped by hand and then riveted together. To make a section of chain the length of a fingertip requires seventy-five or more individual links and rivets; a complete fusee chain can be the length of your hand. One book on watchmaking calls it 'the worst job in the world'. Cox, however, saw it as perfect labour for the little hands of children and, when the Christchurch and Bournemouth Union Workhouse opened in 1764 down the road from him to provide accommodation for the town's poor, he knew where to go looking. At its peak, Cox's factory employed around forty to fifty children, some as young as nine, under the pretext of preventing them from being a financial burden. Their wages, some-times less than a shilling a week (around £3 today), were paid directly to their workhouse. Days were long and, although they appear to have had some kind of magnification to use, the work could cause headaches and permanent damage to their eyesight. Cox's factory was followed by others, and Christchurch, this otherwise obscure market town on the south coast, would go on to become Britain's leading manufacturer of fusee chains right up until the outbreak of the First World War in 1914.

The damage industrial working attitudes to time caused to poor working communities was very real. The combination of long hours of hard labour, in often dangerous and heavily polluted environ-ments, with disease and malnutrition caused by abject poverty, was toxic. Life expectancy in some of the most intensive manufacturing areas of Britain was incredibly low. An 1841 census of the Black Country parish of Dudley in the West Midlands found that the average was just sixteen years and seven months.

As many of us familiar with the Sunday-night blues can attest, the rhythm of the working week dictates people's perception of time, whether they are working or not. The 1937 Worktown Project, a groundbreaking study of life in the Lancashire town of Bolton, observed that workers would 'anxiously anticipate' the end of time off as much as they looked forward to the end of the working week itself. 'Workers were always looking forward to the end of any prescribed period,' they observed, and could not 'escape . . . from time' even when on their summer holiday.*

In 1954 Philip Larkin railed: 'Why should I let the Toad *work* / squat on my life . . .' Like Larkin, I've never been a happy employee. Unlike factory workers, I have had the liberty to follow my enthusiasms, and have been supported by intelligent and generous people over the course of my career. But there have also been low moments. The truth is, I've always found working for others stressful. Even when the work itself wasn't backbreaking or repetitive, I struggled with the unreadable codes and compromises of the workplace – whether it was the requirement to perform as decorative entertainment to watch collectors at a high-profile auction house or negotiating the hidden social rules of the world's richest people. But more than that, it was the feeling of working *under* someone – the sense of their power over my time, even when they weren't aware of it. Bad management and constantly shifting goals meant that both the labour and the worry of work monopolised my non-working life too. My time felt out of my control, and that brought me to breaking point.

In 2012, I woke up one morning and burst into tears. I was shaking, couldn't breathe, couldn't speak, couldn't move. It felt like someone was squeezing my heart in their fist. It was my first anxiety attack. I knew I couldn't go on. I was an anomaly in the watchmaking world, a square peg that couldn't fit, and the effort was destroying me. I was signed off for stress. Craig, who hated seeing me in

---

* Much of the research was conducted in the pub, where, it was noted, workers drank more quickly on a Friday and Saturday – a trait that was put down not only to the fact that Friday was payday, but also to a desire to stretch leisure time to its utmost.

such a state, said it was time to quit. But he didn't want to see me throw my career away. He had another idea. He had been self-employed before, so perhaps it wasn't such a leap for him to suggest getting a business loan and going off to do our own thing. Being our own bosses. The appeal was instant – the chance to stop handing my day over to someone else; to operate according to my own 'logic of need', as E.P. Thompson put it.

Starting the business was my way of combining work life and home life. This combination isn't always easy, I admit. All small businesses are an extension of their owners, which makes separating them from one's personal life challenging to the point of impossible – particularly when you're working with your significant other. It has been one of the hardest things I've ever done, and on several occasions has nearly cost us everything. But nothing has damaged my health more than subjecting myself to the whims of others. I've learned that, when it comes down to it, I need my time to belong to me.

Controlling time played a fundamental role in making empires possible. We are, even now, living by a Christian timetable, regardless of our religion or lack thereof. The year in which you are reading this has been calculated from the birth of Jesus Christ and defined as AD (standing for *anno domini*, meaning 'in the year of the lord'). Colonialists imposed this Christian concept of time on the people they conquered, seeking to regulate the day just as the first religious orders had once called people to prayer with the toll of the church clock.

The anthropologist Edward T. Hall in the late 1950s coined the term 'chronemics' for the study of the perception of time relating to different cultures. According to Hall, Western nations, especially the United States and northern Europe, are largely 'monochronic' societies, characterised by concentrating on one task and by linear processes. They value punctuality and meeting deadlines, are future-orientated and abhor waiting. They are individualistic. 'Polychronic' cultures, by contrast, such as those in Asia, Latin America, sub-Saharan Africa or the Middle East, tend more to multitasking, are more relationship- than task-focused, present- or even past-oriented (e.g. India, China or Egypt). They may not (as in the Sioux language) actually have a word for waiting. When a monochronic society encounters a polychronic society, the result is often a culture clash. These disparities are even present in the ways we greet each other: while a British or American person might simply say, 'Hi, how are you?' a Mongolian person could spend ten minutes asking about how you slept last night and the health of your family. Globalisation has more recently blurred the differences between national behaviours, with smartphones making fast-tapping, fast-talking multitaskers of us all. But there is no question that in previous centuries these distinctions fuelled racial stereotyping and prejudice.

In the United States, colonisers considered Native Americans to be 'savages'. A major reason for this assessment was that their way of working was still closely entwined with the natural world and they seemed unwilling or unable to embrace the Western system of time. They were 'grounded in their violation of the divine imperative to appropriate the world through mixing their labor with nature'. This happened all over the world. Western observers in the 1800s viewed Mexican miners as 'indolent and child-like people' and noted their 'lack of initiative, inability to save, absences while celebrating too many holidays, willingness to work only three or four days a week if that paid for necessities, insatiable desire for alcohol – all were pointed out as proof of a natural inferiority'. Desire for alcohol aside, in an age where so many of us have completely lost our work–life balance, you could argue that our nineteenth-century Mexican miner was the one who really had his head screwed on. Similar accounts can be found about people from Africa, the Middle East and Protestant-ruled Catholic nations such as Ireland.

This was white, male, European time, designed to favour those who determined the means of dividing it at the expense of those who fell under their control. Placing European time culture at the forefront of our social evolution allowed for the inference that all 'others' were 'behind' in their development. Scholars describe this as temporal othering. In the words of international relations scholar Andrew Hom, it meant that Anglo-European time values were perceived as 'mature, adult, and forward-looking; while other cultures become immature, child-like, and backward'. These kinds of stereotypes went on to underpin Western colonialism in its attempt to reform the world in its own image.

For those who fell under the rule of the new industrial age, any rebellion by the workforce was met with dismissal at best and violence at worst. The overriding message was that those unable to adhere to twelve-hour working days, six days a week, from the age of four, in often extremely dangerous and unimaginably unpleasant conditions, were showing their 'natural inferiority'. Needless to say, the leisure-loving wealthy masters and industrial entrepreneurs were not minded to work in the same way.

Time is a commodity – it's something we possess and can sell. Any job we do is a transaction – we're selling (or maybe leasing) a portion of our time to an employer. If an employer tries to make use of our time without paying for it, we feel rightly cheated. At its most extreme, this arrangement can become enslavement, stripping us of the basic human right of liberty.

In Britain the trade unions, legalised in 1824, understood that time lay at the heart of workers' rights. The first fruits of their struggle came in the Factories Act of 1847, when working hours for women and children were capped at ten hours a day, ceding to demands for the 'three eights' – eight hours' labour, eight hours' recreation, eight hours' rest (a description that makes me think of King Alfred's candle clock).* The next success of the trade union

---

* Holidays with Pay Act of 1938 marked the moment that people could holiday without taking unpaid leave. It wasn't until 1998 that the right to a forty-eight-hour week was enshrined in law.

movement was the Factory Act of 1850, which recommended (it was still ultimately the choice of the manufacturer) that all work should stop at 2 p.m. on Saturday and thus ushered in the modern idea of the weekend.

Traditionally, workers had seized their own time off, taking 'Saint Mondays', an unofficial day of rest taken to sleep off the excesses of Saturday and Sunday night. A hangover (to pardon the pun) from when artisans worked six-day weeks, Monday through to Saturday, Saint Mondays persisted right through to the 1870s and 1880s, despite being much disliked by employers. In 1842 the Early Closing Association, supported by temperance societies, persuaded employers that they would see reduced absenteeism and increased productivity if they allowed workers to finish early on Saturdays. This was promoted as an afternoon for wholesome pleasures and 'rational recreation' – a walk in the countryside, gardening or any other pursuit that required daylight – and boosted a burgeoning leisure industry. Theatres and music halls, which once catered to Monday audiences, began to open on Saturdays, and football clubs, first started by churches to stop workers hitting the alehouse too early, played their matches on Saturday afternoons.*

Railway expansion led to an explosion in the popularity of days out. With the advent of the steam engine radically reducing travel times, day-trippers could get further, quicker, meaning outdoor adventures from picnics and hikes to boating or a trip to the circus were possible. Trips to the coast for health reasons had been popular since Georgian physicians first started touting the holistic benefits of briny waters and fresh salty sea air a century prior, and now there was a peak in Victorian

---

* The Victorian era saw the birth of organised sports. The rules of rugby were published in 1845, the rules of football in 1863. Train travel allowed for local cricket, rugby or football teams to travel for away matches, allowing for larger national competitions, like the FA Cup (1871), to be formed. And it wasn't just the teams that could travel; spectators were able to travel too. Older sporting venues like Epsom enjoyed a boom in visitors as people were now able to travel cross-country to watch the races.

beach-holiday escapades. By the end of the nineteenth century the popular seafronts of cities like Brighton and Blackpool were heaving with tourists from increasingly varied walks of life.

For working-class women, 'time off' never seemed to arrive. The typical ten- to twelve-hour working day was just the beginning for mothers and wives, who were expected to continue their labour when they returned home in the evenings to care for their families. Writing as early as 1739, Mary Collier, a washerwoman from Hampshire, lamented:

> . . . when Home we are come,
> Alas! We find our Work has just begun;
> So many Things for our Attendance call,
> Had we ten Hands, we could employ them all.
> Our Children put to Bed, with greatest Care
> We all Things for your coming Home prepare:
> You sup, and go to Bed without delay,
> And rest yourselves till the ensuing Day;
> While we, alas! But Sleep can have,
> Because our froward Children cry and rave . . .
> In ev'ry Work (we) take our proper Share;
> And from the Time that Harvest doth begin
> Until the Corn be cut and carry'd in,
> Our Toil and Labour's daily so extreme,
> That we've hardly ever *Time to dream*.

A similar eighteenth-century Scottish ballad called 'Answer to Nae Luck about the House' tells a story about a man called John who takes on his wife's chores thinking they'll be easier – only to discover how hard her work actually is. He expresses great relief when his wife at last returns home.

It wasn't simply that women had more work to do, often on top of a day's waged labour (many working-class Victorian women were forced back into work as soon after childbirth as they were able), but that their work in the home didn't adhere to the linear result-oriented pattern of the formal workplace. As most of us know too well, domestic chores – washing, cooking, cleaning – are Sisyphean tasks that are never completed. A meal prepared, eaten and tidied away simply creates space to cook another.

Even now, the work of raising children thwarts our capitalist notion of productivity. New parents working in the home have to adjust to being what the psychotherapist Naomi Stadlen calls 'instantly interruptible', whereby the goal of finishing something, anything, is inevitably thwarted by the more clamorous demands of their offspring. Meanwhile the task of raising a child happens imperceptibly, in an accumulation of mealtimes, bedtimes and cuddles (and later, arguments over screen time). Time passes slowly, at toddler pace, stopping every few steps to look at the ants clustering on a wall at the side of the pavement, and then fast, when we look up to find we suddenly have a hairy teenager.

Why does time seem to speed up as we get older? Although parents witness huge changes in their young children, they themselves remain relatively static.* We tend to remember novelty more sharply, experience it as 'slower' in time. We also, instinctively, use the sharpness of a memory as a guide to its recency. The psychologist Norman Bradbury in 1987 described this as the 'clarity of memory' hypothesis. If a memory is unclear we assume it happened further back, whereas those memories that are magical and life-changing – such as the arrival of a child into our lives – will always seem to have happened barely yesterday.

---

* A recent study by researchers from the Institute for Frontier Areas of Psychology and Mental Health in Freiburg, Germany, and Geneva University has now found that parents do actually perceive time to go faster than non-parents.

As Britain's industrial age progressed, the industry that helped the factories run on time was itself in decline. From the giddy heights of its golden age, watchmaking became one of very few British industries that spectacularly failed to industrialise. The economic damage started early on in the nineteenth century, when the French Revolution and subsequent Napoleonic Wars (1803–1815), combined with competition from Dutch forgeries and the inability of British watchmakers to modernise production methods, dealt a devastating blow to the watch- and clockmaking trades. The British watch industry went from being the centre of the horological world in the eighteenth century to the brink of ruin by 1817.

Thousands of watchmakers were now unemployed and facing destitution. One of the organisers of the Worshipful Company of Clockmakers' relief fund in 1817 described going to visit a former London watchmaker's home, where he found the family in a terrible state with:

> hardly a rag to cover them, and children without shoes or stockings, and in want of bread . . . he had a wife and five children. I found the wife and children in a room without a fire, in the month of January last. Rolled up, in one corner of the room, was something in the shape of a bed on the floor; I believe only a bundle of straw in a cloth without sheets, and a thin sort of cotton covering, which was all the whole seven had to sleep on.

Any trade in luxury goods will suffer during times of war and recession. A luxury watch, as one trader put it to Parliament in 1817, is 'the first article put off in times of distress, and the last put on again when distress is removing'. An 1830s report by the Clockmakers' Company bemoaned cheap and cheerful continental watches, like the Dutch forgeries, being commonly found in jewellers', haberdashers', milliners', dressmakers', perfumers', 'French Toy-shops,' and just 'hawked about the streets.'

Competition led to the declining wages of watchmakers. By the mid-nineteenth century, life for apprentices working in Prescot in

Lancashire was 'mostly hell' and journeymen cutters were known by the unflattering sobriquet of 'poverty knockers'. Unlike Swiss and French watchmakers, the British had been very stubborn about scaling up production to produce more affordable watches. These proud master craftspeople, accustomed to making things of such high quality that they were once regarded as some of the finest toolmakers in the world, had been reluctant to start cutting corners to make lower-quality goods. They had resisted manufactories and even employing women.

That reluctance was not felt in the USA, where a slow start in watchmaking accelerated with the adoption of a mechanised version of Swiss *établissage* techniques. Manufacturers like the Waltham Watch Company, founded in 1850 by the American industrial pioneer Aaron Lufkin Dennison, were able to perfect a standardised, machine-built *ébauches* in very large quantities. They were like the ready-made cake mix of the watch world, almost ready to go, just needing a few final additions before they were ready for the oven.

Historically, even *établissage* watches had a great deal of natural variation, because they were assembled by hand. But in the nineteenth century, American watchmaking really found its footing by combining mass production with standardisation through the use of machines. It meant that by the second half of the century the parts, dials and cases of watches could be made in different locations, allowing company owners to make the most of regional skills and even international differences in metal costs. It also meant that, for the first time, parts could be swapped, so, instead of needing someone like me to make a new balance staff to replace the old one on your watch, a replacement could be ordered from a catalogue of parts. This made them cheaper to put together, cheaper to buy and cheaper to maintain.

In 1896, a New York mail-order business, Ingersoll Watch Company, released the cheapest pocket watch so far, marketed for just $1, the price of a day's wage for the average working American.*

---

* The very cheapest English watches, no more accurate than the Yankee, would have been around $12.50 at the time.

It was called the 'Yankee'. Suddenly people from all walks of life – from servants and factory and rail workers to farmers, cowboys, street traders and even their children – were able to access accurate time whenever they liked. In the twenty years that followed their release, Ingersoll sold forty million Yankee pocket watches, enough to supply well over half the population of the United States at the time. Their slogan was 'The watch that made the dollar famous!'*

Ingersoll watches are technically unremarkable, but are incredible watches nevertheless. They're no-frills pocket watches, in budget nickel-plated cases that feign the appearance of silver, with dials made from printed and pressed paper to imitate white enamel. Their movements are bulky and lack finesse. Some of their parts were stamped for speedy manufacture, leaving rough and rounded edges. They look like they should barely function, but function they did, with great success. Ingersoll marketed their Yankee with a one-year guarantee, promising to repair or replace any watch that failed to keep 'perfect time' free of charge. I've handled some of these watches and, remarkably, they're still just about repairable to this day. They can be dismantled and cleaned; repairs can be made to worn or damaged parts. Budget goods today are designed to be thrown in the bin when they stop working. And yet, this watch that cost just $1 could and still can be serviced just like any other mechanical watch of its era.

If Dutch forgeries had left the British industry walking wounded, the sheer scale and immaculate organisation of American mass production delt the death blow. In 1878 one unnamed 'leading London watchmaker' was quoted as predicting 'that Americans would manufacture common watches for the millions, for this would leave British watchmakers to make aristocratic watches for the hundreds'. It was a prediction that proved all too true – only it failed to anticipate the damage that would wreak on what was left

---

* On a trip to Africa in 1910, 24th President of the United States of America, Theodore Roosevelt, proudly described himself as 'the man from the country where Ingersoll are made'.

of the crippled British industry. In the 1870s and 1880s machinery to create standardised watch movements en masse was shipped to the UK from the USA in a last-ditch attempt to catch up. But it was already too late. By the end of the nineteenth century, the once-thriving community of British watchmakers had dwindled to a few workshops. The last manufacturer to produce watches in Britain on a commercial scale was Smiths, who founded their watchmaking division in 1851 and finally ceased production in 1980. Today there are only a few dozen watchmakers left in the UK who have the knowledge and ability to make timepieces in their entirety using traditional methods. Craig and I are among them. Together, we British watchmakers now produce considerably fewer than 100 watches a year.

8

The Watch of Action

'I refused to take no for an answer.'

Bessie Coleman, aviator, 1920s

Some people just hate to be told something can't be done. I'm one of them. I was the first person in my watchmaking class to tackle a verge watch precisely because my tutor thought it would be too advanced for me. In my final year, instead of a clock I decided to make a pendant watch in the form of a dragonfly. In 2011 I met George Daniels – the world's most famous living watchmaker. He wanted to know why I was working at an auction house and not making watches. Fair question, I thought. He asked whether I wanted to make my own watches some day; I said I did. He laughed loudly, and told me he looked forward to seeing them. It took more than ten years to meet George's challenge. Unfortunately, he's no longer around to see what I'm up to.

This trait goes way back. The first novel I ever read was a response to a similar gauntlet, when an unkind teacher (the type who would lock you in a store cupboard or rip up your work in front of the class because of a spelling mistake) told me I could never read Jules Verne's *Around the World in Eighty Days* because it was too long and hard for me. I was eight, and up to that point had preferred science books to novels, but the story – which was full of daring and adventure – piqued my interest. It's ironic that *Around the World*, published in 1872, is itself the story of a wager: the unbelievable (at that time) idea that it might be possible to circumnavigate the globe in eighty days. Phileas Fogg was prepared to bet (and spend) his fortune defying his doubters as he travelled the world by boat, train, camel and sledge, accompanied by his trusty manservant Passepartout (desperately trying to keep on schedule with the aid of his great-grandfather's watch) and tracked by the doubting Detective Fix.

Phileas Fogg's wager wasn't actually that outlandish. The late-nineteenth-century world was, in some respects, considerably smaller than it had been at the start of the century. Since Richard Trevithick invented the first steam locomotive in 1804, railway fever had spread through Western nations, hurtling people and commodities from one place to another faster than ever before. Railways even shrank the vast nation of America – at least tempo-rally. At the start of the century, it would have taken as long as three months for a single letter to be delivered from, say, New York to New Orleans, and then another three months for the reply to come back. By the 1850s, thanks to the railroad, the whole exchange took just two weeks. Improvements in steam-driven ships and the opening of shipping lanes and canals shortened marine voyages: by 1900, the journey from England to Australia took thirty-five to forty days instead of four months. Meanwhile, the arrival of telegraphs (invented by Samuel Morse in 1844) and telephones (invented by either Antonio Meucci in 1854 or Alexander Graham Bell in 1873 depending on who you believe, as evidence suggests Bell plagiarised parts of his design) allowed people to communicate with friends and family across the world in not months or weeks but moments.

The astonishing aviation advances of the Wright brothers, who first took to the air in 1903, made travel faster still. In 1909 Louis Blériot crossed the English Channel (another wager, with a reward of £1,000 from the *Daily Mail*) in just thirty-six minutes and thirty seconds. It was a period of extraordinary possibility. For all kinds of adventurers (and it seems like there were a lot of them at the time!), once-impossible expeditions now seemed eminently achievable. Men and women travelled to the furthest limits of the Earth, drawn by the mystery of the unknown, the draw of getting there first and the thrill of achieving the impossible. But none of them could have achieved these remarkable feats without watches.

When people travelled the world by foot and horse, the fact that noon shifted as you went east or west barely registered. But as steam trains made crossing a country in a morning possible, the complications became all too apparent – even on a small island like our own. Imagine you've agreed to take the train and meet your cousin, a punctual fellow, at Bristol station on Thursday at two in the afternoon. As you prepare to catch your train in London, you check your watch's accuracy against the town hall clock. Reaching Bristol, you fish it out of your pocket again. It is keeping time nicely: two o'clock exactly. But no sign of your cousin. When he does turn up ten minutes later, he's not flustered and makes no apology. It's not until you see the clock at the Bristol Corn Exchange that you understand: Bristol is actually eleven minutes behind London.* That's because up until the mid-nineteenth century time was regulated by the position of the sun in the sky at noon, when the sun was at its highest.

As travel became faster, it became apparent that these small time differences had big – and potentially dangerous – consequences. Many trains still ran on single tracks, so disparities in an agreed time could lead trains heading in opposite directions straight into each other. To solve the problem, standard national time was introduced on British railways on 1 December 1847, and became law in 1880.

The larger the country, the more extreme the challenges of multiple local times. In a nation as vast as the United States, local time, calculated by the position of the sun, could vary by several hours. Before the adoption of railway time in 1883, it contained over 300

---

* Bristol eventually converted to standard time in 1852, five years after it had been introduced.

Entering service in 1938, *Mallard* broke the world speed record for steam locomotive the same year after reaching a whopping 126 miles per hour (203 kilometres per hour) – a record that still stands to this day.

local time zones. Although the new standard time was quick to catch on in cities, those expected to make the greatest leap from local to national time were reluctant to make the change: some rebellious regions refused to adopt national time for many decades. Dual time observations were common in the US until 1918, and it was only in 1967 that the 'local option' in time system observance officially ended.

The next step was a standardised time that could unite the world. On 22 November 1884, the Greenwich Mean Time system – known as GMT – was agreed at an international conference in Washington, DC. It divided the globe into twenty-four zones, each representing 15 degrees of longitude, and an hour of the day. There are a number of reasons why Greenwich, London, was chosen as the meridian for this standard. It was, and still is, the site of the Royal Observatory, which was one of the world's most important locations for the study of astronomy and horology. It was also chosen because it allows the date line, on the opposite side of the world to the meridian, to run through the Pacific Ocean – the only part of the globe that could be divided without causing residents in the same country to be living across two different days. Travellers now had to start doing what to us is second nature: adjusting their watches to a different country's time zone. It's this that provides the final joke in *Around the World*. Passepartout has consistently refused to change his great-grandfather's watch from London time, insisting that the moon and stars are at fault. He is delighted – and vindicated – when twice on the journey his watch is exactly right. Then, as they arrive back in London, Fogg and Passepartout are crestfallen to discover that their journey has taken eighty-one days. But then they realise that they have travelled eastwards. Thanks to GMT, they have won an extra day and, by extension, the wager.

Jules Verne's story is accurate in another respect: by the late nineteenth century, timekeepers no longer had to be expensive to do as good a job for travellers as the intricate chronometers of the eighteenth century. In 1895, just one year before Ingersoll introduced its famous 'Yankee', Joshua Slocum set off on a journey that would make him the first person to circumnavigate the world single-handedly. He'd chosen to leave his original chronometer at home as it needed repair; he'd been quoted $15 for the job and felt it wasn't worth the expense. Ever the frugal adventurer, he purchased in Yarmouth, Nova Scotia, a cheap alternative timekeeper for the voyage. He described it as his 'famous tin clock, the only timekeeper I carried on the whole voyage. The price of it was a dollar and a half, but on account of the dial being smashed the merchant let me have it for a dollar.' Slocum complained how, 'In our newfangled notions of navigation, it is supposed that a mariner cannot find his way without one [a marine chronometer].' But after three years at sea, and a journey of more than 46,000 miles, Slocum returned triumphant. His one-dollar tin clock had served him loyally throughout.

A quiet village in Nottinghamshire is home to one of the most significant watches in the history of exploration. The Museum of

Timekeeping's collection, housed in Upton Hall, consists of thousands of watches and clocks made over the centuries, donated by all kinds of people. The result is an irregular cabinet of horological curiosities that, to me, offers far more interesting insights into our intimate relationship with our timekeepers than you would find in a more strategically curated collection. Exceptionally valuable and rare chronometers and longcase clocks sit alongside cases of mass-produced 1940s Metamec electric bedside clocks, whose vivid shades of bright orange, retro brown pearlescent and duck-egg blue Lucite acrylic take me back to childhood visits to my grandparents' house.

Housed in a glass cabinet in the museum's watch gallery is an unassuming and rather dishevelled pocket alarm watch that dates back to the early twentieth century. Its dark steel case, the colour of gunmetal, is pockmarked with old reddish-brown rust. The glass that protected its dial is missing, along with the hour, minute and small seconds hand. The only hand that remains indicates that the alarm had been set to chime at about twenty past eleven. Its white enamel dial is still comparatively bright: glass enamel, though brittle and susceptible to cracking if the watch is knocked or dropped, never tarnishes or fades, and keeps its vivid lustre even in extreme conditions. Likewise the black Arabic numerals are as clear as the day they were fired into the enamel. Only the dabs of once-luminous paint that sit on the minute track, marking each hour, have matured from their initial glowing green to a flat dirty brown. Instead of an Albert chain, the bow of the watch is secured to a worn woven bootstring, with a rusty safety pin at the other end. The movement of this watch hasn't run since soon after Thursday, 29 March 1912, the date its owner, Captain Robert Falcon Scott, wrote his last diary entry before he and his remaining crew succumbed to the Antarctic weather, just 12.5 miles from the camp that would have provided them sanctuary.

It's been said that this pocket watch was an essential tool in preventing Captain Scott and his team from sleeping for too long and freezing to death. While I was researching this book, I was fortunate enough to be able to speak to polar explorer Mollie Hughes, who made the trip to the South Pole at a similar time of year to Scott. She said the twenty-four-hour sunlight of the Antarctic summer meant that, once inside the tent, the air was warm enough to sleep in base layers and dry out wet clothes (although her gear was considerably more advanced than Scott's knitted jumpers and gabardines). The greater danger, Hughes found, was the risk of accidentally overexerting yourself by not keeping track of time, because there was no nightfall to indicate the day's end. We use the sun to regulate our days more than we realise. Our biological clocks are geared to wake us up in sunlight and put us to sleep after dark. Without the celestial trigger of darkness, it's harder for our brains to tell us when the day is over and it's time to set up camp. Mollie said the most dangerous point in her trip was not the two-week-long storm she endured at the beginning but when she walked for too long each day trying to make up subsequent lost time. If she was too exhausted to set up camp properly and fell asleep exposed to the harsh Antarctic winds, the consequences would have been fatal.

I scoured Scott's journal, which was recovered along with his watch when his body was found in November 1912, looking for clues that he used his watch to limit his hours of sleep to keep everyone from freezing to death, but he made no mention of it. I did, however, find regular evidence of an almost obsessive regulation of his group's daily routines. His meticulous journal details the time of virtually everything. He marks the time they wake up in the morning, when they start preparing breakfast and how long it takes for all the party to finish eating. Like Mollie Hughes, Scott is mindful of the dangers of disorientation in twenty-four-hour sunlight. He imposes structure on the men's time wherever he can, with a strict regime of travel times and meals, so it's likely he used his pocket alarm watch to chime out when the day's walking had finished and then they would set up camp, ringing the bell for supper, waking the crew in the morning and telling them it was time for breakfast.

He even schedules a service on Sundays, as well as a half-hour slot before it in which that week's hymns were selected, and arranges a series of lectures. It's hard to imagine these explorers – who were, after all, on a life-and-death journey into the unknown – settling down for a pleasant illustrated talk on 'Antarctic Flying Birds', but these activities not only kept spirits high but created a link back to the temporal rhythms of normal life: a sense of normality in the white expanse of nothingness.

When I held Scott's watch in my hands, I felt humbled. It seemed as if his whole life, his hopes, his ambitions, his fears, even the loved ones he left behind, were somehow contained in that otherwise ordinary object. This little machine ventured into the unknown with him and served him until the very end. The mere sight of it conjured up a vivid mental diorama: I imagined the places it had been, the things it had seen, the conversations it had eavesdropped on while discreetly hidden in Scott's pocket. Mechanical watches like this need to be wound to function. With no one around to keep them going, they stop, falling silent with their keeper. They also succumb to the elements: their oil congeals in the cold, their case lets in damp causing ferrous parts to slowly rust, their wheel train gradually seizes solid. To me the museum's decision not to restore this watch back to working order feels like the right one. It would feel somehow disrespectful to Captain Scott to wake his loyal companion from its Antarctic-induced sleep.

As the nineteenth century rolled into the twentieth, watches became indispensable allies of action, and action changed how watches were worn. In the second half of the nineteenth century, soldiers involved in Britain's various overseas campaigns reported the benefits of strapping their fob watches to their wrists: this way they could quickly and easily tell the time without having to fumble around in their pockets in the heat of battle. These wrist-worn watches may have evolved from sweetheart's watches, which young women gave to their lovers as they departed for war. To keep these watches safe, men started making leather pouches, known as wristlets, which held the watch securely while strapped to the wrist. Photographs of British soldiers stationed in northern India show them wearing watches in these 'wristlets' at the time of the Third Burma War in 1885. This was a highly important development: in my view it marks the birth of the commercial mass-market wristwatch as we know it today.

A leather cup strap holding a fob watch,
allowing it to be worn on the wrist.

What started as an improvised trend was quickly capitalised upon by manufacturers. In 1902 Mappin & Webb produced the 'Campaign' watch, an Omega fob watch fitted inside a leather cup wrist-strap. The advertisement promised a 'Small compact watch in absolutely dust and damp-proof oxidised steel case. Reliable timekeeper under the roughest conditions. Complete as illustrated £2.5s. Delivered at the Front. Duty and Postage Free.' The front in this instance was the Second Boer War, fought between Britain and the Boer republics from 1899 to 1902. While Britain's scorched-earth tactics and brutal internment camps have led many to label it our most shameful hour,

Omega cheerfully reported that its Campaign watch had proved 'life-saving'.*

At home the wristlet was pressed into a different kind of action with the cycling craze of the 1890s, which the *New York Tribune* claimed was more important to humankind 'than all the victories and defeats of Napoleon, with the First and Second Punic Wars . . . thrown in'. In 1893, an advert placed by London retailer Henry Wood claimed that the specialist 'cyclist watch wristlet' was 'the only way to carry a watch on a Cycle without injuring it, and always to hand'. Another, from 1901, describes them as being perfect for 'the tourist, the bicyclist, the soldier'. One of the advantages of marketing these watches for cyclists was that women, who had traditionally worn bracelet watches, were as enthusiastic about the sport as men. That advert features an illustration of a frilly-cuffed feminine hand with a watch round its elegant wrist.

Cyclists weren't the only people of action who needed to see their watches while their hands were occupied. For the world's first pilots, an accurate timekeeper was a life-saving essential, arguably even more so than chronometers were to ships' captains. Pilots didn't simply use them for navigation but to calculate fuel consumption, airspeed and lift capacity. The first dedicated aviation timepiece or 'pilot watch' is said to be the 'Santos', designed by Louis Cartier for the Brazilian aviator Alberto Santos-Dumont in 1904. It came about after Santos complained that he spent far too much time fumbling around in his pocket for his watch when he really needed to be keeping his hands on the plane's controls. Cartier's design had an unusual square case and dial with highly legible black Roman numerals – tough, masculine and perfect for quick reading. The Santos-Dumont watch went into commercial production in 1911 and is still being produced today, more than a century later.

---

* The internment of some 150,000 refugees in British-run concentration camps led to the deaths of over 15,000 native Africans and around 28,000 Boers – three-quarters of whom were children.

Another Swiss watch company, Longines, was, like Cartier, quick to throw its weight behind the wristwatch, as developing a watch suited to the demands of aviation became the longitude challenge of its day. Amelia Earhart wore her Longines chronograph for two of her Atlantic crossings. Another early American aviator, Elinor Smith, set a plethora of records in the late 1920s and 1930s for solo endurance, speed and altitude flights with the assistance of her Longines timekeepers, visible on her wrist in almost every image of her. At sixteen, she was the youngest person in the world to become a certified pilot. Shortly after, she made her name flying under four of New York City's bridges, a challenge she took up after a pilot who had tried the same stunt and failed told her she would never be able to do it. The *New York Times* were so convinced that she too would fail that they prepared her obituary – eighty years before she eventually passed at the ripe old age of ninety-eight. I like to think we'd have got on.

One of the great perils of air navigation was the lack of visible markers to which you could orient yourself: sometimes time was the only marker you had. Philip Van Horn Weems, one of the great pioneers of air navigation, remarked:

> There is no disgrace in being lost in the air. This happens to the best navigators. The important thing is to reduce the periods of being lost or uncertain of position to the lowest limit humanly possible.

It was to Weems that Charles Lindbergh turned when he wanted to learn the art of celestial navigation shortly after completing the first non-stop Atlantic flight, from New York to Paris, in the *Spirit of St Louis* at the age of twenty-five. Together, Lindbergh and Weems invented the Longines 'Hour Angle' watch, with the first calibrated rotating bezel in wristwatch history. The bezel allowed its owner to calculate the angle of the sun relative to the Greenwich meridian: the so-called hour angle that gives the watch its name. Everything about it was adapted to early aviators, from its extra-long strap (for wearing over a bulky flying jacket) to its supersized crown, so that the watch could be wound while wearing flying gloves.

When mountaineer Conrad Anker scaled Mount Everest in May 1999, his goal wasn't to reach the top. Anker was instead trying to work out a mystery. At around 700 metres below the summit and 8,157 metres above sea level, he solved it:

> I was curious, I stopped, turned around, and there was a patch of white. It wasn't snow, it was matt, a light-absorbing colour, like marble. As I got closer, I realised this was the body of one of the pioneering English climbers, frozen onto the mountainside.

The body was male. His right leg was fractured, his arms were outstretched, his clothes had degraded, and the exposed skin on his back was a sun-bleached shade of milky white. Within the remaining tatters of the body's weather-beaten gabardine jacket, Anker found a tag embroidered in crimson thread – 'G. Leigh Mallory'.

Seventy-five years earlier, in 1924, George Mallory and his climbing partner, Andrew Irvine, had gone missing close to the top of Everest as they attempted to become the first to scale the mountain. Until this moment, Mallory's fate had remained one of the great mysteries of mountaineering. To this day, we are still not certain whether he met his end on his journey to, or from, the summit.

Though Mallory's body, which was too difficult to remove, remains exactly where he fell a century ago, a number of his personal possessions were retrieved and are now held by the Royal Geographical Society in London. They include his broken altimeter, snow goggles, knife, matchbox and a silver watch. The watch seems frozen in time. The hands have rusted to dust, leaving nothing but a burnt ochre shadow on the bright white vitreous enamel dial, suggesting the watch last ticked at approximately seven minutes past five, or possibly twenty-five minutes past one, depending on which shadow you read to be which hand. The black outlines of the Arabic numerals still contain a little of the radioactive luminous paint that would have made it easier for Mallory to read his watch at night or in the low light of a snowstorm.

Surprisingly, the watch wasn't found on Mallory's wrist, but in his pocket. There were also no traces of the glass (or crystal) that protected the delicate dial and hands. It's been suggested that it became lost, so Mallory placed the watch in his jacket pocket to protect it from further damage. It's a theory that certainly holds horological weight. In the days before plastic gaskets, which create a friction fit between the crystal and the case of most modern watches, or specialist adhesives that cure under ultraviolet light, watch crystals were commonly fitted using thermal expansion. The bezel, or uppermost band of the case in which the crystal sits, was heated, causing it to expand ever so slightly. The cooler crystal was introduced and, as the bezel cooled and returned to room temperature, it shrank onto the crystal, embracing it firmly. This made the crystal particularly susceptible to extreme changes in temperature. In fact, Mallory's watch has survived in surprisingly good condition for a metal object with ferrous composition that spent seventy-five years near the summit of the world's highest mountain – a testament, perhaps, to the no-nonsense design and robust build quality of these watches.

I made my own journey to Everest – Base Camp rather than summit – in 2011. It was an awe-inspiring experience, of course: how can you not feel insignificant compared to the magnitude of a mountain? Everything took longer because of the altitude, a little like moving through treacle. It gave me an appreciation of why timing is so important for the climbers higher up. Low on oxygen and only able to move very slowly, they often need to be up before dawn in order to pack up camp and make good progress before the ice starts to thaw, increasing the risk of avalanches.

As I made my way up I entered the village of Namche Bazaar, built into the side of the mountain (in Nepal, most things seem to be built on the side of a mountain), which offered the last experience of something close to civilisation for those headed to the summit of Everest. I recall seeing streets packed with market stalls selling sweets, water or hiking gear for those who had damaged or lost theirs en route. Here what the Sherpas referred to as 'North Fakes', imported from China, were a reminder of how near we were to the border. As I looked closer, I started to notice older items, the kind you might find at an antiques market: a rusty pair of iron crampons, wooden-handled ice axes, like those an explorer might have used in the 1920s or 1930s. Then I spotted smaller, more personal items: a pair of spectacles, a wallet. I asked our Sherpas where these objects came from. They told me that, a few years ago, in response to the amount of man-made rubbish now cluttering the sides of Everest, Nepal's government had instructed Sherpas to collect the scattered detritus. At first they were paid by weight, but when that fast became too expensive, the government switched to the budget option of paying by the day. To make back the loss, the Sherpas started to sell the more interesting objects they found as a lucrative side income. The original owners of these items were likely dead, and we would probably never know their stories. I couldn't stop thinking: what if Mallory's watch had turned up here?

At Chukla Lare, a memorial that climbers pass through as they approach Everest from Nepal, Buddhist prayer flags and mounds of stones create a place to pause and reflect on those who have lost their life on the mountain. There are estimated to be over 100 bodies on Everest, left there as they are too heavy and dangerous to retrieve. It's accepted by mountaineers that, if they succumb to the mountain, there's every chance they will be left to become a part of it.

Mallory's watch, like Scott's before him, would have been a crucial tool in determining where he was both in time and space. But you can't take these objects' accuracy for granted. Watches of this sort have a typical running duration of thirty hours and require regular winding by hand. Mallory would have had to remember, every single day without fail, regardless of the weather, his exhaustion and countless other distractions, to devote some moments to winding his watch. If he faltered, it would falter. If an explorer like Mallory wanted to survive, the first, and last, thing they needed to do was keep their watches alive.

Modern explorers can take their timepieces for granted, as we all do. Whether we use a traditional watch, our computer or a mobile phone, we trust the time to be there whenever we want it. But for those who set out on adventures at the start of the twentieth century, this was not the case. The world might have shrunk but once they had embarked, they were alone and their watch was their only way of ascertaining their whereabouts on this vast, lonely planet. The explorer was reliant on their wits, and their watch.

9

Accelerated Time

They leave their trenches, going over the top,
While time ticks blank and busy on their wrists,
And hope, with furtive eyes and grappling fists,
Flounders in mud.

Siegfried Sassoon, Attack, 1918

One day in May 1905 a twenty-six-year-old patent clerk was travelling home from work through central Bern in Switzerland when he heard the town's famous medieval clock – the Zytglogge – chiming out the time. When he looked up at its vast and elaborate clock face he was struck by a curious thought. What would happen if he was sitting in a streetcar travelling away from the Zytglogge at the speed of light? His watch (a Swiss silver pocket watch dating back to about 1900) would carry on ticking out the time as usual, but if he looked back at the clocktower the time would have appeared to stop.

A few months later, the patent clerk – who happened to be called Albert Einstein – published a paper, 'On the Electrodynamics of Moving Bodies', in the German journal *Annalen der Physik*. It would go on to fundamentally alter our understanding of time, the world and the universe. Einstein's theory of relativity argued that time was not an absolute, unchanging ticking clock, as Isaac Newton had claimed centuries earlier, but a flexible dimension that could be stretched and distorted by space, gravity and even personal experience. Einstein demonstrated that time appeared to travel more slowly as gravitational force increased and that it was similarly distorted by the speed of its observer.[*] Time slows down if you're close to a massive object or if you're travelling at high speed, which has the astonishing result that your clock would run faster at the top of a skyscraper than on the ground floor and slower in a moving car than in a stationary one. While this may not affect the kind of watches I make, it needs to be taken into consideration for GPS systems, which are inside satellites whizzing around at huge heights about the Earth at tremendous speed. For Einstein, time and space were *relative*. 'Time and space,' he summarised, 'are modes by which we think and not conditions in which we live.' He even went so far as to claim that the fundamental concepts of past, present and future are little more than illusions.[†]

Einstein's revolutionary work on time theory occurred while time-keepers were entering their most rapid state of development. While Einstein was developing his theory of relativity, some 40 miles away on the other side of Lake Neuchâtel, another ambitious young man was developing his own plans. Hans Wilsdorf began his career as an interpreter and clerk for an exporting firm in the Swiss watchmaking centre of La Chaux-de-Fonds, but in 1903, the twenty-four-year-old German moved to England. He established himself in Hatton Garden, Edwardian London's jewellery centre, just a stone's throw from the capital's old watchmaking centre of Clerkenwell. Wilsdorf

---

[*] It's worth noting that the subject of relativity itself was not new; it was Einstein's in-depth exploration and rationalisation of it that was revolutionary. Physicists such as Galileo and Lorentz experimented with relativistic mechanics and, much earlier, Polynesian navigators had employed a system whereby their vessel was imagined as a stationary object while the world moved beneath it.

[†] As Einstein said in a letter to his friend, the engineer Michele Besso, 'For us convinced physicists the distinction between the past, the present, and the future is only an illusion, albeit a persistent one.'

had a plan for a new kind of watch business that was to make him one of the most influential horological entrepreneurs in history. Wilsdorf had read reports of soldiers who had served in the Boer Wars wearing their pocket watches on their wrists. He was convinced that this was the gentleman's watch of the future. Hardly a theory of relativity, you might think – but in 1905 the pocket watch had reigned unchallenged for nearly four centuries, longer even than Newton's theory of gravity.

If an eighteenth-century watch turned up on my workbench, I would genuinely struggle to tell whether it had belonged to a man or a woman. But in the century that followed, the differences became pronounced. As women were increasingly cast as frail and emotionally temperamental, their watches become correspondingly delicate. The pocket watch shrank down to a 'fob watch', worn on a short ornamental chain or pinned like a brooch, while small timepieces mounted on a bracelet or cuff became all the rage. These bangle watches were as much jewellery as they were functioning watches. They were often gold and embellished with vividly coloured enamels, diamonds, split pearls, and gemstones like sapphires, rubies and emeralds. In the nineteenth century, wristwatches were for women.

One of my favourite watches I have ever restored was one such timekeeper. It was made in around 1830 in a lovely warm-coloured gold – a shade you only see in watches and jewellery made before the mid-twentieth century, when higher amounts of copper were used in the alloy. The dial is nestled in a wide bangle formed of gold snakes, decorated with brilliant white and jet-black enamel and gleaming red and green garnet eyes. It reminded me of a warrior's cuff, perfect for Wonder Woman if she ever needed to attend an extravagant nineteenth-century ball, but not the sort of thing you would throw on to pop to the local shop. Practical it was not.

Watches held in wristlets might have been useful for men in action, but civilian men at the turn of the twentieth century considered them effete. Newspaper cartoonists lampooned the new fashion, and men caught wearing them risked being called a sissy. Manly men wore pocket watches – even cowboys. When Levi

Strauss introduced his iconic 501 jeans in 1873, the small inner pocket at their front right was designed to hold a pocket watch – a quirk that persists in jeans to this day. One account from 1900 documents how a trial shipment of wristwatches from Switzerland to the United States was returned by the retailer on the grounds that they were 'unsaleable in the States'. In one 1915 edition of the American comic strip *Mutt and Jeff*, Mutt shows Jeff his new wristwatch. Jeff scoffs: 'Wait a minute – I'll go and get your powder puff.' And yet canny Wilsdorf had a hunch that, with the right marketing, men could be turned on to wristwatches, and that wristwatches would become the defining timepiece of the future.

Wilsdorf didn't have an easy start in life – he was orphaned at the age of twelve and packed off to boarding school in Coburg – but it left him with an independent attitude as well as a small inheritance. He had managed to save a bit from his early employment in La Chaux-de-Fonds, but to get his idea off the ground he needed another investor. His solicitor then introduced him to an Englishman called Alfred Davis. Wilsdorf lost no time in persuading him of the brilliance of his idea: to buy watch movements in volume from a Swiss manufacturer and couple them with pre-made cases to supply the English market.

In 1905 Wilsdorf and Davis began importing movements from the Rebberg district of Bienne, or Biel, in Switzerland from a factory owned by Jean Aegler. These were shipped into the UK and fitted into cases. Some of the cases were made in Switzerland, and some were made by companies like Dennison (founded, as we have seen, by Aaron Lufkin Dennison of Waltham fame), who had a manufactory in Birmingham. The watches were sold under the name Wilsdorf & Davis, with the initials 'W&D' appearing inside the watch cases, while the movements were marked 'Rebberg'. The design was the same utilitarian style chosen by George Mallory for his expedition to the summit of Everest.

Dennison, Wilsdorf & Davis's case manufacturer, had relocated from America to Birmingham in search of a very specific Brummie skillset. Birmingham was the gold and silversmithing capital of the world. At the turn of the century, around 30,000 specialist crafts-people were employed in the trade in and around the Jewellery Quarter. Harnessing their skills meant Dennison was able to establish a factory that became one of the world's most prolific makers of watch cases. By the early twentieth century the Dennison Watch Case Company, located within walking distance of my workshop today, was exporting throughout the USA, and producing cases for Waltham, Elgin and Ingersoll, as well as some of the Swiss greats like Longines, Omega, Jaeger-LeCoultre and now Wilsdorf & Davis.

Ultimately, the Dennison Watch Case Company would succumb to the same fate as the rest of Britain's industry, winding down production before finally closing the factory doors in 1967. A few years ago, Craig and I decided to go and have a look for any last traces of what was once one of the greatest names in the British watch trade. Armed with some 1980s photos of the factory and an old map, we headed to its likely location. All we found was a tarmac expanse of nothing – the building had been replaced by an NHS car park. Deflated, we were about to return home when at the far end of the car park we spotted some grass and an old brick wall. As we approached, a booming voice crackled through unseen speakers. 'You are on CCTV, leave this area immediately! You are on CCTV, leave this area immediately!' By unspoken agreement, we ignored the voice, and Craig gave me a leg-up so I could peer over. The main building had been demolished, but there, next to a crumbling ivy-covered section of factory wall, were the rusting remains of some green industrial rolling mills. They would once have been used to thin down sheets of metal like a rolling pin pressing on a sheet of puff pastry. Behind them I could see a wing of the factory, just one small room and, through the long-since-shattered glass in the square metal window frames, a hint of the dark and overgrown workshop within.

Wilsdorf & Davis had barely got going when storm-clouds started to gather over Europe. The outbreak of the First World War generated a huge level of anti-German sentiment in Britain. The Aliens Restriction Act of 1914 meant Germans in England had to register with police and were forbidden to move more than 5 miles. German businesses were closed down. There were anti-German riots in the streets, and homes were attacked. Manufacturers were keen to avoid any association with the country. The London retailer for Swiss-based manufacturer Stauffer, Son & Co. was forced to issue advertisements to remind the public 'that all the Watch Bracelets supplied by them are British Made as Messrs. S., S. & Co. have never Stocked German Bracelets'. Wilsdorf himself was married to a British woman – Alfred Davis's younger sister – and was a proud Anglophile, but he knew that his conspicuously German name was going to be bad for business. In 1908 Wilsdorf & Davis registered a new name for the company, though it was not until 1915 that they officially started to call themselves the Rolex Watch Company Ltd.

When Craig and I started our workshop we weren't (unlike Wilsdorf and Davis) thinking big. Our small business loan of £15,000 was just enough to secure the rental of our first tiny single-room workshop, buy a couple of old desks (proper watchmaking benches were too expensive, so we built platforms to raise vintage office desks up to the correct height for watchmaking), stock up on some essential hand tools and buy our cleaning and timing machines* (which we still have!). The money ran out almost as soon as it arrived and the first few years were a phenomenal struggle as we figured out how – the irony is not lost on us – to charge for our time, a struggle shared by most creative people. We lived below the poverty threshold for the first eighteen months and regularly ended up selling

---

* A timing machine will listen to the tick of an escapement and tell you, by tracing a line on a graph, whether the mechanism is running fast or slow.

things on eBay to make our rent. We couldn't afford to heat our home in the first winter and ice lined the insides of the walls. On the coldest nights we slept fully clothed in hats and gloves with our cat. Finding the energy to pull sixty- to seventy-hour weeks when you're freezing and living off cheese and bargain pasta is something I never want to experience again. It took seven years before we were on a regular wage.

Craig had developed an infatuation with these very early Rolex Rebbergs at his previous employer, and his reputation for working on them had stuck now we were self-employed, to the extent that clients actively hunted him down. Although early Rolex adverts emphasised the watches' accuracy, announcing them as 'Rolex Watches of Precision' and boasting that they held 'twenty-five world records for accuracy', this was not what drew Craig to them. In fact, in his experience, Rebbergs are not particularly high-quality movements. On close inspection, the edges are rough. Flaws in the design mean they wear themselves out and often need new replacement custom-made components to compensate for their increasingly baggy bearings. The majority have only fifteen jewels in their movements, leaving some bearings subject to unnecessary friction, while the design of the winding stem causes wear to the plates. Sometimes they would have been through the mill with other repairers over the decades and important elements, such as the balance staff, had been replaced with a badly made or adapted part. In early Rolex Rebbergs issues like a later replacement balance wheel that hasn't been set up and adjusted correctly, or the wrong hairspring, both of which cause huge problems with timekeeping, are common. But from the start there was an irresistible aesthetic charm to the Rolex. They are a long way from perfect – but they have an appealing sturdiness to them. Craig describes them as tractors – but they are tractors with an undeniable romance.

Wars – like necessities – are the mothers of invention. Conflict always generates intense periods of investment and innovation in science and technology, because better equipment offers significant advantages on the battlefield. But it also creates unexpected

A Rolex Rebberg watch movement dating from around 1920. Craig derived almost as much joy from drawing this as he does from working on them.

inventions, solutions to unforeseen problems. The First World War – which Lenin once called 'the mighty accelerator' – thus gave us blood banks, stainless steel, tanks and drones – but it also gave us the commercial wristwatch.

One thing the war didn't accelerate was Einstein's theory of relativity. The conflict shut down European scientific collaboration, and his theory was only confirmed, by the British scientist Arthur Eddington, in 1919. But in another respect, the war was a perfect fulfilment of his ideas. The conflict was fought on numerous fronts simultaneously, with traditional cycles of time – day and night, even the seasons – ripped to shreds by the perpetual destruction of trench warfare. Technological advances, meanwhile made communication increasingly instantaneous, while the experience of the war created vast gulfs in space – between one side of no-man's-land and the other, between front line and home front – and indeed time, between the prelapsarian pre-war era and the nightmare that followed it. In some ways, the war itself was the theory of relativity made real: a zone in which time and space were destroyed and remade at vertiginous speed.

And yet timekeepers played a vital role in its execution. Fighting on the Western Front was characterised by trench warfare and synchronised attacks. Tactics such as the 'creeping artillery barrage' employed at the Battle of the Somme, which involved sustained artillery fire according to a precise schedule to allow troops to creep closer to the enemy, were highly time-sensitive. As thunderous shelling drowned out orders, timed signals replaced audible calls. Multiple units communicated by telegraph to mobilise at an arranged time. Crawling through the trenches it was near impossible to reach for a pocket watch, so Allied soldiers embraced the wristwatch. Demand was such that it was no longer adapted, as it had been in the Second Boer War, but designed for the purpose. The 'trench watch', as it became known, was fitted with wire lugs to hold the watch directly onto the wrist and could be fitted with a 'shrapnel guard' to protect the fragile glass in battle.

In the Boer War soldiers had had to procure wristwatches for themselves. Now, however, they were retailed in bulk to the military and could be issued to soldiers as kit, along with their uniform, rifle and bayonet, or bought at a discount through Army & Navy stores. Adverts, meanwhile, helped counter any lingering resistance to the wristwatch's historic femininity: 'If HE is fighting at the Front or on the Sea, this Wristlet Waltham will tell him the right time,' proclaimed one in 1914. 'Specially made to withstand rough wear and keep good time under the most trying conditions.' The more affluent officers would often buy fancier versions simply because they could. Gold versions of trench watches are referred to as 'Officers' trench watches'.

Most of the early trench watches were made in Switzerland, where watchmakers had benefited from buying American machine production equipment to boost production. Brands such as International Watch Company (IWC), Omega, Longines and of course Rolex were, unsurprisingly, the first to get in on the action. Their trench watches were simple and functional, with cases formed almost like the sort of pebble you could skim across water and usually made in nickel or brass. These were watches made in volume, created to accompany their owners into the most dangerous of circumstances, whether that be in battle against a human adversary or in Earth's most extreme climates. One 1916 Rolex Rebberg that Craig restored had been bought by the client's grandfather from a man who had served in the Persian Gulf. It came to us tarnished, scratched and dented, missing its bezel and glass. But even with no shock setting or waterproofing, this watch had seen its owner through desert combat and had been worn daily for decades after. It had done what it was designed for.

In the gloom of the trenches at night, soldiers depended on the glow of their watches to read the time. The dials of a trench watch were usually enamel, often bright white, with the numerals applied in luminous radium paint. The hands were skeletonised, pierced out to create hollow spaces down their arm and at their tip, so they too could be filled with glow-in-the-dark paint. The pioneering physicist Marie Curie, together with her husband Pierre, had discovered radium from the uranium-rich and radioactive ore uraninite in 1898. It had rapidly developed a public reputation as a kind of super-element. Thanks to its success in treating cancer, it was promoted as a treatment for everything from hay fever to constipation.* For the watch industry, it was one of radium's decay processes that proved most compelling: when radium was mixed with a phosphor, a radio-luminescent chemical, its action on the phosphor produced a ghostly pale-green fluorescence. The use of radium paint to illuminate the dials of timepieces and scientific instruments quickly caught on. By 1926, US manufacturer Westclox alone was making 1.5 million luminous watches each year. The demand for glow-in-the-dark watch and clock dials, aeroplane instruments, gunsights and ships' compasses was predictably enormous. By the end of 1918, a year after America joined the war, one in six American soldiers owned a luminous watch.

---

* By the early 1900s, factories monetising this new element's glowing properties had sprung up around the world. It was marketed as a health product and incorporated in everything from edibles, like radium-fortified butter and milk, to toothpaste ('for teeth so clean they glow!') and cosmetics. It was even impregnated in clothing and worked into lingerie and jockstraps. The fact that, around the home, radium-laced fly spray was touted for its ability to exterminate pests was not a connection anyone was prepared to make. Radium was a big-dollar business, and those marketing it were keen to suppress any negative press.

Dial factories sprang up across the US and in Switzerland and the UK, with thousands of young women employed to hand-paint the numerals on tens of millions of luminous watch dials.* Radium painting was a prestigious and coveted job – the painters were considered to be skilled artists. Each and every dial went through stringent quality control in a dark room to check its accuracy. Too many mistakes would result in the sack. Out of respect for their talent, wages for radium painters were unusually high, particularly for women of that era. Employers believed that women were the perfect artisans for dial painting, as their small hands were well suited to this intricate and detailed work. It might also be that women were seen as more expendable than men. Male radium

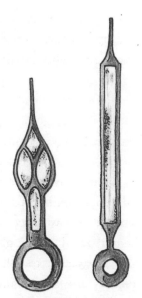

Skeletonised watch hands in-filled with luminous paint.

workers were given protective lead aprons but female radium workers weren't. But women were delighted to take on the work. They were paid by piece, rather than a salary, meaning the more dials they painted each day the more they earned. Some earned as much as three times the average factory worker, taking home more money than their fathers. On average, women could earn the modern equivalent of $370 per week, while the fastest could earn up to $40,000 a year. As many of the factories were established in poor working communities, these wages were life-changing, enabling women to support their families and save for their futures. Although adverts asked for over-eighteens, lax enforcement meant girls much younger managed to gain employment, some as young as eleven.

Radioluminescent paint, given alluring brand names like Undark and Luna, was dubbed 'liquid sunshine'. Dial painters had to mix

---

* The work and its harrowing consequences in the USA are the subject of Kate Moore's brilliant exposé book, *The Radium Girls*.

their own paint in a crucible using a small dab of powdered radium, a drop of water and gum arabic adhesive. In Switzerland, they applied the paint with glass sticks; in France, they used sticks with cotton wadding on one end; in other countries they used sharpened wood or metal needles; while in the US they used an incredibly fine camel-hair brush to apply the paint in lines as fine as 1 millimetre in thickness. The brushes were so delicate they had a propensity to spread, and so a technique called lip pointing was used, introduced by women who had previously worked painting china ceramics. The 'lip, dip, paint' process involved the painter using their lips to form the brush into a fine point, before dipping it in the radium solution and painting it onto the dial. Although radium paint was equally dangerous wherever it was used, it was this application process that proved so deadly in America.

Initially there were no concerns over its safety – quite the contrary. The dial painters were led to believe that radium was good for them – after all, this was a health product, used in expensive face creams and make-up. Not only were they consuming radium as they formed their brush with their lips, they ended up covered in the radium powder that filled the air of the factories. The workers could be seen glowing with that eerie green light as they headed home in the evening twilight. Local residents remarked that they looked like ghosts. The factories were generally happy places, and the women who worked there felt privileged to do so, playing their part helping soldiers on the front line. Some of the painters were known to play a game of sending secret messages to soldiers by scratching their name and address on the back of the watch case and waiting to see if its eventual owner would write to them. Sometimes they did.

From very early on in its use, management knew there were hidden dangers. Sabin Arnold von Sochocky, the paint's inventor, who eventually died from the long-term effects of radiation exposure, had worked with the Curies, who, by this time, had many radium burns themselves. He himself had amputated the end of his own left index finger after radium became embedded there. While it is true that radium has the power to destroy cancer, it has no ability to differentiate between healthy tissue and a cancerous tumour – it destroys everything in its path. Workshop managers and company executives reassured themselves (comforted no doubt by the huge profits they were making from this booming business) that the quantities of radium the painters consumed were too minute to harm them. But with some girls lip-pointing as much as two times for each numeral, and the fastest workers completing up to 250 dials a day, these amounts added up. The body, mistaking radium for calcium (to which it has a similar chemical nature), delivers it to the bones, where it proceeds to eat away at them – slowly, like a ticking bomb.

This cumulative exposure caused the bones of its victims to rot. Most of the painters' symptoms began with their teeth, which became sore, then loose, then fell out or had to be removed. The gaping holes in their gums rarely healed, causing ulcers and infections that left the bone exposed. As the disease progressed and necrosis of the jaw set in, fragments of bone broke away. The affected women were able to pull chunks of their jawbone out of their mouths.

For those who survived the risks of sepsis and haemorrhaging, the radium was still wreaking havoc in their skeletons, riddling their bones with sponge-like holes until they broke or disintegrated. The pain was excruciating. It left these young women in their teens, twenties and thirties crippled. Cancer was another killer. Rare sarcomas, many starting in the bones, would appear in the years that followed the exposure. Doctors, including those employed by the very companies that manufactured radium dials, repeatedly reassured them that the radium girls' problems were due to feminine nerves, hormones and hysteria – despite tests revealing that these women were now, literally, radioactive. The first fatality, Mollie Maggia from New Jersey, who died in September 1922 at the age of twenty-four, was at first believed to have died from syphilis, on no sounder basis than that she was a young single woman who lived alone.

I once heard about a package of old military watches being stopped by airport security. Decades after their manufacture, they still registered as radioactive. However, the amount of radium in one watch dial is so small it can be disposed of in normal refuse, and many restorers still have drawers and drawers full of movements with old radium-painted dials or 'new old stock' radium-filled hands.* One of Craig's many talents is replicating vintage radium paint with safe modern alternatives. He blends Humbrol model paints to match the colour and uses fine sand or grit to replicate the heavier texture of the original paint, which would fluff up from the surface of the dial like a freshly baked muffin. I once bought a load of old tools at auction and found, lurking among them, a small glass bottle of white dust with the word 'RADIUM' handwritten in ink on an age-browned label, like a sinister version of Alice in Wonderland's 'Drink Me' shrinking potion. I fearfully googled 'how to safely dispose of radioactive material', half expecting some sort of alert to sound and armed police in hazmat suits to burst through the workshop door.

I eventually put the powder in oil to stop it becoming airborne and thought no more of it. The radium girls' lives were blighted. More than fifty had died by the end of the 1920s, though it is impossible to know how many were ultimately affected. Their contribution to the First World War effort was immeasurable, but the price they paid for it was criminal.

Their tragic deaths were not entirely in vain, however. Many of these women, both those who survived and those who didn't, through their consent and that of their families assisted in the research into radiation exposure in the second half of the century. Their bodies helped put a stop to nuclear testing in areas where the fallout could end up in the food chain. The radium girls were the earliest examples of our potential future in an increasingly nuclear world.

---

* 'New old stock' is old stock that's never been used, a bit like things on eBay being sold 'brand new with tags' (only these are 100 years old with tags).

By the end of the First World War it was unusual for a man *not* to own a wristwatch: for many they were a badge of bravery. In the years that followed, trench watches became the foundation of a new era of wristwatch design and innovation. Watch manufacturers created a huge range of new designs and styles including elegant, long, rectangular aerodynamic cases that paid tribute to the emergence of the Art Deco movement. Square watches, whose sides curved outwards (to me they look like a plumped-up sofa cushion), became all the rage too. Where the original trench watches looked much like a small pocket watch with wire lugs soldered on, watches now bore lugs that were integral to the case, like shoulders that extended out and hugged the top of the strap.

Post-war advances in metallurgy and materials science allowed for techniques like gold plating to replace the thicker 'rolled' or 'capped' gold in watch cases. Plating reduced the amount of actual gold needed, bringing the status symbol of a 'gold watch' within reach of a broader group of people. Better materials for base metal cases, like stainless steel, were introduced too, aided by improved equipment that made harder metals easier to machine and finish. Stainless-steel cases had none of chrome and nickel's issues with horrible allergic reactions.

The mechanisms that allowed wristwatches to be automatically wound were also refined, making them more efficient and cheaper to make. They used a weight pivoted at the centre of the mechanism that allowed a wearer to wind the watch simply by moving their wrist. The weight's motion is like that of a wooden rattle swung by a football fan at a match. The spinning weight turns a series of gears, which wind up the mainspring in its barrel, removing the problem of looking at your watch at a critical moment and finding it has stopped.

Complications increased in number and affordability. There were now wrist chronographs, which could function as a stopwatch while maintaining the time, and alarms that could buzz to wake you up in the morning. This steady gain in the popularity of the wristwatch also allowed for profits to be invested into development, improving the quality and accuracy of the movements.

It is this innovative interwar period that both Craig and I were drawn to when we started learning the art of watchmaking – it was one of our early shared enthusiasms. There are so many weird and wonderful designs, some very successful, and many more that were such failures only a masochist would willingly attempt to repair them. I've seen watches from this period that barely functioned when new, let alone after seventy, eighty or a hundred years of wear. Automatic winding, for example, went through many iterations before we arrived at the highly efficient systems we use today. There was 'wig-wag' roller winding, which swings the whole delicate movement up and down its oblong case, and the 'Autorist' with its articulated lugs that flex as the wearer moves their wrist to trigger the winding. We have an Autorist in our collection and, having tried wearing it, I can think of very few publicly appropriate wrist actions vigorous enough to make it reach even half-wind. They're all reminders of our very human urge to test our ingenuity and keep inventing, as well as our need to own the latest exciting technical innovations.

Rolex, more than any other brand, rode this wave of development. In 1919 Wilsdorf and Davis had moved Rolex to Switzerland. After the war ended the British government had imposed heavy import duties on imported watch cases as part of their efforts to fill the treasury again. Rolex kept an office in London until 1931; shortly afterwards the British government abandoned the Gold Standard, gold prices crashed, and Wilsdorf moved to Geneva with the company. Their watches improved vastly in quality over the years, but it was Wilsdorf's genius for marketing that cemented their status. Even the name he had come up with in 1908 – Rolex – had a ring to it. It sounded . . . regal – an association bolstered by model names such as the Prince, Princess, Oyster and, later, Royal, as well as a sister brand called Tudor. Wilsdorf said that the Rolex logo with its five-pointed crown was inspired by the five fingers of a human hand, a nod to the handcraft that goes into every Rolex watch. It still appears on the winder of every Rolex made to this day. These associations have all helped Rolex watches to ooze understated luxury, status and wealth.

Wilsdorf also pounced on publicity opportunities. When in 1927 they developed their first waterproof watch design, the Oyster – so called because the case clamped together like an oyster's shell – they didn't just release it in shops and post a few adverts in newspapers and magazines but sent it across the sea. Mercedes Gleitze wore it round her neck as she became the first British woman to swim the Channel. After her ten hours in the water, the watch was inspected and found to be in perfect working order. Gleitze became Rolex's first ambassador, her famous face promoting the brand, vouching for its integrity and reliability. If today we are used to seeing celebrities and sportspeople advertising watches, Rolex pioneered the practice.

Rolex watches became uniquely identified with the daring achievements of their wearers. In 1933, members of the RAF were the first to fly over Mount Everest, wearing Rolex watches. Rolex printed a series of adverts with the title 'Time flies'. In 1935 an Oyster watch travelled at 272 miles per hour down Daytona Beach with Sir Malcolm Campbell in his legendary car, the Campbell-Railton Bluebird, during one of his many speed record attempts. In the 1950s Rolex launched the Explorer, which was worn by Edmund Hillary and claimed to be the first watch to have reached the summit of the world (although Hillary later stated he was wearing a Smiths watch at the summit). These days you might associate Rolex with Wimbledon, equestrianism, gold at the Masters Tournament, or Formula 1. They sponsor arts festivals around the world. It would be hard to find someone who has never heard of the name Rolex or couldn't identify its iconic crown logo. Rolex, in my mind, were the first watchmakers to make a brand name more prolific than the timekeepers themselves.

The interwar years generated the most rapid evolution of the watch in its 500-year history, a pace of innovation that has continued to the present day. By the time the Second World War broke out in 1939, watches were much better adapted to the adverse and extreme conditions of war than they had been twenty years earlier. Huge pocket-watch-sized wristwatches were supplied to aviators, designed to be read at night with ease and worn over bulky flying gear. Divers' watches accompanied navy frogmen on aquatic missions to raid enemy bases from the sea or plant limpet mines on the outside of an enemy ship's hull. These supercharged wristwatches could be shockproof and waterproof, and their cases could even incorporate a shell that protected them from magnetic fields. Calibrated bezels designed to help early aviators calculate their position mid-air were increasingly used by fighter pilots.

During the First World War, Switzerland's neutrality had been a boon to the watch industry. Swiss watch manufacturers had not lost 50 per cent of their workforce to fighting on the front line, and the country's economy was spared the double blow of reduced productivity and investment in the war effort. Swiss manufacturers didn't simply supply the world's military with completed watches, but also exported kits of movements, cases and dials ready for retail under all sorts of different brand names. In the Second World War, however, trading became more difficult. Switzerland was cut off from trade with the Allies by the Vichy invasion. It was now completely surrounded by Axis powers, posing a quandary for Swiss watch companies, who did not want to deal with Axis powers but were struggling to survive. Early on in the war, Rolex supplied movements to the Italian firm Panerai for use by their Italian navy divers, but it wasn't enough to replace the loss of its main market in Britain.

Eventually, like other watch companies, Rolex started shipping watches to the UK using neutral countries and flagged ships and planes from nations who weren't at war with the UK, such as Spain and Portugal. They were greeted with a built-up demand, particularly among RAF pilots, for whom, since Alex Henshaw became the first person to fly from London to Cape Town wearing a Rolex in 1939, they were practically de rigueur. Wilsdorf went to tremendous lengths to maintain sales with the Allies. He famously offered his watches to British officers who were prisoners of war, to replace those that had been confiscated.* The watches were ordered and then sent via the International Red Cross on the understanding that they would be paid for when the war ended. His gentleman's agreement was a boost for morale, demonstrating that Wilsdorf was confident the POWs would make it through, the war would end and the Allies would win. In one German prisoner camp, Oflag VII-B camp, more than 3,000 watches were ordered by British POWs.

Flight Lieutenant Gerald Imeson, who was interned at Stalag Luft III, 100 miles south-east of Berlin, ordered a Rolex 3525, a top-of-the-range Oyster Chronograph. Imeson used his watch, which had a waterproof oyster-shell case and radium hands and numerals, to light his way in the 334-foot-long tunnel, nicknamed Harry,† which he and his fellow inmates dug as part of a daring escape mission. Imeson was one of 200 men who planned to escape, and had been employed in the run-up as a 'penguin', hiding some of the many tonnes of soil dug up from the tunnels under a bulky coat, in order to redistribute it on the camp grounds.

On the night of the escape, it is possible Imeson used his sophisticated watch to calculate how often the guards patrolled the camp, how long it took each man to crawl through the tunnel, and how many men could enter the tunnel an hour (it worked out at ten). At 1 a.m. the tunnel partially collapsed, slowing the men down. At 4.55 a.m., the seventy-seventh man was noticed by a guard. Those who had made it through ran for their lives. Seventy-three were caught after a manhunt, fifty of whom were executed on the order of Hitler to set an example. Only three managed to get to safety. Imeson was not among them. He returned to the camp and, after being moved

---

* Military-issue watches were often confiscated in POW camps on suspicion that they might contain a compass or tool for escape.
† Harry was one of three tunnels. 'Tom' had been discovered and dynamited; 'Dick' was abandoned when the spot where they planned to surface was built over.

to another POW camp, was liberated in 1945. He treasured his Rolex 3525 for the rest of his life.*

By the end of the Second World War, Rolex was in a strong position but Britain and much of Europe were about to plummet into a recession. The brand focused its attention on the US market, and with some savvy marketing and new designs it managed to compete with home-grown American watch brands like Waltham and Hamilton. Key to their success was their partnership with the world's largest advertising agency at the time, J. Walter Thompson – a relationship that endured for decades.

Thanks to advertising, Wilsdorf's watches were not just time-keepers but storytellers. By associating them with the daring of extreme sports, the precision of extreme timekeeping and the luxury of extreme wealth, the watch proclaimed as much about who you wanted to be as who you were. There is a contemporary marketing term that refers to goods that give the impression of exclusivity and luxury but are intended for the mass market. They are referred to as *masstige* – an amalgamation of the words *mass* (-produced) and *prestige*. This, to my mind, was Hans Wilsdorf's genius.

Hans Wilsdorf and Albert Einstein embarked on very different careers: one was a canny businessman, the other a brilliant theorist. But they had a surprising amount in common. They were born in Germany just two years apart, then both moved to Switzerland, where both of them initially worked as clerks. But for me the most intriguing connection between these two remarkable men was that they both altered our relationship with time. While Einstein overturned centuries of received wisdom about the nature of time itself, Wilsdorf overhauled centuries of assumptions about what a time-keeper could be. We are still living with their legacy today.

---

* This extraordinary prison break was to become the inspiration for the film *The Great Escape.*

IO

Man and Machine

'We are always getting away from the present moment.'

H.G. Wells, *The Time Machine*, 1895

As midday approached on 9 June 1940 the three-man crew of L9323, a light bomber Bristol Blenheim Mk IV aircraft, was returning to base. They had just completed a successful mission bombing a German armoured convoy near Poix-de-Picardie in the Somme area of northern France. The German army was fast advancing on the Allied forces, who were trapped between them and the coast. The Dunkirk evacuations had ended. The crew of L9323 were part of Operation Aerial, whose role was to slow the enemy's progress and give their comrades who had missed the main evacuation as much time as possible to escape across the Channel.

As they passed over Normandy on their return to base, the crew came under enemy fire from a Flak anti-aircraft cannon. Their twenty-five-year-old pilot, Flying Officer Charles Powell Bomford, was killed instantly. The plane's observer, Sergeant Robert Anthony Bowman, pushed his fallen friend's body aside and grabbed the centre stick. Robert had never flown a plane before, but he knew that if he allowed it to freefall to the ground, he and their gunner, Pilot Officer Francis Edward Frayn, would be killed on impact. Wrestling with the controls as the plane came down, he managed to slow their descent enough for them to survive the initial crash. But the impact caused the nose of the plane to push back into the cockpit, pinning Robert against the steering column. He was trapped. Francis rushed to his aid, but was unable to free him. The leaking fuel ignited, and the explosion threw Francis from the plane. Robert died in the fire.

For a long time after, Francis's fate was unknown. Aviation records suggest that he survived and might have been taken as a prisoner of war, though no record of his internment was found. The truth turns out to be far more remarkable.

Francis was stranded – too badly injured to move and aware that the enemy wasn't far away. Lying on the ground where he fell, he heard the marching boots of soldiers approaching. I can only imagine his relief when he heard their voices and realised they had Scottish accents. It was the 51st (Highland) Infantry Division, the last large Allied force left in the area as they too fled to the coast. They rescued Francis, carefully folding his flight jacket beneath his head as a pillow and carrying him by stretcher for two days to the hospital in Saint-Valery-en-Caux from where they all hoped to be evacuated. In the days that followed the 'Miracle of Dunkirk', the last troops – nearly 200,000 Allied personnel and injured soldiers – were boated to safety. Francis was among them. The ship he escaped on was the last to depart before the port fell under enemy fire. It was captained by the esteemed naturalist and British naval officer Sir Peter Scott, the only child of the explorer Robert Falcon Scott.

Francis's rescuers, the 51st (Highland) Infantry Division, were not so lucky. The plan had been to return to collect them, but thick fog made the journey impossible. As dawn arrived on 12 June, the men knew that no further ships would be coming back to rescue them. Trapped, exhausted and depleted, they surrendered that morning.

We know Francis's incredible story because he shared it in detail with his son, who shared it with me, as now I have a tiny part to play in his story too.

At the hospital in France, as Francis's injuries were attended to, a nurse took the battered flying jacket that had been folded under his head and shook it out ready to hang by the side of his bed. There was a metallic *clunk* as a small silvery object hit the floor. She bent down and picked it up. It was Francis's watch, which had somehow also survived the crash. Chunks of metal had been knocked out of the case, the strap had broken, and the rotating bezel he'd used to measure bombing intervals was long gone, but both watch and its owner had made it – battle-scarred but alive and ticking.

And now here it is, sitting before me on my bench. Francis kept the watch after the war and left it to his son, who brought it to me in a Jiffy bag. The dial, once a rich ivory like the colour of full-fat milk, is now patinated with dark speckles. I call it 'foxing' as it reminds me of the dots that appear on the pages of ageing antique books. It's not an official horological term, but one that makes sense to me as someone who loves both old watches and old books. The section of the dial below twelve o'clock bears the remains of what was once the brand name 'MOVADO'.

Movado is a Swiss firm that was founded in La Chaux-de-Fonds in 1905, the same year as Rolex. The founder, nineteen-year-old Achilles Ditesheim, must have been excited by the potential of Esperanto, the artificial international language created by Polish oculist L.L. Zamenhof in 1887, as Movado is Esperanto for 'always in motion'. This model, called a 'Weems' after Lindbergh's navigation expert, Philip Van Horn Weems,* and featuring a movable outer bezel to calculate longitude, was issued to RAF pilots and navigators at the outbreak of hostilities. There were only 2,500 ever made and, looking at it now, I wonder how many are still in existence.

Unusually, the 'MOVADO' on Francis's watch has almost completely, and very carefully, been scratched out – almost as if it was done with the tip of a pin. The only letter remaining visible is the central 'V'. Francis never told his son why the dial had been damaged or what it might mean. Perhaps Francis made the modification himself, to leave the 'V' for Victory. Whatever the reason, it's an important part of the history of

---

* It was in fact a version of the Longines watch that Lieutenant Commander Weems developed with Charles Lindbergh. Longines licensed the design to a number of other companies, Movado included, when they could not keep up with wartime demand.

this little watch, so I will leave it as it is. My primary goal is to get the watch itself back up and running so that Francis's descendants can continue to wear it and remember its story.

Pilot Officer Francis Edward Frayn's Movado Weems after restoration. With projects like this we're careful to make sure our repairs can be undone to return the watch to its original state in the future if desired. The replacement bezel can be removed, and the dents and chunks knocked out of the case were left untouched.

All watches have stories, but those of the twentieth century feel much closer to us. We learn them directly from their owners, or those related to them, rather than in the pages of books or letters in archives. The watches of the Second World War – whether standard issue or cherished purchase – are imbued with their owners' experiences. Not all of them saw fighting. Some were destined for use by the military's vast number of administrative staff. The marks on military watches, usually printed on their dial and engraved into the back of their case, give us the reference numbers and codes that can tell us which branch they were issued to, in which nation, and in what year. For example, British military watches carry a broad arrow, nicknamed a 'crow's foot' as its three joined lines look a little like a bird's footprint in the sand. There were various codes: *AM* indicated it was issued to the Air Ministry, *ATP* stood for Army Time Piece, and *W.W.W.* meant waterproof wristwatch (the code's literal translation is Watches, Wristlet, Waterproof). The initials R.C.A.F. signify a watch issued to the Royal Canadian Air Force. Other countries had their own systems and this no-nonsense approach to the marking of military-issued pieces usually (there are always exceptions) makes them straightforward to identify and date.

I have also seen watches that were issued to the Nazi military during the 1930s and 1940s, emblazoned with the swastika and eagle of the Kriegsmarine, the *F.L.* Flieger number of a Luftwaffe-issued timepiece, or the *D.H.* property mark that stood for *Deutsches Heer*, meaning 'German army'. On the rare occasions they make their way into our workshop, I swiftly hand them to Craig. His attitude towards them is more clinical than mine. He points out that a Nazi military watch might never have left the stores; or it could have been issued to a junior clerk of little importance, or been traded for some cigarettes by one of the many soldiers who ended up in the Allied prison camps. Equally, any watch that ends up in front of us could, unless its full provenance is known for certain, have been witness to any number of atrocities, which I try not to contemplate. To Craig they're just inanimate objects. You can't blame an inanimate object for the actions of its owner or creator.

Guiltless as an inanimate lump of metal, enamel and glass might be, one can legitimately question the intentions of their collectors. I know owners who see them as nothing more than pieces of history, and whose general interest in twentieth-century military history means they collect a wide range of paraphernalia from both sides. There is, however, another market for Nazi memorabilia that seeks to glorify an appalling phase in our history, and that poses an enduring dilemma for auction houses and dealers. Very recently, a wristwatch claimed to be Hitler's own, a 1933 Huber, came up for sale at auction. Thirty Jewish leaders wrote to the auction house to object to the sale, but it fetched $1.1 million on the first day, and apparently went to a European Jew. A similar discussion raged in 2021, when a watch that had been issued to Chinese soldiers by their government to commend their participation in the June 1989 Tiananmen Square massacre came up for auction in the UK. On the dial, underneath an image of a soldier in a green uniform and helmet, the script reads '89.6 to commemorate the quelling of the rebellion' – a quelling that killed between 300 to 3,000 people, depending on whether you use the official government figures or those of external observers. The auction house initially took the line that this was an 'object of international interest', that its sale wasn't a statement of support, and that the owner had nothing to do with the People's Liberation Army. Yet after the anonymous vendor was subjected to death threats through the auction house's social media and website, the lot was later withdrawn. Objects play an important role in ensuring that the past is never forgotten. But what should we do with artefacts from the darker periods of history? Should they be kept in museums, either on public display or hidden away in a storeroom? Should they be destroyed? Once they reach the open market, we have no way of knowing where they will end up, or how they will be used. There are no easy answers to these questions.

To me, every watch carries the traces of those who have worn it. When the Nazis rounded up Jews for 'resettlement in the east', many believed they were just being relocated. They had little time to pack and limited luggage allowance, so they grabbed their most precious portable possessions. Watches, along with money, clothes, glasses and artificial limbs, were among the first valuables to be confiscated when they arrived at the concentration camps. When the camps were finally liberated, the watches were found heaped in their thousands. Individually, these watches had stories that would never be passed on; collectively they bore witness to humanity's most shameful hour.

The watches gathered after the American bombing of Hiroshima in Japan now form a poignant display at the Hiroshima Peace Memorial Park. When the 9,700-pound (4,600-kilogram) 'Little Boy' atomic warhead landed on the morning of 6 August 1945, 80,000 people were killed outright. The bomb created pressure waves that travelled faster than the speed of sound. Later, it was discovered that each one of the watches caught up in the blast was frozen forever at the time of the detonation: 8.15 a.m.

Through two world wars, wristwatches accompanied men and women through battle and imprisonment, espionage and escape. In the post-war period, watches continued to build on this heroic inheritance. Even after men returned home to civilian life, feats of bravery and endurance were used to sell the watch. The reliability of a watch in all sorts of extreme situations was promoted to the modern man, even if now he was more likely to be called upon to mow the lawn than fight.

Watches sought to outdo each other in chronometry and technical precision. Improvements in water-resistant cases allowed for diving watches to descend to greater depths. In 1960, Rolex strapped its ultimate diving watch, the Deep Sea Special, to the outside of the bathyscaphe submersible *Trieste* and plunged it to a depth of 10,911 metres in the Mariana Trench. It returned to the surface in full working order, which is more than could be said of any human being if they'd gone with it (the current world record for the deepest scuba dive is 'just' 332.35 metres.) In 1969, Omega sent their chronograph the Speedmaster to the moon on the wrists of Buzz Aldrin and Neil Armstrong after it surpassed all competition in its ability to function throughout extreme variations in temperature, changes in pressure, shocks, vibration and acoustic noise that it would encounter on the voyage. Watches for women, by contrast, were more for ornamentation than function, and became more delicate, their dials smaller and smaller. They were almost a reminder that women, after the wartime expansion in their roles, should be back in an apron ready for the most important hour of the day, the return of their husband from work.

Today utility or sports watches are still one of the watch industry's most popular sectors. You might not be able to survive the same conditions as your watch, but at least it will be in one piece to leave to your next of kin.

The science that had led to such devastation in Hiroshima and Nagasaki found a new direction in the period after the Second World

War. It was to change our relationship with our timekeepers forever. As early as the 1930s, Isidor Rabi, a physics professor at Columbia University, had started work on an atomic clock, building on the research of Danish physicist Niels Bohr, who had developed a theory of the structure of the atom.* Bohr had observed that electrons orbit the atomic nucleus with remarkable regularity and that an increase in energy can cause electrons to jump to a higher orbit. As the electrons jump, they emit energy at a specific oscillation frequency. Timekeeping generally depends on things that oscillate, from the pendulum to the balance wheel and hairspring, but the atom's emitted frequency, which Rabi eventually harnessed to produce the first atomic clock in 1945, was found to be more precise and more stable than anything that had come before. It was swiftly followed by further incarnations working with the caesium atom from the National Institute of Standards and Technology (NIST) in Colorado and the National Physical Laboratory in London. By 1967, the General Conference on Weights and Measures had redefined a second as 9,192,631,770 oscillations of the caesium 133 atom. In the years to come, atomic time would make GPS, the internet and space probes possible.

Atomic time was a major breakthrough, but for now it remained locked away in scientific institutes across the world, housed in machines the size of trucks. Meanwhile, other scientific and technological developments were making their way through to civilian time. Electronic clocks, powered by electric impulse rather than swinging pendulum or mainspring, had been around since the 1920s, but now Swiss and American inventors raced to make the technology work for the watch. The first battery-powered wristwatch to make it through the gate was the 1957 Hamilton Ventura, instantly recognisable with its triangular dial and Art Deco-style stepped golden case. Elvis Presley made it covetable by wearing it in the movie *Blue Hawaii*. But the Ventura was launched in haste; its short battery life meant that once sold, many watches were swiftly returned to their retailers, many of

---

* In the 1940s both Bohr and Rabi played a role in the Manhattan Project, contributing to the development of the atomic bomb.

whose repair staff were untrained in the new technology. By the time Hamilton sorted out the Ventura's teething problems, competitors had made up the distance, including the revolutionary Accutron – the name was a selective conflation of *accuracy* and elec*tronic* – released by the American company Bulova in 1960.

The Accutron kept time with a tuning fork activated by an electronic circuit with a single transistor powered by a small battery. The electronic oscillator assisted the tuning fork to vibrate at a consistent frequency, exactly 360 times a second, which in turn regulated the timekeeping of the watch (Bulova claimed to within +/- two seconds a day), rendering the balance wheel obsolete. To wear, the Accutron both looked and sounded like an object transported from the future. The flagship Accutron Spaceview had no dial, meaning you could see right through to the electronic circuitry of its movement. Pinned to its turquoise-green board were two copper wire coils, supplying the magnetic field to the tuning fork; the rotation of the wheels was accomplished by indexing a tiny wheel with 300 teeth that, in turn, powered a series of gears that drive the rotation of the hands. The tiny tuning fork created a constant audible hum that emanated from the watch surprisingly loudly. (Adverts for the Accutron tried to turn this into a selling point – 'Have you heard the new sound of accuracy? It's the hushed hum of Accutron.') I once slept with an Accutron on my bedside table; it was like sharing a room with a very rowdy bee trapped under a glass. The design of the Accutron Spaceview was a celebration of the very latest in miniature electronics. The transparent design of the dial provided more than a window into the movement; it was a glimpse into the future.

But the watch that decisively overturned the traditional mechanical watch came from Japan. On Christmas Day 1969, Japanese watchmakers Seiko released the Astron, the world's first commercial quartz watch. Instead of the tuning fork, this new invention, the brainchild of Kazunari Sasaki, focused on using piezoelectricity – the process, discovered by Pierre and Jacques Curie in 1880, of using crystals to convert mechanical into electrical energy. When subjected to pressure (the derivation of piezoelectric is from *piezin*, the Greek for 'to squeeze'), a crystal will emit a small electrical pulse which can be used to derive a remarkably stable frequency. This was used to regulate the turning of a magnetic rotor that performed a similar function as an escapement in a mechanical watch. A fashioned piece of quartz can vibrate millions of times a second, compared to 18,000 times an hour in a comparative mechanical watch of the era. The brand-new quartz watch was advertised as being 100 times more accurate than its mechanical rivals.

The Astron was not cheap – only 100 were made initially and sold at Y450,000 (around £10,000 in today's money) – but that didn't remain the case for long. Through massive investment in technology, streamlining production and increasing automation, quartz watch movements became more and more affordable. Today, you can buy a perfectly functioning quartz watch movement for just a few pounds.

It was this speed that took the Swiss by surprise. At the time of the arrival of the Astron, a consortium of Swiss watch companies had, like those in the US, been working on their own versions of the quartz movement for years but, protected by the fixed global exchange rates after the war, the industry itself had failed to innovate and restructure. The Swiss watch industry was still fragmented, with small manufactories, not unlike those that had given it the edge in John Wilter's time, scattered in every town and village of the Jura. Quartz technology required a totally different set of skills – electronics, rather than traditional mechanical engineering – and Japan and Hong Kong were better placed to exploit it than Switzerland and the US.

It's no surprise that the quartz revolution both started and gained pace fastest in the Far East. Japan and Hong Kong had already emerged as world leaders in electronics more generally, with Canon, Panasonic and Mitsubishi becoming hugely successful, and now developed watch companies of their own. Hong Kong had a reputation for producing cheap watches and watch parts for other companies; Japan had brands like Citizen, Seiko and Casio. For the first time in its history, watches were completely machine made, and

At around 44,000 years old, the Lebombo Bone – carved from a finger-sized piece of baboon fibula – is thought to be the earliest potential timepiece discovered to date. The thirty spaces divided by twenty-nine notches average to a lunar month. Found in South Africa in Border Cave, it is clear evidence of calculation, and the wear patterns indicate this object was regularly used.

The mechanism of a small drum-shaped clock made in Germany some time between 1525 and 1550. The maker's identity is unknown, as it was not unusual at that time for watch- and clockmakers not to sign their work. Made from iron, it was probably created by a locksmith or armourer as the skills required were very similar to those of a watchmaker. These small table clocks are the transitional timepieces that gave rise to the first watches as they were portable and small enough to hold in your hand.

The mechanism is housed in a gilded and engraved drum-shaped canister which measures just under 7 centimetres in diameter and less than 5 centimetres in height. The raised beads that mark every hour were used to tell the time by touch in the dark. It has a single hour hand – this might be, in part, because small clocks and watches of this era weren't accurate enough to warrant one that measured minutes or seconds. It also suggests that more accurate time division wasn't as important to typical owners at the time.

Form watches, so-called as they are, quite literally, made in the form of something, were popular in the mid-seventeenth century. This example – a tiny silver lion that could sit in the palm of your hand – was made in Geneva by watchmaker Jean-Baptiste Duboule in around 1635. To read the time, the lion's belly snaps open and reveals its dial. As Geneva was a Calvinist state in the seventeenth century and banned decorative items like this, the watch would have been intended for export, possibly to the Ottoman Empire.

The mechanism of the watch sits inside the main body of the lion and can be swung out once the lid is opened. The plates are made from gilded and engraved brass. Some of the steel work had been blued (the process of changing the external colour of steel with oxidation caused by heat), a decorative process still used today.

A watch mechanism dating from around 1770, made under the pseudonym 'John Wilter' and proclaiming to have been made in London. Wilter became an obsession of mine after I discovered one of his watches at an auction house. There is no evidence a watchmaker by this name ever existed and the style is not typical of a London-made watch. So-called 'Dutch forgeries' would forever change the dynamic of the watch industry and were the first step in the journey towards making watches affordable to all.

Repoussé watch cases like this one were very popular in the mid- to late eighteenth century. They were stamped or hammered to form a three-dimensional relief which was then engraved, typically with classical or biblical scenes. The technique lent itself to the outer cases of pair-cased watches, with the first of the pair housing the movement and the second protecting the inner.

'Tarts, London' was another notorious fictional watchmaker's name found on eighteenth-century forgeries. This is the front of the case pictured previously. The dial is arcaded, meaning the minute track is scalloped between the hour numerals, which was a popular style among Dutch clockmakers. The watch is fitted to a chatelaine which would have allowed it to be worn on the hip, suspended from a waist band.

An early example of one of Abraham-Louis Breguet's *perpetuelle*, or 'self-winding' watches made in Paris in 1783. Breguet is considered to be among the greatest watchmakers of all time, with a number of his inventions still used in watches today. His *perpetuelle* was the first automatic watch, capable of winding itself using its wearer's movement. The hand and retrograde scale to the top left of the dial show how many hours of wind the mechanism is storing.

The shield-shaped weight swings back and forth as the wearer moves, triggering a series of wheels to wind the spring and power the watch. Breguet also fitted this watch with another of his inventions: steel wires, finely tuned like piano strings, run around the outside of the mechanism and are struck by hammers to chime the hours and quarters. Breguet's wire gongs replaced earlier bells that sat inside the case, allowing repeating watches to become much slimmer, which suited the fashion of the day.

This pocket alarm watch belonged to the explorer Captain Robert Falcon Scott, known as 'Scott of the Antarctic'. It accompanied him on his ill-fated *Terra Nova* expedition (1910–1913) to the South Pole and was found on Scott's body, alongside his companions Edward Adrian Wilson and Henry Robertson Bowers, by a search party eight months after they succumbed to the elements. This watch, along with records and other personal effects, was recovered and returned to the UK.

The Rolex Rebberg, an early mechanism made in the Rebberg manufactory for the Rolex Watch Company in around 1920. They were used in the days before Rolex started making their own movements and are one of our favourite vintage watches to restore. Although mostly machine-made, they were finished and built by hand, creating slight variations between parts that are no longer seen in mass machine-made watches.

One of our favourite, and rarest, early Rolex Rebbergs we've seen was also one of Wilsdorf's first patents: the design of a single-lugged fob watch, allowing it to be suspended from a ribbon or band on a nurse's uniform. It was made in around 1924. The devices on each side are 'vibrating tools' that we use to time in the spring controlling the speed a mechanical watch runs at.

A fingerprint from around 1780 found baked into the counter enamel of a watch dial Counter enamel (on the back of the piece) is used to prevent the dial warping when the visible decorative enamel is applied to the front. This was not designed to be seen, and it is highly unlikely the person who left the print would have been aware that it was unique and how special it would be to us today. It was, almost certainly, accidental. Dial enamelling is a separate art to watchmaking. This watch dial was commissioned by James Wilson, a watchmaker on King Street, London, whose name is marked in ink on the back of the dial in the same way that a tailor might put a customer's name on a bespoke suit to identify it on the rack.

no longer required skilled artisans to assist in their production. By 1977 Seiko was the largest watch company in the world in terms of revenue.

Meanwhile the watch industry in Switzerland was sleepwalking towards a cliff edge. Swiss watchmakers, just like British watchmakers a century earlier, were too wedded to a belief in mechanical excellence to move with the times. They were slower to invest in new technology and increasingly had to source parts from overseas. This, combined with the rising value of the Swiss franc, priced them out of the low-priced market. By the early 1980s, the Swiss watch industry was in a catastrophic state of decline with mass redundancies and hundreds of companies collapsing, causing recessions in the old watchmaking world.

And if that wasn't enough, hard on the heels of the 'quartz crisis', as it became known in the trade, came another threat: the digital watch.

I can still remember the schoolgirl envy I had for my classmate Victoria's Casio G-shock 'Baby-G'. It was my first year at secondary school. We'd been taken to an Outward Bound centre, supposedly for a bit of team building, and on our first night we were treated to a 'potholing experience'. In reality this was thirty eleven- to twelve-year-old girls crammed into the outer border of the building's loft space, which had been boxed in to create total darkness and filled with various obstacles. We were supposed to circuit the loft in the dark like moles, which would have worked had it not been for the eerie green glow of Victoria's Baby-G digital watch. One touch on the light-up display and we were following it through the darkness. I loved the way it lit up at the push of a button. How I wanted one of my own! But unfortunately it was beyond my frugal parents' means; I would have to wait.

It's incredible to think that the technology that lay behind Victoria's digital watch sprang in part from research undertaken at NASA. The very first digital watch was American, the Hamilton

Pulsar, released in 1972, and used LED technology developed at the Space Agency. It was advertised as 'The ultimate in reliability – no moving parts. No balance wheel, gears, motors, springs, tuning forks, hands, stems or knobs to wind up, run down or wear out!' But the Pulsar, like the Ventura and the Accutron before it, still couldn't compete with the price of products from the Far East. The Japanese Seiko LCD in 1973 and Casio in 1974 left it in the dust.

You might think 'no moving parts' was my idea of hell, but I still genuinely love digital watches. In fact, I have a collection of Casios and they remain my watch of choice for working in the workshop. I suppose I've always been drawn to opposites (I like cats and dogs equally, and my first two albums as a kid were Holst's *The Planets* and *Cyndi Lauper's Greatest Hits*). I may practise and study the traditional craft of watchmaking, relishing the application of skills that are centuries old, but I delight in the plasticky resilience of my Casio. There's something immensely reassuring about wearing a watch that could survive being dropped from the top of a block of flats (or so the adverts say; I haven't tried) when your day job involves applying infinite precision and care to make its high-grade mechanical ancestors run again.

A late-1970s digital chronograph watch by Seiko.

A Casio doesn't need my love. I don't have to worry about scratching the plastic with sharp metal swarf or bashing it on a milling machine. Nor do I need to repair it: it just goes. If it needs a battery exchange, it's a quick and easy job. And when, one day, it stops and a new battery isn't enough to do the trick, it's not the end of the world: it only cost £30 in the first place.

The salvation of the Swiss watch industry came largely in the form of one man – Nicolas Hayek. A Swiss entrepreneur of Lebanese descent, Hayek was approached by banks to oversee the liquidation of two Swiss watchmaking firms that had been forced out of business by the quartz crisis. But rather than close the businesses altogether, he believed that with substantial restructuring there was a way ahead. He realised that if the Swiss watch industry was going to be saved, it needed to evolve fast, incorporating, rather than rejecting, the competitive quartz technologies, cutting retail prices and presenting something new to the market. So Hayek came up with the idea of producing affordable quartz watches from cheap materials like plastics and resins, in a huge range of bold, fashionable technicolour designs. He called his new brand Swatch.

Swatch quickly cornered the fashion market and single-handedly reinvigorated the analogue watch. In 1985 the *LA Times* called them 'the hottest new fashion accessory on the market'. Swatch watches were so attractive and affordable – Cheryl Chung, then product development manager for Swatch USA, called them 'cheap chic' – that they changed the way people bought and used watches in general. Lanny Mayotte, marketing director for US rival firm Armitron, put it well: 'People today have a wardrobe of watches . . . Years ago you bought a watch for graduation, and it was handed down to the children . . . why not have a fun watch rather than a boring, old, expansion-band watch?'

Five hundred years earlier, a watch had been one of the most expensive personal luxuries money could buy. Now you could purchase one in every colour from your local department store. And if fashions changed? Throw it away and buy a new one. The Swatch watch transformed our relationship with portable time. Yet, ironically, it also saved the mechanical watch. Swatch went on to become such a formidable success that the profits allowed Hayek to buy up faltering historic watch brands and inject new capital into them. The Swatch Group is now one of the largest conglomerates of luxury brands in the world, and owns celebrated marques like Omega, Longines, Tissot and the House of Breguet. The cheap and cheerful watch had rescued the Swiss mechanical watch industry from oblivion.

For American firms and their global outposts there was sadly no such saviour. The strike at US-founded Timex's factory in Dundee, Scotland, in 1993 is among the most notorious incidents caused by the quartz crisis. The picket-line violence has been described as the worst since the miners' strike of 1984. At its peak in the 1970s, Timex was a major employer in the city and the factory employed 7,000 people. By the time it closed, its workforce was reduced to just seventy employees. The trouble arose in 1993 from a dispute over proposed lay-offs, a wage freeze and a reduction in fringe benefits, the result of competition from the Far East. By 1993, Hong Kong was exporting 592 million watches a year and manufacturers could supply large-quantity batch orders in as little as twenty-five days. Timex in Dundee could not compete. Union workers voted decisively to back strike action rather than face a reduction in workforce or wages. Failing negotiations resulted in the striking workers being locked out of the factory and strikebreakers, or 'scabs' as they were known by the protestors, being bussed in to replace them. One protestor described how: 'Cans of Coke and coffee were thrown over cars and there were incidents of vandalism . . . I always carried a pickaxe handle in the back of my car.' The majority of the Timex workers were women and the strike politicised them. In an interview with the *Scotsman*, one female assembly-line worker said they had changed 'from lambs to lions'. 'There were women that were wee tiny creatures and all of a sudden they were unrecognisable . . . They were fighting for their livelihood. The majority were like me. We had worked there since we were young lassies.'

In the end, after six months of violent unrest, the factory was permanently closed, bringing its forty-seven years as an employer in the city to an end. Though the Timex brand prevailed, and the company still has offices around the world (the bulk of its production is now in Switzerland and the Far East), the closure cast a long shadow. In 2019, when the BBC produced a documentary on 'Scotland's last full-blooded strike', it was clear that the wounds of losing the employer from the city were still raw.

For the watchmaker, one of the greatest changes of the last century was the rapid and at times total shift from craftsperson to machine. The quartz crisis, price warring and budget cuts meant that from the 1970s to the 1990s there was little room left for the skill of the master craftsperson. Humans cost more than machines, and so the more a watch could be made by machine the better. This extended to the way watches were maintained. The Ingersoll Yankee watch, which cost $1, still had the capacity be serviced and repaired. The 1980s saw an increasing number of hermetically sealed watches, whose cases are impossible to open,* meaning the moment they stop working you have little choice but to throw them away and buy another. Most Swatch watches today have a similar construction.

Many watches were no longer expected to last. Plastics, being much softer than metal, wear much faster and infiltrate the delicate mechanisms of watches, further reducing their life expectancy. While a mechanical movement can be repaired by a watchmaker, a circuit board like that found in a battery-operated watch can't: the mechanism is often fused to the case so that when you open it, the components fall apart like a glitter bomb. When one part stops working, the whole watch goes with it. In the decades that followed, built-in obsolescence would become part of the way cars, computers and software were designed.

---

* I've tried but had to resort to breaking the case.

This was a poignant moment in the long history of our craft. Craig and I are inspired by the era before the quartz crisis, not least because it represents a time when humans and machines worked in unison. Machines improved watch production, efficiency and accuracy but they still needed us to operate them. A 1940s milling machine needs a skilled operator. It requires someone to set it up, monitor it and operate it by hand. It speeds up our work and is beautifully accurate, but it can't be left to its own devices. By contrast, CNC – computer numerical control – will do all the work for you once it's been set up and inputted with a design programme, machining whole components to a near-finished state. It can even be left to run overnight, allowing you to return to your workshop the following morning to find your work has been completed for you while you slept.

It's a surreal thought for a heritage watchmaker. All of our tools and machines are old. Having started our business with a very small loan, around half the price tag of a new Swiss-made lathe, we had no choice but to start buying old equipment that we could restore and tailor for the jobs required. But working with them is one of the most enjoyable aspects of my job. Once you get to know them, you discover that each of our machines has a distinctive personality – so much so that we give them names. Alongside Helga the lathe is her sister, Heidi, another 1950s East German 8-millimetre lathe; they arrived together in a box from Bulgaria. We customised Helga to be a wheel-cutting lathe using the photograph on a cover of a book (ironically called *The Watchmaker and His Lathe*) as reference – she cuts out every tiny tooth on each wheel and pinion within a watch – while Heidi cuts the tiny arbors and stems using a hardened steel graver. There is George the pillar drill, made in the 1960s by the British Ideal Machine and Tool Company (IME), who can drill holes as small as 0.1 millimetres. He sits next to Albert the milling machine, made by Wolf Jahn in around 1900, who works like a drill but whose bed (which holds the work) moves from side to side, allowing him to cut recesses and long trenches into metal. Our smallest lathe, made by Lorch in the 1940s, was found, as one of a pair, in a box of bits on a friend's workshop floor. We named her Maus, or 'mouse' in her native German. Her sister, who we named Spitzmaus ('shrew'), is now an uprighting tool, aligning the tiny holes drilled in different metal plates. These machines need us as much as we need them. We think of them as our colleagues.

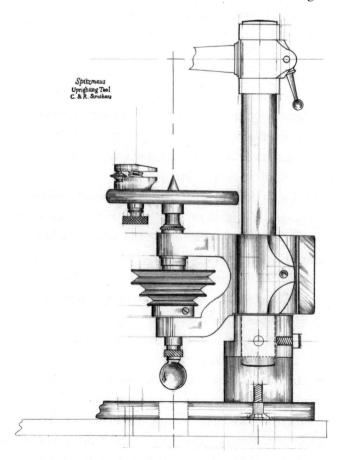

Spitzmaus
Uprighting Tool
C. & R. Struthers

Our uprighting tool, Spitzmaus, adapted for purpose after arriving in a box from our friend's workshop floor.

When Craig and I were training, if I was ever short of a watch to restore for my portfolio he'd let me rifle through his 'tin of movements'. He would beam with glee over his Family Circle biscuit box full of hundreds of old watches and mechanisms, mostly dating from the first half of the twentieth century. While the majority of the other students focused their energy on modern servicing in pursuit of a job at one of the major watch brands' many servicing centres, we'd work on 1920s or 1930s stuff that was often missing parts, cases and case backs, often with no brand name or financial value. Our tutor, Paul Thurlby, a former watchmaker for Omega, would despair: 'Why', he would ask, 'do you only work on old crap?'

For us the appeal is that they offer a trace of human hands. Even though these watches were made at a time when production was becoming more automated, machines were nowhere near as accurate as they are today so some of the parts were very much hand-finished. All the adjustments and fitting were done by hand. When you look at an old watch you can clearly see, if not necessarily errors, the idiosyncrasies of another craftsperson. This always leads me to think about the moment someone first saw that watch in a shop window and decided to buy it, and how precious it then became to them. Did they wear it into the ground? Was it sold back to a jeweller and did it then embark on a new story with someone else? Was it left in a drawer and forgotten about for decades? Or perhaps left to a family member who didn't appreciate its dated style? So many mechanical watches suffered this fate in the 1970s and 1980s when they fell out of fashion. And now here they are, in this Family Circle biscuit tin.

Even earlier trench watches, which are quite small compared to their modern equivalents, regularly met this fate, including those by Wilsdorf & Davis. Their thin wire lugs, often made from soft metals like silver, didn't age well, either physically or in terms of fashion. It's only recently that people have started to realise that they're a century old and worth restoring. We are finally beginning to see the biscuit-tin survivors being repaired and worn again. During the quartz crisis, watchmakers weren't paid much to repair watches and were put under huge pressure to turn work round. We've spoken to now-retired watchmakers who've told us that sometimes they were given as little as half an hour to repair a watch. Craig and I can spend a day at the very least, and sometimes several weeks, working on a single mechanism; we're currently working on a restoration that has taken the best part of two years. Back then, the goal was simply to get them ticking as quickly and cheaply as possible. This resulted in a lot of unintended damage. I don't like to hear people knock the work of these watchmakers, who were working in terrible, unsupportive conditions. I've heard of highly skilled watchmakers making more money changing batteries than restoring vintage watches. The quartz crisis was a real low point for traditional watchmaking in so many ways.

Now we are in a new phase again. Technology has overtaken even the quartz watch. The current Apple Watch has more complications than Breguet could have imagined. It's not only a timepiece accurate to within 50 milliseconds, but a phone, internet browser, email provider, car key and fitness tracker, and can even offer ECG and oxygen-level readings – multiple technologies contained in one small package. Although the watch is still powered by a battery, the moving parts have been replaced by circuit boards, and the time readings they rely on are beamed by satellite and adjusted by atomic clocks. Each time a phone or smartwatch assesses our location, it draws on at least three satellite readings of nanosecond time from space. To tell us where we are, GPS will make adjustments for the relativity between their readings in a way that would have delighted Einstein. Without that adjustment we would register always slightly off, in much the same way that misreading longitude proved fatal for Cloudesley Shovell and his fleet in 1707.

For me, personally, the smartwatch is an advance too far – and somehow invasive. I've never owned a smartwatch. I already feel like my phone and laptop follow me everywhere and that's enough in itself. I love nothing more than getting away from phone signals and Wi-Fi and tracking cookies. I fear that technology will disconnect me completely from the world around me.

When e-books first appeared, everyone claimed the book was

dead. Who needed a book, a bookshop, even a bookshelf, when it could all be accessed on an e-reader? How odd, then, that there was a surprise resurgence in exquisite handmade books, which reminded people of the tactile pleasures of reading. Something similar is happening to watches. Things are starting to come round again. Prices of vintage watches have skyrocketed in recent years. Restoration is valued. Repairs you would have struggled to charge more than a few pounds for in the 1970s are now quoted at tens of thousands by some of the big brands. Today, watches like the Rebberg are increasingly collectable and retailing for good prices, which, in turn, allows for craftspeople to restore them properly. Along with the custom-made larger winding stem to fit their worn plates, we often increase the power of their mainspring to give them every bit of help they can get. Even a tiny difference in the thickness of the spring, maybe just 0.05 millimetres extra, is enough to help improve their accuracy and reliability. There is no spare parts supply, so we hand-make new balance staffs to replace broken ones (another common fault in watches made before shock settings were common-place), which we turn from steel in our lathe. The delicate pivots that support the entire balance assembly as it oscillates are less than half a millimetre in hight and smaller in thickness. Some of the ones we replace were made quickly and cheaply by one of the quartz crisis repairers, so we make a proper one, which we fit and then 'poise'. Poising is the process of carefully removing minuscule shavings of metal, smaller than a grain of sand, from the balance wheel to make sure its weight is perfectly evenly distributed. It's the miniature equivalent of having your car's tyres balanced.

On a Rebberg, the fine hairspring, made from softer metal than we use in watches today, almost always needs manipulation as, having been through those repair shops, it has usually been bent out of shape. To do this, we use tweezers so fine that their tips are like needles to gently return the hairspring to its perfect spiral. Rebbergs were finished with a human touch. You might have three or four movements that are exactly the same calibre, but you couldn't mix and match the parts. They'd need re-jewelling or modifying, as each movement is coach-built – designed to function only with the parts it has been set up with. Sometimes you'll see that the watchmaker has marked the main components of each movement with hidden dots or scratched numbers to keep them together as they finished and assembled them piece by piece.

In mechanical watches made today, swapping parts is possible with precision CNC engineering. I can't fault CNC: it's an incredible technology and without it many of today's most complicated and accurate watches would be unconceivable. But we'll never embrace it in our workshop. As Craig says, 'I'd rather spend hours messing around with bits that don't fit than have everything finished for me.' That's what it means to restore and rescue something that old; it might take a long time but the process – and the outcome – has soul.

One day there will be watches in deep space. For more than twenty years NASA's Jet Propulsion Laboratory has been developing an atomic clock small enough to take on space exploration missions beyond the reach of GPS.* At present, in order to gain their navigational coordinates, a spacecraft must send a signal from a location to an atomic clock,† and wait for orders – a process that, given the distances involved, can take hours. Atomic clocks need updating several times a day to maintain their phenomenal accuracy. The Deep Space Atomic Clock is currently described as 'about the size of a four-slice toaster', but they are working on making it smaller. It uses mercury-ion trap technology that can maintain a deviation of less than two nanoseconds a day (0.000 000 002 seconds) – accurate enough to enable astronauts in deep space to make navigational

---

* GPS can guide spacecraft navigationally up to an altitude of approximately 1,860 miles from Earth.
† The clock calculates the position of the spacecraft by measuring the time it takes for electromagnetic waves travelling at the speed of light to travel between the spacecraft and a known location, such as a satellite or antenna.

decisions on their own. Looking for a timepiece small enough and accurate enough that it can guide us as we venture into unknown territories feels like Harrison all over again – the sci-fi sequel.

But Earth time is only relative to Earth, so the further we travel in space, the less relevant it will become. And even though we might now regulate our clocks by the atomic second – the world's most accurate clock (although there is some debate about this!),* at JILA in Colorado (formerly the Joint Institute for Laboratory Astrophysics), is accurate to one second in 15 billion years, roughly the duration of the known universe† – we still live by the circadian rhythm. Habits, as the French Revolution's decimal time showed, are hard to break. But Mars has a similar day duration to ours, so maybe we needn't change just yet. In 2002 a sundial travelled aboard NASA's Mars lander *Surveyor 2001*, and was placed on Mars. Inscribed with the motto 'TWO WORLDS, ONE SUN', it will record the shadows and diurnal rounds of a new planet. It's like starting all over again.

Meanwhile, here on Earth, the internet has transformed our relationship with time once more. Accurate timekeeping now has to be global rather than local, and accurate to within a millionth of a second. Global air travel, telephone networks, banking, broadcasting, all rely on an extraordinary level of time accuracy. Where once we might have asked someone to wait a minute, now we expect everything in a nanosecond, the time it takes light to travel 30 centimetres.

The modern world is terrifyingly fast. I like slow.

The number of beats per second in a modern watch has become a status symbol. The faster a watch ticks, the more accurate it is, as

---

* The optical lattice clocks at the Observatoire de Paris and the strontium clock at the National Physical Laboratory in Teddington are also strong contenders.
† A standard atomic clock loses a second every 100 million years. They are so accurate that relativity becomes a problem. They can detect the relativistic impact on gravity if they are raised as little as a centimetre.

it becomes less susceptible to changes in position. A sixteenth-century verge watch might tick or beat at most 10,000 times an hour, compared to rates of 18,000 to 28,800 an hour in a modern watch. The fastest escapement in a mechanical watch today can run at a frequency of 129,600 ticks an hour, and at this speed the distinguishable sound of each tick is lost to a consistent hum. Of course, whether it's a slow clunk or high-tech hum, a second or a nanosecond, the pace of the time measured remains the same. But, for me, time can somehow seem more expansive when accompanied by the unhurriedly reassuring *clunk* that verge escapements make every time a tooth from the crown wheel drags itself across a flag and drops onto the next; back and forth, over and over. It's a reassuring sound, like a metronome ticking on top of a piano.

In spacetime, a second might be the difference between landing on Mars and landing tens of thousands of miles away. Here on Earth, I like to remember that the difference in accuracy between the fanciest modern timekeeper and one from the eighteenth century is only a moment – just a few minutes, sometimes seconds, over the course of a day. I can live with that; I've never been one to measure my life by the nanosecond.

11

Eleventh Hour

Though leaves are many, the root is one;
Through all the lying days of my youth
I swayed my leaves and flowers in the sun;
Now I may wither into the truth.

W.B. Yeats, 'The Coming of Wisdom With Time', 1916

I make my living from time – whether it's by making devices that measure it or studying its vast history. It can be overwhelming. Writing this book has been a huge process of bringing together so much that I have learned. Even though I handle timepieces every day of my life, it's only by taking the long view that I've realised just how daunting time can be. It's vast, measured by the movement of distant stars in an infinite universe, and it's also tiny and incredibly intimate, affecting the cells of our bodies right now as you read these words. How we spend our time is personal and also cultural; watches and humans alike are creatures of context.

As a watchmaker first and foremost, I find my place in time through making watches. Leaving work behind in metal feels like my way of leaving a small legacy, something that will live on beyond me, a mechanical ghost to haunt the Earth after I'm gone. Of course, there's no way of knowing how my work will be cared for in the future, if at all. Perhaps, in a few hundred years, the watches we make will sit behind toughened glass in a museum cabinet. Perhaps they will be family heirlooms. Perhaps they'll end up in an old biscuit tin waiting for someone to rescue them. The future is unknown. The past, however, has already unfolded. Considering the history of my subject helps me root myself and my making practice.

One day Craig and I realised that, over the years, we had wound up making virtually every component of a watch in our quest to repair another artisan's work. The time had come to set ourselves the challenge of making every component of a watch of our own. We nicknamed this watch Project 248 as a nod to the way it would be made – by two people, their four hands, and a traditional 8-millimetre watchmaker's lathe. As we surveyed our stable of old tools and repurposed machines, it felt natural to look to the late nineteenth century for our inspiration, to go back to the point in British watchmaking when our home industry was on its last gasp and pick up where our horological ancestors left off.

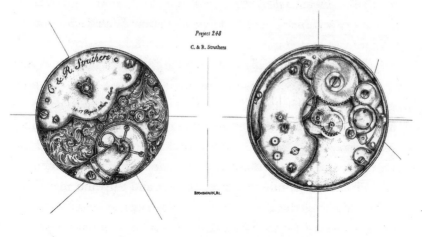

Craig's original concept illustration of Project 248.

We took our design cue from a machine-made fob watch created in the manufactory of a Coventry maker called Thomas Hill in the 1880s. The mechanics and materials of our wristwatch, however, pay tribute to a wider range of watchmakers and companies we admire from across the centuries. It's our little version of al-Jazari's elephant clock, an homage to the international contributions that have made our work possible. As Thomas Hill's pocket watch had no shock protection, we made our own in the style of Abraham-Louis Breguet's 'parachute'. The plates were cut from a metal called German silver, which is a base alloy made of copper, zinc and nickel. The colour, as the name suggests, is silvery in tone with a subtle green-grey hue. It develops a beautiful patina and warmth with age and has been popular among South German watchmakers for over 150 years. The balance, or ticking heart, of the watch was made in the style of one of the greatest watchmakers in living memory, George Daniels, the man who once asked me whether, and when, I would make a watch of my own. Daniels, who lived and worked on the Isle of Man until his death in 2011, was a one-off and an inspiration to Craig and me, not least because as a one-man band he made all the components he needed from scratch, using heritage tools, machines and processes.

Making Project 248 with Craig was the culmination of our combined experience – nearly forty years between us. Ever since I met Craig and started training as a watchmaker, he has had my back. At college he was a vital ally in a workshop where the unusual presence of a girl created a constant source of interest, and rarely in a good way. Now we encourage and motivate each other, something always important in both business and personal relationships, but even more so when you share both. Where we lack confidence in ourselves, it's made up by the unwavering confidence we have in each other. We drive each other on to push our boundaries and test our capabilities.

Bringing our skills and influences together in a single object was a fun, and frequently challenging, mission. From seedling concept to finished object, the whole build took nearly seven years and covered life-changing events for us both. This is how I know that what appears to be a straightforward object can be symbolic of so much more than its form or function. Every element bears the imprint of our hands, our skills and the hours we have taken to work on it. Every part, every stage in the process, is connected with our memories, of moments spanning those seven years.

Of all the tools we might use, none is more incredible than our hands. People often think that because I am a watchmaker I must have very delicate hands. In fact, as everything we do uses fine twee-zers, the physical make-up of your hand means very little. I have watched in awe as a man with hands the size of baseball mitts deli-cately manipulates and restores a balance spring barely thicker than a human hair. What matters is your sensitivity of touch and your understanding of the tolerances of the materials you are working with.

A few years ago, I attended a talk by cardiovascular surgeon Professor Roger Kneebone. I was fascinated by his descriptions of surgery: the years of experience it takes for surgeons to develop the haptic understanding of the body that enables them to predict how the tissues of a child will respond compared to those of an elderly patient, say, or those of a healthy adult compared to an unhealthy one. He described how the vessels in young people are strong and tough, like rubber, whereas when we reach old age they can be as delicate as tissue paper. In watchmaking you learn something simi-lar: brass that is 200 years old will have a different malleability to brass that is new. Steel in a watch made in the sixteenth century will react differently to heat than steel in a watch made in 2020. Ironically, in a watch, older materials, although marked by age and wear, tend to be better quality than new.

Interestingly, a disproportionate number of our clients over the years have been surgeons. In fact, one of the very first customers to entrust us with his precious watches was an orthopaedic surgeon who specialises in hands. Craig and I found it beautiful to listen to him talk about his love of hands, how, 'when opened', the structure beneath the skin is both robotic and alive – almost biomechanoid – operated by tendons, nerves, arteries and bones, with very little soft tissue. Hands are more than objects of human biology, and orthopaedic surgeons work with the physics of our joints. Like watchmakers, they work with saws and files and drills. He says a surgeon should be 'master craftsmen first and foremost',* and

---

* If you had to choose between a surgeon with brilliant theoretical knowledge and one with a track record of dexterity, he would urge you to choose the latter.

describes our hands as 'medical marvels', akin in intricacy to the most complicated of watches.

It is the human hand in 'handmade' that makes a watch unique. When I buff a watch case to achieve a form that is comfortable to wear, I hold a cloth loaded with polishing compound rather than use harder buff sticks or paper. It allows the metal to take its shape according to the curve of my palm, until it is irresistibly tactile. I could hold a machine-made and a handmade watch blindfolded and be able to tell you which is which; until we develop truly sentient artificial intelligence, no machine will be able to replicate the degree of variation you find in a handmade watch.

Handmaking something takes time. When I finish a watch, the last thing I do before it goes in the case is regulate it, adjusting the length of the hairspring to make the time run accurately. In old English watches the index, the pivoted lever we use to adjust the hairspring, is marked 'Fast' and 'Slow' – we included one in the earliest prototype of Project 248.* When I think of making a watch by hand, it is like turning the lever of time to 'Slow'. Indeed, part of the joy of being a craftsperson is that creating something – anything – has its own timeline. Each stage takes as long as it takes, and we have no choice but to give ourselves over to it. Yesterday I spent the whole day filing back the sides of an octagonal band so that it would fit perfectly into an octagonal case. It was only out by a tenth of a millimetre, but it took me nearly eight hours. But now that watch contains the time I have devoted to it. In a fast-moving world, I think there's a generosity to that. Watches not only measure time, they are a manifestation of time – signifiers of the most precious thing we have.

---

* In the end, we gave the 248 a free-sprung mechanism, in which the index is omitted and the hairspring is timed using the mass of the balance. It's a more precise means of regulation.

I have always had a peripheral awareness of the preciousness of time – in a workshop full of watches, how could I not? – but it has taken events in my own life for me to realise it acutely. In June 2017 I woke up one morning with intense pins and needles in my left leg and severe hypersensitivity down the left of my torso. The pain was so severe it reminded me of the time I cracked my ribs when I slipped climbing out of a window as a teenager. I went to the doctor but, as there were no visible marks, he reassured me it was probably just stress. After a few weeks the symptoms stopped. Then, in September that year, I suddenly lost partial sight in my right eye, a symptom accompanied by searing pain. It felt like I'd been punched in the face by a boxer. But again, there were no marks, no swelling. There was nothing noticeably wrong from the outside. This time too, the doctors I first saw tried to pass it off as stress, but I was becoming convinced there was something very wrong, and so fought for further tests. A few months later, after multiple referrals and an MRI scan, I was diagnosed with multiple sclerosis.

I will never forget those months undergoing tests. No one had taken the time to explain the possibilities to me, and so I feared it was a brain tumour, the only diagnosis I could rationalise out of my weird neurological snags. I'd never thought I was immortal, but the sudden intimation that I might be facing death a lot sooner than I'd anticipated changed the way I experienced the world. I remember that winter there was a sudden heavy snowstorm. The house we lived in backed onto a park, and when the snow settled a hill near the foot of our garden provided endless entertainment for children out sledging. I insisted on dragging Craig, who hates the cold, out on a long lap of the park, detouring from what would have been a straightforward trip to the supermarket at the end of our road, because it was rare to see snow that heavy in Birmingham and I wanted to experience it while I had the chance. That memory – of the ice-cold breeze biting my face, the way the light bounced off the white ground and trees, the sounds of children playing and excited dogs barking muffled against the soft snow, is as vivid today as that day years ago. I wanted to soak up the world while I still could.

It wasn't death that scared me so much as how I had spent my time. I had worked and worked – for what? What did I have to show for it? I had spent years being stressed, anxious and exhausted, but I hadn't allowed for happiness, and now I might be out of time.

As it turned out, I was lucky: life expectancy for people with multiple sclerosis has improved in recent decades. It's not pleasant, walking around with what feels like an undetonated Second World War bomb in your head, which may sleep peacefully for the remainder of my life, or go off at any moment, but things are far more promising than they used to be. I am incredibly fortunate to be living at a time when treatment options for MS have improved. I stand a good chance of living what I refer to with my neurologist as a 'long and blissfully uneventful life'. I'm also glad to live in a country that affords me access to one of the finest publicly funded healthcare systems in the world. I have a lot to be grateful for, and I just wish it hadn't taken preparing myself for a much worse diagnosis to realise that.

The French composer Louis-Hector Berlioz (1803–1869) famously declared: 'Time is a great teacher. Unfortunately, it kills all its pupils.' We are all students of time. Certainly, the lesson I've learned is how precious time really is.

My diagnosis has permanently changed how I want to live these seconds, hours, days I've been given. Those first doctors were right in one respect: a lot of my condition can be blamed on stress. Every single one of my relapses has followed shortly after an anxiety attack, and I've not had an anxiety attack that didn't precede a relapse. I used to see stress as a badge of honour, an internal scar I wore that proved how brave I'd been, throwing myself into high-pressure situations, surviving bullying and discrimination, and coming out still fighting. I operated at a fever-pitch of anxiety. Now I realise tolerating stress is not something to be celebrated. It takes just as much strength and resilience to say no, and to know when to walk away. Knowing how ill stress can make me, I avoid it like the plague. I'm a reformed workaholic.

Sadly, I can't claim that I now lead a flawlessly calm life, at peace with the universe. I still get wound up over stupid things, like energy companies failing to manage my account properly or my painfully slow-loading computer. But I am far more relaxed about life than I have been in the past. I focus my energy on the people in my life who support me; those who don't aren't worth my time. I allocate more time for taking in and appreciating the world around me.

The truth is, of course, all of us have limited time, and the amount we have is not within any of our control. Yet how we spend it is, and our experiences can alter our perception of time beyond the minutes logged on our watches. That day when Craig and I walked in the snow is lodged in my mind as if it were yesterday, defying clock time. Recent studies in neuroscience support the theory that our experience of time is tethered to the quality of our experiences.

Philosophers have known for centuries that there is a difference between having lived for a long amount of time and having lived an eventful, active and rich life. Old age isn't always a sign of experience or wisdom. There's a wonderful quote in *On the Shortness of Life*, written in around 49 AD by the Stoic philosopher Seneca, where he sums up the contrast between time lived and experience gained through the metaphor of a storm:

> you must not think a man has lived long because he has white hair and wrinkles: he has not lived long, just existed long. For suppose you should think that a man had had a long voyage who had been caught in a raging storm as he left the harbour, and carried hither and thither and driven round and round in a circle by the rage of opposing winds? He did not have a long voyage, just a long tossing about.

The experience of living a full life might leave us feeling as if our lives are flying past us, but making vivid memories means that in years to come we will look back on a life that feels long.

We all measure our lives in moments of time, and the memories that accompany them. Watches, which tell the time for us as they did for our relatives before us, provide a constant in those memories. People who would not consider themselves 'into watches' still want the treasured pocket watch they've inherited from their great-grandfather restored, even if they know they'll never use it. Looking at the dial of an old watch, we see the same hands passing the hours and minutes that our parents, grandparents or great-grandparents saw, hear the same ticking sound that they heard as they measured the passing moments of their lives. And then, if we are lucky, we wind it up and carry it with us as we go through ours.

Fast and slow, an English regulator.

How to Repair a Watch

A brief (and personal) guide

Every watch is unique. Even those made in vast quantities in contemporary factories. Once a watch has been worn, it will pick up traces of its owner's life: the adventures they've been on together, the daily wear, the special occasions it's been brought out for, or even the time it's been sat in its box. This is why, when a watch arrives in our workshop, I start a systematic process of inspections to make sure I capture any and all of the faults it has picked up over the years.

First, I don my watchmaker's loupe, a small magnifying glass that nestles against my eye and enlarges the watch in front of me to three times its size. I carefully check the overall appearance of the watch, looking for marks on the case and signs of water damage on the dial, and whether the winding button (or crown) is worn or shows a telltale flat impact mark, indicating the watch has been dropped on its crown, possibly causing further impact trauma inside. Tiny little scratches on the dial can be a clue that it has been to a less careful watchmaker in the past and might have other repair damage hidden inside. I check that the winding works, that the movement is still running, and that the winding button pulls out and turns the hands freely but not too sloppily; there needs to be some friction, but not too much. By the time I open the case back, I usually have a fairly good idea of what to expect.

The inside of watches' case backs is the prime location to spot the marks of past repairers. Sometimes they're scratched into the metal or scribed in permanent ink. There can be an identifiable name and date, or a code that means nothing to anyone other than the person who left it there. Several marks can indicate that a watch has been regularly sent for servicing over the years, like taking a car for its MOT, and so has been well cared for. Sometimes, however, they forewarn of a string of unresolved issues that it's now my job to unravel.

What happens from here varies from watchmaker to watchmaker, and depends on the calibre of the movement. There is near-infinite variation in the faults I encounter inside a watch, including some surprises I might see just once in my whole career. What follows is how I would approach a very typical manually wound wristwatch from the mid-twentieth century. It can't pretend to be a comprehensive guide, but it's an introduction to the process.

Assuming the watch is running at all, I start by checking it on our timing machine to get a basic indication of performance. Then I remove the bezel, the ring that holds the crystal protecting the dial, allowing me to access it. I pull the winding button into handset and set the time to twelve o'clock, or another time where the hands sit directly over each other. This provides the optimum position to lever the hands up and off without damaging them. To protect the dial, as well as taking great care, I use a thin sheet of plastic while I lift the hands. I repeat the process with the small seconds hand that sits just over six o'clock. With the hands off, I replace the bezel to protect the dial again and turn the watch over.

I slightly loosen the small screw just to the side of the stem of the winding button, allowing me to slide the button and attached stem out of the case. Two screws with large polished heads, slightly domed like the back of a ladybird, hold the movement in its case. I remove these two screws and remove the bezel again to release the movement from the case band, dial first. Putting the case to one side ready for cleaning, I remove the dial by loosening two tiny parallel grub screws that travel horizontally into the sides of the movement, engaging with the feet of the dial hidden inside. Once loose, the dial lifts off, and I place it carefully in a little sealed box along with the hands, to keep them safe while I turn my attention to the mechanism.

The movement is now fully undressed, so I fit it into an adjustable holder, which has sides that can be screwed inwards to hold the edges tightly. Then I lift off the motion work, the wheels that control the rotation of the hands, that has been sandwiched between the movement and dial. I place them to one side in my dust cover. A dust cover is like a horological bento box, with low-walled divisions separating areas in a tray to house the various collections of parts that make up each mechanism within a watch. It is covered by a clear plastic dome with a little handle on the top, a bit like a serving dish, which protects the parts from dust or rolling off my bench.

Now I fit the winding stem, with the button still attached, back into the watch, tightening the little screw to hold it in place again. This makes it easier to handle the movement and allows me to double-check the winding and keyless work now I can see it all moving.

Next I start the mechanical checks. I make sure the fine spiral hairspring is sitting perfectly over the balance wheel, that it's flat and the coils are not bunching to one side. I check the index, which controls the rate of the watch by lengthening or shortening the active length of the hairspring and has settings for 'slow' and 'fast'. It should be aligned in the middle. If the index is cranked all the way over to 'slow' (often marked 'R' for the French 'Retard' meaning 'Delay') it's a clue that the delicate spring might have been replaced or broken and re-pinned in the past and is now too short, making the watch run fast.

I check the whole movement for rust or oil build-up. I may have already spotted both of these faults before I removed the dial. Rust can tint the dial a reddish brown and is usually paired with water staining, while too much oil left by a previous repairer can begin to ooze through the hole in the middle of the dial, leaving a greenish residue and sometimes causing the paint to start lifting. I check for missing parts, which can work loose and end up getting jammed elsewhere in the movement, or are sometimes missing altogether. I switch to a closer magnifying loupe, which enables me to see the tiny components at twenty times their size, and look for cracked ruby jewel bearings that might be wearing the pivots rotating in them. Sometimes, if the watch has been dropped, these pivots break completely, causing the horizontal wheel to flop to one side. I remove any loose chunks of dirt or old jumper fluff that have worked their way into the winding with my tweezers. For general jobs I use my no. 3s, which are as fine at their tip as a sharpened pencil.

Next, I start checking for the faults that are so tiny it's easier to *feel* them than attempt to see them but are nevertheless so significant they could stop a watch altogether. Each wheel inside the watch, from the balance to the train wheels and mainspring barrel, needs to have the right amount of something we call 'shake'. This is the wiggle room that each part needs to function efficiently. Too much wiggle and the depthing (distance between the parts) will be wrong, causing unnecessary wear, variation in timekeeping and wheel teeth skipping through out of control. Too little wiggle and the mechanism will seize up and lock altogether. The right shake is often measured in hundredths, sometimes thousandths, of a millimetre. We check it by holding the part in a pair of finely pointed watchmaker's tweezers and giving it a gentle wobble. It takes a lot of practice but eventually you can feel instinctively whether the shake is right or wrong.

I work through this part of the process in stages. First, I check the shake of the balance. If I'm happy, I remove the balance cock screw and lift it out of the movement, the balance suspended on its hairspring below. I check the balance over, making sure the pivots aren't worn. I check the underside of the wheel for little marks that indicate the staff has been replaced and whether any weight has been filed away to poise it. I remove the jewels and check whether they've started to wear and need to be replaced. By my calculations, a typical watch train running for eighty years would vibrate 12,614,400,000 times. Even something as tough as synthetic ruby can start to erode against the steel staff after this much friction.

Next, I check the depth of the pallets in the teeth of the escape wheel, gently flicking the lever across with the tip of my oiler to make sure the locking is just right. Watchmakers' oilers are like very tiny spatulas made from steel, and are the thickness of a stiff paintbrush bristle. They have a little olive-shaped tip on the end that you dip in oil, the form of the olive holds it in place with surface tension until it's touched against your target. Even without the oil, they are a useful and precise tiny tool and sometimes more gentle than using tweezers. There is a mathematical formula for the perfect angles of depth between the pallets and the escape wheel teeth ('depthing') but, again, tables and charts mean very little when what you're working on is so small. It's easier to go by eye and feel. Depthing is bigger than the shakes, so you can usually see it as well as feel it.*

I'm careful to let any power out of the mainspring before removing the pallets as, once the movement is palletless, there's nothing to stop remaining power whizzing through the train, which in a dirty movement can potentially cause more wear and damage. I pull back the click that engages with the ratchet wheel and stops the mainspring unravelling. Holding the winding button carefully, I slowly let it reverse between my fingers under the releasing power of the spring, allowing the mainspring to unfurl and the power to seep away. I check the shakes of the pallet pivots and remove the bridge securing them. I check the ruby faces of the pallets, to make sure they haven't worn or chipped. I put the bridge, bridge screws and pallets in my dust cover to join the other parts I've placed there so far.

Working through the train, I check it's free and that the shakes are all acceptable. I remove the bridge holding the tops of the

---

* A way of illustrating depthing is if you interlock your right and left hand with straightened fingers, so the middle knuckles align. Give your hands a wiggle, and you'll find you are able to move your fingers, but they still stay firmly locked (correct depthing). However, if you lace your fingers together at their last joint, near the tip, your fingers might slip apart (depthing that is too shallow). If you lace your fingers together at their base you will struggle to wiggle your hands at all (depthing that is too deep).

pivots, which allows me to thoroughly inspect the pivots for wear. I then remove the last and largest bridge to reveal the barrel, removing the ratchet wheel and click to check the shakes. Then, once the barrel is free and in my hand, I check the shakes of the arbor inside, before finally popping the lid off. I remove the arbor and spring and check they are intact and in good condition. In old watches, the mainspring has nearly always fatigued with age and use and needs replacing. Replacing mainsprings can be a balancing act. They need to be the correct height and thickness to give the right power, which is very individual to the watch. Quite often, if you consult the official old charts and use like-for-like measurements the mainspring ends up exerting too much power, as modern springs are made from more efficient steels. Conversely, heavily worn tractors like our old Rebbergs might need all the help with power they can get, requiring a stronger mainspring now than when new. It can involve a bit of trial and error, testing a few springs to make sure you use the right one for the watch rather than the right one by the book. That's something we can only test once the watch is clean and back together, so for now I set it to one side. I remove the winding stem for the last time before cleaning, allowing the wheel and sliding clutch that control the interaction between handset and winding to drop out with it. These are enclosed in a pocket under the barrel bridge so, once exposed, the only thing holding them in the mechanism is the stem they thread onto.

The last set of components to be removed is the keyless work – the parts that allow the watch to be wound without a key and control the action between winding and setting the hands. I flip over the bare movement in its holder, to access the side usually hidden under the dial. The piece that was holding the stem as well as a small wheel, a lever (called a return bar) and a spring all hide under what we call rather literally the 'keyless cover'. I remove the screw and carefully lift this cover, making sure the spring doesn't ping out across the workshop. Return bar springs, often shaped like a shepherd's crook, are a little thicker than the hairspring but still tiny enough to make them a bugger to find! I remove the last parts with my tweezers, storing them away, and set about cleaning out the ruby jewels and bearings by hand with a sharpened piece of wood to make sure any dried old oil is removed. I then put all of the parts I've been carefully storing in my dust cover through our specialist watch cleaner.

Watch cleaners are a bit like dishwashers: you get the best finish if you make sure most of the mess is removed first. So I hand-clean the parts with solvent to remove excess grease and oil before loading them into little steel baskets that clip into the harness of the cleaner. I generally sort the brass parts together in one mini basket and the steel in another (to stop them scratching the brass as they spin in the cleaner) and store any very delicate parts in more individual mini baskets. The harness is like a robotic arm that lowers and lifts the baskets in and out of a series of cleaning solutions or rinses. The last receptacle conceals a heater, which evaporates any remaining rinse out of the movement and leaves the baskets hot to the touch.

While the movement is being cleaned, I turn my attention to the case, cleaning away any dirt and sometimes brightening it with a little polish. If a case is seventy years old or more, it will inevitably look like it's lived a little. I don't try to make cases look 'like new', unless their owner has requested it. I'm sensitive to the fact that those marks can be stories to their owner and they might want them left alone.

Once the parts of the movement are sparkling clean, I reassemble them. I start with the train and barrel, recoiling the mainspring with a special tool and fitting it back in place, laced with a little fresh grease. A slight bit of pressure to the barrel should send the whole train spinning, so it's easy to check that everything is moving freely. If something's locking, there's no point continuing until you've resolved the issue.

Now I reassemble the keyless work, carefully loading the little spring in place before capturing it under the cover before it pings away. It's a bit like trying to catch a grasshopper under cupped hands in the garden. I apply a little grease or oil to all the jewelled bearing surfaces and points of friction, careful to use enough to do the job but not so much it oozes onto the dial or anywhere else it shouldn't be. I check the keyless work is all functional, clicking between handset and winding. When I try winding the watch the train should now whizz through freely.

I lock the train with the pallets, retuning them to their position engaging with the escape wheel teeth. I wind the watch, turning the button once or twice, and watch the lever suddenly jump into position, held by the pressure now backed up in the train. I check the lever is flicking across in a lively fashion before moving on to the final stage.

I chance winding the mainspring right the way up to full power and return the assembled balance with its freshly oiled jewels to the movement. This is the moment when the watch starts to tick again. It's particularly poignant after the long restoration of a watch that has been silent for many years. It's when the watch comes back to life, its little escapement starting to beat and its hairspring methodically breathing in and out again.

I can tell by sight how healthy a balance is by the degree of its oscillations. In a watch of this age and style, I like it to be between 280 degrees and 300 degrees. Any greater and the balance risks swinging right the way round and knocking the wrong side of the lever, creating a tick that sounds a bit like a galloping horse. This can happen if your new mainspring is too strong. If the angle is too small, the watch is susceptible to positional errors, gaining or losing with the movement of the wearer. I wait until I can tell by eye that it's in the right ballpark before putting the watch anywhere near our timing machine, which reads the ticks to tell us exactly how the watch is performing.

The timing machine allows us to check the poising – that the weight of the balance wheel is evenly distributed. If one spot is too heavy, the reading on the machine will show a loss when the balance is on its side (the position it would be in if you were wearing it with your arm by your side), and a gain when the movement is rotated the other way. This impacts timekeeping. If the balance is out of poise, I remove it and roll the wheel on a set of sharp ruby jaws that allow it to roll freely until the heavy spot drops to the bottom. Once the culprit is identified, I remove a tiny scraping of metal with a cutter (these are the marks you can see if someone has done this before) to reduce the weight, and try rolling it again. And again. And again. Until it rolls free and comes to a slow stop without swinging. Sometimes I get this right first time, and sometimes it takes hours of patience to achieve.

When the power is off, the ruby pin (known as an impulse jewel) under the balance that flicks the pallet lever back and forth should rest centrally in the fork of the lever; we call this being in beat. With the other checks completed, the movement is ready for casing up. I replace the motion work and dial, securing those little grub screws to hold the dial securely. I replace the hands, checking that they align perfectly at the hour. I return the move- ment to the case band, removing the stem one last time and thread- ing it back through the hole in the case before securing it tightly. I replace the last two screws hold- ing the movement in the band and replace the back.

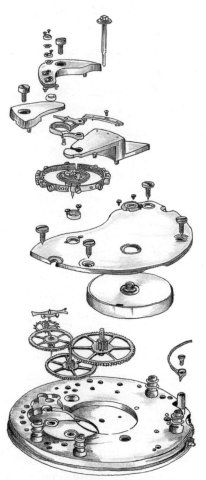

An exploded pocket-
watch movement.

The last test is a practical one, or as close to practical as we can achieve in the workshop. We put the watch onto the 'wrist' of a machine that rotates it through all positions as it runs, checking it continues to perform while being moved around like it will be in wear. We test it in movement during the day, and rest it in one position at night, over the course of seven days, ticking away non-stop. Only when we're satisfied that the watch is behaving itself do we reunite it with its owner.

Glossary

Arbor is a rod or parallel spindle onto which a part that needs to rotate or pivot is mounted, for example a watch train wheel.

Automatic, also called 'self-winding'. A watch with a mechanism that winds the mainspring using the movement of the wearer, usually using the swinging or rotating motion of a weight.

Atomic time is a high-precision form of timekeeping generated by atomic clocks. An increase in energy can cause electrons orbiting the atomic nucleus to jump to a higher orbit. As the electrons fall back, they emit energy at a specific oscillation frequency that can be used for timekeeping.

Balance spring (see hairspring)

Balance bridge is the name of the component fitted to the top plate of a watch with a verge escapement (occasionally early lever and cylinder escapements), designed to hold the top pivot of the balance staff. It consists of a round plate, or table, which is often decorated with piercing and/or engraving and attached to the top plate by two feet secured by two screws forming the shape of a bridge.

A pierced and engraved balance cock with Green Man design, found in an English watch from the 1760s

Balance cock is as above, only held by a single foot and screw.

Balance staff is the central component, or backbone, of the balance assembly. The balance wheel, impulse roller and collet in the middle of the hairspring are all friction-fitted to this central staff, which oscillates on pivots at the top and bottom. Variations of the balance staff form the basis of every single watch balance assembly invented.

Balance is an oscillating wheel in the movement of a watch, similar in appearance to the design of a steering wheel, and is responsible for regulating the release of power from the mainspring. It is friction-mounted and then riveted to the balance staff.

Bezel is the front part of a watch case in which the transparent glass (or mineral crystal or sapphire) is secured to reveal the dial underneath.

Blued steel is steel that has been heat treated, traditionally using the flame of a spirit lamp, to create an oxide layer that makes it appear blue in colour. As the steel is heated it changes colour, first turning a straw-yellow shade before going deep purple, then midnight blue. This blue gradually lightens in colour to a bright, almost electric blue before fading back to a grey steel colour. The oxide layer offers some protection from rust and tempers hard steel, but in watch-making it is also performed purely for aesthetic reasons. Watchmakers choose their preferred shade and stop the colour transition by removing the steel from the heat.

A glass watchmaker's spirit lamp, used for jobs like heat-bluing steel.

Champlevé is a style of solid silver or gold dial, often with engraved numerals in-filled with black enamel or wax, and typically decorated with ornate engraving, piercing and chasing. The modern term also refers to dials that are engraved, then covered with translucent vitreous enamel.

Chronometer is the term given to a watch or clock capable of operating to the very highest standards of accuracy. The title is awarded by national independent governing bodies who test them. Historically, this occurred in observatories such as Kew (UK), Besançon (France), and Neuchâtel and Geneva (Switzerland). The current main assessment centre is the Contrôle Officiel Suisse des Chronomètres (or COSC). Accuracy is measured over many days (fifteen at COSC; Kew used to have a forty-four-day trial) and through a range of different temperatures and positions. Only watches that meet these exacting standards are allowed to be referred to as chronometers.

Chronograph is a timepiece that also has a stopwatch function that can start, stop and reset without interfering with the timekeeping element of the watch.

Complications are functions that go beyond any device used to improve timekeeping, such as a chronograph, calendar display or repeating work.

Crown, also known as the button, is the part on the side of the case that can be turned to wind up or set the time on a watch.

Depthing is the positional relationship between two interacting components. Optimal depthing ensures the parts work efficiently together without slipping past each other (which means the depthing is too shallow) or locking or generating too much friction (which means the depthing is too deep).

Ébauche is the term given to a standardised movement supplied to order in a complete but unfinished state with the intention that the purchaser can customise, finish and sign it accordingly before adding their own case, dial and hands. *Ébauche* manufacture was perfected in the mid-nineteenth century.

Engine turning is a style of engraving performed on a rose or straight-line engine machine. The finished result is geometric in form, resembling a design a spirograph might create, or a weave like the side of a wicker basket.

English lever is a type of escapement invented in England and used there from the second half of the eighteenth century to the beginning of the twentieth century. It differs from the Swiss lever in the shape of the teeth of the escape wheel, which are long, fine and pointed.

Equation of time is the difference between mean solar time and apparent solar time, meaning the difference between the time shown on a watch or clock and the time shown on a sun or ring dial.

Escapement is the collective name given to the group of components in a watch responsible for controlling the release of power from the mainspring and reducing the speed the train wheels rotate to one useable for timekeeping.

Établissage is the early production-line manufacturing of non-standardised blank watch movements (evolved into the *ébauche*), which was developed in Switzerland from the start of the eighteenth century.

Form watch is the term given to a late-sixteenth or early-seventeenth-century watch that is, quite literally, created in the form of something else. Popular forms included flowers, animals, skulls and religious iconography.

Fusee is a device used to achieve uniform torque from the mainspring. When the mainspring is at full wind it exerts a stronger force than when the watch is nearly unwound. The force is evened out using a gut line (or later chain), which transmits the power to a graduated barrel known as the fusee. When the mainspring is at full wind, the line turns the smallest diameter of the fusee barrel, reducing its power. The line works its way down the graduation, inverting the mainspring's power against the fusee diameter.

Gilding is the predecessor to gold plating, also referred to as 'fire' or 'wash' gilding. An amalgam of gold and other metals and chemicals in the form of a paste would be painted onto the surface of silver, or a base metal like copper or brass. The liquid was burned off with a flame, leaving the gold baked on to the surface. The fumes created were incredibly dangerous as the process uses ingredients such as mercury, ammonia and nitric acid. Gilding was replaced by electroplating.

Greenwich meridian is the imaginary line that runs through Greenwich in London, UK, opposite to the international date line.

Hairspring, also known as a balance spring, is the fine spring formed in a spiral that regulates the speed at which the balance oscillates. When shortened the watch runs faster and when lengthened it runs more slowly. It is secured to the balance staff by a collet at its centre. In early watches it was made of a hog's-hair bristle.

Jewels are synthetic corundum (ruby or sapphire) pivot bearings. Corundum is used for its exceptionally hard-wearing properties. These replaced brass bushes and start to appear in the eighteenth century.

Keyless work is the group of components that work together to allow a watch to be wound and control the transition between winding and setting the hands (by pulling out the crown, on modern watches). During handset, the keyless work engages with the motion work.

Latitude is a position calculated using the series of imaginary lines on a map or globe that run from east to west, parallel to the equator.

Longitude is a position calculated using the series of imaginary vertical lines that run at 15-degree intervals from the North to South Pole on a globe and are counted from the Greenwich meridian.

Loupe is a magnifying lens that, in watchmaking, is worn against the eye. For general work a three-times magnification is used, although for more detailed inspections a range of loupes are used, generally up to around twenty-times magnification.

A loupe.

Mainspring is the power source in a mechanical watch. It is the name given to the coiled ribbon-like spring contained within a barrel (the mainspring barrel) that can be wound up. The wound spring creates a rotating force as it releases, which is transferred down a series of toothed wheels to the escapement, where the speed of release is regulated.

Masstige is an amalgamation of the terms 'mass produced' and 'prestige', used to refer to the commercialisation of luxury objects.

Motion work is the series of wheels and pinions that control the movement of the watch hands. The hour wheel is geared to turn once every twelve hours, and the canon pinion once every sixty minutes. In normal operation, their movement is controlled by the train wheels. In handset, the keyless work overrides this to allow manual handsetting.

Oscillation is the regular back-and-forth movement, or rhythm, of an object such as a pendulum or wheel.

Pallets (entrance and exit) are the parts of the escapement that allow the catch and release of one tooth at a time of the escape wheel. This regular staggered catch/release controls the release of power from the mainspring at a rate that can be used for timekeeping.

Pare-chute, or parachute, is the name of the first type of shock absorber introduced in a watch mechanism in 1790. It was invented in Paris by Abraham-Louis Breguet and helped protect the delicate pivots of the balance staff from breaking in the event that the watch was knocked or dropped.

Pair-cases is the name given to the design of watch in which the movement is housed in an inner case, which is then protected within a further outer case.

Piezoelectricity, from the Greek for pressure or push, in this context refers to the small electrical pulse produced by a quartz crystal when it is subjected to mechanical stress.

Pivots are the narrow terminal points of an arbor, which are smaller in diameter, and highly polished to reduce friction as much as possible. They are the points the arbor rotates or pivots on.

Positional error is the variation in timekeeping caused by changes in the direction of gravitational pull as the watch moves through a range of positions, as it does during wear.

Quartz crystal regulates the electronic oscillator in a quartz-battery-powered wristwatch using its piezoelectric properties.

Raising is a process in which metal is repeatedly hammered and then annealed (softened by heating) to stretch and form it into a shape such as a bowl, cup or watch case.

Repeater is a mechanism in a timepiece that chimes hours and quarter-hours, or hours, minutes and quarter-hours, on gongs made from wire or small bells.

Louis Audemars Quarter Repeater

C. Struthers
Birmingham, B1

A quarter-repeating pocket watch movement made in Switzerland by Louis-Benjamin Audemars in around 1860.

Repoussé is a technique used in silversmithing where a design is punched into the back of a piece of metal to create a relief, before being engraved and chased from the front to refine the detail. Commonly used on the outer case of pair-cased watches in the eighteenth century.

Shake is the amount of end and sideways movement a pivot has in its bearing. Too much movement will interfere with depthing and too little will cause the part to seize.

Shock setting is the general term for the components that help to protect pivots from breaking in the event of an impact. This includes the *pare-chute*, the concept of which has been refined and redesigned over the centuries to include a wide range of systems.

Staff (see balance staff)

Temperature compensation refers to the range of different inventions developed to compensate for the variations caused by thermal expansion and contraction.

Tourbillon is a device invented by Abraham-Louis Breguet in 1801 that reduces positional error by keeping the escapement and balance assembly constantly rotating through 360 degrees.

Train wheels are the series of toothed wheels and pinions that interconnect in a watch movement to gear down the power from the mainspring to rotation speeds compatible with timekeeping.

Verge is the name of the earliest escapement found in watches. First used in clocks and phased out at the turn of the nineteenth century, the mechanism consists of a balance staff with two flags positioned at right angles. The staff, secured to the oscillating balance, rotates back and forth, allowing the pallets to release one tooth at a time of an engaging wheel, referred to as the escape or crown wheel.

Acknowledgements

This book brings together so much of my life, career and education that I could write a list of equal length for every person who has helped me on my way. But I'll try to keep it short.

For introducing me to the world of watchmaking I'd like to thank my tutors Paul Thurlby and Jim Kynes. As a jeweller I'm indebted to the tuition of Peter Ślusarczyk and Eimear Conyard. As a historian, my studies wouldn't have been possible without the support of Dr Lawrence Green and Professor Kenneth Quickenden.

This book came about at a moment when, after many submissions and rejections, I'd given up all hope of publishing it. Thanks to Kirsty McLachlan for finding me and encouraging me to give it another go, to my agent David Godwin, and to my editor Kirty Topiwala for commissioning this book, along with her wonderful team at Hodder: Rebecca Mundy, Jacqui Lewis, Tom Atkins, Helen Flood and my picture editor Jane Smith. Holly Ovenden designed the beautiful cover. Roma Agrawal has been my informal mentor, keeping me on the straight and narrow. I must also take a moment to honour the long-suffering Victoria Millar who helped me to weave together the multitude of strands of this story. Thank you for your late nights and cheer-up cat photos when things have been a struggle.

For assisting with my research enquiries and helping me check the details of this vast subject, thanks to (in no particular order) Justin Koullapis, Alom Shaha, Dr Michelle Bastian, Professor Kevin

Birth, Michael Clerizo, Mike Cardew, Dr Richard Hoptroff, Dr Stephanie Davies, David Goodchild, Dr Jim Beveridge, Karen Bennett, Chantal Bristow, Professor Francesco D'Errico, Dr Katie Russell-Friel, Mr Ronald Mifsud, Anna Rolls, Elizabeth Doerr, Seth Kennedy, David Barrie, Mollie Hughes, Mike Frayn and Dr James Fox. Thanks also to Professor Joe Smith and Professor Renata Tyszczuk from Smith of Derby for showing me around their epic workshop.

Thank you to the ultra-talented Andy Pilsbury who photographed all of the images in this book as well as the majority of our photographs over the years. We're lucky to have such skilled friends. Andy welcomed his first child, Poppy Pilsbury, in 2022 so I'm going to take this moment to immortalise her in print and say welcome to the world, Poppy! I'm also grateful to Jen O'Shaugnessy for her photo-editing skills, and to the owners and caretakers of the watches and objects included whose support of the book made these images happen.

Much of my research would not have been possible without the support of museums and their curators. I particularly want to thank David Thompson, Paul Buck, Oliver Cooke and Laura Turner of the British Museum; Alan Midleton, Alex Bond, Izzy Davidson, Dr Robert Finnegan and Dave Ellis from the Museum of Timekeeping; Anna Rolls from the Worshipful Company of Clockmakers; Dr David Morris from the McGregor Museum; and Amy Taylor from the Ashmolean Museum. The thorough list of museums with horological objects and collections all over the world was greatly enriched by ideas from my amazing followers on Instagram; thank you all for your kind suggestions.

There have been a number of educational bodies and charities that have supported us and our workshop over the years. I mention them here both to thank them, and in case anyone reading this would also like to find out more about the traditional crafts described in this book or learn about the art of watchmaking: the British Horological Institute, Queen Elizabeth Scholarship Trust, Heritage Crafts and the Association of Heritage Engineers.

I'd like to thank the creative cluster of fellow makers with whom I've had the pleasure of collaborating, and who make what we do possible: Henry Deakin, The Wizard (a.k.a. Steve Crump), Dave Fellows, Andrew Black, Anita Taylor, Liam Cole, Sally Morrison, Lewis Heath, Florian Güllert, Mike Couser, Neil Vasey, Anousca Hume, Gabi Gucci, May Moorhead, Callum Robinson and Marisa Giannasi. And special thanks to our exceedingly patient friend and director Jan Lawson for having the astute business mind that Craig and I lack. I doubt we'd still have a workshop to write about without you.

And for keeping me well and able to continue watchmaking, my thanks to Sharon Letissier and Dr Niraj Mistry.

To my soulmate, inspiration and illustrator Craig Struthers, thank you for your unwavering support while I've been writing this book, and over the last two decades. And my four-legged family of waifs and strays, my short and furries, for keeping me on my toes and being there for a cuddle whenever I've needed one: watch dog Archie, cats Alabama and Isla, and our mouse Morrissey. And, of course, to my parents and family, I am who I am today because of you, so when I'm a pain in the arse you have no one to blame but yourselves.

This book is dedicated to the memory of two great losses that occurred during its writing. Adam Phillips was the last independent watch casemaker in the UK who took Craig on for an informal apprenticeship in 2017 to share his skills. His generosity, knowledge and kindness will not be forgotten. Neither will Craig forget the excuse Adam provided him with for always needing new (old) tools. One can 'never be over-lathed'. As a fellow cat-lover, I hope Adam would have been happy to share his dedication with my loyal old friend Indy, who sat on my lap for most of the time I was writing this book. She passed away the day after I submitted my manuscript. She was a very fine cat, a very fine cat indeed.

Picture credits

Bibliography

Abulafia, D. (2019). *The Boundless Sea: A Human History of the Oceans.* Allen Lane, London.

Albert, H. (2020). 'Zoned out on timezones.' *Maize*, 30 January. Available at: https://www.maize.io/magazine/timezones-extreme-jet-laggers (Accessed 12 May 2021).

Álvarez, V.P. (2015). 'The Role of the Mechanical Clock in Medieval Science'. *Endeavour*, 39 (1), pp. 63–8.

Anon (1772). *A View of Real Grievances, with Remedies Proposed for Redressing Them.* London.

Anon (1898). *The Reign of Terror, a Collection of Authentic Narratives of the Horrors Committed By the Revolutionary Government of France Under Marat and Robespierre.* J.B. Lippincott Company, Philadephia.

Antiquorum (1991). *The Art of Breguet.* Habsburg Fine Art Auctioneers. Sale catalogue 14 April. Schudeldruck, Geneva.

Baker, A. (2012). '"Precision", "Perfection", and the Reality of British Scientific Instruments on the Move during the 18th Century'. *Material Culture Review*, 74–5 (Spring), pp. 14–28.

Baker, S.M. and Kennedy, P.F. (1994). 'Death by Nostalgia: A Diagnosis of Context-Specific Cases'. *NA – Advances in Consumer Research*, vol. 21, eds. Chris T. Allen and Deborah Roedder John, Provo, UT: Association for Consumer Research, pp. 169–74.

Balmer, R.T. (1978). 'The Operation of Sand Clocks and Their Medieval Development'. *Technology and Culture*, 19 (4), pp. 615–32.

Barrell, J. (1980). *The Dark Side of the Landscape: The Rural Poor in English Painting*. Cambridge University Press, Cambridge.

Barrie, D. (2015). *Sextant: A Voyage Guided by the Stars and the Men Who Mapped the World's Oceans*. William Collins, London.

Barrie, D. (2019). *Incredible Journeys: Exploring the Wonders of Animal Navigation*. Hodder & Stoughton, London.

Bartky, I. (1989). 'The Adoption of Standard Time'. *Technology and Culture*, 30 (1), pp. 25–56.

Baxter, R. (1673). *A Christian directory, or, A summ of practical theologie and cases of conscience directing Christians how to use their knowledge and faith, how to improve all helps and means, and to perform all duties, how to overcome temptations, and to escape or mortifie every sin: in four parts . . . / by Richard Baxter*. Printed by Robert White for Nevill Simmons, London.

Beck, J. (2013). 'When Nostalgia Was a Disease'. *The Atlantic*, 14 August. Available at: https://www.theatlantic.com/health/archive/2013/08/when-nostalgia-was-a-disease/278648 (Accessed 14 May 2021).

Betts, J. (2020). *Harrison*. National Maritime Museum, London.

Birth, K. (2014). 'Breguet's Decimal Clock'. The Frick Collection, *Members' Magazine*, Winter.

Breguet (2021). *'Grande Complication' pocket watch number*. Available at: https://www.breguet.com/en/house-breguet/manufacture/marie-antoinette-pocket-watch (Accessed 18 May 2021).

Breguet (2021). *1810, The First Wristwatch*. Available at: https://www.breguet.com/en/history/inventions/first-wristwatch (Accessed 18 May 2021).

Breguet, C. (1962). *Horologer*. Translated by W.A.H. Brown. E.L. Lee, Middlesex.

Centre, J.I. (2021). 'Bacteria Can Tell the Time with Internal Biological Clocks'. *Science Daily*, 8 January. Available at: https://scitechdaily.com/bacteria-can-tell-the-time-with-internal-biological-clocks (Accessed 22 April 2021).

Chapuis, A. and Jaquet, E. (1956). *The History of the Self-Winding Watch 1770–1931*. B.T. Batsford Ltd, London.

Chapuis, A. and Jaquet, E. (1970). *Technique and History of the Swiss Watch*. Translated ed. Hamlyn Publishing Group Limited, Middlesex.

Chevalier, J. and Gheerbrant, A. (1996). *Dictionary of Symbols*. Translated 2nd ed. Penguin, London.

Church, R.A. (1975). 'Nineteenth-Century Clock Technology in Britain, the United States, and Switzerland'. *Economic History Review*, New Series, 28[4].

Clarke, A. (1995). *The Struggle of the Breeches: Gender and the Making of the British Working Class*. University of California Press, Berkley.

Clarke, A. (2020). 'Edinburgh's iconic Balmoral Hotel clock will not change time at New Year'. *Edinburgh Live*, 29 December. Available at: https://www.edinburghlive.co.uk/news/edinburgh-news/edinburghs-iconic-balmoral-hotel-clock-19532113?utm_source=facebook.com&utm_medium=social&utm_campaign=sharebar&fbclid=IwAR0HxWdnV5H4VrQT51OofOkUMWs_kXaHMo_h4LvHCu2Fr1PFsLTgfl6Qono (Accessed 5 May 2021).

Clayton (1755). *Friendly Advice to the Poor; written and published at the request of the late and present Officers of the Town of Manchester*.

Corder, J. (2019). 'A look at the new $36,000 1969 Seiko Astron'. *Esquire*, 6 November. Available at: https://www.esquireme.com/content/40676-a-look-at-the-new-36000-1969-seiko-astron-draft (Accessed 14 May 2021).

Cummings, G. (2010). *How the Watch Was Worn: A Fashion for 500 Years*. The Antique Collectors' Club, Suffolk.

Cummings, N. and Gráda, C.Ó. (2019). 'Artisanal Skills, Watchmaking, and the Industrial Revolution: Prescot and Beyond'. Competitive Advantage in the Global Economy (CAGE) Online Working Paper Series 440. Available at: https://ideas.repec.org/p/cge/wacage/440.html (Accessed 8 April 2021).

Daniels, G. (1975). *The Art of Breguet*. Sotheby's Publications, London.

Darling, D. (2004). *The Universal Book of Mathematics: From Algebra to Zeno's Paradoxes*. John Wiley & Sons, New Jersey.

Davidson, H. (2021). 'Tiananmen Square watch withdrawn from sale by auction house'. *Guardian*, 1 April. Available at: https://

www.theguardian.com/world/2021/apr/01/tiananmen-square-watch-given-chinese-troops-withdrawn-from-sale-fellows-auction-house (Accessed 14 May 2021).

Davie, L. (2020). 'Border Cave finds confirm cultural practices'. *The Heritage Portal*. Available at: http://www.theheritageportal.co.za/article/border-cave-finds-confirm-cultural-practices (Accessed 6 July 2020).

Davis, A.C. (2016). 'Swiss Watches, Tariffs and Smuggling with Dogs'. *Antiquarian Horology*, 37 (3), pp. 377–83.

D'Errico, F.; Backwell, L.; Villaa, P.; Deganog, I.; Lucejkog, J.J.; Bamford, M.K.; Highamh, T.F.G.; Colombinig, M.P. and Beaumonti, P.B. (2012). 'Early Evidence of San Material Culture Represented by Organic Artifacts from Border Cave, South Africa'. *Proceedings of the National Academy of Sciences of the United States of America*, 14 August, 109 (33), pp. 13, 214–13, 219.

D'Errico, F., Doyon, L., Colagé, I., Queffelec, A., Le Vraux, E., Giacobini, G., Vandermeersch, B., Maureille, B. (2017). 'From Number Sense to Number Symbols. An Archaeological Perspective'. *Philosophical Transactions of the Royal Society*. B 373: 20160518.

De Solla Price, D. (1974). 'Gears from the Greeks: The Antikythera Mechanism – A Calendar Computer from ca. 80 B.C.' *Transactions of the American Philosophical Society*, 64 Pt. 6. Philadelphia.

Dickinson, H.W. (1937). *Matthew Boulton*. Cambridge University Press, Cambridge.

Diop, C.A. (1974). *The African Origin of Civilization: Myth or Reality*. Chicago Review Press, Chicago.

Dohrn-van Rossum, G. (1996). *History of the Hour: Clocks and Modern Temporal Orders*. Translated ed. The University of Chicago Press, Chicago.

Dowling, J. and Hess, J.P. (2013). *The Best of Time: Rolex Wristwatches: An Unauthorised History*. 3rd ed. Schiffer Publishing Ltd, Pennsylvania.

Dyke, H. (2020). *Our Experience of Time in the Time of Coronavirus*

Lockdown, Cambridge Blog. Available at: http://www.cambridge-blog.org/2020/05/our-experience-of-time-in-the-time-of-coronavirus-lockdown (Accessed 11 February 2021).

Erickson, A.L. (Unpublished). *Clockmakers, Milliners and Mistresses: Women Trading in the City of London Companies 1700–1750*. Available at: https://www.campop.geog.cam.ac.uk/research/occupations/outputs/preliminary/paper16.pdf

Evers, L. (2013). *It's About Time: From Calendars and Clocks to Moon Cycles and Light Years – A History*. Michael O'Mara Books Ltd, London.

Falk, D. (2008). *In Search of Time: The Science of a Curious Dimension*. St. Martin's Press, New York.

Forster, J. and Sigmond, A. (2020). *Accutron: From the Space Age to the Digital Age*. Assouline Collaboration.

Forsyth, H. (2013). *London's Lost Jewels: The Cheapside Hoard*. Philip Wilson Publishers Ltd, London

Forty, A. (1986). *Objects of Desire: Design and Society since 1750*. Cameron Books, Dumfriesshire.

Foulkes, N. (2019). 'The Independent Artisans Changing the Face of Watchmaking'. *Financial Times*, How to Spend It, 12 October.

Foulkes, N. (2019). *Time Tamed: The Remarkable Story of Humanity's Quest to Measure Time*. Simon & Schuster, London.

Fraser, A. (2018). *Mary, Queen of Scots*. Fiftieth-anniversary ed. Weidenfeld & Nicolson, London.

Freeman, S. (2021). 'Parents find time passes more quickly, researchers reveal.' *The Times*, 22 February. Available at: https://www.thetimes.co.uk/article/parents-find-time-passes-more-quickly-researchers-reveal-sqvvod65v (Accessed 22 June 2022)

Fullwood, S. and Allnutt, G. (2017–present). The AHS *Women and Horology* Project. Available at: https://www.ahsoc.org/resources/women-and-horology/ (Accessed 18 May 2021).

Ganev, R. (2009). *Songs of Protest, Songs of Love: Popular Ballads in Eighteenth Century Britain*. Manchester University Press, Manchester.

Geffen, Anthony (director) (2010). *The Wildest Dream* (film). United States, Altitude Films with Atlantic Productions.

Glasmeier, A.K. (2000). *Manufacturing Time: Global Competition in the Watch Industry, 1795–2000*. The Guilford Press, London.

Glennie, P. & Thrift, N. (2009). *Shaping the Day: A History of Timekeeping in England and Wales 1300–1800*. Oxford University Press, Oxford.

Good, R. (1965). 'The Mudge Marine Timekeeper'. *Pioneers of Precision Timekeeping: A Symposium*. Antiquatian Horological Society, London.

Gould, J.L. (2008). 'Animal Navigation: The Longitude Problem'. *Current Biology*, 18 (5), pp. 214–216.

Guye, S. and Michel, H. (1971). *Time & Space: Measuring Instruments from the 15th to the 19th Century*. Pall Mall Press, London.

Gwynne, R. (1998). *The Huguenots of London*. The Alpha Press, Brighton.

Hadanny, A.; Daniel-Kotovsky, M.; Suzin, G.; Boussi-Gross, R.; Catalogna, M.; Dagan, K.; Hachmo, Y.; Abu Hamed, R.; Sasson, E.; Fishlev, G.; Lang, E.; Polak, N.; Doenyas, K. et al. (2020). 'Cognitive Enhancement of Healthy Older Adults Using Hyperbaric Oxygen: A Randomized Controlled Trial'. *Aging* (Albany, NY), 12 (13), pp. 13740–13761.

Häfker, N. S.; Meyer, B.; Last, K.S.; Pond, D.W.; Hüppe, L.; Teschke, M. (2017). 'Circadian Clock Involvement in Zooplankton Diel Vertical Migration'. *Current Biology*, 27 (14), (24 July), pp. 2194–2201.

Heaton, H. (1920). *The Yorkshire Woollen and Worsted Industries, from the Earliest Times up to the Industrial Revolution*. Clarendon Press, Oxford.

Helfrich-Förster, C., Monecke, S., Spiousas, I., Hovestadt, T., Mitesser, O. and Wehr, T.A. (2021). 'Women Temporarily Synchronize Their Menstrual Cycles with the Luminance and Gravimetric Cycles of the Moon'. *Science Advances*, 7, eabe1358.

Hom, A. (2020). *International Relations and the Problem of Time*. Oxford University Press, Oxford.

House of Commons (1817). *Report from the Committee on the Petitions of Watchmakers of Coventry*. London, 11 July.

House of Commons (1818). *Report from the Select Committee Appointed to Consider the Laws Relating to Watchmakers.* London, 18 March.

James, G.M. (2017). *Stolen Legacy: The Egyptian Origins of Western Philosophy.* Reprint ed., Allegro Editions.

Jones, A.R. and Stallybrass, P. (2000). *Renaissance Clothing and the Materials of Memory.* Cambridge University Press, Cambridge.

Jones, M. (1990). *Fake? The Art of Deception.* British Museum Publications, London.

Jones, P.M. (2008). *Industrial Enlightenment: Science, Technology and Culture in Birmingham and the West Midlands 1760–1820.* Manchester University Press, Manchester.

Keats, A.V. (1993). 'Chess in Jewish History and Hebrew Literature'. University College, University of London, PhD thesis.

Klein, M. (2016). 'How to Set Your Apple Watch a Few Minutes Fast'. *How-To Geek.* Available at: https://www.howtogeek. com/237944/how-to-set-your-apple-watch-so-it-displays-the-time-ahead (Accessed 8 February 2021).

Landes, D. (1983). *Revolution in Time: Clocks and the Making of the Modern World.* Harvard University Press, Massachusetts.

Lardner, D. (1855). *The Museum of Science and Art, Vol. 6,* Walton & Maberly, London.

Lester, K. and Oerke, B.V. (2004). *Accessories of Dress: An Illustrated Encyclopaedia.* Dover Publications, New York.

Locklyer, J.N. (2006). *The Dawn of Astronomy: A Study of Temple Worship and Mythology of the Ancient Egyptians.* Dover Edition. Dover Publications, New York.

Lum, T. (2017). 'Building Time Through Temporal Illusions of Perception and Action: Sensory & Motor Lag Adaption and Temporal Order Reversals'. Vassar College, thesis, p. 6. Available at: https://s3.us-east-2.amazonaws.com/tomlum/Building+Tim e+Through+Temporal+Illusions+of+Perception+and+Action. pdf (Accessed 19 April 2021).

Marshack, A. (1971). *The Roots of Civilization.* McGraw-Hill, New York.

Masood, E. (2009). *Science & Islam: A History.* Icon Books Ltd, London.

Mathius, P. (1957). 'The Social Structure in the Eighteenth Century: A Calculation by Joseph Massie'. *Economic History Review* (Second Series), X (1) pp. 30–45.

Matthes, D. (2015). 'A Watch by Peter Henlein in London?' *Antiquarian Horology*, 36 [2] (June 2012), pp. 183–94.

Matthes, D. and Sánchez-Barrios, R. (2017). 'Mechanical Clocks and the Advent of Scientific Astronomy'. *Antiquarian Horology*, 38 (3), pp. 328–42.

May, W.E. (1973). *A History of Marine Navigation*. G.T. Foulis, London.

Mills, C. (2020). 'The Chronopolitics of Racial Time'. *Time & Society*, 29 (2), pp. 297–317.

Moore, K. (2016). *The Radium Girls*. Simon & Schuster, London.

Morus, I.W. (ed.) (2017). *The Oxford Illustrated History of Science*. Oxford University Press, Oxford.

Mudge, T. (1799). *A Description with Plates of the Time-keeper Invented by the Late Mr. Thomas Mudge*. London.

Murdoch, T.V. (1985). *The Quiet Conquest: The Huguenots, 1685 to 1985*. Museum of London, London.

Murdoch, T.V. (2022). *Europe Divided: Huguenot Refugee Art and Culture*. V&A, London.

Myles, J. (1850). *Chapters in the Life of a Dundee Factory Boy, an Autobiography*. Adam & Charles Black, Edinburgh.

Neal, J.A. (1999). *Joseph and Thomas Windmills: Clock and Watch Makers; 1671–1737*. St Edmundsbury Press, Suffolk.

Newberry, P.E. (1928). 'The Pig and the Cult-Animal of Set'. *The Journal of Egyptian Archaeology*, 14 (3/4), 211–225.

Newman, S. (2010). *The Christchurch Fusee Chain Gang*. Amberley Publishing, Stroud.

Oestmann, G. (2020). 'Designing a Model of the Cosmos'. In *Material Histories of Time: Objects and Practices, 14th–19th Centuries*. Bernasconi, G. and Thürigen, S. (eds.). Walter de Gruyter, Berlin, pp. 41–54.

Payne, E. (2021). 'Morbid Curiosity? Painting the Tribunale della Vicaria in Seicento Naples' (lecture, Courtauld Research Forum, 3 February 2021.)

Peek, S. (2016). 'Knocker Uppers: Waking up the Workers in Industrial Britain'. BBC, 27 March. Available at: https://www.bbc.co.uk/news/uk-england-35840393 (Accessed 10 January 2021).

Popova, M. (2014). 'Why Time Slows Down When We're Afraid, Speeds Up as We Age, and Gets Warped on Vacation'. *The Marginalian*. 15 July. Available at: https://www.themarginalian. org/2013/07/15/time-warped-claudia-hammond (Accessed 16 September 2022)

Quickenden, K. and Kover, A.J. (2007). 'Did Boulton Sell Silver Plate to the Middle Class? A Quantitative Study of Luxury Marketing in Late Eighteenth-Century Britain.' *Journal of Macromarketing*, 27 (1), pp. 51–64.

Rameka, L. (2016). 'Kia whakatōmuri te haere whakamua: I walk Backwards into the Future with My Eyes Fixed on My Past.' *Contemporary Issues in Early Childhood*, 17 (4), pp. 387–98.

Ramirez, A. (2020). *The Alchemy of Us: How Humans and Matter Transformed One Another*. The MIT Press, Cambridge, Massachusetts.

Rees, A. (ed.) (1820). *The Cyclopaedia, or Universal Dictionary, Vol. 2*. Longman, Hurst, Rees, Orme, and Brown, London.

Ribero, A. (2003). *Dress and Morality*. B.T. Batsford, London.

Roe, J.W. (1916). *English and American Tool Builders: Henry Maudslay*. McGraw-Hill, New York.

Rolex (2011). *Perpetual Spirit: Special Issue – Exploration*. Rolex SA, Geneva.

Rooney, D. (2008). *Ruth Belville: The Greenwich Time Lady*. National Maritime Museum, London.

Rossum, G.D.v. (2020). 'Clocks, Clock Time and Time Consciousness in the Visual Arts.' *Material Histories of Time: Objects and Practices, 14th–19th Centuries*. Bernasconi, G. and Thürigen, S. (eds.). Walter Gruyter, Berlin, pp. 71–88.

Saliba, G. (2011). *Islamic Science and the Making of the European Renaissance*. MIT Press, Massachusetts.

Salomons, D.L. (2021). *Breguet 1747–1823*. Reprint by Alpha Editions.

Sandoz, C. (1904). *Les Horloges et les Maîtres Horologeurs à Besançon; du XVᵉ Siècle a la Révolution Française.* J. Millot et Cie, Besançon.

Scarsbrick, D. (1994). *Jewellery in Britain 1066–1837: A Documentary, Social, Literary and Artistic Survey.* Michael Russell (Publishing) Ltd, Norwich.

Scott, R.F. (1911–12). *Scott's Last Expedition.* (1941 ed.) John Murray, London.

Seneca, L.A. (c. 49 AD). *On the Shortness of Life.* Penguin, London.

Shaw, M. (2011). *Time and the French Revolution.* The Boydell Press, Suffolk.

Snir, A., Nadel, D., Groman-Yaroslavski, I., Melamed, Y., Sternberg, M., Bar-Yosef, O. et al. (2015). 'The Origin of Cultivation and Proto-Weeds, Long Before Neolithic Farming'. *PLoS ONE,* 10 (7). Available at: https://www.sciencedaily.com/releases/2015/07/150722144709.htm (Accessed 10 August 2020).

Sobel, D. and Andrewes, W.J.H. (1995). *The Illustrated Longitude: The True Story of a Lone Genius Who Solved the Greatest Scientific Problem of His Time.* Fourth Estate, London.

Sobel, D. (2005). *Longitude: The True Story of a Lone Genius Who Solved the Greatest Scientific Problem of His Time.* Walker & Company, New York.

Stadlen, N. (2004). *What Mothers Do (Especially When It Looks Like Nothing).* Piatkus Books, London.

Steiner, S. (2012). 'Top Five Regrets of the Dying'. *Guardian,* 1 February. Available at: https://www.theguardian.com/lifeand-style/2012/feb/01/top-five-regrets-of-the-dying (Accessed 23 July 2020).

Stern, T. (2015). 'Time for Shakespeare: Hourglasses, Sundials, Clocks, and Early Modern Theatre'. *Journal of the British Academy,* vol. 3, 1–33 (19 March).

Stubberu, S.C.; Kramer, K A. and Stubberud, A.R. (2017). 'Image Navigation Using a Tracking-Based Approach'. *Advances in Science, Technology and Engineering Systems Journal,* 2 (3), pp. 1478–86.

Sullivan, W. (1972). 'The Einstein Papers. A Man of Many Parts'.

New York Times, 29 March. Available at: https://www.nytimes. com/1972/03/29/archives/the-einstein-papers-a-man-of-many-parts-the-einstein-papers-man-of.html (Accessed 14 May 2021).

Tann, J. (2015). 'Borrowing Brilliance: Technology Transfer across Sectors in the Early Industrial Revolution'. *International Journal for the History of Engineering and Technology*, 85 (1), pp. 94–114.

Taylor, J. and Prince, S. (2020). 'Temporalities, Ritual, and Drinking in Mass Observation's Worktown'. *The Historical Journal*. Cambridge University Press, pp. 1–22.

Thompson, A. (1842). *Time and Timekeepers*. T. & W. Boone, London.

Thompson, E.P. (1967). 'Time, Work-Discipline, and Industrial Capitalism'. *Past & Present*, 38, (December), pp. 56–97.

Thompson, D. (2007). *Watches in the Ashmolean Museum*. Ashmolean Handbooks. Ashmolean Museum, Oxford.

Thompson, D. (2014). *Watches*. British Museum Press, London.

Thompson, W.I. (2008). *The Time Falling Bodies Take to Light: Mythology, Sexuality and the Origins of Culture*. Digital printed ed. St. Martin's Press, New York.

Unknown Author (2019). 'BBC documentary examines the deep scars left from Dundee Timex closure, 26 years on'. *Evening Telegraph*, 15 October. Available at: https://www.eveningtele-graph.co.uk/fp/bbc-documentary-examines-the-deep-scars-left-from-dundee-timex-closure-26-years-on (Accessed 14 May 2021).

Various (1967). *Pioneers of Precision Timekeeping*. A symposium published by the Antiquarian Horological Society as Monograph No. 3.

Verhoeven, G. (2020). 'Time Technologies'. *Material Histories of Time: Objects and Practices, 14th–19th Centuries*. Bernasconi, G. and Thürigen, S. (eds). Walter de Gruyter, Berlin, pp. 103–115.

Wadley, L. (2020). *Early Humans in South Africa Used Grass to Create Bedding, 200,000 years ago*. YouTube Video. Available at: https://www.youtube.com/watch?v=AzUui4eZI2I (Accessed 8 November 2020).

Walker, R. (2013). *Blacks and Science Volume One: Ancient Egyptian*

Contributions to Science and Technology and the Mysterious Sciences of the Great Pyramid. Reklaw Education Ltd, London.

Weiss, A. (2010). 'Why Mexicans celebrate the Day of the Dead.' *Guardian*, 2 November. Available at: https://www.theguardian.com/commentisfree/belief/2010/nov/02/mexican-celebrate-day-of-dead (Accessed 2 September 2020).

Weiss, L. (1982). *Watch-making in England, 1760–1820*. Robert Hale Ltd, London.

Wesolowski, Z.M. (1996). *A Concise Guide to Military Timepieces 1880–1990*. Reprint. The Crowood Press, Wiltshire.

Whitehouse, D. (2003). '"Oldest sky chart" found'. BBC, 21 January. Available at: http://news.bbc.co.uk/1/hi/sci/tech/2679675.stm (Accessed 12 June 2020).

Wilkinson, C. (2009). *British Logbooks in UK Archives 17th–19th Centuries. A Survey of the Range, Selection and Suitability of British Logbooks and Related Documents for Climatic Research* [online].

Wragg Sykes, R. (2020). *Kindred: Neanderthal Life, Love, Death and Art*. Bloomsbury Sigma, London.

Yazid, M.; Akmal, A.; Salleh, M.; Fahmi, M.; Ruskam, A. (2014). 'The Mechanical Engineer: Abu'l –'Izz Badi'u'z – Zaman Ismail ibnu'r – Razzaz al Jazari' (seminar on Religion and Science: Muslim Contributions Semester 1 2014/2015, 9 December, Skudai, Johor, Malaysia.)

Yoshihara, N. (1985). '"Cheap Chic" Timekeepers: Swatch Watches Offer Many Scents, Patterns'. *Los Angeles Times*, 21 June. Available at: https://www.latimes.com/archives/la-xpm-1985-06-21-fi-11660-story.html (Accessed 14 May 2021).

Zaimeche, S. (2005). *Toledo*. Foundation for Science Technology and Civilisation. June 2005. Pub. ID 4092.

Zaslavsky, C. (1992). 'Women as the First Mathematicians'. *International Study Group on Ethnomathematics Newsletter*, 7 (1), January.

Zaslavsky, C. (1999). *Africa Counts: Number and Pattern in African Cultures*. 3rd ed. Lawrence Hill Books, Chicago.

Further Resources

Object-led histories are only possible if you can physically see and examine the object of your study. This book would not have been possible without support from some of the many horological collections in museums and art galleries around the world.

If you'd like to see some of the timekeepers I describe in this book, and other examples like them, the following is a list of museums with collections of watches and clocks on public view. Some horological collections are dispersed within wider exhibitions of art and design, and might require a bit of leg work to see them all. Some of these museums are small and open part-time; some offer guided tours and behind-the-scenes visits if you book in advance. I recommend contacting them before your visit to try to find out what objects are currently on display or whether a curator has availability to show you things that aren't!

EUROPE

United Kingdom
Bury St Edmunds: Moyse's Hall Museum
Coventry: Coventry Watch Museum
London: Clockmakers' Company Collection, Science Museum
London: British Museum
London: Royal Observatory

London: Wallace Collection
London: Victoria & Albert Museum
Newark: Museum of Timekeeping in Newark
Oxford: Ashmolean Museum
Oxford: History of Science Museum

Austria
Karlstein: Uhrenmuseum
Vienna: Uhrenmuseum of the Wien Museum

Belgium
Mechelen: Horlogeriemuseum

Denmark
Aarhus, Den Gamle By: The Danish Museum of Clocks and Watches

Finland
Espoo: Finnish Museum of Horology

France
Besançon: Musée du Temps
Cluses: Musée de l'Horlogerie et du Décolletage
Paris: Conservatoire National des Arts et Métiers
Paris: Musée des Arts et Métiers
Paris: Breguet Museum
Saint-Nicolas d'Aliermont: Musée de l'Horlogerie

Germany
Albstadt: Philipp-Matthäus-Hahn-Museum
Furtwangen: Deutsche Uhrenmuseum
Glashütte: Deutsches Uhrenmuseum
Harz, Bad Grund: Uhrenmuseum
Nuremberg: Uhrensammlung Karl Gebhardt
Pforzheim: Technisches Museum der Pforzheimer, Schmuck und Uhrenindustrie
Schramberg: Junghans Terrassenbau Museum

Italy

Bardino Nuovo: Museo dell'Orologio di Tovo S. Giacomo
Milan: Museo Nazionale della Scienza e della Tecnologia Leonardo da Vinci

Netherlands

Franeker: Eise Eisinga Planetarium
Joure: Museum Joure
Zaandam: Museum Zaanse Tijd

Romania

St Ploiesti Prahova: Nicolae Simache Clock Museum

Russia

Moscow: Museum Collection
Siberia, Angarsk: Angarsk Clock Museum
St Petersburg: The State Hermitage Museum

Spain

Madrid: Museo del Reloj Antiguo

Switzerland

Basel: Haus zum Kirschgarten – Historisches Museum
Fleurier: L.U.CUEM – Traces of Time
Geneva: Musée d'Art et d'Histoire
Geneva: Patek Philippe Museum
La Chaux-de-Fonds: Musée International d'Horlogerie (MIH)
Le Locle: Château des Monts
Vallée de Joux: Espace Horloger
Zurich: Beyer Clock and Watch Museum

AFRICA

South Africa
Kimberley: McGregor Museum

ASIA

China
Beijing: The Palace Museum, Forbidden City
Macau: Macau Timepiece Museum
Yantai: Polaris Heritage Museum of Clock and Watches

Japan
Hiroshima: The Hiroshima Peace Memorial Museum
Nagano: Gishodo
Tokyo: National Museum of Nature and Science
Tokyo: The Seiko Museum
Tokyo: Daimyo Clock Museum

Thailand
Bangkok: Antique Clock Museum

AUSTRALASIA

Australia
Melbourne: Museums Victoria

New Zealand
Whangarei: Claphams Clock Museum

MIDDLE EAST

Israel
Jerusalem: The Salomons Collection, Meyer Museum of Islamic Art.

Turkey
Istanbul: Topkapı Palace

NORTH AMERICA

Canada
Alberta, Peace River: The Alberta Museum of Chinese Horology
Ontario, Deep River: The Canadian Clock Museum

Mexico
Mexico City: Museo del Tiempo
Puebla: Museo de Relojeria

USA
California, San Francisco: The Interval – The Long Now Foundation
Connecticut, Bristol: American Watch & Clock Museum
District of Columbia, Washington: National Air and Space Museum
Illinois, Evanston: Halim Time & Glass Museum
Maryland, Baltimore: B&O Railroad Museum
Massachusetts, North Grafton: The Willard House and Clock Museum
Massachusetts, Waltham: Charles River Museum
New York, New York: Metropolitan Museum of Art
New York, New York: The Frick Collection
Ohio, Harrison: Orville R. Hagans History of Time Museum (AWCI)
Pennsylvania, Columbia: National Watch & Clock Museum (NAWCC)
Pennsylvania, Philadelphia: Philadelphia Museum of Art
Texas, Lockhart: Southwest Museum of Clocks and Watches

SOUTH AMERICA

Brazil
São Paulo: Museu do Relógio (Professor Dimas de Melo Pimenta)

There are many more museums with horological collections than those I've listed here.

Notes

A full list of works referenced can be found in the Bibliography on p. 221.

A Backward-facing Foreword

p. xiv 'In fact, the word *time* . . .' According to the BBC, cited by Hom, A. (2020). The top ten nouns include two further time words: 'year' and 'day'.

Chapter 1: Facing the Sun

p. 4   'Archaeologists found more . . .' Wadley, L. (2020).

p. 5   'If our ancestors rotated . . .' Walker, R. (2013), p. 89.

p. 5   'A recent study found no . . .' Helfrich-Förster, C.; Monecke, S.; Spiousas, I.; Hovestadt, T.; Mitesser, O. and Wehr, T.A. (2021).

p. 7   'In other words, they too . . .' Häfker, N.S.; Meyer, B.; Last, K.S.; Pond, D.W.; Hüppe, L.; Teschke, M. (2017), p. 2194.

p. 7   'If anything, our ability to . . .' Popova, M. (2013).

p. 7   'The Lebombo Bone suggests . . .' The Pradelles hyena bone is another example of a bone that has been marked with regular parallel incisions that look completely unlike anything you would expect from butchery, and dates to a similar time as the Lebombo Bone. Where it differs is that the Pradelles bone was carved by our cousins, the Neanderthals. The marks may well be decoration or are possibly evidence of some level of numeracy. See Wragg Sykes, R. (2020), p. 254.

p. 7 'In the Democratic Republic . . .' The Ishango Bone currently resides on permanent public display at the Royal Belgian Institute of Natural Sciences in Brussels, Belgium.

p. 7 'As the philosopher William Irwin Thompson . . .' Thompson, W.I. (2008), p. 95.

p. 8 'This site is the earliest . . .' Snir, A.; Nadel, D.; Groman-Yaroslavski, I.; Melamed, Y.; Sternberg, M.; Bar-Yosef, O.; et al. (2015).

p. 9 'This is how the Kenyan-born . . .' Zaslavsky, C. (1999), p. 62.

p. 9 'For some coastal Aboriginal . . .' Wragg Sykes, R. (2020), pp. 278–9.

p. 10 'For them, a meaningful . . .' Rameka, L. (2016), p. 387.

p. 11 'The Sumerians, the first . . .' Zaslavsky, C. (1999), p. 23.

p. 11 'About 5,000 years ago . . .' Locklyer, N.J. (2006), p. 110. Dedicated astronomers that they were, the Egyptians may even have pioneered heliocentric theory – the theory that the Earth and other planets revolve around the sun – though for centuries Ptolemy of Alexandria's Earth-centred model of the second century CE held sway.

p. 12 'In the hours of darkness . . .' Ibid., p. 343.

p. 12 'Astronomers identified . . .' Walker, R. (2013), p. 16.

p. 13 'They also knew of the planets . . .' Ibid., pp. 18–19.

p. 14 'In ninth-century England . . .' The description was made in the accounts of King Alfred's royal biographer, Bishop Asser, which was written in 893 CE.

p. 14 'In his *Documenti d'Amore* . . .' Balmer, R.T. (1978), p. 616.

p. 14 'In the late fifteenth century . . .' May, W.E. (1973), p. 110.

Chapter 2: Ingenious Devices

p. 20 'In ancient Mesopotamia . . .' Masood, E. (2009), p. 163.

p. 20 'We don't know exactly . . .' Ibid., p. 74.

p. 21 'Al-Zarqali managed . . .' Zaimeche, S. (2005), p. 10.

p. 22   'Al-Zarqali's clock remained . . .' Masood, E. (2009), p. 74.

p. 22   'Nearly 4,000 miles east . . .' Foulkes, N. (2019), p. 64.

p. 22   'The brief for Su Song's . . .' Morus, I.W. (2017), p. 108, as cited by Foulkes, N. (2019), p. 65.

p. 22   'With its bronze . . .' Foulkes, N. (2019), p. 65. Sadly, Su Song's clock has been lost in history. It was taken during the Tatar invasion of China in 1127 but – similar to the fate of Al-Zarqali's clock in Toledo – the invaders' scholars were unable to put it together and make it run again. Today, the closest example we have is a fully functioning scale replica that stands outside the Gishodo Suwako Watch and Clock Museum in the Suwa area, one of the great horological centres of Japan, and a short drive from the famous watch company Seiko.

p. 23   'They include automata . . .' Yazid, M.; Akmal, A.; Salleh, M.; Fahmi, M.; Ruskam, A. (2014).

p. 23   'A mahout, or elephant trainer . . .' Masood, E. (2009), p. 163.

p. 26   'The Elizabethan dramatist . . .' Stern, T. (2015), p. 18, citing a seventeenth-century manuscript compiled by a Richard Smith and quoted in Bedini, Doggett & Quinones (1986), p. 65.

p. 26   'In late medieval . . .' Glennie, P. & Thrift, N. (2009), p. 24.

p. 26   'In an age when . . .' Ibid.

p. 28   'Medieval church clocks . . .' Oestmann, G. (2020), p. 42.

p. 28   'Devices like lunar volvelles . . .' An example of a lunar volvelle can be found at Bodleian Libraries, University of Oxford, accession number MS. Savile 39, fol. 7r, https://www.cabinet. ox.ac.uk/lunar-tool#/media=8135_(accessed 19 April 2021).

p. 29   'Astronomers drove clocks . . .' Baker, A. (2012), p. 16.

p. 29   'The regimented nature . . .' Álvarez, V.P. (2015), p. 64.

p. 29   'The story goes . . .' Johnson, S. (2014), p. 137.

p. 31   'Over the course . . .' Álvarez, V.P. (2015), p. 65.

p. 34   'In 1511, the scholar . . .' Lester, K. & Oerke, B. V. (2004), p. 376. Although the claim of 'forty-eight hours' is dubious as watches of this era rarely run longer than a day.

Chapter 3: Tempus Fugit

p. 36   'On the forehead . . .' Horace: '*Pallida mors æquo pulsat pede pauperum tabernas Regumque turres*' ('Pale Death, with impartial foot, knocks at the cottages of the poor and the palaces of kings').

p. 38   'At the end of . . .' Thompson, A. (1842), pp. 53–54.

p. 38   'But in the early 1980s . . .' Jagger found evidence that a skull watch – presumed to have been Mary's – was known to have existed in Salisbury in 1822, and this, in turn, could have led to the creation of a further two in the nineteenth century. A letter written in 1863 by Queen Victoria's clockmaker in Scotland describes a 'death's head watch, formerly belonging to Mary Queen of Scots', which had come into the family of Sir John Dick Lauder through Catherine Seton, sister of Mary Seton, 'to whom the unfortunate Mary gave it before her execution'. Yet an 1895 transactions ledger from the St Paul's Ecclesiological Society lists the sale of 'a devotional watch, with a very sweet tones bell, on which the hours are struck, set in a case in the form of a skull, covered with engravings . . . It very closely resembles the watch belonging to Sir Thomas W. Dick-Lauder . . . which was by him stated to be one of the twelve presented by Mary to her favourite ladies of honour'.

p. 38   'It would have had . . .' An inventory made while Mary was at Holyrood in 1562 lists an impressive range of items that included sixty gowns – many heavily embroidered – in her favourite colour, white, as well as jet black, crimson red, and orange with silver detail. There are cloths of gold, silver, velvet, satin and silk waiting to be made up into clothes. She had fourteen cloaks, mantles (similar to a cloak only without sleeves) of purple velvet and ermine, and thirty-four vasquines, or corsets.

p. 40   'Her letters and the quotes . . .' The quote '*En ma Fin gît mon Commencement*' ('In my end is my beginning') was said to have been embroidered by Mary during her imprisonment.

p. 40   'When they stripped . . .' Fraser, A. (2018), p. 669.

p. 41   'In a portrait of King Henry . . .' There are a number of Holbein portraits that decisively depict the watches and clocks of his wealthy patrons, including those of the French ambassador Charles de Solier (1480–1552), the Swiss merchant Jörg Gisze (1497–1562), and the group portrait of lawyer Sir Thomas More (1478–1535) and his family.

p. 41   'The list, drawn up . . .' Cummings, G. (2010), p. 14.

p. 41   'One was more spherical . . .' Incense was important not only to help sixteenth-century Europeans with refined noses navigate the olfactory challenges of living in a city without a proper sewerage system, but it was also believed to ward off diseases like the plague.

p. 42   'After carefully removing . . .' As things stand, we still don't know who buried the hoard, when, or why. The treasure's custodian, Hazel Forsyth, has made the pursuit of these questions into a significant body of work. Her theories are that it could have been a goldsmith heading off to fight in the Civil War, which started in 1642, from which he never returned. It's also possible that he was fleeing that conflict to seek safety abroad. It's likely it belonged to a Jacobean goldsmith who had been resident at that site, but whoever he was, it is clear he wasn't around for long enough to reclaim his buried treasure.

p. 46   'Over the next two decades . . .' For atrocities on such a large scale that happened nearly five hundred years ago, with little attempt to officially document them, it is understandable that estimates for the number of dead and displaced vary greatly between sources. For these figures I have referenced Murdoch, T.V. (1985), p. 32.

p. 49   'Flamboyant dress was . . .' Ribero, A. (2003), p. 65.

p. 49   'Even the powdered wigs . . .' Ibid., p. 73.

p. 49   'Puritans dressed modestly . . .' Watches were already such invaluable tools that even strict Puritans couldn't give them up. There is a watch in the collection of the British Museum with provenance linking it, albeit somewhat questionably, to Cromwell, while another example that also has claims to once serving the Puritan leader appeared at auction in Carlisle, Cumbria, in 2019.

p. 50   'They believed that to prosper . . .' The Huguenots in particular were renowned for their industrious working habits. In 1708 Edward Wortley described to the House of Commons how work for them was 'the practical exercise of a calling appointed by God'. Watchmaker David Bouguet was a devout Protestant – he served as an elder at the French Church on Threadneedle Street four times – and would likely have held the view that not to apply his God-given gifts to the best of his ability would have been sinful. To those of Calvinist faith, time is an example of God's plan manifested in nature and its observance was a matter of religious importance.

p. 50   'In the Puritan day . . .' Rossum, G.D.v. (2020), p. 85.

p. 50   'It was even argued . . .' Richardson, S. (1734) as cited by Rossum, G.D.v. (2020), p. 85.

p. 50   '"Time being man's opportunity . . ."' Baxter, R. (1673).

p. 51   'As his reign progressed . . .' Fraser, A. (1979) as cited by Rossum, G.D.v. (2020), p. 74.

p. 52   'In the years that followed . . .' Murdoch, T.V. (1985), p. 51.

p. 52   'While the majority . . .' Ibid.

p. 52   'These Huguenot migrants . . .' Thompson, W.I. (2008), p. 40.

Chapter 4: The Golden Age

p. 57   '"So home and late . . ."' Samuel Pepys' diary entry from Tuesday 22 August 1665, https://www.pepysdiary.com/diary/1665/08/22/.

p. 59   'When I take apart . . .' House of Commons (1818), p. 4. Any object which has required significant human effort to create becomes authored and will exhibit some degree of unique personalisation both in the subtleties of the finishing and the obvious fingerprints of the craftsman (such as signatures and maker's marks). With a trained eye, these marks can be read like a text.

p. 60   'Their watches were charged . . .' Cummings, N. & Gráda, C.Ó. (2019), pp. 11–12.

p. 65   'In 1777, watchmaker . . .' Dickinson, H.W. (1937), p. 96; Tann, J. (2015) as cited by Cummings, N. & Gráda, C.Ó. (2019), p. 19.

p. 65   'According to a 1798 . . .' Thompson, E.P. (1967), p. 65.

p. 66   'Some of the stars . . .' Stubberu, S.C.; Kramer, K.A.; Stubberud, A.R. (2017), p. 1478.

p. 67   'He was also able . . .' Abulafia (2019), pp. 17, 812–13.

p. 67   'And nor, in his opinion . . .' Sobel, D. (1995), p. 52.

p. 71   'By the time she returned . . .' Robert FitzRoy as cited in Barrie, D. (2014), p. 227.

p. 71   'They are liable . . .' Henry Raper as cited in Barrie, D. (2014), p. 89.

p. 71   'Most early-modern navigational . . .' Baker, A. (2012), p. 15.

p. 72   'Shortly after, a key . . .' Baker, A. (2012), pp. 23–24.

p. 72   '"Trim took a fancy . . ."' National Maritime Museum, Flinders' Papers FLI/11 as cited in Barrie, D. (2014), p. 204.

p. 74   'And although the Board of Longitude . . .' Wilkinson, C. (2009), p. 37.

p. 74   'In 1802, thirty years . . .' Rodger, N.A. (2005), pp. 382–3 as cited in Barrie, D. (2014), p. 115.

p. 75   'It is the descendant . . .' Good, R. (1965), p. 44.

Chapter 5: Forging Time

p. 81   'All of them, like Wilter's . . .' Although Mudge had introduced his lever escapement in 1767, it was to the average pocket watch of the day as the Zenith Defy (currently proclaiming itself to be the most accurate mechanical watch in the market) is to a simple Timex. The verge escapement was still the standard.

p. 82   'Watches were complicated . . .' Chapuis & Jaquet (1970), pp. 80–82.

p. 82   'One worker in the north . . .' Heaton, H. (1920), pp. 306–11 as cited by Cummings, N. & Gráda, C.Ó. (2019), p. 6.

p. 83   'One observer noted . . .' Ganev, R. (2009), pp. 110–11.

p. 83   'Some local manufacturers . . .'Cummings, N. & Gráda, C.Ó. (2019), p. 6.

p. 83   'Artisan culture was almost . . .' Clarke, A. (1995) as cited by Ganev, R. (2009), p. 5.

p. 83   'The number of women serving . . .' Erickson, A.L., p. 2.

p. 83   'One ongoing study . . .' The list is frequently updated and the latest figure can be viewed at https://www.ahsoc.org/resources/women-and-horology/.

p. 87   'As a result of *établissage* . . .' Landes, D. (1983), p. 442.

p. 88   'Renowned father–son watchmakers . . .' Neal, J.A. (1999), p. 109.

p. 89   'He justified it . . .' Pitt introduced a plethora of taxes over this time, including Income Tax, to compensate for the financial burden of the French Revolutionary and inevitable Napoleonic Wars.

p. 90   'The closer Moll gets . . .' Rossum, G.D.v. (2020), p. 78.

p. 90   'The proliferation of watches . . .' Ibid., p. 73.

p. 90   'Timepieces were highly prized . . .' Ibid., p. 86.

p. 90   'Records from the Old Bailey . . .' Styles (2007) as cited by Verhoeven, G. (2020), p. 111.

p. 91   'Time awareness was slowly . . .' Verhoeven, G. (2020), p. 105. Increases in clock ownership also contributed to a rise in time awareness. In 1675, barely 11 per cent of London households had a clock; thirty years later the figure stood at 57 per cent. In the 1770s, more than 10 per cent of all cases that went through the Old Bailey related to the theft of clocks. While this was going on, their average prices decreased, falling by as much as 75 per cent over the century. Nonetheless, over half the reports of clock theft from the late eighteenth and early nineteenth centuries were from affluent owners, despite their being a minority of the population.

p. 92   '"[He] introduced the making . . ."' House of Commons (1812), p. 67.

Chapter 6: Revolution Time

p. 97   'The barrister and horology buff . . .' Salomons, D.L. (1921), p. 5.

p. 98   'By moving the dial . . .' Credit for this analogy must be attributed to the wonderfully talented watch restorer and rose engine turner Seth Kennedy.

p. 100   'It was Breguet's stepfather . . .' Researching Breguet's personal life is challenging. He was a celebrity in his day, and has been commonly written about ever since, with little by way of hard facts for anything outside his watchmaking endeavours. There are books about his life that make no mention of his family at all; even dates for the completion of his works can vary. I have done my best to tread a middle ground.

p. 100   'He lost his father at . . .' Depending on the source, Breguet's age at the loss of his father has been quoted as ten, eleven or twelve years old.

p. 103   '"This devil, Breguet . . ."' Breguet, C. (1962), p. 5.

p. 103   'Breguet "had the power..."' Ibid., p. 6.

p. 103   'In the months that followed . . .' Also referred to as 'the Terror' and 'the Reign of Terror'.

p. 103   'Although the most famous . . .' Ironically, the guillotine had been invented as a more humane method of execution by a medical doctor opposed to execution, Dr Joseph-Ignace Guillotin. The machine was based on previous sliding axe designs used in Italy and Scotland, improved and refined to ensure a swift and clean despatch.

p. 103   'The accounts of the survivors . . .' These accounts and many more like them can be found in Anon (1772).

p. 104   'In April 1793 . . .' Antiquorum (1991).

p. 104   'Two months later . . .' Daniels, G. (2021), p. 6.

p. 105   'So important was this . . .' The watch mentioned went under the hammer at Sotheby's on 14 July 2020, selling for £1,575,000. My description was aided by their cataloguing which can be read in full online at https://www.sothebys.com/en/buy/

auction/2020/the-collection-of-a-connoisseur/breguet-
retailed-by-recordon-london-a-highly.

p. 106   'As Jamaican philosopher . . .' Mills, C. (2020), p. 301.

p. 106   'Decimal time, as it was known . . .' Shaw (2011).

p. 106   'This new calendar was . . .' To balance leap years, every fourth
year there would be an additional 'Festival of the Revolution'.

p. 108   'His cover story . . .' Daniels, G. (2021), p. 7.

p. 108   'Breguet agreed . . .' Ibid., p. 9.

p. 109   'In his last years . . .' Salomons, D.L. (2019), pp. 11–12.

p. 110   'Stendhal declared Breguet's watches . . .' Breguet, C. (1962),
p. 10.

Chapter 7: Working to the Clock

p. 112   'Before industrialisation, Britain . . .' Published in 1967, British
social historian and political campaigner E.P. Thompson's
seminal essay 'Time, Work-Discipline and Industrial Capitalism'
is an exceptional source for discovering how heavily time
transformed from a force of nature to a force of organisational
control. If you find this section of interest, I strongly advise
tracking down the original essay and those it inspired.

p. 113   'This change was less about . . .' Thompson, E.P. (1967), p. 61.
The idea of a working day and of paying an hourly rate dates
back to the sixteenth century: see Glennie and Thift (2009),
p. 220.

p. 114   'In 1757, the Irish statesman . . .' Edmund Burke, cited by Ganev,
R. (2009), p. 125.

p. 115   'He describes stepping . . .' Myles, J. (1850), p. 12. Chapters in
the Life of a Dundee Factory Boy, an autobiography, James
Myles, 1850

p. 116   'Their services became more . . .' Peek, S. (2016). Knocker-
uppers were such an important part of London life that Charles
Dickens mentions them in Great Expectations, with Mr Whopsle
being woken by a knocker-upper. The job was far from short-
lived and continued until relatively recently: it wasn't until the

1970s that the UK's last knocker-upper hung up their pea shooter.

p. 116 'An account from a mill worker . . .' Alfred, S. K. (1857) quoted by Thompson, E.P. (1967), p. 86.

p. 117 'As one industrialist explained . . .' Rev. J. Clayton's *Friendly Advice to the Poor* (1755), quoted by Thompson, E.P. (1967), p. 83.

p. 117 'In 1770 English cleric . . .' Temple (1739–1796) was from Berwick-upon-Tweed in Northumberland and had been educated at the University of Edinburgh. He published a number of essays expressing his views on religion, power and morality.

p. 117 'In 1772, in an essay . . .' Anon. (1772).

p. 117 'The watch industry had . . .' Newman, S. (2010), p. 124.

p. 120 'The year in which you are reading . . .' Mills, C. (2020), p. 300.

p. 121 'They were "grounded in ..."' Mills, C. (2020), p. 308.

p. 121 'Western observers in the 1800s . . .' Thompson, E.P. (1967), pp. 91–92.

p. 122 'These kinds of stereotypes . . .' Hom, A. (2020), p. 210.

p. 122 'Needless to say, it was not . . .' Thompson, E.P. (1967), pp. 56–97.

p. 124 '"when Home we are come . . ."' Collier, M. (1739), pp. 10–11. 'The Woman's Labour: an Epistle to Mr. Stephen Duck; in Answer to his late Poem, called The Thresher's Labour' (1739), pp. 10–11.

p. 124 'A similar eighteenth-century . . .' Ganev, R. (2009), p. 120.

p. 125 'New parents working in the home . . .' Stadlen, N. (2004), p. 86. One of the new mothers Stadlen interviews declares, 'Clock time doesn't mean anything any more.'

p. 125 'If a memory is unclear . . .' Sophie Freeman, 'Parents find time passes more quickly, researchers reveal', *The Times*, 22 February 2021; Popova, M. (2013).

p. 126 'Thousands of watchmakers . . .' House of Commons (1817), p. 15.

p. 126 '"hardly a rag to cover them..."' House of Commons (1817), p. 5.

p. 126 'By the mid-nineteenth century . . .' Hoult, J., 'Prescot Watch-making in the xviii Century', *Transactions of the Historic Society of Lancashire and Cheshire*, LXXVII (1926), p. 42, as cited by Cummings, N. & Gráda, C.Ó. (2019), p. 27.

p. 128 'In 1878 one unnamed . . .' Church, R.A. (1975), p. 625 as cited by Cummings, N. & Gráda, C.Ó. (2019), p. 24.

Chapter 8: The Watch of Action

p. 131 'Railways even shrank . . .' Ramirez, A. (2020), p. 49.

p. 131 'By the 1850s . . .' Ibid., figs 18–19.

p. 132 'It's not until you see . . .' 'Corn Exchange Dual-Time Clock', Atlas Obscura, https://www.atlasobscura.com/places/corn-exchange-dualtime-clock.

p. 133 'Dual time observations . . .' Bartky, I. (1989), p. 26.

p. 134 'He described it as his . . .' Slocum as cited in Barrie, D. (2014), p. 245.

p. 135 'Likewise the black . . .' The dangers of luminous paint were brought to mainstream attention through the tragic harm they caused the Radium Girls, who painted these dials by hand with a fine brush in the 1910s and 1920s, and whose story we will return to in the next chapter.

p. 137 'He even schedules . . .' Scott, R.F. (1911), p. 235.

p. 137 'It's hard to imagine . . .' Ibid., p. 210.

p. 138 'While Britain's scorched . . .' 'South African concentration camps', New Zealand History, https://nzhistory.govt.nz/media/photo/south-african-concentration-camps (accessed 23 November 2022).

p. 139 'At home the wristlet . . .' 'The History of the Nato Watch Strap', A. F. 0210, https://afo210strap.com/the-history-of-the-nato-watch-strap-nato-straps-in-the-great-war-wwi-era/ (accessed 12 January 2023).

p. 139 'In 1893, an advert . . .' Ibid.

p. 139 'Another, from 1901 . . .' Ibid.

p. 141 'At around 700 metres . . .' Geffen (2010).

p. 141 'Within the remaining . . .' Gabardine is a tough and durable tightly woven cloth, typically made from wool or cotton, which was commonly used in the production of uniforms, coats and outdoor wear.

Chapter 9: Accelerated Time

p. 148   'One account from 1900 . . .' Gohl, A. (1977), p. 587 as cited in Glasmeier, A. (2000), p. 142.

p. 148   'In one 1915 edition . . .' Cartoonist M.C. Fisher cited in Cummings, G. (2010), p. 232.

p. 150   'In 1908 Wilsdorf . . .' Dowling, J.M. & Hess, J.P. (2013), p. 11.

p. 154   'By 1926, US manufacturer . . .' Moore, K. (2016), p. 171.

p. 154   'By the end of 1918 . . .' Ibid., p. 25.

p. 155   'Each and every dial . . .' Ibid., p. 9.

p. 155   'On average, women . . .' Ibid., p. 11.

p. 155   'Although adverts asked . . .' Ibid., p. 45.

p. 155   'Radioluminescent paint . . .' Ibid., p. 8.

p. 155   Dial painters had to . . .' Ibid., pp. 7–8.

p. 156   'In Switzerland . . .' Ibid., p. 10.

p. 156   'The "lip, dip, paint" . . .' Ibid., pp. 9–10.

p. 156   'Some of the painters . . .' Ibid., p. 16.

p. 156   'He himself had amputated . . .' Ibid., pp. 18–19.

p. 157   'The body, mistaking . . .' Ibid., p. 111.

p. 157   'Doctors, including those . . .' Ibid., p. 224.

Chapter 10: Man and Machine

p. 170   'The auction house initially . . .' Davidson, H. (2021).

p. 173   '(Adverts for the Accutron . . .)' 'Reinventing Time: The Original Accutron', Hodinkee, https://www.hodinkee.com/articles/reinventing-time-original-bulova-accutron (accessed 23 November 2022).

p. 175   'By the early 1980s . . .' Glasmeier, A. (2000), p. 243.

p. 176   'It was advertised as . . .' Finlay Renwick, 'The Digital Watch Turns 50: A Definitive History', Esquire, 18 November 2020, https://www.esquire.com/uk/watches/a34711480/digital-watch-history/.

p. 177 'In 1985 the *LA Times* . . .' Yoshihara, N. (1985).

p. 178 'By the time it closed . . .' Unknown Author (2019).

p. 178 'By 1993, Hong Kong . . .' *South China Post* (1993), p. 3; Hong Kong Trade and Development Council (1998) as cited in Glasmeier, A. (2000), p. 231.

p. 178 'manufacturers could supply . . .' Glasmeier, A. (2000), p. 233.

p. 178 'One protester described . . .' Unknown Author (2019).

p. 185 'The Deep Space Atomic Clock . . .' 'Deep Space Atomic Clock', NASA, https://www.nasa.gov/mission_pages/tdm/clock/index.html (accessed 23 November 2022).

Chapter 11: Eleventh Hour

p. 195 '"you must not think a man . . ."' Seneca, L.A. (c. 49 AD), p. 11.

Index

THE
BJORKLUND
LEGACY

*Philanth
at 25*

ABOUT THE AUTHOR

Betty Ladd Halliwell graduated from UCLA as a Phi Beta Kappa with Honors in English and earned the M.A. in English from the University of Chicago as a Woodrow Wilson Fellow. After studying in the Theology and Literature program of the University of Chicago Divinity School she received the M.A. and Ph.D. in sociology from UCLA, specializing in content analysis of the mass media. As a professor at several universities she has taught a wide range of courses and focussed on global and social problems from an interdisciplinary perspective.

Dr. Halliwell has been active in politics at all levels and divides her time between Southern California and Washington. In 1980 she began writing screenplays on political themes, and one of these became the basis for *The Bjorklund Legacy*. More novels in the series are planned.

THE
BJORKLUND
LEGACY

Philanth at 25

FIRST IN A SERIES OF NOVELS

by

BETTY LADD
HALLIWELL, Ph.D.

The Pribiloff Press, Santa Monica, California

Grateful acknowledgment is given to the following
for permission to reprint:
the excerpt from "A Fellow Needs A Girl"
by Richard Rodgers and Oscar Hammerstein II, on page 9.
Copyright © 1947 by Richard Rodgers and Oscar Hammerstein II.
Williamson Music, owner of publication and allied rights throughout the world.
Copyright Renewed. International Copyright Secured.
Used by Permission. All Rights Reserved.
six lines from "Shine, Perishing Republic", from *The Selected Poetry of
Robinson Jeffers* by Robinson Jeffers, on page 279.
Copyright © 1925 and renewed 1953 by Robinson Jeffers.
Reprinted by permission of Random House, Inc.

The passage quoted on pages 191-192 is from Chapter IV (authored by William
A. Hance) of *Africa: From Mystery to Maze*, edited by Helen Kitchen, Volume
XI of a series published by Lexington Books in 1976 for a Commission on
Critical Choices for Americans brought together by Nelson Rockefeller.
Thanks to Helen Kitchen for her assistance regarding this source.

All other quotations and references in this work
have been judged to be in the public domain or to constitute
fair use. For additional copyright acknowledgments
please contact The Pribiloff Press.

INFORMATION ON OTHER NOVELS IN THIS SERIES
WILL BE FOUND IN THE BACK OF THIS BOOK.

CONTENTS

This book is dedicated

to the men and women of Peace Corps

and to Professor Michael John Halliwell,

developer of Carole Bjorklund's

secret files.

PART ONE:

WASHINGTON

Chapter 1

THE RUINS IN WASHINGTON

Nothing told him where he was; he had no sense of where he was going. Everything was familiar, yet nothing made sense. Washington had been designed to be confusing, but this went beyond its normal craziness!

He couldn't pause to study the situation. He was in a hurry, getting sick with anxiety, and the traffic was swift and merciless.

Everybody was rushing toward the center of power or away from it. But right now he couldn't even be sure which was which.

Desperate to find some—*any*—point of reference, he swung the wheel and left the main thoroughfare. Looking around, he let the car lose momentum. He squinted, trying to bring the next street sign into focus.

But when he reached the intersection he could find no signposts. He continued to let the car slow down.

He wasn't really driving, anyway. He was merely drifting, carried along by other people's momentum. Now that he had left the forceful flow he could wander in the peaceful residential side streets for the rest of his life and never get anywhere at all.

Still searching for a sign, he doubled back to the wide avenue he had left. It seemed right to turn left, but as he re-entered the heavy traffic he felt more lost than ever.

Carole usually routed him to his destinations. His wife sat beside him as if lost in her own thoughts.

"Carole," he began, trying to keep his voice level, "I,

3

uh, need some directions . . . ."

She would think he was losing his mind. He drove and she fed him instructions from a map only when they were venturing into unfamiliar territory. Now he was only trying to get from his home to his office. But he had forgotten the way, his mind wiped blank by panic.

She didn't seem to have heard his request.

He was in a hurry. He had no time to waste pretending he didn't have a problem.

"Carole," he repeated more loudly, and an angry horn blared beside and then behind him, diminishing as he was swept along.

Cold fear swept through him. He had almost sideswiped a truck! He was going to crash if he couldn't decide what he was supposed to be doing.

Was this Pennsylvania Avenue, or was he still on M Street? And how could he not know that? Realizing he did not know made him frantic.

"Carole," he practically screeched, "why won't you help me? Carole? *Ca*—"

He jerked forward with the effort of trying to cry her name and found the steering wheel becoming a fold of cloth in his hands. He was lying on his back in his bed, breathing quickly, his heart pounding.

Thank God, it was only a dream! What a relief. Even the real world was better than *that*.

Forcing down his fear—and trying to ignore the despair underlying it—he strove to bring his breathing under control as he looked around the peaceful, orderly bedroom his wife had created. In the soft dawn light the blue-and-brown room looked soothingly immutable.

But he was alone in the double bed. Carole was not there. He was alone in silence.

No. Birds were twittering. And a siren wailed.

4

This was Georgetown, and it was Monday morning again, so that siren probably was just some big shot going to the office.

His nightmare resurfaced: he had been trying to get to his office. —No, to a meeting at his former office, across the street from the Capitol.

But he had gotten lost—hopelessly lost—in tidy little *Georgetown*, where he had lived for more than seven years! And Carole, who had patiently guided him for so many hard years, had been unaccountably deaf to his pleas for directions.

He had been anxious to get to a meeting, an important meeting, the sort one would never be late to: a chance to present himself to some party fat cats interested in his prospective bid for the Senate. These kingmakers had been waiting for him, drumming their fingers and giving each other looks while his administrative assistant tried to distract them. He had been frantic at being delayed in getting to them.

Apparently the prospect of his first statewide race had him scared half to death. With reason.

And Carole, who had found him in storefront law practice and had helped him get into Congress and had always hoped he might become president someday, hadn't even cared enough about him any more to tell him how to find his way to the Capitol! No wonder his fear had waked him up.

For if Carole had really lost faith in him at this most precarious stage of his career, all they had built with their lives could come to nothing. She had helped him begin a bridge to power; now it was a fragile, perishable structure extending into mid-air. He couldn't complete it without her.

He couldn't even want to try.

For fourteen years they had labored on it together,

pressing the limits of their endurance. Always they had comforted each other when they were abraded and exhausted in body, mind, and spirit; and together they had celebrated the victories which had been essential to their continuing. Fourteen years of discussing, arguing, and working out their consensus, but always admiring and praising and loving each other in good times and bad.

They had been so married. Out of their union had come a steady stream of speeches and articles, as much an investment of their hopes as children.

And they had done so much. It was mind-boggling, looking back on all that work.

Yet virtually all of that toil and all of that accomplishment would be wasted, thrown away, if they did not continue to press hard now: to complete the slender bridge which could take him on to the next firm ground rather than collapsing under him. For most of what they had done had had little discernible effect except gaining him recognition. A career in politics required enormous maintenance just to prevent its disintegration.

Looking out over the bustling city on the Potomac, the most political city in the world, he saw not the glowing, spotlighted public buildings of the "city of many dreams", but its blocks and blocks of boarded-up and burned-out ruins. He saw the wreckage left from other men's ambitions.

Looking across the wide country, he saw a landscape littered with such ruins. Structures assembled with great hopes were being quietly re-absorbed by the environments out of which they had been molded with such soul-destroying efforts.

Oh, they often started out bravely and full of promise, these careers in politics. But they could end abruptly, in defeats as undeserved as a rainstorm on

Election Day. Or they were abandoned in disgust, disillusionment, or disgrace. Often the builders simply died too soon.

But to end in nothingness, all those efforts and sacrifices and humiliations wasted, when he was only entering his forties! No wonder panic nibbled at his sleeping mind.

Meanwhile, outside the obsessions of any one ego, the world's real needs continued to feed on each other. While others looked at recent historic breakthroughs and saw the promises of peace and plenty, he took a broader, deeper view and perceived that the world was in a dangerous time. It was an awakened lion, thrashing about in a net of constraints it did not understand. As it became more frustrated it could grow more savage. If it lashed out too wildly before it could be slowly and patiently freed from the most painful binds of the tangles it had gotten into, it could kill its would-be liberators; all that had been gained could be lost. There was no time to waste. He had no time or energy to spare for fear or self-doubt.

—Or for pain, the pain of knowing how much of all the human suffering politicians were supposed to "*do* something about" was indeed unnecessary. Or for frustration and despair because trying to point out the obvious avenues away from all that unnecessary suffering—measures so simple as letting go of greed or grievances, achieving balance between revenues and costs, or producing only children one could bring up to be productive citizens—was politically dangerous. People seemed so wedded to their follies that sometimes it was tempting to feel that the human race wasn't worth saving from itself.

That feeling brought on a greater fear. It was not mere fear of personal failure, but the fear of giving

7

up: fear of letting go of the one fundamental value ever deeply cared about. It was the fear of moral suicide.

For him that surrender was in part what the pundits called "compassion fatigue": giving up on the struggle to do something about poverty or refugees or malnutrition or any other problem that would have been so simple to deal with, had human nature been more rational and less self-centered than it was.

The temptation to give up was a disease. The spectacle of just one arrogant, self-aggrandizing individual weakened other people's resistance to it, for one successful egoist made all the giving people look futile: their very refusal to be corrupted kept them from power. So each little sin may make the human prospect more problematical.

Was humanity able to survive despite all its inherent weaknesses and the constant betrayals of those who should know better? Could even he become one of those who stopped making an honest effort to help it survive because the costs and burdens of caring had become more than he could bear? And if even *he* was tempted to give up, how could there be hope?

Yet he could not shut out even the anguish on the other side of the world by concentrating on his own work, "cultivating his garden", as most people could. He could not evade awareness by focussing on his private life, for aside from his career-oriented marriage, he had none. Moreover, he was mentally a citizen of the global village, and the field in which he had chosen to spend his life included all of the world's problems.

Since it was Carole who had given him the courage to enter politics, it was to Carole he turned when his courage faltered. In his senior year of high school he had played "the hero" in a musical, and his role had

8

imprinted one song on him:

> A fellow needs a girl to hold in his arms
> When the rest of the world goes wrong;
> To hold in his arms and know that she
>   believes
> That her fellow is wise and strong.

But of course not just any girl would do: he needed a woman he admired and respected so much that she could give him assurance that he was entitled to his dismay, even entitled to her respect and affection because he struggled with monsters which others could not see or could choose to ignore.

So what really sapped him these days was Carole's strange behavior. He had been going through a bad time in his new job this past year, and he was afraid that his wife had at last realized the full extent of his frailties and lost respect for him.

Without respect there could be no love, not as he understood it. Her love had given way to a secret contempt, and naturally desire had fled.

What else was a man to think when he was by all accounts attractive and knew himself to be clean in his habits and unswervingly faithful, but his wife wouldn't let him touch her and wouldn't tell him why? She had withdrawn even on the level which mattered most because it lent its passion and happiness and certainty to all the others.

She had built him up with her faith and then withdrawn it.

Yet he couldn't stop working, whether Carole cared or not, not for any reason whatsoever. He had an obligation. He mustn't stop pushing himself. He must push until he dropped, or until he couldn't endure living any more. —Especially since he was

9

underway now, after all the early struggles, and he still had a bit of momentum left. He had to use it to keep driving himself forward.

Driving, driving, driving himself as he had all these years. But running on empty now, pretending that everything was fine while he prayed for a refill that did not come. Because even if he were already a has-been at forty, if he admitted it to himself he might as well lie down and die.

As it was, he too often had to fend off the feeling that he was inching along with his face in the dirt. Despair was his intimate enemy; it prowled outside his gate, nosing about for its next chance to rush in and pull him down. As he grew older and found his limitations it was coming to look more menacing.

When weariness and the depression it brought on did not impair his objectivity he understood that this personal monster drew its power from the same aspect of his central nature which had determined the moral and professional commitments of his life. Carole, ever the psychologist, had once shown him an article in a women's magazine which she said "put its finger" on the reason he was subject to depression: like many women with the same problem, he was "an excessive giver": he assumed the burden of feeling he was responsible for other people's well-being, instead of only thinking that he would like to do something about their problems.

Now he had attained what might be the one position for which he was best and worst suited: director of the Peace Corps. When his defenses were lowered,—say, when he was worn out after an exhausting day,—a stack of articles and reports he felt he had to study could appear horrible to him, quivering with the sufferings of all the real people represented by each piece of paper laden with abstractions.

10

Clean water supplies. Birth rates, child immunization, health education. Overcropping, overgrazing, deforestation, erosion, malnutrition. Urbanization, crumbling infrastructures, social breakdown, political instability. All of these and many more, woven together in the framework of the old Malthusian equation of overpopulation equals poverty, famine, disease, and war.

Yet he welcomed the excuse this job gave him for learning all that he could about poor countries and their problems, and the depth of his caring spurred his imagination. There was so much to do! If there had been ten of him, a hundred of him, they all would have been pushing themselves to their limits.

And running themselves into the ground.

They all would have been frustrated, for in his view the responsibilities of his position were infinite, while its powers were trivial. Continually trying to do more with less within idiotic constraints imposed by the federal bureaucracy was exasperating.

When it all started to get to him he needed Carole. In fact, he needed her so it would not get to him in the first place.

He had to do something about that. It was stupid to lie there feeling sorry for himself. He swung his long, narrow legs out of bed, and without pausing to put a robe over his pajamas he padded barefoot on the polished hardwood floor into the hall and over to the little second bedroom they had called "the study".

He stopped, then, and stood resting a forearm along the paint-softened old door frame, watching his wife sleep.

Carole was plain. Her dull, brown hair was worn short and straight to express her no-nonsense personality; she despised her image in the glass and considered efforts to improve it a foolish waste.

Whereas he had been told all of his life, in one way or another, what a handsome blond he was. Fans of old movies compared him to Leslie Howard.

He could see the facial resemblance: high, wide forehead, deep-set eyes, V-shaped jaw, and thin mouth. As for manner and voice and temperament, he was afraid people who mentioned that resemblance were thinking of the overcivilized Ashley Wilkes in the movie of *Gone with the Wind*; so whenever the comparison was made he struck a dashing pose and retorted, "You mean, like in *The Scarlet Pimpernel*?" That usually got a laugh.

Okay, so women regarded him as exceptionally handsome. His mother had always said, "Handsome is as handsome does," and he really liked looking at his wife, seeing her strength and sense of humor even when, as now, she still looked tired at the end of a night's rest. For her spirit seemed tireless.

She was bolder, more flexible, more cunning than he was. But he had to be the straight arrow anyway, for he was the one who always had to step forward and deal with whatever anybody chose to throw at him. He didn't mind that, accepting the fact that constant carping was more intrinsic to public service than adulation.

And his wife's job always had been helping him. It was what she had married him for, she had said so! She was always planning ahead and making arrangements for him. She was wonderful at managing: time, money, people.

Free of concerns about when and where and how, he could concentrate on doing each delicate interpersonal thing as gracefully and yet as effectively as he could. He seldom had to be harsh, and he saw no reason to hurt or offend anyone unnecessarily. He need only murmur, "I'm sorry, but my most trusted adviser has persuaded me that . . . ." or something less transparent. Being able to muster enough presence of mind to sense what to say and how to say it in order to get what he wanted, he could leave people charmed and impressed even when Carole had been sure they would be mad at him.

It was only a matter of natural gifts, he would say to her, meaning that it was nothing for which he deserved praise. (On the contrary, it was culpable not to use them, for the general welfare. He believed that in the depths of his religion, though it would have been insufferable to say so.)

But it was Carole's opinion that besides his voice and manner and looks, he also had sensitivities rarely so highly developed in a man. She had noticed that quality the first time they met, when she interviewed him for a story in the *Los Angeles Times* because he had won a David-and-Goliath legal battle to protect the California coastline.

In her role as inquiring reporter she had quizzed him thoroughly, and she had made up her mind: "This is the man I want." Then it had been only a matter of showing him that she recognized his "enormous potential" and that she was the helpmeet he needed if he was to fulfill it.

It had been a courtship as funny and full of surprises as it was short. Carole had known that being denied the right to take the initiative was withering to the male libido, but she had a massive intolerance of wasting time. She had never in her life tried to flirt, much less seduce, and she made it clear that he was the exceptional case: he was not to encourage her unless and until he had become as serious about her as she was about him. Then she had offered him the book of herself and invited him to decide whether he wanted to buy it.

13

Once before, in law school, he had gotten serious about a woman because her mind delighted him. But that mind had been too personally ambitious to accept marriage to him. This time he had been a pushover.

Just to prove to each other how rational they were, they had spent a year confirming their initial impressions before formalizing their decisions. All the same, it had given his ego a whale of a boost to be so decisively coveted by such a tough-minded and intelligent woman when he had fallen into the habit of denigrating himself as an ineffectual idealist.

He had never doubted that he had personal assets. Nevertheless, using them did require constant fine-tuning. He could always use the wrong word at the wrong moment and come off worse than he should have been able to.

Especially since he was forever hurrying on, pretty much making everything up as he went along. He lived like a rat running a maze, seldom with time even to really look at his wife.

Small wonder if the years of counseling and supporting him had made her look tired. For always she moved around the edges of the maze, working to get the overall picture and warning him when he had to make an extra effort to be sharp or keep on schedule; watching for openings up ahead, feeding him directions to help him avoid making a wrong move: "Wear your raincoat, you're due at the *Washington* Hilton at two o'clock, and here are the details. Try to get whoever introduces you sweetened up beforehand, especially if he seems inclined to be jealous of you. Oh yes, and this speech has to be adapted here, here, and here. See? Where I've marked." A live-in, lifelong aide with no salary, yet worth a fortune.

That was particularly so because as a public man

14

he could not afford even one slip of judgment which could later come back to haunt him—yet he had to move fast. Other rats, many of them far richer and just as handsome and all the rest of it, were running the maze to reach the same goal. In each man's life the timer would buzz, and no man knew when.

He wondered how many of those others also lived with the fear that if *they* had not made it to the goal with years to spare, there could be terrible consequences for millions. Every passing minute brought more hungry babies into the world; more want, and more ruin of the land and air and water by people who made their own wants paramount. Following upon these damages would come still more, as destruction of resources bred more want; want bred more unrest; and conflict and destruction went on with their self-perpetuating cycles. Ignorance went on breeding more ignorance, pressing for more devastation of the life-sustaining powers of the Earth, more loss of faith in the future—which activated more selfishness, which further threatened that future.

More shriveled babies with bloated bellies; more women already old, worn out by childbearing, at thirty. More bewildered children whose limbs had been shattered by the toys of war. More bodies of brave young men sprawled in heaps, rotting and turning into drifts of clean-picked bones. More elderly people huddled among the ruins of what had held their entire lives. Silent.

Silent. The powerless and the helpless, the great throngs of Earth's ever-increasing billions who asked only to be left in peace to toil and eat and rear their children and occasionally enjoy a sunset! They all would be silent with the silence of Auschwitz and Buchenwald after the killing finally stopped.

One day there *would* be silence over all the world,

the ultimate solution to all human problems, the ultimate peace of universal death: the extinction of the only species known to be able to think about eternity. Only moving beyond dependence upon their own sun could give that species the physical immortality for which it yearned so deeply. Only politics could muster the resources to help technology take the long series of steps which could make that eventual escape from extinction possible. Yet millions demanded those same resources for the current needs of populations which continued to grow . . . and grow . . . and grow.

In the midst of this crisis men like himself, polished men, sat at polished tables, arguing strategies and weighing alternative scenarios, sacrificing essential truths to protection of careers and egos. Meanwhile the world, the beautiful Earth, cradle of infinite varieties of life, spun on toward death.

He shouldn't be just standing here! He was starting out tired. He must at least get what he needed to run the gauntlet of another day.

So hurry, little mouse in the maze! Begin by waking her up and making love to her so you can be sure that the world still holds some trust and honor and tenderness. In such goodness we can find enough sanity and hope to make the effort required just to muddle through.

But it was still so early. He hated to wake her up before it was necessary. Her lean face looked so peaceful and so weary. She had been keeping long hours, researching her new "pet project", a series of articles she wanted to publish in a women's magazine. In addition to helping him as usual—well, not "as usual", now that he was no longer a congressman, but still keeping his home running smoothly.

She had turned out to be more wife than he could

have dared to wish for. Indeed, she had concealed her family's money from him until after they were married. He had had to forgive her for that, since she had used this resource to help him meet the costs of his desire for public service.

Unable to resist the urge to touch her, he bent and moved a strand of her tousled hair. Greater than his desire was his tenderness, and greater still his gratitude.

But much as he cherished her, right now he needed her awake. Another demanding week was starting, and he needed the comfort which only this one woman in all the world could give him. He eased down onto the edge of the twin bed, watching as her eyelids fluttered open.

Seeing him as she awakened, she automatically started to smile, and for a moment his fears wavered. "What is it, hon?" she mumbled.

If he bent and started nuzzling her she would shrink from his touch, and he couldn't stand any more of that. Besides, after so many months of her rejections it wouldn't even be honest to imply that he felt amorous. First he had to feel relaxed and happy from being close to her again.

But how could he honestly explain what he needed, without sounding even weaker than he must already seem to her?

"I want you."

"No, you don't," she answered dryly, yawning and starting to turn away.

"You mean, you don't want me." He reached across her to rest his hand on the bed. "What have I done? What's changed?" His tone demanded an answer.

She closed her eyes. She didn't respond, just shifted a little as though mentally digging in her heels. She could be as stubborn as occasion required, and the worst tactic anyone could use on her was to seem to bully her.

He made his tone reasonable. "You remember the last time, just before my trip to Puerto Rico and Florida? Last May, and now it's September! You said afterward it was 'spectacular', that I'd made you feel like a Roman candle about to go through the wall. You went on and on—"

"*Please*, Charles, *not now*."

He tried to jest: "'If not now, when?'"

Again she did not answer.

If he could only hold her, feel she was there for him! "Let's at least cuddle up," he urged, trying to edge closer in order to nudge her over against the wall.

She didn't give way. Instead her voice became alert. "I'll meet you in the big bed. We need to talk. There's a matter you've got to deal with."

He groaned, for that sounded like some no-win decision regarding major repairs. He sat still.

"Carole, why won't you sleep with me?" It was time to be blunt.

"Another time, Charles." She would not look him in the eye, but her voice was hard.

"No, now! Don't you have any idea what it does to me every time I reach out and you draw back? Do I really repel you so much? *Why?*"

She bit off each word: "I don't want to talk about it."

"Carole, love, give me some sort of explanation! Anything is better than not knowing!"

He tacked on a bribe: "Then we'll talk about whatever you want me to 'deal with'."

"But I . . . can't discuss it with you right now, Charles." She had turned plaintive.

"And not at any other time either, apparently." He

18

captured one of her hands and caressed it. "I've been patient, haven't I?"

"You've been an angel." She withdrew her hand. "But I have to ask you to wait a while longer. I'll explain it when I can."

"Tell me when!"

"I hope when you get back from circumnavigating the globe the problem will be under control."

That trip would last through the first week of December, but any promise was better than all the stonewalling he had been getting, so he straightened, brightening. He waited for more, something specific. But she simply lay there staring at nothing.

"You mean we can get back to normal after you've had a couple more months?" he asked, but his flash of hope already was fading as he watched her.

"We can talk about it then."

"But why can't we talk about it now?" His voice rose. "What's the problem? Don't I at least have the right to be told?"

She met his eyes at last, but only for a second. "Yes. But I have the right not to tell you." There was no hint of wavering, but he thought he heard a note of regret.

So he went on trying. "Carole, let me understand! It can't be so bad as to justify all this to-do!"

"You're making all the to-do," she complained, "and I'd hoped you wouldn't.

"It's not something anybody can explain to you now. It's not a matter of talking things out. It's a matter for . . . changing circumstances to resolve."

Not quite keeping his face straight, he shot at her, "Are you having an affair? You want to tell me only when it's over?"

She ignored the note of laughter in his voice. "I know it's hard on you, but believe me, I don't want

19

you to suffer any more than is necessary. I have to get the problem worked out in my own way. So just set it aside. I'm sorry, but this is how it has to be, for a while longer."

"But you say after the trip you'll . . .?"

"Yes, just give me until after the trip." Now she met his gaze squarely. "Trust me. Time will resolve it."

When something was complicated but not worth explaining to him she would sometimes just say dryly, "Trust me." She never said it as a joke, and over the years he had learned to do as she asked. That ability to rely on her judgment was a significant factor in what success he had achieved.

He bent and kissed her cheek. He lingered in spite of his recent efforts to avoid trying to arouse her lest it cause her some sort of discomfort. As had become her practice, she shied from such intimate contact.

He drew back, saying, "Time." His tone tried to make it a pact even though she was not party to it.

He attempted a lighter tone. "I'll be so frustrated by then, we'll have a wild second honeymoon."

She gave him a look he could not read, but it was not encouraging. Nevertheless she pressed his arm, and he saw that whatever the problem was, she was not proud of herself.

He got up, sighing but giving her hair a gentle ruffling to say, Whatever it is, I love you.

After taking turns in the aged bathroom they re-convened in the front bedroom of the townhouse.

She arranged the blue sheet and bedspread over her chest and folded her arms above her waist. Then she unfolded her arms to tug down the cuffs of her pajama top as though preparing to address a staff meeting in one of his congressional offices.

"Marjorie won't be going with you on your world

tour. We've agreed Philanth Devon will fill the position nicely."

Bolting upright in bed, he squawked, "What?!"

"She knew you'd be a little upset, ..."

"What the hell d'you mean, she 'won't be going'? Why the hell not?"

"Really, Charles, you know better than to indulge in such language. It's not like you."

"Tell me! Why isn't Marjorie . . . ?"

"She's too experienced to be gaga over the idea of travel as such, and she doesn't find what she's heard about travel in Peace Corps countries appealing. Also, she keeps hearing about what a heavy schedule it will be, considering the conditions. She thinks a younger person would bear up under the stresses much better. Philanth is the obvious choice."

"Is she? Wasn't this girl brought in as a temporary only about a month ago?" Carole nodded. "Well, she can't *possibly* handle it! The Peace Corps bureaucracy alone takes months to learn, and we've got all these outside parties to coordinate with too. Not the least of which is the State Department! And it's a trip involving sensitive matters. You must be out of your mind to even suggest such a thing."

"Now, now," she soothed. "Your office will function much better while you're gone if Marjorie's left in place there; and it'll be valuable for you to have such a reliable main point of contact with the headquarters. Philanth will have another month of intensive training. I'm sure it'll work out fine."

"Out of the question! Aside from all the practical considerations, Philanth's too young for me to travel with. Marjorie's obviously old enough to be my mother, whereas Philanth . . . ."

"Looks young enough to be your daughter."

"She *looks* like a *Boopsie*!"

21

"Well, but she acts like the sweet, old-fashioned girl she is, and she still wears her wedding ring."

"I'll just talk Marjorie around."

"Don't even try."

"If she won't budge, then I'll take Jeff."

"I've discussed your alternative choices with Marjorie. She's talked with Jeff's wife Annie. Annie has already vetoed Jeff's going. She's had a difficult pregnancy, and now he's excited at being a father. They've had their fill of travel for quite a while."

A careful end run. Now he would be accused of stirring up trouble between husband and wife if he asked Jeff. Outflanked by women, he pondered.

"Then I'll ask Jeff to name somebody with all the necessary skills and experience."

"Charles, Philanth *has* 'the necessary skills and experience'. She can get along in French and Arabic, and it's *her job* to be familiar with all the arrangements for that trip. The procedures of your office go on around her all day. She's had ample secretarial experience, and she's excellent at 'word processing'."

"That's all very well, but I have my career to think about. This trip is going to be heavily photographed, and we're going to be scrutinized by a lot of interested parties with pipelines back to Washington. So I'll simply have to take either Marjorie or a man."

"You're being paranoid."

"I don't think so. You're proposing the sort of situation that can blow up and destroy a public man. Anyone in my position excites jealousy. He makes enemies if only because of the fierce competition for the jobs he has the power to fill, and the gossip of some of those disappointed parties can be ugly.

"Besides, they'd have something to work with: Marjorie and I are slated to have connecting bedrooms because of all the important international

22

phone calls and the language problems involving switchboard operators. The per diem regulations wouldn't allow us a suite so we could have an intervening room for my aide.

"And it's a five-week trip! A politician noted for his 'looks' and 'manner' is a sitting duck for speculation about his personal life anyway; he'd have to be crazy to take a long overseas trip on official business with a young female companion!"

"A politician's reputation is his only protection in dealing with women, and surely yours is impeccable.

"You've always tried to be known for giving opportunities to qualified women, and Philanth shows lots of promise. She's an excellent administrator, able beyond her years. I have it directly from her previous employer before her Peace Corps service . . . ."

Carole paused, then hurried on.

"Furthermore, she has a knack for getting things done pleasantly, which is more than Marjorie can say for most of the people brought in under your predecessor." Carole sounded determined to get through a rehearsed sales pitch. "*They* have the Washington syndrome of acting as if working for somebody important puts them at the right hand of God—and an insufferably arrogant god, at that."

"Yes, I'm going to have to do something about those people—as soon as I can get around to figuring out what. At present I'm stuck with them because of my peculiar relationship with the White House."

"At least you don't have to take one of *them*."

"Thank heaven for that. But there's been no jockeying for the chance only because it's always been a foregone conclusion that Marjorie goes.

"It's mighty odd, Marjorie's springing this on me *now*," he grumbled. "This trip has been in the planning stages for most of this year."

"Well, Marjorie didn't think about not going herself until she saw how sensible and intelligent and pleasant to work with Philanth is."

"Who originated this idea, Marjorie or Philanth?"

"I did. I could see that Marjorie wasn't overjoyed by the prospect—after all, she got several trips with you during your years in Congress—and I knew Philanth would jump up and down at such an opportunity."

"Philanth knows about this proposition?"

"Of course not. I thought you should be the one to broach it."

"If I did she'd see that I don't want her, and if she's as sensible as you claim, she'd beg off." He folded his arms and plopped back down. "So I won't."

"You have an alternative?"

"Of course! I'll just talk Marjorie round. She'll never have to leave the hotels, except to go to and from the airports."

"Oh, she won't be taken in by that! She knows the facilities can be unreliable in those countries. Not to mention the possible illnesses. And she's a Southern Californian, remember? She despises the *Washington* climate, let alone a jaunt back and forth along the Equator!"

"She'd do it for me."

"Don't bet on it! I'll talk to Marjorie if you do, and I guarantee she won't go. In fact, by the time you and I get through pulling on her we'll be lucky if she doesn't just quit. She's eligible for retirement, and she's looking forward to getting off the nine-to-five treadmill."

Carole's tone told him that she meant business, and he knew she could do what she threatened. Carole accepted no obstacles when she had made up her mind to something.

24

"Five weeks," he mused, seriously considering the preposterous notion for the first time. "That's a long period on the road with somebody I scarcely know well enough to speak to."

Resettling herself, Carole replied smugly, "I'm her best friend. What would you like to know?"

"Never mind. I'll think about it." He glanced at the bedside clock and moved to get up, thinking about breakfast and getting ready for work.

Carole grabbed his arm. "Where have I heard *that* before! You can't dodge this, Charles Bjorklund. The paperwork has to get underway—immunizations too, if you don't take somebody who already has them. Marjorie told me the visas alone are a huge problem for this trip, because you're visiting so many countries. And they're countries that can pose all sorts of problems for anyone with such a complicated and inflexible itinerary."

"I have to be locked in with Philanth or Marjorie or whoever right away," he mumbled as it dawned on him, "to the extent that I want to avoid exacerbating the already horrendous paperwork."

"I'm sorry, Charles, but Philanth came back from North Africa in mid-August, only about six weeks ago. I had Marjorie start her through the Personnel procedures as soon as the doctor let her out of bed. Her letters to me and my visits with her in the hospital made me suspect she had the qualifications to be a good assistant for you, and now her work in the office has confirmed it."

"But it's me that's making the trip and has to work with her! My tasks will be delicate, and extremely important! It'll be demanding work, highly visible, and under severe time pressure. I can't afford even *minor* problems between me and my closest staff! It has to be my choice."

"Then get to know her better."

"Let's see," he sighed. "This girl is special assistant to Marjorie for coordinating the arrangements for the trip, right?"

"In practice that means she also works for Jeff and Les."

He already had seen that Jeff liked Philanth and Les didn't, but that told him nothing except what he already knew about the two men: Jeff had risen through the ranks of Peace Corps and seemed to like everybody, whereas Les was a political appointee and probably didn't even like himself.

"I'll pay more attention to what she's doing."

Carole nodded, looking satisfied. "What would you like for breakfast: waffles, French toast, or biscuits?" She was "rewarding" him by offering "sinful" treats he loved, and it wasn't even a holiday.

"All of them," he growled, getting up. "The waffles today, the French toast tomorrow, and the biscuits on Wednesday."

Carole chuckled. "Frozen waffles tomorrow," she amended, "because they're so fast, and you go in early on Tuesdays." He agreed.

He had back-to-back meetings all morning. After an interview and some appointments with staff he went over to the State Department for briefings and meetings which took up the rest of the afternoon. But between the morning meetings he found time to order and look over the official file on Philanth Devon, and after lunch he obtained an authoritative, confidential explanation of its more arcane notations.

He came in late from a dinner at a hotel. He had made a presentation to a convention of veterinarians, explaining Peace Corps' program for mid-career professionals interested in a few months of service in developing countries. He had outlined the crying need for veterinarians in Peace Corps countries, showing slides and giving statistics which must have made it hard to digest the good meal everyone had just enjoyed.

Praying that he could recruit at least a few of these people who were routinely giving better medical care to pets than most of the world's poor ever received in their entire lives, he had lingered for an hour after the program ended, trying to talk to prospective volunteers about livestock problems in tropical rural areas. However, they were leery of the discomforts and inconveniences and their lack of relevant experience, and he had come away feeling that he probably had wasted his time and the time of the earnest staff who had put together his materials.

He greeted Carole and started preparing for bed. But when he was in his pajamas he returned to "the study", which Carole was starting to refer to as "my bedroom". Carole was working at the computer.

As he sat down on the bed he remarked, "I didn't realize Philanth had served less than a year of her tour of duty as a Peace Corps Volunteer."

Without taking her eyes from the screen Carole replied, "You knew she was medevacked."

"I suppose I did. But I wasn't paying attention. To me she was just another of the young Returned Volunteers in the outer offices.

"Now—even after studying her file and talking to a staffer who specializes in medevac cases—I still don't understand why Philanth wasn't sent back overseas after she got well. The staffer just kept saying it was the decision of the medical officer in the field, and he may have sent in an incomplete account out of concern about confidentiality. She says if I get him on the horn he may be reluctant to say more, even to me. It doesn't feel right."

"So? Why tell me?"

With two quick strokes Carole divided a paragraph in the text on the screen before her. It seemed to be a case history of a "Susan W.", presumably someone Carole had interviewed for her study.

"Has she discussed the circumstances with you?"

"Yes." Carole bent to consult notes in a steno pad.

"Well?"

Her hands returning to the keyboard, Carole said absently, "I don't see why you're so concerned."

"Then I'll tell you! Instead of being returned to her post or reassigned so she could apply the training Peace Corps has invested in her, she's shuffling papers up in the director's office. And *you* encouraged her to apply for the job she has now. This job was wired anyway, because of your influence with Marjorie. But it also looks like she could have used her friendship with you to wriggle out of returning to her responsibilities in the field."

"Impossible." Finally appearing to take an interest in the conversation, Carole turned from the screen. She turned back to the keyboard long enough to instruct the computer to save her work, then darkened the screen.

"Philanth had agreed to being separated from Peace Corps before she came back to the United States. She didn't know anything about the headquarters job until she was here and I visited her in the hospital.

"I even have a witness. Her previous boss before Peace Corps was visiting her when I came into her room one day. He was trying to recruit her back, and she was wavering. I knew he was a sadistic bastard but she was emotionally involved with him. She was also weak and depressed and hard up for a job. So I jumped in with *my* offer."

"Just the same, somebody interested in making trouble for me could ask why she was medically separated from Peace Corps when she needed only a couple of weeks in bed to get well and was less than halfway through her assignment."

Carole dropped her gaze. Smoothing her lounge robe against her thighs, she began carefully, "There are some things not in the files."

"That's what I've just been saying—I thought." Thinking that it had been a long day, he beckoned with both hands. "So give!"

"Well, the Peace Corps medical officer responsible for her well-being apparently was overcautious, and other people in the pipeline went along. You know how it is—how it must be, in a global network of government bureaucracy with a million things going on at once."

"But there are safeguards against overly hasty decisions."

"Yes." Carole spoke disinterestedly. "Philanth had appeal procedures and other personnel protections available to her. She knew her rights. She chose not to avail herself of them. It's water over the dam, now." Carole started to turn back to her work.

"Don't trivialize it. Failure to complete a tour of duty reflects on the viability of our whole program, especially in the eyes of the host-country nationals closest to a project. It's also a damaging statistic that affects the data we use in the appropriations process. Besides, a lot of work is required to fit a Volunteer into a slot, since we're responding to requests from host governments. Early termination is a serious matter."

"But it wasn't her fault."

"Why not?"

"She just hadn't been taking proper care of herself

when she got dysentery."

"That could be another way of saying she refused to follow reasonable medical instructions. Refusal to follow orders is grounds for termination."

"You're suggesting *Philanth* could have been guilty of *insubordination*? That's ridiculous! Her inclination is to err in the opposite direction. She actually went through a period of preparing to take vows of obedience, et cetera. Any personal assertiveness on her part takes conscious effort."

"Then what *were* the circumstances?"

"The hospital was grossly overburdened. The staff were overworked. The conditions were appalling, especially to someone who hadn't had time to become case-hardened. Philanth didn't want to shift her work onto others who were already taxed to their limits, so she fudged on the prescribed bed rest. She shouldn't have done that, of course, and one thing led to another . . . and they had her shipped home."

His voice rose. "But *why* was she *terminated* rather than being sent back to work when she recovered? Come on, Carole, I haven't got all night!"

Carole began choosing her words carefully. "There was some question about her psychological ability to adapt to field conditions. That allegation threatened her career plans. So rather than try to appeal it, she kept it out of the record by accepting a medical separation."

"She couldn't stop working long enough to recover her health, so they threatened her with a psychiatric termination," he repeated to make sure he had it right. Carole nodded. "And so she plea-bargained for her medical separation!"

"Something like that."

"Oh, tell me another!"

"Threatened with a psychiatric stigma in your

30

official government employment record, what would *you* have done?"

"I'd have stayed in bed in the first place!"

"I dare say you would.

"Well, so she's too conscientious. Some young people feel the world is depending on them much more than it is. I hate to think where we'd be without them."

He shrugged, obliged to concede that point at least.

Carole continued, "Just bear in mind that she wasn't guilty of anything except overwork and not taking adequate care of herself. Dealing with an endless succession of dying babies would bother a *lot* of people, and Philanth is unusually maternal."

"The Peace Corps is a secular institution. We don't glorify people who make themselves martyrs due to lack of good sense. It looks like she's neurotic."

"She's altruistic, and altruism shouldn't be dismissed as neuroticism."

"Look, I have to be careful. On the surface this looks suspect; and when you start picking it apart it looks like it must have been worse than you're making out, because the alleged circumstances are not adequate to explain the outcome."

"Don't cross-examine *me*, Counselor!" Again she half turned back to her work. "If you want more information you should talk to Philanth."

"I'm trying to get to the bottom of it without approaching her because you've proposed that I take her with me as my special assistant on the most difficult and important trip I've ever made."

"I can assure you that there's nothing for you to worry about.

"It's true that her sense of self-preservation is weakened by her ideology, but try to understand *her* point of view: she can see that being a woman of

31

altruistic rather than personal ambition, she's probably going to be consistently blocked from advancement into positions where she could make a significant contribution to society. So she may sometimes be too ready to risk herself in a good cause.

"But she *was* sufficiently in touch with reality to slip through the Peace Corps screening by emphasizing her career plans and relevant practical experience while concealing her intense love of helping people."

Seeing that he wasn't reassured, Carole added, "A lot of the people you're used to dealing with are self-aggrandizing, and that's made you suspicious. I expect you'll find that Philanth is unlike anyone you've ever known before.

"Such a strong . . . well, . . . selfless love as she has is rare, nowadays, since it's not supported by the culture even to the extent that it used to be. That can explain why Philanth got bounced out of today's Peace Corps: she developed independent of the cultural mainstream. In fact, she feels little obligation to fit the current norms.

"But despite her unusual degree of altruism, she has a serene, well-balanced personality which makes people like having her around."

"Yah." Seeing that he hadn't succeeded with his major assault, he switched to this aspect of his actual objection. "As to that, they like her too much. I got her involved in a staff meeting I held this morning, and it looked to me like she had 'a special relationship' with half the men in the room!"

"Don't women seem to like her too?"

"Yes," he admitted. "Except for the self-important bitches I can't get rid of because they were sent to us by the White House."

"Well, then you're just trying to find fault because you felt safer traveling with Marjorie."

32

"Yes! I don't understand why you won't admit there's a problem! I'm too likely to be accused of 'poor judgment' if I travel around the world on a thoroughly publicized tour with a girl who reminds people of the mindless bimbos they see on t.v."

"I know her appearance is all some people will have to go on, but they certainly won't go far on that basis alone." Her tone lacked certainty.

"What about her personality?"

Carole frowned inquiringly.

"She's too cheerful. She acts like she hasn't a care in the world. She comes across as brainless."

"That doesn't make her a bimbo either. What you may have seen of her personality is the result of her desire to be on good terms with everybody. St. Paul said to try to do that, and she respects his ideas. Besides, it's important to her to make people feel good by converting any hostility or grouchiness she encounters into warmth, or at least melting it into civility. And aside from her rationales, she's compulsively 'nice': it's a matter of breeding and temperament."

"That's all very well, but she overdoes it. Men like her too much. She's a—a curvy little blonde, and . . . I see how they look at her. I don't need to make it any clearer than that."

"She doesn't issue come-ons."

"Doesn't she?"

"No! She's told me she thinks of herself as happily married to her late husband. She won't even consider the idea of a replacement."

"That's a good line until she finds one. And how long can a girl of twenty-five have been a widow?"

"That doesn't matter. She's as dedicated to continuing her celibate life as any nun. In fact, she cultivates techniques for living a full life alone."

"That sounds neurotic too."

"But it's only because she's completely dedicated to her work. Ken's been dating her ever since she got out of the hospital last month, and he's told me that she arranges her life so it completely serves her ability to work."

"Your *brother* is dating Philanth? You gave him *her* number?"

Carole's mouth twisted into a cynical smile.

"You must've thought she'd be good for Little Brother, then, 'cause she certainly doesn't sound like what he was looking for . . . even if she looks it."

"He seems satisfied. He complained to me that she's working so hard, she spares him only Friday and Saturday evenings. And then they do only the things she'd do anyway alone: buying groceries, cooking and eating, and exercise."

"He must like her 'cause she's a cheap date."

"You're determined to underrate her. I think it's because people with a low opinion of humanity tend to assume that a compulsion to please everyone must reflect weak character or mediocre intelligence."

"Hmmph."

"Philanth's I.Q. places her in the top two percent of the population. She was amazed when I told her. She only knew what she'd scored and didn't think much of it.

"And she has a strong character. Her personality is pliant, but she has deeply rooted principles and well-developed self-control. She regards every bit of her behavior as mandated by a religion founded on a logical obligation to love humanity."

He had no interest in Philanth's religion. Her character was not the issue. And Washington collected intelligent people as a magnet collects iron filings.

Seeing that he remained quite unimpressed, Carole

34

added archly, "And she won't feel mistreated if you make infinitely expandable demands and never give her time off for sightseeing or shopping. She's used to working for people who have no compunction about exploiting her, and as long as it's in a good cause, she doesn't object."

"That's the Peace Corps spirit."

"In quite a pure form. So she'd be an ideal assistant for you during a trip that would offer any normal person a lot of temptations."

"Okay, okay. I'll check her out some more."

He was resolving to disqualify Philanth by more direct methods. He absolutely would not be seen traveling on that five-week trip with a well-endowed, cherub-faced young blonde who was so transparently eager to please. It would make him a laughing-stock.

But Carole added, "Charles, I mentioned that Philanth shows promise. Before she was interviewed for the job in your office I explained to Marjorie why I favored Philanth: it was because I had further plans for her if she did well.

"Marjorie wants to go back to California and enjoy her golden years near her grandchildren, not slave for you through yet another election campaign—especially a much bigger and rougher one than you've ever had before. But she wouldn't want to feel she was deserting you when the worst ordeal of your life is looming before you. So she doesn't have any objection to helping select and groom someone to take over her position with you.

"I told Marjorie, 'I want you to observe every breath Philanth takes.' Now Marjorie thinks very highly of her."

"But even if her looks and personality weren't against her, it's too complicated a job for such a newcomer." He couldn't stop arguing, since it was clear that Carole would have to be talked out of the idea eventually anyway.

She sighed emphatically. "It's not as though Philanth had to pre-arrange every detail single-handed. I'm sure you've got the in-country and headquarters staff doing their best, and you mentioned that there are liaisons helping them in the State Department as well as our embassies. Philanth only has to learn enough about the priorities while the plans are being finalized to be able to help you complete your agendas on a day-to-day basis."

"It's too much for anyone to learn so fast."

"She's easy to train." Carole's manner became an embrace of her subject. "She's like a hungry young bird waiting with open mouth for everything anyone will tell her. She soaks up other people's reports of their experience like a sponge. It's not just a matter of a quick and retentive mind, it's her . . . nature.

"So allow for the fact that she's not Marjorie by remembering that though Marjorie's more experienced in working with you, she's also forty years older. Whether Philanth will be the best possible choice for this particular task or not, we've got to look to the future, not take the path that's easiest right now." Carole was invoking his basic theme as a politician.

"Even if she has potential, I don't have to take her with me on this important trip." He thought that was so cogent a point as to give him the final word.

Yet Carole's urgent but reasonable tone implicitly dismissed it. "Give her a chance. Get to know her. I'm not making this proposal lightly."

Chapter 2

SACRIFICIAL VIRGINS
AND DEVOURING MONSTERS

The next morning he arrived at the office a little after seven to go through the weekly reports which had been due several hours after he had left for the State Department the previous day. It was important that he check every detail in them before the meeting which would begin at ten.

The offices were virtually empty at that hour, so he was startled to see Philanth already busy at her desk in the work area she shared with Marjorie just outside his private office. Though she appeared to be absorbed in studying some papers, she reminded him of Alice in Wonderland: she wore a head-band of the same rosebuds-on-white material as her blouse, and her long, blonde hair hung down her back.

Seeing him pausing to notice her, she shyly wished him a smiling "Good morning."

Going over to her, he asked, "Why're *you* so early?" His voice came out sounding hostile.

"Several reasons—but basically because there was lots waiting to be done here, and I saw no reason to stay away." Her furry voice sounded light-hearted.

"If you're a grind, you're just the item we ordered."

As he started to move away his eye caught the mangled condition of the dictionary lying open on her desk. Besides not being the standard government issue, it looked out of keeping with its immaculate formal surrondings: it appeared to have played football on moist soil without protective gear.

"Why do you have that dictionary?"

"I brought it from home."

"A junk dealer wouldn't touch it."

"I admit it looks out of place." She shut the book and tucked it among the loose-leaf binders in a metal rack on one corner of her tidy desk. "There. It has to keep a low profile. Everybody and everything in Washington is expected to look ideally beautiful, since we're all exhibits in a museum for the world."

"Get rid of it. I have V.I.P.'s coming through here sometimes. If they happened to pause in this area they might notice it just as I did."

Her hands covered it protectively. "But it's my friend! And it would be unemployed at home: there I use an old dictionary that belonged to my father." She stroked the dictionary. "It's *very sincere*."

"Very battle-fatigued, you mean."

"Yes. It's a war veteran. And it's government policy to give preference to veterans."

"Be serious."

"I am. It's earned its scars: my last boss before Peace Corps threw it out a picture window—along with everything else in my office he could pick up and hurl.

"I had to have the window replaced. An expensive self-indulgence we could ill afford."

"Why'd he do that?"

"I wasn't too big for him to pick up, but I wouldn't hurl well." Even a grim expression brought out some of her uncountable dimples. "I really aggravated him. I was always trying to improve his character."

"But what made him *that* angry?"

"It doesn't matter." Glancing up and seeing his expression, she added, "Really, it's not important."

Carole was pressuring him to spend five critically important weeks depending on this girl, in more or less constant contact under what could be very trying circumstances. Moreover, once they were overseas there would be no replacing her without truly crippling effects on his far-reaching objectives.

"Is there any chance of your ever aggravating . . . someone else that much?"

"Not much. I'd just talked him out of raping me."

"*Ra*—"

These bizarre revelations might lead him to some excuse he could use to appease Carole. He beckoned the girl into his office and shut the door behind her.

As they both sat down he asked, "In your *office*?"

"It was in an isolated building, formerly the slaughterhouse. The Hunger Foundation estate had been a farm. There was no one else around."

"Weren't you frightened?"

"I was preoccupied with trying to teach him something." She kept her eyes lowered.

"You don't look out for yourself very well, do you?"

That made her look up at once. "I could have kept it from being an unbearable violation of selfhood by controlling my feelings toward him. And I'd been steeped in the traditional culture, which conditions women to suffer and endure in silence."

"Oh—what would 'condition' you to 'suffer in silence' an action like that!"

"When I was little I read a great deal: old-fashioned inspirational poetry, chivalric romances, mythology, stories from operas, watered-down versions of the classics. Most of our *traditional* entertainment depicts women as admirable if they're selfless and long-suffering. Women are forever at men's mercy, and usually all 'good' women do about the horrible consequences is be meek and bravely dignified."

"Yes, I guess that's so."

"You can bet your life on it. The archetypal ideal woman of western culture is the sacrificial virgin: she

39

may be allowed grief, but no self-assertiveness. Without a brave, strong man to protect her interests she has no choice but to endure the depredations of the devouring monster—whatever form it may take: a lascivious boss or landlord or stepfather or teacher, a howling mob, or a powerful man who threatens her sweetheart, husband, father or brother. Whether it's an opera or pulp fiction, the pattern is the same. It's not more noticed because of the great exceptions."

"You're not saying you'd been conditioned to accept that model for *yourself*! Not when it's been stoutly rejected by the women's liberation movement!"

"They would *say* they reject it, but it's not that easy to avoid the traditional role-playing when those same expectations are still being enforced in the marketplace: women are required to conform to them in order to get favorable evaluations that allow them to get and keep most of the jobs readily available to them. Most men still enforce the same standards in mate selection. Parents reinforce them in their efforts to make their female children socially acceptable. And the model is still held sacred.

"The ideal persists because it's functional: there are social benefits from the 'feminine' virtues. A pliant, soft-spoken person is easy to work with; and patient, gentle people make good teachers and caregivers of the young, the old, and the disabled. Self sacrifice is needed by society: self interest doesn't begin to get done all the jobs that need doing."

"That's true." His eyes narrowed. "But I wonder why such stories had so much appeal for *you*."

"I know what you're thinking, but it's not the masochism found in a socially subordinate group. It's just that a character motivated by the simple-mindedness of selfish goals isn't nearly as interesting as one with a larger commitment. Compared with the scope of the motives of archetypal sacrificial figures like Prometheus or Christ, most modern entertainment seems squalid; only violent action or perversion or some threat of major proportions saves it from being insipid." She added with that engaging smile which came so readily, "Since I wasn't a pathological personality, I didn't identify with characters whose motives were squalid or perverse."

He too had to smile a bit at that.

She went on, "And I don't accept the assumption that people don't want heroes any more. People have enjoyed identifying with heroic figures for thousands of years. They've just been conditioned by *our* popular culture to believe that 'people aren't really like that'. Allan Bloom's *The Closing of the American Mind* points out how pitifully empty people are when they have no heroes, no enthusiasm for great virtue, because of 'the perversion of the democratic principle that denies greatness'."

"Yes, it's demoralizing to believe in nothing."

"It's also socially dysfunctional—if only because assuming that people can't really be heroic is a self-fulfilling prophecy.

"You can see the resulting cultural demoralization even in Peace Corps: because cynics can't distinguish between durable altruistic motivation and naïve expectations which lead to burnout, the new 'realism' about motives discourages idealism among prospective Volunteers. That smothers the joy of giving and kills the soul of Peace Corps. The result is only a different kind of morale deficiency in the field, one that's worse than disillusionment.

"Our culture's 'realism' is mistaken to deny our spirituality. People need high goals and uplifting visions, and not just in order to tackle great obstacles. If they're denied permission to love widely

41

they wither emotionally. Promising them 'career opportunities' won't keep them from feeling they're marking time on the way to the grave."

She pursed her lips and folded her hands. "Sorry to lecture you, but I deplore a lot of what's going on around here; and how many chances do *I* have to talk to the Peace Corps director about basic policy?

"Besides, our entire society is in trouble because of this error. Cynicism is our cultural cancer. It invites every form of sloth and social breakdown."

"Indeed!" So she was opinionated, was she? Good. A chatterbox on a long trip would be intolerable.

"*Yes*. The fact remains that people can be virtuous. A lot of people are. But they're pressured to conceal their virtues and flaunt their vices because of the inversion of values by popular culture: cynicism is treated as sophistication, and advertising openly works to validate shameless self-indulgence. Everyone is assured that they're as good as anyone else.

"That's sick when it's taken to the extremes we see now. A healthy culture doesn't tell everybody to feel good about themselves without constructive effort.

"So let's look at our entertainments and ask questions. At what level of people's psychology are they being asked to identify with characters: at the level of their aggression and self-indulgence, or the level of their ability to care about others? When you look at all the nihilism and self-absorption in what passes for literature these days, it's no wonder so few people care to read it. And much of the entertainment being produced for adolescents is appalling.

"The anti-heroic fashion in art is culturally suicidal. You can see the same fashion in Germany in the 1930's: nihilism and the degradation of persons was 'chic' because the society itself was demoralized. Then Nazism offered glorification."

"Why did you get interested in all this, Philanth? It wasn't just from the reading you did 'when you were little'."

"My parents were scholars deeply interested in values. And in sociology courses I learned about the functional nature of religion and about analyzing the value content of the mass media. Those insights meant a lot to me because I had the feeling I should try to carry on for my parents in some way, since they both had been wiped out by an excrescence of the mass culture they deplored."

Before he could ask about that strange statement she went on, "Once I understood that people have always imparted meanings to existence even when they were uncritically receiving those meanings from others, I saw that the negativism in contemporary art and literature is not sophisticated, as so many people assume. It's actually sophomoric in the conclusions it draws from scientific materialism: it understands that there are no absolute and universally recognizable values from the supernatural, but it doesn't move beyond that to deliberately re-imbue existence with meanings which give it value.

"The scientific perspective only explains how things work; it doesn't assign value to *anything*. By not taking the additional step of conscious affirmation of some specific value, and hence some purpose for human life, culture fails to provide intellectual content for the will to live. Deprived of self-transcending purpose, a culture sinks into hedonism and anarchy. That's self-destructive for both the society and its members."

"I can see that. People whose religious faith isn't able to counteract the demoralizing effects of scientific realism's view of man can become very despairing about the worthwhileness of existence."

"Yes." She looked into his eyes, making him fear that she understood more of his comment than he had intended.

To distract her he commented, "Your parents were into some pretty heavy stuff."

"Well, uh, they helped me to lay the groundwork and gain conceptual tools, but my dad was essentially orthodox, and my mom worshipped beauty as a model for order in human life. Don't hold them responsible for my do-it-yourself philosophy."

"Okay. Tell me about *your* philosophy." Her unpretentious cheeriness was softening him.

"It's simple. Being alive is like having been given a million dollars, except it's your decision whether this money has any buying power. For your life to have value you must assign it some purpose.

"In order to have purpose in life I chose to affirm the value of human potential. That's easy when you see what some individuals have achieved.

"I find that value well served by many of the principles for living found in Christian ethics. As I've indicated, many of the models for conduct in that tradition are socially functional, as well. And what's good for society is good for people in general."

"Very tidy. Do you deliver scholarly discourses like this often?"

"Only when I'm lucky enough to find a qualified listener."

He smiled wryly, half accepting her sly flattery.

She went on, "It's 'no big deal'. Having traditional altruistic values is personally liberating."

At last he could really object: "*I* haven't found it so. Devoting your life to others is about as liberating as carrying the world on your shoulders . . . like one of the archetypal figures you were talking about, Atlas." Carole had said that he had "an Atlas complex".

44

"But that's a magnificent opportunity to develop the ol' biceps. Why would you prefer to become ensnared in the petty concerns and hostilities that are bred by an egocentric orientation? What would be the sum of your life when it came time to die?"

"That's true. But one does get discouraged by human nature."

"Yes. But all you have to do to escape that despair is find some example of great moral beauty."

"Is it still possible to find such examples? Haven't even saints been discredited by the psychoanalytical perspective?"

"One doesn't discredit moral beauty by explaining its psychological bases any more than one takes away the loveliness of a flower by explaining the chemistry by which it is derived from the rotten matter which has enriched its soil."

"But now all the flowers appear malformed, if not blighted."

"That's why we need to try to perfect our own conduct: when we set aside the view of reality which made us perceive each other in moral rather than psychological terms, we must find new reasons to maintain our faith that human beings are more than apes in trousers. So we ourselves are needed to provide examples, to serve as Christ *to each other*. We have to encourage each other, like lonely Volunteers out in the field. Because we're surrounded by people who—like ourselves—need to be shown that a great deal more is possible than our current mass culture implies.

"Otherwise the shallow cynicism generating that cultural content may indeed contain the seeds of our destruction as a civilization. That's especially so because the 'right to be selfish' bred during a period of privatized affluence must give way to altruistic norms which help people tolerate increasing deprivations. For populations grow, expectations of poorer nations and classes rise due to global mass communication, and at the same time non-renewable resources become scarcer and more expensive."

"Oh, you too see that we're heading into a long-term global crisis—one we can get through intact only by lowering our expectations for ourselves in order to provide for a decent future."

"Yes." Their eyes met, and she smiled and said softly, "Your thinking helped crystallize mine. The first time I heard you give a speech, I practically danced out of the hall."

Her eyes and voice dropped. "So that's why I prefer the old-fashioned models of behavior, even though their theological premises about the nature of human nature are being displaced by social-scientific thinking: they still demonstrate our potential beauty and dignity as human beings; they demonstrate patterns of conduct which will help us preserve our vitality and happiness and sanity as individuals and as a society. They preserve values which make surviving worth the struggle, even when they're eclectically reinterpreted within a new philosophical framework, a functionalist rationale dependent only upon a humanistic affirmation."

"Good Lord," he murmured, and began clapping. "Bravo, bravo." She tipped back her head and laughed quietly, as though she were accustomed to such reactions but did not think she had done anything spectacular.

"How do you keep producing statements like that? Even if the material is thoroughly familiar, ..."

"I have a sort of typewriter in my head that helps me keep track of what I'm saying. I probably developed it by growing up with my nose in a book. It's

46

certainly no more remarkable than the mind of a concert pianist who can go through a whole symphony without any sheet music in front of him."

He nodded.

"But back to the subject." She hesitated. "Do you think my thesis exaggerates the importance of the models for behavior provided by our commercially produced entertainments?" She required another substantive comment.

Not being at all prepared to be dismissed as a dunce by a girl of twenty-five, he made one more effort to keep up with her.

"Certainly not. Some ancient Greek said, 'Let me but write the songs of a people, and I care not who writes their laws.' What you've said cuts through a lot of debate over whether antisocial behavior may be inspired by entertainment or 'artistic expression'.

"But to get back to the *original* subject, were you saying that the traditional glorification of submissiveness in women requires acceptance of rape?"

"Of course not! If human dignity were truly held sacred, rape would be a capital offense. I could go on and on about how blasphemous it is.

"But you have to find some way to accept something terrible before it can drive you crazy. People avoid the self-destructiveness of helpless rage by perceiving themselves as martyrs because it's one way of enduring the unendurable."

"That's all very well," he said cautiously, "but I'm concerned about anybody who's so convinced she has an obligation to be walked on."

"I don't accept any such obligation. I'm too philosophically liberated to follow *any* rules without a practical reason which serves my values. I do generally let people walk on me, often because I want to have a particular edifying effect on them.

"In general, I find there are lots of benefits in living according to the old rules. Another is that it's educational to try to find ways to survive and accomplish things while adhering to them. For example, I'm now more articulate on certain subjects as a result of the conflict with my last employer. That was a period of madness. But I learned from it how to be more sane in the future."

She looked at her watch, so he looked at his.

"You're right," he said, "it's getting late.

"Anyway, I like your ideas about entertainment providing models that reinforce the better facets of human nature. Demoralization feeds on itself, and that *is* dangerous."

She nodded approvingly, and he felt that he had acquitted himself reasonably well.

But he had not found anything he could use to justify rejecting her as his assistant during the trip.

However, he had to go through the stack of reports Marjorie had left in the middle of his desk. He and the International Operations staff making arrangements for his November tour were moving into the day-to-day scheduling phase of what had turned out to be an exceptionally difficult coordination and planning operation. The Travel Office would be unhappy if there were any late requests for changes in the carefully constructed sequence of reservations required by his long itinerary, for the ambitiousness of his trip had presented them with unusual difficulties.

"I have to get busy now, Philanth. But I've enjoyed talking with you."

Getting up, she answered pleasantly, "It's been an honor, Mr. Bjorklund."

"Oh, by the way. You must be aware of the planning-and-review meetings held every Tuesday morning to prepare for my November trip."

"Indeed I am, sir."

"Well, why aren't you attending them? You of all people should always be there!"

"I thought that was pretty obvious, but Les said no, it 'would be inappropriate'. So I had to accept the idea that I was too unimportant, and not wanted."

"Well, Les seems to feel that way about all women."

"I'll be delighted to attend from now on."

"Then I'll see you at ten o'clock."

She nodded and slipped out, closing his door.

He could see, now, how she might have gotten into trouble during her Peace Corps service: her beliefs regarding self-sacrifice and perfection of personal conduct could have gotten out of hand when her mind had been impaired by illness.

Her mind surely wasn't impaired now. He liked the modest self-assurance with which she addressed him when he drew her out of her formal role as a low-level temporary clerk interacting with a presidential appointee. Indeed, she had taken on the manner of an academician holding forth about a favorite thesis. This amused him, for she must have acquired that behavior pattern from her scholarly parents.

It might actually be pleasant to have such a cheerful, thoughtful, and verbally gifted assistant, after she had become experienced in working for him. And Carole had a point: Marjorie was going to retire one day, and he would be badly inconvenienced if she had not thoroughly trained a replacement for herself.

However, regardless of her qualifications, he couldn't take her on the trip, for the problem of appearances remained. He had his reputation to consider. His career depended on it.

Early that afternoon he encountered a problem which gave him an idea. He told Marjorie to send in Philanth, then laid a folder in front of her.

"My speech writers haven't given me what I want in this. Try to come up with something fresh and effective."

She gave him a look of pleased surprise as she picked up the folder. She opened it to study the heading of the top page.

"Yes, sir. What time frame?"

The speech was not to be given for more than a week. But he had another sort of deadline built into the workings of the Travel Office downstairs, the State Department, and more than thirty foreign embassies representing nations celebrated for overbureaucratization.

"Right away."

She nodded and left.

About an hour later, while he was between appointments, Marjorie entered and wordlessly laid before him the folder containing the speech draft which he had assigned to Philanth. She left without waiting for any reaction from him. Since she and Carole were in frequent contact and she had brought this to him as if it were a top-priority matter, she must have surmised what he was up to.

He quickly found that his speech to a group of new Volunteers about to depart for in-country training had been nicely restructured and reworded so that it said much more effectively what the first writer probably had had in mind.

A page of new material had been inserted, bearing a little stick-on note to call it to his attention:

Another problem you may have is loneliness. When you confront the overwhelming mass of seemingly intractable, mutually reinforcing problems you are being sent to combat, you will be far from settings and people which

50

have always given you unnoticed comfort because of their familiarity. You may experience culture shock compounded by loneliness and frustration. You may begin to despair.

Let me suggest a way of avoiding this, or at least coping with it: keep in mind that all over the world there are people working alone, as you will be at times, for a better future for the human family. Each of us is part of that great network of hope and purpose. We can draw from each other the strength and courage we need in order to do our best work and find satisfaction in it.

So stay plugged into that network. Don't try to 'run on your own batteries' any more than you have to, because during prolonged hardships your own reserves can become dangerously depleted. Keep them charged as much as you can. Make as many new friends as you can, but also write to old friends and to other Volunteers, to whomever you feel close to, as often and as freely as you can. That's important. Never let yourself feel alone and powerless, because you are not. You are always part of something quietly magnificent.

"Schmaltz," he scribbled along the margin of this page. "Forget it."

He summoned Philanth and handed her the folder. "Any comments on my comments?"

She quickly scanned the pages. She looked sober, but she said apologetically, "I was *afraid* my advice was too motherly to sound right coming from you.

"But even if there's nothing in the new passage that can be useful to you, I think the clearer structure I imposed strengthens the rest of it. If you still

51

don't like it, I'd be more than happy to make a fresh start. If you could give me a little more guidance as to what you want?" She held out the folder.

"No, it'll do," he said, taking it and laying it aside. "Sit down, Miss Devon."

She obeyed, demonstrating how it was possible to sit down while remaining at attention.

"Please, sir, it's 'Ms.' or 'Mrs.', if you don't adopt the common usage of first names around here."

She was being tactful in saying "if" instead of "since". Probably thanks to Marjorie, it seemed to be common knowledge all over the headquarters that he disliked being addressed by his first name unless the speaker had an intimate relationship with him. He also did not want to be obliged to tolerate being called "Chuck" by a lot of people in their twenties. His preference for formality was a violation of Peace Corps tradition.

But that was not Philanth's point. "Oh yes, you're a widow," he said as if the notion were patently ridiculous. Anyone looking only at her face would doubt that she was old enough to be legally married.

He went on, "I'm considering you for a more visible position. Wouldn't you say you could stand to lose some weight?" That was hitting practically any woman where it hurt, and if he made a habit of that sort of talk it could get him into trouble. But with a deferential little thing like Philanth he felt he could get away with it just this once.

Philanth looked nonplussed. "Uh, no, sir," she answered faintly. "It's my fat cheeks that give that impression." She spread her fingers over the offending features. "And, uh . . ." She hung her head and made her hands into a ball in her lap. "And a, uh, a body suited t-t-to childbearing."

She rushed on, "I lost weight in Africa, below what

52

I weighed in high school, and when one of the girls in high school saw me in the locker room she said, 'Why, you've got a *cute* figure. You should show it off.' But I prefer to dress modestly.

"I've only now gotten my strength back after being sick, and I haven't gained enough to have excess around the middle. I just can't look skinny like it's fashionable to be. My—my bones aren't shaped right." She hung her head so that her pale hair draped forward over her shoulders. Her color had begun to rise; she looked rigid with embarrassment.

He rubbed his mouth to conceal his incredulous amusement, for he felt as though he had suddenly found himself dealing with an extraordinarily sheltered young virgin.

"Very well." He started to add, You're right, it's better to look fat than distractingly female. Instead he said, "You're right, it's better to be modest."

"I suppose this light color makes me look fat too," she added, smoothing her pink linen skirt, "but I'm trying to stretch my spring-summer office wardrobe through September because my wardrobe suitable for fall-winter office work in Washington is sparse; and I don't want to acquire clothes I won't need when I go back to Africa to live, probably in just a few months."

"All right.

"Now, . . . about your hair. I think it looks unprofessional to wear it so loose like that."

"I know I should wear it up, and I usually do," she answered quickly, lifting it off her shoulders with both hands, "but I was interrupted just as I was about to start dressing this morning; then my landlord's wife offered to drop me off at the Metro station if I could leave right away, and it was raining, so I threw myself together in order to accept her offer. I promise not to wear it this way again."

"What interrupted you?" He wanted to know about anything in her personal life which might interfere with her job performance.

Philanth began smiling irrepressibly. "My landlord's oldest child—she's thirteen, and very bossy—came to me to 'make' her little sister get out of bed. Her sister had been disciplined last night, and she couldn't bring herself to face the world. So I got her started. I'm like an honorary nanny: the children love me, and I love them."

"Think you'd like to have some like them, someday?"

Philanth's eyes danced. "I don't think I could manage that. Both their parents are black."

"Your landlord is a black man?"

"Yes, it's an all-black neighborhood. Except for me."

"Oh. Well, of course I meant, do you want some little girls of your own someday?"

"No. There are far too many children being produced as it is. It's the nurturing that's in short supply—*terribly* short supply. So I've always believed I should devote myself to making things better for *other people's* children."

"Uh." This was not going as he wanted it to. He toyed with a pen and studied her with distaste, thinking that all that pink and white made her look like she belonged in a dish in an ice cream parlor rather than in an office.

She simply waited, rigidly erect with hands clasped on her lap, the image of respectful docility.

He looked at his watch and consulted his day's schedule. He had so many important things to do, and he was wasting time on this! Not letting Philanth know why he was suddenly showing so much interest in her was putting him in a bad light, too, and that was always risky.

Suddenly running out of patience with the whole situation, he decided he really was too busy to take the business of disqualifying Philanth any further under these conditions. There would be a cost to her feelings when she was rejected, but letting her know that she was being tested would allow him to put the process on a basis which would give him protection as well as being more efficient.

"Ms. Devon, how do you feel about taking on a lot more work?"

"I'd be glad to do whatever I can."

"You've been helping with preparations for the 'world tour' I'll be on during November. Marjorie doesn't want to go, so I'm looking at possible substitutions. Carole has suggested that I might take you."

To his surprise, her face fell in disbelief. "I-I might go 'round the world with you?" she stammered.

He nodded, closing his eyes to agree that the idea took some getting used to.

"Ohhh . . . . Why doesn't Marjorie want to go?"

"Too hard a trip for her."

"So Carole suggested me? Well, . . . I guess she must've had her reasons, ..."

He laughed, pleased by her lack of presumption. "You don't seem to think they can have been very good ones."

"Well, I guess the idea would be to take the cheapest person available who knows what it's all about, and leave the experienced staff in place to take care of matters properly during your long absence."

"Right.

"Just the same, I'm sure that I'm going to make other arrangements. I need someone much more seasoned. But I have to raise the possibility with you so you know that you're being evaluated." This would get him off the hook with Carole. "*Do* you want to be considered? Are you physically up to it?"

"Oh," she breathed, bobbing on the chair, "yes, yes, yes!"

"Are you quite sure you're up to a lot of Third World travel? You've had recent health problems."

"I'll be stronger by November, and I'm a hard worker, sir, everyone has always said so. And I enjoy a challenge. And—"

"Okay, fine. Now, if you'll excuse me."

She jumped to her feet. "Yes, sir. Thank you, sir. And I promise not to get my hopes up."

Philanth started for the door, but the fact that she was taking some sensitive new information out of his private office made him say, "Wait." He motioned for her to come back.

"You must understand, Ms. Devon, that until I announce to the staff that someone else is going with me instead of Marjorie, only those directly involved are to know that a substitution is even under consideration. It has to be handled carefully, for your sake. Because of your inexperience I must remain free to make other arrangements. Is that quite clear?"

"Yes, sir," she chirped.

"I suppose you think you're going to be working for an ogre," he remarked, "—if you go."

"I don't mind a demanding person as long as I'm not encouraging sadism by acceding to him," she replied artlessly. "*You* could never be guilty of that, never let your own unhappiness twist you so."

He answered dryly, "I appreciate the assurance."

Chapter 3

TONY

It was Sunday morning. He had slept late and awakened naturally. He thought, But I'm still tired.

He felt that he could stay in bed all day, and it wouldn't help much. I'm getting old, he thought, that's what it is.

He had turned forty the previous month, but he felt much older than that in many ways. He found changes in his ways of feeling and perceiving: more and more there seemed to be a colorless sameness in the world, a repetitiveness and empty futility.

What really disturbed him, though, was that in his interior world of ideas, which still mattered, there was more and more a leeching away of vitality. He was tired in his soul; tired all the way through.

He did not need an article in *The Washingtonian* to tell him that this was a symptom of burnout. But it was not burnout from his work. It was burnout from life.

He blamed himself. His easy youth had given him a grateful desire to help "the less fortunate", but it had given him too little understanding of how difficult it would be to do that so as to have wide or lasting impact. He was, as Carole had put it, too easily dismayed by overwhelming obstacles.

He had replied that he had a bad habit of looking at "the big picture". Her "advice" for protecting his drive to succeed had been to cultivate mindlessness.

At least he did not regret his choice of profession. If he had wanted to build or renovate houses he would have trained as an architect. Because he wanted to do the same sort of work on the structures of society he had studied law.

But altering by even a minuscule amount the operations through which government shaped society required enormous resources. Gaining access to those resources without the quick start of a well-connected family or a large, inherited fortune required time—daunting amounts of time, for anyone who understood that constructive ends normally cannot justify destructive means because means have consequences of their own. Undertaken on such scrupulous terms, even the longest life was terrifyingly short.

Those obstacles were discouraging enough, when one saw so much urgent business remaining on the human agenda. What made them really demoralizing was that even as he worked to gain access to resources more rich in the ever-dangling promise of "making a difference", his faith in action itself was being dissipated: as he gained useful experience he became weighed down by his awareness of the intractability of both things and people. He understood too much now, and that made him hope too little.

Soon, perhaps, it really would become possible for him to do significant things. Time was catching up with the California senator whose term would expire in two years. That seat would be winnable; if the senator had the sense not to seek re-election it would be up for grabs.

But he too was aging, losing his youthful stamina. Already he was moving down the other side of the hill, being dragged toward the end of all endeavor.

Yet his life thus far had only been preparation!

That irony mocked him like a grinning skull. "Accomplishment", it laughed. "You glimpse it first just as you absorb the reality of your own aging; as you draw closer you attain to the prematurely paralyzing vision of its elusiveness; and just as you seem about to grasp it, the trap door opens under you. This is mortality. Relax. Accept."

But "As you get older, being good-looking becomes ridiculous unless you've ignored your looks and piled up solid accomplishment. That takes everything you've got—plus hard work." His father had given him that news in a note tucked around a new wristwatch on his fourteenth birthday.

He had never forgotten it. Now, seeing his "good looks" beginning to be creased by decades of effort and caring, he had to ignore the grinning skull's counsel and listen to his father: since he was getting older, he had to try harder than ever to feel hardworking in order to avoid feeling ridiculous. A man could at least choose which way to be absurd.

He got up, resolved to spend his Sunday getting through a lot of paperwork. He liked to spend his Sundays reading, and he had plenty of studying to do before his travels in November. Some of that study would be his reward once he had done the routine paperwork which had been piling up at the office.

The reasons he had fallen behind were depressing. On Wednesday, for example, he had been obliged to mediate a tiff between the deputy director and the chief of staff over which of them had priority when they wanted to use the Peace Corps car and driver at the same time. It had been inordinately time-consuming because both men had such ego problems.

On Thursday he had had a long meeting with Sandy, his fresh-faced young ombudsman. In addition to other problems discussed, Sandy had given him a demoralizing report regarding problems in the staff of Public Relations.

The stupidity and bitchiness of the political appointee who was Public Relations director was a constant source of trouble for him anyway: of extra work because Courtney Simmons didn't really know much about media, and of heartaches because that office could have done so much more.

On Thursday Sandy had reported that the director had fired a seasoned newspaperwoman, apparently just for not being a toady. This was a significant loss to Peace Corps even though Courtney had wasted her on mindless chores which should have been given to an intern or not done at all. He asked Sandy to let the journalist know that he would serve as her reference. Being past fifty, she would need a strong one.

Recruitment for the position was already underway, and Sandy had learned of the firing through another complaint: he had received an anonymous tip that the director had rejected an outstanding potential replacement on the ground that she was overqualified. The director refused to acknowledge the obvious fact that because of its unique character Peace Corps routinely got excellent service from lots of people who were overqualified for their jobs.

It wasn't hard to figure out why Courtney took that position: one only had to look at her own qualifications, or rather the lack of them. The rejected applicant could have quickly worked her way up to become an outstanding director of Public Affairs if that position had not been filled by political appointment; so the holder of that position had seen to it that the applicant never had a chance when she showed up for the interview stage. Again there had been a significant loss to Peace Corps.

But he was in no position to initiate open conflict with Courtney, since she held her plum job because she and her husband enjoyed the favor of the White House Personnel Office thanks to political and/or personal connections.

Also on Thursday there had been a meeting about Peace Corps' wrenching current budget crisis.

In the afternoon he had attended another funeral for a Volunteer. This time there had been a car accident on a mountain road. Two other Volunteers had been seriously injured.

On Friday afternoon he had had to make a phone call to inform a Volunteer's parents of her death. It had been even more difficult than usual. The young woman had been an only child, and her mother had taken the news so hard that he wished he could have arranged to tell the parents in person. Their daughter had been stationed in a remote village, but she had gone into Manila on routine errands and gotten shot when a bus was robbed. He could still hear the broken voice of the woman at the other end of the line, telling him what an outstanding student her daughter had been and how she had always wanted to do things for others ever since she had been a child; and how much she had loved the Peace Corps.

Then, after quickly apologizing because her husband was unavailable to talk to him and thanking him for calling, the woman had scarcely managed to articulate a "Good-by" before hanging up.

Her heartbreak had stayed with him as he left Washington for a twenty-four-hour trip to Vermont. After addressing a reunion of Returned Volunteers at the School for International Training and fulfilling a few media engagements while in the area, he had returned to Washington the previous evening.

Upon returning home he had been notified that a Peace Corps country headquarters had been bombed by someone who accused Peace Corps of being a front for the CIA. The explosion had occurred when the offices had been unoccupied because of the weekend, so there had been no injuries. But they had just received their new little desk-top computer from Peace Corps' Information Resources Management office, and the country director was distraught over its loss. The computer had been donated by the manufacturer, but the IRM office had not really had the funds available to ship it, much less replace it.

Only two days earlier IRM had been among the offices where Les had been making the office directors and deputy directors jump through hoops of budgetary craziness. Those offices had not been allowed to ask for enough funding to meet their fixed minimum operating expenses to begin with, so they had been spending huge amounts of time trying to cope with all kinds of difficulties throughout the fiscal year.

It was fascinating to compare what little Les had casually told him about that episode with the hair-raising accounts Sandy had gotten from the financial analysts whose office heads had been put on the hot seat. But as usual he had not had any time to react to what he had learned.

The coming week promised to be just as busy, so he was anxious to catch up on his backlog of reports, memos, drafts, and correspondence. However, as he ate breakfast and settled down to work, Carole kept exclaiming over the "perfect October weather". She finally twisted his arm so hard that he agreed to "take a few minutes off just to whiz over to Rock Creek Park for a picnic" since he had to eat lunch "anyway", and he really ought to "take time to smell the flowers, or rather to look at the leaves."

He had no sooner given in on that than Carole suggested it would be a good chance for him to "get better acquainted with Philanth in the less formal sort of circumstances you'll have during the trip."

Carole was being deliberately obtuse: she had to know that no amount of getting better acquainted would end his opposition to the idea of taking that girl on his long trip. Besides, he needed peace and quiet on his one precious day at home.

"It's pretty short notice. She'll probably have other plans. Besides, you said we'd just take a few minutes extra to—"

"No," Carole answered blithely, "I told her to keep the time uncommitted, just in case. She's crazy about autumn foliage, and she's never been into Rock Creek Park, since she doesn't have a car. So she was trying not to sound too hopeful that it would be possible for us to go, and take her along. She'll bring her share of the food. She insisted. So it's less work for me if we take her, and we'll get a better meal."

"Oh." Outmaneuvered again.

Carole phoned and told Philanth they would pick her up at her place, but Philanth insisted on meeting them near the Woodley Park Metro Station, which they could go past on their way into the park. As Carole hung up she said, "Philanth doesn't want you to see where she lives. And if she thinks she can spare somebody else ten minutes of inconvenience she'll go to twice that amount of trouble to do it."

"Wonderful." He was emphatically uninterested in Philanth's shining virtues.

After dealing with a number of fairly straightforward matters he paused to contemplate a memo he had had prepared for his signature by a team in Financial Management. It notified all associate directors and office heads that stringent budget conditions would continue, and indeed that some overseas posts would lose money already given to them. It noted that the associate directors would meet on the following Friday to see whether any money could be redistributed.

He was wondering whether the memo wasn't too

cryptic. He had suggested that this memo include some explanation of the reasons for the budgetary problems, but the experts in Financial Management had objected that that could invite trouble. They had preferred to send someone around to brief the senior staff of each office at their regular meetings.

The briefing was to explain that one cause of the current severe hardship was that Peace Corps was on a long-term expansionary program mandated by Congress, but Congress' appropriation for the current fiscal year had not accommodated that expansion. Thus Peace Corps had placed in the field the numbers of Volunteers called for by the expansion plan before receiving an appropriation nine million dollars less than requested in its budget.

Congress' recurring difficulties in getting any budget package passed in a timely fashion aggravated the difficulty of avoiding this sort of problem. It would have arisen anyway because most Volunteers were put into the field in late summer, which was late in the fiscal year, so the full impact of their costs was first felt in the next fiscal year. For example, partly due to the net increase in Volunteers, unemployment and social security allotments had increased by $900,000 in one year.

Since the Volunteers in the field had to be supported, the cuts to support the Volunteers had been taken mostly out of Management. Thus Management had had to cut its budget more than $700,000 below its level for the previous year.

IRM, within Management but having the support of Volunteers in the field as one of its major functions, was particularly hard hit. That office had to meet ongoing service commitments regardless of how much money it was told it would be allowed to spend. Aside from the fixed costs of maintaining the mainframe computer and keeping the desk-top computers all over the headquarters functioning and providing software specially adapted to myriad tasks, there was no point in getting new desk-top computers donated by a manufacturer and then not spending the money to ship them to the country posts and send someone from post to post to install them and teach the local staff how to use them.

Meanwhile the older models had to be kept operational, for a post needed its own desk-top computer for all the record-keeping which the federal government required the country staff to do regarding every aspect of Peace Corps activities in their country. Everything had to be properly accounted for, since there was always somebody who thought it was cute to turn U.S. government property to his own use.

The fiscal starvation was much more difficult to cope with because of the extreme degree of budgetary compartmentalization and inflexibility within Peace Corps. There had to be something fundamentally wrong with a budget system which itself caused so many grotesque difficulties.

As for the next fiscal year, Peace Corps had been given what OMB said was a maintenance rate, not a growth rate, and had been told not to appeal. So this Mad Hatter's Tea Party would go on and on.

He had to put up with this collective insanity until and unless he could get the entire budgetary apparatus put onto a more manageable footing. He was not sure that was do-able, since there seemed to be no resources of appropriate expertise available to Peace Corps which could realistically undertake such a herculean task. Peace Corps existed in a perpetual Catch-22. The career bureaucrats at Peace Corps with whom he had discussed the quagmire had no solutions to offer, saying only that over the generations

Congress had created an impenetrable thicket of regulations which had to be followed.

His minimizing the costs of his November trip by making lots of stopovers could not ease the fiscal distress at all, because his expenses were budgetarily separated from all the operations which were going bankrupt. This protection of his own expenditures was not remarkable, since the congressional oversight committees urged him to travel, travel, travel.

Thus—free from the constraints which had obliterated the tiny travel budget of IRM—he was preparing to go off on another tour which should promote international good will for Peace Corps and by extension the United States. Congress generously supported publicity while letting underlying realities starve. It also carefully protected its own claims regarding the educational value of congressmen's travels at taxpayers' expense.

"Time to go, dear," Carole was cheerfully calling up the stairs to him.

He reluctantly initialed the memo from Financial Management and laid it on the pile for his "out" box.

He picked up the next item for a quick look, in case it was something he could be mulling over during this silly picnic. It was a compendium of proposals for improving recruitment of Volunteers with minority-group ethnicity and/or technical skills. He had requested that it be drawn up because Volunteers in these categories were most needed and because there had been all sorts of interesting, untried ideas bouncing around on this for years among Peace Corps people and oversight committee staff on the Hill. The report brought together a lot of them and tried to explore feasibility and methods of implementation. He dashed off a note instructing Marjorie to have a memo drafted to express his approval of the effort and to ask that the report be circulated for comments of interested parties.

"CHAR-ules!" Carole was hollering from the foot of the stairs, not angrily but teasingly.

"Coming, coming!" He got up without letting go of the paper. In a moment of desperation he thrust it into the "out" pile, leaving Marjorie to decide which recipients of the memo he had had in mind.

No, she would have to ask him, he thought as he hurried downstairs. He would have the authors of the report decide. That might seem stupid, but he badly needed to delegate anything he could.

At the foot of the stairs Carole started unbuttoning his plaid flannel shirt and made him replace it with a beige polo sweater and brown tweed jacket. He thought that was very silly, but this was one of those issues on which it made more sense to comply than to argue. At least she had not tried to get him to shave just for Philanth's benefit. Carole was adamant about his "image", and that meant appropriate dress for all occasions. He didn't even *have* a polo sweater or tweed jacket in Los Angeles, where they might be thought pretentious.

As Carole helped him into the jacket he decided he would try to develop some ideas for his speech to a conference on international development at some university at the end of the week. He had the impression that it was in Florida. Yes, with a stopover for a speech or something in Atlanta. Wherever his major engagements took him, he insisted that Public Affairs always have lined up some local media events to flesh out a worth-while trip.

Carole handed him his wallet and car keys, then led the way through the dining area into the small kitchen which filled the back of the narrow town house. He picked up the carton box she indicated

67

and followed her outside. The expensive picnic basket someone had given them probably was languishing in the basement because Carole couldn't be bothered to locate it.

As he waited for her to lock the back door he smelled warm baked beans, and his mouth watered. Like syrup and biscuits, this was another of his favorites which his dietician normally refused to serve. Carole's list of sinful treats was interminable.

The World Food Day symposium was this week. It would be at the Department of Agriculture.

He had to "say a few words" at a late-afternoon ceremony at a hotel, an awards program sponsored by US AID to coincide with World Food Day. The Presidential End Hunger Award was going to go to Peace Corps again. There also would be a convocation at National Cathedral and exhibits at Peace Corps headquarters as well as at the Department of Agriculture.

But the item of concern for him was the symposium. He must see how that speech was coming along.

He put his box next to Carole's packages in the trunk, shut it, and got into the driver's seat.

If the speech was not good enough he might be able to borrow some of the speech for the conference in Florida, because his best writer was working on that one. And he must check on who had the speech for accepting the AID award. The writer handling that had quit to go to work for a news service.

He certainly needed another speech writer. The speeches he was getting were really uninspired. One day he had read one cold to a large gathering of staff, and it had been so ridiculous that he had interrupted himself to exclaim, "Who wrote this, anyway?"

His writers must have become used up because he spoke to so many different groups. Committees on the Hill, graduating classes, meetings for families of Volunteers, business clubs, senior-citizen organizations, civic groups . . . . There was no end to it. He had to avoid saying the same things too often, since such events were written up for the newspapers. There also were all the articles and letters for publications and personal notes and materials to feed interviewers that he was responsible for producing.

Carole had wanted this post for him so he would get more good national exposure, and he certainly was getting it—in small doses, but almost all of it pretty favorable. So far. But surely it took a genuine dunce to blow the public-relations aspect of being Peace Corps director.

Carole had taken her place beside him and pressed the garage door opener, so he started the car.

But the object of all that publicity was to recruit and promote. Recruit, recruit, promote, promote—even though we have to turn away qualified applicants by the thousands and turn down countries' sound requests for new programs because we don't have the money.

Because we've got to keep Peace Corps from being starved into oblivion and then killed with the claim that "there's no interest in it anymore". Returned Volunteers, you're supposed to be an informed and growing constituency, and we've had trouble even scrounging the money to keep in touch with you so you can help us tell Congress to let us promote the goals for which you gave years of your lives!

"Turn right," Carole said crossly. He put the car into gear and drove out of the garage into the alley. Carole pressed the garage door control unit, and they were on their way: turn right, turn left.

Philanth seemed prepared to be delighted by everything that day: she was ecstatic over the crisp, sunny

69

weather and leaf-scented air and autumnal scenery, she loved Rock Creek Park's steeply sloping wooded hillsides and long, winding drives, she was even pleased by the "thoughtfulness" with which a full range of picnic facilities had been provided in the meadow where he finally chose a parking space. It was no wonder he had at first taken her for an idiot: no sensible person could be that happy. Carole, beside him in the front seat and turning to smile back at Philanth, caught him making a face to himself caricaturing the girl's buoyancy and surreptitiously punched him.

As Philanth and Carole laid out the lunch Carole remarked that Philanth "shouldn't have gone to so much trouble" over the fruit salad she had brought as part of her share of the menu, and Philanth gushed over how much she appreciated being invited. They all settled down to eat.

"Have you met Charles' bodyguard yet?" Carole asked Philanth.

"Don't call him that!" he snapped.

"You mean the aide/photographer who'll be accompanying him on the five-week trip?" Philanth asked, and Carole nodded, chewing. "He started working with me last Monday. I knew him a little bit before that because he's been working in IO, International Operations, in the Africa Region. His name's Stuart Smith."

"D'you think you and he will work well together under the conditions of the long trip?"

"Oh," Philanth said, chuckling, "I'll—I'd have to be careful we didn't get along *too* well. I'm haunted by his grin, and we've had only superficial contacts. It's a wonder he wasn't stolen when he was a baby, he's *so cute*."

Carole laughed.

Philanth rattled on, "At first he seemed very somber and uncommunicative, but he's really charming once you get to know him: funny, and very smart, and unpretentious in spite of having tremendous natural dignity. Quite sophisticated, yet without conceit. And as I say, I find his grin irresistible.

"He has a white girl friend who also works at headquarters. She's obviously 'gone on him', and it's not hard to figure out why. I suspect he's standoffish toward her because of his race, or she'd have gotten him to the altar by now."

"So they did get a black man for the position?" Philanth nodded. "That's good."

Carole began quizzing Philanth about her work on the travel arrangements. Carole persisted in referring to the traveling party as "you" while Philanth tried to talk in terms of "they", with the result that Philanth kept getting tangled up in her pronouns.

When the pattern was clear Carole turned to him and said, "Give her credit for taking nothing for granted, Charles. But not knowing whether you're really taking her is wasting effort needed on other things."

"Double-think isn't hard," Philanth protested, "and I'm doing the best job I can regardless of who gets to go."

"Of course," Carole said sympathetically, "but you'll be disappointed if you're left behind, won't you?"

"I don't hope more than a little, so I'd be only a little disappointed."

"But how many times in your life will you have a chance to take a trip around the world, stopping in so many interesting places and being guided and looked after and staying in comfortable hotels?"

"Never, of course! But if I don't get it someone else will. The benefits would be less for someone

71

who's traveled more than I have, but that person might be more helpful to Mr. Bjorklund because of their greater prior experience."

The girl finished with forced cheerfulness, "And I have to admit I may not really *need* all that wonderful experience of other places if I'm settling into the local problems in North Africa."

To change the subject he interjected, "You're committed to going back to North Africa?"

"As soon as I can find enough of a job to get the necessary visa." Philanth made no secret of her anxiety on this point. "With my languages and experience it seems the best choice. I'm not cut out for an equatorial climate, and I'm reasonably able to fit into a woman's role in a Muslim culture. I'm a retiring person."

"Why are you so willing to recede into the background?" Carole asked.

"I got so much attention when I was in school that by the time I got to college I just wanted to lose myself in the crowd. I was tired of the resentment of others who felt I made them look bad."

"Fear of jealousy is hardly adequate reason for underachievement," Carole objected.

"A person can do a great deal of good without ever being regarded as an important person," Philanth countered. "Of course, being respected does give a fantastic boost to your ability to get things done, but the moment you poke your head above the level of the crowd there's always someone wanting to cut you down. So you waste time by being on the defensive."

"Maybe so," Carole said, "but you have to learn to accept being pecked at because of your achievements." Turning to him, she added, "Isn't that right, Congressman?"

"That's what you have to do," he agreed.

"How?" Philanth asked him.

He put his arm around Carole's shoulders as he answered, "It helps to be with someone who thinks you're wonderful."

Carole looked displeased and shrugged off the contact with a movement too slight for Philanth to see. For a few moments no one said anything.

He had had enough of this chatter. "I hate to break this up, Carole, but you did say I could get right back to work, ..."

Carole gave him a reproachful scowl, but she began repackaging the uneaten food and collecting the trash.

Philanth began hurriedly trying to scrape the last traces of her yogurt out of its little carton.

"Sorry, I didn't realize you hadn't finished," he apologized.

"Oh, that's all right. I always try to taste every molecule I eat, and that makes me slow. It bothers me when I don't get the most I can out of everything." She threw him a smile which flicked on her largest dimples. "As long as you're alive you may as well try to make the most of every minute of it, right?"

Philanth quickly replaced the lid on her carton, but when Carole tried to take it from her for the sack of trash Philanth withheld it, saying, "I save these for packing my lunch and various uses around the house."

"Oh, fine," Carole said smoothly, starting to fish through her bag of trash. "You'll want the one I emptied too?"

"Thank you, if it's no trouble," Philanth replied meekly. "I usually buy the bigger cartons because it's cheaper, so I don't have enough of the little ones yet." The assorted flavors of fruit yogurt had been one of Philanth's contributions.

Philanth took the three empty cartons and replaced them in a ratty-looking zippered canvas-and-vinyl bag covered with an atrocious red-and-yellow floral pattern.

Seeing him staring at Philanth's bag, Carole remarked to her, "That, uh, valise doesn't look like something reflecting *your* taste. For one thing, you like peacock colors."

"I got it second-hand, very cheap. It doesn't mind having food spilled inside, and it was easy to carry on the Metro. Giving aesthetics high priority wastes resources needed for other things."

Carole nodded, flicking him a glance, and finished repacking their picnic gear.

"That was a virtuous lunch." His tone was neutral, but Carole would understand that he was complaining.

Carole lifted one eyebrow and reached for her handbag. From it she drew a small gold candy box with a stylishly flattened red cloth bow on it.

"Here's some sin to top it off."

He lifted the lid and set aside a chart which lay under it. The box was full of the largest, most strangely decorated truffles he had ever seen—except for the box like this one which Carole's brother had given them the previous Christmas.

"From Ken?"

Carole nodded, her eyes twinkling. He took the huge, perfectly formed truffle decorated with only a dot of pink icing which made it suggest a woman's breast in the most opulent sex fantasy.

Remembering his manners too late, he passed the box to Philanth.

"What was the occasion this time?" he asked before starting to suck on his chocolate.

"A peace offering." Carole's tone was sardonic. "He'd bawled me out before he left town, but by the time he—"

"What for?"

"—got back to San Francisco he'd cooled off enough to realize that his little disaster wasn't entirely my doing."

"What 'little disaster'?"

"Well, ..." Carole flicked a look in Philanth's direction. "I suppose it's confidential."

He too glanced at Philanth, who seemed absorbed in comparing the candy and the candy chart. "Oh."

He dismissed the matter and carefully bit into his truffle.

"What's Grand Marnier?" Philanth murmured. "Well," she went on as she extracted a piece gingerly, "if this is as good as the same-flavored petit four I once had, I'll have to take all afternoon to do it justice."

She passed the box to Carole with thanks, then added, "That petit four gave a new meaning to the word 'cake': I carried it around until the exquisite bittersweet chocolate icing melted in the warmth of my hand and I had to lick it off the napkin. I studied the contents, and I'll never forget it. There was even a layer of brownie in it, as well as one of Bavarian cream filling. And on top it had 'Grand Marnier' in Gothic lettering on an icing medallion that looked just like fancy printed paper. Sampling the foods of the elite certainly opens up new horizons."

"How did you get this once-in-a-lifetime experience of the foods of the elite?" Carole appeared to be enjoying Philanth as if she were a precocious child.

"At a luncheon for lobbyists sponsored by the World Hunger Foundation, at a big hotel just off Embassy Row."

75

Philanth carefully took a dainty bite off one edge of her chocolate. Despite her caution, the viscous filling came away after her mouth, the parting as sensuously prolonged as slowed-action film.

"Oh, dear," Philanth murmured, and when she saw that he and Carole were watching her with knowing smirks, she dimpled and dropped her eyes even as she carefully licked her lips.

"I'd never let anyone photograph me eating one of these," Carole remarked, picking out a truffle covered with tiny pink and green swirls. They all ate slowly.

"You're closing your eyes in order to concentrate," Carole observed.

"Sorry!" Philanth chuckled embarrassedly. "Knowing you've eaten something isn't much, so you have to get everything out of a sensual pleasure at the moment it's happening. It's always over so quickly."

"Yes." Carole kept her eyes on Philanth's face as she added, "I guess everyone has trouble following that principle even when making love, though. The mind always wanders."

Philanth leaned over to put the lid back on the candy box, then took it off again to replace the chart and closed it again. "The mind is supposed to be able to stay focussed, if given special training. Religious adepts work at that."

"I don't suppose it's much of a problem for *you*," Carole said, still watching her young friend. "You surely believe in getting the most out of that activity as well."

Philanth's mouth tightened, but still her lips trembled. She answered distantly, "Getting the most out of every moment of life, that's the ideal."

"Mm. You're a sensuous woman, too."

Philanth's mouth began working nervously. She threw a focussed look at Carole and then glanced

76

around the meadow as though seeking a path of escape.

"I've written poetry celebrating the lovely things of life," Philanth managed, gazing into the red and yellow foliage of the nearest trees. "Like rain and snow and the effects on plants as seasons change. I wrote a poem about autumn—"

Carole laughed and interrupted, "Why won't you talk about sex in front of Charles? He's old enough."

"*Please*, Carole!" Philanth burst out.

"My dear girl, grownups are relaxed enough about it to have a little fun with it. As witness that chocolate I knew Charles would take."

"I took it to get it out of sight, same as you took that weird one to get it out of its misery," he protested, but Carole only laughed at him.

"A lot of such fooling around degrades people, especially women," Philanth objected, "and degrades something so special it's interwoven with great religious traditions, the Judaeo-Christian being a notable exception."

"Pooh. You're candid enough about it at other times. Aren't you being hypocritical?"

Philanth looked reproachful. "I'm trying to avoid compelling your husband to listen to things he has no wish to hear. If you really think I'm at fault, tell me how."

Carole waved away the challenge. "I have yet to find fault with you," she told Philanth warmly. "But you admit you have a sensual nature."

Philanth looked pained.

"Don't you?"

"Yes, yes! I'll admit to anything, only please show some regard for my professional relationship with Mr. Bjorklund!"

"You're too easy to bait, Philanth. You should

stand up for yourself more."

"I don't mind being mistreated as long as I can meet someone's needs in a way that could have salubrious long-term effects."

"What needs did you hope to meet just now?"

"Whenever *you* poke at me I know it's in the course of scientific inquiry. You're always studying people. It no doubt contributes to your formal work in all sorts of ways. And I know you mean no harm."

"I should think not!" Carole agreed tartly, and reached over to return the candy box to her purse.

After they had let Philanth out at the same intersection where they had picked her up Carole said encouragingly, "Well? What do you think of her?"

"She talks too much."

"Charles," Carole said sharply, "That's thoughtless and unreasonable. She's not a moron; she's aware that you need to get to know her as quickly as possible before you commit yourself to taking her. So she was doing her sweet little best to cooperate. You certainly made no effort to make her task easier."

"Okay, so she's not stupid. But she *is* a *prig*."

"Then you can feel perfectly safe in traveling with her, can't you!"

That must be why Carole had subjected him and Philanth to that tasteless grilling about Philanth's "sensuality": to bring out that side of her personality. If Carole would go that far she must be determined indeed. He wasn't giving up, but he would have to keep from being worn down while he watched for an "out".

Two days later his pink "While you were out" slips included a message from Stuart Smith: "Arm broken while practicing for State Department's security training course. Hopes there'll be no problem in lining up a replacement for such a desirable opportunity to go around the world." He cursed. He had Marjorie get Stuart on the line, confirmed that Stuart could not possibly perform his duties during the trip, and conveyed his heartfelt regrets.

The moment he hung up he summoned Marjorie to set the wheels in motion to find a replacement. He was plenty upset, for he certainly wanted a black man traveling all over Africa with him, and it was quite important that the man be both a former Peace Corps Volunteer and handy with a camera. Of course, he also had to be personable. And State wanted the aide to have had combat experience or some equivalent evidence of fitness for security training. Stuart's deficiency in that regard had caused official displeasure, which could be the cause of the unsupervised practice that had caused the accident.

"One of the other applicants interviewed for the position was also a black man," Marjorie told him under her breath.

"Hallelujah!" He had tinkered with the wording of the position posting in the hope of encouraging black men to apply.

Having been forced by scheduling pressures to delegate to Jeff and his choice of panelists the tedious and exacting business of interviewing all of the candidates selected by Personnel, he got a qualified "OK" from him and told Marjorie to tell Personnel to offer the job to that candidate at once. They would be lucky indeed if the man were still available.

Just as he left for the day Marjorie told him that the offer had been accepted a few minutes earlier. However, due to some sort of procedural requirements in Personnel the new man could not report for work until the following Monday morning. He barely managed not to swear out loud as he turned to go.

When they were introduced nearly six days later he was pleased to see that Tony Hall was even better-looking than Stuart Smith. Besides being clean-cut, he was tall, broad-shouldered, and muscular. He would look just fine in black tie.

When he had a little free time at the end of his afternoon schedule he sent for Tony so they could become acquainted. As soon as they had gotten past some male-bonding rituals involving Tony's having played fullback for the University of Illinois he sat Tony down and asked whether he had any questions about the job.

"Philanth's been explaining things to me all day," Tony said in his deep, genial voice, "but there's one question only you have the answer to: Why are you still keeping her dangling about whether she's going or not? I know it's hush-hush that Marjorie wants Philanth to go in her place, but what's the problem?"

"I'm not sure Philanth's equal to the responsibilities. She's too inexperienced."

"That's what she told me. But what other choice have you got?"

"I think Marjorie will do it if I really want her to. It's not something you should be concerned about."

"Well, I *am* concerned, because if this trip's going to be as hard as everbody keeps saying, it'll sure be better if Philanth's along."

"Not if she isn't on top of every detail well enough to prevent foul-ups!"

"No, look at it this way: there're *going* to be foul-ups. It's in the nature of the trip. So who'd you rather have around to help you cope with the aggravations, Philanth or Marjorie?"

"Marjorie's a very nice lady."

"Sure she is. But let me put it to you like this: if you've had a really rotten day, which would you want to have around: a sweet li'l former Volunteer who's ready for anything, or your maiden aunt, who didn't want to be on this safari in the first place . . . and thinks johns that don't flush are 'an abomination'?"

He shook his head, gathering his dislike of this line of argument.

"Think about it! Imagine you been on the road with rough conditions for weeks. You're exhausted and stressed out. The weather's like a sauna, and there doesn't seem to be good air conditioning *any*-where. You're also sick from the local bacteria, and I mean *miserable*. Then you get trapped for hours in some suffocating metal shed called an air terminal by some delay nobody could've anticipated. The phones don't work, so there's no way to talk to the people at the other end of your flight; so your schedule full of foreign V.I.P.'s gets blown to hell.

"Now, Marjorie's 'experience' is in clean, air-conditioned offices where she's treated like the power behind the throne. What are weeks of hard travel going to do to *her* disposition? Some 'nice' people get super cranky when they're denied their comforts for a while, if they're not used to rough conditions. And, man! Havin somebody like that around when *nobo*-dy's exactly havin fun can drive everbody else up the wall!"

He had to respect Tony's quick grasp of an aspect of his plans he had not considered. But he had no wish to address this attack on his fiercely loyal long-time secretary.

"Could it be that *you'd* rather make the trip with Philanth rather than Marjorie?"

Tony snickered and sat back, lifting one ankle onto the other knee, as though deciding that the director was a regular guy after all. "Actually, yes, I expect we'll have some good times together when we're off duty. Philanth says we're going to be in some capital cities with lots to offer: interesting food, shopping, local sights . . . ."

He didn't consider a bodyguard with multiple other responsibilities ever off-duty. He began to appreciate Carole's assurance that Philanth could be counted on for total dedication to her own responsibilities during the trip.

He let the silence extend until Tony added, "It's sort of inevitable, isn't it? Philanth and I are about the same age, unattached, . . . "

Still he said nothing, only shifted while regarding the young man thoughtfully.

A little unsettled, Tony finished, "Anything wrong with that?"

"Yes. For one thing, Philanth is probably just as strait-laced as Marjorie."

Tony exploded with ridicule. "Naw! I know the type. She's the nurse in the bush clinic, the teacher in the thatched-roof school for pore little black chillun. 'Long as you say, 'Yes, ma'am' and 'No, ma'am' and dust the erasers when she wants you to, she'll wind up doin anything you want. We'll have fun."

He homed in with careful blandness. "I don't understand. If she seems that dedicated, how can you expect to 'have fun' with her?"

"Oh, but she's not limited the ways women that type usu'ly are." Tony was unbending as though his expertise on women were a favorite topic.

The young man's speech and manner were also shifting in a way typical of upwardly mobile blacks whose ghetto origins had given them a dialect and body language which they shed when occasion required. Stuart Smith's personality, he realized regretfully, was not just thoroughly middle-class, but

82

distinctly on the up-tight side.

Tony was continuing, "I can tell jus' from talkin to her a while that her mind is . . . adventurous. For esample, her and I were talkin about my bein an econ major in college, and she tol me how much she admires the 'intellectual style' of Thomas Sowell. She loves iconoclasts, she says.

"And she's not cold with defenses. Even when she's just met you, there's instant contact. Know what I mean?"

"She does have warmth and charm."

"More'n that. She's intense. She reminds me of a gal I met on a plane, once. She was jus startin her vacation, and she was comin on to everbody, just out of esitement. So . . . ."

Apparently sensing that he had been led on, Tony shrugged and refused to say more.

"So?" he prompted reasonably.

"So we can do our jobs well. But if we have a few laughs on the side you'll find you've always got a team in good spirits, ready for the starting lineup." Tony finally had realized that it might be time for a little damage control.

"No chance of emotional involvement, you think?"

"Oh, no. She's a smart lady."

"Oh, good." He didn't care whether this youngster sensed his sarcasm or not.

"Look," he said, leaning forward for emphasis, "I have to be sure there's no misunderstanding here. Because there won't be any tolerance in our regimen for interpersonal problems."

"I'm easy to get along with, and I'm sure Philanth is too," Tony replied in his big voice.

"You've *got* to be."

However, he would have no way to be sure until it was too late! He was beginning to feel stupid for not having simply insisted upon taking Marjorie and Jeff.

But when he overruled Carole's considered advice she always got angry and said something like, "Then what do you need *me* for?" Besides, Carole generally had sound judgment and was persistent in presenting her views only when she felt she had a strong case. She always dealt with issues on the merits: ordinarily, if he brought out a stronger argument she would accept it at once and that was that. So over the years he had become accustomed to yielding when she kept pressing him about something.

And Tony had a point about not taking Marjorie. Hot, humid weather could be particularly hard on older people. He didn't need even the slightest additional burdens during that trip.

Also, Jeff was white, and he really needed a black man with him on this trip, since he was going to be spending most of his time in Africa, being photographed and stared at by Africans.

Moreover, it *was* wise to leave his most reliable staff in place at headquarters when he was going to be away for such an extended period. A previous director had gotten into publicized trouble because of events which had transpired while she had been in Africa for three or four weeks, and this trip was even longer.

However, because of these considerations he was facing an unanticipated risk: it really would be great, wouldn't it, if Tony the jock started fooling around with sexy little Philanth after they were overseas?

"We'll be almost constantly on show, representing Peace Corps to all sorts of people."

"I'll keep my shoes shined."

"But you also mustn't ever be found in a compromising situation—by anyone, including hotel staff. You'll have to curb any temptations to flirt."

That gave Tony pause. "No fraternization to relieve all the stress everybody keeps talkin about?

"Look, Mr. Bjorklund, don't you think we should be frank and realistic about this? —With all respect, sir."

"A person can put up with difficult conditions for the length of time we're talking about."

"Sure, but is he doin his best work while he's concentratin on just hangin in there til it's over? Are *you* willing to settle for sleeping in hostels full of damp and heat and bugs insteada international hotels jus cause you could 'put up with it', or are you goin over there to try to get some good work done?"

"What's on your mind?"

"Look, you need Philanth 'cause she's a little sweetheart, . . ."

"Do I?"

"Yes! I know that awready because I been watching her all day while she was talkin to staff from downstairs. People like her—unless they're real soreheads.

"So whoever else you take with you, they're gonna like her too. But you gotta take a guy, a big guy, cause there'll be a lot o' bags to tote and venal clerks and porters and taxi drivers to stand up to; not to mention possible terrorists and kidnappers of prominent Americans to rassle to the ground.

"Now, sooner or later you're gonna see this guy comin outa her bedroom. Put her looks and personality together, and she's like an open box o' fresh, yummy pastry a guy can't walk past. So you're gonna start to wonder what's goin on when those two are alone together in her bedroom. Now, how ya gonna deal with that?"

"By assuming she's been working with him, coordinating their duties, of course!"

"Right," Tony said, nodding but looking at him as if he were a fool.

Rather than let himself in for tensions between them later, he opted for more candor.

Leaning forward on his desk, he said, "If you've got a more realistic approach, or think you do, let's have it. Any differences in perception have got to be ironed out fast."

Tony seemed to ponder, and he suspected that Tony was trying to weigh just how badly this whitey needed him.

"It's not that we're not going to be working hard, Mr. Bjorklund. Or that we can't control ourselves. But being together morning, noon, and night does things to people. Just a few days together in a car can get the juices started even when people have nothing in common; and we're talkin about five weeks, staying in nice hotels, in exotic places. With sleeping quarters close together because of the phone and security considerations and all the running in and out coordinatin' everthing . . . .

"And a rough day goes a lot better if you know there's somethin nice waitin for you at the end of it."

Tony paused, and his voice hardened a bit. "Philanth's a widow. She's been alone for a number of years. And she acts like she loves *ever*body. Her eyes and mouth flirt all by themselves, she can't help it. And she's as full o' bounce as a new rubber ball.

"Africa's been hit hard by AIDS, and if you value your life that keeps you careful. Since I'm just back from a full tour o' duty there, I guess I see Philanth like a man comin in off the desert sees a lush oasis."

"I'm sorry," he said, shaking his head, "you'll have to accept the fact that this 'oasis' is a mirage. My wife's assessment of Philanth is utterly different from yours, and it's based on much longer and closer association."

Tony failed to repress a smile. "I'm glad your wife feels that way. But she isn't looking at Philanth the way a man would. Any man with . . . ." Tony at least had enough sense not to finish that.

They both were silent for a moment.

"It's not going to be a problem," he declared. "You yourself were just talking about all the discomforts and inconveniences; and with virtually unlimited numbers of things to be done, there'll be no energy left over for romance. At the end of a long, hard day, each of us will be only too happy to take a shower and fall into bed . . . and be alone."

"I got lotsa stamina. And your job will carry a lot more stress than mine."

"I'm not thinking just in terms of my own role. I asked the Travel director to produce the first draft of your job description for the position posting, and when we discussed it she remarked in passing that whoever got the job would be 'very busy, keeping everybody happy'."

"But as long as we do our jobs well, how we get recharged is our own business, right?"

They weren't getting anywhere. "You need to become better acquainted with your duties. And with Philanth. This will be a big job, with satisfactions of its own if it's done well. Forget anything else." He added coolly, "Thanks for coming in."

Tony rose with obvious reluctance. "Just remember," the black man admonished him, "I been tryin to improve African agriculture for the las' two years. After that, anything you can lay on me has got to be a piece o' cake. And my next stop is gonna be the best graduate school of business I can get into. Harvard and places like that are lookin for blacks. So I'm not about to mess up."

He nodded.

But as Carole sat down to dinner with him at home that night he remarked, "I've got a jinx on the traveling gofer job. Stuart Smith bunged himself up trying to become a bodyguard, and the only other black man we've gotten from Personnel isn't suitable. I'm getting backed into a corner where I'll have to draft somebody from International Operations on short notice."

Carole looked alarmed. "What's wrong with the replacement who isn't 'suitable'?"

"He had the incredible brass to suggest he considers Philanth the main attraction of the trip."

Carole smirked. "Maybe she is, you're so determined not to have a good time."

"I guess I haven't made myself clear. This big jock informs me that Philanth is a hot little number, and if I had any . . . anything like normal manhood I'd see it for myself."

Snickering, Carole said, "Now, I'm sure he didn't say *that*."

"It's amazing what unseasoned people will say if you just make them go on talking. He hung himself."

Carole smiled some more. When she commenced eating and didn't comment he got angry.

"You gave me a sexpot as my assistant," he complained, "so now I've got a problem with the bodyguard—and I haven't even left the country yet. What'm I going to do?"

"Oh, Charles, Philanth would be hurt—if she took that seriously! 'Sexpot', indeed!"

"She *is*."

"Well, you're still just talking about looks and manner. There's nothing wrong with her morals; and the stupid, dirty-minded prejudices that force women to give up their femininity in their struggle for equality have got to be faced down.

"Besides, Tony sounds like somebody she can wind around her little finger."

"How'd you know his name?"

"I talked with Philanth this afternoon."

"And?"

"Well, he may be a bit of a jock, but that's fine. The more subtle, controlled types would take more of Philanth's time to figure out how to handle."

"You assume any man I take with her would have to be 'handled'? Com'on, Carole!"

"If somebody doesn't like her it means there's something wrong with them. So assuming Philanth is going, it's best to also take a man who can be relied on to keep their interaction casual. If you want a superficial relationship pick a superficial person."

"I'd love to know what was in your mind when you paired her with Ken. Were you trying to punish him for coming to town with naughty intentions? Did he blow up at you because she wouldn't 'put out'?"

Carole snickered again and moved on to her sliced carrots in mashed cauliflower sauce.

After a moment she said, "You need someone with the types of specialized expertise Philanth says Tony has. And Philanth says he's charming. You should be counting your blessings."

"Philanth seems to find *all* men charming 'once she gets to know them'. It looks like it's the other way around: they *become* charming to *her*."

"Then aren't you going to be lucky to have her working for you in that position!"

"Very clever. Do I want all the top staff in the entire Peace Corps to decide I've selected both members of my traveling party on the basis of 'least likely'?"

Carole looked annoyed and ate in silence for a minute. Then she laid down her fork.

"I know what the problem really is for you, Charles. It's not that Tony or Philanth won't project the proper Peace Corps image overseas. It's not that they won't be equal to their jobs. It's that Philanth makes you uncomfortable because she's full of youthful spirit, and you don't like Tony because he's too crude for your Beverly Hills taste.

"But Tony may turn out to be better adapted to the trip than you or most of the people you'd feel comfortable with. And Philanth is a treasure: a blend of intelligence and empathy and genuine selflessness, and conscientiousness overlaid by easygoing humor and flexibility, patience, and tact."

"However did you manage to steal her from Heaven?" he snarled.

"She was drawn to *you*, if you remember: she came to hear you speak on foreign aid and population policy at Town Hall. I merely provided her with the bibliography she requested afterwards. She wrote to thank me, and we started a correspondence."

They fell silent.

"Tony seems to be an excellent candidate with one little flaw," Carole summarized. "A flaw which has no bearing on his function in your mission, since it makes him not Philanth's type, so she'll find it easy to get along with him. You worry too much."

He hesitated. Then, seeing that he had not kept the matter in perspective, he nodded. He was worn out after his typically full day of running around trying to be wise and witty and all the rest of it.

"I'm sorry, I did get carried away. I'm just anxious for things to go well."

"You're trying too hard, as usual."

90

Chapter 4

JUST ANOTHER PUBLIC TROUGH

"A reporter from the Washington *Post* is asking to speak to you. She's preparing a story about an article in the new issue of *Common Cause Magazine*. It's very critical of Peace Corps operations."

"Put her through."

When they had run through the preliminaries the reporter explained, "I didn't accept being referred to your Public Affairs office because their director is one focus of the criticism—besides, I thought you'd want to give us your comment for tomorrow's edition.

"The author is using Peace Corps as a case study showing that the country can no longer afford the incompetent management of government agencies by political appointees."

Since he himself was a political appointee, he really was being put on the spot. "I'll be glad to take a look at it."

She continued, "Peace Corps was singled out for the attack with the argument that if things are that bad there, where the organizational objectives are uniquely idealistic and one would expect many of the staff to be particularly able and anxious to achieve those objectives, what must conditions be like in other government agencies?"

Only a fool would have commented extemporaneously.

"That's very interesting, but I'd better defer discussing the allegations until I've seen the article."

"I'll fax you a copy. I'll also send an old *Post* article from our files that I'm using as background.

Items involving the competence of appointees appear pretty regularly through the years, so you can have more from that file if you're interested."

She seemed to be assuming he had been born yesterday, but since taking this job he had gotten used to that from people who didn't know his background.

"Thank you." They exchanged fax numbers. "How much time do I have?"

"The earlier this afternoon the better."

The moment he hung up he realized that here was the opportunity he had been looking for: Philanth certainly wouldn't be able to handle *this*. He flipped on the intercom. "Send in that dizzy blonde who's been sharing your work area. What's her name? Devon."

He hardly had time to draw a breath before his prospective assistant entered. She announced solemnly, "One dizzy blonde reporting, sir."

He was not amused. "Go get the fax copies of two articles the Washington *Post* is sending me. Prepare whatever response you consider appropriate for me to give the *Post* for its forthcoming article about the *Common Cause* article. I'll expect your written report by noon."

Philanth did a double-take and frowned at her watch. It was almost 10:30.

"Any problem with that?"

"N—No, sir."

"Then, get going!"

"Yes, sir," she breathed. Light-footed as a ballet dancer, she turned and sped out, pausing to close the door carefully behind her.

He speed-dialed his political adviser, whom he had brought with him from the Hill and given a desk down the hall in Congressional Relations. He tersely explained the situation. "Get copies of the articles from Philanth, and let her know that you're going to be working on a rebuttal also."

As soon as he hung up he studied his schedule for the rest of the day. He told Marjorie to reschedule several appointments which had been made for the first half of the afternoon.

Then he started pushing paper. If he didn't keep up with his paper flow there would be all sorts of additional problems, and he already had more than he could handle.

After five minutes or so Marjorie brought in still more papers. She delivered them to his "in" box, then emptied his "out" box.

As she started to turn to leave she paused to ask, "Do you really consider Philanth a dizzy blonde?" He saw that her tending of his boxes must have been a pretext for this query.

"Don't you?"

"Are all young blondes dizzy, in your opinion? Or only those who fly to do your bidding?"

"Get outa here, Marjorie," he replied good-naturedly.

But his confidential assistant didn't obey. She stood frowning at him through the tops of her bifocals and tapping her bunch of papers and folders against her hip. Marge had hawk-like features and gray hair worn in a chignon; since coming to Washington from his congressional field office she had enhanced her look of authority with tailored suits and lighter makeup.

"You'd've gotten Philanth into trouble with Les if she hadn't taken a bundle of work home with her last night," Marjorie reproved him. "Les isn't a tolerant supervisor. But she crawled into bed after supper, then got up at 5 a.m. and got Les's chore done."

"She's been complaining!" he surmised, a note of triumph in his voice.

"On the contrary. She was just explaining to me why it was okay for you to dump this project on her: how she'd already managed to get done the work I'd seen Les give her late yesterday, even though she's been busy with Jeff's people all morning. Les had said he wanted it done as soon as possible, so she'd delivered it to his desk on her way in."

"It still sounds like she was showing you what a martyr she is."

"She's *never* complained!" Marjorie was incensed. "Though she *has* commented that the 'wage-slave-style payroll system' is demeaning to people who've risked their lives in the field, serving Peace Corps. She says it's 'axiomatic' that overregulation of highly motivated workers, which we definitely attract in great numbers, alienates them."

Here was an opportunity to draw Marjorie out regarding Philanth, so he gestured for her to sit down. Marjorie laid down his "out" papers and eased herself with arthritic caution into the chair he indicated.

"Philanth sets herself up as some kind of expert?"

"She took a course in college."

He smiled his contempt.

Marjorie went on with a grandmotherly chuckle, "But you should hear her and Sandy chattering at each other. They agree they speak the same language, only Philanth calls it 'sociologese' and Sandy calls it 'managerese'." Sandy had a degree in management.

"Anyway, Philanth says the fact that she's not paid as a professional doesn't mean she shouldn't be allowed to work like one."

"Who says otherwise?"

"I guess you don't understand. Under our payroll system, a 'professional' is anybody who's paid by the day. He or she works until a job is done, one hour or twenty. Most staff are paid by the hour, to be on duty during certain periods. So Philanth is saying that she should be allowed to do unpaid overtime if she wants to, even though she's not professional staff."

"Oh."

"She doesn't in the least mind all the extra work she's already been doing at home. She doesn't even report her extra hours in the office to earn compensatory time off. She claims she's still just learning her job. And Les signs her time sheets without objection."

"Is all that extra time she's spending justified?"

"Certainly. This trip is too heavily loaded for preparations to be less than meticulous."

"I meant, is there pay-off in the quality of what she does?"

"Jeff is impressed. I think Les is too, though he won't admit it. She's being careful about details and showing initiative. By my reckoning she's doing GS-8 work for a GS-6 salary.

"She'd do still more advanced work if it were given to her. I think it's fine if you're starting to give her more demanding things to do."

"Glad you approve," he replied sardonically.

"The problem is, she's already overworked. *She* doesn't mind, but *I* mind. The young are used to being able to recover from their excesses. But they'll pay for them later on."

Marjorie gave him another direct look. "So if you're going to keep laying extra things like this on her whenever you feel like it, ..." He saw that Marjorie did not want to bring up her reason for suspecting that he had started testing Philanth's abilities for himself.

He accepted her challenge. A lot of Philanth's work could have been handled down on the seventh floor in International Operations but for the fact that the point of her temporary position was to keep the right hand aware of what the left hand was doing as the plans for his major trip were finalized.

"If there's a conflict because I want to give Philanth things to do, talk to Les. He can always ask Jeff to line her up a little help from IO, maybe from among the assistant country desk officers."

"'Talk to Les'?" Marjorie snapped. "A lot of good *that* would do! Why not just tell me to 'stick it'?"

He affected shock. "Marjorie! Your language!"

"Well, Les's contempt for other people is . . . ." She glanced around at the door, but it was closed.

He slumped back in his big executive chair.

"Marge, if even you can't deal with Les, I'm worse off than I'd realized. He can undermine everything I try to do, if only by being the kind of person he is."

"Yes." She leaned forward and lowered her voice even though the private office was quite soundproof. "Arrogant bastard. Can't you get rid of him?"

He lifted his hands helplessly. "Ordinarily the chief of staff would leave with the director who brought him. This is an unusual situation."

"It certainly is!" There was a pause.

Marjorie went on, "I remember that during your confirmation hearing one of the senators commented on the fact that Les had 'graciously consented to stay on during the transition', and what an aid that would be. But the transition is over."

"He probably has feelers out, since there'll continue to be some shifting around of executive personnel even if the President wins re-election."

Marjorie pulled a face. "You think Les is prepared to hang on where he is, on into next year?"

"Much longer, if he can't line up anything more desirable. And with his erstwhile mentor in disgrace Les may have too little pull of his own and too few friends to lobby for his desire to move up.

"Ordinarily, though, after his present position Les could have expected to go on to some *really* plum job, like a sinecure in Geneva. It must be hard for him to lower his expectations."

"You can't just invite him to move on?"

"He's a protected species: an appointee approved through White House Personnel. He's one of the party regulars who'll do anything for whatever fool or scoundrel his party nominates. Whereas I'm barely tolerated by the loyalists—they just needed a 'Mr. Clean' to clear out the bad smell in this office.

"So I don't really have the authority to replace him. I wasn't in any position to haggle with the White House over the terms of my appointment, and by the time I realized that the political appointees still in place here were doing much more damaging things than insulting women, it was too late to even raise the question."

He knotted his hands on the desk top as he went on, "Besides, how'm I going to get replacements so clearly qualified that the White House would accept them over the appointees put in here with their blessing? A person like Les—Sandy has discovered that the director of Public Affairs is an even more flagrant example—makes sure that nobody who might learn his job well enough to move into it gets a position anywhere near it. You know how he felt about my creating the ombudsman position."

"But Sandy's no threat to Les's job!"

"Sandy's position chips at the edges of Les's power.

It's instinctive, with Les. Les rules by fear, and an ombudsman's role is . . . presumably . . . to help officialdom temper justice with mercy."

"Yeah." Marjorie smiled conspiratorily. "Since Peace Corps already had an Employee Relations office, you had to do some fancy explaining." Marjorie knew that Sandy's real job was gathering intelligence among lower-level staff so that Charles Bjorklund would be one Peace Corps director who did not have to depend on a lot of bootlickers to tell him what was going on inside his own organization.

Ignoring her digression, he started thinking out loud. "Anyway, Sandy's new to us, and fresh out of graduate school. I'd *love* to promote somebody who's already seasoned in the complex technical details of Peace Corps operations—yet not so beaten down by experience that flexibility and initiative have been lost. But ordinarily one political appointee from outside is always brought in to replace another, leaving the staff who know what they're doing frozen in the subordinate positions."

He rubbed his forehead as he added, "Being chief of staff is no easy job. Anybody who'd really want it might be exactly the wrong person for it."

"Maybe you're overawed by it yourself. Philanth makes a fairly simple job significant. Maybe it could work the other way around. With so many people having a piece of the picture, the main qualification for that central position might be having the right attitude."

"And frame of reference and communication skills.

"But everybody'd have to pull together, or there could be much worse problems than we've already got. We can't risk that.

"And considering what Sandy's learning about the morale around this place, it's really hard to imagine everybody pitching in to help a new chief of staff do his job properly. —Especially all the other top appointees. If I had somehow pushed Les out it's more likely they'd leak our problems to the press."

"Maybe I've seen *Mr. Smith Goes to Washington* too many times, but I should think the experienced rank-and-file could be activated to make it work."

"Sandy says they're so used to being at the mercy of dictatorial and incompetent associate directors that critical thinking about the larger context of their tasks feels like 'thoughtcrime'. So how can they tell us what we need to know?"

Marjorie shook her head and said nothing.

He concluded, "So I need a new chief of staff who's brilliantly qualified to begin with. But if he's been spending recent years learning Peace Corps operations it's hardly likely that he's playing a significant role in the current presidential election campaign; so how could I get him past White House Personnel? Which is where we came in."

"I see the problem." Gripping the arms of her chair to hoist herself up, Marjorie concluded, "Well, all I can say is, 'The world is not coming to an end, so you must suffer along and learn to cope'." That old *New Yorker* cartoon had seemed funny to him, once. His confidential assistant took up her papers and left.

He went on pondering the dilemma. Marjorie's suggestion that the chief-of-staff position could be scaled down implied she shared his feeling that Sandy was his only hope. It made sense in that Sandy moved around all over the headquarters, and thanks to the graceful and personal yet professional way in which he had made himself and his new position known to the five-hundred-person headquarters staff when he first "came on board", he was finding out where the bodies were buried. Given several more months of solid work, would Sandy still be too inexperienced to at least take the title of chief of staff?

The possibility that White House Personnel would let him elevate Sandy to the post was made extremely unlikely by the fact that since the President was up for re-election, in a couple of months there would be a whole new flood of job-seekers entering the pipeline from the current campaign. Yet he could not even think of any way to move Les out of the position without getting flak from the White House and from right-wing columnists happy to have dirt about Peace Corps.

His position was especially delicate because of his uneasy relationship with the deputy director, who also was an appointee who had come in under his predecessor. Just the previous week there had been a wine-and-cheese reception for him and the deputy director at Decatur House, and on their way over there Archie had pressured him about being given more to do. But badly as he needed to share his workload, he could not delegate to the deputy director: Archie's zeal for applying the workings of private enterprise to a federal bureaucracy made him a menace to Peace Corps operations at all levels. Archie also welcomed opportunities to make public appearances, but he was an unctuous little man with a big stomach and not much presence as a speaker, so everybody strongly preferred appearances by the tall, handsome, debonair director.

Soon after taking the helm at Peace Corps he had fast-talked Archie into spending his time cultivating contacts for development of private-sector involvement in projects. That was working out well except that the deputy director kept pressing for "more scope for his talents". In the car going to Decatur

100

House he had once again told Archie how much Peace Corps needed all the corporate donations Archie could solicit.

He had to keep Archie pacified. A falling-out between director and deputy director during the Reagan administration had gotten into the newspapers and been nastily exploited by the right-wing press, and the scuttlebutt was that for most of those eight years that deputy director had gone on collecting a good salary while the director allowed him to do nothing but "sharpen pencils".

This deputy director might be so full of inapplicable convictions derived from his success in the business world that he was more trouble to work with than he was worth, but at least little Archibald wasn't pushy. Those women running Public Affairs and Management were another matter: needlessly aggressive and all too full of themselves. Stylish, and charming to him, but overimpressed by their own opinions. Unprepared to admit how little they understood Peace Corps and eager to apply irrelevant experience, they had become petty autocrats endlessly destructive of Peace Corps' interests.

Sandy was invaluable because he could report on exactly what damage these associate directors were doing. Sandy was also a very imposing-looking but prudent and pleasant young man. He was getting the impression that Sandy was probably a dyed-in-the-wool conservative Republican—Sandy was from Utah, after all—but it might be valuable for a Democratic senator to have such an intelligent, likable conservative Republican on his staff, if only to keep his other staff in contact with that point of view.

However, that was for the future.

As for the present, talking candidly with Marjorie

and Sandy was helpful, and he trusted their loyalty and discretion, but sharing his frustrations with these subordinates was unprofessional. He wished he had a confidant who was close to this particular situation and whose experience in federal agency administration he trusted. But his closest contacts were on the Hill, where Peace Corps' oversight committees held forth. Moreover, it would be hard to find anyone who could be counted on not to enjoy passing on some inside gossip about his problems.

Now that Carole was no longer working with him and had become wrapped up in her magazine-writing project he didn't even have a decent sounding-board any more. He had a small-calibre case of "Lonely at the Top."

Realizing that he had chewed this over enough, he leaned forward to consult his schedule. He saw that he had a little time before his luncheon meeting in which to review and initial papers.

He had almost finished getting the latest arrivals into his "out" box when Marjorie buzzed him to say that Philanth's report was ready. It was noon. He put on his suit jacket and went out. He stopped by Philanth's desk to get her manila folder and tell her he would see her soon after he got back.

He trekked down the long, narrow, blue-carpeted hall, nodding in response to obsequious salutations from those he met. He reminded himself not to look grim, for Marjorie had commented a month or so earlier that that seemed to have become his natural expression.

In the softly lighted reception area at the center of the building a bronzed, muscular young man in shorts and a beard, a heavy-looking backpack at his feet, was asking for directions to the office for Returned Volunteers. Such apparitions from another world, so incongruous in their rugged vitality that they made the carefully groomed and formally dressed Washington office workers look effete, were a common sight on this floor.

Though pressed for time and feeling oppressed, he felt morally obligated to stop and introduce himself, shake hands, and go through the expressions of personal interest and appreciation which he believed he owed to every Volunteer. The ritual took a few minutes, but it always had to be done with sincerity and presence of mind, for otherwise one did better not to attempt such gestures at all. He only had to remember that all of his glory came from their sweat, and his resulting charisma made his heartfelt thanks a token they seemed to appreciate.

As he concluded the exchange he was turning right to go through the heavy glass doors opening out into the bare, bright elevator lobby. Staff recognizing him hastily made space for him in one of the six elevators, which were busy with the noon rush.

The elevator contained mouth-watering aromas, probably from take-out food which had just ridden up in it from one of the many ethnic establishments in the neighborhood. Five floors below the five floors of Peace Corps offices, fragrant food smells were also coming from the various eateries in the two-level mall which ran from K to I Streets. The foot traffic was heavy, and as he climbed the steps to the exit on the Twentieth-Street side of the building he looked down at a long line of people waiting to get into the cafeteria.

Once he was outside he had only a short walk to a dark but reasonably decorous bar-and-grill. He moved briskly; but his friend, a nondescript little man from one of the New England states, was waiting. The congressman had graciously come to him, "just to get out of the House for a while", but he held out a menu, for they both were men with no time to waste.

Having eaten there several times before, he quickly made his selection. While his companion continued to study his own menu he took a peek inside Philanth's folder.

The first item in it was the faxed copy of the magazine article. He winced as he saw that the Peace Corps article was the cover story. The lower part of the cover presented a cartoon drawing of a trough lined with small, dark ovals—they had to be plums—and surrounded by eager pigs happily taking up the plums in their mouths. Most of the rest of the cover shouted in caps of three sizes, "PEACE CORPS—Just another public trough—Effects of the spoils system on one federal agency, *page 21*". The validity of this image hit him and made him shudder.

After the magazine article came the old *Post* article, which had some passages underlined. He lingered over it, studying the photograph of Edmund Muskie, Paul Volcker, Elliot Richardson, and Elmer Staats. The caption which listed their names began, "Members of National Commission on the Public Service discuss recommendations to restrain political appointments." The huge headline below read, "Too Many 'Plums' for Political Picking?" Below this a smaller headline explained, "Panel Urges Fewer Appointees, More Opportunity for Careerists". The column, by Judith Havemann, had been cut from page A21, the Federal Page, of the *Post* of October 20, 1988.

He experienced a bitter internal chuckle. He was now in the position of being obliged to try to parry this respectable panel's previous attack upon the political spoils system as well as answer the current attack upon Peace Corps as an example of the effects of the system. It was a bit much to bear, considering all the time and effort he expended on wrestling with the consequences of using Peace Corps as a harmless playpen for opinionated long-time fans of the President.

They wanted the pleasure of throwing their weight around and getting access to Washington's goodies and glamor, all in the sacred name of "public service". Honest, ordinary jobs in the private sector, doing things they truly understood, would have been far more of a public service than mucking about in the vast and finely tuned mechanisms of the federal bureaucracy.

"Chuck?" said his friend. The waiter was at his elbow. He quickly gave his order, and they were left alone. He apologized for having become engrossed in "Just something somebody handed me as I was going out: the crisis du jour."

His friend waved aside the slight. "I understand.

"Now, about the President's veto. You wanted to discuss the possibility of mustering an override after the election next month, and whether there's anything you can do to help."

At once he had no thought for anything but the resuscitation of the package of bills which he had come to regard as his only meaningful accomplishment during his seven years of hard work in Congress.

When he returned to his office he told Marjorie to hold his calls and sat down to study the contents of Philanth's folder. The ball was in his court now, and the clock was running.

First he went back to the old *Post* article, since that antedated everything else. Its key passages had been underlined:

Four-time Cabinet secretary Elliot Richardson urged the next president yesterday to go easy on rewarding political supporters with top government jobs and to consider career civil servants for some of them, not only for the sake of the workers but for the good of democracy.

Former secretary of state Edmund Muskie said "the use of political appointees has dipped down into the middle level and the lower level of management in government service to the point that . . . it closes off opportunities for advancement to career people. Opportunity for advancement in government service is as important as it is in the private sector."

More important, he said, "too often people are not qualified on the basis of their private experience are appointed to positions in the government that ought to be filled with people whose experience is relevant."

He called for more active and effective screening of appointees.

Volcker, Richardson, and Muskie, members of the public service commission, called for restraint on political appointments as the transition to a new administration begins in less than three weeks.

The 5,342 jobs listed in the last "plum book" —the list of available presidential appointments—should be compared, Volcker said, to figures of about 50 in the French government, 25 in the Japanese government and 100 in the British government.

Those 5,000 jobs were only the officially designated appointive positions, he thought. Judging by what quietly went on all the time at Peace Corps, in practice there had to be many times that number: as many more thousands of jobs as those with the requisite clout and nerve desired to create. His year at Peace Corps had shown him all too clearly that the spoils system was not only placing grossly unqualified people in the top and middle levels of federal administration, it was constantly being unofficially expanded.

—And not just by dreaming up new positions in which an aggressive political hack might do little harm until the next election year. Political slots were being created also by displacing experienced civil-service employees—including sharp, highly dedicated young Returned Volunteers, at Peace Corps—by bouncing them down into other jobs vacated by attrition. The damage to morale was incalculable, and this was hardly following through on the official policy of encouraging high-quality people to pursue careers in government service.

Those who were aware of what went on and were not among the beneficiaries were in no position to protest. As one brave civil servant had pointed out in a signed letter to the editor of the *Post*, "A 'Tight-lipped' Bureaucrat Is an Employed Bureaucrat". The bureaucrats' silence was like that of the twenty women raped by the same doctor: since all were afraid to complain publicly, nobody knew about any of the other cases.

Moreover, Sandy had told him that these abuses were treated by everyone who knew about them as part of the regular routines: because of time-honored norms, civil servants felt it was part of their professionalism to quietly endure gross misjudgment and misconduct by the political appointees who were always their superiors. People immersed in an ongoing situation could rationalize anything when they had a commitment to stay and saw everyone around them tolerating what was going on.

Finishing with the *Post* article, he turned it over and found one more item: a computer-printed breakdown of the argument of the *Common Cause* article, with comments: Philanth's original work, a preliminary to the "response" he had asked for.

He left that at the back and turned to the article itself.

It was five full pages of text, preceded by a full-page cartoon drawing. The drawing depicted Peace Corps headquarters as a five-tiered variant of Orwell's *Animal Farm*: some pigs were gaily trotting off with travel stickers all over their suitcases; others were lolling at big desks, each attended by several sheep with the accoutrements of secretaries; and countless horses with dismal expressions were collapsing under bales of documents. The artist must have warmed to his theme as he worked, for it was an exuberantly detailed depiction of exploitation, waste, and mismanagement.

The fax was in black-and-white, but the shadings indicated that the original was in color. Visualizing how effective the page would be in the glossy, full-color original which was being mailed out to Common Cause members all over the country, he groaned.

He turned to the text. Across the top of the two-page spread, joining the full-page cartoon and the first page of text, ran a huge title: "EVEN AT PEACE CORPS!" The first page featured a large-type blurb:

> With so many of the decision-making positions filled by political appointees, a sophisticated structure of dedicated, specialized professionals is rendered a government of the

competents by the incompetents and for the incompetents.

He skimmed through the columns of text. The author built credibility for his thesis by roasting the day-to-day operations of the headquarters and regional offices and proceeding to illustrate effects on the field operations. It was apparent that unhappy staff had not reserved all of their inside information for Sandy. Some of the lunacies and abuses Sandy had discovered were there, and more besides: the incredibly detailed safeguards over pennies combined with the lack of control over self-indulgent waste by those who could get away with it; the smothering tangle of paperwork too often done by people who had not been adequately trained for it; and basic flaws in the structuring of tasks, especially in the entire systems of disbursement and budgetary allotment.

The author quoted a former Peace Corps bureaucrat to point out that complex administrative decision-making in a large government bureaucracy was not the sort of work that the hard-nosed political ideologues who wage election battles are likely to perform well. Moreover, "There is no Republican or Democratic way to ensure that technical tasks such as ensuring computer security or managing programs for training fish farmers in a different culture are competently performed." Yet it was a matter of public record that a Peace Corps director appointed at the beginning of a new administration could expect to have to fight White House Personnel over every appointment of someone whose political credentials had not been approved through "proper channels".

As he reached the conclusion of the article he heaved a dismal sigh. He had no choice but to try to rebut this excellent job of investigative reporting, for as even the remainder of the old *Post* article showed, political appointees typically went rigid at any suggestion that the President did not need and deserve the power to fill the top positions of his administration with those who had demonstrated their loyalty by helping him get elected. It would be folly for him to cripple himself politically over such a routine matter as "reacting" to a published complaint.

He turned without hope to Philanth's analytical breakdown with comments. This proved to be well done, especially considering the conditions under which she had been working. She had concluded by noting that she would continue to work on how to respond to the criticisms. He had glimpsed a banana peel and an open jar of peanut butter among the papers on her desk as he came in, and she had not glanced up from her computer screen to notice his return.

He buzzed Marjorie. After a brief delay to have her computer print the pages she had been drafting, Philanth came in. She was clutching another manila folder to her bosom as though it might afford her some physical protection from him.

Chapter 5

DARING PROPOSALS

She stood before him like a model schoolgirl sub-missively awaiting discipline in the principal's office for some nonexistent transgression. This impression was reinforced by her plain white blouse and navy-blue pleated skirt. Presumably to keep herself limber for combat, she had omitted that essential of the Washington dress code, her suit jacket.

"For God's sake, girl, sit down," he exclaimed, and she sat down, stiffly erect. She certainly didn't take any liberties on the basis of their previous conversations.

"Now, what do you propose I say to the *Post* for their article about this forthcoming 'exposé'?"

"So far I haven't found any weak spots in it that you'd be able to use. I didn't have time to do more than one spot-check of the accuracy of the reporting, and I found out at once how sensitive the issues are."

"Which issues?"

"The associate director for Management's awful judgment regarding the security system and her taking regular personnel away from their desks to serve as rotating receptionists on all the floors except ours; and her refusal to bring in someone who can actually make all the 'smart' phones work the way they're supposed to. I was told that she has flatly ordered everyone not to discuss these decisions any more.

"Anyway, we can't expect to be given space to rebut the data even if it's possible. So I can only suggest a response that side-steps the attack."

"Let me have whatever you've developed since you

gave me this," he said resignedly, nodding at the papers on his desk. "Maybe it'll help me get started."

She opened her folder, remarking, "It's fortunate we don't have to address his implications regarding our overseas programs."

"But the overseas programs are what we're all about! His argument that decision-making at headquarters is faulty shouldn't be allowed to impugn our work in the field!"

"Yes, it should." Philanth flipped through some loose sheets of paper. She found a page, crossed her legs, and began reading aloud. "Because of trendy program development, the Peace Corps' original purposes have in fact been distorted by U.S. domestic politics and lack of realistic, social-scientific perspective on the long-term, macro-level effects of programs. One could go much further than this writer does by pointing out that the premises of present development efforts are questionable for intrinsic reasons, aside from the question of the qualifications of those who instituted them.

"For example, it can be argued that much of our work now only bolsters corrupt host-country governments, fosters more hunger, and fails to address other causes of political instability. This in turn discourages foreign investment, reinforces dependency, and perpetuates the isolation of the developing countries from the emerging global economic community."

"Whoa! What's your source for that?"

She replied apologetically, "I just dashed that off to show why we can't try to address the author's attack on our programs. It's my normal style, and I know it's very general, but I could of course elaborate and defend every—"

"You're out of your depth, Philanth."

To his surprise, she smiled. "Not nearly as far as a lot of policy-makers are," she responded politely, "including most presidents.

"Even if they know better, politicians don't tell the truth. They know they'll incur the displeasure of the uninformed if they endorse 'doomsayers'' analyses of long-term trends instead of mouthing popular shibboleths.

"—Excuse me, sir, I know you're exceptional—as a congressman from a highly educated district you were able to be.

"I'm not a politician at all, and the publications of the World Watch Institute, the works of the Ehrlichs, and others of that school of thought are my natural meat. The op-ed columnists' efforts to refute them are always pathetic in their ignorance and ridiculous in their reasoning."

Seeing that he had nothing to say to that, she resumed reading. "Even with technological innovations, economic development is limited by the nonrenewable resources of the planet and the carrying capacity of its ecosystem unless we've developed systems which cost-effectively exploit the infinite extraterrestrial resources. Yet we don't even attempt to look beyond the goal of enabling Africa to feed itself, though that's manifestly impossible without dramatically reducing birthrates. Global scientific perspectives could guide Peace Corps and other international development efforts toward a time when the new technologies—"

"We've got a deadline." He was deliberately being unfair, switching signals after drawing her out.

"Uh, yes, sir," she responded meekly. She turned back several pages, saying, "Anyway, be thankful you *don't* have to defend the programs you're preparing to examine with a view to instituting overhauls.

There'd be the devil to pay, if you were candid about *your own* concerns."

Her casual presumption took his breath away. Premature statements about possible changes in programs obviously *could* cause all sorts of unnecessary problems, but her remark also seemed to imply that she had been deducing more about his hidden agendas for his November trip than he would have imagined possible.

She must have been sensitized by her prior awareness of his "concerns" about foreign aid policy before he had left Congress. Nevertheless she must be an astute guesser as she saw from her unique vantage point the details of all the arrangements for his "world tour" falling into place. Remarkable.

But perhaps he was only projecting his own thinking onto her. He alone knew that his hidden agenda for the trip had become so extensive and detailed that he had developed a personal short-hand code so he could keep track of his tasks during the journey without risking an international brouhaha if his notes should fall into unfriendly hands.

Philanth went on, "I asked Sandy to glance over the charges against our operations, and he agrees that logically and factually, this article nails us to the wall. It's sure to receive other press attention across the country. Therefore your response to the *Post* is likely to be cited and commented on as well, in columns and letters to editors, by some sharp people who care deeply about these matters. It follows that your logic must be sound and on target."

"You've just told me its argument is unassailable. What do you suggest?"

"We have to go outside the boundaries of this attack: approach this whole subject from a different angle than the author does."

"We can talk about the fiscal starvation at Peace Corps. That's a strong defense: our funding is allocated to give maximum support to our service in the field, so our headquarters staff are consistently overworked. And we can't keep our best staff because of our salaries plus the five-year rule. So we're plagued by high turnover. The technical specialists who keep things running around here are predominantly middle-level people who must have the patience of Job."

"Nobody will pay any attention to our horror stories about our fiscal problems. They've heard similar complaints too many times. It doesn't matter that a situation is crazy and desperate if no one will listen when you talk about it."

He had not expected such perceptiveness in one so young. She was showing him that he would be perceived not as a former congressman still issuing his clarion call for more intelligent foreign aid, but as a bureaucrat mouthing the same pleadings as countless other bureaucrats, each competing for the same tax dollars to help his own fiefdom.

"So what do you suggest?" he asked again.

"If we don't go with my daring proposal there's nothing you can say that won't sound flimsy, banal, and/or evasive. Everything else I've come up with is things we can't say, for practical or political or other strategic reasons."

"What's the 'daring proposal'?"

"First develop the answer you'd have to give, without it. Otherwise you won't properly appreciate it."

"I haven't time to play that game, Philanth. I've asked you to advise me as to what I should do."

She mimed her regret. Then she said, "You must address the central problem: the infliction of amateur administrators upon a complex technocracy by a

115

spoils system left over from times when government functions were primitive in their simplicity."

"How?"

"Advocate getting rid of it, of course!"

"You think *I* should try to change that system?" he asked wearily. "It would have to be done by a congress able to override a presidential veto. And can you imagine the arm-twisting its members would be subjected to if the measure seemed to have a chance?

"Even if that congress by some miracle stood up to the president and overrode his veto, the measure would go to the Supreme Court and the Court would kill it, invoking theorists and hacks who say we need patronage to keep the party system alive, as a countervailing force to the power of media-driven politics."

"That's a stupid argument," she declared, startling him. "It clings to a nineteenth-century conception of how things have to work. Meritocracy and technocracy, strengthened by genuine campaign reform, can create a more effective basis for a responsive and competent system of government and politics in the type of society now emerging."

"Perhaps." Achieving the enlightened society she envisioned would require such widespread changes in priorities regarding childbearing and education as to make the question academic. He would have enjoyed discussing it with her anyway if there had been time, but the need for quality rather than quantity of children was too politically explosive for a Democratic politician to discuss with anyone but his wife.

"Be that as it may," he went on quickly, "to curb this remainder of the spoils system would require a second revolution of the sort which created the civil service in the first place."

"Fine, so why not do it! The American people

could develop a higher opinion of their government if all the hard work of career specialists in the civil service weren't constantly being trashed by their politically appointed superiors."

"I grant you that the present system of filling key posts with reliable deputies is abused. But where can one draw the line?"

"I refer you to the old *Post* article: it mentions numerical proportions we need to tighten up. Strict, detailed wording is needed to draw the lines clearly."

"But the fight wouldn't be over numbers. Whenever this comes up the argument never gets beyond the basic principle that a new administration needs loyal supporters to carry out the policies for which it was put into office."

"You're too pessimistic. It's becoming more and more apparent that that rationale has glaring flaws— most notably, the fact that our presidential elections seldom truly are mandates on more than one or two mindless issues. Also, an administration full of loyalists lacks critical thinking among insiders."

He smiled wryly at her. Then he pondered. "Compose a careful statement to the effect that I don't endorse appointing people to jobs when their qualifications for those positions are questionable. That's always an acceptable public pronouncement."

Her mouth registering disdain, she leaned forward to pick up the dictaphone on his desk and offer it to him, thumbing the switch. As she began writing he repeated his statement without his cynical comment.

He went on dictating. "But it takes time to make informed personnel decisions when one comes to a top position in an organization from outside. As for the idea that top people should work their way up, that too can have drawbacks: it tends to entrench the status quo, while 'a new broom sweeps clean'."

"That's balderdash," Philanth muttered.

He added, "We can only start from wherever we find ourselves and try to make a situation better."

He handed back the dictaphone, switching it off, and her mouth twisted scornfully as she took it.

"All right, you have my pitiful fall-back! What's the proposal I'm now supposed to be sufficiently desperate to jump at?"

With an understated flourish she lifted away the top papers in her folder. "You might make headlines of your own if you used this opportunity to propose congressional study of a proposal that's been around Peace Corps for years. Here's my summary." She handed him a few pages. "I also have the proposal itself." She lifted by its stapled corner the document she was keeping. He settled back to read.

> Because of its unique nature Peace Corps should be removed from the regular government bureaucracy by being made a quasi-autonomous organization like the Atomic Energy Commission or TVA. Thus it would have authority to run itself as an independent organization. Freed of the hobbles kept upon our foreign policy by the general public's shortsightedness regarding the world beyond our borders, it could grow by its own strengths into a really effective scale of operation, instead of being kept a struggling token effort by the overall federal budget squeeze. As a private foundation it would receive grants from both government and the private sector.
>
> It would cooperate fully with similar organizations in other countries, using cross-training programs, an expert loaner program, a staff exchange program, and joint projects.

It would be exempt from all civil service rules, and its managers would have the right to hire and fire their subordinates. It would be run by a board of directors made up of the most respected, creative, and active experts in the fields related to Peace Corps' work. They would select their own chief executive. They

He broke off reading to squint at her. "How did you come across this?"

"I found it in the Peace Corps library, buried in a file. It was done in the 1980's as part of the self-assessment and long-term planning exercises for Peace Corps' silver anniversary."

"Why were you looking into *that* file?"

"I noticed it while I was looking up coverage of previous directors' travels overseas to help me get perspective on my new job. I was on my own time because I hadn't officially started working here yet.

"While you were at lunch today I went downstairs and got the librarian to help me find it again. It took her only a minute.

"By the way, she's *very good*, and she's being forced out by the five-year rule. She can't get an extension because the library is being cut back to a staff of one. That's outrageous."

"Is the library under Management?"

"I guess so." She quickly prepared to write in her steno pad. "Shall I confirm that? Maybe it could go under program support."

"No. We have to cut headquarters staff to the bone because of the major discrepancy between current congressional appropriations and prior congressionally mandated commitments in the field. Since the cutting *can't* come out of Volunteer support it has to come out of Management."

119

"So we leave it under Management in order to be able to cut it? But running that wonderful library with only one librarian is stupid! There's so much record-keeping required. Its operation could at least be merged with the Information Collection and Exchange office, which isn't even on the same floor."

"That's an interesting idea, but I can't deal with it at the moment. Besides, any change to save money would cost money, and we *simply haven't got it*.

"Now, as to *this* proposal, . . . "

"Right," she agreed quickly. "Naturally your assignment reminded me of it. One of the major issues identified in the document is . . ." She consulted an inner page of the document before her. ". . . how to separate Peace Corps from the 'wasteful, stifling, depressing, resource-devouring requirements of the government bureaucracy'.

"What makes that phrase remarkable is that it— and this whole proposal—was written by one of the early Peace Corps Volunteers who'd become an experienced government bureaucrat and was working here at headquarters. I found that out from Personnel.

"By the way, I also brought a bit of supporting material from another file I'd noticed." She was handing him a taped-together photocopy of a large newspaper clipping. "The file contains news items about criticism of previous Peace Corps directors' management. My proposal could put a stop to the sort of chronic misconduct involved here by changing the way 'senior officials' are selected to run Peace Corps."

Again he was looking at an item from the *Post*'s Federal Page, this time page A25 for April 24, 1992: "Peace Corps Torn Over Internal Report: Officials Accused of Reprisals Against IG's Office After Critical

Review of Management". He skimmed the article.

It noted that the agency inspector general's office had "long been critical of the agency's management". Now an internal report by the Peace Corps inspector general had indicted personnel practices and poor management and criticized individual officials. The draft of the report said "the personnel office had 'an established reputation for inefficiency and for unfair and manipulative personnel practices,' a 'rigged' selection process and 'the worst kind of bureaucracy.'"

"Senior agency officials" had tried to get the report altered and suppressed and had escalated from pressures to threats to various forms of retaliation. But an inspector general has a legal mandate to be independent. Congressman John Conyers "accused the agency of 'cronyism and mismanagement' and said the 'attempted manipulation and suppression of official inspector general documentation by senior management'" probably violated the law. Both he and Senator John Glen had "expressed growing concern in recent months".

He noted that the article repeatedly referred to "senior officials" and "senior managers" as the villains. Obviously these were members of the ruling caste, political appointees who—then as now—were manipulating Peace Corps' personnel procedures for their own selfish purposes. The article noted that the previous Peace Corps director had brought in his aides at clearly inflated salaries, but this was only a useful example because it required little explanation.

There was lots more about whether the inspector general's independence was being protected, but he had read enough. He returned the photocopy to Philanth.

She knew better than to comment. Instead she smirked at him.

He leaned back, letting the chair carry him as he eyed the summary on his lap. He must be working too hard, because he had to admit that he was tempted to lash out against this ongoing mess of exploitation and mismanagement by adopting her proposal. It would be a fantastic response to that article in *Common Cause*, and the editors at the *Post* would see how significant it was. It would make a splash, all right.

Too much of one.

His voice gentle with regret, he said, "I couldn't broach a proposal like this overnight in a newspaper, Philanth. It would sink me."

"But it'll prove viable! It only needs people with the brains and commitment to make it work."

He turned to the last page of her summary.

My Comments

The most important consideration is that it would truly be taken out of politics and put on a professional basis. It wouldn't be run by political types who can't find Togo on a map, much less micro-manage a complex set of offices dealing with all sorts of technical questions. It also wouldn't respond to requests based on political agendas of host-country strong men or our top politicians' willingness to pander to domestic pressure groups promoting otherworldly religious agendas which don't take into account the economic, demographic, political, cultural, and ecological realities involved.

That means it could *effectively* pursue its goal of working for a peaceful world society by reducing causes of instability and repression instead of trying to explain itself as a nice chance for Americans to get acquainted with other cultures and make friends with some villagers. For it could focus on people's real needs as identified by the professionals experienced in working directly with the people asking for our assistance.

This would give Peace Corps an entirely new modus operandi. We wouldn't have to go on working against ourselves by doing everything *except* the realistic and (therefore!) politically sensitive job of helping people bring down the birth rates which overload everything we do manage to accomplish. We could do *intelligent* things, like the Pathfinder Fund and all sorts of other non-political, non-profit humane organizations. We could *work with* other big, independent international organizations already doing these things, like the Boy Scouts.

He looked up, chuckling. "The *Boy Scouts*?"

"Sure, their Family Life Education merit badge program in Africa has gotten international recognition. It began by reaching twenty-nine countries in subSaharan Africa.

"All *sorts* of international groups are working together on *effective*, *meaningful* approaches to poverty, independent of intimidation by racist chauvinists and pious ignoramuses. Peace Corps should be able to work with them."

He nodded, thinking that here again was the problem which U.S. foreign aid for underdeveloped countries was continually facing in the appropriations process: trying to squeeze family-planning funds past congressmen afraid of America's anti-sex league.

Even in central Africa archbishops were still preaching abstinence rather than condoms as the means to stop the spread of AIDS. It was wonderful to be ignorant enough to be able to fly in the face of reality that blindly, to ignore the fact that it was not rational to base a public policy on the assumption that people could be expected to behave responsibly even though they had not done so in the recent past.

The Cold War between economic systems might be over, but there remained the global conflict of ideas between two other totally different world views. Couched in Malthusian terms, it was an argument over whether it was right for human beings to multiply to the limits of the support capacity of their environment and then suffer population crashes like dumb animals in a state of nature. Capitalism had won its contest with Communism, but unfortunately birth control lacked the self-replicating dynamics of capitalism as a force for social change: while capitalism fed on self-interest, birth control was blocked by religious traditions and male dominance.

The afternoon was passing, and Marjorie's pink slips requesting return calls were undoubtedly piling up outside his door. So were papers for his "in" box. In the business of probing Philanth's mentality for flaws he kept forgetting the journalist preparing her story for tomorrow's *Post*. He resumed reading.

He had reached the final paragraph.

> And *then* would we get results! Then governments that didn't care to tackle family planning would see the economic progress of their own neighbors which did, and they'd climb on board. Everybody would finally recognize a key reason why Japan got rich in one generation: it switched to small families after World War II.

He looked up and saw her watching him. "It's everything you promised," he conceded. "But even a president gets in trouble when he pops off about a bright idea instead of going through the proper preliminaries."

"'Popping off about a bright idea' was how the Peace Corps was born: Kennedy was overtired from campaigning, and . . . . You must know the story."

He smiled at her, but he dipped forward to spread his palms on his desk as though that contact would bring him back to reality. "What I'm trying to say is, I answer to some people who are very aware of their prerogatives. So I'd have to go through channels: maybe talk to a senator who's a former Volunteer and on one of our congressional oversight committees, and then *if* he took time to look into it and liked the idea I could suggest he might like to sponsor a bill."

Seeing her dubious expression, he added, "It's a major proposal. It would have to be fully staffed out and developed in legislative terms. People would have to be 'brought on board'. Peace Corps is a popular agency, but it's always been battered from both the left and the right. There'd have to be hearings, to bring in different points of view. The actual provisions would have to be hammered out, and all the stages of revision required by the process of enactment would have to be gone through—probably several times, by different congresses, as alliances built and shifted and outside pressure was mobilized. All that would take years."

"Since it wouldn't have the constituency that pushed through the TVA legislation, you're saying it would never happen."

"It might if the right man were in the White House to push it."

He handed the summary back to her, saying kindly, "Make me a copy to look at when I get back from my November trip."

Her mouth turned down in disappointment, but she seemed unsurprised. She placed the summary on top of the other papers and closed the folder before she spoke. "Yes, sir. I understand."

She gathered together the folder, her pad and pencil, and the dictaphone, adding heavily, "We go with your fall-back response."

"Draft whatever you can in a few minutes. Then see Mark, in Congressional Relations, and work out a consensus between you. When you've got that, come in together and we'll polish it. Get back here quickly, because it really must be finished very soon."

She nodded and rose. She moved to the door, then paused with her hand on the knob to say, "It's waited this long, I guess the world in its misery can wait a little longer." She slipped out.

He was left alone in his big, climate-controlled corner office with its framed color photographs of the powerful and its humble handcrafted "folk" artifacts from all over the world and its dazzling expanses of windows. He looked out along both I and Twentieth from the eighth floor, untroubled by the heavy traffic in this expensive "business district" a short walk from the White House.

The girl's gratuitous phrase seemed to echo: "the world in its misery". Her caring was larger, or at least closer to the surface, than the compassion of most people. She was like him in that.

He turned to stare thoughtfully at the door through which she had gone. She was more than he had expected not only in her knowledge but also in the ways in which she applied it. Her apparent simplicity was not due to simple-mindedness, as he had assumed. He had kept expecting her to be vapid because she had that baby-faced innocence in looks plus that childlike gaiety of manner. But she was not childish, only lively and unassuming; and her vivacity was a result of her intensity.

He leaned back to lift one foot up onto the edge of his desk, crossed his legs, and reflected. He was not accustomed to staff people who were that forthright with him, but her deferential manner had told him that she knew what she was doing. In hoping that he would not be annoyed she had complimented him more than any of the staff who told him only what they thought he would be pleased to hear. Her brave honesty pleased him more than any flattery.

But was he now willing to take her with him on his big trip in spite of appearances? *No.*

Indeed, he now perceived aspects of her which could make her seem more attractive than he had realized. For example, when her husky voice relaxed it took on a furry timbre which was sexy in a more mature way than the girlish sticky-sweetness of her normal alert tone. The sexiness of both levels of her voice was not important when she was face-to-face because of her wholesome appearance and deferential manner. But how would the men he dealt with react when they heard it over the phone and then *he* immediately came on the line, speaking from some hotel room in Nairobi or Katmandu?

Anyway, her appearance-cum-personality remained a problem. Besides being full-figured, she had that ready, engaging smile and those uncountable dimples which continually came and went as she spoke or reacted to someone else. She even had a large dimple below the left corner of her mouth.

His initial impression of her had been wrong partly because she was a study in contrasts. Her manner

127

strove for simplicity, but her speech was full of nuances. Her clothing was businesslike, but her movements and personality were ingratiatingly feminine. Her face was that of an innocent little girl, and she had a peaches-and-cream complexion. But with her long, blonde hair put up in a sophisticated style and with makeup and clothes which enhanced her other physical assets, plus her self-assured manner and obvious intelligence, she might be . . . quite something.

Something quite different from any of the types of women he was willing to be seen traveling with for five weeks with connecting bedrooms in foreign countries. Philanth had the looks and mannerisms of a worldly man's plaything.

Carole, because of her feminist convictions and her lack of the male point of view, would continue to refuse to accept this objection. So he would have to keep looking for some excuse to reject Philanth, even though trying to prove her inability to do the job seemed to be a non-starter.

The competence required for the job included a high level of ability to work well with others. Philanth seemed to be satisfactory in that regard, but he could try further.

He got his message slips from Marjorie and returned calls until Philanth returned with Mark. Philanth put her diskette containing their combined draft into his desktop computer, and he and Mark sat on either side while she typed until they all were reasonably satisfied.

He noted that she was easily the best word smith of the three of them. Mark even got a bit put out because he could not see the point of some of her alterations, but she knew her stuff and politely insisted on "getting it right".

When they were finished Mark left, and Philanth went out and came back a few minutes later with the laser-printed final version. He checked it and signed it, and she went off to fax his reply and tell the *Post* reporter that it was coming.

Five minutes later she was back. "Uh, sir . . . ?"

"What is it," he said impatiently.

"May I ask what you, uh, plan to do about that *Common Cause* article in terms of internal adminis- tration?" As though aware that that was a pretty presumptuous question, she added as an excuse, "It might be helpful in days to come if you'd gotten the comments of various staff before you have to make any follow-up statements in response to further press inquiries."

"Right. Whip out a memo from me to all the associate directors, asking them . . . ."

"And office heads?" she ventured, slipping into a chair as she flipped open her steno pad.

Now, that was a loaded question! "Um." He paused. "Uh, yes," he answered reluctantly, "and office heads. Note that submissions may be anony- mous." Boy, the associate director for Management was going to be in a snit. But what could she say?

"Ask everyone to comment in writing as soon as they can. Reply to my office, Attention: Marjorie. Enclose a copy of the article with each memo, of course, and you may as well wait to enclose a copy of the article coming out in the *Post*."

She wrote quickly. "May I pass this to Marjorie to deal with in the usual way?" Her high voice indica- ted that she still felt on dangerous ground. "I'm not trained to prepare a memo for your initials, even if it doesn't go through or from several other people."

"*What?*" The girl was a flaming genius and— "You can't do a simple little memo like *that?*"

129

She wasn't rattled. "There are numerous procedures to be followed, and all the items in the folder have to be arranged correctly, with the plastic oversheet over the original, before you can sign it and I can have it duplicated.

"I know there are little bits of colored cardboard to be clipped on at certain points, but I forget where one of them goes or what it means, and I don't know how to tell the computer to put the string of initials, et cetera along the bottom of the yellow confirmation copy that has to be printed out along with the original so it can be initialed in the little boxes along the right-hand margin. And I don't know which copy to photocopy when, or how to file it. And I don't understand the distribution code that tells the Mail Room—"

"Oh, come off it, Philanth!"

"I'm not making any of this up! It's really simpler to let someone do it who won't forget to not date it so the rubber stamp can be used at the proper stage. And the format has to conform to the correct pattern. If the heading and the closing aren't just so it'll have to be sent back to be done over.

"Sometimes a memo goes back and forth for weeks until it's all done right, especially since in the meantime someone may decide it's to be sent out in somebody else's name or thrown out altogether, or maybe completely rewritten. And it can be lost for weeks and people have to keep hounding other people to look for it, if the computer tracking system breaks down because somebody forgets to log any stage of the procedure for a given document. Ordinarily if something gets buried on somebody's desk the other secretaries can—"

"Tell Ginny Blake to see me about the memo-writing procedures!"

"It's the same for letters."

"All right. Letters too."

"You'll be taking on a lot," she warned. "Those procedures exist because people all over the headquarters are being required to generate not just authorized drafts but final copies ready to be signed in any one of several offices above them in the chain of command. The requisite handbook of instructions is unwieldy, unreadable, and out-of-date; and trying to streamline the process while maintaining adequate control is difficult because relations between secretaries at different levels can be poor.

"That's inevitable, when their bosses don't like each other because the higher-level boss is an incompetent autocrat, and her government-career secretary can become a bully because she knows the ropes and it's part of her job to enforce every little detail.

"She's justified, of course, by the duty to teach everybody to follow the procedures correctly despite the constant high turnover in staff.

"But the little games of correcting things unnecessarily so a lower-ranked secretary has to get her boss to sign or initial her work all over again can get rather nasty. Especially since it takes years to learn all the little details that various offices deal with, and they're constantly changing, and only one person in the whole Peace Corps may have the knowledge and the authority to say whether a form or memo or letter has been done correctly."

"How do you know all this?"

"I share Marjorie's office, and Ginny Blake comes in with stuff for Marjorie after it's cleared all the other hurdles. Ginny's been 'educating' Marjorie, 'cause Marjorie doesn't like the system much, either: things take too long."

"I should think so!"

131

"I also pick up tidbits in the rest-rooms—especially early in the morning, when women are fixing themselves up after the trip to work. I'm very interested in organizational dysfunction because of a sociology course on bureaucracy I had in college; and thanks to the course and my previous work in offices, I know how to elicit information rather than make people freeze up.

"Of course, I never tell anyone I work in the director's office. I say I'm just a temporary working with IO. That's so big they can't check it."

"I see."

Marjorie had mentioned that Philanth and Sandy "spoke the same language." He had learned to act on ideas as they came to him if possible, for it saved a great deal of time as well as keeping important matters from sliding and small matters from being lost. So on an impulse he asked, "After your temporary job coordinating arrangements for my big trip is over, would you be interested in working for the ombudsman?"

"Uh, no, thanks." She was gracious but definite. "It sounds like a very interesting and worthwhile job, and I'm sure Sandy could use a female assistant, but I have other plans. Sorry."

"At least take *a minute* to think about it," he chided. Having taken a risk in even mentioning the idea to her, he felt such an offer deserved more of her attention.

"I appreciate the thought. But I'd rather work on bigger problems than paper flows or ego conflicts."

That stung. "Find a job where you don't have to do that, and I'll apply for it myself!"

She looked troubled, though she smiled placatingly. "I should have said, more *basic* problems. I only meant that . . . ."

This could be a chance to damage the chemistry of their relationship, maybe even pick a quarrel which he could escalate into a "personality conflict". Lacing his fingers, he sat back to watch her try to extricate herself and gave his voice an unfriendly edge. "Yes?"

"Let me explain it this way," she began. "Sandy's job is important, but it's only making an existing mechanism work better. I'm interested also in *innovation*: creating new mechanisms, if necessary, to deal with fundamental needs still not being addressed. To be specific, there's still much new ground to be broken in helping poor women gain control over their own childbearing." Her direct but deferential manner emphasized the analytical distinctions which retracted her unintended insult.

She added pointedly, "One could wish that the organizational environment of Peace Corps would allow *it* to focus on such a fundamental goal, using the organizational apparatus already in place. But apparently it isn't allowed to because too many Americans still can't deal with sex rationally—not even in behalf of vague masses of dark-skinned, illiterate people who live a million miles away. They'd rather have rock concerts to give them a few handouts during famines, just enough to enable some of the wretched people to live to watch their children die of the diseases associated with malnutrition."

"I see." He added deprecatingly, "Well, I don't even know that Sandy'd *want* an assistant. I just noticed that you two seem to be similar types."

"I'm aware of the honor you did me by suggesting it, Mr. Bjorklund." Her sweet voice carried an apology it was impossible not to accept. So much for trying to create personal conflict with Philanth!

"As to our being similar types, it's true that both Sandy and I have studied sociology. And it looks like

133

Uncle Sam needs a sociologist in every office, judging by how much there is to be done around *here*." She dimpled as if she had said something naughty.

She started to get up but paused. "Regarding the scope of *your* position, Mr. Bjorklund: you really shouldn't get into the nitty-gritty of improving the correspondence flow. Could the deputy director handle it? Or the ombudsman, just because he's obviously qualified?" He noted that she did not suggest the obvious choice, the chief of staff.

"I think it's simpler if I just try to improve the situation with a new policy directive.

"In fact, you draft it for me: say that the office of the person who is to sign a document is responsible for the correctness of its final form, not the office which drafts it. The authorized draft should be transmitted with a diskette for quick polishing. And while you're at it, have Sandy check around regarding other needed changes along the same line."

Philanth looked dubious. After a moment she said, "I applaud your move, but it might be wise to put these changes into effect while you're here, say, after the first of the year."

"Carole was right: you show promise."

She smiled dismissively. "I'll talk to Sandy."

Chapter 6

FINAL PREPARATIONS

The following Tuesday morning Marjorie gave him a lecture about the problems which were being caused by his delay in making official the "fact" that Philanth was to travel with him: handling the visas and shots on the present basis was sticky, and fudging in communications with all the people involved in the project was creating more and more awkward situations. This was completely unfair to Philanth, since she was being forced to act as if an important element in the arrangements for accommodations were an established fact when she had Marjorie's assurance that it was definitely untrue.

Moreover, it was touch-and-go as to whether Tony could keep his mouth shut, since he had recent intensive experience of the conditions in equatorial Africa and strongly endorsed Marjorie's wish to be replaced by Philanth.

He was about to go into another planning-review meeting, so in the spirit of one has to take his first dive off the high board he walked into the meeting and began it by making the announcement.

In explaining the change of plans he did not spare Philanth's feelings. He made it clear that he had been very reluctant to accept her as a substitute for Marjorie and had subjected her to considerable testing before deciding to go along with Marjorie's recommendation. He even said that he was willing to take her only because he needed all of the more *experienced* staff keeping things on an even keel during his absence. He made serving as his assistant during this trip sound like a job that only an idiot would want; he kept his tone so negative that by the time he was finished he felt that Philanth must have the sympathy of everyone present.

Just before adjourning the meeting he emphasized to everyone that Philanth was to be immediately and thoroughly informed of all changes in the plans from then until they returned, noting dourly that she was "going to need all the help she can get." As the meeting broke up, Philanth actually got hugs from two of the other female staff present, and Jeff jokingly offered his condolences for "having to spend five weeks with that slave-driver". Tony barely managed to keep silent about his satisfaction with the substitution; his lack of surprise was evident.

As he watched the scene he hid his satisfaction at having created a protective attitude toward Philanth. His unconcealed dissatisfaction with her would also help to protect him from any suspicion of improper behavior in all those connecting hotel rooms.

As he climbed into bed that night he remarked, "I announced to the staff today that Philanth is making the trip with me."

Carole was opening and closing dresser drawers. "I know," she replied sourly. "You do postpone things you don't want to deal with."

"It went well. Not much chance she'll be hampered by the jealousy of people she'll have to depend on in the crunches."

Carole sat down on the corner of the bed. "Speaking of Philanth, any objection if she helps Kate at the party in L.A.?"

"Wha-at?" He rose onto his elbows to look around at her. "*What* party?"

"*Char-ules*! The fund-raiser for your war chest, at that huge house you like so much in Sherman Oaks, looking out over the whole San Fernando Valley! We're hoping to clear seventy-five thousand that night, if the parking problem can be worked out for a reasonable amount.

"We've arranged for shuttle service from a church parking lot near the bottom of the hill, but—"

Rolling over to face her, he repeated, "*What* party?"

Carole's tone expressed an effort to be patient. "When we decided on a twenty-four-hour layover in L.A. for you to rest before coming on back to D.C., I naturally checked with Marty about tacking on some political contacts. He urged me to agree to a party.

"Marty's handling the mailing of the invitations. He's put together some very good lists, and he's using the occasion to start lining up your old stand-bys for other events later, especially in Beverly Hills."

"First *I* heard."

"Oh, I *must've* mentioned it."

There was no point in arguing about it, so he scrunched his face back down into his pillow and said nothing. These things happened. Even Carole was not infallible, though she sometimes complained that he seemed to expect that of her.

"The invitations are beautiful," Carole said brightly. "Silver and royal blue. Want to see one?"

"Nope." He wondered how many of those swank mailing pieces he had seen in the last twelve years.

Then he thought of all the "SPEECH"es dotting the detailed five-week schedule which led him at last to the refuge of his little apartment in his old district.

"Oh. Gosh. Guess I'll have to make a speech."

Carole burst out laughing. "Yes," she answered firmly, "and it had better be as good as any that month or the preceding one. And without any help from the briefing materials about desertification and livestock marketing they're putting together for you at headquarters." Carole seemed to make a point of demonstrating how well Marjorie and Philanth kept her informed about his work.

She resumed, "Shall I talk to Philanth and Kate about Philanth's putting on a maid's uniform and handing around hot hors d'oeuvres? The caterers won't mind if we don't spring it on them at the last minute, and I think Philanth would like the chance to see how we do things in California, since she put together black-tie affairs for the Hunger Foundation."

He could see Philanth now, taking the ladies' furs and bobbing a curtsy. "Okay by me. May as well keep her working."

"I'll see that Marty sends her a check afterwards.

"Now, there's one problem: she has no place to stay that night, because Marjorie would have stayed with her son and his family. And Philanth tries so hard to be frugal that I hate to see her use her money on a motel. It wouldn't be justified as a Peace Corps expense, because she could've flown on to Washington from Los Angeles while you stopped off for your personal business. Besides, there'd be the transportation down the mountain late that evening.

"So would you mind if I asked Kate and Ned to put you both up after the party? Then next morning you could simply drive her back down to the apartment so Gus can take you both back to the airport."

"That sounds best . . . if Kate and Ned wouldn't mind. They've certainly got enough room." That hillside home with its three big floors plus a separate apartment could have housed four families, but Ned and Kate lived there alone. Ned worked at a nearby movie studio and had won an Oscar in special effects.

"They'll like telling people you stayed with them.

"That leaves the afternoon you arrive at LAX. You'll want Gus to take you straight to the condo, of course. Now, what becomes of Philanth?"

"*I* don't know," he snarled. He had been ready to fall asleep when he hit the bed, and then all this business had started. Carole's need to clear such little details with him could be a nuisance.

"Well, she'll be pretty tired too," Carole observed. "It would've been cruel to schedule her to fly right on to Washington: to be sure of making the flight after coming all the way from Manila she probably would have had to languish a long time at the airport, too tired to do anything. And you agreed you'd prefer she just completed the trip with you so you could go on giving her things to do up to the time you got back to your office.

"So I thought, why not let her rest that afternoon on our sofa bed, while you do your telephoning and take a nap in the bedroom? Just let Gus know you've kept her with you so you could continue to work with her all the way back to Washington. It's simplest in terms of logistics, and safer and quieter for her than a motel, as well as cheaper."

"No big deal," he agreed sleepily, "she can snooze or work in the apartment that afternoon."

"I'll talk to Kate and Philanth."

"Fine, fine." At last she let him sleep.

*　　*　　*　　*　　*

The Saturday before their Monday departure he mentally wallowed in the luxury of working at home. He had the house to himself and was in an upbeat frame of mind. He had completed his shots and knew what pills he had to take when in order to minimize the chances of illness; his briefings from the State Department were completed. All of the final preparations seemed to be falling into place.

He began his day's work by turning on the computer he would take with him and glancing through the contents of the computer diskettes which contained his briefing materials for the trip. He told himself he wanted to see whether he could spot gaps which the staff could fill by sending him supplements before he reached the pertinent countries: additional program or project data, staff performance evaluations, officials' backgrounds, or whatever. Actually he just wanted to "get acquainted" with the material like a child with a new set of toys. Les had personally and pompously delivered the diskettes and accompanying folders to him the previous afternoon, and he had been too busy or tired to examine them until now.

Each diskette's label had been painstakingly filled so that its contents were clear. There were several "program background" diskettes to help him master technical information which might be useful in any country with a program of the given type: maternal and child health or forestry, for example. But the diskettes in which he was most interested now had been organized so that each contained a group of countries representing one section of his itinerary. Each of the country-by-country documents contained sections and subsections, always in the same order.

First came the "Travel Notes", which had also been printed out and compiled in an appropriately labeled folder. These pages had been formatted for folding into a pocket of a man's suit jacket, and each one began with the information he could give others about how he was to be contacted most directly during each successive period of his stay in a given country. After noting any time-zone change upon arrival, the rest of both sides of each page described his movements during a given interval between international plane flights, noting transport, eating, and hotel arrangements plus names and backgrounds of key people who would meet him at each juncture. For example, if his driver from an airport was to be a country director with whom he particularly wished to talk one-on-one, he had better have that aspect of his agenda in mind when he stepped into that car.

The next section, of variable length but also duplicated in a folder of independent sheets of paper for carrying around in his pockets, provided his anno-'tated day-to-day schedule for that country. This included descriptions of "social" and ceremonial events and key individuals involved in each event, "talking points" for specific occasions, and meeting agendas.

Then came the country and country-program briefings and the country-post personnel information. Even the program briefing papers had been laid out so that they filled pages without spillovers, to facilitate quick location of sections while he was on the road and inserts of new material when the first drafts of his findings were to be assembled after his return. This dense technical material was made accessible also by a consistently applied set of bold-face and underlined subtitles.

A fair amount of push must have been required to get those country and program briefing papers chiseled into such perfect form, considering that they all had had to come from different authors or groups of authors who would not willingly suffer any alterations in content to produce the beautifully practical consistent format which had been achieved. He was glad Les was a chief of staff who knew how to crack the whip, even though he didn't like him.

The overall organization plus the format made it easy for him to find his way around all of the material quickly. Moreover, the whole thing had been so well structured and edited that it was actually a pleasure to read despite also being succinct.

Quite a piece of work. It set a new standard for his future travels. It was even more impressive because he had not been consulted: it had simply been taken for granted that he would be provided with briefing book and schedule as usual, then decided that he would take the risks of traveling through all those countries with a computer in order to travel light with voluminous briefing material and expedite production of his findings after the trip. Philanth would use the daily diplomatic pouch connecting all the resources of Washington to the U.S. embassies to maintain regularly updated backups in case misfortune befell his computer or his diskettes.

The material was important to him not only for its practical value but also because it brought together the results of many people's work over a period of many months. Including his own. During the year since he had been confirmed as the new director and in the name of learning his new job, he had spent many hours closeted with the program specialists who worked at the other end of the eighth floor and many more hours on the phone and at the Agency for International Development and the State Department. He had interviewed practically everybody in the upper layers of International Operations except newcomers. He had called in miscellaneous personnel from all over the headquarters: a nurse, the head of overseas computer operations, and many more.

In the course of these careful, sometimes confidential consultations he had developed well-defined objectives for this trip. These included getting his own answers to questions regarding specific Peace Corps personnel and host-country officials, as well as about programs, projects, policies, and procedures.

Out of this process he had developed his wish list

of countries for this trip. He had sent the list downstairs to the director of the Travel Office, indicating the minimum time he wished to spend on each country. With her astonishing computer and expertise she had given his list a working over, then briefed him. After studying the tough choices his planners would face because of the vagaries of the commercial airline services involved, he had set decision-making guidelines providing his priorities regarding countries, individuals, and types of activities.

Applying these had reduced his "ideal schedule" to little more than a series of meetings at air terminals. But the air terminals in these countries were often totally inappropriate settings for meetings. Which countries he actually favored by going into their capitals for a night at a hotel—and attendant receptions, dinners, breakfasts, luncheons, and meetings—was also determined in part by where larger gaps fell in the airline connections and by his regular need for a shower and a decent night's rest. There had been extra hassle over whether he was really required to use U.S. carriers where a far more convenient flight was available, though the safety of those other airlines had also been a consideration.

This model was complicated by several other sets of considerations.

One was diplomatic. Less "hands-on" administrators in his position had followed the far simpler and more comfortable procedure of flying off to one or a few countries at a time, making a few congratulatory speeches and "factory tours", accepting ceremonial presents and getting their pictures taken, and coming home. That was all very glamorous, and it made them visible but non-controversial and hence "successful" to those whose favorable impressions affected their fortunes when they left Peace Corps. However, this practice apparently had fostered development of ordinary hospitality obligations into local *de facto* protocols: when a Peace Corps director came to town his hosts assumed it was *necessary* to hold a reception for him here, a dinner and reception there.

But for him to fulfill his role as director basically by just going around talking about how wonderful Peace Corps was and how much it deserved more support wasn't good enough, to his way of thinking. Because of his years of involvement in foreign aid issues in Congress he was deeply interested in how Peace Corps' programs might be improved to meet changing priorities and more realistic conceptions of enduring needs. Therefore he could not have proceeded the way some people felt he must and still have attempted the searching review of Peace Corps operations which he believed was needed at this time.

In-country and international recognition gained by the news media's coverage of the ceremonial aspects of a Peace Corps director's visit to a country was no doubt conducive to Peace Corps' effectiveness, but even the best "image" was not enough to feed and educate people. The best long-term investment in Peace Corps programs was promotion of sound practical results, not public-relations blurbs which plastered over the problems.

Moreover, as in any other agency, operations had to be monitored as wisely and carefully as possible. Otherwise any deterioration would feed upon itself, and eventually the resulting waste and demoralization would be leaked to journalistic watchdogs.

So this trip was serious work. Yet even country directors were accustomed to a tradition of Peace Corps directors whose qualifications for the position consisted of having influential friends and/or having been involved in a winning presidential campaign, and some of them had had difficulty in grasping the fact that he was not interested in playing the starring role in mob scenes which they aspired to direct. He had had to avoid offending or hurting them, but he had had to be firm, since yielding to such expectations and desires in each country could have turned every stop on his long itinerary into a series of just the sort of ceremonial affairs he sought to avoid.

However, some of those parties with conflicting scenarios for him were not his subordinates. They were other significant players in the international scene. He could only try to balance their agendas against his, plead scheduling problems when there seemed to be an irreconcilable conflict in priorities, and then have his staff work to make the best of the compromises. Thus the development of his schedule had required some stroking and hand-holding via overseas calls from him personally, and now there were ongoing attempts to encourage officials on the guest lists for the "social" affairs in his honor to come prepared for substantive discussions. Those with cultural backgrounds similar to his had welcomed the suggestion, being as loath to squander an opportunity for substantive face-to-face discussion as he was.

Another source of complications lay in the magnitude combined with the sensitive nature of his trip: because he had tasks unique to each country and because the countries differed widely, he had to accord very different treatments to various countries for reasons which must not become apparent. This situation raised the possibility of damage to in-country staff morale from invidious comparisons. For example, while he was to spend two days and nights in a very small country, staff serving a nearby sparsely populated country in a remote city would have to travel to meet him in the capital of a smaller country.

Such decisions had been implemented so as to avoid imposing excessive hardships or otherwise causing hard feelings: asking a country staff to travel had been made a reasonably attractive and cost-effective opportunity for them to exchange information and receive in-service training with their counterparts in another host country with the same regional technical problems. This process had required determining whether a country's Peace Corps staff would receive enough benefit so that they would (honestly, in terms of personal agendas) regard the proposal as an opportunity and what would be the cost and quality of facilities available for travel, lodgings, and meetings. Planning for dovetailing his agenda for his trip with this typical type of program-support activity had been the province of OTAPS, the Office of Training and Program Support, but of course every aspect of this undertaking had had to be carefully coordinated.

He had confined such multi-country conferences to cases where the airline schedules and routes made visits to particular little countries excessively difficult in relation to his overall trip. In an instance where it had been determined that such a conference could not be justified in terms of benefits to the staff involved, he had agonized over all of the factors in each case and then regretfully dropped two little countries from his list.

Since he had thoroughly warned his planning team at the outset against leaks which might reach country posts through the country desk officers in International Operations, staff in those countries should not know what they had missed. The standard line was that the sixty-some Peace Corps countries he would not visit had no reason to feel slighted, since they could always hope for a visit from him later on.

Besides the possibility that posts would feel they

were competing for the honor of his visit, he had the opposite problem: they might learn the truth. Because of the constant communication between headquarters and the posts he had to prevent scuttlebutt which revealed that a country he had chosen to visit must have problems he wanted to address personally.

His visit would then seem to reflect negatively on in-country staff even if they were being heroic in an impossible situation. If on the other hand he let it be rumored that he felt some of the country staffs' problems seemed to call for his personal intervention in their behalf, which was another potential aspect of his trip, even that could make his tasks more difficult. As for those posts where personnel might in fact be part of the problem, it went without saying that he had power to advance the careers of people under him, so the possibility that his visit might actually threaten anybody should be left implicit to prevent unnecessary distortion of his findings.

Thus how he actually made the final decisions as to which countries to involve on what basis required a great deal of discretion even in his dealings with the Peace Corps staff working most closely with him. Ideally this trip was to be regarded as just his own way of trying to prepare himself to fulfill his responsibility of accounting to Congress for Peace Corps activities during spring budget hearings.

Keeping anyone from becoming concerned about his intentions was not as difficult as might have been expected, since still other variables had obscured the decision-making process. Some of these were quite non-quantifiable and even speculative factors: local conditions which might make travel more difficult; the nature (as contrasted to the numerical size) of Peace Corps involvement in a country (past and prospective as well as current); and the often dramatic difference between a country's geographic extent and its population. Most useful were the diplomatic considerations such as the character and stability of the host government: since this factor involved impressionistic and classified information it provided him with a particularly useful cover for many of the decisions regarding this trip which were ultimately his alone.

To be sure he always had the right questions and information ready at the right moment regardless of his physical and psychological circumstances he himself had to insert the sensitive material he had been accumulating where he would see it in his briefing material at the appropriate times. He would do that during his plane flights in the earlier part of the trip. After he had left each group of countries and had inserted notes of his findings he would pouch the sealed diskettes for those countries to Marjorie for safekeeping.

He was eager to get started on the final phase of preparing his briefing material. But he must put first things first. He shut off the computer and set the diskettes aside, promising himself he would give them as much attention as the weekend allowed after he had done his duty on the more immediate matters. Resignedly he turned to the unusually large stack of Friday-afternoon deposits in his "in" box.

There was an internal memorandum explaining the proposed reorganization of the New York regional office and the Puerto Rico training operation.

There was a memo from Sandy, confirming horror stories regarding the hardships of the staff responsible for getting office supplies from the Supply Room. There weren't enough staff to keep it open even during the few hours a week when it was supposed to be open. Often there were long waits in line, and the erratic flow of basic supplies to Peace Corps headquarters created scarcities which likewise made getting supplies evocative of grocery shopping in one of the old Communist dictatorships.

Les or the deputy director or the associate director for Management should have dealt with this. But Sandy knew that all three had opposed the appointment of an ombudsman, so he had sent this memo to him. Oh, when was he going to find the time to find ways to get rid of those three, much less replace them, if he was continually swamped by problems they should have been dealing with?

Maybe there *was* no way he could get rid of them. Maybe he could not afford all the bad press and problems with the White House they could cause. He was trapped between his own ideological enemies and those who really believed that no one who had not made an important contribution to the winning of the last presidential election should have a significant position in the federal government.

Right-wing columnists would enjoy adding to the scars Peace Corps had received from the sexual harassment scandal which had precipitated the departure of his predecessor. They could construe his getting rid of any of the President's loyal supporters as proof of his disloyalty to the President who had appointed him. Such firings could also be used to suggest personal inadequacies on *his* part, especially if he got rid of more than one individual.

Bad press would arouse the virtuous indignation of his party's True Believers. They would feel that he of all people had no right to fire any party loyalists, regardless of how they did their jobs. But for publication they could wring their hands over his poor handling of Peace Corps' interests.

He could see the headlines now. The Heritage

Foundation had published a broadside assault on one of his predecessors, calling her "an annoying thorn in Reagan's side" and noting that making personnel appointments without political clearance by the White House had been one of her sins. Evans and Novak had run a column, "Liberal to the Corps", because she had retained a Democrat as one of her administrators for a while. The discussion in the press had then been kept alive by revelations about her struggle to remove her deputy director and by manufacture of mini-scandals when some of her appointments turned out badly. She had even been treated as an "uppity woman" for working at Peace Corps instead of helping her husband campaign for the Senate back in Wisconsin. Having violated the taboos which protected the politically sensitive spoils system, she could not put a foot wrong without her enemies in the press spilling ink over her.

Now the reverse situation applied: he already had enemies, and he needed to get rid of administrators who had been hired only because they *had* been given "White House clearance".

He knew that his relationship with the White House was delicate because the President had "jokingly" gone out of his way to extract his promise to "behave yourself, for a change" if he were given the appointment. The injunction had been a clear warning that he was in danger of being fatally stigmatized by party insiders because he had not "always voted at his party's call and never thought of thinking for himself at all". The President's acting as if appointing him were a big favor had in fact been hypocritical: because of his outspoken criticisms of sensitive aspects of current foreign policy, the President must have been as glad to get him out of Congress and into the bureaucracy as he had been glad to go.

He was brooding again. He turned back to his "in" pile. He wrote a note on Sandy's memo, asking him to obtain background information which could be used to frame solutions to the problems associated with the Supply Room. Into an inter-office envelope and into the "out" pile.

There was a letter from one of the three senators who really did their oversight job on Peace Corps, asking for his comments on the report provided to her at her request by the Peace Corps Congressional Relations Office. The report dealt with personnel turnover data and exceptions to the five-year rule, and she had questions about it: was the five-year rule exacerbating turnover of all the highly technical staff to the extent that the efficiency of operations was excessively impaired? And were exceptions granted too erratically?

The senator must have received and paid attention to a letter from someone with inside information, for as far as he knew, this was the first official recognition that inside favoritism or capriciousness affected granting of exceptions to the five-year rule. Moreover, the tenor of this inquiry was the opposite of the position expressed by senators during his confirmation: the five-year rule had seemed to be held sacred, up until now. At least there had been no recognition that it caused turnover which was much more disruptive than the framers of the rule could have intended.

He wanted to postpone this matter, but there would be no time for it later either. Besides, inquiries from Congress had to be given high priority. So he pondered the matter for a while, then wrote a note for Marjorie to turn into a letter he could sign before he left the office on Monday.

In it he suggested that the rule should be refined to ensure an impartial review process which would grant extensions on a case-by-case basis whenever the interests of Peace Corps would be served. He noted that government salaries made it impossible to ever replace some of the invaluable people who were forced to leave Peace Corps only because the five-year rule was used to trim staff due to short-term budget pressures. He was sure that other inequities and miscarriages abounded, even though the goal of preventing bureaucratic stagnation was laudable.

There was a memo regarding another death of a Volunteer. He had attended the funeral at St. Alban's the previous afternoon. Either this memo had gotten shuffled out of order, or he was more behind on his paperwork than he had thought.

There was a memo from the director of Public Affairs reminding everyone that Peace Corps staff did not cooperate with writers doing articles about Peace Corps for magazines "which have a primary focus on sexual exploitation".

There was a memo from himself to all staff regarding the unpleasant headlines given Peace Corps by the *Washington Times* two days earlier and a draft of his reply column for the op-ed section of that newspaper. He made changes in the draft and initialed it and the memo.

There was a draft of a letter from himself to the National Council of Returned Peace Corps Volunteers Newsletter, in response to the letter which was attached to it. He noted objections for a rewrite.

Next came a folder regarding the Volunteer of the Year. He set that aside to deal with on the plane on Monday evening. He could pouch it back to Marjorie from Morocco.

The conference facilitator for the Inter-American Operations Country Directors' Conference had submitted a first draft of the plans being developed for the conference. He put it into the stack for the plane.

There was a thick, stapled sheaf of Weekly Activity Reports, the distillation of write-ups of what practically everybody in the headquarters had accomplished during the previous week. For the plane.

Yes, even the WARs were going to follow him around the world. He was going to try to keep up with his basic office duties throughout his extended absence from Washington. He had no intention of letting "his" deputy do anything as acting director if he could prevent it. Every day he was gone, he and the headquarters would exchange packets of material via the State Department's regular embassy pouch, with the assistance of whichever country post he was dealing with on that day; and he would stay as available by phone as possible. Philanth would have plenty to keep her busy when she was not staying on top of the paperwork and logistics for his moment-to-moment in-country activities and travel.

The next item in the stack was a preliminary report on the study which was being done of what cuts would have to be made in which programs if Peace Corps did not get specified amounts of increased funding for the next fiscal year. This was only an early phase of preparation for the appropriations process which would require him to testify on the Hill in early spring, but he wanted to study the alternative projections carefully and make note of his questions. He put the report into the stack to be dealt with on the plane.

It had been pointed out several years back that Peace Corps was receiving less federal money than the government spent on military marching bands. All sorts of such comparisons could be made. But no matter how one analyzed the numbers, the results in

153

human terms were heart-rending—not to mention incredibly stupid in terms of the country's interests.

He decided he was going to have to spend more time going around the country in public appearances and do more to stimulate Peace Corps' efforts to involve the private sector. He wanted to numb himself to the pain of presiding over more and more rejection of qualified potential Volunteers and more and more rejection of proposals for badly needed projects. They *had* to get more money from somewhere! He made a note telling himself to call in the deputy director and staff specialists to talk about that on Monday, to spur them to make progress in getting help from corporations which they could report to him when he returned.

The phone was trilling. Carole had gone shopping, so he had better answer it. Glancing at the clock as he picked up the receiver, he saw that it would be permissible to break for lunch.

The call was from their favorite catering firm. The lady wanted to speak to Carole "right away", so he explained Carole's unavailability and asked whether there was something he could help her with.

"She engaged us for your bon voyage party tonight, but we never did receive the flyer with the map and directions she promised to mail us, and she said we'd need them. I'll be in your area, so I'd like to come by and get a copy of it. If I may?"

He could see the problem. The site of the party was in an area of Fairfax County where one wrong guess could send a hapless driver careening through sylvan glades for half an hour or more before he could even find some road sign which confirmed that he was in trouble. Truly erratic streets and unlighted, unconscionably poor signs made detailed instructions for finding any place for the first time a folkway of the Washington area anyhow; traveling through the dark countryside without detailed directions and a specially prepared map, the caterer could be in for a nightmare.

It wouldn't do for the caterer to be touring Northern Virginia while his bon voyage party was mobbed with people expecting attractive delicacies to eat, especially when the caterer had been arranged for by the wife of the guest of honor. He told her, "Of course you can come by, and I apologize for the inconvenience." They agreed that Carole hardly ever slipped up on anything, so they would blame the Postal Service.

Carole was an orderly person who had never given up on trying to get him to remember where she kept things, so he did not have much trouble locating their copy of the flyer which had been distributed at the office.

However, the hand-drawn map was quite detailed, with brief written instructions to supplement it, and if he gave it away he himself might have trouble in making it to the party in his honor. He would have to make a copy for the caterer.

When he turned on their desktop copier it started flashing a red light at him. He swore under his breath. It looked like Carole had finished a copying job just as the machine ran low on ink. But as with every other maintenance area except the car, keeping the copier operational was Carole's responsibility, so he couldn't be sure what was wrong, much less fix it. He did know that when a red light was flashing he could do very expensive damage if he tried to use the machine anyway. So no copier. He switched it off, swearing again at the helplessness that usually got him out of the sort of tasks with which he didn't want to be bothered.

Over the years full of meetings and social gatherings this problem had presented itself before on occasions when he had not had access to a copier, and his solution had been to use a carbon paper. He felt quite clever at having come up with that idea. He stepped over to the small metal chest by the typewriter and took a sheet of carbon from the appropriately labeled drawer, reflecting that it was a very good thing to teach little girls to be orderly, for otherwise the men who depended on them later in life would go crazy.

Hey. This carbon had been used once before, and it was the heavy blue kind which was supposed to be used only once, for all the ink had come off wherever pressure had been applied.

They had had such carbons around the house for years because of some long-forgotten political project, and these once-only papers had gotten mingled with the regular carbons. Carole must have absentmindedly put it back like a regular carbon instead of throwing it away.

At first glance he could see that it had been used to reproduce a carefully neat and closely spaced letter which Carole had written by hand. This was strange, since Carole seemed to use her typewriter or their desktop computer for practically every written communication except the grocery list. This must be some special personal document.

The backwards script said, "My dear Philanth." That was strange too, since Carole and Philanth were in frequent contact by phone and presumably were together at that moment, having a fine time in some department store. The date in the corner indicated that the letter had been written that very morning, yet their shopping date had been made some time in advance.

Mystified, he took the carbon paper to the mirror in their bedroom. He and Carole had no secrets from each other, so he felt confident that she would not mind his reading it.

In the mirror the letter was perfectly legible.

My dear Philanth,

During this trip Charles' time will be especially precious, and you know that it will be your job to see that he doesn't waste a minute if that can be avoided. You likewise must do your best to use all of your time well.

During this trip you will be sharing Charles' burdens, and because you and he are so much alike in the most important ways, I know that before long you both will begin to desire intimacy. Neither of you should waste your valuable time fighting these impulses, because I love both of you, and I have no objection.

His hand jerked, dropping the paper as if it had caught fire, and he grunted. He bent and fumbled before he could manage to pick it up. For a moment his eyes wouldn't focus, he was so stunned. He could not believe what he had just read.

He saw his bewilderment reflected in the glass as he tried to position the paper so he could read the opening section again. At last he steadied the carbon.

Yes, that was what it said. He wasn't dreaming. He read on, a sick feeling beginning in his throat.

You may object that you feel you must keep your distance from this beautiful, loving, and charismatic man. Instead you should set aside any fear for your own well-being, trusting in my assurance that you may continue to give him whatever he needs from you after you return to Washington and that you will never receive anything from us in return other than gentleness and caring. Because you both are wise, there need be no pain for either of you, only enrichment from mutual giving.

I know you are dedicated to a code of loving abstinence. But I am sure you are selfless enough to put aside your agenda for yourself if I ask you to defer to my fuller understanding of how your ultimate values, which that code is only designed to promote, can best be served.

I know what a devoted husband Charles is, so when he needs to be assured that I want this for him, show him this. Please do this also for me, to help him in our great work, for his task is such that he needs every form of help he can get.
Love,
Carole

He rubbed the flimsy sheet between his fingers, but it still did not seem real. He sat down on the bed, trying to make the whole thing sink in.

It couldn't mean what it said. But it could hardly be a joke. It had to be some kind of trick. No, it simply made no sense at all.

He made himself believe that it was real. Assuming it was, how ought he to react? His first reaction had to be embarrassment. Then anger.

Did Carole really have such a low opinion of his fidelity? For her to imply that he would be seriously tempted by another woman, let alone during a separation of merely five weeks, was just plain insulting!

But whatever Carole intended with this incredible prank, what really made him scarcely able to see straight was the fact that his wife had deliberately undercut his dignity by giving such an outrageous letter to his new, young assistant! Maybe he could be called a little strait-laced even by other people who took responsible, mature views of the proper place of sex in adult life, but he had no doubt whatsoever that this absolutely exceeded all bounds of decency.

As well as of good sense. He wondered whether Carole's mind were somehow becoming impaired.

Had she become frigid, and was she handing him a concubine like some barren wife in the Old Testament? Had he complained too much about her refusal to accept his advances?

Actually he had hardly complained at all. He had been too bewildered, and stretched too thin by his attempts to do right by the Peace Corps without abandoning his commitments involving his own political future.

He could not imagine how to confront Carole about this. He did not want to show this much anger toward her when they had only one more day together before their long separation. It would be better, perhaps, to simply pretend he had not seen this and let his normal behavior speak for itself during the weeks he would be away.

But she had often commented, sometimes in exasperation, about his inclination to postpone anything unpleasant. Was she counting on that now?

No, because she could not know that he had found this letter.

This would gnaw at him until he had it out with her. This was not a matter of risking a quarrel, it was something washing around inside him which had to be stabilized.

How could she have spent most of her adult life advancing his career and then taken the risk that something like this could fall into the hands of someone who would sell it to a tabloid? Why, she had even made a copy. Counting this carbon and the original, *three* copies!

Without using the desktop copier. Had she been in too much of a hurry to replace the ink supply, and therefore whipped out a carbon? Not likely. The copier must be waiting for a visit from a service man.

Anyway, how could she do this, in terms of their relationship? He was mystified.

One thing he knew: this had to be destroyed.

He tore up the carbon paper and threw it into the wastebasket; retrieved the pieces of it; and took it downstairs. He found some matches and burned it in a corner of the tiny, bricked-over back yard. He collected the ashes and returned to the second floor to flush them down the toilet. A politician couldn't be too careful.

Damn. The caterer already was clacking the knocker at the front door. He quickly washed his hands and went down to let her in.

He laid on lots of charm to excuse the delay, and then, leaving the poor woman to cool her heels in the sitting room, he went back upstairs. He returned with a fresh sheet of black carbon paper, the map, and a sheet of typing paper.

Feigning what he liked to call "typical male ineptitude" when he wanted to be excused from something routine, he let her create her own map and copy the instructions which supplemented it. While she did so he went to the kitchen and assembled his lunch. She left, thanking him and apologizing for the inconvenience.

Chapter 7

A GUIDED MISSILE

Carole called at five o'clock to say she was taking Philanth to supper and tell him what to fix for himself. She arrived home nearly two hours later, making it clear that she had barely enough time in which to gift-wrap two packages, shower, and dress. He didn't want any "vibrations" between them while they were the center of attention at the party, so he postponed the scene over the letter.

When they arrived at the deputy director's house and rang the bell the door was opened by Philanth. She was wearing a cream-colored dress cut exactly like a man's long-sleeved shirt except that its thick, stiff fabric kept going all the way down to her ankles.

"I like your dress," Carole said.

"It cost seven-and-a-half dollars," Philanth told her. "The same thrift shop where I found my coat." She touched one of the gold-painted buttons which marched down the front and fastened each cuff as she added, "I'd replace the buttons with something less cheap-looking, but they'd cost more than the dress. Besides, I *wanted* to look unprepossessing."

"It's very modest," Carole commended her. It certainly was. It also made her look plump.

"Well, the last thing I'd want is to look sexy." Her fetching hairstyle, a crown of braid softened by curling tendrils, made the statement suspect.

"Heaven forbid," Carole replied with amusement.

He was bewildered by the straightforward way both women played the exchange. He could only conclude that Carole had decided not to give Philanth

161

the letter after all.

Expressing the hope that they would "Enjoy the party," Philanth went to join a couple of men who had their heads together. Meanwhile he and Carole were being descended upon by their host and hostess, who were expressing their appreciation for Philanth's "faithful service as our doorperson".

As he and Carole were escorted toward the nearest refreshments he passed Philanth's group and recognized the dark-haired man facing her as Kingsley Merriweather, one of the area specialists from the State Department who had briefed him for his trip.

"Look at the blond," Carole was whispering in his ear. She indicated the tall man beside Philanth. "That must be her date. He's CIA."

He turned to her. "Really?"

Carole just nodded, her hands preparing to slip their hostess the brown paper bag containing her gift-wrapped packages.

He knew he was going to be kept busy for the next couple of hours. Like a robot in a movie, he could switch to the appropriate program and start shaking hands and exchanging pleasantries in genial tones.

However, as soon as he could speak in Carole's ear he asked, "What's a CIA man doing *here*?"

"She asked me today whether I thought you should be asked or told about it. I said I'd tell you, and she shouldn't tell anybody else; and she should ask her friend to avoid being introduced."

Later, while Carole stood beside him helping Tony resolve his problems in packing for a man's overseas business travel, he found himself watching Philanth's group still holding forth somewhat removed from the crowd. It had grown to a closed circle of five, including three area experts from the State Department.

The presence of someone identified with the CIA

bothered him. It was essential to the Peace Corps mission that Peace Corps always stay clear of that agency. Indeed, as the recent bombing of the Peace Corps office in Central America illustrated, preserving that separation in the view of others was a matter of life and death for Volunteers.

The young man looked like a professor in his thirties. The large lenses of his dark-rimmed glasses followed his eyebrows below a broad forehead which was emphasized by fluffiness in his well-trimmed blond hair. The face tapered down in short, straight lines to a firm chin; and from unwavering eyes to slit-like mouth to snug white collar and tie, it projected intense seriousness.

One of the three men from State addressed the CIA man directly, and the tight lips of the blond opened in an almost boyish half-smile as he briefly responded. The young man was good-looking, though the dominant impression he gave was of a strong, passionless intellect.

Noticing him watching them, another of the area specialists apparently remembered his manners, for soon all three came over to pay their respects.

Philanth and her "date" hung back, and he was surprised to see the young man flirting with her as he offered her some tidbit from a tray which was being carried around. She let him put it directly into her mouth, and for a moment they appeared completely absorbed in each other.

As the men from State prepared to move away, saying polite things about hoping to be able to assist him during the journey, he told them to stop working long enough to get some nourishment. They obligingly went off toward the buffet.

At once he went over to Philanth, took her elbow as he said to her companion, "Please excuse us for a moment," and drew her to the nearest space where they need not be overheard amid the racket.

He bent to murmur into her ear, but he was biting off his words as he asked, "What the hell d'you think you're doing, bringing someone from the Central Intelligence Agency to a Peace Corps function?"

Her wide blue eyes and blonde braid made her look the quintessential innocent. She gave him a stare as though she didn't know what he was talking about. Then she gave him an amused grin.

"Don't you suppose he can guess what you dragged me away from him to talk about?" Her manner suggested she was saying something facetious. "Maybe he doesn't belong here, but no reason to be rude."

"Don't play with *me*! What's going on?"

"It's too complicated to explain right now. But it's going well."

"What is?"

"Just a—a getting-acquainted . . . so to speak."

"Why here?"

"It seemed convenient.

"Really, it's okay. Stop making a fuss. You'll only cause trouble."

Seeing that he was not reassured, she added in a tone which made it practically an order, "*Just buzz off.* Please."

Shocked by her impertinence, he let go of her. She returned to her escort, who had been watching them. He turned away quickly to hide his anger. Assuming someone in the State Department was behind this, how dare they treat him this way! But he wouldn't have time to do anything about it before his trip.

The climax of the party, the formal presentation of going-away presents, came off as was to be expected. After the gifts had been unwrapped and exclaimed and joked over, the crowd began to drift back into cocktail-party mode.

He saw Carole go to Philanth and slip her an envelope, murmuring something in her ear and squeezing her shoulder. Philanth smiled, murmuring a response, and gave Carole a peck on the cheek.

Suddenly he very much wanted to be there with them. But in the time it took him to make his way over through the crowd, Philanth moved off and Carole was snagged in conversation with someone else. He had to wait for a break in the discussion so he could pry her away.

While he stood poised, watching for his opening, someone offered him a piece of cake, and he ate it. He needn't have bothered: he could scarcely taste it.

As soon as he could he bent to murmur in his wife's ear, "Wha'd you give Philanth in the envelope?"

"A little something for souvenirs. She won't indulge herself, usually, unless it's forced on her."

He wasn't satisfied, but he couldn't risk a scene. He needed to get her outside and away from the house. He grasped her elbow. "Time to go home."

"Don't grab your present and run," she chided him. She took his empty plate from him and set it aside. "Surely there's somebody here you need to talk to."

Having lost track of Philanth in the crowd, he gave up. "Ten minutes?"

She nodded, then added in his ear, "Don't look so stern. A rising politician is supposed to act dashing and charming and on top of the world, not go around as if he carried it on his shoulders."

He found the deputy director and took care of the business he had wanted to discuss on Monday. His last morning at the office would be too crowded without that meeting anyway.

While he waited for Carole to wind up her conversation with Mark and a couple of hangers-on he stood and worried. He hardly ever erred on the side of impulsiveness, and thus he made few obvious blunders. But this time he might have missed a major opportunity. Sometimes erring on the side of caution could prove to have been a catastrophic mistake.

However, he could hardly have rushed over to Philanth and asked her to let him examine the contents of the envelope only to find that it contained a little money.

On the other hand, even such an embarrassing and inexplicable *faux pas* as that would have been better than risking letting the letter pass into Philanth's hands.

But that had not been an option anyway. He had not been close enough to grab the envelope before Philanth opened it.

But he might possibly have kept her from reading it, even if he had left her permanently bewildered and thinking he must be a little crazy.

He was mentally kicking himself for not having done *something* to prevent what might have just transpired. He could have pretended to want to be sure Carole had given enough, then added to it. Ah, hindsight! He would have looked ridiculous, anyway.

As soon as he could extricate Carole from the useless chat about Peace Corps appropriations he said into her ear, "I've earned my bed."

She hesitated, nodded as though recognizing that she had pushed him as far as she could, and let him lead the way through the crowd to their hostess.

As they made their way out to their car Carole remarked, "That was fun." He grunted.

As they started home he was working on how to bring up the cause of his funk. "How sure are you that Philanth really won't object if I work her hard the whole time we're gone?"

"Quite sure. If she's not allowed enough time to sleep she'll probably let you know and leave it to you to decide what's best."

"But will she really not *mind*?"

"She really won't. She feels it's petty to think in terms of her 'rights'. Her vow of obedience is in remarkably good working order."

Philanth was strong on 'the feminine virtues'. He ground his teeth at his image of Philanth visualizing herself as some archetypal sacrificial maiden while climbing into bed with him in some foreign capital.

"Besides," Carole added, "she's very aware that she owes me her job. And she respects you enormously. So she'll do whatever either of us asks of her."

"Including sleep with me?"

"Excuse me?"

Keeping his voice as calm as possible, he told Carole how he had found out about her letter. "Did you give it to her tonight, in that envelope?"

"Yes."

He hit the steering wheel with both hands. "Damn! I would've said something when you finally got home this evening, late as it was, if I'd had any idea that you might not've given it to her yet! How stupid! My God! I must still have been numb from the shock!

"Anyway, I still can't believe you would do such a thing. It's . . . terrible."

"Let's wait until we're home," she replied calmly. "We don't want to get lost or have an accident."

He concentrated on getting them home as quickly as possible. Carole provided her usual assistance.

When they reached their bedroom Carole began straightening the disorder left by her hasty departure as though everything were business-as-usual.

"Well?" he thundered, standing in the middle of the room with his fists on his hips. "What about that letter?"

"Relax, honey!" Carole didn't even pause in her work. "I just wanted to find out how she'll react."

He unbelievingly repeated what she had just said. "*Why?* You gave me a song and dance about how reliable she is!"

"Did you believe it? You did *not*. But I have reason to think we can believe in her whole system of commitments; and if so, I want to know it."

"Jealous?" he jibed without humor.

"Of course not." Carole sat down on the bed. "But there would be considerable satisfaction in knowing we have such an idealistic person working for us."

"Surely there has to be more to it than that!" He sat down in the valet chair, facing her.

"Imagine the peace of mind in relying on her consistency and steadfastness. If she's as locked into her entire personal code as she seems, and if we get her committed to a career with you, she won't leave you because some young man wants her for *his* purposes.

"In that case you may have an assistant who's worth considerable training.

"Remember that Marjorie is ready for retirement. Philanth's so young she might go on being your good right hand until *you* retire. An able assistant who's been with you a long time is a great asset."

"Well, this is not exactly a test of whether she's likely to quit and get married!"

"It's a multidimensional experiment. For example, if we find out that Philanth is as flexible as I suggest in the letter that she should be,—so easily led and so easy to manipulate in terms of her ideals,—we need to know that too. Subordinates that pliant are tools which have to be used carefully, in Washington."

"Speaking of being careful, what if that letter falls into the wrong hands? What if she's unstable? It could ruin me!"

"She's got a psychiatric stigma in her Peace Corps record, even though it's not in writing. Whereas your reputation as a 'straight arrow' is secure."

"You're not worried she might . . ." He choked. "Actually offer herself to me? 'Hell hath no fury . . . .'"

"Nonsense. Philanth's incapable of such a reaction."

"Anyway, it seems to be what you'd call a fishing expedition. And it surely violates the ethics of behavioral research, which you should know as well as anybody. It's sinful to—"

"Yes, yes, yes. But Philanth sets lofty standards for herself. For example, instead of being personally ambitious commensurate with her intelligence, she lays on herself commensurate obligation to set an example of right conduct for those less equipped to reason things out.

"Yet her code is her own work; what she consciously made she can unmake or amend as she chooses. We have important practical reasons for needing to know whether she'll stick by her principles or make adjustments when she's under pressure.

"It may be that she's merely caught up in youthful idealism which she'll outgrow. But she seems more likely to *develop around* her high principles as she matures—yet somehow adapt effectively to the real world. As you did."

He noted her use of the past tense. He surely was not adapting effectively now.

Carole was structured in such a way that the question did not apply: she had once said that she had grown up a cynic, but she wanted to see something decent triumph for once, and he was it.

Not sure whether she still really wanted to help

169

him or might have become so frustrated that she wouldn't mind doing him in, he found himself wringing his hands. "You realize this letter is like a bomb."

"Philanth will be careful. My copy is well hidden."

"My embarrassment with Philanth is a heavy price for me to pay, all by itself."

"You needn't discuss it with her at all. And I believe it'll be worth it."

"But what can this really *prove*?"

"We shall see." Carole got up. "Good night."

Since the argument had begun to resemble a dog chasing its tail, he didn't try to detain her. Carole was unlikely to yield an inch from the position she had taken, and there seemed to be no way he could undo the damage anyway. The best course seemed to be to let the matter drop and devoutly hope that it never resurfaced.

However, he followed her as far as the study door. "Promise me you'll destroy your copy."

"Oh, very well."

"And tell Philanth to destroy hers."

"That would abort the experiment. Tell her yourself. *That* should prove interesting." She turned away.

He stomped back into the master bedroom and shut the door. He was violating a cardinal rule of their marriage, Never go to bed mad or sad. It couldn't be helped.

The next afternoon he worked steadily on inserting his self-briefing notes in the travel diskettes while Carole did her usual careful job of packing for him. Several times she snapped back at him when he snarled at her for making him decide whether he wanted to take this or that. Their tempers were only mildly improved by the latest polls showing that the President could expect to be re-elected on Tuesday, even though that meant his position at Peace Corps probably was safe.

Carole finally pointed out that she was coping with an unusually tough job, considering the range of conditions and occasions for which he would have to be properly dressed and the baggage-weight limits for international flights. "Philanth warned me that there might be an occasion when you all have to be able to carry your belongings at a dead run or miss a flight."

"I'm sorry to be cranky. But it's because Philanth has that letter. God only knows what she's thinking! She *probably* thinks *you're nuts*."

Carole did not respond.

As Carole cleared the dinner table that evening she remarked, "I hope that made you feel better, 'cause I want to talk to you."

"What now?"

Carole resumed her seat opposite him. "I know you're still annoyed with me because of that letter I gave Philanth last night."

"'Annoyed' scarcely begins to cover it!"

"Well, before you go off with her tomorrow there are a few more things you should know about her. They'll help you understand that I do have valid reasons for having given her that letter."

"I don't care how able and idealistic she is, there can't be *any* justification for this 'test'!"

"Don't be so sure."

"Well, I am! For one thing, it's not as though we were going to have to decide whether to ask her to sign a life-time contract as soon as I get back from the trip. I could just offer to make her Marjorie's permanent assistant and take things as they come."

"That might be a reasonable objection regarding anybody else. But Philanth was hired only to help with this trip. She wants to spend the rest of her life working in poor countries. Her Peace Corps service didn't give her the start she needed, so she's begun applying to other agencies. If we decide we want her to stay on after your trip is over we've got to dissuade her from pursuing that goal."

His tone heaped ridicule: "You're going to try to convince me that she's so special we should try to persuade her to change her plans for her entire life in order to work for somebody with *my* shaky prospects?"

"Charles, convincing you of that is partly what the *test* is for. Because if *I* tried at this point to tell you how special she is, you wouldn't be . . . prepared to . . . accept it. You need time to see for yourself.

"Meanwhile you've got to be preparing to persuade her to give up her resolve to make her life overseas. That may not be easy, because of the way she prioritizes her objectives. But you may decide it's well worth the effort."

"Okay," he sighed. "I'll see about it."

"There are some other aspects of this situation you need to know about."

"All right, go on."

Carole took a deep breath and looked him in the eye. "Philanth thinks you're the most wonderful man she's ever met. Only a woman certain of the purity of her feelings could be so candid about her admiration of a man to that man's wife.

"And she fell in love with her last boss before she joined Peace Corps, even though she says herself that he has 'the soul of a reptile'." Carole raised her eyebrows as if waiting for him to add two and two.

"But—But that's ridiculous! Considering the conditions of this trip, by the time it's over we'll be lucky if we can still stand each other!"

"Philanth is a very loving person. She's idealistic. And she has a vigorous sex drive."

"Now, how do you know that?" he exclaimed.

"It's in her journal. I forced her to let me read it when she was in the hospital. She remarked later that if she'd remembered what was in it she would've managed to resist even *my* insistence."

"And knowing that, you're sending her with me? With that letter? What on earth am I getting into? Won't I have enough problems without coping with her . . . uh . . . ?"

"Don't worry. Philanth says in her journal that since the appetite is boundless anyway, it doesn't matter that she'll never make love. She compares it to an addiction: the craving has to be lived with one day at a time."

"That sounds horrible. Not to mention unnatural."

"She regards it as a physical impairment caused by an accident for which she was partly responsible. Her last boss deliberately tormented her, knowing she was in love with him and trying to get her to . . . submit to him. She went around in physical pain for extended periods. Sometimes it hits her suddenly; she compares it to a gigantic mallet that comes out of nowhere and leaves her vibrating. I hadn't known lust could be so excruciating.

"But she believed she had a moral obligation to endure that situation rather than leave it: running away from her tormentor only to save herself would have discredited her—and her efforts to reform his views. That would have left him more cynical than ever about the concept of 'love'. At any rate, that was her reasoning.

"When she knew she was risking a nervous breakdown due to lack of sleep she finally allowed herself to start masturbating. She climaxes easily. It was all that saved her."

"She 'allowed herself'? She has a problem with that too?"

"No. She's basically scientific in her thinking, so she's objective about the need to relieve sexual tension. She knows she could simply accept and enjoy orgasm as a free, harmless source of refreshment and relaxation that's available whenever it's needed.

"Yet she feels she should yield to the need only when it interferes with her obligation to get on with significant matters. She'd prefer to use the tension as energy for additional effort in her work."

"My God! And I thought *I* was dedicated!"

"Besides, it's a tension she still wants to divert into her spiritual development. That's a carryover from her study of mysticism. She regards every orgasm she gives herself as a turning aside from the quest for further emotional or psychological development: self control, understanding of other people's problems, or religious experience which can make her a more effective personality. She works with her sexual need in her journal, trying to draw lessons for herself from it—including lessons in humility. Her ascetic convictions are always at war with her sensual nature: not trying to repress it, but to harness it."

He shook his head.

But he saw a chance to make a point of his own. "People can also be spiritually enriched by making love," he said, giving Carole a direct look.

Avoiding meeting his glance, Carole asked, "But can fulfillment teach as much that's of practical value as deprivation can, if the deprivation is undertaken voluntarily? It's said that we learn nothing from *success*, ..."

"I never considered the question," he replied, and

his tone said that he also did not intend to.

"Well, Philanth is sure that it can't, not when the deprivation is undertaken *willingly*. Whether it's sexual or material, she embraces it as her greatest teacher.

"You see how much sense of inner abundance her character was built on. But you can also see how inexperienced she is. Life hasn't stretched her until she found her limits; hasn't taken all she had to offer until she felt the need to defend what she must have for her own survival . . . except during periods she felt she could endure because she could recover from them later. She still has a complete commitment to giving; and it's reinforced by strong feminine sexual impulses.

"And this is important, Charles: because her system of beliefs was consciously constructed, it's an integrated whole. This is why parts of it that cause difficulties for her don't flake away as she gains experience. Instead she works at adapting *to* it.

"So she's becoming locked into a character structure that puts her at risk. Unless someone or something intervenes to protect her from the excesses of her own altruism, she *is* potentially self-destructive, just as the Peace Corps medical officer responsible for her suggested to her.

"That's where you come in. Working with you will give her the protection she needs, because you're too decent to exploit her to the extent some men would.

"And it *won't* cause problems for you, because she's taught herself to love without desiring to possess.

"In fact, unrequited love is her natural state. She's never experienced any other kind, not even with her husband."

"Then why did he marry her?"

"They saw it as a matter of economic necessity:

175

'Two can live as cheaply as one.'

"But that's off the subject.

"I expect her to fall in love with you. I also expect her to behave quite properly, in spite of my letter."

"And you expect me to pretend to ignore that, and try my luck at recruiting her to work for me long-term?"

"I also expect you to start feeling at least a little affection for her. She's a nice girl. It would be only natural.

"But your acting too fond of her would make her uncomfortable. And if you did anything that seemed improper to her she would not accept it."

"I trust she regards our marriage as a 'given'."

"That's not important. If your marriage seemed weak she would try to strengthen it, even if you were propositioning her at the time. It's her policy in such cases. She says married men always appreciate her more than single—"

Carole broke off because of his laughter. "You make her sound like a model character out of . . ." He paused as he realized what he was about to say: "an old-fashioned storybook."

"Philanth seems to me to be as perfect as humanly possible. And yet I find her lovable. We'll see what *you* think, five weeks from tonight."

"But what if I decide she's too out-of-the-ordinary for me to use, and in the meantime she's decided she's in love with me? What happens then?"

"Why, nothing! She'll go on her way whenever you want her to. She of herself is nothing, so her emotions are inconsequential.

"Besides, emotions pass, like a headache or the flu. Principles, on the other hand, can be eternal. Principles define all that's most important about a person. Principles are everything.

"That's true for her because she's developed a set of principles consciously chosen to give life value, starting from complete nihilism."

"She *truly* believes in *nothing*? So she might be induced to . . . ."

"Use the right argument on her, and she might be induced to do anything."

"*Anything?*"

"Anything she's persuaded is necessary to best serve the value she has chosen to treat as fundamental: fulfillment of the highest spiritual and intellectual potential of humanity, with an endless future for our species.

"So I've devised the sort of test which should be most difficult for her. The one consideration which could induce her to do anything is the one I've invoked: I argued in my letter that by making love with you she can help you in *your* service to *humanity*."

"Talk about exploiting my position!"

Carole went on as if he had not spoken. "That's why this is in part a test of her common sense as well as her reliability: how does she juggle her priorities when she's constantly seeing the enormity of what you're undertaking during this trip, and when hardships have worn down her judgment?

"I envy you this trip, Charles, if you make good use of this opportunity I'm giving you."

That grated on his sense of ethics—yet he couldn't help thinking, This opportunity to spend five weeks of nights in connecting hotel rooms with a sexy, engaging little blonde who's passionate and smart and loving and eager to please and thinks I'm the most wonderful man she's ever met?

Or this opportunity to play head games with her, which might be interesting too—but the Devil's work, especially given an innocent victim.

"So you mean for me to play head games with her." He spoke softly, but it was an accusation.

He reverted to practicality. "I won't have time."

"With so much hard work to do, your mind will have to take time off, occasionally. This will provide it with constructive recreation."

"That's a perverse notion of recreation."

"You misunderstand. I expect it will be pleasant to watch how Philanth conducts herself when all her internal pressures on herself are combined with the stresses of the trip, plus the strain of this 'test'."

"It won't be any problem for her. She seems like a sensible girl."

"But constantly with a man she finds immensely attractive, morally and intellectually and physically."

"If you're right about her sexual feelings, then it's cruel."

"Remember that she welcomes a moral ordeal as a learning experience."

"Carole, do you take pleasure in imagining her in the sort of suffering you're trying to set her up for?"

"There's something in all of us, I believe, that takes pleasure in seeing how much a 'good' person is willing to suffer patiently. Philanth would say that it's a way of exploring our own capabilities.

"There *are* darker gratifications. But I'm really not playing with her. I hope to prove that you should find a way to persuade her to stay on with you."

"I couldn't work with a woman who was in love with me. —Unless of course she was my wife."

"Ordinarily that would be a reasonable position. But as I keep telling you, Philanth won't cause you any problems if it happens."

"I don't see how she could help it!"

"In the first place, she's an undemanding person.

She wouldn't expect the slightest regard for her feelings. And she's analytical, so she wouldn't allow herself to behave irrationally.

"Most important, she's very controlled: she's loving even to people she fears or despises, so you wouldn't notice any changes in her behavior. That may be hard to believe, but wait and see.

"Furthermore, she has powerful incentives to exercise that control. She regards loving as a great blessing sufficient unto itself. She would consider herself very fortunate to be working for a man she loves. She would consider it imperative not to spoil the situation by behaving 'improperly'.

"She's been through it before, and in her journal she spelled out what she had wanted to happen.

"Loving even a married man in the selfless way she loves isn't wrong, provided of course that she doesn't let the man get any wrong ideas about it. If you did, she would want to 'educate' you about Christian love as she tried to educate her previous boss.

"Unfortunately like lots of other people, he had never learned to separate love from sex. Since he had hired her with the intention of debauching her, he found her resolve to set an edifying example for him an irresistible challenge."

"I'll bet he did. And it's easy to see what went wrong. Her theories about sex and love are the sort of conceptions a very sheltered and idealistic young woman might develop. In practice she couldn't avoid signaling conflicting impulses, and that must have inflamed him."

"If he had only treated her decently she could have kept erotic overtones out of their relationship—even though he was loaded with animal magnetism."

"How could she have that much control?"

"She's conditioned herself to behave entirely in

terms of the interests of others. She doesn't have the conventional sort of morality, which can break down under pressure from within, from the ego. Philanth has removed that central basis for motivation, as some religions teach people to do. At the center of her is only her altruistic commitment."

"She's some kind of saint?"

"Applying that concept would be very misleading. She deals with questions of right conduct in what she calls 'sociological terms'."

"I don't understand."

"Start with the basics: society, not God, has created her. So she serves it in glad repayment for the privilege of existence.

"As part of her duty to maximize her ability to serve others as well as she can, she believes she has an obligation to keep her inner life orderly. Therefore she rejects sexual relations outside of marriage."

"But she's really committed to a *life* of celibacy? Despite her sexual feelings?"

"Marriage is for other people to enjoy. Because of her inner wealth it's her role to help make other people's happiness more possible: not just by 'being nice' in personal contacts, but by improving basic conditions. Everything she does is oriented to that."

"She *is* a saint, if all of this is true."

"No. What we might call her saintliness is to her only a means to an end. And it's not the only means to the end. She doesn't rule out the possibility that one might in some instance be of greater service by doing something that's widely regarded as wrong, even atrocious."

"Oh, *dear!*" Carole's test was undeniably interesting.

"*Yes.* She knows that's a dangerous position. But she's willing to admit it into her system, with great caution regarding the facts of a particular case. She understands ethical relativism.

"The freedom to abandon conventions at will is incalculable. That alone could justify this experiment."

"But this is all just theorizing. She still sounds too good to be real."

"It's true that I have little to go on except her own words. You can watch her in action and decide what her reality is for you.

"However! Bear in mind that you can do that only through your own present ability to perceive and interpret.

"Though she may want to extend even *your* capacities. She feels compelled to try to improve people, to help them fulfill their potential."

"*C'est an enfant terrible.*" It was one of the few French sayings he knew.

"No. Just morally ambitious. She believes each of us is needed to provide models of conduct in an unheroic age. Otherwise the low standards set by our mass culture could lead to the ruin of our civilization."

"She's spoken of such ideas to me. But you really must have studied her!"

"She let me dissect her while she was lying around in the hospital. Her character gives her secure self-esteem, so she doesn't mind enduring what she calls 'exercises in humility'. In conversation as well as in writing, she's let me know things she would never want to reveal to anyone."

"Why did she let even you know such things?"

"She'll do anything she can to oblige me. It's hard for her to find friends: men let sex get in the way, and insecure women feel competitive. She feels especially indebted to me for being her friend because she respects my 'superior status'.

"Also, because of my graduate degree in psychology, she trusted me not to be put off by her candor. She saw that I wanted to study her as a psychological phenomenon, and it was obvious that I shared her non-judgmental perspective. People not familiar with the analytical framework of social science probably wouldn't tolerate a person's self revelations that seem to imply, 'See how good I am', even though she doesn't regard herself in moral terms.

"Of course, she assumed my interest was *merely* scientific. And she could hope that like a therapist drawing her out, I might help her to know herself."

"It must've taken considerable time."

"Not really, since she was so cooperative. I was only asking a technician to describe the instrument he has been laboring for years to perfect.

"Of course, her journal was a mammoth short-cut to my asking pertinent questions. And I had her letters to draw on. I even taped my interviews, without her knowing. The tapes no longer exist."

"You invested that much effort with a view to replacing Marjorie?"

"In politics one can never have too many capable, experienced, and devoted staff."

"I don't need 'devoted staff' so badly I have to be sexual bait for them. Your letter is really just a way for you to indulge in sexual torture of her while I watch.

"As for working for a man she was in love with, she'd be better off going her own way and letting time and distance return her emotions to normal. I think your whole experiment stinks."

"*I care about Philanth much more than you do!* I want her to be happy and achieve her basic goals.

"She tries to achieve beatitude by living immersed in the perfection she calls 'God'; her character was formed around a genuine love for Christ. When the Protestant faith she'd been brought up in began to fail her she took instruction from a priest, hoping to become a nun. Having lost all ability to believe that there is any divine companionship available, and unable to find a kindred soul who doesn't want to co-opt her in ways she can't accept, she's very alone.

"Yet she has unlimited courage when it comes to doing whatever she believes is right. She has that courage because she's not afraid of anything. She's not afraid of death, since her religion transcends it. She's not afraid of life, since she has no personal ambition and no fear for herself.

"If she pleases you as she does me, I want her working for us: for your career. So stop being so ethically superior and . . .

"What I started to say was, I love Philanth, and I want what's best for her too. The lonely life she's got planned for herself would be one of quiet desperation, because she has a powerful need to express love in a personal way, not just through her work.

"She could be very happy as your slave. She could lay her sexuality at your feet and never let you suspect. Because your goals are hers. You're even her ideal kind of man as to physical type: tall, blond, ascetic, and 'as spiritually handsome as the young Mikhail Baryshnikov', as she put it. She asked for a campaign photo of you because of your 'beautiful face'."

"Did you give her one?" he growled.

"No. Because if she gets one she'll have that to idolize instead of the original, and that'll make it easier for her to leave you. She regrets having given away her statue of St. Francis. She's ready to settle on a new idol. It's hard for her not to worship *something*, when she was formed by growing up in the

Church."

"Well, she can't 'worship' someone she'll be as close to as she will be to me during the next five weeks!"

"Wanna bet? How much?"

Her playfulness made him frown.

"Don't let it bother you, Charles. Feel free to ignore her. She'd genuinely prefer it. Her basic impulse when she deeply loves a man has always been to try to avoid his notice so he won't be tempted to reciprocate."

"How strange."

"Not to her. She believes that the sort of man she would love unreservedly shouldn't be distracted from his high mission in life by love for a woman. Just as a woman with a great mission shouldn't be turned aside from it by love for a man."

"That's genuine self-abnegation," he murmured.

"But you resent my having foisted her on you. If you continue to take your resentment out on her when you're traveling, she'll continue to let you— even if by deliberate nastiness you drive her to secret tears because she's so committed to pleasing you. And Marjorie won't be around to tell me.

"Philanth is especially prepared to accept mistreatment from someone she respects and admires the way she does you. She would pity what she took to be your unhappiness over some unrelated matter. Even when someone hurts her deeply she never lets them know, since that would be 'self-important'. She'll fight for an important principle or an altruistic objective, but where her own interests are concerned, she's a doormat."

"I'll be fair to her. I really didn't need to hear about her sex life."

"I'm sorry, but I have to insist that it's relevant."

"Because it makes her easier for us to exploit?"

Again Carole exploded. "No! Because it's a central element in her self-sacrificing nature! And because of that nature, Philanth is going through life *inviting* people to exploit her. It's only a question of who gets to her first, and for what purposes; and whether she will be wasted or destroyed, or fulfilled and content.

"Do you want to let one more altruistic young woman be raped, tortured, and murdered by the anti-American faction in some stupid little civil war, one more innocent martyr to our ignorant, myopic foreign policies and misdirected benevolence?

"Well, she prefers to risk that rather than become household drudge and round-the-clock Girl Friday to some young hotshot who fancies himself God's gift to American politics."

"So you want her to become Girl Friday to some *not*-so-young hotshot who fancies himself, and so on."

"*Yes.* Her combination of ability and selflessness would make her priceless to anyone she might decide to devote herself to."

"I never knew you could be this cold-blooded."

"I'm only putting together the facts in a way that shows how to prevent Philanth's character from taking her into danger. What's in her best interest may be in our best interest too."

"Does considering ourselves altruistic in our personal ambitions give us the right to try to co-opt someone else's life?"

"I never knew you to be such a moral wimp! Since when do you take the pious, mush-minded position that it's 'wrong to play God', so irresponsible inaction is better?"

"So she's a missile that may self-destruct eventually anyway, and all I have to do is re-program its guidance system so it can fulfill its purpose and mine too."

"That's the idea." Carole sounded relieved.

"Well, bear in mind it's an assignment I don't need on top of a load of more immediate concerns. And my career may not be going anywhere that would enable me to keep Philanth interested. So if it doesn't work out it won't be so awful to let Philanth go ahead and 'do her own thing' as she sees fit."

"*Hon*estly, Charles," Carole exclaimed, slamming her hand on the table and getting up, "sometimes I wonder why I bother with you."

Trying to assume that she was being facetious, he started to laugh, but she turned and stalked out of the dining room. The water in the sink was immediately turned on full blast, but above it he could hear Carole rattling dishes and tossing silverware and banging pans. Even her footsteps sounded angry.

A remark like that from Carole, especially if made in earnest, was hauling out a shillelagh and lowering it onto his head. Carole had made her position clear: if he had an ounce of ambition, if he respected all the hard work his wife had done for him over their fourteen years together, he would give careful attention to the potentialities in Philanth Devon and try to recruit her to his long-term service if she seemed worth his effort.

He walked into the kitchen, came up behind his wife, and started to slide his arms around her waist. Her hands in the dishwater, she used her elbows to shove his hands away with a force which killed any hope of negotiation.

"I'm sorry, honey," he said, his lips close to her neck. "You're still my personnel director, and if you want me to evaluate somebody for a career-track position, I'll do my best."

Carole nodded, saying, "Fine!" as though satisfied, and went on doing the dishes.

PART TWO:

BYZANTIUM

Chapter 8

THE SPLENDORS OF BYZANTIUM

After they had been airborne for a while he finished going through the newspapers Philanth had bought for him at the airport and traded them for her news magazines as she had suggested. She dove into them. After a few minutes she remarked, "Your Los Angeles paper is substantial, as such things go."

He smirked at the qualification and agreed.

She began cutting out something.

"Don't do that. You'd have to lug it around the world."

"I can't resist 'think pieces' with enduring value. I'll mail a package home to myself when they become burdensome." She went on whacking at the newspaper.

When she stowed it under the seat he commented, "You seem to have gotten a lot out of it—literally."

"Yes. It's social history. If you put just one issue of any major American newspaper in a time capsule, and it were exhumed by extraterrestrials trying to find out why our planet was lifeless when it once had held an inconceivably rich civilization, they would decipher it and say, in effect, 'Aha. Now we understand everything.'"

"What would they understand?"

She hauled out the paper and began flipping pages. "See for yourself: side-by-side with articles about poverty and ignorance and oppression you find ads for high-tech adult toys and extravagant personal adornments: a blouse for over three hundred dollars, a coat for thousands. I got a lovely coat for twelve.

"Here's where I cut out an article about the vice president's signing an agreement to give Mali some aid, and it goes on and on about all the poverty and hardship there. Right under the article they put an ad for a crystal decanter and matching glasses."

"You're implying that the money being spent on personal luxuries could keep civilization from destroying itself. That's simplistic."

"Don't reduce what I implied to redistributionism. *Judiciously* used, the cost of one evening dress could transform the lives of a thousand people, with changes that would carry on into subsequent generations—if you only started with vasectomies and bitubal ligations."

He smiled reluctantly.

"You surely have seen statistics about all the women in poor countries who want no more children," she reproved him. "Unwanted children are a fundamental reason poverty is perpetuated from generation to generation. Yet most of what's being done to combat poverty and all the evils it breeds is wasted because basic causes go untreated."

"It's a matter of political feasibility."

"Invoking 'political feasibility' capitulates to ignorance. Ignorance is an enemy that should be assaulted at every opportunity. The human race may die of ignorance."

He didn't feel like pursuing the matter, but he was glad that Philanth was starting the trip with such matters on her mind. She probably had decided he knew nothing about Carole's letter. Early that morning at the office she had greeted him cautiously. He had behaved normally, and gradually she had returned to the vivacity which was normal for her.

He worked on office business until dinner was served, but after that he tried to relax. At the end of the nine-hour flight he had a full day of meetings, and when it was 9 a.m. in Paris he was afraid his body would know it was 3 a.m. in Washington. He hoped his prepared statement at the plenary opening session would prove apropos in the context of whatever preceded it, but he would need to be sharp enough to ad lib gracefully when the time came.

But the bends in his reclining chair made sleep impossible; and for the first time he had leisure to contemplate the tens of thousands of people who would be affected by how well he did his job during this trip. So instead of getting sleepy he kept hauling out his loaded briefcase to check on various matters.

Soon he started a list of items he wanted sent to him in his first daily pouch. He started another list of items he must be sure to receive at various points later on. The journey ahead began to open out before him, proliferating details in all directions.

Finally he knew he must stop this. He decided to concentrate on the conference the next day, in Paris.

Program development and training specialists from the OTAPS section had flown over at the end of the previous week in order to have the weekend in Paris at their own expense. They would conduct the substantive business after he had gone on to Africa.

However, he needed to be as well-briefed as possible. He reviewed the preliminary papers. The theme of the group of conferences was, "An End to 'Random Intervention'!" To explain this term the notice of the conference carried a quotation from *Africa: From Mystery to Maze*:

> "It is remarkable and tragic that several of the major 'benefits' brought by both colonial and independent governments have had adverse effects on the people and the land. More

often than not, those effects have occurred because development efforts were focused on one or two specific features, and no attempt was made either to view existing problems holistically or to prepare for the new problems the innovations would create.

"In one sense the lowering of death rates through disease control is an example, for with no attempt to lower birthrates or to provide for an increasing population, this was an invitation to tragedy. Of course it was not wrong to improve health standards and lower death rates, but there must be wider recognition that modern medicines can in the long run take more lives than they save if their introduction is not accompanied by other modernizing programs."

Though still a novice in international development, he could think of other examples: fostering of cash crop production and drilling of wells in dry areas where the resulting concentration of population and live-stock destroyed the soil and thus furthered the spread of the desert.

With its promising theme and more than a year of preparation, the conference was expected to bring together more than five hundred professionals in international development. The U.N., the Common Market countries, the World Bank, and several Asian and Pacific Rim countries were sending representatives of their technical assistance organizations. The American Friends Service Committee, Oxfam, AID, CARE, UNICEF, the Overseas Development Council, and the U.S. Ambassadors' Self-Help Fund also would be represented. There would be delegations from privately sponsored programs for overseas development, and the U.S. alone had more than a hundred private voluntary organizations involved in overseas relief, development, or rehabilitation programs. This should be a major opportunity for pooling ideas, information, and organizational efforts.

He put away his briefcase and tried to relax into that optimistic thought.

He would have liked very much to remain for the three days of technical meetings, just to educate himself. He wanted to be everywhere, doing everything, because there was so much to be done.

Eventually he decided that he must have gotten a little sleep. But there was an unhappy baby on board, and someone was sneaking cigarettes despite the prohibition; it was a long, uncomfortable night.

He was getting too old for red-eye flights. This one was his own fault because he had felt he must have one last Monday morning in Washington to wind up loose ends as much as possible. Just the same, by the time the stewardesses brought around croissants and fruit cups and beverages he was morose. He *hated* trying to sleep on planes. He was a busy executive, and he needed more rest when he traveled, not less. Moreover, he liked his comfort.

He liked his comfort so much that he should never have gone into politics.

But he wondered, as he always did when his thoughts turned in that direction, what else he could have done that would have made life so interesting, so intensely *real*, so worth paying attention to. There could be nothing more absorbing than trying to improve the future, carrying on the work of those who had brought humanity out of the caves and given them libraries, electricity, ...

But why couldn't he do his share in making life better for more people at a more comfortable pace?

He knew the answer, of course: "We must take the current when it serves." His present job was a priceless opportunity to fulfill the dreams of a lifetime. Once one got into such a position, there were never enough hours in the day.

However, on the mere external level his day at the meeting of directors of the International Secretariat for Volunteer Service consisted mostly of a second Continental breakfast, a rich lunch, and endless talk. His whole life seemed to consist of talk.

After lunch his difficult night caught up with him, and as the afternoon and the translators droned on he drank so much coffee that his stomach rebelled. He became not only muzzy-headed but faintly nauseated by the smoke. No smoking was permitted in Peace Corps headquarters, so he had gotten spoiled.

By the time the closing session of the afternoon broke up he was developing a headache. He told Philanth he wanted only a quick supper and an early bedtime. She endorsed showing respect for jet lag.

Tony had had the day free since seeing him safely to the big convention hotel that morning. He met them as promised in the reception lounge. Philanth had the taxi leave them at a modest café near their own hotel and ordered for them.

While they waited for their food Tony began showing Philanth some maps and brochures he had collected during his day at leisure. As she went through them Philanth got so excited she started mixing French and English. Their simple orders were served promptly, but she pushed her plate aside and went on debriefing Tony.

Finally he picked up her fork, cut a piece of her ham omelet, and stuck it into her mouth. "Eat!"

"Yes, sir.

"Tony," she went on with the food in her mouth,

194

"does your map show the produce market? A friend of mine at the Hunger Foundation, the chef who helped me with my French, told me about it."

"Philanth, forget all these other places. This is your only chance to go to the Paris night clubs with me! You must've brought a dress, for the black-tie things later on."

"I'm sorry, Tony," she said, finally starting to eat under her own power, "but I'm going to Sacré Coeur if I don't get to see anything else."

Tony groaned so loudly that the customers at a nearby table looked over at them.

Hoping to halt an incipient argument, he asked, "What's so special about Sacré Coeur?"

"It's Byzantine," Philanth began, and he noted that she laid down her fork. "Both it and the Capitol remind me of Yeats' poem about the holy city of Byzantium, where power and religion blended to produce beauty.

"There are so many parallels! Each dome is sacred to its respective religion: Christianity and Americanism. And each dome dominates its respective city: the Capitol psychologically, Sacré Coeur visually. And these white domes dominate their surrounding cities as their cities dominate their worlds: the world of power and the world of beauty. For both power and beauty we have instinctive attraction."

Tony was consulting a section torn from a book. "My guidebook says Sacré Coeur is 'ugly'," Tony informed them, "and its decorations are 'meretricious'."

"That was written by a snob," Philanth declared. "Anyway, what better adjectives could there be to describe night clubs?"

Tony made a moue, raising his elbow to block an imaginary blow, and tackled the big French bun

195

which was his *sandwich jambon*.

After washing down a mouthful with Coke Tony growled, "Don't ask me to be your bodyguard, then."

"No," she agreed reluctantly.

Tony's churlishness aroused his sympathy for her. "Is this really so important to you that you must go tonight, even alone?"

Tony had slept well on the plane, but he could not understand Philanth's not wanting to go to bed early almost as much as he himself did. They both had had a wearing day, and they had to catch an early-morning flight, and they would need to be rested in order to work efficiently on the plane to Morocco.

Philanth's face lit up as she again tried to explain. "Just in going to Sacré Coeur you become a pilgrim. When you ascend Montmartre you're climbing a natural hill, toiling up cobblestone streets through centuries of the pursuit of pleasure and art to a pinnacle of ascetic mysticism."

Suddenly self-consciousness made her duck her head, dimpling shyly.

"That's very lofty," he said, trying to count her dimples.

"There's a more personal reason," she admitted. "I love beautiful old churches. Their beauty is eternal. I'd like to stay in one forever, if only I felt I had the right."

"You must know Montmartre is a red-light district. Suppose you're walking alone through it tonight and some drunk assumes you're a lady of the evening and then makes an issue of your right to refuse him your services? Suppose some pimp decides you're a newcomer trying to free-lance? Suppose he tries some maneuver that can get you hauled in on drug charges if you refuse to cooperate with him?"

Philanth looked shaken. "I—I thought my French

would be good enough to . . . ."

"It's not enough to be able to prove your innocence. The Peace Corps must avoid any unsavory incidents that could be reported in the newspapers."

She nodded, her mouth becoming an inverted "U" flanked by dimples. The excitement went out of her like helium released from a bouncing balloon.

She picked up her big, fresh croissant, tore it carefully, and studied the convoluted layers of tissue-thin pastry inside each half of it. She carefully nibbled at the first half, pausing to recapture the delicate, golden flakes which fell to her plate no matter how carefully she bit into the torn fragments. It was like watching Ken's five-year-old.

"I *hate* being a woman," she burst out, putting her fist to her temple. "And the worst thing about being a woman is having to put up with men."

Tony took her small, white hand in his large brown one. "It's no big loss, Philanth. Even if it's pretty, so what?"

She reclaimed her hand. "It's not merely idle pleasure. I can justify it in utilitarian terms: enjoying lovely things can add to your supply of inner wealth, if you go about it right. In the manner of your action you can share your inner wealth, to the extent that others are open to receive it. That's just one element in a complete, happy way of life in which you are nourished at little cost and become able to feed others from the abundance you can generate in yourself."

She caught his eye and again became self-conscious. "It's just a way of doing things I've developed," she mumbled.

After a moment she began eating. She finished her omelet, which had to be cold by now, while cherishing her croissant down to the last flake.

As she addressed herself to the last of her salad she said without looking up, "I should go to bed early, too. Mr. Bjorklund has the right idea."

"In *Paris*," Tony cried, gesturing as though calling upon all the gods to witness. "Our *one night in Paris!*"

He watched her face as he said, "Actually, I'm thinking I might pay a quick visit to a Byzantine church before I turn in. The fresh air would do me good."

He chewed the last fragment of unconscionably hard bun which he had previously abandoned as he enjoyed the dawn of hope in her eyes, the tremor of hope in her lips.

"In fact, I definitely will, if there's anyone interested in going with me."

Her hands fluttered, and he thought she had the most expressive mouth he had ever seen. "You're a good man. I know you'd much rather go to bed. I'll try to make it up to you."

Tony reacted with a heh-heh-heh. "You mustn't say things like that, Philanth."

She glanced at Tony, and her mouth became an inverted "U" again.

She clasped her hands in her lap and bowed her head. "I'm sorry, Mr. Bjorklund. It's something I hate so much I ignore it sometimes. I'll try to be more careful."

He tried to behave as he would have if this were the simple situation it should have been. "I don't want to embarrass you, Philanth, but Tony does have a point. We mustn't make any mistakes when other people are around, so we have to stay in practice." He sounded more gentle than he should have.

Her mouth went into twists and curls of distress, but she said in a normal voice, "You're quite right.

The fact I'd rather ignore such ugliness is no excuse."

She leaned her plump cheek on her fist. "I prefer to act as if people were going to react the way they should. When they don't, I always hope that afterwards they realize they've fallen short, so they might try to behave a little better in future."

"Philanth," Tony said in a peculiar voice, "why aren't you wearing your ring?" The black man was scrutinizing her fist.

"Oh." She took her hand down from her cheek to spread it and study the third finger on all sides. "It's almost healed. I get raw fruit juice or thinner for correction fluid under the ring sometimes, and I don't know it's happened until the skin gets inflamed. I think it'll be okay to put it back on tomorrow." She covered her hand. "It makes me nervous not to wear it, I'm so afraid of losing it."

She smiled bashfully at him. "Not that it's so valuable to anyone but me. The engraving inside it cost more than the ring itself." She picked up her glass and coaxed a final drop of milk from it.

Tony was eyeing her. "I hope it's in a safe place."

"I have it pinned. It's always on me, one way or another." She moved her fork over her plate, found a trace of egg or pastry, and put it into her mouth.

Scooting back his chair, Tony said, "No use my hanging around if you two are going to church."

Tony abruptly stopped moving, looking at him. "Uh-oh. I got to go with you. Don't I? I don't remember any instructions for this sort of situation. It's your official schedule that puts you at risk."

He didn't want Tony tagging along on this: his attitude would ruin it for Philanth. He gestured dismissively. "I'm in no danger. Run along."

Tony needed no urging. He unfolded his subway map and quickly taught Philanth how to use it to get to her destination. After giving her the necessary coins Tony took their two briefcases, paid their tab with Philanth serving as translator, and left for the hotel to get ready for his one night in Paris.

Montmartre was easy to reach by Metro. They climbed out of the subway to find themselves in a crowd which ranged from men in heavy-duty overalls to women dressed as though for the Academy Awards at the Los Angeles Music Center. Philanth's pleased, impressed comment on "so many beautiful women" told him she did not realize she was admiring common prostitutes. He did not enlighten her, reasoning that she would figure it out soon enough.

Soon they began climbing, passing ageless shops and little bistros already crowded and noisy. A door opened and accordion music came out, accompanied by a jumble of male voices and laughter. It was Paris unconcernedly imitating herself, refusing to change anything merely because she was being as predictable as an old movie.

He stole a look at Philanth. She was looking about her with the expression of an enthralled cherub.

All in cornflower blue, and with her tailored denim jacket concealed by a big, hand-knitted shawl above a long, full skirt, she could have been a runaway country girl. In the soft, shifting light she looked more like sixteen than twenty-five. Men out for a good time in a district famous for its streetwalkers might go out of their way to engage such a fresh-looking girl. Such men could become very interested in finding out to exactly what extent she was as innocent and sweet-natured as she looked.

Passers-by must think he was going with her to her shabby *pension* for a quick one. She was too young to be his wife, and girls that appealing in that neighborhood at that hour would not be expected to be anything but commodities. He felt like an aging lecher about to exploit a poor, young girl who, being as light-haired as himself, might have been his daughter.

In spite of her jacket and shawl, she was hugging herself.

"A wool jacket would have been warmer," he remarked, indicating her shawl, "and not excessive, for the higher altitudes in the countries where we're going."

"But this packs beautifully, and it has multiple uses. And it was absurdly inexpensive, since I was teaching knitting classes anyway. The North African yarn was wonderful, after the flimsy American stuff.

"Besides, I didn't have a warm jacket that was light enough and nice enough to take on this trip—only sweaters I couldn't wear with my suit jacket."

Buying a jacket apparently had been unthinkable.

As they continued their climb, a woman near his own age winked and tossed her head to beckon him. Her smile said, "You don't want that clumsy little girl, Dearie; you want a woman with *finesse, savoir faire.*" She wore a rhinestone tiara in her curly black hair: a queen of lower Montmartre. He looked away, pretending not to have noticed—

And saw a skinny redhead lounging in a doorway. Seeing that he had noticed her, the girl gave him a direct smile as she struck an inviting pose.

Missing Philanth beside him, he turned and saw that she had paused to peer into a shop window. At once she saw him turning back, dimpled apologetically, and scampered to catch up with him.

"It's so exciting," Philanth confided breathlessly. "It's possible that Degas, Renoir, and Monet have all walked these same pavements. Think of it!

"How lovely to have lived before technology became

201

a threat to the whole ecosystem, when it seemed the world would always be safe enough for the enjoyment of beauty and the infinite variations in human experience."

He didn't think to try to respond to her awed enthusiasm until it was too late, and perhaps that was why she said no more.

It was quite dark when they reached the church. Philanth went over to look down the steep hillside, but the lights turned upward onto the white basilica were too blinding to let them see much. "Paris at our feet, and we can't see it," she lamented.

He had had the same disappointment on the Eiffel Tower, in his own youth. Maybe that was a part of losing youth: climbing a height only to look down and discover that there was really not much worth seeing, after all.

Now, after uncounted journeys overseas and innumerable crisscrossings of the United States, he had decided that farm land looked about the same in Belgium as in Nebraska, every city had street corners very much like some in Los Angeles, and even every tourist attraction fell into one of a few basic types. There no longer was anything worth his attention except what was related to his work, and his work was the sole justification for his existence.

As they turned back toward the main entrance Philanth took a long, white scarf from a pocket of her skirt and arranged it over her hair. "That's not necessary, is it?" he remarked, watching her toss the ends of it back over her shoulders with a flourish so graceful it was incongruous with her plebeian clothes.

"I prefer it." Her furry voice was gentle. "For me the old custom expresses respect for the concept of eternal perfection.

"There's nothing wrong with ritualizing female

modesty, either. Modesty is culture's answer to biology."

She smiled up at him with nothing in her manner to suggest that this was other than a casual academic discussion. "You can see I've thought about it."

"There seems to be a pattern in the topics you've 'thought about'".

"Yes," she admitted, suddenly shy, "it's all part of a whole."

"And your 'do-it-yourself philosophy' seems quite important to you."

"I come by it honestly," she replied lightly. "My dad was a German-American who taught Systematic Theology. The final exam was to 'build your own' theology, just like Paul Tillich."

"Really!" he said, smiling a little.

"Uh-huh." She had paused with her thin little hand on one of the massive doors of the church. "But Dad didn't just teach theology. He was a dean, and he taught other religion courses. He was primarily interested in the study of ethics."

"What did your mother do?"

"Mom's specialty was religious poetry. She had done her dissertation on Gerard Manley Hopkins. But she taught a range of literature courses. Her greatest public recognition came from a comparative study of the presentation of women in pulp fiction and other commercial entertainment."

He smiled to himself as he reached to open the door for her, but she caught his look and put out her arm to block his entrance.

"You think they were 'irrelevant'," she accused him.

"No, no, I was just . . . I guess I was thinking that 'The fruit never falls far from the tree'."

"I wouldn't want to anyway," she replied loyally.

"They collected all sorts of material and talked about it and wrote about it, and I read some of their work, so I know how important it was.

"Their study convinced me that our whole civilization is in danger because of deterioration at its motivational core, in its central values and the norms for behavior arising from them. There's tremendous need for a new foundation for human values which incorporates the cosmological view of our emerging global technological society but *doesn't* fail to grow beyond philosophically adolescent negativism . . . .

"Sorry. I forgot I'd talked to you about that before. But it *is* terribly important."

"There are lots of important things in the world that very few people care about. Why is this so important to *you*?"

She hesitated. Then she answered shyly, "Because I want to know how to deal with the roots of the troubles in the world—and at this stage, thanks to technology, the fundamental obstacle to progress seems to lie in the values systems in the affluent nations. In the simplest terms, I want to 'save the world', just as you do."

"Hm!" He turned, reaching for the handle of the big door of the church, but this time she stepped in front of him in direct confrontation.

"What does 'Hm!' mean?"

"It means I'm too old to admit to that."

"It doesn't make you crazy or a fool, Mr. Bjorklund. You only have to find some way to cope psychologically with the larger picture—difficult as that is."

"It's true. But why do we get drawn into something so pretentious—and so crushing? Why do *you* want to save the world, Philanth?"

"Same reason Buddha did." She slipped off her shoulder bag and eased it to her feet as she turned to

204

face out into the night. "When you have all you want for yourself the only happiness still to be attained is seeing the happiness of those you care about.

"Now ask why I care about the world. Because the greatest wealth is love. And what greater—at least, what more vast and challenging—object of love can there be?"

"What about God?"

"Should I love a deity for which I find no evidence outside of the minds of human beings and the orderly processes in the universe? Or should I love a species capable of conceiving of such wondrous gods in order to transcend awareness of its own helplessness and ugliness?

"As for first beginnings, do you explain them better by saying that intelligence, will, energy, or matter came first or always was and always will be? More importantly, what difference must that make? *I* cherish a species with the ability and courage to raise questions for which we can find no answers beyond those we can create within ourselves."

She paused, glancing at him, but he said nothing.

"Almost everything is imperfect and incomplete," she went on, "and always becoming something else. So I honor the potential and the image of perfection. Since I find myself in an impersonal universe, what else can I make the measure of every other value if not the human mind, with its demonstrated potential for what Plato called 'the good'?"

Her voice was playful even in challenge: "Ask me something hard."

He could not pass up such an opening for the assignment Carole had given him. As he leaned against the building he remembered a key point in Carole's briefing the previous evening. He molded it into what he believed was the hardest question of all.

"If you understand even a little how difficult it is, how can you be sure you'll go on wanting to 'save the world' *for the rest of your life?*"

"What alternative do I have?"

"Hedonism, of course! Me-ism! When things get utterly discouraging or you feel used up, there are constant temptations to relax and enjoy yourself; to focus on material things; to concentrate on looking out for your own interests. It's become a cliché: 'Living well is the best revenge.'"

"Oh." She leaned against the wall beside him. "But it's a person's values that define his goals, not his tastes or desires. It must feel awful to have revised your values downward just so you could achieve the goals they set. It's far better to take care how you define 'success' in terms of the values you truly find worth serving.

"I learned when I was a child that the majority are often wrong. So I resist the pressure to define 'success' in terms of other people's standards."

"But how can you resist all that pressure? How can you embrace what the world regards as failure?"

"It must be terribly hard for *you*, since you deal with the socioeconomic elite, and they tend to avoid unwanted contacts by measuring others in terms of the standards which protect and define their own positions. It's easy for me, since I've dropped out of that competition: all I had to do was consciously choose what to value, understanding the psychological and philosophical foundations for my choices."

"You're never tempted by the pleasures of materialism?"

"Constantly. No matter how irrational its appetites, the body remains a sensualist.

"But the stupidity of the values that pressure me to strive for socially damaging forms of 'success' like

206

wasteful consumption give me a special obligation not to internalize those pressures. Society depends for its survival upon those who do what they believe is right regardless of immediate personal consequences. We're the yeast that leavens the whole loaf, the core who keep things functioning when otherwise they'd fly apart due to self-interestedness."

"So you *dare* not change?"

"You need not *want* to change, if you've made sound basic choices and organized yourself according-ly."

Carole's briefing was proving useful. "You must change as circumstances do, or as you learn more."

"Not the broad foundation of your way of living and your choice of a reason for living—not if your basic purpose is not to extend your existence, but to make it part of something which will let you regard even your death as a secondary matter. On the contrary: your ways of doing things and your very nature keep rebuilding themselves around clearer definitions of your fundamental purpose. You make your own satisfaction out of your consciousness of the worthwhileness of your purpose. That way you aren't tempted to junk your purpose to seek the sat-isfactions of achieving some less worthwhile purpose."

He thought that over. "So if you're doing your best for a worthwhile goal," he said slowly, "even if you don't make much headway, you're a failure in moral terms only if you have a failure of will itself."

"Yes. And you should be able to maintain your emotional commitment to your intellectual convic-tions if you avoid emotional malnutrition."

"How do you do that?"

"How do you find adequate emotional nourishment, when the traditional sources are obsolete and the contemporary culture is debased? If your taste is as selective as mine, you may have to find some way to 'grow your own'. That takes time you may not want to spare. But there *has* to be *some* reliable source of love and order and beauty in your life, if you're to go on offering them to others. That's certain."

Seeing that he had no rejoinder, she straightened and moved away. He followed, raising his eyebrows to himself only after she had turned her back. Together they opened the door.

The church was dim and empty. But the floor, walls, and ceiling shouted with decoration, and the lighted altar blazed with white and gold. Its brilliance beat them back like intense heat; wading into deep light, they approached slowly.

Then she surprised him: she dropped almost to one knee and quickly crossed herself.

As she straightened she tilted her head back to drink in the spectacle: the lavish altar beneath its mosaic, unprettified Christ; the sweeping lines of the converging arches in the high ceiling; the racing lines of mosaic in the shining floor. The hard, bright, ornate surfaces were like the inside of a Fabergé Easter egg, but this shell enclosed silent, soaring space filled with soft light and hovering peace. Here was a sanctuary indeed.

An empty church at night had an altogether different quality than in the daytime. For the first time he wished that he and Carole had been married in the evening.

Philanth would be a bride for evening: serene, quiet, glowing.

Tony's words came back to him: "Something nice waiting for you at the end of the day."

Philanth pressed his arm to keep him where he was, then went toward the back. He turned and saw her kneel, again genuflecting. She shut her eyes, folded her hands, and began a soundless monologue.

The prayer lasted for a minute or so. A third time she genuflected, but this time with great deliberation. She rose and returned to him, eyes on the floor and lips compressed. As she drew close to him her mouth unwillingly revealed that she was deeply moved.

They slowly walked around the sanctuary. When they reached the large double doors she paused to look back at everything, then nodded to him and led the way outside.

He remarked cautiously, "You were acting as if you were Catholic."

"I tried hard to become a convert after my husband died. I wanted very much to become a nun. I even wrote to the Dominican motherhouse where I wanted to become a postulant."

"What went wrong?"

"I couldn't swallow the whole package of doctrines. Once you've learned to think for yourself it's hard to give it up."

He managed not to laugh. Apparently it would not occur to Philanth to go through the motions of accepting the totality of a creed in order to get what she wanted from the institution which sponsored it. Everything had to be part of the coherent whole. That need for consistency of policy reflected a seriousness about life, a commitment to making every moment count toward something whole-heartedly believed in, which he found refreshing.

She went on, "Later, by the time I'd finished college, I found the Church's positions on birth control and abortion unspeakably stupid. And there are too many other conflicts between faith and science in the theology, as well. I came to find it unnecessary to acknowledge any conflict between faith and science at all: nothing faith affirms should be contradicted by reason regarding empirical reality. So to me it's not worth all the intellectual acrobatics to believe in the orthodox conceptions of God."

"Yet you worship as a Catholic."

"I worship everywhere," she answered distantly.

"I see."

She turned to look searchingly up into his eyes. "Do you really? Then you must be looking forward to this trip as eagerly as I am. The Buddhist temples, the Hindu statuary . . . ."

"Ah, well, I . . . I hope you don't imagine you'll find time for visiting temples and so on during the *rest* of this trip. I admit it was nice just this once."

"Oh. No." She bit her lips and turned away. After a moment she said in a gentle, half-strangled voice, "Of course not."

She wandered away for a little, looking at the church.

Finally she returned. "It's strange, isn't it? So far from Constantinople.

"Of course, like Washington, Paris and Byzantium are primarily states of mind.

"Nevertheless, what an extraordinary civilization it must have been, for its time."

"Yes."

She began walking away very slowly, head bowed, so he joined her.

She went on, "Some things are so beautiful I'd like to spend my whole life living in their shadow. That's how my mother said she felt about an unutterably magnificent rose window she saw in one of the European cathedrals: Rheims, I think she said."

Philanth sighed a little as she added wistfully, "She and Dad were going to take me there, in fact all over Europe, as soon as I graduated from college . . . .

"But there are too many urgent problems in the

world to allow the self-indulgence of the contemplative life. More than ever, since the proliferation of threats to the whole life-support system of the planet, beauty has 'its fingers always at its lips, bidding adieu'. So we must learn how to enjoy it on the run or not at all."

"That's from a poem?"

"Yes. I have to be careful not to quote poetry too often. My mother taught it—and lived it."

Her voice became sad. "She wrote it a little, too. But in it she said truly happy people seldom write poetry because they're too busy living it. I found it after she died.

"Sometimes she would recite it just for the pleasure of the sounds and the turn of the polished phrases, when we were working around the house or going somewhere.

"I remember one time she ran through a whole poem, and then she chuckled and said, 'That's nonsense, of course, but it reflects the thinking of that time. And it still sounds good, so people don't notice there's really little comfort in its meaning.'

"But she wasn't a debunker, she just looked at what was really there in front of her, as she'd been trained to do.

"It seemed to me that she had a snatch of poetry for every thought. She took pleasure in sharing it with me. She told me in later years that I had been enjoying the rhythms of Theodore Roethke's lovely nonsense with my first milk bottle. For the first seventeen years of my life I was fed poetry.

"The cadences still creep into my speech, if I'm not careful—especially when I'm inspired about something.

"Ah, she was a lovely woman! I don't mean her looks—in what she was. Other girls hated their mothers, but she had so many fine qualities . . . ."

Philanth stopped and swung around to look back. "Good-by, Church," she called with soft longing. She stood looking up.

He waited until she turned to move on and again fell into step beside her.

After a few minutes he began, "Philanth, I've been wanting to talk to you privately. I've been too preoccupied until now. I want to know more about that CIA man you brought to the party Saturday night."

"I suppose nobody has bothered to explain it to you. They owe you that much, but I suppose they thought you'd let it pass, or that I'd explain."

"Then please do."

"There's nothing wrong with a little unofficial conversation at a party, is there? When people who have things to talk about 'just happen' to bump into each other?"

"You mean between the men from State and the man from CIA? They set it up, with you as cover?"

She shrugged.

"But why use a Peace Corps party? State and CIA . . . ." He paused. "Some sort of back-channels communication? But again, why use *us*?"

"I . . . can't answer that, Mr. Bjorklund.

"My role in the situation was that of 'bubble-headed tomato'. Yours is that of the innocent administrator who finds out too late about anything irregular."

"My child, you have a lot to learn, and your attempt at sophistication in this matter ill becomes you."

"I was only speaking a bit ironically. I assumed *you* would understand that.

"But to speak seriously: you should take my word for it that it was only an informal meeting in which nothing harmful occurred."

"Don't I have a choice?"

"You shouldn't embarrass yourself by asking State about it. After all, 'It was only a party.'"

"All right," he said, letting her see that he was still annoyed, "let me ask you this: why would anyone from State ask you to pretend to be 'dating' a CIA man at a Peace Corps party, especially when spouses or dates were not invited?"

"Nobody can know everybody at headquarters, so nobody knew he didn't work there except our little group.

"And I wasn't pretending, I was actually dating him—in a sense. It wasn't as though I'd never met him before. I'm actually a little in love with Petya."

"'Petya'? What's his—uh—background?"

"Russian.

"We met at an embassy reception more than a year ago.

"That was back when I had to attend such functions with my boss at the Hunger Foundation, because he went in order to lobby people like you.

"We met you at one of them, in fact. I suppose you don't remember how Rod dragged me away when I started agreeing with you about the wrong issue. You were wearing a velveteen-trimmed tuxedo; you were all in black and white with gold studs and golden hair—"

"Don't try to change the subject.

"You're in love with a Russian? Who is he?"

"He's a very interesting man."

"Philanth," he ground out, "what was going on, Saturday night?"

"Between him and me, only a flirtation," she said placatingly. "We've been together only a total of about five hours, and almost always with lots of other people around. We're only *attracted* to each other.

"Though we have a lot in common, I think, in our characters and values."

"I doubt that."

"No use in arguing the point."

He gave up on trying to get her to stop evading or deliberately misconstruing his questions. He decided to try an indirect approach. "Do you plan to go on seeing him?"

"I hadn't seen him between the time I applied to Peace Corps and this past Saturday night. I just saw an opportunity to bring people together because of my passing acquaintance with the men from State."

"*Why?*"

She appeared not to hear the question. "We've never *really* 'dated'."

She added regretfully, "We never will. Petya's extremely ambitious."

"I'm vastly reassured."

"That's not a strike against him," she added quickly. "It's quite natural. After all, he's handsome, tall, blond, extremely intelligent and quick, cold-blooded, obsessed with power, thoroughly pragmatic. He finds me interesting only because of my blend of contradictions; and he's curious about American women.

"I think he may go far. And at the center of him there's a small flame of aspiration for the welfare of his country and peace between the world's peoples."

"Uh, whose country?"

"His native country. He's lived in lots of places, though.

"I think you'd enjoy talking with him too, Mr. Bjorklund. His English is impressive, and he can be charming in a detached way which has no analog in my experience. And he has a great deal of sense. He always weighs his words carefully and goes straight

214

to the point—they're a literal-minded people, you know. The Russians.

"I might also note that he's not macho like the Russians one hears about. He's quite acculturated, I mean cosmopolitan, and he's psychologically beyond treating women as second-class citizens—at least, in a superficial relationship."

"All right, all right, you can stop singing his praises."

"I'm a connoisseur of people, Mr. Bjorklund. And my chances to meet the extraordinary types that come together in Washington have been limited. With Petya I got lucky."

They strolled in silence for a few moments. "Could you ever come to want to marry him?"

"Oh, *no*! I'd be just a *wife*. *He'd* get to do all the *interesting* things. Besides, intercultural marriages have poor prospects of success."

She added demurely, "But it has been a worthwhile social experiment. And the men from the State Department were pleased to make a little use of it, with Petya's consent."

Someone at the State Department had wanted an off-the-record meeting. That indicated that Petya's tie to CIA was not official; the episode began to look like cloak-and-dagger stuff.

"Philanth, is there anything else at all you can tell me about 'Petya'?"

"He's opportunistic, yet of refined sensibility."

"Sounds like someone in a spy movie."

"Then maybe I'd better put it in context.

"When he took me home Saturday night he asked for permission to come in for a nightcap. I said I was afraid that would disturb my neighbors. He promised he'd be very quiet. Then he admitted that he . . . wanted to make love to me.

"It seems nowadays some men feel they have to ask for the whole nine yards just to see what happens. That must ruin a lot of budding relationships.

"Anyway, I said I was sorry, but even if there were no other considerations, I couldn't agree to that for personal reasons."

"And he took it in stride?"

"You underestimate him.

"I expect any man I'd have time for in the first place to be mollified if I refuse him my body but promise to love him forever—even if I make a half joke of it because we aren't well acquainted.

"Petya's response was that he hoped we'd meet again someday. He'd already told me he would be gone when I got back from my trip with you. He said if we met again, he'd 'try again'.

"He was trying to be romantic, but I said I could never marry him, so his repeating his suggestion would offend me. That seemed to make him thoughtful, and he kissed my hand. We said our good-bys, and he saw me safely inside my front door."

"A real paragon."

"The point is that he made his pass verbally, rather than insulting my intelligence by taking the ordinary physical approach. He acted seductive only after accepting my refusal, when he kissed me. So he was relating to me as mind to mind, and he was respecting the fact that even actions taken in passion have consequences which should be weighed in advance. That makes him a very special kind of man. I adore men with sensitivity and restraint."

"Hm."

Chapter 9

THE LETTER

They continued to walk slowly, following the street as it sloped and wound gently down the hill.

She *was* a nice girl, he realized, just as Carole had said. He had to admit that it was not her place to decide what the State Department should or should not tell him. He had become irritated because she was too logical to agonize over the conflict between the position in which she had been placed by others whose authority she respected and his claim on her loyalty. Moreover, she *had* made a gesture toward getting his permission for the apparent violation of policy while keeping him protected by his ignorance. It had been petty of him to be annoyed with her.

Philanth kept a secret better than he had wanted her to when it was somebody else's, but he should be glad to find that she did not violate such a trust. Even her refusal to admit that there was much of a secret was a laudable protection of it.

But because of her conscientious nature she probably had brought Carole's letter with her while she tried to figure out what to make of it. Their constant travel made the risks of her losing it unacceptable. He had to be sure it was destroyed.

But that would require letting her know that he knew about it.

Well, it had to be done. The practical risk outweighed any psychological hazards.

Just the same, he needed to take into account the effects of bringing the matter into the open. He would have to proceed carefully.

"Let me buy you a drink. Not in one of these noisy bars, but when we get back to the hotel."

"Oh, that's not necessary . . . ."

"Of course not, but . . . "

"I may as well tell you, Mr. Bjorklund: I don't drink. When we're in situations where it seems obligatory, I'm good at pretending. Now that you know, I won't have to work so hard at it, because you won't give me away—or create difficulties."

"Of course not. But a drink can be a pretext for conversation. Don't take such invitations literally."

"Please don't be kind to me, Mr. Bjorklund. It makes me sad. I guess because you remind me of my father." She added hastily, "You're not nearly that old, of course, but he was a wonderful man, wise and kind, and—I miss him."

In even greater embarrassment she went on, "Anyway, let's do our best to stay businesslike. Now that we're not even going to be doing any more sightseeing, ..."

"I appreciate what you're saying, but it can't be like that," he began, and her head jerked up and turned toward him as she slowed her steps.

Her stare made him say hurriedly, "I mean, obviously if we're working closely together all the time it helps a lot if we can be friends. Right?"

"Oh, right, of course!"

He had stopped walking, so she had stopped too, still peering up at his face in the semi-darkness.

For a moment they simply stood there. He did not want to go through with this.

Then the strain he had just heard in her voice dispersed his conflicting impulses.

But it was too chilly to talk it over where they were. "Let's hustle on back to our rooms. There's something we need to discuss so you won't worry."

He started to walk quickly.

"*Please* tell me what it is! Now, so I won't have to wonder all the way back."

He stopped again and stood looking down at her.

Finally he said, "Before we left home Carole gave you a letter."

She drew back, and even in the poor light he could see that now she looked frankly scared. Almost inaudibly she said, "Yes, sir."

"I . . . Of course I'm very embarrassed about it, and I want you to know that . . . ."

"That Carole meant well," she supplied quickly, "but in her effort to be generous to you she, uh . . . ignored her better judgment?"

"Um, that's a nice way of putting it."

"It's a tribute to you that she feels so secure that she's not possessive," Philanth hurried on. "And that she loves you so much." Though gulping with tension, she was actually attempting to put *him* at ease. "And that you have a wife who's so rational and objective in her thinking. I admire her for trying to anticipate a problem and, uh, advancing to meet it. I tend to do that myself."

"That's admirable in many contexts. But . . . ."

"But of course quite unnecessary here. We'll simply ignore the idea. It's not as though I'm suggestible. I make up my own mind about things."

"I'm sure you do." He resumed walking.

Philanth went on with forced briskness, but she had to keep gasping for breath as she almost trotted in order to keep up with his long, downhill strides on the very uneven pavement. "She just applied—the wrong concepts. That's the problem with— theoretical models: you have to—pick one that really fits the—specific circumstances of—reality well enough to—help more than it impedes—your coping ability."

He started to laugh, but after catching her breath a little, she went on.

"In this instance—Carole was trying so hard to—ensure your short-run practical productivity—that she felt it best that—you not be obliged to call upon—any psychological reserves . . . to deal with instinctual deprivation . . . during a long, critically important period."

He reacted with a short burst of nervous laughter. These two women flattered his total supply of energy, never mind his sex drive.

Seeing that she was not going to beg him to slow down, he did so. "This is not something I want to discuss now. Wait until we're back in our rooms."

She mutely accompanied him back to the hotel.

When they reached their floor she accompanied him into his room. As he shut the door she slipped off her shoulder bag and took off her shawl. She sat down on a chair and immediately started talking.

"Your wife's wish was loving, but it ignored the permanent, indirect costs of conserving one's psychological reserves by depleting one's basic operating capital in one's moral self-esteem. It even ignored the effects of stress as a depressant." She obviously meant, on the libido.

Sitting down on the corner of his bed, he replied, "That's a very selfless analysis. Don't you resent her treating you like a slave expected to be available for the master's pleasure?"

She looked surprised and amused. "No, ... That doesn't bother me at all."

"Because the idea appeals to you?" His harshness dared her to admit it.

Her tone was matter-of-fact. "The idea is simply unthinkable."

"You ought to be repelled, insulted, resentful."

"I don't resent it because the request doesn't threaten me: I have nothing to lose by ignoring it. I can't believe that Carole will be seriously displeased because I declined to honor her request.

"It doesn't insult me, since I'm familiar with the traditional relationship between low-status women and high-status men. It would be absurd to get worked up over such a universal fact of life as differential power . . . and its sexual implications even in *our* species. One can only combat its effects.

"I don't feel affronted by this attempt to thrust me into that primitive formula, since I respect you and the basis of your status. In a culture where treating women as chattel was the norm, a woman like me—without property or family—could rejoice at being a gift to a kind and prosperous man like you. She would be delighted to bear your children, even if other women were doing so too.

"However, we're not in such a culture. So Carole seems to be asking me to oblige her. I do try to oblige people, but this isn't a question of being obliging: this is a serious matter. If only because of practical considerations, the idea is a non-starter."

He relaxed, seeing that in this matter too, Carole had had her facts in order. To put the final stamp on his sense of relief he asked, "Did I do right to bring the matter into the open?"

"I think so," she replied comfortably. "It's good that we know that we agree on how absurd it is."

"I *beg* your pardon," he said, smiling a little. "Is it *absurd*?"

"Begging *your* pardon, sir," she replied, also beginning to smile, "the idea I could want to go to bed with you *is* absurd, because it's so far removed from the way I feel about you."

"Oh? I remind you of your father, right?"

"To use the cliché because it's so apt, 'I respect you too much.'"

"I'm very reassured," he said, and she laughed softly. But she blinked away tears of stress.

"You've taken this very well!" Gratitude made him confide, "I was afraid of making the matter worse if I let you know that I knew about it."

"You were worried, unsure of me. And your public position makes you terribly vulnerable to sick or petty women. But you overcame your fear, out of consideration for my peace of mind."

"That's another too-charitable construction.

"However, it's true that I couldn't just keep silent and watch you try to deal with the situation alone. That was too corrupting for me. I had to either openly disavow Carole's scheme or become part of it."

"What 'scheme' are you talking about?"

Relief had loosened his tongue, which had promptly betrayed him. He hesitated; and he saw that her eyes did not miss his sudden concentration.

"Oh," he said with an offhanded wave, "it was—the whole idea of the letter was just Carole's idea of a test to see how you'd react. I'll tell her you passed with flying colors."

He started to get up, but she spread her hands to ask him to stay where he was.

"The letter was a test? It's a strange sort of test! Why would Carole want to test me in *that* way?"

"She wanted me to see how fine you are, so I'd treat you better," he said, still trying to act casual.

"So the letter wasn't what it purported to be?"

He shook his head.

She threw herself back in her chair. "What a relief! I suspected that, but oh, I'm so glad to have it confirmed! Now it does make *some* sort of sense."

She tried to smile at him. "It's pretty scary when

a flower enjoying the sun is told it must cease to keep its proper distance. It starts to tremble in uncertainty as to what may happen next."

"What a lovely image. And how very flattering."

She reproached him gently, "You should never have worried about how I'd react to that letter, Mr. Bjorklund. I do see you as set as far above me as you see yourself."

That touched his conscience. It was true that he was at the top of the pyramid at Peace Corps and she was near the bottom, but she had never deserved his treating her as less than his moral equal.

He decided to be entirely direct with her, to exorcise the demon Carole wanted to trouble this girl before it had any chance to gain a footing in her.

Making the concept sound as frivolous as it was, he asked, "You're sure you'd never be so silly as to believe you'd 'fallen in love' with me?"

The question seemed to amuse her, but she lowered her eyes. "It's no big deal, one way or the other. A man with your responsibilities shouldn't be bothered by such matters."

"I quite agree. But please explain why you think 'It's no big deal, one way or the other.'"

"You're so good, Mr. Bjorklund, and so . . . so polished and commanding and yet caring that you must get used to being around women who are in love with you," she informed him gently. "It's merely instinct, and it can't be allowed to have any practical significance. Sensible people understand that, and I like to think I'm a sensible person."

"You don't reassure me."

She shrugged. "Then give it a little time, and you'll see that I'm not as young as I look."

"Philanth, that doesn't reassure me, either."

Again she smiled, but she acted embarrassed.

"Whether I have the passions of a mature woman is no concern of yours, as long as I have the self-control of a mature adult."

Her candor made him stare, and she gestured helplessly. "Please, Mr. Bjorklund, I'm going to start blushing any moment now. I haven't implied anything that should disturb you. I've simply gone to the heart of the matter by pointing out that I recognize that such matters are of no importance—not when one considers the needs of the world."

"I'm concerned because it could affect your work."

"I understand that! But I'm trying to show you that I view the question realistically!"

"I think it's open to debate which of us is being more realistic," he decided, and she nodded.

"However we reason about my feelings," she added, "you should rest assured as to my intentions. Causing problems for you would be the last thing I would ever be willing to do."

"I believe that. I'm concerned about the effort of self control if your feelings should start to interfere with your intentions. I know I must sound conceited, ..." It was his turn to be embarrassed. "I wouldn't bring this up if Carole didn't believe you're at risk."

Philanth simply nodded again, folding her hands. "But one has to calibrate the nuances of one's interaction all the time anyway," she said as though it were self-evident. "I'm *always* calculating how much expression of affection I can get away with.

"So I can't see what your concern is. —I mean, I can, but it's inapplicable in my case. My being madly in love with you wouldn't make any real difference in how I go about things. I speak from experience of just such a relationship in the past.

"Indeed, I like to believe I seldom fit the stereotypes of the silly female in *any* situation. I noticed how foolish such behavior is when I was a preadolescent observing teenagers."

"All right," he said as a sense of hopelessness descended upon him. Carole indeed seemed to have mastered the workings of Philanth's inner life as a clockmaker comprehends a clock.

He added threateningly, "But if I do find out you're emotionally involved with me, I'll fire you. I won't work with a woman who's doing that much 'calibrating of the nuances of her interaction' when she should be concentrating on her work for me. Is that clear?"

"Of course! And it's no problem," Philanth replied equably, picking up her shawl and shoulder bag as she rose. "I was expecting to leave Peace Corps by Christmas anyhow."

He imagined how Carole would react to that exchange and felt like swearing. But in the next moment Philanth threw him an impish look. It said that she was conceding nothing, either about her emotions or about the validity of his position.

He would do well to leave the whole matter alone while this discussion faded into oblivion.

But there was one bit of unfinished business first: the object of this whole discussion. As she set the strap of her bag on her shoulder and started for the connecting door to her own room he asked, "Did you destroy that crazy letter?"

"I made a copy that deletes anything which might compromise you. Then I destroyed the original."

"Why keep it at all?"

"Oh, maybe now I don't need to. I wanted to be able to study it while I figured out what to say to Carole when I got back—and to you, if you knew about it and brought it up. But if it's only a test so you'll see I'm a good kid and stop being mad at having to bring me on this trip, ..."

She added as though thinking aloud, "Now perhaps I can figure out why Carole offered me those irrelevant inducements on the second page."

He flew at her and grabbed her shoulders, yelping, "The *second* page? *What* second page?"

She gaped at him. When she could close her mouth enough to speak she stammered, "You—you didn't know . . . about the second page?"

She sagged, automatically lifting a hand to secure the strap of her bag as it started to slip. "Oh, Lord, then I guess—maybe—you weren't meant to? How can that—"

He gave her a little shake. "*What does it say?*"

She twisted free of his grasp, exclaiming, "I promise you, you don't want to know!"

"I must, and you know it." He started to go around her. "Where is it? In your room?"

"I have it on me, for greater security." As he spun back to face her she put her hand on her midriff. "In my money belt, under my blouse. Carole said when she handed it to me, 'Here's something to keep in your new money belt, where it'll be safe.'"

"Let's have it!"

She turned her back and began working at her clothing. Finally she swung around and handed him a strip of folded onionskin stationery.

"Page Two of the original begins with the 'P.S.' on the lower part of this page," she explained, jumping out of his path as he strode over to sit down where he could study it under the bedside lamp.

The handwritten page of which he had found the carbon had been transcribed into small, single-spaced type with narrow margins. Date, salutation, and closing had been omitted. The lower portion of the page read,

P.S.: You will find him an admirable lover. Nothing is withheld or given conditionally; there is no selfish hidden agenda; he is completely present in his giving and his sharing.

He is like you in these ways. So as soon as you have come to know each other the two of you can find completion and profound happiness together.

What is more remarkable in light of the above is that C. understands all that a woman wants, even that spirit of artistry which has the self-distancing required for calculation yet expresses intense caring. Your pleasure becomes his masterpiece, achieved with constant attention to detail—yet always with honest, unqualified devotion.

I tell you this quite sternly, though in a motherly way: you must not deny yourself the benefits of sharing yourself with this man. Life is too short for such an enriching experience to be deliberately passed by. You must accept the fact that in this context, in giving you cannot help but receive; certainly not when you give to a deeply loving person.

Also bear in mind how much a loving person needs to express love. You would wrong him and hurt him if because of some misapplied principles of ethics you forced him to try to repress all the tenderness he assuredly will need to give you when he has truly come to know you.

With a heartfelt groan he thrust the letter back at her.

"I'm sorry," she exclaimed as she took it.

Slipping the letter into a pocket of her skirt, she added tightly, "But it doesn't change anything. I suppose Carole didn't let you know about the postscript just because you'd be even more embarrassed.

"But it means nothing to me, Mr. Bjorklund. When a conclusion ignores highly pertinent facts, the question of the validity of its *stated* premises doesn't even arise."

"Meaning?" he snapped.

She hesitated. "Meaning that this 'P.S.' can have no persuasive power for me at all."

"Because you don't *care* what I'm like as a lover," he finished dryly, and Philanth looked shocked by the bluntness of his translation.

"It's such a simple test that she tried to make it harder—just in case I'm not quite the sort of person she has every reason to believe I am. That's all there is to it."

He did not want to worry her, but the facts didn't add up. Just as a logical matter, why would he be told to evaluate her reaction when he had not been given a complete picture of what she was reacting to? So why had he found the carbon for the first page but not the second? Had he been intended to see that the first one was used and reach for the second in the drawer before noticing the peculiarities of both of them?

But come to think of it, hadn't he gone upstairs and gotten a second carbon after the caterer arrived? He surely would have noticed if it had been like the one he had taken out first! So what had Carole done with the carbon of the "P.S." page?

Apparently he *had* been intended to find the carbon of the letter, but not the carbon of the postscript. As Philanth surmised, it must have been done that way to avoid embarrassing him—and making him very angry indeed.

He began to feel that he was being adroitly manipulated; that he could be sure of nothing.

His long silence made Philanth say uncomfortably, "Don't fret about it, Mr. Bjorklund! It should have no bearing on our conduct. We both know what's sensible, honorable behavior."

"Right," he said bleakly.

"I can see you're ready to call it a day," she said, gathering up her shawl and bag again. "Good night, and thank you so much for the lovely evening. It was a privilege, seeing Sacré Coeur with you."

"Good night, Philanth," he returned, scarcely glancing at her. "My pleasure.

"—Philanth."

"Sir."

"Destroy that paper. At once. I would've taken the liberty of tearing it up when I had it, but I was too appalled to do anything but push it away."

"Oh, sir!" she protested in a heartfelt tone. "It's a beautiful thing, and there's nothing in it that identifies you. It might as well have been written about someone who never existed, like an imaginary mistress addressed in a bit of sixteenth-century verse."

That did make him turn to look at her. "You regard it as a—a piece of literature? Like a poem?"

"Yes, sir."

He turned his back on her. "Burn it, Philanth. For your own sake, as well as mine."

"I certainly needn't burn it for my sake, Mr. Bjorklund! To me it's like a beautiful sculpture a woman created in the likeness of her husband: *her* husband, who could never conceivably be mine! My knowing the artist and her subject has no significance: an artistic creation stands independent of such considerations. It's about her love and their love, and that makes the personal aspect of the subject matter inalienably theirs. Yet at the same time the beauty of its sentiment belongs to anyone who can be touched by it."

"Nevertheless, if you want beauty you must find it in some other object."

"It only embarrasses you. I respect your feelings.

"But Mr. Bjorklund, you must see that it's not at all like being an actor who has enough sensibility to resent being confused with his roles by women—"

"Don't argue, just do it!"

"I'll do it only because you're paranoid about your reputation." Her tone made the statement a challenge, and it was followed by a silence.

After a minute the connecting door closed quietly. She had given up on waiting for him to relent.

He took off his jacket and threw himself face down on his bed. He tried to make his mind blank rather than think about that little metal drawer full of carbon papers in Carole's study.

But it would have been easy for Carole to set him up to find that letter, her knowledge of his reaction patterns was so thorough. She could have deliberately not sent the caterer the flyer so the caterer would have to come to him that day while she was gone, forcing him to make a copy without the copier.

Carole's elaborate spiels about her plan to "test" Philanth could have been a cover story to mollify him, perhaps to gain time.

For what purpose? He did not want to think beyond that point. He went to bed.

Chapter 10

KEN

Laden with briefcases, computer, and luggage, they hurried down the corridor toward the hotel elevators. A uniformed maid was coming toward them carrying a stack of linens. As she drew close to them Philanth began speaking what he took to be Arabic in a soft, friendly voice. Turning, he saw that Philanth was setting down her bags as she continued to talk to the girl. The maid responded shyly, but within moments both girls were chattering.

He and Tony stood by, halted in their rush to meet their driver and get to the airport. Tony's glance at him said he didn't have any idea as to what was going on, either.

Philanth turned to him and Tony. "Mr. Bjorklund, would you and Tony please shake hands with my friend?"

He and Tony complied, Tony grinning at the chambermaid's discomfiture.

After more soft Arabic the two girls embraced warmly, the maid with the towels still clutched in her arm. Philanth bent to gather up her shoulder bag, briefcase, and suitcase, the other girl shyly trying to help her.

With more Arabic flung over her shoulder and an answer in about the same words from the girl, who seemed genuinely sorry to see her new friend leave, Philanth got underway again. He and Tony did likewise.

Considering how many Senegalese and U.S. embassy officials and Peace Corps staff were anxiously awaiting their arrival in Dakar, he was beginning to be annoyed by the delay, though the whole scene had taken only a minute.

"What was all that about?" he asked as they neared the elevator hall.

"I was only saying good-by."

"You seemed to be good friends with her," Tony commented with his usual jovial humor.

"We got to talking yesterday when she came to clean my room, about how she could increase her earning power by improving her French. She seemed to welcome my suggestions.

"I think it was actually my personal interest in her that she appreciated. And my being able to speak her language. She feels rather alone, working here, but her family needs the money. So she's venturing out into the big, bad world. Considering her upbringing, it's difficult for her."

"You should've introduced me sooner," Tony jibed.

"You're much too late. Her marriage was arranged for her when she was in her early teens. If she hadn't gained access to contraceptives when she married she couldn't be earning money outside the home now.

"The indirect consequences of this kind of change in one person's life are far-reaching: the economy, education, and eventually all other aspects of society are affected." Having worked in family planning in this region, Philanth could be expected to feel a sense of personal achievement in this state of affairs.

"She certainly seemed to have taken to you," he observed as an elevator opened and they went in.

"Yes. I wanted to start exchanging letters with her, but she said no. I think because the postage would have been too much of a luxury for her. It's so sad to be that limited in your choices!

"I really wanted to tip her, but under the circumstances I felt that would hurt her feelings.

"I know! I'll write to the manager here about what a motivated worker she is."

The elevator opened, and they hurried through the lobby and out into the bright sunshine to find their driver waiting to help with their bags.

The young Peace Corpsman had brought a package which had arrived for them in the daily pouch to the Moroccan embassy. But being a member of the country headquarters staff, the young man was eager to discuss the problems of Peace Corps/Morocco all the way to the airport. So neither he nor Philanth opened the package. Finally Tony took it because he was anxious to be sure it contained his next installment of press kits from Public Affairs.

Their seats on the plane were two across, and Tony got a pair of seats to himself several rows behind them. When they were belted in he watched as Philanth opened the remaining packet, which had been directed to her from Marjorie. Suddenly she paused, looking at a letter envelope with consternation. It was addressed to her and marked, "Personal—PLEASE FORWARD." The return address told him that it had come from Carole's brother in San Francisco.

Philanth fidgeted, then bent double to slip it deep into her bag without completely removing the bag from under the seat in front of her.

She opened a manila envelope marked "IN BOX" and took out its contents, but he saw that she had lost the ability to deal with them. She looked stricken.

"What's my brother-in-law been up to, to make his unopened letter get such a reaction from you?"

"I . . . don't think I should discuss it."

"Carole said you'd been dating Ken while he was in town this past summer."

She nodded.

"You're obviously disturbed. It'll save me trouble if I don't have to tackle Ken about it. —Especially since I'd have to keep reminding myself to do it until I got a chance."

"It was a shock, getting a letter from him—particularly *here*."

"But why was it so upsetting?"

"He . . . tried to . . . change my mind about getting married."

"What! My brother-in-law proposed marriage to you?"

"I was surprised too," Philanth answered wryly.

"I take it you refused."

She dipped her head.

"He took your refusal badly?"

She nodded vigorously.

"Well, he'll get over it."

"But it hit him hard because . . . of his divorce."

"I understand." Ken's wife had unceremoniously dumped him for someone else.

"It was so unfair. He got wrapped up in earning tenure so he wouldn't lose his job,—'publish or perish', you know—and his wife got . . . to feeling neglected."

"Yes." That would be Philanth's charitable expurgation of Ken's self-serving version of the story. But during their last Christmas family reunion Ken's wife had angrily told Ken in front of everyone that he was "such a jerk!" merely because he had made an unkind and pseudo-sophisticated wisecrack which in fact missed the point of someone else's statement.

Philanth went on, "But he's a dedicated teacher and a conscientious scholar, so he had a double handicap in getting promoted. The system rewards professors who woo students rather than teach them and grind out quantities of unoriginal trivia."

"Does it, now."

"I'd heard my mother complain about the same things, and my father and their friends agreed. So I understood his frustrations.

"And I understand why he was hit hard by his wife's defection: it felt like one injustice too many. And taking his little children to live with another man!

"Then I made it worse: I caused him to feel rejected again. If I hadn't been blinded by my own selfishness I'd have realized that the more we enjoyed each other's friendship, the more I risked his wanting more. I worry that I may have made him even more soured on women."

"Yes, he *was* soured when you met him, wasn't he."

"That's why I was taken aback by his proposing! Carole had written me in June that he'd come to Washington determined not to get seriously involved with *any*one."

Ken had read about all the unattached women in Washington and had not expressed his plans for the summer in such polite terms.

"He's cutting himself off from so much," Philanth lamented. "Surely he'll change his mind again. A tenured professor is a good 'catch'; and considering the generous offer he made me, lots of women would be delighted to marry him."

"Really? What did he offer you, if I may ask?"

"He promised I could go back to school for the Ph.D. I'd always wanted. And have a full-time servant, and he wouldn't insist on having a child with me if I insisted on our being foster parents instead.

"It's a splendid offer, especially from someone paying child support, plus extras, from a professor's salary. And he says San Francisco's an expensive place to live."

"Sounds like he's prepared to take up moonlighting, now that he's got tenure."

"He said he'd wind up the scholarly work he was doing in the Library of Congress this past summer and concentrate on making money if I'd marry him. I said money wasn't an issue."

"No, if you don't love him, you don't love him."

"Now, how do *you* know I don't love him? He's sensitive and caring."

"He's pretty messed up by the divorce and the loss of his children, though, isn't he?"

"He has wounds. But wounds heal clean, treated properly."

Her voice softened. "I would have enjoyed healing him. I felt I already had made a start.

"It must be wonderful to watch someone become whole and grow and become strong and happy. He could have blossomed, over the years.

"I feel so bad because what I offered ultimately did more harm than good!

"But that happened because he wanted too much—more, at least, than I was prepared to give."

"You weren't in love with him. You talk about him as if he were an injured child." Carole found her younger brother's life-long penchant for self-pity hard to put up with and suspected that Ken's wife had come to feel the same.

"Well, *of course* he was too damaged to manifest much of the psychological strength which arouses instinctive female desire—but surely what one woman could destroy, another could rebuild."

"He never had it, Philanth."

"Then he was beginning to have it with me."

"If so, I can see why he asked you to marry him."

"We didn't have to get married. I could've made it a productive friendship."

"He must've thought there was much more to your relationship than *that*!"

"He was overly impressed by our compatibility. I could've married *any*body, on *that* basis! I've always made it a point of honor to get along well with anybody, no matter how impossible. Like the sun in the fable about the contest between the sun and the wind, I just keep up the warmth until a person sheds his hostilities. —Of course, 'warmth' as metaphor may include tact or reasonableness or whatever it takes."

"You don't really think you could get along well with any sort of *husband*!"

"Traditionally that's how pliant and adaptable 'a good woman' is *supposed* to be."

"Sometimes I have trouble deciding whether you're serious."

"Depends upon which level I'm operating on pleases you. Truth is multi-layered. If you want to be amused, I'm not serious. If you won't be offended, I am. If both, then I'm both at once. I'm not obligated to operate on only one level, and really try to tell people truths which will only offend them because they're too ignorant or irrational to recognize that they're true."

He saw what she meant. "One reason it's so hard to find an honest man is that it's hard to find people who aren't hostile to those who tell an unpleasant truth."

"Yes. And people don't deserve truth if they're going to turn against you for telling it."

"So you have to be tactful.

"But surely a woman of your sensibilities couldn't be a good wife to *any* man!"

"Suppose someone pointed out a brutish-looking stranger and said to me, 'Girl, here is your husband.' I can control most of my actions, so I could at once become a loving and submissive wife to him. Even if he were as beastly as he looked, I could find some qualities to like in him. Women have had to be able to do that for thousands of years."

"It's monstrous."

"I'm very thankful to live in a society which produces men who would say so.

"But over time he also probably would become kinder to me, within the limits of his capacities. I usually manage to have that effect on people, given enough close association—insecure or intensely competitive people being exceptions."

"So you felt you could have had a happy marriage with Ken?"

"If I'd wanted to. But I couldn't feel my life was properly spent if I didn't feel fully engaged in doing something that was truly needed."

"But it seems Ken does need you, rather badly."

"He does need a good wife. But lots of other women could fill his needs in that regard. If they were generous and compassionate, they'd be glad to make the effort: he just needs more petting and stroking than some men do. After what he's been through, he might respond with considerable loyalty and gratitude."

Now Ken sounded like a homeless dog. However, Philanth seemed to have understood him pretty well.

"But the world's needs supersede those of any individual," she concluded.

"I hadn't realized the world situation was so desperate as to morally preclude marriage!"

His joke elicited only an impatient glance. "The world situation needs people who choose to find their happiness in doing the things that need doing most, especially through the ways they support themselves; people who enjoy minimizing their wants and meeting their needs as economically as possible, while trying to give as much as they can in their personal and professional lives.

"Ken merely leads a conventional middle-class American life, with the usual consumption habits. That's not nearly good enough: it perpetuates so much that's wrong in the world! Someone from a poor country would regard him as rich and insouciantly wasteful and self-indulgent."

"Your alternative must be pretty grim."

"It's happy and peaceful and liberating.

"For one thing, it lifts you out of the struggle to earn and keep the superfluous things people spend so much of their lives on.

"And it must be difficult to truly enjoy living for your own pleasure if you're aware that everything we care about may be destroyed unnecessarily and without renewal, by the hand of man."

"So you reason that doing all you possibly can for the world's needs is a sound basis for being 'happy'? I find that incredible. Perhaps I could understand if you said 'content', but 'happy'?"

"Yes, I'm happy. I seem to be the happiest person I know. Why do you find that incredible?"

"I think you've made a realistic appraisal of what the world needs, in view of the large-scale, long-range trends. It's very good, in fact: we do have to conserve and redirect our efforts in order to be able to afford to reach out for the unlimited resources beyond our planet. And of course doing what is needed is satisfying.

"But what about *personal* happiness?"

"You seem to divide happiness into 'personal' and 'impersonal', between what you want for yourself and what you want for others. Why not just get rid of the first category, the things that enslave the self with petty concerns? Happiness is simply having what you truly want and need, and knowing it."

"But there's so much you must want and can't have, to the extent that you care about the whole world at large!"

"Yes. But I divide wanting into two widely separated kinds. I'm virtually always happy on the level that needs only the minimal physical conditions for health in order to be glad to be alive. On the other level, I'm virtually always in despair because I can't remake the world as a perfect place. The first level is a solid platform of a sense of personal security and well-being on which to stand while I strive for the second. The second gives me something to keep reaching toward so I have the satisfaction of growth and action and purpose, as well as the urgency of fear for the world's future.

"I couldn't have the sense of security and well-being which allows me to focus my attention outward if I let myself want much in terms of conventional pleasures. So I have to reject as entanglements, frivolous distractions, all the ingredients of 'happiness' in between the minimum personal and the maximum global levels: all the degrees and kinds of gratification involving nice clothes and housing and transportation and food, and also fame or prestige. That's not difficult when you have a wealth of inner resources."

"But how do you *sustain* those inner resources? I asked you that at Sacré Coeur, and you said a person might have to 'grow his own'."

"People find all sorts of ways to do that. But it's easiest and most productive if you find nourishment in your work. Loving someone is important, too.

"Someone did a study which confirmed that the two major factors in happiness are love and work. Having work I believe in and through which I can express love is all I need to be happy."

"But there's no guarantee that you'll continue to find work that's that satisfying to you. Lots of people never do."

"I don't need much income. I minimize my personal wants, and I have no dependents. And I'm very frugal: I *like* making use of other people's castoffs, for example, because I cherish my freedom more than luxury; and I don't feel the need to prove anything through my possessions. And I don't object to working under harsh conditions. Those factors greatly reduce the competition for the jobs I'd like to have.

"And if you're competently performing worthwhile tasks nobody else wants, you can feel significant even in a world where hundreds of people compete for any job that offers the *conventional* kinds of rewards. Work done in obscurity can demand just as much skill and commitment as the work of people who receive recognition.

"It's not even necessarily unimportant. For example, we can afford to have mediocre people clinging ferociously to their positions near the top only to the extent that we have excellent people toiling conscientiously at the lower levels. Look at Peace Corps headquarters, for inst . . . ."

She smiled and put her fingers over her lips. "Present company not included."

"Those are arguments to console yourself for not achieving success."

"You mean, for not achieving *recognition*—like I explained at Sacré Coeur. And a lot of people need consolation for not achieving that, because their work will never be given the honor it deserves. It's a complex, interconnected world, so lots of invaluable work goes unrecognized forever.

"Meanwhile useless and even counterproductive 'achievements' are being recognized and rewarded extravagantly, so those who waste their time on them are accounted big 'successes'. As I was trying to say at Sacré Coeur, we mustn't get pressured away from valid values by the irresponsible foolishness of mass culture. Many of the people who run after 'success' defined in those terms are a menace: to the ecosystem, the culture, the country, and themselves.

"It's far better to do only what needs to be done, hoping that somebody, somewhere, sometime, will benefit as intended."

"But what about love? You mentioned the importance of love to happiness, but love must have an object. Even if you are able to love your work because it's needed, you—one needs a personal relationship."

She bent to pull her shoulder bag from under the seat and took out her black vinyl wallet. She opened it to show him some photographs. "My mother; my father; my husband. I didn't stop loving them just because they're dead. I believe they have to live through me, now, so they're within me. You're talking about an intimate relationship, and that's pretty intimate."

He took the wallet, studying the pictures.

The facing pair of tinted studio-portrait photographs of her parents were beautiful. He looked from them to her face, picking out family resemblances: an impish, dimply smile and fair coloring from her mother, a strong jaw from her benignly dignified father.

Her father looked so much older than her mother that they could have been father and daughter. "Were your parents' pictures made at the same time?"

"Yes. Their fifteenth anniversary," she answered sadly.

"They look like fine people."

She didn't bother to respond to that.

"Are they really companionship for you?"

"Yes."

He flipped her father's picture over and found a stark contrast: her husband had been a scarecrow who looked impatient at being asked to pose, and the amateur snapshot did not do the boy any favors.

To the right of the mug shot of her husband was a representation of Christ on the Cross as viewed from directly above His head. "This is exquisite," he said, showing it to her. "Where'd you get it?"

"It came from the Glasgow Museum, where they have the original, by Salvadore Dali: 'Christ of St. John of the Cross'. It was my mom's."

He flipped it over and found the collection ended with a vibrantly beautiful Madonna, probably by a contemporary calendar-art painter. Turning back to the beginning of the set, he found that the back of her mother's portrait was covered by an illuminated Prayer of St. Francis.

That was the full extent of her wallet's collection of photos and memorabilia. He handed it back. Such simple piety in such an intellectual young woman.

"A person needs someone who's still alive: someone who can share everything on an equal basis. That sharing provides crucial support when the prospects of dealing with our global problems seem just about hopeless because of human nature. To the extent you feel the urgency of the problems, you'll need that."

"Lots of people never have that source of strength, so one has to be able to live well without it, in order to help the others."

"Why must you set yourself to help in that way?"

Her tone was stubborn. "Because it's needed and I think I can do it. I believe in the imitation of Christ."

He studied her. "D'you think Christ was lonely, Philanth?"

She nodded, not looking up. "I'm used to it. When it gets bad I remind myself that it's better than a drug addiction."

He adopted a practical tone. "Well, I don't see how you expect to go very far without the real, living companionship I'm talking about. I've often marveled at how so many people function at all without the psychological balance happily married people have. It's no wonder to Carole or me that they so often fail to do it *well*."

"There are millions of refugees in the world today," she replied. "*Millions!* Some are in camps, yet still subjected to starvation, disease, and terrorism. Some are trying to make their way to some form of relief through areas made wastelands by war: in rags, helpless to save their children from dying before their eyes, often dying themselves. All their livestock are gone,—butchered, starved, or stolen,—and the land has been ruined because they abused it in their efforts to support their growing families. Sources of aid we take for granted are nonexistent.

"In the swiftly growing shanty towns around the cities in poor countries there are ten-year-old prostitutes who regard having v.d. the way we regard having a cold. Boys of twelve are scavenging for scrap and hiding small coins to buy their first pair of shoes so they won't have to walk on the excrement that washes across the ground and into their huts every time it rains. They're literally trying to figure out where their next meal is coming from; they're sniffing cobbler's glue because it dulls the taste of the rotten food they find in other people's garbage.

"D'you think many of those people are fretting because there's not enough 'sharing' in their lives, enough romance or understanding or affection?"

He did not attempt to answer. He was glad he would never have to go up against her in a public debate, for her position would enable her to go straight for the soft underbellies of a lot of cherished assumptions he knew were unfounded.

"If *they* can't afford such luxuries," she resumed, "what right do *I* have to concern myself about such matters?"

But her position was too pure to be lived in. "The fact that you can't touch or see something doesn't make it a luxury. And the fact that others are starving doesn't mean you should try to teach yourself to function effectively when you're weak from lack of proper food. And different purposes in life, different kinds of activity, generate different genuine needs. Besides, we're talking about necessities you could have whether others have them or not."

"I might have lots of things that would help me function better: a car; servants; a private plane. Where does one draw the line? So why should I accept your idea of what constitutes a legitimate necessity for me?"

"You're an able and educated person, Philanth. You shouldn't limit your ability to accomplish good things by more self-denial than is practical."

"I have to test the limits of what I can do without."

"Why?"

"So I know what they are, of course. And to strengthen myself; to learn ways of doing without while still being effective; to maximize my independence.

"And because I'm not even a tiny bit more deserving than some poor person just because I'm healthy and educated and protected by an orderly society. I don't have more entitlements. I have only the obligation which accompanies knowledge and the ability to make choices.

"Did you ever see the movie, *Soylent Green*?" At his shrug she went on, "It starred Charlton Heston and Edward G. Robinson, and Joseph Cotton appears briefly. Self-indulgent people seem to hate it, but it was a wonderful film: a remarkably credible projection of what our own country could be like in the middle of the twenty-first century if our population growth and political irresponsibility continue.

"At the end the Charlton Heston character tells his girl friend he wants her to stay with the nasty, self-indulgent man she's dependent on because he himself has nothing to offer her. He understands the awful truth about what's happening to the world, and because of that knowledge there's no place in his life for her—no time for pleasure or sharing, no means to protect or provide for her. He now has something he has to try to do even if it kills him, and it nearly has already.

"I don't want to be melodramatic about something so ordinary, Mr. Bjorklund. But you oblige me to spell it out.

"Different rules apply to you. You're like a general: positioned where you can do a great deal without being on the front lines.

"You've been fortunate all your life, haven't you?"

He nodded.

"And you've made good use of your advantages.

"People can have comfortable lives behind the lines. They can even buy homes and bring up children. I'd enjoy that, too—I admit it. But the rules are different out where the guns fire real bullets.

"I don't mean for a moment to suggest I'm morally superior. I'm just called to a different lot in life by my circumstances than you have been by yours."

"But Philanth, believing that you should be able to get along without something isn't the same as being able to do it. Is your track record in this area really so good?"

Her mouth tightened. "It's a fair question. But a person can't set track records while he's still learning to walk.

"When I'm depressed I remind myself that doing things for others is the most important part of sharing, and that depends only on myself. You don't need any particular person in order to engage in it, and there are plenty of needy people around—especially children, who need so very much . . . and don't always have inhibitions about accepting what is offered."

"Are you deliberately being obtuse, Philanth, pretending you don't really understand what we're talking about? Do you secretly believe you're some special type of being who's naturally above the need for ordinary loving?"

She looked appropriately hurt. "I understand 'ordinary loving' well enough: it's basically appreciation for needs met. It's essentially selfish. I'm not interested in that, nor in a conventional kind of marriage, which is founded on that: on reciprocity of met needs."

He clicked his tongue over such a misapprehension.

"Selfishness doesn't *enter into* what I'm talking about! Loving and being loved completely is wholeness. The concept of selfishness fades into irrelevance." He had veered too close to a concept implicit in Carole's damnable letter, but it couldn't be helped. "The good of either is quite literally the good of both, and they both feel that it is."

He asked more gently, "You do understand that, don't you?"

"'Wholeness'," she repeated hesitantly as though recognizing that it was necessary to say *some*thing.

Remembering that Carole had told him Philanth probably had never experienced reciprocated sexual love, he suspected he had gotten into an argument about music with someone who had never heard a symphony orchestra.

"The wholeness of spiritual union is a perfection the saints write about," she ventured.

"Put aside all you've read for a moment—if you can. Try to answer from your own experience."

She threw him a troubled glance. There was a long silence.

She said reluctantly, "One has to have someone to share it with."

"And you don't think you can find someone who meets your standards? There are lots of good people around."

"But if someone's really special it's best to keep one's distance, because one wouldn't want to get in their way. They have their own agendas, which it would be wrong to interfere with. Either they're married, or they can't marry because of the nature of their work."

"Why not?"

"For example, they might be obliged to move around a lot; they may even be forced out of this place or that when conditions deteriorate too far. I lack the training to have anything to offer as a partner of someone like that. And I certainly wouldn't want to make things more difficult for such a person."

"Not even by letting them know you loved them?"

"Especially not that way."

"They're not all unavailable. Not in your age group."

"I knew someone like that even in high school," she said so quietly that he had to read her lips because of the vibration of the plane. "But people that beautiful are in great demand. Everybody seeks them out, wants to be near them as much as possible. If I can stay out of their way because I can get along without, they'll be available for someone else, who . . . won't hesitate to go after them and . . . marry them."

He nodded deeply, catching his lower lip between his teeth, for he could imagine, now, what Philanth had been like at sixteen. She hadn't changed much, except to develop an elaborate set of rationalizations for acting according to her nature.

"What made you that selfless, Philanth?"

"Going after something or someone for yourself seems petty to me.

"Besides, living for yourself can't really make you happy, because with all your self-centered efforts in 'the pursuit of happiness' you're only serving a small, greedy animal that's going to die. Whereas living for others frees you from that burden. And in the service of something great you grow toward greatness. That process transcends the conventional concepts of happiness."

"Humanity needs all the help it can get, but I don't find it 'great', nor an inspiration to greatness. On the contrary: as a whole it's a discouraging spectacle."

"It's all in how you choose to interpret the facts. If there *ever* was *anyone* morally beautiful, learned yet wisely humble, capable of loving deeply or creating great beauty, why must his or her imperfections cancel out those qualities? And—as I said when we talked early that morning at headquarters—the examples provided by individuals prove the potential of the species."

"That's pretty abstract stuff for getting through a life of loneliness."

"When I do crave male companionship I like to think about some other woman being happy because I went without. As I was just implying, any man I'd like to spend my life with would be so good that lots of other women would appreciate him too, even if he were ugly and awkward and had very little money."

"I give up!" he exclaimed half-facetiously.

Then, in the back of his mind, Carole raised her shillelagh over his head. He must have married her to be sure he would never give up and then despise himself. He renewed his effort.

"You really don't have closeness with anyone special who's alive?"

She shook her head, then added shyly, "Since I don't, I'm free to be married to my work. And I feel that my home is wherever my work is, since 'Home is where the heart is.'"

"Surely you have *some* family."

"The human family." She was very shy, now. "And that's what my work is for."

He frowned and shook his head.

"As my superior, you should be *glad* to know that I live only for my work. Carole said *she* was."

"Oh, did she now!"

The girl nodded, looking puzzled by his anger. "She only meant she was glad I was happy."

"Well," he said, "you'll surely meet someone who can persuade you to get married again."

"I'd never intended to marry even once. There were special circumstances which can never apply again. But I can't explain that, so people never believe me."

"You can't expect anyone to."

"There's one kind of person who would believe it, once I explained my reasons: the sort of person who could share them.

"I don't expect to meet anyone like that, though. They're probably mostly in religious orders and other out-of-the-way places."

Carole's briefing had been thorough. "I think you've got the *Bells of St. Mary's* syndrome. How many real people in religious orders have you ever known?"

"Can't you believe that there really are nuns who think and feel and act the way they're ideally supposed to?"

"Yes," he said, looking at her, "I can believe there are people like that. But I wouldn't look for them in out-of-the-way religious houses. I think there's more of the misery they need in order to stay 'poor in spirit' to be found in a life of trying to help people outside, in the real world."

"Oh, *thank* you!"

"For nothing!"

Seeing his hostility to the idea he himself had expressed, she looked disconcerted, and they were quiet for a moment.

"If you live only for your work you're going to be emotionally frustrated. You must see that that could make you less effective."

"I find emotional intimacy in relationships where I can help people develop. F'rinstance, I made it a 'project' to improve my boss at the Hunger Foundation, and . . ." Her voice tightened with shyness. "At present I'm focussing on Tony. Even if his selfishness is just a macho front, he should grow enough to shed it. People who espouse the fashionable view that it's smart to be selfish need to develop their capacity for caring so they don't hurt others or live shallow or parasitic lives. It's important to me to try to help them if I'm presented with an opportunity."

"How can you?"

"It's not hard to demonstrate what love is *really* like, when a man is attracted to me and we have a great deal of close contact."

"Surely you've realized how dangerous that can be! Your 'project' at the Hunger Foundation could've gotten you not just raped, but murdered!"

"Oh, Tony wouldn't hurt me. Like my boss at the Hunger Foundation, he's too secure about his appeal for women to get that angry over a rejection."

"Surely you can find a more 'safe and sane' emotional outlet!"

"I choose not to take the conventional route. Having a great deal of love to give but no one to give it to who's easy to love makes the love build up and become concentrated until it seeks expression in the more difficult, more needed, more *practical* ways of loving. You can understand that, can't you?"

"I suppose. Though I'm not sure it's true."

"It's one of the many things which are true if one makes them true within oneself."

"But once an altruistic motivation like that has formed your character you're free to live a normal, fulfilled adult life. And you need to. Because you shouldn't expect loneliness to be a permanent source of the spirit of giving. Once you outgrow the emotional excesses of adolescence and the altruism you've accumulated gets used up by the world's orneriness, your solitude engenders a despair that can destroy you."

She looked startled, then cried, "How can *you* know that? You're amazing."

"I didn't know it until I said it. But I already knew that in public service one has to be either insensitive or well-shielded, financially or emotionally, to withstand the abuse. Grabby people and power junkies seem to be exceptions. But if celibacy were required too I don't want to think about all the decent people who would find it intolerable."

"So you're thinking of public service at the high levels where *you* operate, where people are subjected to impossible expectations. We were discussing my very different circumstances. For me, being unattached provides a freedom to find the greatest need that isn't being met which I am qualified to meet and give it all I've got. There's not much competition or criticism or pressure in work that almost nobody else particularly wants to do."

"Nevertheless, you'd have more to give people if you were married."

"More than I have?" She was chuckling bitterly. "Would that be bearable, when they already want less than I need to give them?"

"Wait a few years, Philanth, and things will look different to you. Believe me, the world will use up *anybody's* altruism quickly if they don't have a strong source of renewal. You haven't convinced me that you've got one."

She responded firmly, not looking at him, "Then I'll be sure to acquire one. But I won't have to get married to do it."

"But marriage can meet the need so beautifully! And I must completely disagree with your idea that marriage precludes doing things that really need doing."

"It does for a woman," she shot back, "considering what's expected of women in marriage!"

"Ah. But you wouldn't mind marriage to a man with the same priorities you have?"

She gave an exasperated laugh. "Show me a man like that: who didn't want me to bear his children in an overpopulated world, keep his home looking like a magazine photo to make a favorable impression on thoughtless materialists, and all the other superfluous tasks married women are wasted on.

"You needn't worry about my emotional life, Mr. Bjorklund. I plan to do volunteer work with children in my spare time."

"Oh, great."

"I love children, talking with them and cuddling them and telling them things they want to know and doing things with them, just watching them and studying their reactions. And they love me."

"Volunteer work isn't enough."

"How do *you* know? Do you know how it feels when a little girl says she thinks you're 'nice'? Or a little boy decides he 'likes' you? A child is saying so much in a few words that sometimes it can break your heart, when you know what that child's frame of reference is. And when a child wants more and more of your time, I think that can feel as good as winning a Nobel Prize. And when she keeps giving you little presents from her childish treasures . . . .

"Have you ever watched a child you knew well when she or he was asleep after a hard day? You sit there looking at all that moral simplicity and fresh physical beauty, and you know so well how little they imagine what life will bring to them, how terribly hard it can be sometimes, and how much, how very much they'll need everything you can give them, show them, teach them in the here and now."

She had become passionate. "How can people be so heartless as to rush around making money for unnecessary *things* while neglecting a *child*?"

He couldn't answer. He had never experienced any of that, really. He had not even thought about wanting children for many years, because one should not look back on options deliberately forgone.

That had been one of his sacrifices for a career in politics: children looked good in a campaign photo in a brochure or on a Christmas card, but bringing them up well required a great deal of time—not to mention money. Getting anywhere in politics required forty-hour days and a large personal fortune. He probably would inherit a trust fund from his parents, but he was not a wealthy man. As for the time children required, the alternative uses Carole had made of her fourteen years with him had amply vindicated their decision.

He wanted to return to the attack. Out of his own personal conviction but perhaps also in defiance of Carole's cold-blooded design for Philanth's life, he wanted to change Philanth's mind about marriage.

"Adults need closeness to other adults. There's only so much you can get out of giving to a ten-year-old."

"All my needs will become manageable once I'm back in a job that takes everything I've got, both emotionally and intellectually." She added dreamily, "If I could've just gone on with my Peace Corps duty, and taken courses so I could move up into more responsibility, that would've been perfect. Teaching and clinical work, gradually moving into administration, are an ideal blend for me. Birth control is such an exciting field, there's so much to be done!"

"You think every problem you've got is going to be solved by making a career of the destitute."

"If I'm mistaken in my vocation that's my problem. We shouldn't waste your time arguing about it.

"And you can't advise me on the basis of your own situation, Mr. Bjorklund. *I'm* not going to be holding any important public office; *I* don't expect to have any enormous responsibilities in which things I can't control go wrong and I'm widely blamed. I'm not sufficiently personally aggressive to rise that high."

That gave him exactly the sort of opening he was supposed to be working for. "You'd *like* to do significant things, wouldn't you? Of course you would! I'll bet you'd take *my* job, if it were offered."

Her abashed smile was sufficient answer.

"Well," he continued, "Don't you see that *you're setting out to be a failure—that is, to do far less than you could—*by refusing to arm yourself enough to take on big enemies, by establishing a sustaining personal relationship?"

She frowned in alarm. "If you're alone and you're taking a beating, sometimes you have to do as the old Scottish proverb says: lie down and rest a while, and then get up and fight again."

"That might suffice if you only want to swing a club like a zombie! I'm talking about using your creativity and initiative, and sparking the best efforts of other people! That takes a lot more than the will to drag yourself from day to day, take it from me!"

Her mouth and chin showed that he had at last struck home. "I admit depression can be a problem. But it's a sign you need more rest, or a reorganization or restructuring of some kind. I find it helps to keep a journal, to work things out of your system. And I try to find a satisfying project to lose myself in. Eventually you get . . . better."

"That whole approach is pitiful! You're pretending you can do just fine with only one leg when you could have two. You're cultivating the art of making crutches."

"Yes, that's what I'm doing. Lots of people can't get that good second leg, and they may find a use for the crutches I may be able to offer them because I'm continuously having to make my own. Every insight born of deprivation is another potential aid in getting down the road.

"But you can't help someone very much if you can't convince them that you understand the problem. If they can identify with you, they can believe that what worked for you might work for them."

"Why do you want to live that way, Philanth? Why the compulsion for emotional and material poverty?"

"It's better than feeling guilty for wasting while others want."

"But why must you feel that guilty? Other people obviously don't."

"They don't feel other people's need partly because they can't afford to let themselves feel: they're too compelled to cling to material things because they don't have great *internal* wealth. *I* don't *need* pretty things to make me feel good.

"More importantly, it's simpler not to have a lot of nice things to protect and take care of. It's more important to focus on taking care of people. A neglected child is more important than the most beautifully furnished room in the world."

Her voice became tinged with impatience. "I don't mind being open with you so you can accept me well enough to be comfortable working with me, but you're going 'way beyond that; and I don't see the point of being required to spell out things that should be obvious to you. You make me feel like a silly child, with you the adult telling me I'll 'change my mind when I'm older'.

"It's not exactly a novel set of values we're talking about! Through the centuries many great minds have agreed on its basics. Either you know how to appreciate them or you don't, and I can't believe *you're* one of those who can't."

"I'm sorry, I do appreciate it. I just want you to have everything you need in order to do all the good you can with your life."

She said gently, "That's kind of you. But you must take my word for it that you mustn't worry about it any more. You may feel responsible for the world, but that doesn't make you responsible for me individually."

"But I can't leave it at that. I want you to have a normal, fulfilled life."

"Mr. Bjorklund, if you'd met Christ when he was my age, would you have tried to persuade him to 'marry and settle down' so he could have 'a normal, fulfilled life'? For some people that just isn't good enough. It isn't good enough for *you*; you don't exactly flatter me by implying it should be good enough for *me*."

"Point taken." He had made absolutely no progress in trying to protect her from herself.

But at least he was now convinced that his fears about her becoming emotionally involved with him had indeed been groundless. She had too well-defined and generous a character and too broad a frame of reference to allow her loving to become a problem for anyone. The peace of mind which that conviction gave him was worth the time they had spent arguing.

Unfortunately he also had made no progress in his assignment from Carole, and now he wanted to. But he could not imagine how.

Philanth said kindly, "I appreciate your fatherly desire to set me straight about the necessities of life, Mr. Bjorklund. But I have to try to extend myself. A person makes effective use of his opportunities to help people according to the kind of person he is, as well as by how much he knows."

"I understand."

She touched his hand with her fingertips and smiled into his eyes, saying, "I think you do."

As she withdrew her hand he understood how she could instantly bind people to her. There need not be a trace of sexuality involved, any more than when she had hugged the Moroccan chambermaid. Just a few words, a look, and a touch: that was how an American politician also might try to do it, but she made the conventional uses of the technique look grotesquely clumsy. Christ too had touched people— literally.

After a moment she said, "It really needn't be such a hard life as you make out. Contentment is a matter of one's expectations. And other people's burdens *are* always easier to bear than one's own. Even the personal burdens you can't escape—illness or people who hurt you, for example—feel smaller after you've tried to help others with their much worse problems for a while."

"I'm afraid I don't have such a positive attitude. My whole life has been taken over by work on other people's problems, and to me it's simply a burden I wouldn't know how to let go of without losing all respect for myself."

"Then something's wrong with the way you're taking hold of things," she replied compassionately. "If you're willing to discuss it I might be able to help."

"I think you're getting a little too personal."

Her mouth became a thin line, causing a lot of dimpling, and he saw that she felt put down harder than he had intended. In fact he had not considered her reaction at all.

But she said politely, "I'm sorry, Mr. Bjorklund. I forgot myself. Please forgive me."

"Forgive *me*, Philanth. I obviously set up a double standard."

"You carry something that gives you pain. It's large, and I don't wonder that you pushed me away when I offered to touch it."

"Philanth," he said curiously, "have you no fear at all?"

She seemed astonished. "Of what? Your firing me for not keeping my place? Of your anger? No. If I can help you, in this time we have together, I want to.

"Carole didn't have to ask me to do whatever I could for you. I always want to help people, if I can.

"There's a controlled anger in you that was not always there. I know it wasn't because it's inconsistent with your basic habits of kindness and caring. So it's something that can be dealt with.

"Perhaps it's not anger. Perhaps it's cognitive dissonance that hasn't been relieved by growing into acceptance. But a pain, a loss of something."

"Suppose it's just a loss of the illusions of youth?"

"Oh," she said, brightening, "I do know ways to deal with that."

Then he did get angry. "You think a girl your age can help a man my age sort things out as though his soul were an untidy attic?"

She drew back, licking her lips nervously. "Obviously not."

"You mean, because I'm too proud to admit such a thing could be possible."

"Yes." She slightly bent her head like a cat flattening its ears in anticipation of being cuffed.

After a little he said, "It's nothing to do with you, Philanth. It's real, and there's no cure for it, and no one could claim any virtue by accepting it."

"You might be mistaken. Humanity has been working on ways of dealing with the anguish caused by its boundless imagination for a very long time."

"I don't want platitudes!"

"There are morally and intellectually respectable ways of *redefining* things so it's possible to overcome one's personal demons, at least temporarily. A person who claims to be irreconcilably bitter is clasping to himself something that's killing him, no matter what dignified name he gives it."

He didn't answer. She was right. But since the difference between them lay in the distance between his ambition and her humility, he was hardly interested in trading his kind of frustration for her kind of peace.

"If there's ever anything I can do I'd be grateful if you'd let me know."

Nastily he flung at her, "Aside from giving me the comfort of your body, as per your instructions?"

"I might also demur if you asked me to stitch up a doll of somebody and stick pins into it, but aside from such inappropriate measures, my offer stands."

"Ha."

He brooded over her temerity. "Sometimes you're too candid, Philanth, and too articulate. You hope to elicit tolerance by adopting a meek manner, but you must make people uncomfortable. How have you never learned to keep your light under a bushel?"

"Surely you don't suppose I talk to just anyone as I talk to *you*! Never in my life! —Except to Carole. You're both special cases.

261

"But when I sense that someone might be hungry for a little truth and strong enough to appreciate it and use it I do reject the easy, cowardly course of keeping silent—or following social conventions for merely slipping through a situation with minimal contact.

"Sometimes my effort is rejected and I get put down, sometimes painfully hard. But other times people open to me quickly if I only offer a bit of insight: it comforts them, quickens them, maybe even delights them. For a few moments they are more fully alive, and I leave them smiling.

"If I seem brazen to you, if I seem lacking in the slavish deference everyone shows you because of your position, it's because I believe *you've* got the sense to tolerate what I say and take it for whatever it's worth to you.

"Thank goodness you're not an insecure woman, who'd feel threatened because I'm a woman too. Then I *would* have to be careful! Just the same, there's one way I'm willing to risk getting in trouble, and that's in trying to help people.

"Why do I keep trying? Because when I was a child I saw a movie that made a lasting impression on me. It was about a nun who was pressured to waste her talents for service to others in order to cultivate 'humility'. Such perfectionism can be moral narcissism, and enforcing that version of 'humility' in others is a weapon by which the shallow and petty can protect their precarious self-esteem.

"So rather than waste myself by being too 'humble' I sometimes choose to risk being shot down because small-minded people are so quick to consider a happy and articulate person conceited, or because dull people think I'm showing off when I'm only relaxing and enjoying a good conversation. And I'm likewise willing to risk offending older people, like you, by seeming to presume to operate on their levels when in fact I'm only trying to go about my business."

He stared at her. After a moment she said, "Maybe I lost my temper a little bit just then. If there ever was a sore subject with me, you've found it. I'm getting *tired* of apologizing for what I happen to have read or figured out. And 'hiding my light under a bushel' is so crippling! —Worse than foot-binding!"

She was only getting angrier as she went on, "Our 'democratic society' is like the isolated village in the famous short story: In the city of the blind the one-eyed man isn't king at all: he has to have himself blinded in order to 'fit in'. It's no wonder we keep electing fools and then paying for our folly for generations afterward, in lives and dollars and irreparable damage to the world."

"All right, all right," he soothed. She made a disgusted face but subsided.

After a moment she added, "One thing I really love about Washington is that I don't have to apologize so often for inadvertently intimidating someone. There are so many people who've come there basically just because they've got their brains humming under full power that I fit right in. The men don't even mind that I'm a woman."

His mouth twisted in amusement.

"It was simpler for you, wasn't it. And not only because you're a man."

"It's true. The slots have always been there for me to move into."

He looked at the assortment of papers still lying on her lap and realized that they had not gotten far in dealing with the mail. And Ken seemed to be hiding in her shoulder bag, waiting to jump out at her. That was not going to help her do her best work.

"Why don't you go ahead and open Ken's letter? Then you can put it behind you."

"Thank you."

She fished out her letter, and he took over the job of going through the papers from Marjorie. Soon he was absorbed in the new batch of work to be done.

After a few minutes he noticed her stillness. He looked up and saw her eyes narrowed above the fists folded together in front of her mouth.

There was no use in beating about the bush. "May I see?" He indicated the letter on her lap.

She hesitated, then nodded. "Oh, Mr. Bjorklund," she mumbled, and as he took the letter she unseeingly turned to rest her forehead against the side of his shoulder. "I've exploited him!"

Before he could decide how to react to this contact her breath caught, and she straightened. "Oh, I—I beg your pardon!" She shrank back against the wall of the cabin.

He unfolded the letter.

My dearest Philanth,

I'll never stop hating myself for having been so stupid. I can't go back and undo the harm that was done.

But I want to spend the rest of my life trying. I'm ready to do anything, Philanth. I'm not begging, I'm just promising.

"What's he talking about? What harm did he do? What did he do that was so stupid?"

She was hugging herself. "He . . . got mad when I rejected his proposal. And . . . wouldn't let him make love to me. To . . . change my mind.

"But that isn't as bad as it sounds. He'd been giving me lessons in ballroom dancing, in his apartment, and . . . it seems we developed different interpretations of the situation."

"What was *your* interpretation?"

Her voice was laden with remorse. "I was eager to learn to dance, it was my first opportunity, and he was delighted to teach me. We'd agreed on our first date that we weren't interested in getting married again, and when I said I didn't care for extramarital sex either, he said that was fine.

"So I thought we had a beautiful friendship. We had stimulating conversations about literature, and it had been so many years since I'd known somebody I could talk to on that level.

"I assumed he understood that the physical attraction which developed was just a natural consequence of the circumstances—it didn't mean anything."

He nodded and went back to the letter.

> I want to spend my life with you; and I think that given the tasks you set for yourself you would find it very comforting, perhaps even necessary, to have one man's tenderness and devotion and admiration.

He did not want to read any more. This was already starting to repel him, for over the years he had come to despise the man who was abasing himself in this shameless fashion.

Ken offended him deeply. His brother-in-law's habit of denigrating politics and anyone connected with it could have been dismissed as conformity to a vapid convention of American culture, but Ken was one of the great number of Americans who actually preened themselves on their political illiteracy. Regarding ignorance as a kind of moral purity, they never realized that in fact they were moral imbeciles compared with those who were prepared to endure the hardships of the political arena to combat those who exploited government for their personal benefit.

It was not the stupid self-righteousness of such people which antagonized him. It was the practical effects of their refusal to "get involved". In the life-long cocoons of their personal lives they gained so little experience of what happens to human nature in the crucible of the struggle for power that a dictator who had ordinary civilians tortured or slaughtered to maintain his rule by terror was no more real to them than Darth Vader. They watched the evening news on television with little sense of its reality, much less of moral participation.

They also remained too ignorant of the political process to know how to pick their battles and act effectively. Yet only by fighting for years did people become sufficiently organized and knowledgeable to win. Thus "decent people"'s refusal to get their hands dirty allowed those who were not so fastidious to perpetuate their own privileges at the expense of everything else.

Moreover, because as members of the middle class they legitimated their attitudes and dispersed them through the general population, they made life miserable for any public-spirited person who ran for office and for those conscientious citizens who worked to help him or her get elected simply because they believed in their candidate. So much knee-jerk contempt from the very people one hoped to serve helped cure many an honest and able individual of his or her desire for public office. People like Ken, with their self-serving political cynicism and apathy, were as much a threat to the future of their country as the corrupt officials they despised, since those politicians counted on that apathy to keep them in power.

What was especially unforgivable in the case of affluent people like Ken was the fact that such people benefitted all their lives from the protections others had suffered and died to make possible for them, yet they would not part with any of the time and money their privileges afforded them in order to do their share of the work of maintaining the system which protected them.

His resentment of Ken's attitude of moral superiority to all politicians was sharpened because Ken exploited his privileged position as a relative: for a long time Ken had seemed to sense that his smug and boorish put-downs had to be endured out of consideration for the person who linked them in a family relationship. That was contemptible.

He decided he had no right to read this, anyway. It was an invasion of his brother-in-law's privacy.

But when he glanced over at Philanth he knew he couldn't opt out. She was punishing herself for having been so "selfish" as to enjoy friendship with a man who was coming to want to marry her. To be practical about it, he did not want his assistant to be distraught because of his tiresome brother-in-law's posturings when he had to depend on her to keep him from slipping up or falling behind in his complicated agenda. He had to know what was in the letter in order to help her get over it quickly.

> I've barely been able to do the minimum required of me since I saw you last. All I want to do is stay in bed and court oblivion. I'm glad I've never used drugs or alcohol, because now they would finish me. I understand now why some soldiers have stood up in their trenches and let themselves be mowed down by enemy gunfire; why some people step off a

ledge without any preliminary. Sometimes there simply isn't any further desire to go on living. There isn't any will to do anything, no will even to figure out a way to kill oneself, because that would require some drive, some impetus to act, and sometimes there is none. None.

I'm not threatening to do something drastic if you won't marry me. You know I'm not such a child as that.

I'm only letting you know what a void there is for me now that knowing you has cleaned out all the motives which kept me functioning before. Without in the least intending it, you've shown me how small and ordinary I was, in my self-absorption and self-indulgences. Now all of that is gone, wiped away.

But how am I to live the rest of my life if I can't be with you? I can't see how to carry the light you give without the candle which bears it.

And I lack your mastery of despair. In the depths of the quiet, clear, sunny-seeming lake of your personality you have explored the dark caverns of meaninglessness. You must have been there at some time, not merely dealt with it at the level of abstract thought where you now display your understanding of the roots of the meaning of everything. Because you have achieved a silent victory over all those shadows.

So you must know what I'm talking about. You can understand that I'm dead to myself and waiting to be born again. I'm an empty vessel waiting to be filled more fully than I ever dreamed possible. I'm waiting to be filled by what you are in the unassuming ways in which you go about the business of daily living so that it acquires a timeless loveliness for those who understand it.

Philanth, I admit I exceed my own past nature in writing this. You probably think I've gotten flowery from too many years of poring over the works of Dostoevski. But please let me have the chance to go on growing in the ways in which your thought and feeling move.

I know you want—any woman wants—a man to share with, not a child to nurture, when she marries. Maybe I couldn't offer you that before, but I can now. Could we possibly make a fresh start?

He let the letter resume its original folds. Such groveling would not have been so repugnant if he had not known that he was going to have to go on politely interacting with the man who had written this, from time to time, for the indefinite future. Yet now in order to help Philanth he was required to act sympathetic toward this puerile academician, this spoiled son of a real estate speculator who considered himself superior to a mere politician who was perpetually begging for votes or money.

He handed the letter back to Philanth, glad to get rid of it. He was not even sure that he had finished reading it, but he had done his duty.

He had to nudge her with it to get her to open her eyes and take it. She unfolded it again and ran her eyes over it as though to be sure that it said what she remembered. She slipped it back into the envelope, stuffed it back into her bag, and resumed her tightly self-contained posture.

His tone carried his disgust. "A man should never put a woman in such a position."

"You're a little old-fashioned, Mr. Bjorklund. Sex roles should have nothing to do with it."

"Well, the man seems to have no sense of what a woman wants in a man. No self respect."

"Suffice it to say that he's past caring. Anybody in 'the dark night of the soul' has the right to cry out for what he needs in order to keep from slipping into the abyss. Anybody who's embarrassed by the spectacle should be ashamed of themselves."

"Ah," he exclaimed.

She cringed. "Excuse me."

After a moment she said, "A couple of times in my life I've wanted to borrow a certain book so badly that I was ready to get down on my knees to the person who had control of it. But I couldn't have it because I didn't have the right kind of library card.

"Isn't that ridiculous? Nobody could have believed how badly I needed those books, just for psychological reasons.

"How can you ask the world to bend its little rules—about library cards, or the etiquette between men and women—when it's shameful even to admit that your soul is hanging by its fingernails and every minute you're slipping further down, with no glimmering of an idea of where the bottom might be, only a mindless fear when you think of it . . . .

"But the rules *must* be ignored, when you're the one in the position to do it and you recognize a need that justifies it.

"Mr. Bjorklund, what can I do for him? I can't marry him, but I do have to accept responsibility for the situation he's in."

"A person is responsible for his own emotional problems."

"Even if that's so, it doesn't relieve others of any obligation to help if they can."

"Well, God knows I sympathize with the man, but . . . ."

They were quiet for a little.

"I've got to do something.   He could be pre-suicidal."

"Is that a term?

"Anyway, I don't believe he's as desperate as he sounds."

"You don't?"

"There's a certain amount of histrionics in Ken's makeup.   It's probably part of what drew him to teaching."

"But even if it's as superficial as you think, I mustn't let him decide to revert to what he was before he knew me. He's caught a vision of something and confused it with the person who helped him see it. If he fails to move forward he may become worse off than before."

"I don't see what you can do about that. A person has to work through these things for himself . . . if he's without a close friend able to help him.

"And I wouldn't think *you'd* want to even see him again."

"I don't. He's become so volatile that that could get messy and make the situation worse.

"But I could write to him. I *must* write to him.

"But it's going to sound cold if I tell him what he has to do. What he needs in order to get through this is to go through with what he thinks he's already done: restructure his values so that he's no longer self-centered. He can escape all of his anguish just by moving out from under all of the burdens of ego. The self stripped of all its self-absorption can slip through the dark and narrow places, even where mortality and meaninglessness converge to crush the hope out of us.

"But he has to be able to re-orient himself outside himself, on the genuine burdens of others. When the self learns to do that it truly dies and is reborn.

"If he doesn't manage that, he could escape the burdens of self the other way, the way he mentions. Being willing to die to yourself can be dangerous, when you're seeking to escape pain for yourself but you lack genuine caring for others."

He sighed, thinking it was unfortunate that such an able mind as hers should have felt it necessary to spend so much time on such matters—and also that a person with his responsibilities didn't have time for this kind of problem. Philanth herself did not want to become heavily involved in trying to help one grown man to outgrow his psychological immaturity.

However, Philanth was a caring person, aware that one is not justified in kicking a beggar out of the path on the way to try to save millions.

"Write him a letter," he agreed. "I know you'll give him the sympathy he needs, but he's got to face the fact that he must learn to stand on his own feet: to catch his own fish, not depend on you for handouts, the way he wants to."

The tip of her tongue moved nervously between her lips. "All right."

She added, "But I think you just undermined all your arguments about why *I* shouldn't try to go it alone. You can see in that letter how terrible being alone can become when you develop enough emotional vitality to need intimacy at a deep level. Yet you tell me to help him learn to endure all that loneliness. I couldn't hope to be able to do that if I weren't in the same situation."

Chapter 11

BANJUL

His visit to The Gambia was to be unique in that
it had been shaped by plans for a video camera to
follow him around making tapes for promotional purposes. Since most of the rest of the trip would be
devoted to business meetings and conventional embassy and hotel affairs, the taping would not extend
beyond that one country.

This visit had its own country-specific aspects, as
well. Most notably, he was interested in cultivating
a certain host-country official who not only was crucial to Peace Corps activities but also was reported to
be respected by his colleagues and counterparts in
the region. He had maneuvered the arrangements so
as to give himself a day with this minister of
agriculture and "operation video". As in the other
countries on his tour, the incountry Peace Corps staff
would remain in their offices rather than accompanying his party; thus his contacts would be kept personal and unpretentious.

He could not have refused the invitation to dinner
at the prime minister's home, but that had left him
with no evening to spare for a reception-dinner at
Banjul's U.S. embassy. A luncheon-reception had
been substituted.

For reasons best understood by those at the scene,
the reception had been "moved" to the British embassy while he had been in transit between Morocco and
Senegal. Due to the practical problems officials in
developing countries encountered in meeting visitors'
standards, he had agreed with Philanth that it was

best not to inquire deeply about such irregularities. Philanth remarked that at one time the U.S. embassy in Banjul had been a motel room, and the British embassy could be expected to be at least as good as the American one for filming purposes.

In cooperation with the firm policy of making the most of every minute of his time, the U.S. ambassador to The Gambia had arranged to be in Dakar at the same time he was visiting Senegal and to share his short flight south to Banjul. This was regarded as no hardship, since Dakar was the center for Senegambia and the ambassador went back and forth.

En route to The Gambia they were supposed to have a productive discussion about the relationships between AID and Peace Corps in Senegambia and the local workings of the "Ambassador's Self-Help Fund". He also hoped for some insights into the continuing obstacles to cooperation between The Gambia and the country surrounding it, Senegal, and the continuing conflicts between Senegal and Mauritania, which was yet another Peace Corps country.

Before they could get into any of these topics, however, the ambassador informed him that the Peace Corps country director for The Gambia had been flown to Dakar for medical attention and sent his regrets that he could not meet with him. He tried to feel the ambassador out regarding the country director's performance, but soon the ambassador was giving him advice about keeping healthy in the tropics which his scheduled activities would make impossible to follow. This led off into a discourse about the food, climate, and daily life of West Africa.

Finally noticing how the time had flown, the ambassador gave him some briefing papers on the subjects they had been scheduled to discuss during the flight. They agreed to try to address them when they got together again at lunch. The ambassador pointed out the lush vegetation fed by the Gambia River, the extensive beach front near the wide mouth of the river, and the port of Banjul before their descent to a very unpretentious air field.

It was a bumpy ride. "Welcome to The Gambia," the ambassador shouted above the noise, and he forced a smile as he gripped the armrests of his seat.

The ambassador had made sure that his party would be first out of the plane. He followed his host, with Philanth and Tony close behind him.

But at the top of the metal steps he paused, for at the foot of the stairs was a cluster of bright, gold-braided uniforms, men with cameras pointed at him, and Africans in long, flowing robes. Beyond them were polished black cars with fluttering pennants and waiting drivers. The brilliant morning sun and the abrupt transition momentarily immobilized him. Then, realizing that he was the center of attention, he waved and showed his teeth for the cameras.

But still he lingered for a moment, looking over the scene and letting the significance of it sink in: all of this was for him! And for the U.S. ambassador, of course, and for the Peace Corps, and for the United States of America. And for whatever a tiny bit of U.S. foreign aid had done for this tiny, poor, peanut-growing country, especially in the years since Peace Corps Volunteers had begun work here in 1967.

Just the same, he was not yet accustomed to receiving this level of VIP treatment, and the symbolism of this little gathering at this improbable-looking "international airport" gave him a lift. The pleasure reminded him of his formless desires at moments when the drums and trumpets of "Ruffles and Flourishes" seemed to stir every cell in his body as they announced the arrival of the President. Maybe he was not so different after all from men who went rabid from the taste of power.

Philanth was at his elbow, pressing a little because he had stopped and she kept expecting him to move forward. He threw her a bemused look over his shoulder. "Enjoy," she mouthed, and smiled at him.

Feeling that she was seeing straight into his mind, he hastened to debark. The band began to play.

He recognized the tune. Daring to be different, the Gambians had opted not to play "The Star-Spangled Banner," and he approved their choice. He stood at attention, there near the top of the movable metal steps. The ambassador, now at the foot of the steps, glanced up at him and did the same.

> O beautiful for spacious skies,
>   For amber waves of grain,
> For purple mountain majesties
>   Above the fruited plain!
>   America! America!
>   God shed his grace on thee
> And crown thy good with brotherhood
>   From sea to shining sea!

He realized that this was a good summing-up of natural advantages which his country had and The Gambia did not.

However, it was so unaffected and appropriate a welcome that though he had stepped out into warm sunshine, a chill ran up his side.

There was a pause; for a moment there was only the sun blazing down on him like the spotlight of some cosmic theatre. Only the breeze stirred, wafting scents of sun-warmed earth and vegetation.

Then the African officials began greeting the ambassador, and he descended the steps to join them.

But as if to make sure they had gotten their point across, the band began to play the tune a second time. They were trying to play softly, but they were too close at hand. Voices introducing people to him were rendered unintelligible; and as he nodded and smiled, returning handshakes and mouthing acknowledgments, another verse was running through his mind:

> O beautiful, for Pilgrims' feet
>   Whose stern impassioned stress
> A thoroughfare for freedom beat
>   Across the wilderness!
>   America! America!
>   God mend thine every flaw!
> Confirm they goal with self control,
>   Thy liberty with law!

Beautiful words, he was thinking, even if not entirely coherent, and never more timely than in these latter days of our might and self-indulgence. Some lines from Robinson Jeffers which he once had dared to use in a speech also ran through his mind and heart:

> And you, America, that passion made you.
>   You were not born to prosperity, you were
>   born to love freedom.
>   . . . .
> The states of the next age will no doubt
>   remember you, and edge their love of
>   freedom with contempt of luxury.

Africa speaks to the American: look at your own past more closely, remember your own strivings and imperfections, for this can help you to understand us even though we seem to occupy a separate planet because we are so different from you, so vast, so multiform. Thank you, Gambia, for saying it so kindly, so very diplomatically.

The introductions were being concluded, and it had been natural for him to acknowledge such a reception with warmth. After completing the protocol he shook hands with the Gambian musicians and their leader, thanking each one individually and dispensing compliments. They seemed genuinely pleased, and he was impressed by the dignity and unaffected friendliness of their personal welcomes. After France, Morocco, and Senegal he was glad to be in a country where English was the *lingua franca*.

Cameras had been activated afresh during this unexpected exercise in international good will, those of a European and an African as well as Tony's. He recognized the Peace Corps "media specialist" from Public Affairs, who had arrived before him with the minicam. That most imposing camera had its inscrutable eye on him also.

He caught a glimpse of Philanth smiling at the proceedings from a position behind the cameras.

One of the Gambian officials he had just met introduced the newsmen to him. The local man was from Banjul's major paper. The European photographer was with an international wire service and, the Gambian noted, one of many Swedes who came to The Gambia for winter holidays.

The Swede hurriedly explained that he needed a story to accompany his photos and asked him to answer a few quick questions. Tony hastily stepped forward to offer the journalists the appropriate tailor-made press kits, and these they gladly accepted.

He started to turn to the waiting dignitaries, eager to get started with his substantive business.

However, the Swede wanted to know what part of Sweden had been the home of the ancestors of "the distinguished Charles Bjorklund". He said he was sorry, he didn't know and asked to be excused.

Before he could turn and speak to the African ministers the Swede started trying to interview Philanth, who had come up beside him. She denied any knowledge of Swedish ancestry and sweetly but unequivocally refused to be included in any picture-taking. She claimed to be "only Mr. Bjorklund's typist", and of course he did not contradict her.

However, he was starting to worry about whether her minimizing her practical importance to him might be counterproductive when this awkward little skirmish was cut short by the ambassador. With apologies because "Dr. Ablebody"—indicating the African in red and yellow draperies and traditional red fez who had introduced the newsmen—had "a full schedule for our guests before their luncheon at the British embassy", he cut the newsman off.

Because of the band music he had caught no one's name during the introductions, but the name rang a bell: this was the politician he particularly wanted to get to know and who had agreed to serve as his host.

At once Ablebody smilingly took charge and with a hospitable gesture motioned him toward one of the waiting cars. The ambassador told them, "See you at lunch" and moved off toward another of the cars.

However, the Banjul newspaperman tagged after them, begging Ablebody for "jus' one more," this time including the Peace Corps director's "assistant"—indicating Tony—as he and Tony were officially welcomed to The Gambia by Ablebody. This Ablebody could not refuse, so the three of them posed with happy grins, the one white face in the middle. According to plan.

At the open door of the car the other Gambian ministers shook hands with him again, assuring him at some length that they looked forward to seeing him again later in the day. Leaving Tony and a local Peace Corps staffer named Cory Watson to deal with their baggage and travel documents and deliver these to their hotel in the Peace Corps Jeep, he and Philanth rode off behind Ablebody and his driver.

It had been quite a reception. He caught a glimpse of other travelers from the flight lingering outside the terminal building to gawk at his car. The sign behind them said, "Welcome to Yundum International Airport".

Philanth murmured to him, "What should I have done, Mr. Bjorklund? I shouldn't have been so unfriendly to the press, but . . . ."

"It isn't always possible to oblige news people who consider bad manners part of their professional role."

Leaving the dismal-looking terminal area, they began passing through scrubby open country. After routine pleasantries Ablebody broke off to point out the poor condition of a side road they were passing.

"Motor cars"—Ablebody's African-British English had no "r"'s, so he pronounced it "Motah cahs"—"have sho't, ha'd lieves on owah mos'lee unpaeved rohdz." Almost none of Ablebody's vowel sounds were used in the same ways as in standard American English, and final consonants also were often altered if not lost. So he had to pay careful attention in order to avoid missing so much of what Ablebody said that he failed to piece together his meaning.

Ablebody went on, "For trips like the one we make tomorrow, when we can't use waterways because so much iss to be done inland, we try to take two cahs. 'Specially wise because we can nevah have enough mechanics."

He and Ablebody got into a discussion of the problem of attitudes toward work with one's hands. It was even worse in countries where everyone with some education aspired to work for the government. "But my tribe dominates the government," Ablebody said, "and even for us there are not enough jobs."

"I suppose colonialism is blamed for *several* aspects of that situation," he commented politely.

"Blaming colonialism can be national pastime, like baseball. These do not produce rice, cattle, or ground nuts—what you call peanuts. Not even manufactured goods can be produced jus' by talking."

He laughed. "May I ask where you studied, Dr. Ablebody?"

"London."

"And how many groups of do-gooders like ours do you personally entertain in a year?"

"Do not disparage yourself. You do not ruin with hand-outs as has been done in some cases. UNICEF, AID, CARE, and the U.S. Embassy's self-help fund have supported Peace Corps with good effect."

"But you do have help from other countries."

"This year I have talked with Japanese, several groups from European Common Market, and two missionary societies from your country."

Ablebody's rich voice took on a sing-song quality. "More Japanese go come small time; have plan for catchee many fishee." Ablebody was stooping to pidgin! "Many people is finding out where is Gambia. Want to know is him joined with Senegal yet." Ablebody was reverting to normal tones as he went on, "Not think we mind learning French when we still trying to teach everybody English so we can at least communicate among our own tribes."

Another typical problem inherited from the colonial powers: they had drawn the boundaries between

281

their colonies without regard for tribal regions. The Gambia was one of the smallest, poorest, and least developed countries in the world, but its handicaps were typical of subSaharan Africa.

Despite Ablebody's ridicule of foreigners' ignorance about Africa and the bitterness of Ablebody's allusion to The Gambia's having been created by the British along the banks of a river deep inside what had been a French colony, he considered this man's degree of detachment regarding the humiliations of colonialism a remarkable personal achievement. It probably also reflected the magnanimity of a man secure in his own intellectual gifts.

Such breadth and flexibility of perspective was a tremendous help to cross-cultural communication. Indeed, trying to talk with a diplomat who lacked it was often a waste of effort. He could see why a regional specialist at the State Department had recommended Ablebody as a West African with whom he "might want to spend some time".

Ablebody continued, "People also want to know what British ever saw in this country, to build fort here. Oil? No. Gold? No. Diamonds? No, not found yet. Sure around here somewhere, though.

"After big slave trade out of Dakar ended, British must have been looking for the White Man's Burden. Found it here, oh, yes. Still carrying it, oh, yes."

"Four hundred years is a long time. And maybe they can't afford it anymore." Even in this tiny country, foreign debt from import of manufactured goods was making outside support of the economy an increasingly burdensome proposition.

Ablebody sighed quietly, shifting to stare out the window at the fields they were passing.

It was not an inspiring sight. The natural vegetation was only scrub, and the tidiness imposed by mechanized farming and extensive irrigation was absent. Moreover, Africa had never had the benefit of the glacial movements which had so enriched the agriculture of vast areas of the Northern Hemisphere, and The Gambia's soil was poor and sandy.

"We are not without options," Ablebody said. "Someone has suggested that we sell our beach front to the Hilton chain and the rest of the country, the strip along the river, to Disney Enterprises."

At his laugh Ablebody went on with mock defensiveness, "Make very fine Disney ride for grownups. Already Swedish Cruises makes regular winter trips upriver to visit 'native villages' and view wildlife from rubber rafts. And to see 'African Stonehenges' marking mass burial sites.

"Fine school for study of African culture. Take steamboat ride to visit village of Alex Haley's *Roots*." Ablebody's tone was less than enthusiastic. "All very quaint, very authentic: winnowing by hand, pounding millet at dawn. Peace Corps project has designed 'village tourism infrastructure' for 'village live-in' tourist experience."

"Like the Reservations in Aldous Huxley's *Brave New World*," Philanth muttered, and Ablebody looked at her as if she were a parrot which had suddenly broken silence.

Whether or not he understood her devastating comment, Ablebody became completely serious. "We are hospitable people. And with Peace Corps help we have improved production of many traditional handcrafts for export and sale to tourists. Tourism is important to many countries in this great continent, and ours is no exception."

To help along what was surely a canned speech, he said, "The ambassador pointed out the unspoiled beaches as we flew in."

"This could be another Miami," the minister continued. "It is seven hours from New York to Dakar, same as for many Western European cities. Closest point to South America. Winter holidays at beach front hotels are inexpensive. Americans and British come ashore from cruise ships and take buses to villages, sightseeing and shopping. We are now included in package tours, besides the Swedish charter flights. The available tours cater for the full range of tastes, from beach hotels with European meals to camping trips through the bush."

Ablebody's cultivated conversational tone again took on an ironic edge: "Or we may adopt another solution someone has suggested: become refuge for wealthy deposed African leaders. Why should they prefer Paris? Much more peaceful here. More than six hundred and fifty species of birds, plenty of fish, ..."

Ablebody broke off and turned to look out, for the car had stopped. They all sat and watched a very lean, long-horned cow taking a remarkably long time to leave the road.

Off to the left was a family compound: tall, woven fencing linked together several thatched-roof mud huts. In this vicinity the scrub had given way to cultivated fields. Nearby was a palm tree.

Philanth touched his sleeve and pointed to a big, dark tree which looked like it belonged in a children's fantasy of evil enchantments. "A baobab tree. I recognize it from photographs."

He was not much interested in the local flora. The red dust of the road was being carried into the car through the open windows, into his nose and throat and onto his pale-blue poplin suit and white shirt. The dust-bearing breeze brought little relief from the heat which was rapidly building up in the car, and he was starting to feel trickles of perspiration which made him twitchy.

The late spring of The Gambia was sultry. Close to the equator, someone had warned him, there was not much seasonal variation in temperatures. He prayed for the relief of strong hotel air conditioning. He dreaded the moment when that perpetual promise of relief was not honored because of one of the many mechanical breakdowns he had heard about.

This car surely must have had an air conditioning system at one time. But presumably it wasn't working now, due to the difficulty of getting spare parts from overseas. He couldn't possibly ask.

The driver gently prodded the cow with the bumper of the car. Time moved slowly, here. The people had incredible patience. Often they had little choice.

With one finger he discreetly wiped his forehead just below the hairline. He mentally debated taking off his tropical-weight suit jacket. But Ablebody seemed comfortable in his voluminous robes, so he decided that to someone of the British heritage his shirt sleeves might not seem adequately respectful of his host's position.

The cow at last grudgingly yielded the right-of-way. The car crawled forward, limping over the bumps and tripping into the potholes. This climate was hard on man-made things, and during the rainy season pieces of a road could disappear altogether.

He took out his handkerchief and carefully mopped the tickling wetness from his forehead. The day, begun prematurely by clatter in the street outside his hotel room in Dakar, stretched into shimmering distances before him, an obstacle course of social, political, and—if he were lucky—technical niceties.

In the afternoon he would be meeting with the in-country Peace Corps staff. That would be a time when he could hope to get recharged a little, if their offices weren't too warm.

But at the end of the day loomed a formal dinner of strange, unappealing foods with too much spice and grease and starch; a large group of African officials who would have come to study his every move; and speeches in more or less heavily accented English—or translated. He too would have to give a speech, using his American English slowly and clearly without appearing to consciously do so.

Also, as the guest of honor he would be required to communicate effectively but diplomatically on literally any subject anybody brought up. The chasms of cultural differences he must bridge were daunting.

He was quickly wilting in the stifling car, and he knew from experience that by the time he was called upon to speak and field questions this evening he would be functioning distinctly below the level he expected of himself. He always regretted for years afterwards such imperfect use of his opportunities.

But first things first: he must not let himself be recorded the way he felt right now when he met the unforgiving eye of the camera at his next stop. The skill, judgment, and motivation of the cameraman could do only so much, yet he would have to appear "picture perfect" from any angle and at any range, like an actor in the carefully edited final "takes".

If he failed in any way at any moment, that much valuable footage would be wasted. And if he looked to the minicam like he felt right now, all of the tape to be shot at the peanut processing plant—the only "factory tour" in his five-week program—would be laughed at by a few insiders at headquarters and then regretfully thrown away. But good videotape was needed for t.v. promotionals and documentaries, for presentations to past and prospective Volunteers and for distribution to schools and civic groups.

He was lucky that his beard was blond, because the camera always exaggerated a dark beard by "seeing" it through the skin, unless a natural look was achieved by using makeup.

Forget t.v. makeup—he couldn't even expect to have any chance to freshen up before the filming began. He again took out his handkerchief and mopped his face, this time trying to spread the sweat around to get a cooling effect. He brushed at the sleeve of his suit jacket, examined his cuffs, and imagined the red dust being rubbed into his collar.

Now he saw a new significance in something a senator had said at his confirmation hearing: he had been told to "put on your grubbies" and get out into the field, because "nobody ever ran the Peace Corps from headquarters". Busy doubting the senator's main point and questioning his motives, he had overlooked the fact that casual clothes did not look ridiculous when they met sultry heat and dirt the way business attire did. He should have dressed like "one of the guys", overcoming his prejudice against dressing down for a fake effect.

But should he have planned to wear a sports shirt to lunch and a diplomatic reception at the British embassy? Of course not. And there was no provision in his schedule for stopping at the hotel to change. He would just take off his suit jacket at the peanut factory. What about his tie?

Ridiculous as his problems were, he must honorably bear his part, his central role in all of this ceremony and consultation, for his country's sake. He reminded himself how privileged he was. This *was* better than sitting on a congressional oversight committee, cross-examining experts.

But might not all this be an exercise in futility?

"Seriously," he said to Ablebody, "our reports say that the country's overall economic situation has been steadily deteriorating. We don't deny that that's partly due to foreign governments' trade policies, or that those could be altered if their domestic political pressures allowed it. But do you think that if we in the industrialized countries *could* mollify the countervailing special interests in our domestic politics, we could help agrarian countries get on their feet through trade development?" His follow-up would be the crucial question: If so, how?

Ablebody nodded and started to speak, but again he was distracted. The car had stopped again, this time because a dog had gone to sleep or died curled comfortably in the middle of their path. A truck was coming toward them, so they had to wait before going around the obstacle.

Several children were waving to them from the side of the road. One little boy was waving while he urinated against a tree. Philanth put her hand outside her window and waved back to them.

A couple of goats started across the road, and a boy who seemed to have been herding them began yelling at the goats in unintelligible expostulations. One of the goats seemed to bleat a rejoinder.

Behind them the other cars which had formed their motorcade from the air terminal likewise waited in the dusty, shimmering heat.

The truck also stopped. Its driver climbed out and with his sandaled feet gently pushed the dog off the road. The animal seemed to be alive but sick.

A uniformed cyclist came past, slowing to appraise the situation. The rider then waved back to them and moved on. He leaned forward to ask their host, "What's the uniform that cyclist is wearing?"

"Our police ride bicycles."

The driver returned to the truck. It soon moved past them, its open sides shaking and rattling. The two drivers waved to each other, and their car moved forward again.

Farther down the road they passed a group of mostly bulky women in long, loud-colored dresses. The entire back of one of the ankle-length dresses was printed with a huge portrait of a man, probably a leading politician. Several others had babies bound to their backs by lengths of cloth. The women turned to watch their group of cars, and several of them clapped, ignoring the baskets resting on their head wrappings.

"Why're they clapping?"

"Old custom to greet people in cars. Also, they may hear you are coming and know my car. Was in *Gambia News*, also on Radio Gambia. We welcome visitors sincerely."

"How many hours of daily broadcast do you have, now?" Philanth asked.

"It varies."

"I asked because with illiteracy high but with transistor radios now being used even throughout the bush, such a powerful tool for social change should be well utilized from morning until night.

"With foreign aid as available to you as it is, surely the necessary technical expertise should be imported as much as necessary to produce more high-quality educational programming, in spite of nationalistic sentiment against hiring outsiders.

"There's a surplus of writers in our country. It's disgraceful what some of them spend their time on. And it can be extremely difficult for them to get their work produced or published. After they'd studied your problems and culture they could work with local people who wanted to participate as writers or script advisers or actors. With technicians imported from Los Angeles and New York and London you could start an on-the-job training school in drama and media production; draw in people from adjoining areas to study here; and soon you'd have a program that could serve the entire West African region, or even the whole subSaharan region. They could start with the Anglophone countries, then begin parallel productions for the Francophone countries in cooperation with Senegal.

"Of course, I'm talking about entertaining shows like we have on public television in the U.S. There's tremendous potential in drama and comedy as vehicles for practical information. One popular series, using basic types of characters the audience finds believable, could be exported simply by translating the scripts. They have series like that on Indian television, to modernize attitudes toward health practices. Just one soap-opera series in Mexico has drastically improved the family-planning practices there, and it's been exported all over Latin America.

"And with satellite transmission and inexpensive village television receivers like in India to overcome conventional broadcasting problems, you could then expand into t.v. production, . . ."

Ablebody was staring at her.

"Excuse me," she said. "I forgot myself."

"Don't apologize. It's an interesting idea. Where did you find it?"

"It mostly just came to me as I said it. I have no idea what your local programming is at present."

"You say you're 'only Mr. Bjorklund's typist'?" Ablebody exclaimed, laughing. "Your country must indeed be oversupplied with talent ripe for export!"

"It certainly is in the nation's capital," she replied tartly. "Countless people are being wasted in jobs they hate because they're overqualified for them, just because they want to be where interesting things are being done. But interesting things can be done *here*!"

"Yes, indeed."

"Just the same, I should've kept my mouth shut."

"Why?"

"I—I'm not a man," she stammered.

"You think women 'should be seen and not heard' in this country?"

"Women are allowed to haggle in the market and gossip with other women."

"You are our guest. Be at home, Miss Devon."

Philanth gave Ablebody her warmest smile. "Please call me Philanth." She laid her left hand on the back of the driver's seat. "And I'm married."

"I beg your pardon. Your husband must miss you."

"Thank you."

At once she began telling Ablebody about a "marvelous" article she had read. It explained in detail how countries could set up instructional satellite systems. When she cited the source, Ablebody nodded as if impressed. She promised to send him a copy of the article as soon as she returned home.

While she was making a notation in her steno pad he asked, "How did you come across *that* book?"

She smiled knowingly at him. "I found it in your bookshelves while I was visiting Carole. With my interest in educational problems in developing countries I just naturally gravitated to it."

"It was influential in my efforts to launch a new aspect of our foreign aid program that would be integrated with our orbiting satellite programs. That developed into a package of bills, the most important legislation I sponsored in all my years in Congress."

"You got it enacted?"

"I'm still waiting for the final vote."

291

Ablebody was watching them. He quickly leaned back, away from Philanth. The exchange had assumed an intimate tone because it had been an improper aside from their conversation with Ablebody. But Ablebody might not understand that nuance of manners and could have gotten a wrong impression about the nature of his relationship with Philanth.

Philanth addressed Ablebody. "I have read that everyone in The Gambia prefers to be easygoing. Forgive, please, if I seem rude." She was slipping into Ablebody's verbal style as though unaware of it, even blurring the differences between her normal accent and his. "I am eager to know what plans you have for us."

She was trying to check Ablebody's plans against their own, to make sure they tallied. This sort of constant verification was a central aspect of her responsibilities throughout the journey.

"We begin with a drive around points of interest in Banjul and a call at Government House."

"I suggest we leave Philanth at the hotel before going to Government House," he said.

Ablebody bowed acquiescence. Then he asked Philanth, "But you would like to see something of our little capital city?" She nodded eagerly.

"Then if you have no objection, we will stop soon at a peanut-oil processing plant." Ablebody seemed unaware that they already had this noted in the printed schedules disseminated to many interested parties, including the cameraman who presumably was dogging his dusty trail at that moment.

"Oh, I want to see that," Philanth was exclaiming, bouncing on the seat.

"You do?" He detected a note of amused condescension in Ablebody's tone.

"Yes, and would it be very much out of our way to

visit Perseverance Street? I mention it only because I saw a photo of it in an old book."

"It showed ditches and dust and structures made out of scraps?"

She lowered her head like a guilty dog. Their host was silent for a moment.

"We'll show you several areas, some like that and some rather different."

She responded with a nod which accepted his judgment without question.

But she asked timidly, "And the Albert Market?"

Ablebody gave her a penetrating stare, and his mouth began a slow upward curl.

She added quickly, "If it's not out of the way. I've never seen a map of Banjul."

"But if you had, you would know where everything is, yes?"

She began dimpling irrepressibly at the black man, and Ablebody was smiling too, looking into her eyes in a way which seemed to cross international boundaries. Her gift for dissolving interpersonal barriers began to alarm him.

"Yes, the open-air market where ordinary people do their shopping. Onions and plastic sandals and patent medicines. And folk pharmacopoeia.

"But also you must go inside the American-style supermarket in Bakao, the suburb where most embassies are situated. Then we will leave you at your hotel, and I will take Mr. Bjorklund to the British embassy for lunch. Will that be satisfactory?"

"Oh, that will be wonderful!" she exclaimed, clapping her hands. "I'd like to buy some things for snacks in my room." She added anxiously, "I'll only take a minute."

"Since Peace Corps has aided development of tourism and handicrafts, perhaps you also should visit the Tourists' Market: for postal cards and a look at Gambian silver jewelry and batik and wood carvings and leather purses and sandals." Ablebody's sly smile said he knew what women liked.

"Oh, Mr. Bjorklund! . . . Uh, that's not possible, is it? I was hoping to get to the National Museum, but . . ."

He did not share her instant enthusiasm. "Can she get to the market from our hotel?"

"By taxi it will be no problem."

"I—I'd b-better not," she told Ablebody reluctantly. "I have to stay by the phone . . . ."

Ablebody's eyes flicked back and forth between them. "As you wish." There was a brief pause.

Ablebody began describing the dinner to be held that evening at the home of the prime minister. "I apologize because we did not arrange for a European-style affair," Ablebody told Philanth. "I hope you will understand that we did not plan to include women."

"I do understand. Where even boys are lucky to be taught to read, women can have nothing to say in the highest councils of government."

"I understand that the other member of your party will be spending the evening with the friends he made while working in this area," Ablebody said. "You will be obliged to dine alone at the hotel?"

"I don't mind. I hope to meet somebody interesting. There must be world travelers passing through."

"That might be awkward, in the dining room. I suggest you try to meet some of the women on the beach before dinner."

"No. On second thought, I'll dine in my room."

Hearing the edge which had suddenly come into her voice, he diverted Ablebody's attention by asking exactly who would be at their dinner and what topics concerning Peace Corps it would be appropriate to discuss then and which would have to be discussed at breakfast on the morning he left Banjul.

"Not to worry. We will be happy to just get better acquainted. You may mention any topic you wish."

That degree of good manners depressed him. It suggested that these Africans thought he was only a figurehead who had been rewarded for service to his chief with a plum job and was traveling for pleasure and status. It warned him that even if they had not considered his position that deeply they probably felt it would be beneath their dignity to get into specifics with someone of his rank.

Hadn't they read their mail carefully, hadn't they really heard what the American officials stationed there had told them about the nature of his trip? Couldn't people who regarded themselves as experienced nevertheless take statements at face value?

It was a matter of cultural blind spots: the traditional attitude was that it was good manners to act as though practical matters should be left to underlings. Previous directors undoubtedly had fostered this attitude by making their trips largely ceremonial. Thus his trying to *really* be a "hands-on" administrator while "out in the field" might be a matter of swimming against the current all the way.

It was an impractical attitude they had. It must be changed. He couldn't let cultural barriers stand in the way when what he wanted to do was so important. Much that was wrong in government agencies had to be changed from the top, since only there could the necessary authority and influence be found. Yet the knowledge of *what* needed to be changed was dispersed everywhere *except* the top. That was the impasse he was struggling to overcome.

Ablebody broke the silence. "Has either of you read a novel by John Updike about a *coup d'etat* in a fictitious African country?"

He shook his head. Philanth threw him a guarded look and then ventured, "I have."

At once Ablebody began questioning her about her reaction to it, and the two of them got into a heavy discussion about the book as literature and as political comment. They were soon involved in subjects so sensitive they made him cringe: the defensibility of CIA efforts to overthrow grossly incompetent and oppressive governments, the industrialized countries' export of economic dependency to agrarian societies in the form of consumer luxuries, and what (if anything) the exporters' and importers' governments should do about this crippling of national economies which instead needed outside investment to process their own raw materials, diversify, and produce more of the goods they needed for themselves.

On the latter issues Philanth felt that "national leaders should not merely attack the toys of decadence, as Updike's dictator does, but inspire their people to make sound values the basis of their choices as consumers." She had this prospective prime minister's absorbed attention.

It irked him that Ablebody preferred this social chat with his "typist" to the high-level policymaking approach to the interrelated problems of international trade, debt, and monetary policies which he had tried to discuss earlier. Of course the sort of home-grown remedy Philanth was advocating would have more appeal than his and Ablebody's trying to educate each other on matters neither of them could do much about even if they became heads of state.

Ablebody went back to asking probing questions about Philanth's reactions to the novel. Philanth responded politely, but again she made him nervous. What particularly distressed him was the constant confirmation that Philanth had what Tony had called "an adventurous mind": as Carole had warned, she was ready to consider *anything*. She was even willing to discuss it with a major foreign official.

After a momentary pause in the conversation Ablebody asked, "Did you consider the book realistic?"

"That's a more personal question than you might suppose. Even if there's a consensus that the reality depicted is congruent with the reality a given group experiences, there's still some degree of solipsism involved, in that others will have different impressions of the nature of reality.

"Besides, the illusion that 'realism' has been achieved is only a minor aspect of art. Art is a striving to reflect truth which not only requires distortion in order to bring some aspect of experience into the foreground, in so doing it in fact creates or reveals new truth. Therefore to some extent it can *change* 'reality': can help reshape it by changing our perceptions in the process of pretending to imitate it. Striving for realism by emphasizing negative aspects of reality has fed back into culture and degraded it."

Ablebody smiled, then glanced at him before again focussing on Philanth. Ablebody must think she was functioning as his unofficial alter ego so it was all right to pursue a discussion with her while he sat by.

"The book has countless nuggets of insight," she went on, "and they can be found by mining it at different levels. But it's more realistic to call it 'mythic' than to debate its realism, even though it was carefully researched. Perhaps its greatest genius is in the ways it *plays with* realism."

"An interesting observation. But do you believe its basic points about African development are valid?"

"Dr. Ablebody, I would rather not examine the realism of the book with you in those terms. If we did I'm afraid we would be obliged to confront religious differences." Finally there was an area she would not discuss.

She added, "As with any other cross-cultural economic undertaking, we should stick to identifying specific goals we can agree to work together on, not get off into the quagmire of differences in underlying world views. Those can eventually resolve themselves in a cosmopolitan culture."

He groaned inwardly. Even if she could charm Ablebody to such an extent that she could be this frank without giving offense, he wished he could have been somewhere else rather than be a party to such a high-risk and unnecessary discussion.

"Ah," Ablebody said. "You see obstacles for development in the reactionary elements of Islam?"

Now he was ready to smother Philanth.

Ablebody went on, "That may be an understandable view for a Westerner in view of what's happened in other parts of the Islamic world, but in subSaharan Africa we have other variations."

"It goes far beyond any particular religion," Philanth replied, "despite the causal relationships between different major religions and economic development demonstrated by Max Weber. My own reading of that novel's interpretation of the dilemmas of a developing nation tends to reinforce my own view regarding the dysfunctions of organized religion in pluralistic society. Since nations are being woven into a global economic and technological community, we are certainly talking about pluralistic society."

"Would you care to elaborate?"

Philanth hesitated. "Institutionalized religion can be very functional among the less educated and nurtured, for example in encouraging the habits of deferred gratification without which no high culture is possible. Its codes of conduct also strengthen the fabric of society.

"But I reject the view that it's necessary because of the limitations of human nature, or because of the limitations of human reason regarding ethics." She could not resist going forward where even she had seen that she must not.

"Really!" the African said.

Ablebody turned to him, his dark eyes twinkling. "Mr. Bjorklund, you have a remarkable 'typist'."

Clearly Ablebody now regarded this characterization of Philanth's relationship to him as a wonderful private joke. Imagining what Ablebody concluded about the actual nature of their relationship made him even more upset. He barely managed a smile.

"Another illustration of how subtly intelligent the book is," Philanth said quickly, "is the parallels between the background of the dictator in the novel and the biography of Ho Chi Minh. The American novelist creates a man who has studied in the United States and learned to loathe the vulgarity of its mass culture. Ho Chi Minh emerged from the brutal colonialism of the French in Asia."

This change of subject let her launch into an invective against most "histories" as "stories of carnage and negotiations in which men of all nations take such piously qualified pride, when it's basically the work of vicious little boys." Such men were "the puppets of egoism and economics, even though the poor fools who slaughter each other have been conditioned to perceive their behavior in altruistic terms of nationalism and/or religion."

"So you think we would do well to put both nationalism and religion behind us?" asked the African.

"I'm afraid we're ignoring how to get from here to there, and we should get down to the sort of specifics which brought Mr. Bjorklund here. I'm truly sorry to have broken into your discussion." At last she had come to her senses.

"Continue for as long as you please," he told Able-body politely. They seemed to be entering the out-skirts of the city, and he was praying that they soon would reach the first stop on their tour.

Philanth shot him a look which seemed to mean, "Did you have to say that?" Ablebody smiled and turned to survey the ramshackle commercial-indus-trial area through which they were passing.

Philanth made an apologetic moue at him, and he shook his head at her.

"We should have talked about *The Zin-Zin Road*," she muttered.

Ablebody turned to her at once and said cheerfully, "A less complicated novel about West Africa and help-ful Americans."

"It was perhaps Knebel's best work, by my own standards," Philanth replied.

"Why do you say, 'perhaps'?"

"I haven't read everything Kneble wrote."

"Of course," was the smiling answer. Ablebody regarded Philanth with a warmth which surprised him, considering what outrageous things she had been saying. He could only conclude that Ablebody entertained few foreign visitors whose conversation he found so interesting—or at any rate, so candid.

Glancing out again, Ablebody said, "We are close to our first stop. I should let you look around."

He had absently been doing so while being left out of the conversation. He kept his expression neutral to conceal his dismay. This was such *endemic* pover-ty! Usually even the poorest countries had at least a tiny elite who amidst wide-spread destitution would raise a few ostentatious edifices: churches and palatial homes in earlier times, but now public buildings and unrealistically grand commercial or apartment buildings which typically began to fall apart soon after completion. Here he saw little except struggling impoverishment.

To say something positive he commended the apparent absence of showy but economically ill-advised structures. Philanth replied that most Gambians had been practical about the limitations imposed by their isolation within a former French colony and by their poor natural resources.

"We do have some structures in the British colonial style," Ablebody said, "the hospital and public library and the Parliament buildings."

The car slowed. "But first, here is the peanut-processing plant, updated and expanded with Peace Corps assistance." His first impression was of a tall, oddly shaped wooden building hugged along one side by huge mounds of peanut hulls. Two small children cavorting in the mounds paused to stare at them as their car drew up and stopped.

A group of people began emerging from a low, cinder-block building adjoining the plant. As he opened his door, reluctant to step out into the full force of the sun, a second car pulled up beside him. From it emerged the two news photographers and the Peace Corps cameraman.

The cast were assembled. Another performance was about to begin. He had come a very long way to be the star, and he had better make it good. He took out his handkerchief and once again mopped his face and neck, then checked his hair. He felt ready to collapse under the smothering blanket of heat, and he heartily wished that he had insisted upon more than a mere Continental breakfast earlier that morning, during his final meeting in Dakar.

As Ablebody went through the introductions and the Gambians offered their traditional extended welcomes and the still cameras recorded each grouping as they greeted him he was mentally giving himself a pep talk.

They were about to tour a piece of tangible evidence of industrial development in a time when much of Peace Corps' work had become too organizational and technical to be captured in interesting images. Americans were not inspired by photos of people planting tiny trees to save land from becoming desert or wading around in artificial fish ponds to save people from malnutrition, and the various types of picturesquely primitive instructional settings had been overworked. Photos of dark-skinned "natives" horsing around with their American friends during periods of simple recreation did not show the "work" of Peace Corps as most Americans understood "work": the achievement in such friendships could not be appreciated by those who had not experienced the difficulties of cross-cultural cooperation. But this processing plant dealt with a product of this area's major economic activity, its one cash crop to be weighed in on the international market against the innumerable finished products the country needed to import. The hulling machinery embodied power and efficacy, and using it as props was legitimate.

Go to it, Bjorklund. Pay your dues for all the honor you receive merely because of all you now represent. You're an embodiment, a symbol; and even human symbols have to be perfect. He grimly made up his mind to look alert, interested, and inspired by what he would be seeing.

Chapter 12

TRAVELING COMPANION

Ablebody had done yeoman service in getting him through the banquet. As the driver brought them back to the hotel he said, "I want to express my thanks for all your help this evening. Can I offer you some refreshments in the hotel lounge?"

"I was glad to be able to assist. Take the time to rest before our early start tomorrow.

"Perhaps you might ask your pretty assistant to join you." In the late afternoon Ablebody must have reviewed his documents, since he had started using Philanth's correct title. "After being on her own all evening she should be happy to have your company. A pleasant way for you to relax."

"I expect she's asleep. I'll follow her example."

They shook hands and said good night with more thanks and compliments which expressed his feelings as well as being in accord with the custom of the country. Official gifts were a sticky matter for him, but he was resolving to have something special delivered to Ablebody through diplomatic channels.

Nevertheless as he went up to his room he had a queasy feeling about what Ablebody might have had in mind in suggesting that he "relax" with Philanth. People saw what they expected to see, and viewed from the perspective of traditions in this part of the world, the nature of his relationship with a female traveling companion who looked, acted, and talked like Philanth must seem beyond question. Ablebody could imagine that in comparison with a fresh dish

303

like Philanth a wife his own age must seem dull and unattractive. Such a traditional wife of a powerful man would be contenting herself with her position while he pursued his own pleasures.

Moreover, to the extent that he was not culturally African, Ablebody was Anglo-cosmopolitan, not American. And the global distribution of American films was notorious for giving foreigners exaggerated notions of American women's sexual availability. Having become the appendage of a man with his perquisites, Philanth might be married in name only. All of these influences considered, he could hardly assume that Ablebody correctly guessed his strict principles regarding marital fidelity.

This was exactly the sort of possibility he had feared when he had fought Carole's insistence that he take Philanth on his travels instead of Marjorie. It was worse, in fact, because Philanth was achieving a much higher profile than he had anticipated. Indeed, she had had "a wonderful time" their second evening out, carrying on in three languages at the reception for him in Morocco.

Moreover, Carole had chided him about how hard Philanth had been working during weekends on preparations for the trip. If Philanth had done her homework on the other countries they were to visit as she had on The Gambia she was going to be noticed by other officials in host countries.

Wives could ease the boredom of their insular lives by making sly remarks about "Bjorklund's little traveling companion" after they were gone. Because of his leading-man looks and austere bearing—especially juxtaposed with his unique official position—it was choice cocktail-party material.

In a time when a good topical joke could leap across a continent the day it was conceived, snide innuendoes about his lapse of judgment were too likely to travel the international diplomatic circuit back to Washington. Those who had helped him obtain this appointment would hardly be pleased.

Those who were jealous of him or disliked him because of his political views plus his outspokenness as a congressman *would* be pleased, in a twisted way, and would start looking for the best means to exploit his situation while it remained current. They could easily find out that he was on an extended journey and imagine what a high, old time he was having with his cute, little buxom-blonde assistant in all those foreign hotels while his plain, no-nonsense wife remained in Washington. And they would lose no time in devising ways to inspire some columnist to begin the public crucifixion.

Considering his looks and the national attention he had earned as a congressman, he would be lucky if nobody sold the "story" to one of the major tabloids. He could see it clearly, his flamboyant sex life in foreign capitals at Peace Corps expense shouted in headlines to everybody waiting in a grocery-store check-out line throughout the country.

Anyone "developing" the story would surely discover at once that the Peace Corps director's young assistant was a former Peace Corps Volunteer. This was to be expected; a bit more effort would reveal that she also had been a school teacher in Ohio and had worked for the World Hunger Foundation before her Peace Corps service at a hospital in Africa. Whereas he was a lawyer and a politician, and too handsome to be expected to have developed much self restraint regarding women. Philanth looked as innocent as a Cupie doll, and their respective ages would be duly noted. He would look depraved, a seducer of children.

It would not matter how many people actually bought and read those millions of tabloids and gossip magazines. One full-color, front-page photo of him, juxtaposed with a picture of Philanth and a headline, and the sensation-seekers in political journalism would pick up the scent. Since the apparatus for publicizing this trip in every capital he visited was already in operation it would be easy for photojournalists to extend that coverage: to begin stalking Philanth and taking more pictures.

Those additional pictures would "confirm" the "basis" for the "stories". Then, by one pretext or another, newspapers could keep this interesting little sidelight on his trip in circulation for months: the fact that other publications had created a story was made a story itself. He knew from long observation how American journalists felt "obliged" to perpetuate political sex scandals, despite their pious disclaimers.

And women had incredibly long memories for that sort of thing: sexploitation by a politician seemed to be the one abuse of office they could never forgive. As the "story" was dragged on and on he would have to kiss his already fragile political career good-by. All his years of effort and experience would be dismissed, his future blasted.

Oh, God. How had he been so preoccupied with all of his normal work and worries plus all of the complicated preparations for this trip as to let himself get pressured into this situation? But it was no time for lamentation. What was he going to *do*?

Sending Philanth home at this point would only call attention to her. Office gossip could piece together the unusual speed which she had been sent in from the field, put into his office, and then picked for the trip. All that could have been his doing, following a short trip he had made with some congressmen in the general area where she had been working. But people didn't have to work that hard: one look at her, and those so disposed—Les immediately came to mind—could settle upon a sexual explanation, the easiest being that "she wouldn't put out, after all." Or else she was "no good in bed".

Her untimely return would be underscored by the logistical foul-ups which would result if his traveling assistant were yanked out of the picture.

The picture. The tabloid editors might not be so tempted if they couldn't get even one photo of the two of them actually together. The fakey juxtapositions of cropped photos those rags routinely used weren't convincing, and skillfully doctored photos were grounds for protest even by a public figure.

Philanth had had enough sense to stay out of the picture-taking at the airport that morning, and come to think of it, she had hung back during the filming and picture-taking at his factory tour, also. He did not remember her being close to him during any of the picture-taking in Morocco or Dakar. He could not think of any time when she had not avoided being photographed.

They had interacted extensively in the company of a third party only today, in the car with Ablebody. Except for that and the reception in Morocco, she had been inconspicuous or out of sight, tending the phone in her hotel room or whatever. Maybe little risk had been incurred so far, and he could manage to minimize people's awareness of her for the rest of the trip. Just keep her out of sight as much as possible, handling the telecommunications and paperwork at his command post at the various hotels. Only their drivers would see her, usually, and she could be told to keep her mouth shut.

It was the only way he could play it.

While he had been worrying and working through to the way he would deal with the problem he had automatically gotten ready for bed. He found a little note on his pillow: Philanth asked him to be sure that she was awake in time to be ready when Ablebody came for them in the morning, as she had never been able to get that precise information earlier. She apologized for her request and promised "to try harder to avoid letting this happen again."

He requested his wake-up call, set his travel alarm clock as back-up, switched off the light, and fell into bed. The wind was coming straight in off the ocean, rattling the windows and making the colorful African draperies creak like sails. In the midst of the tumult he had a secure haven in which to rest. That hot shower did him so much good at the end of the day.

He was awakened by Philanth's knocking. He had a phone call from Marjorie and Les. Rather than risk losing the call, he stretched the line of Philanth's phone into his room. She came in later and used his phone to order his breakfast from room service. By the time he was through with business at headquarters he had to hurry to shave and get dressed in order to have time for his good British breakfast.

As a matter of fact, however, it was not so good. There was something wrong with everything except the orange marmalade: the toast had been made of the same tasteless, textureless white bread as the watercress sandwiches he had been served the day before, and it was hard and cold. The eggs were too runny, the bacon overdone. Remembering the warnings against dairy products, he did not even touch what looked like butter and cream. The coffee was too strong.

But he crumbled the toast into the eggs and mixed them well and decided he should consider this a good meal compared to what he was likely to get for the rest of the day. Lunch and dinner probably would consist of the locally produced foods which the ordinary people in The Gambia ate, and he would be the victim of the hot spices which had come into favor during millenia of hot climates with no refrigeration.

As he ate alone in his room he skimmed his program briefing material on The Gambia one last time, looking for reminders of concerns he wanted to bring up with Ablebody during the long drive out and making a note when he found one. Since the projects here mostly involved improving health care or producing more food and he had had no involvement in those areas before coming to Peace Corps he was not good at carrying much of this information in his head. He was like a dull student hastily preparing crib notes for a day-long oral exam and knowing that he was hopelessly unprepared.

So many of his days were like that during this trip. And as though everyone else guessed how little he really understood about agriculture or public health or whatever, they never really assumed that he knew much of anything. Thus time was wasted while they took his measure, and they never got into specifics unless he pressed them. He had to question them while trying to keep them from realizing how dependent he was on his briefing material. He had to focus his interpersonal skills on avoiding embarrassing himself or making others embarrassed for him.

A good deal of his work as a congressman had been like that too, of course, but he noticed it now because he did not have Carole and most of the rest of his staff working furiously to be sure that he was fed exactly what he needed when he needed it. There was so much to know about every problem. No one could know all he needed to know, so outside the fields where he had developed special expertise a politician could not be much better than his support staff.

Just the same, he could not stand the thought of being just another politician who spent his whole life on the business of getting elected and re-elected. Therefore he had press himself to learn as much as he possibly could. That was one reason why he was treating what could have been a pretty mindless public-relations job as an extraordinary educational opportunity.

All of his rushing to get ready and worrying about how ignorant he was began to seem a little silly when they arrived at the Bambatenda Ferry. Their car simply went up to the water's edge and stopped. It was a very peaceful, almost static scene. Decorated canoes moved on the wide expanse of the Gambia River, and a few motorboats went past.

Ablebody explained that they had just missed the ferry and that it left whenever it became filled. However, Ablebody added, there would be a number of cars waiting for it on the other side at this time of the morning, so their wait for its return should not be more than an hour.

Ablebody's tone implied that they were fortunate. He was glad that Ablebody, who was driving the car himself this morning, could not see the looks which passed between his aide and his special assistant, who were together in the back seat.

It was a discouraging beginning to the day. This day was bound to be full of unpleasant little novelties for him, since it was the one day in all his five weeks which was devoted entirely to giving him direct experience of conditions in the field as well as to photographing and videotaping him as he demonstrated his willingness to "get out there where it was really happening", "get his hands dirty", and so forth.

The day was too important for him to let it get off on the wrong foot. After a minute of getting his head straight he said, "Philanth, Tony's going to take you window-shopping." He waved toward the little tin-roofed, patched-together open-air booths set up along both sides of the road. He could see displays of clothing, jewelry, and vegetables.

"Tony, here's your chance to put your Mandinka to work again. Volunteers are supposed to be ambassadors, in addition to everything else, so be ambassadors: spread around some international good will. Let people know we were here."

Tony grinned, opening his door. "You got it, Boss. I'll leave 'em laughing."

"But with you, not at you. And try to buy something, a wood carving or whatever.

"Meanwhile Dr. Ablebody may be so kind as to give me a crash course in The Gambia's needs and problems in relation to what the Peace Corps has to offer. By the end of today I expect to have absorbed all the information about the problems in this region that my brain can manage." There were approving nods all around.

As Tony and Philanth climbed out he added, "Keep an eye out for the country director's car—Jeep, excuse me. Let me know when it arrives."

It would contain Max, the cameraman sent by headquarters. In pursuit of briefing for the day ahead, Max had left the prime minister's home the previous night under the wing of the country director's representative, Cory Watson.

His aide and assistant assured him they would do so, for they knew he wanted to discuss the plans for the day with Max.

Noticing Tony's camera case dangling from his neck, he added, "Oh, and take some pictures. Philanth, get some good shots of Tony with the vendors."

She nodded, and he thought her direct look at him seemed to register the one-sidedness of his instruction. He began to suspect that Carole had given Philanth some version of why he had not wanted to take her with him on this trip, and that was why she had been avoiding getting in front of any cameras.

He turned to the African beside him. "I hope you'll honor me by telling me what's on your mind, Dr. Ablebody. Everybody keeps trying to make this a pro forma inspection tour, with everybody spouting off about what a good job the Volunteers are doing. I want much more. I want to hear about the problems, from different points of view. I emphasized throughout the planning of this trip that I wanted small, candid, business-related meetings as much of the time as possible, with the larger groupings where little gets done concentrated into single events.

"I hope I'm not being too blunt. I don't want to offend by seeming to be overly involved in details or in too much of a hurry." He flashed the smile which usually made things go his way. "But being with you this morning is a very precious opportunity for me. Please help me. Tell me what you really would like for me to know."

Ablebody began to talk slowly and carefully, and he took out his pocket notebook with his crib notes and began to ask questions and to write. He forgot the warmth in the car, he didn't let the arrival of the cameraman or the arrival of the ferry stop the flow, and he was sorry when they reached the other side of the river. Then Ablebody had to turn his attention to negotiating the car over washboard ruts and into

312

breath-taking dips without destroying it. Ablebody told funny stories about the need for auto mechanics.

After considerable driving along dusty roads he was taken to pay a courtesy visit to the local headman. The local Volunteers showed up to greet him, and he posed for pictures with everybody and accepted some local artifacts as ceremonial presents. Peace Corps headquarters would find places to display them among all the others: the carved animals and woven hangings, the warrior's shield made of hide with "Peace Corps" burned into it in French.

The only thing he did not expect was the fresh-roasted peanuts, the best he had ever tasted. He and everyone else in his party were given more to take away with them, wrapped in twists of newspaper. Wherever he went people gave him things, no matter how little they had. It was not just because he represented Peace Corps. It was their way.

Next he visited an open-air classroom where a Volunteer was holding forth. The teacher, a retired lady in her sixties, ceremoniously presented everyone in his entourage to the class, saving him for last; then the little boys greeted him in unison, in the cadences of careful rehearsal.

However, they still hadn't mastered his last name, and everyone got a laugh out of that. Meanwhile the camera turned here and there, capturing it all. He believed the tape would convey to future viewers even the warm feelings in the vicinity of his heart.

The next stop was the nearby clinic. While he and the two clinic staffers did their thing for the camera Philanth took down all the information they were being fed about the clinic's caseload and needs—to "save time later, when we're processing all our new materials," she explained when he pointed out that that wasn't necessary because it would be on the tape. He suspected that the truth was that she had a compulsion to keep busy.

After a simple lunch with local Volunteers his afternoon was to be devoted to a tour in the air-conditioned Peace Corps Jeep. As the two vehicles moved on toward a model farm some distance from the area served by the school and clinic he stayed with Ablebody, and his host discussed agricultural problems of the area. When they reached their destination his guides became Bob and Cheryl, the Volunteers who had made the farm a "model" of numerous integrated agricultural innovations.

This young couple, both recent graduates of an agricultural college in Pennsylvania, were finishing their tour of duty as guests of the farmer and his family. Government and Peace Corps policy required that a project be maintained at a high technical level, as a teaching program attached to the educational programs of the area, so the Volunteer couple were to be "replaced" by a single Volunteer who, as part of his work in training agricultural extension agents, would be expected to help the Jaiteh family see that the model was maintained and promoted.

Bob and Cheryl proceeded to make him the beneficiary of the very informative lecture which he had requested. The amount of technical expertise which had been brought to bear here was indeed impressive.

A close view of the basic operations of the farm was an experience for which nothing in his previous life had prepared him. Moreover, the degree of poverty reflected in the lack of manufactured materials for modernization stunned him. The patient, backbreaking work he saw being done in the fields and the poor results yielded in the people's lives seemed to indict him merely for being accustomed to comfort. He was all the more impressed by the good will and hospitality with which everyone received him.

With Tony, Philanth, Cheryl, and Ablebody following them in the car, there were four of them plus Max's camera in the Jeep. But they had to keep leaving the vehicles to get closer views of important new structures such as the methane system and the grain dryers while Bob or Cheryl explained their workings. Moreover, the lecturers frequently paused to invite Dr. Ablebody's comments on "the larger picture". Cory, who did not have to leave the Jeep when everyone else did, loaned him a straw hat; but by mid-afternoon the heat during these extended stops began to make him feel faint. Gradually he stopped asking questions or taking notes or even making intelligent comments.

Finally Philanth took a close look at him and suggested to Cory that they pull into some shade where they could have something to drink. He saw that her face had become flushed even though her umbrella and broad-brimmed hat and long-sleeved blouse and long, full skirt protected her from any direct contact with the sun. Since the air conditioning in the Jeep couldn't do him a lot of good because of the frequent stops he probably did not look much better than Philanth did. Cory took one look at him and announced that he would take them back to the compound.

When they had driven up and stopped, Cory hauled from the back of the Jeep a battered metal cooler full of bottled drinks, remarking that it looked like a good time to break them out.

The one big shade tree near the compound was not much help, so with the permission of the only adult around, Mr. Jaiteh's mother, they all crowded into the largest mud-brick hut. It contained only two cots and a chair, but as they all chose places to sit he insisted on taking a place on the ground. The youngest children of the family had been playing in that room, but as all the visitors crowded in, the old woman took these babies and toddlers away.

He and the rest of his "party" settled down to nurse their cold soft drinks. Cory soon began handing out seconds, and they drank until the supply was exhausted. They were a sweaty lot. Ablebody said the men should remove their wet shirts, and they didn't need much assurance that it wouldn't offend anyone's sensibilities. Max was forcefully told to turn off his camera until further notice.

Philanth promptly collected the used shirts, begged a couple of small towels or rags "for us ladies" from Cheryl, dipped the shirts and rags in water, and passed them around. Following the others' example, he began applying the wet cloth to his head and face and swabbing all of his exposed skin with it. Bob and Cheryl scrambled around the compound to collect the whole family's supply of locally made fans. Soon they all were fanning themselves and each other by turns.

Combined with fanning, the wet cloths were astonishingly refreshing. He realized that he had been starting to get a headache, and applying the cool bundle to his head with persistence had stopped it. He felt foolish at having just rediscovered the principle of the ice pack.

However, this discovery regarding the efficacy of combining ordinary water, cloth, and fans was a relief, for it reduced his constant dread of breakdowns in hotel air conditioning. There was satisfaction to be found in the self-sufficiency provided by "appropriate technology", a long-time Peace Corps principle he had just seen demonstrated in the operations of the farm.

Looking him over, the group half-laughingly decided that he had "seen enough." They didn't want the Peace Corps director collapsing from heat stroke while under their care. Using gestures for want of visual aids, the host Volunteers finished their lecture without the accompanying tour.

When Bob and Cheryl decided they were finished, Max got up and excused himself. Cheryl went with him to locate another hut where the cameraman could freshen the water in his shirt/sponge and try to take a nap "until things pick up again". Despite all the sweating they had been doing, people slipped out to visit the latrine.

In this more relaxed atmosphere the talk became general, and almost at once the conversation turned to the work of the IITA and ILRAD. This was a subject which everyone present seemed to find exciting—even Philanth, though he would have supposed that she knew as little about it as he did. He found out why Philanth was interested when she told him that these were "two of the dozen or so international research facilities around the world that are at the center of efforts to assure the survival of three-fourths of the world's population." Aside from Philanth and himself, everyone else was or had been directly involved in tropical agriculture, so the International Institute for Tropical Agriculture at Ibadan and the International Laboratory for Research on Animal Diseases at Nairobi were involved in all sorts of research which could directly affect their efforts.

Cory and Ablebody got into a discussion of the breeding of cattle tolerant of trypanosomiasis, and Bob jumped in with some opinions on how the farmers *he* knew were likely to feel about switching to an animal with reduced milk capacity.

Cheryl brought up the research on farming techniques which compensated for farmers' inability to afford chemical weed control, and from plant infestation the group moved on to the terrific problems of plant disease in equatorial Africa.

He and Philanth had discovered a shared loathing for plantains, and they exchanged guilty glances when Cheryl talked about the tremendous importance of this banana-like fruit in the diets of sixty million Africans. According to Cheryl, a fungus disease had been accidentally brought to Central Africa from Latin America in the early 1970's and had spread ruin over millions of acres of plantain trees, eventually extending to areas two thousand miles from where it had started.

Cheryl began talking animatedly about the research she and her husband wanted to get into when they went to Nairobi for their graduate work in tropical agriculture. He saw Philanth smiling at Cheryl as though she had found a kindred soul. Here was another woman who knew she would never want for work in Africa. The governments and foundations and charities being as strapped for money as they were, trained people might search in vain for a job that paid enough to live on, but they would not lack for work.

Cory reluctantly tore himself away from the lively discussion to "go collect a few lonely Volunteers" stationed at two other projects within driving range. They were to join his party for dinner. The farmer's wife and daughters returned from their work in the fields to start dinner, so Philanth and Cheryl went outside to help. From outside came the squalling of an infant and the shrieks of other little ones, so he knew that the grandmother had let her charges out of hiding.

His back against the unsoft cool mud-brick wall, he rested, alternately fanning himself and rearranging the drying shirt around his neck or his head or over his chest while the other men went on talking, happily telling each other about the many advantages of improved varieties of cowpeas and cassavas. Nitrogen to the soil and protein to the diet even in the drought-prone areas; providing mulch . . . .

He wished his eyes wouldn't close when his brain started refreshing itself with quick snatches of REM sleep. He speculated about what there was to drink in the compound and longed for the probably unsafe ice which presumably was now tepid water in the bottom of Cory's old ice chest.

There were so many terrible diseases in African waters: cholera, typhoid, river blindness . . . . A government could stand or fall on whether it could promise to bring safe water to all its villagers. He had read somewhere that democracy had once ended in Ghana because the aid for such a program had not been forthcoming and so there had been a coup. That had been years ago . . . .

He roused himself when he and Ablebody and Tony and Bob were joined by Jaiteh and his teen-aged sons and several of Jaiteh's neighbors. Becoming much too large for the windowless hut, the group moved to the shade of the big tree outside the compound, out of the way of the women fixing dinner. As he had the previous night, Ablebody took his place at his right and served as his translator.

He tried to engage the local men in worthwhile conversation about their farms and what they really thought of Bob's effort to develop more cooperation in grazing and related aspects of land management. Since he was just as much an outsider as they felt he was, it was awkward work.

319

The local Volunteers arrived with Cory in the Jeep, their enthusiastic American voices sounding out-of-place in the quiet heat. He gathered from their greetings to him that they had pitched in to augment the monotonous menu of rural West Africa with some prized offerings which would be more familiar to him. Since the newcomers were men, they joined his all-male group under the cottonwood tree. Max reappeared, camera on his shoulder, and captured the gathering of men in ankle-length white shirts sitting around with him before retreating to the compound to capture the women's more picturesque activities around the cooking pot and the toddlers playing underfoot.

Philanth came to squat down and quietly tell him that dinner would be ready soon and that the night insects would begin to become active before they left; so to ensure his protection she had borrowed one of Mr. Jaiteh's fresh shirts and found him a smaller hut in which he could bathe in heated water before "dressing for dinner". She took his half-dry shirt to smooth it on some rocks to catch the last heat of the sun.

He wondered whether his long-suffering white shirt could be rehabilitated to serve another day, back in the world of pristine hotel meeting rooms. He felt like a fool for having worn such a shirt today. But a man who did not even spare time for gardening could not pretend to be any sort of up-country type, and he still felt that his sports shirt would have been too casual for the formal introductions that morning. So he had worn a white one but left his collar open, to look unpretentiously official in all those reels of multipurpose videotape. He should have brought a change. But he had become so accustomed to Carole's deciding what he should do about such details that he was slow in catching on to that aspect of this job.

He had to visit a genuine rural African latrine for the third time, and he hoped that this would be the last. He knew it was a cultural prejudice, but he could not help feeling that Africans were too matter-of-fact about bodily functions.

How would these people react to one of the palatial potties he had seen in his own country? An affluent Washingtonian spent more money remodeling a bath-room than any of these Africans would see during his entire lifetime. Now his own elderly bathroom at home, which he had always despised because it was tiled in tiny blue-and-white hexagons like an antiquated public rest-room, looked just about right.

As he passed through the compound he saw the women sweating over the dinner, which was to include something from a big pot that simmered on the fire in the blazing late-afternoon sun. Containers made of gourds lay here and there in the cooking area. Yet this was described in his briefing as "a prosperous farm family". The number of children and daughters-in-law gave a distinctly African connotation to the concept of "prosperous".

When dinner was ready the neighbors said their good-bys. Jaiteh and his sons and everyone else except the Jaiteh women settled in a circle on woven prayer mats under the make-shift shelter of branches in the middle of the family compound. The ground was as hard and bare as concrete. But he was thankful that he had not come during the rainy season.

Besides groundnut stew, which was hot from the pot and spicy and greasy and seemed to contain a little of everything, the menu included rice and *un*spiced garbanzo beans and roasted chickens. The chickens were stringy. But since Volunteers had been so involved in the preparations, he relaxed and enjoyed the dishes he liked and avoided those he did

321

not. The main event was the fresh fruit—not really because it had been so prettily cut up and arranged by Philanth, but because it was accompanied by custard made from Bird's Custard Powder which had made the journey all the way from Great Britain. He was solemnly assured that it was safe, having been made with powdered milk and boiled water. To Max's camera they would appear to be having a feast.

The role of the farmer's womenfolk as mute serving-maids to the family's menfolk and visitors disturbed him. But he did his best to acknowledge their participation, and dinner was a jolly affair. Due to language difficulties there was much exaggerated gesturing, which contributed to the general good humor. Having worked hard all day, they gradually abandoned any effort to be serious.

The after-dinner entertainment was late in arriving, so while they waited to see whether it would appear they "went around the table" doing "party tricks". Tony proved that he could wiggle his nose three different ways. Philanth did a charade representing the local proverb, "Soflee soflee catchee monkey," providing a imitation of a grimacing, flea-bitten monkey which Cheryl pronounced "darling". He presented one of his best shaggy-dog stories from his years as a campaigning congressman. Mr. Jaiteh and his sons sang a few songs, which the Volunteers were happy to translate and explain. Max had just done his imitations of W.C. Fields and a few recent presidents when the proper entertainers arrived.

They were a group of local dancers and musicians, already in traditional costumes. He didn't like to think about how much trouble they must have gone to just in order to show up. They were strong, handsome people, and he found the women especially pleasant to look at. Soon they were making him smile with their indefatigable sense of humor and vibrate with their ear-shattering exuberance.

Philanth sat across from him, her cornflower-blue skirt spread to completely cover her bare legs and sandaled feet. Ablebody had encouraged her to sit beside the guest of honor, but she had declined, saying, "There's going to be videotaping going on; and in my country it's considered self-serving for someone to be photographed with an important official when they only work for him." So much for his hope that Carole had said anything to Philanth about his reasons for not wanting to take her with him.

Happening to glance around the semi-circle of faces turned toward the performers, he was struck by her rapt expression. She seemed to be drinking in the music and dancing at every pore. She was like that about so many things!

She saw him looking at her and smiled intensely, giving her head a tiny shake to express her delight in what was transpiring.

He smiled and nodded to agree that it was very good. She went back to her avid watching as the incredibly limber dancers shook themselves and patted and teased the naked earth with their bare feet and leapt and tumbled and swayed in unison with a gaiety and joy of life he had never seen before. There were frequent performances of African troupes in Washington, but he had never attended one except for an annual program at Peace Corps headquarters which celebrated Afro-American culture on Martin Luther King's birthday.

Tony and Cory and a couple of the other Volunteers accepted the dancers' invitation to join them in a dance with simple movements and obviously enjoyed this tremendously. He too felt the contagious delight of the movements, but he could not imagine letting go of himself that much. He could not wrestle with the management of a big, complex, important organization day after day, night after night, and suddenly throw it all aside for some light-hearted fun.

He could not remember when he had. It did not bother him that he couldn't. After all, he wasn't young any more.

The video camera was watching the dancing and the watchers of the dancing, too. It had captured much of everything he had seen that day: women carrying water along the road in kerosene cans on their head-wrappings, men sweating in the fields with only hand tools, school children with no books, pencils, or paper. And Volunteers teaching, healing, demonstrating more effective ways to get things done.

He understood now why Peace Corps publications carried the same sort of pictures over and over: those who had experienced the realities which the pictures failed to capture in their flat, little frames could never resist reminders of what he was feeling now. From the outside this do-goodism looked hackneyed; from the inside it had aspects of religious experience. People never tired of that. Having tasted it once, they would always want more.

When the music ended one of the male dancers stepped forward and made a little speech in the local variation of English, addressing all of them but focussing on their guest of honor. It was enthusiastically applauded.

As the applause died down he climbed to his feet and responded in kind, expressing his gratitude and thanks for all the kindness which had been shown him and saying how much he would cherish his new friendship with the people of The Gambia. Then he singled out their hosts, Dr. Ablebody and Mr. Jaiteh, for special thanks and compliments. After expressing his appreciation for the dinner and entertainment to those who had provided them and his hopes for the future he expressed his regret that they must leave so soon in order to begin the journey back to Banjul.

Taking the hint, the party broke up, though the leave-takings went on for some time.

The dinner and entertainment had been scheduled as early as the afternoon sun made feasible in order to minimize their exposure to the night insects. But since complete darkness would soon overtake them on the road and the casual habits of the local children and animals made driving after dark hazardous, Ablebody took the lead car. Max's flight out of Banjul would not leave until late the next day, so the minister proposed to pass the time as he drove by helping the cameraman develop ideas for additional videotapes which might be spliced in as "background". Cory loaned Max his flashlight so he could take notes.

Tony and Cory took the front seat of the Jeep, and he and Philanth took the back. They all rolled up the windows, and before he started out Cory put on some insect repellent and passed it around "one last time".

As they caught up with the bobbing tail lights of Ablebody's car Tony and Cory began trading "war stories" of their experiences in West Africa. Tony's friends at the Agricultural Research Centre had added to his fund of local lore the previous evening, and Cory had friends at the nearby boys' school which had begun as a British project in hen-raising, so the two of them could have talked and laughed all night.

The vibration of the powerful engine muffled the young men's chatter in the front seat, and he relaxed into the lull. Philanth also seemed disposed to quietness. Knowing her as he did now, he enjoyed feeling her silent presence. The insect repellent scented the air between them. Outside there was only darkness. He closed his eyes, tired but content. A good day's work, for once, though he could not take the credit.

He had been half asleep, swaying with the motion, when at last the Jeep ground to a halt.

They had returned to the great river. The little booths along the road on this side were deserted now. One lantern burned, showing that the ferry was there. But it seemed deserted too. There was a bright moon.

Ablebody came back to lean in the driver's window. "Again one has to wait a little. But I will find the ferry captain. His house is nearby. I recommend you keep your windows rolled up and turn off your lights. I regret the delay." Ablebody's light robes faded into the darkness, and Cory hastily rolled his window back up. Cory and Tony started trying to swat a large flying insect which had slipped in, turning on the overhead light to see it.

When it had been dispatched Philanth asked, "Would you mind if I lay down, Mr. Bjorklund?"

The seat was so wide and Philanth was so short that the idea seemed feasible. "I wouldn't mind. I'd have slept all the way here, if I could've lain down and the road hadn't been so rough."

"Same here. If I'd lain down I would've been thrown all over the place."

She pulled her cape-filled pillow out of her bag. The cape had come out and become her seat cushion as they sat on prayer mats in Jaiteh's family compound, and now she set the pillow beside his hip. She removed her sandals and curled up with her bare feet pressed against the door on her side of the vehicle.

"How does it go?" she asked no one in particular. "'How sweetly sleeps the moonlight on this bank! Here let us sit, and tell sad stories of the deaths of kings.'"

Tony snorted. But Cory said defensively, "I think that's *nice*."

After a few moments Tony said, "I need to talk to Max; and I won't see him again before he flies back to Washington. Come with me, Cory."

He noted that the two men did not head for the car ahead of them, but off into the darkness on one side of the road. He could hardly blame them for not waiting for the next gas station.

He and Philanth were left alone. After a moment she asked, "Was the day satisfactory to you, Mr. Bjorklund?"

"Yes. Everybody did a fine job." Then he added, "But it's hard to grasp how little people have, here."

"That's true. Just the same, anybody who believes there's no gold in The Gambia is blind to its human riches. To a teacher, an artist, anybody who deals in direct human experience and wants to get close to life or make it better, it could be a gold mine."

"You really think that?"

"I believe you do too, or you wouldn't be here.

"I know it looks hopeless for lots of countries like this. But we all started from nothing at some time, and look how far we've come: the Sistine Chapel; the desktop computer; the global net of communications satellites.

"And even with all its natural and historical handicaps, just in recent years Africa has produced splendid novelists, statesmen, and artists. People anywhere are like the 'mute, inglorious Miltons' in Gray's elegy: here are a lot of people who could do wonderful things"—for a moment her voice caught—

"if they were only given a chance. You're here to look for ways to help them get that chance."

Suddenly she had gotten choked up. She did not care that this feeling was out of fashion. She made him see the glory of what he was doing, or at least trying to do.

It wasn't so hard, since most of the time his cynicism was directed against himself. Philanth had rejected *any* version of maturity involving cynicism: she had made a considered decision to keep the faith.

Shortly before dinner, when she had come into the windowless thatched mud-brick hut where he was to wash off the heat prickles and grittiness of the day with the blessed efficacy of heated water, she had hardly been able to look at him while he handed her his damp shirt in exchange for the dry one she had borrowed for him. The group of half-naked men to whom she had handed out wet shirts also had been difficult for her to look at.

Nevertheless she had admonished him, "You look wrung out, Mr. Bjorklund. Remember, 'They that wait upon the Lord shall renew their strength. They shall mount up with wings as eagles; they shall run and not be weary; they shall walk and not faint.' Realizing that that's what you're doing lifts you above the tedium of life, and gives your spirit wings."

Without waiting for a reply she had gone back out into the compound. Yes, he had thought, smiling to himself as he set about bathing, he could see her as a nun very well. She would never let the sinful world tell her what not to say or feel.

Lying beside him now in the darkness, she shifted a little and sighed. He looked down at her.

"Philanth. You quoted Scripture to me, this afternoon. In the hut, remember?"

"So I did. But I wasn't endorsing orthodoxy, just

equating 'the Lord' with a satisfying ultimate value. Much religious teaching is valid without any supernatural referent because it reflects realities of human nature."

"Are you . . . an atheist?"

"The concept behind that label is simplistic. Divinity is real, but we know it only as experience in the mind and emotions of the beholder."

"D'you know what you're talking about?"

"Does anybody?" she retorted good-naturedly. "At least I'm clear about my position. Being mushy-minded on the basic questions is like going through life half-dressed.

"It's amazing how many people are like that: ill-equipped to deal with the deepest levels of human experience. That's because our pluralistic culture is so afraid to even talk about religion with any depth that it ignores the fact that the questions religion tries to address are as inescapable as love and death.

"Mr. Bjorklund, you're still a fairly orthodox Christian, aren't you?"

"Something about the context and timing of that question suggests that I seem to you to be 'going through life half-dressed'."

"Sorry." She chuckled.

"But you are, aren't you?"

"How'd you know?"

"The way your mind works."

His voice was warm with indulgent amusement: "Tell me how my mind works."

"Some things you don't try to understand: you simply accept them as mystery around the perimeters of human understanding. Nothing's ever forced you out of the comfortable definitions you were given, but it's never seemed terribly important to be sure you had the right answers to everything. You understand that what works for you doesn't work for everybody. I dare say you pity those who lack your comfort."

"It sounds lazy, put that way."

"No more so than wearing ready-made clothes. Nothing wrong with that if it meets your needs. As long as it's not a system that justifies harm to others."

He got interested in the difficulty of getting political consensus on the definition of "harm to others" and did not answer. So she went on.

"It was the way my father's mind worked. But it didn't serve him well, in the end, even though he had at his fingertips many more of the traditional answers than you do. It's a view for minds that have never truly grasped the horrors of the twentieth century."

"You have quite a darkness in your view." For someone so young and apparently so sheltered. There was in her manner a gentility which made him imagine her spending her youth brousing in her father's book-lined study—in an ankle-length dress with ruffles at the wrists, and with a faithful old servant to bring tea.

"There's no reason why that should bother you. I have seen things which could be permitted only by a god worthy of hatred or contempt. But I have made my own peace in my own way."

Her slight emphasis on "I" reminded him that she had interpreted his expressions of concern about her emotional life as reflections on his own. "And you think I haven't?"

"You're no stranger to despair."

He couldn't deny *that*.

As though his silence had admitted to the more than she had actually stated, she said frankly, "Maybe that's what bothers me about you. A person doesn't *have* to fight two enemies at once, objective evils *and* despair about them. That breeds despair and self-contempt over one's despair! You only need to know enough about the monsters out there in reality to know what to do."

"How can I not despair, when the little I know that I know is rejected by those who obviously know less?"

"You should count yourself lucky: unlike my father, you don't presume to try to accept the—the most terible things as permitted by a divine will. Do you?"

"I don't try to deal with that."

"So much the better. But you *are* conventionally religious, and a ready-made faith not reinforced by your own effort isn't strong enough to bear much strain. So your sort of view often fails to give people the help they need when it's tested by disaster or death."

Her caring but matter-of-fact manner made him imagine her playing under her father's desk while he did pastoral counseling.

She added, "Some people can't handle even the fact that they'll die. That's normally because they have not truly submerged their sense of identity in something greater than themselves. But you seem to be altruistic rather than egocentric, so I don't think it's your own mortality that seriously troubles you."

"You've been working on what it is that does?"

"I think you can't accept not being God so you can put things right."

He chuckled.

"If you're going to be a Christian, Mr. Bjorklund, you need to take the good with the bad: not assume all the responsibilities without really believing the promises."

She was on target. Despair was a deficiency of faith in God's promises. He realized he wasn't such a good Christian: the spirit of the ethics was important to him, but he couldn't imagine himself acting upon any of the theology.

That did not bother him. He felt that she had just freed him from something. He saw that he need not despise himself because he sometimes could hardly bear the weight of his self-imposed burdens. She had taught him a new kind of self-acceptance.

Maybe she was right to want to spend her life as she did, learning how to make crutches for others.

However, he thought that must be, even more than most occupations, a matter of diminishing returns: she would find that the upper reaches of wisdom, like most kinds of knowledge which offered no tangible payoff, were in little demand. She therefore would feel obliged to focus more completely upon the "practical" problems which kept so many souls in shackles.

Could she then—alone—find the strength to effectively assault such hydra-headed monsters as their minds both found, when he with his far greater advantages was faltering? He could not believe it was possible. Like him, she received intellectually from too many directions; moreover, she was too fully developed emotionally. She did not have the shallowness or tunnel vision of some activists who doggedly carried on year after year, never embittered and never giving up. Even though she had taught herself that she should be satisfied to labor without tangible results she was not through with despair.

Could she somehow be sold on the idea of attaching herself to his advantages when he had so little confidence in their adequacy himself? That did not seem likely even if she were ready to recognize that for do-gooders, politics—frustrating as it could be— was ultimately the only game in town. Besides, such a realization could not be rushed. Often it had to be grown into through experience of the alternatives. It was quite a selling job Carole had given him to do.

She was asleep by the time Ablebody returned and Tony and Cory scrambled back inside to move the Jeep onto the ferry. He prevented them from waking her, and they resumed their journey.

In the passenger loading area outside the hotel Max got out of Ablebody's car with all his gear, loudly thanking Ablebody "for everything". Ablebody too left his car and walked over as the Jeep came to a stop.

While they exchanged some words of parting Ablebody could not overlook the fact that Philanth's head was nestled beside the hip of the man she worked for. The minister diplomatically commented upon how soundly she slept, but he could almost hear the African thinking, "Bjorklund's woman is delectable." With a final "Good night" Ablebody moved away from the Jeep.

Tony and Cory had turned to peer down. "Ain't she sweet?" Tony whispered to Cory. "Definitely worth a little effort."

Though chagrinned by the compromising situation, he felt too fatherly to be ungentle in jostling her shoulder. "Philanth."

She made an agreeable "Mmm" in acknowledgment.

"It's worth a fortune to be able to sleep like that. Is it always so easy for you?"

"I feel so safe, with you," she mumbled, still not opening her eyes. She began a dainty yawn and a luxurious stretch that arched her back as she added, "I feel so lucky to be working for such a good man that I can feel secure enough to fall asleep."

Tony snickered.

At once Philanth scrambled to a sitting position, blinking in the light from the hotel entrance.

"That's a mighty sexy stretch you do," Tony said.

"Ohh," she said exasperatedly. "All in your point of view." She dove to the floor for her sandals and set about putting them on.

They collected their belongings and gave Cory their thanks and best wishes. Max wished them luck on "the rest of the world tour" and rode off in the Jeep with Cory. He and Tony and Philanth went into the hotel.

During the night he began experiencing uneasy shifting in his bowels, and as he awakened in the morning he was getting cramps and chills. He had hoped that the daily dose of antibiotic would keep him free of this problem, but he was forced to accept the fact that something had gotten past his defenses.

He regretted having ridiculed Philanth for regarding the prescribed precautions so seriously. Even this early in their journey he had gotten the feeling that she would have nagged him about being careful, if she had dared, out of her instinctive desire to take care of him. Already there had been countless little cues which together told him that in her book a woman always took care of her man, regardless of what job title came with that responsibility.

Actually, come to think of it, she was inclined to want to take care of everybody.

He rested, trying to get comfortable and instead becoming more worried, until the alarm went off. Then he opened the flight bag Philanth and Carole had agreed to designate their "medicine chest" and took the appropriate patent remedy.

The phone rang as he was setting out his clothes for the day. Wondering why the call had not gone to Philanth's room, he answered.

"Mr. Bjorklund?" asked a young male voice.

"Yes."

There was a pause, so he added, "This is Charles Bjorklund."

"It's Cory Watson. I decided you should know: Tony's got a bet with somebody back at headquarters. Five hundred dollars. He'll take Philanth's ring to prove he's scored with her. I didn't want any hassle for you, and she doesn't deserve that. Tony's a fun guy, but . . . I guess he's used to easy women."

"You did the right thing. Thank you."

He hung up and went on about the business of getting ready for the day and worrying about his bowels. A diplomat with the trots: now, there was an inspiring concept.

It occurred to him to wonder why Cory had not just given the message to Philanth, since he must have had to go to extra effort to get his call put through as it had been. He decided Cory had hoped he would quash the matter so that Philanth would never have to know about it. Cory was a decent young man. Over and over he found reasons to be enormously proud of Peace Corps Volunteers.

As a group. Tony was not the right stuff, just as he had tried to persuade Carole. He would wait for the right moment to have a talk with Tony.

That reminded him: he had to have a talk with Philanth about keeping a low profile. Problems, problems, he thought impatiently. He had been right to want to travel with grownups. He decided he could at least make things easier for himself by passing Cory's tip on to Philanth so she could look out for herself.

After his working breakfast with Ablebody and his colleagues they were returned to Yundum Airport for their next flight.

The seats on the plane were in pairs, so he got Tony the luxury of two seats by himself. But he failed to prevent the too-helpful clerk from putting Tony immediately behind him and Philanth.

When he and Philanth settled down to wait for boarding Tony went off to inquire about foreign currency exchange rates and see whether he could get some "pocket money" for their next destination. So he took this opportunity to tell Philanth what Tony must have confided to Cory.

"You mustn't let Tony know that you know," he warned. "His behavior would deteriorate if he knew that we know he's that much of a creep."

"It's disheartening."

"I'm sorry. I wouldn't have told you, but I felt you would need to be on guard—especially since you were hoping to improve his attitudes in that area."

She nodded absently. *"Five hundred dollars.* Think of it! For a plain little ring Larry got for *seventeen* dollars. In Shakespeare's *Cymbeline* the 'proof' was a bracelet. —And discovery of a mole.

"D'you have any idea how it feels to know somebody thinks that way about your body?"

"Try not to let it get you down." She seemed to be getting more depressed and hurt as she thought it over, so he allowed her time to absorb the blow in peace.

But after the flight was underway he gave her the more important admonition he had planned: as she knew, he was on a strict timetable, and he had a heavy agenda. Because of their conception of good manners the foreign nationals he was dealing with were easily diverted from the business he had come to discuss with them. He would be better able to oblige his hosts to get down to official business if they had minimal distractions. If she would confine herself to minimal interaction with these people she would be helping him to accomplish the important business he had come to do. She should stay out of sight as much as possible.

He thought he explained it quite adequately, but she seemed to have trouble taking it in.

"No more of the diplomatic receptions, even if I just keep quiet and watch and listen?" She seemed to have trouble even framing the words. It wasn't like her to be so slow.

"I need you by the phone as much as possible, anyway."

"But in countries where the phones are bad?"

"You've still got to be available, however messages come for me."

Her voice rose a little. "I'm to stay all the time in my hotel room, except for staff meetings and going to and from the airport?"

"And be quiet even then."

"Even if there's only Peace Corps people inside the car?"

"Well, . . ."

She hung her head, taking a shaky breath. "I talked too much with Dr. Ablebody, didn't I."

He took so long to answer that she said, "That means, 'yes'. But—but I promise it won't happen again, Mr. Bjorklund!" He was taken aback by her distress, by her naked pleading. "I can be so much more useful to you if I get to listen to as many of your discussions as possible! And sometimes in the Francophone countries you need a translator."

"I think I can count on the in-country staff to take care of me. If there's an exception I'll let you know."

She bit her lips so hard that tendons emerged in her neck. "Pleeease, give me another chance." Her tone was abject, and she was becoming emotional.

"Philanth, I know you mean well, but your charm and conversational skill are not what's called for

337

here—not even enhanced by all your recent study."

She nodded. "I upstaged you. I'm so sorry." Her voice tightened and cracked. "I forgot . . . my place."

He didn't disagree.

Her voice not quite even, she said, "I know you didn't mind it personally, Mr. Bjorklund, and I see your problem. I'm terribly sorry."

Her lips curled with repressed emotion as she ventured, "But if I were to know in advance what your covert and overt agendas are, I—I could help you."

"You can't know that," he said bluntly, and she cringed. To soften this he added, "It's often a matter of instinct with me, as I feel my way through a situation. If I tried to brief you in advance you might misplay it, and that could do a lot of damage because of the nature of the things I'm doing."

"Because I'm too unfamiliar with your ways of thinking and reacting to specifics. And too inexperienced."

"No real harm in that, as long as you don't seriously try to take over *my* job." He attempted a smile as he said it, but he could see that this only made her feel worse.

Unbuckling her seat belt, she said softly, "Is there anything else, Mr. Bjorklund? I need to go to the rest-room."

He shook his head and let her squeeze past.

Tony's head and shoulders appeared over the top of the empty seat. "You really hafta do that?"

Tony, his attention snagged by Philanth's distraught voice, must have begun making a serious effort to eavesdrop.

When he did not reply Tony went on, "You realize you jus' took away most of what this trip meant to her?"

"Why—why, that's ridiculous."

"No, it isn't!" Tony was indignant. "Don't you know she's starved for the chance to talk to jus' the sorta people you spend all your time with? Man, even the little birds get to peck for the crumbs from somebody else's lunch! First you told her no sightseeing, and now this! What's she got left to look forward to?"

"Spare me, spare me! I should've known better than to bring along two staffers so green they'd let all the possibilities of this trip turn their heads. Next *you'll* be complaining if I tell you to stay out of the hotel casinos full of European 'chicks' playing roulette."

"Well, I hope you're not goin' to forbid me to try out some French restaurants when I get a chance! I'm *tired* o' West African food."

"No, of course not—if you get a chance."

The conversation had taken on an unpleasant tone, so he unbuckled his seat belt. "Excuse me. Might as well go while she's going too." He would have liked to spend the day on the toilet just for the peace of mind that would have given him. Tony sank back into his seat as though giving up on him.

He saw no sign of Philanth either going to or from the tiny washrooms. He resumed his seat and tried to rest, for he was still suffering from chills and urgency and did not feel at all like attempting to read.

After five minutes or so Tony stopped beside his seat and announced, "She's not in any of the restrooms. She's disappeared."

He got up, edging around Tony, and retraced his steps to the washrooms at the rear. Then he checked the ones at the front, as Tony probably had done. He went back and checked the cubicles at the rear

again, in case Philanth had been lingering somewhere between attacks of the kind he was having,
rather than risk having to edge back and forth in
front of him again.

Finally he stopped a stewardess and explained the
problem.

As soon as she understood, she looked knowing and
gestured toward someone reading a newspaper beside
a European man who was leaning against the cabin
wall, dead to the world.

He went to peer over the top of the newspaper.

Philanth was holding up one side of the newspaper
with one hand and using the other to cry into a tissue. The state of her face indicated that she had
been at it for some time. The instant she saw him
looking down at her she raised the newspaper higher,
grabbing it with the other hand to keep it from collapsing onto her head.

He went back to Tony. "I found her. She'll be
along when she's ready."

"Is she okay?"

"She'll be all right."

"What's she doin'?"

"She's having what is known as 'a good cry'."

"Congratulations."

When Philanth finally returned to her seat he said,
"You understand, Philanth, you've done nothing real-
ly wrong. It's just that our job has to come first.
There's no time for personal agendas."

"I do understand, Mr. Bjorklund," she said huskily.
"You can count on me."

"Fine."

But after a few minutes she began to weep quietly.
She took out her big sunglasses and put them on.

"I'm sorry," he said, though his tone said he could
not understand why she was being so emotional.

"It's happened to me before," she managed. "It reopened an old wound. I've been getting shot down for this ever since I was in high school. I should have learned by now."

He did not disagree. She blew her nose.

"I love to hear myself talk," she went on bitterly in a crackly voice. "I'm so clever. It's no wonder some people can't stand me when I start enjoying myself."

He was feeling so lousy he couldn't think straight, but he put his hand on hers. "You can always go back to teaching. Five hours of nonstop talking a day is surely enough for anybody; and proving you're one step ahead of forty adolescents at a time must be no mean feat."

"It's a waste of effort: the mass media make them too cynical and materialistic to respect constructive values. I'll have better prospects with illiterate peasants."

Then she burst into tears.

"What *is* it!"

"My . . . life. I'm . . . so . . . wasted. I ought to—to try to become a man."

"Don't be ridiculous."

"I know." She blew her nose again. "I've considered it thoroughly. I can't change my personality or my bones. So I'm not a good candidate for the procedures. I'd just get messed up and wind up a freak. I'm completely trapped."

"I should be offended because you complain of being 'wasted' just because I can't let you go on acting like a Nobel laureate doing a book tour."

"I was referring to my life."

"Give it a few years, Philanth. You'll find your battle."

"The sort of battle a woman is allowed to fight? Think about that! If I could kill myself and come back as a man like you, I swear, I'd do it in a heart-beat."

"Well, you can't. If you want to fight the sort of battles I'm allowed to fight, you can go on working for me."

"Thank you." For a moment she sounded sarcastic: "Thank you very much, considering the present circumstances."

She went on in a normal tone, "But women who can do what I'm allowed to do for you *now* are a dime a dozen—at least, they are in Washington."

She reached under her sunglasses to wipe her eyes with her fingers. She said no more, only blew her nose a final time and positioned herself to rest.

She had not even pretended to take his suggestion seriously. He could understand her reaction. Her personnel file showed that she had started doing secretarial work the year she left high school. It was hardly surprising if she already had had enough of it.

No position less than that of a senior senatorial staffer could keep her happy on a long-term basis. But Washington attracted so many bright people that it had liquor stores managed by people with multiple graduate degrees. Philanth had come to Washington, seen the situation, and decided to fulfill her altruistic ambitions elsewhere.

When the plane landed she kept her sunglasses on during the introductions and began following his new guidelines to the letter and in spirit, speaking pleasantly but briefly and prosaically when it was necessary to say anything at all.

Chapter 13

A DIVERSION

When he finally left the embassy reception he felt barely able to shower before he collapsed into bed. But just because the evening had been so full of business, his sense of duty took him to Philanth's door. The microcassette in his little hand-held dictaphone and his spare tape had been filled, and besides getting tapes which Philanth had transcribed and erased he wanted to look at some speech material and memos he had given her to work on.

As usual, she greeted him wearing her blue housecoat which became a dress when she added a belt. There were indications that when alone in her room she kept the dress at hand to throw on as she went to open her door. Thus she was always well covered when he showed up wanting to work with her.

Without having to say a word he gave her his full cassettes; she took them to her briefcase.

A glance at the piles of published material on her bed indicated that she had been going through periodicals.

"Goodness, you must read fast," he joked.

"I was just skimming and tearing out the good stuff, as usual. It's a fairly automatic process."

He saw that she had acquired some French- and English-language journals on African mining and agriculture. He looked more closely, fanning them on the bed, and found government or trade publications from several of the countries they had recently visited. It had become apparent that she had recruited Tony, who was always on the go, to collect such items for her whenever he had a chance. Tony was taking his role as her provider fairly seriously.

As she opened her briefcase she was continuing in an as-you-know tone, "The time during this trip is too valuable for in-depth study. We're just grabbing all the information we can get, to digest later."

He looked at the box of torn-out, stapled-together pages. "You're sending all this sort of stuff home?" She nodded. "Isn't that a lot of trouble and expense?"

She took out two microcassettes and shut the case. "Tony mails a packet of them to me in care of Marjorie whenever he's at a post office anyway, shipping home his souvenirs. He'll definitely be doing that in Nairobi, by the way, in case you want to do the same. It's surface mail, and the cost is worth it to me; and it's not time taken from our work, so it's nothing for you to worry about."

"But just out of curiosity, why do you collect stuff about . . ." Sitting down on her bed, he pulled toward him the now-familiar blue-and-white pasteboard box which had started its career full of new manila folders. "The Chinese economy, the ruin of soil by irrigation, chemical dumping in Africa by industrialized nations, air pollution in Mexico City— Oh, I see, they're mostly about ecology.

"But here's one about television programming in West Africa—well, that's understandable, but . . . here's one about palace politics at the White House."

His original assignment from Carole was becoming second nature whenever a new opportunity to "get to know" Philanth occurred. "Why are you saving something about the goings-on inside the White House?"

She did not conceal her amusement. "It fascinates me. I'd love to work there. Wouldn't you?"

He didn't bother to answer. Turning back to look further at the items she felt it worthwhile to ship home, he ran into a group of articles in French but deduced that one dealt with the European space program and one with exports from Ivory Coast. A feature story from an Arabic magazine had a picture of an oil refinery. Her being trilingual increased the quantity of the material she was gathering.

"Is there any one criterion for your ravaging every news periodical you can get your hands on?"

"Anything I save has to be of broad and enduring significance. I don't pay much attention to day-to-day skirmishes or trivial sensationalism. I learn a lot just by going through foreign publications, but when it comes to saving things I'm always looking for information about the basic, long-term patterns."

"Why?" he asked, openly testing her. "It's not relevant to your present job or your plans for the future."

"It isn't?"

"In case you hadn't noticed, Philanth, most people are selective about what they pay attention to. They don't go around collecting information like somebody trying to stuff food into his mouth with both hands, and reaching in all directions."

"I concentrate on fundamental societal and global trends and problems, like I said," she answered plaintively. "I can't help it, I need the overall picture: I'm anxious to see where I can fit into the general scheme of things. It's the secular equivalent of a theological framework, psychologically speaking, though I hadn't thought of it that way before."

He grunted.

"If you take exception to that, I'm sorry. But I'm sure it's not conceit or dilettantism, since I don't believe I really understand anything about how the world works. I'm fascinated because the motivations of the rich and powerful seem so alien to me. And I need lots of background before I feel comfortable with a subject."

"But how did you, with your parents' interests, ever become so involved in—well, in world problems?"

"I've been drawn to policy issues ever since I was eleven years old and Uncle Allie came to live with us. He was a political scientist, and he expressed himself beautifully. I hardly knew what I was eating when he had dinner with us—which was often. He let me ask questions, and afterwards he might go on talking to me for an hour just because I wanted to know all the things he knew. He taught me an awful lot."

He settled himself more comfortably. "Which of your parents had a brother who was a political scientist?"

"I just called him 'Uncle' because I liked him so much. 'Allie' wasn't his real name, either. It was short for 'Alligator'. That was a joke between us because once when I was upset about being persecuted by my classmates he taught me to say, 'Deny the allegations and defy the allegator.'

"He was such a sweet man. He said he'd gone into political science because the field seemed to be overrun by inside dopesters turning out new generations of cynics, and the Republic needed some idealists among its teachers of future lawyers and public officials.

"But back to your question, he'd met my parents at some faculty function, and they'd become friends. He was a paraplegic, and my parents had a specially designed addition built onto the back of the house for him, saying the extra income from a renter would be useful. Uncle Allie stayed with us for about five years. I cried when he got a job somewhere else and left us. I still have his picture and some letters, tucked away somewhere."

"That was a valuable friendship." He heartily approved of "Uncle Allie"'s influence on her, since it made his assignment from Carole easier.

Philanth nodded. She went on shyly, "But I realize now that he was the first man I was ever attracted to. He was a new professor when my parents met him, and he used to get into embarrassing situations at the university because he looked like one of the students. My parents were very protective of me, and I suspect they were afraid I'd . . . become serious about him as I grew older.

"Anyway," she added, glancing at her articles, "Uncle Allie gave me a lasting gift: he showed me that politics is where it all comes together, the values of the masses and the decisionmaking processes of the power elites. I've been 'tuned in' ever since.

"I started systematically cutting articles out of newspapers so I could use them in my teaching. By now it's a compulsion. I think of my brain as the hard drive and my files as the diskettes I can insert as needed; so I don't have to try to retain everything in detail, just become aware that it's out there and be able to retrieve it."

She threw him a self-conscious glance as she added, "Naturally I'm anxious to talk with people who know something about these subjects first-hand every time I get a chance. People speaking from actual experience usually give a different perspective than people writing for publication, and they make the subject matter more real to me."

She ran a caressing hand over her articles. "I don't believe my little hobby of pack-ratting ideas and information is interfering with my duties. There's only so much I can bug people by phone to make sure everything's on track for you before it becomes counterproductive: meaning, they get mad."

"All right. I have some more work to keep you busy."

"For a while. But I'm becoming more efficient as the work becomes more familiar."

"So you're cooped up here without enough to do?"

"I have plenty to read that I could read just as well in my bedroom at home, and I've got the rest of my life to do it." She replaced the microcassette in his dictaphone and gave it to him with the fresh spare cassette.

"I'm sorry," she added, "I didn't mean to sound impertinent. It was hard, missing the reception . . . again. So many people and places one only dreams about, so near and yet so far!"

"I understand.

"Philanth, Carole told me what your I.Q. percentile is."

"You know how many limitations and qualifications there are on the significance of that score!"

"Yes. But—"

"I always thought my I.Q. was classed as only borderline genius."

"So you're only a low-grade genius," he teased.

"And I'll never be recognized even as that because my aptitudes are all over the map. We live in an age of specialists."

"But you're top-of-the-scale on 'verbal'," he surmised.

"Yeah. Probably jus' 'cause I liked to read so much when I was a kid.

"My mom got disgusted 'cause she'd buy me a new hardback I really wanted and I'd polish it off in three hours. So I'd re-read it a few times to make her feel I'd gotten her money's worth. But I've always had good retention."

"You grades in college weren't anything special."

She looked pained. "I—I had to work part-time," she explained weakly. "But I still had to get through at the normal pace so I could start teaching before I ran completely out of money. With a little worse luck I could've gotten trapped indefinitely working full-time for subsistence wages, with no energy left for study. So I did without good grades on the courses that were harder for me. And some were so badly taught that—"

"What was the highest honor you received in high school or college?"

"Outstanding Graduating Senior, I guess—in high school. Merit Scholar, of course, but no money, because both of my parents were working then. The rest were minor. For instance, I was president of four clubs, and student body secretary in my junior year, and I won prizes in essay and speech contests promoting Americanism."

"Ever thought of going into politics?"

"Good heavens, no! When a people find the truth too unpalatable to elect those who tell it, all that's left for their politicians to do is preside over their decline and fall. You don't have to be very bright to be able to do *that*. I'd be wasted."

"It takes real intelligence to say what needs to be said and yet survive. Don't you find that a challenge?"

"I wouldn't survive. I lost the one office I ever really wanted, student body president, first to the captain of the football team and later to the cartoonist of the yearbook. Neither one of them could even conduct a meeting like he had good sense. But that's the sort of people who defeat serious-minded types like me every time. Besides, lots of boys wouldn't vote for a girl, even though I was obviously the best-qualified candidate.

"That sounds conceited, but I wasn't just the smartest girl in school, with far more experience in student organizations than anybody else, I was also terribly idealistic about student government.

"It's no wonder so many of my fellow adolescents couldn't stand me.

"Is the grown-up world so different? No! It's still the 'regular guy' who gets elected; and then it's a lucky fluke if he knows what he's doing more than half the time.

"I did accomplish one lasting service, though. After my second defeat I forced the new student body president to get the student body constitution rewritten, which was what I'd wanted to do in his place, just by threatening to mount a petition campaign for a recall election." She paused to examine her fingernails and then polish them against her dress in the traditional gesture of mock self-congratulation. "That was nicely engineered, if I do say so: I practically lost my voice on the phone that evening, but in twenty-four hours I got him to reverse himself in front of the whole student council!"

He laughed.

"Anyway, I'm not cut out for it in any respect, except for liking to organize things and keep them running smoothly."

"Unquestionably minor considerations. And there's the little matter of your abilities in oral and written communication."

"Yeah. I like to run off at the mouth about ideas."

"And you're good at dealing with people."

"That's just the way I was brought up. I believe very strongly that it's a serious offense against another person's humanity to be rude or arrogant or snobbish. 'There's no excuse for bad manners.'"

"That's fine, but my point stands."

350

"But I'm too retiring for politics." At his incredulous smile she said, "Really, I am! I'll only ride out in defense of a principle. I can't believe the vulgarity of the self-promotion I hear from candidates."

"It doesn't have to be that stupid. They get mindless from too much campaigning."

"Just the same, all that campaigning takes tons of money. Women who make it into national politics usually get started on some man's name and connections or money. Women get to be heads of state in other countries because they have different ways of selecting their leaders."

"Some women do get pretty far on their own."

"And I'm supposed to be one of those rare exceptions everybody makes so much of?

"Besides, women are especially likely to get started on local issues; and most politicians never really rise beyond them. You know the saying, 'All politics is local.' That's talking about human nature: it means few people care much about the big picture." She indicated her clippings. "As you can see, local issues aren't my kind of thing.

"But candidates who don't much care what interest-group particularists want, only about rational solutions in terms of the long-term, overall needs of society, normally get creamed. If they'd rather be right than be president they're sent packing. From the ivory tower they can advise the *real* politicians when their recommendations finally become popular."

"You might overcome the handicap of caring for the common good if you have enough understanding and instinct regarding politics. Your knack for pleasing everybody could be helpful, there."

"No. I'm not ordinary enough for voters to like."

"Being liked is only a matter of personality and image."

"Maybe. What I'm referring to is the fact that voters seem to *want* to be able to despise their leaders. It's healthy not to believe in a 'great man' who'll solve all our problems for us, but this culture has become *so* fiercely 'democratic' that we're being handicapped by our prejudices against intelligence."

"The voters really don't mind if you're sharp. They just want you to have good judgment and character, and it helps a lot if you're likable. You're almost *too* likable, which is why I grounded you."

"Interesting point." Her voice lost its hardness. "I've gotten used to thinking of myself as the Ugly Ducking. I was the one the teachers called on whenever they wanted to show the other students what was possible. Naturally some of the other students would have liked to peck me to death. As for *voting* for me, . . . !

"But in high school everybody was in direct competition on the same scale, so their hostility put me on the defensive. That's not entirely analogous to real life, is it."

"No. But somebody's always going to be hostile toward you if you're doing anything really significant, if only because they're jealous. It's a fiercely competitive world."

"It is in sectors that involve people's lower natures, like the drives for wealth, status or power. But in some fields they at least let you survive if you believe in cooperation rather than competition."

"Sometimes. Sometimes they let you survive. But that's not the same as being allowed to prosper.

"If you want to make a difference, sooner or later you're going to become a threat to *somebody's* vested interests, psychological if not economic. It doesn't matter *what* field you're in."

"Well, be that as it may, I can't imagine why you'd

want to persuade a woman without money or connections to go into politics."

"Couldn't you find satisfaction in doing high-powered work behind the scenes? For somebody whose agenda you really believed in?"

She began to smile in a lopsided way which said that she saw right through him. Considering her personal and intellectual qualities and all the work experience she had listed on her job application, men who wanted her to go on working for them must be nothing new to her.

Then she dropped her eyes and shrugged. "People would line up all around the Ellipse for an interesting job on Capitol Hill. If the job were also morally satisfying they'd probably be willing to kill for it."

He snorted, and she made those cute inverted U's at the corners of her mouth.

She went on matter-of-factly, "And they'd all be better cut out for it than I am. It doesn't matter if they're so addicted to power they can tolerate its chronic misuse. They're willing to sit through all the long-winded banalities at subcommittee hearings and do other idiot work just to be 'a part of the process', while I'd have all I could do not to walk out and go home and take to my bed, I'd be so demoralized by all the posturing. I saw enough of that in my work for the Hunger Foundation."

"Ever worked in a campaign?"

"Yes, I have: to fulfill a course requirement. The men sat around the headquarters talking politics while the women did all the work. Two women staffers got into a fight over which of them I was supposed to be working under: apparently they had a shortage of people willing to count campaign buttons from carton boxes into manila envelopes. The other woman had me manufacturing press releases about

353

local 'Committees for Candidate X' so his supporters could see their names in the newspaper. I can see a great future for myself in politics, answering the phone and typing addresses."

"But suppose you could work for someone you really believed in."

Again she dropped her eyes. "But doing what? The egocentric, competitive men who gravitate into politics treat women like furniture. I've seen it.

"And I saw in high-school club work that if a woman presumes to do anything men consider in their domain the men will only sit around watching her slave away and making snide remarks. I don't see that it's any different in politics, especially since small-minded, insecure men *and* women dislike women who are smarter than they are. All the polite public treatment of women in American politics is cheap veneer for consumption by the credulous."

He shook his head in discouragement.

"I'm honored by your interest, Mr. Bjorklund. But I find the spectacle of Washington politics profoundly disheartening. The physical beauties of the place are outweighed by its psychological squalor.

"That's hardly remarkable, when public officials have to be dumb or dishonest enough to satisfy voters who are so irrational as to assume that being American imparts the right to an affluent lifestyle regardless of marketable skills and number of children. As one cartoonist has put it, there *are* people smart enough to run the country, but they're too smart to go into politics."

"But you can't give up on something so vital as politics! It determines the course of your country, and your country affects the rest of the world!"

"I've given up on my country. It's lost the vision that made it great and gave it so much influence.

It's become so preoccupied with hanging onto what it's got that it's sinking under its own decadence. It's a great might-have-been, perhaps the greatest that will ever be."

"You can't say that! You mustn't!"

"I'm sorry. But trying to persuade most people to share their wealth in effective ways is a massive waste of time. They feel it's a big deal if they give away five bucks or get involved in some silly fund-raising activity that mostly just makes them feel good.

"If you can't hope to persuade enough people to make enough sustained, intelligently directed effort to accomplish something significant you have to go off and do what you can by yourself—or if possible, plug into whatever apparatus is already in place that does address the real needs."

"Like AID or Peace Corps."

"No. Like those *sectors* of those organizations really addressing root problems like high birth rates instead of taking mopping-up-without-turning-off-the-faucet approaches because they're politically safer."

"But any agency of democratic government may be required to serve conflicting goals. It follows that if you're working for an *organization* you may become disenchanted because it's doing so many other things with the resources you need for *your* objectives."

"You're so right."

"So I still say you should consider working for an *individual* whose goals you share. You only have to find someone you trust not to let you down."

"I trust only myself, and not even myself completely."

"But you'd have much more chance of influencing an individual you were working for than an organization."

Again she avoided meeting his eye. "But I can't rely on anyone to stay the course. Even if he wants to he may not be allowed to."

"But you have to trust *some*one, to *some* degree."

"Do I?"

"Yes! Politics is individual people—not causes, not abstractions. And even if you never cast a vote you entrust your interests to officials just by lack of opposition. Nobody can choose to be uninvolved in politics any more than they can choose to be uninvolved in the ecology.

"And women can fight the big battles despite the impediments you've mentioned—the way my wife does, for example." Too late he saw his blunder.

"Yeah," she said, compassionate even in seizing the advantage he had handed her. "Look at her. Sitting at home while you go around the world talking to scores of important people."

That was only one aspect of the inequity. He also was getting coverage in the world press and had accumulated a respectable resumé. What had Carole accomplished, even with her graduate degrees in journalism and psychology? Virtually nothing—just service to him as unpaid adviser and maid-of-all-work. If he came to nothing then she came to nothing too.

How could he reassure Philanth when he was worried about that himself? He had seen this flaw in Carole's plan from the beginning: he could not offer Philanth a ride when he could not be sure that he was going anywhere.

To keep the dialogue going he asked, "Have you got a better alternative?"

"My ultimate ambition would be to become a minister for family planning in some country that had a government strong enough and enlightened enough to get serious about the need. Some of the African governments have begun family planning programs."

"You imagine you'd encounter less sexism in a country where women are still basically chattel? Not to mention racism: being 'European', you'd have to keep as low a profile as possible. And you might also consider the sort of reputation American women have among men from traditional cultures, thanks to our worldwide export of movies and television. It will hardly help if you have no husband."

"Being a woman is like being blind: once you're sure the condition is incurable you're freed to figure out how to do the best you can with what you've got. If I'm kept down because of my gender and race I can at least work among women in ways men don't, in traditional cultures: birth control and education. I'll still find ways to help women lead more productive lives by controlling their reproductive lives.

"I need very little, so surely I can scrounge *some* sort of work that provides me with enough to live on. I've *got* to! Poor women everywhere are bearing childred they can't provide for because I wasn't there to help them get family planning services. 'The fields are white with harvest, and the laborers are few.'"

"But you're obviously a gifted person, Philanth. I don't see why you're prepared to spend your life as a foot soldier on the front lines. Why should you be wasted on mindless work under harsh if not dangerous conditions?"

"It *takes* a certain amount of intelligence and education to see what really needs to be done and preserve the motivation to do it, most especially when the risks are serious and the material rewards are few. D'you think it's the *dummies* of America who keep getting killed being do-gooders in foreign countries? *No.* The *dummies* are *safe at home*, watching t.v. or cruising the local shopping mall."

"You'd soon find it boring to trudge around in the heat and dust, trying to motivate illiterate people to act upon concepts which nothing in their experience has prepared them to understand."

"I don't expect to have to work at such levels forever. Even if I can't move up so far as to become a minister of family planning, I'll make *some* sort of contribution. Because every child not conceived only to die means more resources which can give those who *are* born a better chance to fulfill their potential as human beings. Accomplishments are not less great when they're not measurable."

"But you must see that everything you want to do depends directly upon politics: somebody has to create and fund even the sort of positions you want to hold."

"Not necessarily. Grass-roots, self-help organizations are also possible. And in any case I'd be able to operate on the basis of ability rather than electability."

Before he could tackle that point she added, "When there's so much to be done in family planning and the areas needing birth control worst can least afford salaries for personnel, why should I work in politics, where there are plenty of bright people eager for any job that's available? And that's where we came in."

She looked at her watch. "You're going to be *so tired* in the morning, ..."

He had been hurting his cause by pressing it without having laid necessary groundwork. "All right.

"Let's see, I wanted to look over some things I gave you to work on. . . ."

After he went to bed he resumed thinking about their argument. It had repeatedly made him realize that he could hardly ask Philanth to throw in her fortunes with his when she could not know him as well as he—having been sensitized by Carole's briefings—now knew her.

Moreover, in the normal course of events this situation would not change, since she would not presume to try to get to know him better.

But whenever they were together she was attentive in ways which told him that she was intensely interested in discovering more about "the way his mind worked". To her he must seem to possess the integrative perspective she was laboring to piece together from her reading.

If he satisfied that curiosity she would see the extent of the congruence between their concerns.
The she might accept the idea that her agenda was subsumed by his.

Most important of all, she might come to trust him—not only in terms of political priorities but also at the moral and psychological levels—to "stay the course". Only then could he reasonably hope to interest her in continuing to work for him rather than pursuing an overseas career which would allow her to unleash her energies for the goals she had chosen.

Once she was content with the general terms of their symbiotic relationship she might come to allow his observations to shape some of her still inchoate thinking about specific political questions and adapt her objectives accordingly in those areas as well. That was the ultimate goal which Carole had implicitly set for him.

But reprogramming a missile with such a highly developed guidance system might prove to be a time-consuming operation.

He began to take time to chat with her while they were traveling, and he encouraged her to ask questions. She responded carefully, but her timid eagerness to learn his opinions flattered him.

He began to amuse himself by showing a reciprocal interest in her thinking and discovered that she had a self-perfecting code which required her to answer his every "penny-for-your-thoughts" question thoughtfully and with unsparing honesty. Like Carole, he found her willing to serve as a specimen for analysis.

From then on it was easy for him to get to know Philanth as thoroughly as he wished. For whenever Africa made them wait, sometimes for hours and especially at air terminals where conditions for their work were poor, Philanth was at his side: awaiting his pleasure, humbly honored by any indication of his personal interest, and ready to try to oblige if he seemed to wish to be amused.

Her mental office he already knew. It was full of filing cabinets devoted to most of his own interests, especially to the same tragically factual materials about the world's poor as the briefings he studied. But the parlor of her mind was a cozy, homey place, bric-a-bracked with humor and stocked with a wealth of beloved literature.

When one particularly tedious wait for a long-delayed flight made him irritable she resorted to telling him one of the stories by which Chaucer's pilgrims had entertained each other to pass the time during their journey to Canterbury: a poor peasant girl named Griselda had found favor with a powerful lord; he had had her dressed in beautiful clothes and married her; then he had subjected her to years of cruel testing, and she had accepted it all with humility.

Thanks to their long previous discussion of traditional values Philanth did not need to tell him why this "Parson's Tale" of "The Patient Griselda" was her favorite: she only noted that while Griselda would be denounced as a doormat by modern standards, she had embodied all of the Christian virtues as they were understood in that period. In the manner of her telling the tale he saw that Philanth not only understood Griselda, she felt that she would have been able to emulate her. Her soft words wove a soft net around his heart, for her shy delivery communicated her emotions even as it strove to conceal them.

She must have suspected that he was moved. But she was careful not to acknowledge even implicitly that there might be anything emotionally involving to him about what she was telling him. She left him free to scorn her and her quaint, old tale as much as he pleased. Overtly she was only responding to his request for entertainment less demanding to a tired mind than reading.

She was not afraid of anything which might develop in their personal relationship, since it would be kept as pure as her own intentions.

Because Philanth continued to measure up to the advance billing Carole had given her, ventures into the private chambers of her mind became his reliable diversion in the spare moments which punctuated their constant travel and broke the tension of their work. It was more than an amusing diversion. It was spiritual recreation because of what she had to share.

Tuning into Philanth's inner life allowed him to more fully understand her childlike qualities and mannerisms which at first had been so misleading. He saw that they served as an ingratiating social camouflage. But she retained them also because preserving the wonder and the acute sensations of a child was a duty she embraced: the duty of a sentient being to live fully, rejoicing in the privilege of its own existence, if only out of an obligation to help justify the aeons of suffering and struggle out of which it had emerged. Because her strict moralism and her fundamental independence of the world's conventional threats combined to keep her at peace within, she felt free to keep her consciousness open to the fullest enjoyment of every pleasure of the mind or senses which her work permitted.

Contemplating her readiness to enjoy the simplest pleasures, he caught himself imagining her in the heights of pleasure, moaning with ecstasy from the attentions of a lover. He thought of her loving disposition joined with the intensity of her central nature, investing them in the pliant grace and pyrotechnic responsiveness of a woman's body—and quickly pushed the images away.

Yet they lingered. They began returning at inconvenient moments. He was beginning to react to her in distinctly sexual ways.

This worried him. He wanted to talk about the matter with Philanth, even though it had to be done carefully. Whatever his worry, her perspective could always comfort or reassure him.

As they were concluding one of their regular late-evening work sessions he asked, "Did you ever destroy your copy of the letter Carole gave you?"

She did not look up. "You ordered me to, so I did."

There was a pause. Then she looked at him.

"Mr. Bjorklund, why did she do it?"

In those five words Philanth was dropping into the wastebasket their previous discussion of the subject. His first explanation had not stood up to examination. He was not surprised.

He sighed, raking his fingers through his hair as he realized that he had to be careful. "She just wanted me to see for myself how you'd react." His routine tone reminded her that this was old news. "Then I'd feel you were more worth training."

She appeared to ponder that. Then she said humbly, "I promise to use what you're teaching me as well as I can, wherever my work takes me."

He was a little ashamed. To Philanth, a teacher and the daughter of teachers, knowledge was to be given freely to anyone who desired it, for the benefit of humanity. He did not want to point out that he had spoken in terms of her long-term service *to him*.

After a moment Philanth added, "But there's something wrong here."

To find an appropriate evasive tack he asked, "That explanation still doesn't satisfy you?"

"It's not that I don't believe your explanations. But there has to be more to this than you've told me. It would have been out of character for Carole to create such an extraordinary document for such a routine and unimportant purpose. And I still don't see what it was really supposed to be testing."

"I can't tell you any more. I can't even be sure she would have told *me* anything about that letter, if I hadn't confronted her after finding a copy."

Philanth looked dismayed. "Really? Then how could it possibly have been the test you told me it was?"

She was too sharp, and he was too tired and unprepared. For want of a better answer, he growled, "Use your imagination."

"Oh, dear!" Philanth muttered, collapsing to lean on her knees like Rodin's "Thinker". "Oh, *dear*! Could it be that I was really supposed to be tempted to show you that letter? That's sickening! Even if it was a test of my integrity and judgment, I can't believe that of her!"

"I can't either," he lied.

Philanth's voice rose as she went on, "This could be a trap for both of us! Whatever she intended at first, she could've completely changed her mind! She could be planning to ruin you! Mr. Bjorklund, has there been any indication of mental illness?"

He didn't want to deal with that. "We're jumping at shadows."

"Yes." She took a deep breath and brought her voice under control. "I'm getting hysterical. This is my good friend Carole we're talking about: probably as sensible and loyal and devoted a wife as any congressman ever had. I'm verging on paranoia just because I've seen how completely unscrupulous and vicious some 'nice' women can be, even to their friends who trust them. *I'm no threat to her!*"

He said nothing.

Philanth sighed, shaking her head. "It does seem as though she's playing with our minds. But to what purpose?"

"I'm sure her intentions were good," he said half-heartedly. "She probably intended for me to find out about the letter."

"But you just said almost the opposite," she flared. "Why, then, . . . ?"

He sighed emphatically. Then, seeing that this discussion had gone out of control due to his blunders, he thought quickly. He tried to look at the situation from Philanth's point of view.

He saw that he should have been downplaying, not reiterating, the relationship between Carole's letter and his interest in Philanth's future. For if Philanth examined the clue he had just given her about his being motivated to train her and realized that Carole's ultimate objective in writing the letter had been to get Philanth enlisted in the service of his career she would also realize that he had in fact begun the recruitment effort. The more deeply she considered the matter, the more an appearance of connivance and manipulation could undermine her trust and respect for him. Yet cultivating that trust and respect was essential to the whole enterprise.

The only protection remaining to him was her inability to imagine that he and Carole could stoop to what they were in fact guilty of.

At least Philanth had no way of figuring out a second bit of the information she knew she was missing: she would never suspect that Carole had been refusing to sleep with him. That really would frighten her. It would suggest that Carole's preposterous instructions could be taken at face value, and that put the entire recruitment situation in an ominously different light.

Since he could think only of things she must not learn or suspect or figure out, he would have to let her remain mystified. He was pretty mystified, himself.

He *was* beginning to be certain that Carole had put the carbon paper back into the drawer quite deliberately and therefore that she had tried to set him up to find it by not sending the flyer to the caterer as she had promised and making sure the copier would not be operational that day.

Had she wanted him to stop her before she gave the letter to Philanth? He *couldn't* have stopped her if she hadn't *wanted* him to, because if he *had* confronted her before they went to the party she could have pretended that Philanth already had the letter.

Two alternative hypotheses suggested themselves: first, his wife had become mentally ill. He could not believe that, since he could think of no other irrational elements in her behavior.

Second, Carole had wanted to plant the idea of making love with Philanth but provide a cover story which would keep him off guard while the suggestion worked its way into his mind. If that were so, she had had to actually give Philanth the letter and indirectly let him know about it.

The more he thought about that second alternative the more possible implications he would be able to derive. He did not want to pursue that line of speculation even within himself.

He shook himself. "We're making too much of this. Carole knew I resented having to take somebody on this trip when I knew very little about them. You were her candidate, so she wanted me to develop a good opinion of you. Whether or not I was supposed to find the copy of the letter doesn't matter, since if I hadn't found it she would have had plenty of time to tell me about it after she gave it to you."

"All right. But there's nothing else you can tell me? There *must* be *some*thing else she had in mind. If you don't know what it could've been, please at least tell me *that*!"

Not looking at her, he asked wearily, "What does it matter?"

Philanth's voice was anguished. "I need to know all I can in order to figure out how to handle this situation so as to do the best I can for both of you."

"What's to 'handle'?"

"Oh, spoken like a man! There are all sorts of options."

"I'd be interested in your views."

Her voice and gestures reflected her effort to work it out. "If I failed her test by *following* her instructions, presumably I'd be doing something she had no serious objection to. That surely seems bizarre. But she trusts you, of course, so there'd be no serious—"

"Not necessarily," he interrupted as a new idea struck him. At once he regretted his outburst, for Philanth turned to stare at him.

366

He had meant, "Not necessarily bizarre" because Carole no longer wanted him herself. As for Carole's trust, that might no longer be relevant.

But he could not explain what had just occurred to him, the thought that because of Carole's sexual aversion to him she might actually be willing to let Philanth relieve her of the pressure he had begun to bring on her. Carole could have reasoned that there would be nothing like a guilty conscience to make a man shut up about his conjugal rights. In the letter Carole had even implied that she would countenance his continuing the proposed sexual relationship after he returned home. Therefore it would not be bizarre for Carole not to object to Philanth's following her instructions.

Philanth was waiting for him to explain. Since he wasn't about to enlighten her, he waved the point aside.

Philanth gave him a puzzled frown. But since she could not properly ask why Carole didn't "necessarily" trust him,—the construction she would have put on his interruption,—she obediently returned to her oral meditation.

"*Of course* she trusts you. So she would have no objection if I did make a fool of myself by—by . . . offering myself to you."

She shook her head. "But I still can't believe she could imagine I'd be capable of such a thing. I'm stumped. None of it makes any sense!"

"Well," he said reluctantly, trying to give her *some*thing so she would let go of it, "she knows I've been depressed about various things lately. This was just a diversion, to get my mind off onto some little fantasy that might have suggested itself anyway."

"Ah." She almost smiled, clearly not resenting the implication. "Well, a person *can* get quite low, when they're worked to exhaustion, and sick, and the world seems to be full of wrong-headed people."

Shifting as though about to get up and break off the conversation, but using a bit of the old charm to distract her, he murmured, "Your being around will keep it from seeming *quite* full."

She smiled, laying a hand on his jacket sleeve. This characteristic risk-taking to express innocent affection was under the circumstances particularly daring, though he had seen that at the last moment she had chosen to touch his sleeve rather than his hand.

"Please don't get too depressed, Mr. Bjorklund," she said in her own ingratiatingly intimate way. "I mean, I'm not on the pill, or anything."

That made him chortle—especially the "I mean" and "or anything", which she would have avoided except when intending to be funny. She was deliberately playing the round-heeled, air-headed blonde. She still was so sure of herself as to consider this aspect of the matter a joke.

She hadn't let his suddenly unleashing some medium-voltage blarney fuse her mental circuits, as some women did: she had kept herself cool enough to frame a jest which was for her rather daring. Even in the charm department, she could hold her own with him.

Smilingly he promised, "I'll try."

Chapter 14

A WONDERFUL NURSE

He was gripping the arms of his seat and keeping his eyes shut. The plane shuddered and dipped some more, and for several moments his stomach seemed to remain up where the plane had been a moment before as though it floated unattached in his body.

He had the thought that if he died in the next ten minutes that would at least end the problems in his bowels.

Something soft settled lightly over his left hand, making him visualize a small bird spreading its warm wings protectively over his cold fist. In the next instant he knew it must be Philanth's touch, shy but embracing. His eyes flew open to stare at her in surprise.

Leaning toward him, she breathed, "You okay?"

Afraid the plane's bucking might make him bite his tongue, he answered through clenched jaws. "I've flown a helluva lot, Philanth. The law of averages may catch up with me."

"You know it doesn't work that way." But her hand freshened its light embrace.

"Doesn't this bother *you*?"

"The pilots must be used to the violent weather in these latitudes."

"People die *because* people are 'used to' taking chances." He took her hand and examined it, glad of the distraction from the alarming shaking of the aircraft. "You really don't seem afraid."

"I'm not important, one way or the other. But you are."

"Yeah, I'm so important." The plane dipped, and he hastily released her hand to renew his grip on the armrests. He quoted his favorite cartoon caption: "'Who will look after the world after I'm gone?'"

As he spoke, the plane lurched sickeningly. It righted itself unsteadily and began to climb. The turbulence abated. Under his breath he muttered, "Why didn't he do that in the first place?"

Philanth smiled.

He waited for his stomach to settle, resisting an impulse to press it into place. There still was some roughness, but they still were climbing, and it was getting better.

He was glad Tony had gotten himself a pair of seats elsewhere in the cabin. Irritable from his lapse of dignity, he asked, "Do you enjoy touching me, Philanth? Did you welcome an excuse to touch me while the plane was shaking?"

"I hope you will accept the gesture in the spirit in which it was made."

He started to feel guilty. It had been about a week since he had made her cry during the flight from Banjul. Once chastened, she had become a model of reserve without diminution of her deferential conscientiousness.

Meanwhile they had touched down in four capital cities, and in three of these he had conducted multi-country meetings. Through all of that unremitting work her diligence and polite equanimity had been unflagging.

"I apologize. I'm not feeling well."

"It's all right. I'm sorry you're having problems." Philanth returned to her reading. After a while they descended into yet another modest airport.

He stepped out of the aircraft into the climate of a tropical rain forest. After the usual handshaking, picture-taking, and distribution of press kits he said goodnight to the country director and host-country officials who had formed the welcoming committee.

For the past two days his hours for rest had been squeezed by airline schedules, so his planning committee had granted him this evening in which to recuperate if he happened to be on schedule despite the vagaries of the airlines. Early the next morning he would meet with his hosts to conduct their business. Meanwhile he was fixated on the thought of a hot shower and a good night's sleep. Indeed, he had been pining for them through most of the day.

As he climbed into the car to go to his hotel the driver handed him his latest packet from the daily pouch from Washington. As the car moved away from the group of sheds which served as the country's international airport he flipped through his mail, picking out the interesting-looking items.

He had a letter from Carole; he slipped it into his breast pocket. Tony had an envelope which had come all the way from Chicago and smelled of perfume. Tony soon was absorbed in it.

"Sorry, Philanth," he said, "the postman seems to have passed you by this time."

"I wasn't expecting anything." She was watching the passing landscape, trying as usual to absorb all of her new surroundings at once.

He opened and scanned the cables first, then the memos from Sandy and Courtney. They were not happy with each other. Then the memo from Les. Les was unhappy too. All very minor matters, basically just back-biting due to Courtney's and Les's oversized egos. Then a memo from the deputy director. Damn! The old fool was starting to meddle again. He felt like abandoning his schedule long enough to fly home and knock some heads together.

That would make headlines all right, besides messing up things for a lot of good people who had been trying hard to put this trip together and then keep it from falling apart.

A stench made him look up. They were passing through squalid shanty towns, and he needed only his nose to tell him that the sanitary conditions must be deplorable. He also caught the smells of dust and wood smoke as he hastily rolled up his window. The others did the same even though the car already was stifling.

The driver began muttering apologies and excuses about the delay in getting parts to fix the car's air conditioning. He gathered that the car had been waiting at the air terminal in the late-afternoon sun for the two hours that his plane had been late in taking off.

It was safe to assume that his reception committee had not been able to call the airline ahead of time and find out about the delay. They had been in one country where people stole phone lines out of the phone system, so walkie-talkies, cellular phones, or telegrams were used instead.

Two hours in the baking heat of that tin-roofed shed explained why the Americans greeting him had seemed tired and edgy, despite their obvious determination to make him feel welcome. The country director, a Nordic type like himself, had looked beat.

But nobody in his welcoming committee had complained, just said they were sorry he had been delayed. Delays and hardships were all in a day's work, in a developing country.

Philanth leaned forward to say in a low voice to the driver, "We're coming in from the northwest side of the city, aren't we?

"Yes, ma'am."

"I hope nobody was offended because I questioned putting us in the best part of town instead of one of the planned zones."

"No, we understood your concern. You couldn't tell which hotels were where."

After a moment Philanth went on, "I understand the roads connecting the airports to these capitals normally don't let visitors see what most of the cities are really like. I *thought* this would be an exception."

"It wasn't planned this way!"

"No, all this sprang up because this is the only city in the country with urban amenities. And this side of town was dictated by the geography.

"Which section of town do the Peace Corps staff live in?"

"They're better off than what you see here, that's for sure."

He saw that Philanth had indeed prepared herself for other countries besides The Gambia. Apparently she was hoping that since they were alone with one of their own people and he was busy with his mail, his post-Gambia injunction might not apply. She was keeping her head and voice down in case she was mistaken.

Still leaning forward and in the same confidential tone she went on, "From what I've read, it's a wonder anybody here is still on his feet. With no sewer lines these shanty towns must be filthy during the rainy season."

"Yeah." The driver finally opened up. "Health conditions are worse here than in pre-colonial times. A long list of diseases are endemic, including all kinds of v.d."

"Of course, if the appalling mortality rates were curbed, the population would explode. This wouldn't be the first government to think of that and let nature take its course while it attended to other matters."

"I suppose not."

She went on, "There's so much else that would be necessary in order to institute birth control in tandem with death control.

"Given reasonably adequate medical supplies, a doctor could work herself to death here in short order. I can never understand how doctors in our country can endure knowing how much more needed they are in countries like this."

"The pay is incredibly low, and the conditions are lousy," was the matter-of-fact reply. "As you'd expect, trained personnel are in desperately short supply. Even the government hospital is unsanitary. And away from the towns, brew some herbs or forget it, unless you're near a dispensary run by one of the Christian missions."

"And the diseases of malnutrition are getting worse," Philanth added.

"As well as soil degradation and erosion. Some of the best farmland in Africa has been ruined here by overcropping. Rather than starve, people come to the city and take their chances.

"Have you seen the U.N.'s estimated figures on unemployment and on teenaged unmarried women with children? Among the unemployed urban young there's a lot of hopelessness and growing use of alcohol and drugs. But Americans who say, 'We have the same problems here' don't know what they're talking about! They haven't seen that the spectacular disasters of Africa are only the tip of an iceberg.

"Here, for example, the whole social structure is in desperate trouble: far too much dependence on the government, and a growing parasitic urban population that can topple governments.

"I could go on, but I guess you hear the same sorts of things at some of your other stops. Only, the well-known problems of Africa are all together here."

"We hear that too," Philanth told him gently.

"Well, do you also hear that we could get forced out and all our efforts here come to nothing because our aid is too little too late?"

"No. But that's happened before, in other places.

"Tell me: d'you think if we did put in the needed resources we could turn the tide here?"

"What's the point of asking that, since we won't!"

"Just as a basic policy question," she urged in the just-between-us manner which was drawing the young man out. "Humor me."

"Okay. The capital is only one city in a mess. The country has rich resources."

"Which is more than can be said for countries in less desperate circumstances."

"But . . . ."

"But the country's in a Catch-22," she supplied. "Again, nothing new: unrest impairs ability to attract investment capital on terms that will allow exploitation of the resources so as to develop the nation's economy and benefit its own people; the resulting discontent breeds more unrest, which scares away capital, and so on."

"Healthy natural economic growth from foreign *or* domestic investment is pie-in-the-sky. We can't lay the groundwork for enough stability to attract investment even through *direct aid*."

Philanth leaned closer to the back of the seat, if possible. "Why not? I know that's a complicated question, but where do you see the real difficulty?"

"Personally," the young man said as though confiding a terrible secret, "I think the worst problem is our own country's lack of will. Not trying to find out which countries can be saved by effective intervention, we just putter along and let all of them slide toward deeper indebtedness, worse poverty . . . ."

The car paused for a group of boys who were playfully nudging each other into its path.

The driver looked back over his shoulder as though hoping to find him reading his mail, taking no interest in the chatter of underlings.

"Go on," he urged. "This is what I came for, to hear what everyone really thinks." But there was an awkward pause.

This was hardly the first time he had caught a glimpse of demoralization among country staff. In fact hundreds of people in Peace Corps must be bursting to talk frankly to him. But in spite of all his nice speeches and printed "Messages" in Peace Corps publications, he was a stranger to them.

And they were not just shy or in awe of his position or sure he was too out-of-touch to care about their viewpoints. Sometimes they were afraid.

That shouldn't be. The Peace Corps director was supposed to be every Volunteer's pal. It was part of the tradition.

But the driver was no longer a Volunteer. He was on the in-country staff, and many country staff were likely to want to go on to careers in international development or the Foreign Service. They would feel it was foolish to risk offending anyone in his position; and people tended to be offended by unpleasant truths, especially from people they did not know very well.

Moreover, according to the oral history of Peace Corps' internal affairs which Sandy was picking up, the staff had specific reasons to be afraid. Past directors had made important personnel changes based upon superficial contacts with in-country staff during their travels, and results for Peace Corps had been disastrous.

His visit was supposed to break down such barriers, to let in-country staff know that he was not just another political appointee foolish enough to punish anyone who delivered bad news or promote someone merely for being adept at putting up an impressive front. He refused to treat Peace Corps as if it were entirely an exercise in public relations in which everyone would find it personally expedient to act as if everything were dandy.

"Do you think," Philanth said, "that if our own people *could* be inspired by, say, a charismatic president to put their own wants into a long-term, global perspective and make the kind of commitment it would take to help the *most promising* poor countries become self-sufficient, that policy could possibly cause *this* government to get its act together?"

"I don't know what you mean. A government's 'getting its act together' is a difficult business, when its one urban area is loaded with a young, volatile bunch of welfare cases and it can't even provide them with basic services like water . . . and the military needed to maintain order can always switch sides."

"Right. But take that as an example of what I'm asking: are the military still getting too much competition from the politicians for the best mistresses, so that another coup on that basis might occur?"

A short, bitter laugh burst from the driver. "Your information seems to be better than mine. I don't move in those circles."

"I just found it in a book."

"You mean there's *a book* about this place?"

Philanth ignored the sarcasm. "It was an old book. Books about Africa are hard to find, so I take what I can get. You're on the scene, so you hear things.

"Do you hear complaints about rampant corruption and budgetary crises? Do you hear politicians playing ethnic groups off against each other?"

"The Peace Corps doesn't deal in those matters," the driver reminded her unnecessarily. "And the local press is muzzled, while the international press doesn't bother with places like this except when the people start slaughtering each other. So who knows?

"The Peace Corps briefing paper I've seen talks in terms of 'cordial relations between our two governments' and 'the host government's high regard for Peace Corps' work in the rural areas'. People back at headquarters probably reason that if we find out much about the larger picture one of us might start meddling in the local politics and get Peace Corps thrown out of the country.

"I try to mind my own business. I've still got my hands full mastering Sango. My previous assignment was in another tribal language area, so I'm having to struggle to get by on French.

"Ask me about water catchments and sanitation systems.

"I hope you're all well protected against malaria, by the way."

"I hope so too. I understand a new strain came out that was a killer."

"I trust the current vaccines are equal to it."

"I trust so too."

"Too much hope and trust can get a body into serious trouble."

"Yes. But please," she said in her most ingratiating voice, again cozying up to the back of the driver's seat, "tell me all you know about water catchments and sanitation systems. For starters, how do you—"

Nudging her elbow with his handful of unopened mail, he said, "Before you get into that, Philanth."

"Yessir," she said, taking the bundle. To the driver's ear she said, "Please excuse me, duty calls."

"Sure," was the warm reply.

As Philanth began going through his less-interesting official mail he decided he need not wait until he was alone in his hotel room to read his letter from Carole. He pulled it out of his pocket and opened it, thinking about how he missed her dry humor, her thorough knowledge of him, her quiet, practical companionship. If only he could have kept his simon-pure public image safe while keeping her officially working for him when he moved from Congress to Peace Corps! Then he could have brought her with him as his assistant on this trip. That would have been the ideal arrangement.

But that was a pipe dream. Too many other officials had traveled with wives who were not thoroughly qualified to pull their own weight. Reporters could be so nasty, always trying to prove how smart they were. Yet they could be unbelievably lazy and careless about the facts even when they had plenty of sound information readily available or already in hand. Politics was by no means the only field in dire need of more conscientious practitioners.

The general shortage of ability combined with character explained why Carole was so interested in his recruiting Philanth. In the way Philanth went about her work for him he saw the open-mindedness and systematic thoroughness of a trained scholar, a pattern of thought she must have absorbed from her parents. Her unusual cautiousness in developing opinions about most subjects gave weight to her militance regarding the few—notably overpopulation—in which she felt completely sure of herself.

Carole's letter began with exultant remarks about how the President's re-election had made the gamble of his leaving Congress for the Peace Corps "a viable proposition".

But she had been leading up to breaking some bad news. A slight choked sound escaped him as he grasped the import of the sentence before him.

"Is something the matter?" Philanth asked.

"Nothing, nothing." He didn't want to betray his emotion, especially in the presence of the staffer he had only just met. "Some bad news about some legislation." He could have added that it affected the future of Peace Corps, but that would have opened up a general discussion which he definitely was not up to at the moment.

Philanth asked their driver to take them by a famous monument in the center of the city before driving to their hotel. He regretted that he had not made her aware of his latest preoccupation with getting quickly from close proximity to one toilet to close proximity to another.

The memorial was in a park setting, so he asked the driver whether the park had rest-rooms. The driver took them to a bar favored by Europeans and accompanied him as he slipped in.

After that he was in a somewhat better frame of mind to enter into the spirit of the occasion. The late-afternoon light was failing, but Tony thought his camera might be "smart" enough to compensate. So they got some shots of him paying his respects to the country's founding father.

As they stood before the statue Philanth noted that this statesman's death in a plane crash had been a heavy loss because his weak and divided new nation had been in the process of being born, and this man alone had been the unifying influence needed. Many of the disasters which had plagued the nation since then might have been avoided or reduced if this one man's life had not been cut short at a critical moment.

The lonely figure in a business suit, framed in the simple modernistic structure of the memorial, hastened his plunge into depression. Thanks to his latest news from Washington, he felt all too deeply the tragedy of high aspirations mindlessly destroyed. Looking at the image of his fellow politician in the fading light, he felt his own mortality loom larger and darker.

A little later, when he sat down with Philanth and Tony for a quiet meal in the modest hotel dining room, he at once decided, "I can't eat anything."

"But you must! Please tell us what's wrong," Philanth entreated.

He was perspiring visibly, but he still would not discuss his intestinal problems. "The House failed to override." Even now, when the initial blow was no longer fresh, he had trouble putting the situation into words for anyone who knew nothing about the staggering amounts of effort which had been tossed away.

But it turned out to be possible to explain, since Philanth remembered his referring to it in the car with Ablebody, and Tony wasn't much interested.

He couldn't help mourning openly. "The only real accomplishment I would have had to show for seven years of work, the one truly significant thing I hoped to do for the space program as well as our foreign aid program—" He smacked his open palms together as though crushing and brushing away something which clung. "It's dead." He leaned on the table, his fist against his forehead. "It's like losing a child, a beautiful, wonderfully promising child."

He straightened angrily. "The *work* that went into that thing! Crafting and negotiating ways to mesh the functions of all the relevant branches of the agencies affected; getting a reconciliation among the views of all the people who really knew what would be involved and were ready to testify for it if it satisfied them. A special team of mostly borrowed staff, including some real crackerjacks from research institutions and the best legislative brains I could gather together. I directed them myself during their most productive phases. Months and months *and months* of consultations and negotiations and hearings. Also countless revisions and refinements and the lobbying of other congressmen and senators over the years as we got more of them on board as co-sponsors."

"But what was it going to do?" Tony asked.

"It set up a program which effectively integrated and coordinated aspects of our satellite program with aspects of our foreign aid program. You *can't imagine* the technical and organizational problems that had to be worked out!

"It provided for needs of equatorial countries while also fostering development of our high technologies needed for our trade balance."

After years of practice the familiar prose began to flow again. "It would have set up much better organizational and funding mechanisms which would enable the poorer nations to obtain fuller use of geosynchronous satellites. Those satellites would build up their communications networks, improve educational broadcasting, and enhance local weather prediction. The livestock and property that could have been saved in the monsoon seasons alone could have justified what we put into it!

"And building on data available anyway through UN agencies and the offspring of LANDSAT, it would have aided their governments' planning: for example, in developing agricultural policies. It would have given them more sophisticated ways of deciding what they needed from Peace Corps and Oxfam and all the other international aid and development agencies.

"Don't you see, there was no limit to its potential applications and effects—even just as a precedent for other legislative initiatives later on!"

Frustrated by Tony's vaguely sympathetic expression, he wound down. Philanth was resting her mouth against her fists, her elbows on the dining table. They sat around in silence.

For the past year, as Peace Corps director, he had continued to push for the NASA-AID bills, making submissions to the Senate Foreign Affairs Committee and the House Foreign Relations Committee—both of them, not incidentally, oversight committees for the Peace Corps. For months at a time that effort had been Mark's full-time job; indeed, it had been the main work of the Congressional Relations Office.

He had likewise gone on trying to lobby OMB and anybody else who might be interested, showing how directly these bills' contributions to mass education, crop surveillance, and weather prediction were tied to the Peace Corps' labors in the field and to the United States' long-term interests in the world. The Peace Corps' argument had been that countries able to feed themselves were less likely to succumb to takeovers from the right or the left, and that instability due to corruption and oppression would in the long run be reduced by the increasing sophistication of populations reached by satellite broadcasting. Relevant passages of John Kennedy's speeches had echoed in his mind as he worked on drafts of his memoranda and presentations.

Despite cross-examination by those who thought the governmental apparatus required to administer the new plan was too complex, he had stood his ground; and on both the House and the Senate sides the committees had voted out bills which had gone to conference and at last won final passage.

"The President vetoed the bills?" Tony asked.

"As a campaign gesture. He claimed he was demonstrating his commitment to cut-backs. The truth is that the legislation called for no additional funding.

"And partisan politics crippled the effort to override his veto because it's not a high-visibility issue."

His tone went from despondency to bitterness. "It never had a chance of override, I see that now. Because the press won't ever try to explain anything complicated and abstract, no matter how fundamentally important, unless it's really sensational."

"It's enough to make you sick," Philanth said, watching him. "Enough to send Cincinnatus back to the plow."

"Now I'm no longer in Congress," he went on, not caring if she thought he was being a crybaby. "I can't even act directly to try to resurrect it. I can't even make a public statement." He slammed his fist on the table. "*Damn* that . . . that—" He couldn't say in front of his staff, That idiot, that bastard in the Oval Office. But he knew somebody had given that idiot rotten advice—again.

The idiot had only had the acumen to take a restless, outspoken congressman growing bold and prominent out of indignation over chronic ignorance and stupidity in the White House and convert him into an innocuous bureaucrat.

"That's tough," Tony agreed.

An obsequious voice spoke above his head.

He went on silently cursing the President for his veto, a cynical play for vague support from the ignorant which had irresponsibly prolonged the vast amounts of misery and unrest the aborted program could have helped to reduce.

Meanwhile Philanth was ordering dinner in French for herself and Tony, charming the waiter into helping her make her selections.

She plucked at his sleeve and began cooing solicitously at him. "*May* I order something for you, Mr. Bjorklund? Couldn't you eat something light, at least? Something smells awfully good, and the waiter says—"

"I'm really not interested."

She spoke briefly, and the waiter vanished.

In a low voice Tony began trying to persuade Philanth to go with him to a floor show in a larger hotel nearby. She admitted that she would have liked it very much, but she had something else to do after dinner. Tony kept arguing and cajoling and she kept quietly refusing, saying it couldn't wait.

Losing patience with them, he stood up. "Good night."

"Hey, Chuck," Tony said, "why not come with us? You know you liked the dancing that night in The Gambia, and this'll make you feel better."

Philanth grasped Tony's forearm. "It wouldn't help," she said firmly. "The contrast of moods would only make him feel worse. And when you try to run away from something it only hurts longer, and often more because you hate yourself for having wasted time trying to get away from it. Leave him alone." Tony shrugged.

As he turned to leave, Philanth asked quickly, "Mr. Bjorklund, was your legislation really killed because of the general push for cut-backs?"

"Cutting the cost of government was the reason given. In fact it was only a more effective way of coordinating existing resources," he explained again.

"There was no interest group promoting it at the White House? There were no friends of yours in the staff?"

"The space lobby has its hands full working for appropriations. We only had some lobbying of the staff by agency officials and the bill's co-sponsors."

"Has it occurred to you that potential future rivals or their supporters didn't want you to have this feather in your cap? If staff resented your not being 'loyal to the President' and knew this was your baby and you were regarded by the press as a rising star in the Democratic Party, as Carole's clippings about you show, ..."

He looked at her with new respect. "Actually, I hadn't . . . gotten around to looking at it from that angle. I was so convinced of its merits." It was the sort of thing someone else might have eventually pointed out to him. It made him sick to think about people acting on that basis, but to people for whom his bills were just another program, ...

"I guess if I saw it as a personal achievement others would too," he admitted. "But as something to have done with my life, that was what I wanted."

"You never once thought of referring to it in your future campaigns?"

"Oh, that may have flitted through my mind in a vague way when I needed to keep pushing myself to stay with it in spite of all my usual activities. After all, there were some periods when it was like holding two full-time jobs.

"But it hadn't occurred to me that I'd reached the point of being regarded as so much of a threat that someone would want to stab me in the back by killing that program! I just thought the President doesn't particularly like me, which he doesn't."

"Are you really that pure?"

"I believe the word is 'naïve'."

"Yes," she said, but she seemed unaccountably pleased to find that he could be such a numbskull in his chosen profession.

Turning away, he said, "Good night again."

"Things'll look better in the morning," Tony said.

Taking the hint that his nervous exhaustion was obvious, he went to his room and started unpacking and getting ready to turn in.

He was so low that when he found the shower at the end of his hallway inoperative he felt ready to cry. That told him he was worn down to the breaking point.

Still wearing his raincoat/wrapper and pajama trousers, he went upstairs to try the shower there. He got only a gargling trickle. Then nothing. This time he swore: a much healthier reaction.

There was no water from the basin in the WC, either. Having read about what happened to hotel ceilings when foreigners visiting these countries left the taps open and water service was later restored, he made sure he had shut off every faucet he had touched.

Even that simple fulfillment of an ordinary duty took all the effort of will he could muster. He felt ready to collapse into some sort of massive self-indulgence as the only way to get his own back from the world.

He returned to his room and threw off his wrapper. Feeling he was using up his last spark of rational effort, he opened one of the bottles of mineral water which the local Peace Corps staff always laid in for him in his hotel rooms, along with fresh fruit, small electric fans, and a typewriter for Philanth. With that bit of water he made do.

387

Finally he put a dose of anti-diarrhea medicine into his stomach. That capricious organ was now showing a belated interest in dinner. Mentally throwing a contemptuous obscenity in its direction, he turned the bedcovers all the way down and flopped onto his bed.

His skin prickled all over, demanding the warm shower that was his greatest comfort of each day he spent in these latitudes. The mineral water had only made it want more, and hot. He needed to feel cool and fresh so he could relax and sleep. But if he used up all of the mineral water he would have to call room service for something to drink because his system would need more fluids during the night, and he couldn't speak French, and anyway it was all too much bother.

He turned off the light and tried to relax so he could forget everything in sleep. At least if he rested for a while he might feel like doing something more about the water situation.

But worn out as he was from the past three days,—or had it been three weeks?—he was like a taut rubber band.

And all of the grief came back, worse and worse, marching back and forth in his head and building in his chest.

His friend from the House must have been just holding his hand when talking about an override because it had been obvious that he cared too much about the legislation to see that it was already dead. That embarrassed him and made him feel stupid. He was afraid his inability to face reality was the reason his friend had met him near the Peace Corps offices, rather than be seen with him in or near the Capitol, where he might have told a third party about the subject of their meeting.

He was scarcely comforted by Philanth's speculation about the causes of the disaster. In fact, to remind himself that there were men in Washington who were quite capable of the sort of thing Philanth had suggested hardly helped. It made him feel worse.

At the same time he despised himself for letting that hurt him so much. As though he were a child who expected adults to be fair!

He didn't just want to cry. His emotional turmoil was mixing with his exhaustion and physical miseries and the weakness brought on by persistent diarrhea and the compounded despair his sickness brought on when he thought of how it would affect his performance during the coming days of heavy schedules.

He had less than a day for the problems of this Godforsaken little country, and then it was on to the next country and its problems. And the next. And the next . . .

It had been insane to put all these countries together in one trip. Who did he think he was, the political equivalent of Superman? He was a striving, pretentious nothing.

That thought brought on his general malaise over the course of his life in general and his relationship with Carole in particular.

It was ironic that he missed her so much, considering that she wanted nothing to do with him when they were together. But she had been such a wonderful, warmly supportive companion for so many years. He couldn't stop needing her. He needed her now more than ever. He would go on needing her more and more during the next two years while—no, *if*—he went on building toward trying to win that seat in the Senate.

He tried to tell himself that he had seen this final blow to his legislation coming for a long time. But he had not diverted any of his precious effort into making himself believe that it truly could happen. He had been too preoccupied. Besides, like dying or going blind, this possibility had been too awful to really imagine.

Carole criticized him for his aversion to facing up to really large unpleasant things even while he kept himself feeling conscientious by fretting over more remote or secondary problems.

But a big one had come to him now, and it had hit him at a time when he was already stretched to the limit. Like a small animal crushed under a tanker truck on a busy highway he was suddenly and unceremoniously flattened. He was nothing.

He felt so wasted. He started trying to think of what he had to show for forty years and could not think of a thing, not a thing that had any substance.

A good marriage? He wasn't at all sure any more.

A house in unbelievably expensive Georgetown? Made possible, in effect, by Carole's father's fortune in California real estate and her skill in persuading the old man to see things her way. It was just a small, dilapidated old town house, anyway.

A career. A good job. Ha. In two years he might easily find himself with no job and deeply in debt, so resoundingly defeated in a race so big that he would never be able to build himself back from the stigma of being "a loser". Even if he had the vast financial resources for a comeback, which he did not.

It hardly seemed worth the effort and risk, any more. None of it. He probably shouldn't even try for the Senate, just wait until he was forced out of Peace Corps by the next administration and then go quietly back to private law practice in California.

Washington had come to represent for him all that was cold and ugly in human nature, all that was at once both cynical and irrational. The ethical and intellectual quality of its leaders was the direct result of a national political culture permeated with the obtuseness and apathy peculiar to those who take moral pride in their political illiteracy and refuse to acknowledge the most basic laws of economics. That general civic laziness allowed the wealthy to continue to control the elections process and everything else by extension.

It was not merely the legislative defeat itself which made him look at his career and his life and see only that he was entering middle age without even a secure job on which to hang his sense of personal worth and identity. This crushing of his professional pride and hope had raised into his consciousness an accumulation of personal hurt and anger which could sink him as a politician—and as a human being—unless some way of unloading it could be found.

He felt deep compassion for the many congressmen who had reached burnout as he was doing now even though he had given up his safe seat and taken the Peace Corps job in an effort to escape it. It was no wonder decent people kept getting out and leaving politics to the alley fighters. For there no longer was any doubt in his mind that for someone like him, someone with every asset John Kennedy had had except an unscrupulous and extremely wealthy father, political life was not worth the psychic costs: the irrational abuse, the lack of privacy, the constant looking over his shoulder and fretting about how everything he did could be criticized without regard for the truth . . . and on top of all that, the constant financial pressures!

The irrational abuse, from people who delighted in the discovery that politicians could be used as their own personal whipping-boys, hurt the most. It seemed now that every insult, every rebuff, every snub or sneer that had come his way since the day he began his first campaign for office had left a wound. Now, when he had lost his illusion of having actually accomplished one thing in all those years despite the daunting obstacles, the cumulative effect of all those verbal cuffs and kicks—just for being "a politician"—descended on him at once.

What he really wanted to do now was turn his back on his whole dream of public service in Washington and return to the safety and comfort of private life in Southern California. Every time his plane moved into its landing pattern for Los Angeles International Airport, in daylight or in darkness, he looked down at the miles and miles of relaxed, unpretentious urban sprawl "from the desert to the mountains to the sea" and fell in love with the city of his birth all over again.

But going back there to stay would mean regarding himself for the rest of his life as someone who had aimed too high and fallen back, a man who had somehow failed to make the grade. Certainly it would mean that he had failed as a person to measure up to the unforgiving standards which seemed to permeate every facet of life in Washington.

But he was ready to spend the rest of his days dealing with the conviction that he was a might-have-been if only he could be allowed to escape from what he was feeling now. He thought not even Carole could have done much for his depression this time, for it was she who had inspired him to make all the effort to come to Washington. He would be telling her that he had failed her, that he was ready to renege on the basic commitment for which she had married him. What he felt was readiness for moral suicide.

He should turn tail and go back to his David-versus-Goliath storefront law practice for the little people who were being crushed by "the majesty of the law" as it was shamelessly manipulated by the legal hired guns of the rich and powerful. And he again could get his main pleasure in life from trying to defend the polluted, butchered environment that had been created so rich and lovely: a beautiful, fertile goddess inexorably being made into a wretched old bag lady.

But if he did that, Carole would certainly give up on him. Sure, she had thought his *pro bono* work and his fighting the big law firms over the environment were great stuff when she first met him, but she had—she thought—helped him to move far beyond that now.

He would have let her down, wasted fourteen of the best years of her life, and he did not know how he could live with that. *She* would never be able to forgive him, for she knew the scope of her abilities. And she had seen what he was capable of.

She probably would dump him and try to start over, an embittered man-hater. The thought of her contempt scared him more than the end of his career in politics. Her opinion had become that important to him.

But maybe, having committed the oversight of not becoming at least a millionaire before he entered politics in the Age of Television, he deserved to slip back to what he had gotten by his own efforts.

It hardly helped that he knew he was wallowing in self-pity. He only despised himself that much more. He understood suicide out of pure self-hatred.

He bit his hand to draw off the pain. He didn't have what it took, whatever that was—chutzpa or driving egoism or ideological fanaticism or crass opportunism or great family wealth. Or maybe just unwavering faith in himself. Whatever it took to make it in American politics today, he didn't have it.

And trying to do without it didn't seem worth it anymore. Hell, *nothing* seemed worth it any more. Nothing, nothing, *nothing*. He wanted to die. Just to fall asleep and be done with everything forever.

There was a soft tap on the door which connected his room and Philanth's. He cursed under his breath. "Come in!"

Philanth opened the door but did not come in, only stood with her small personal pillow under her arm, silhouetted against the light from her room. She had changed from her cornflower-blue traveling suit to her smock and trousers of the same deep-blue denim.

"Wha'd'you want?" he growled.

"I've come to pay a sick call," she faltered.

"Go away!" He turned onto his stomach, punching his pillow to make it support his chin.

She spoke again from his bedside. "I know a cure for what you've got," she ventured.

"Oh, come on, Philanth," he snarled, rolling back onto his side to face her. "Surely it's not that stupid letter that's on your mind!"

"What? Oh. No. Nothing like that. It's . . . quite honorable.

"It's not any more effective than you let it be, but it could prove surprisingly helpful. May I try?"

He gave her a baleful look.

When he did not speak she added, "You have only *one day* for the Central African Republic! You've got meetings, speeches, the full treatment, all packed into tomorrow.

"I know you don't feel good. But they need you badly, here. You heard how awful it is for these poor people, you even saw a little of it, coming in from the airport. *Please* let me help you!" She was using a tight, little voice which made it hard to refuse whatever she was asking for.

He gestured his indifference. "Do your worst, it can hardly do any harm."

At once she dropped her pillow by the foot of the bed and went to bring a chair to his bedside. She sat down as close to him as she could get and leaned forward. "Please move toward me a little." Still lying on his side, he scooted closer.

She lifted her hands to spread them flat within a few inches of his chest.

"Now. I'm taking all the pain out of you. It's coming out of you into my hands. See?" She let her eyes meet his wondering stare for only a second. "You're holding back, because you don't believe. Let it all come free.

"Good." She withdrew her hands, holding them high and keeping them spread flat and watching them so he had to watch them too. Still in her little-girl voice she continued, "Now I take it all and hold it here." Slowly, carefully, she turned her hands and laid them, still spread flat, so that they covered the area above her breasts. In the dim light he could see that at the moment her hands settled she started a little, as though from an impact. His attention was held by this astonishing performance.

"A simple transference." Now she was using the tone of a hypnotist, soothing yet professionally detached. "I'm feeling and thinking what you were. It's *very bad*. If you doubt it I'll tell you about it, but that seriously interferes with the cure.

"Now, relax completely, . . . close your eyes, . . . and think about what pleases you most: the place you always want to go back to when you're tired of the rest of the world. What would that be?"

395

He grudgingly obeyed her instructions except to blink at her so he could answer. "Spring mornings in Los Angeles. There's no pain there, the way there is here." For a moment he had forgotten that they were not in Washington. "I mean, there's no bursting out and riotous joy after the winter; only a gentle gladness. No extravagant hope that's bound to be disillusioned later."

"Go on. What else?"

"And winter nights when the Santa Ana blows, like a wild, warm spirit off the desert, and the stars seem to twinkle in the wind. You feel like you could float out of your body and fly away, on that wind. It's the way you felt at special times when you were young."

"Good. Very good."

She got up carefully as though not to jar him out of a trance. She moved toward the foot of his bed, and in the light that came through the connecting door he saw her slowly kneel and then sit on the floor, arranging her pillow under her and using the side of the bed to support her back.

"If it starts to try to return," she warned, "you must look at me so you'll see that it's all still in me. I'll have to stay close enough for you to see me, to be sure the transference is kept complete." She bent her legs to lean comfortably on her knees and was quiet for a little.

Stunned by the strangeness of all this, he simply lay there looking at her in the dimness.

Her soft, reassuring voice resumed. "After a while the pain is no longer fresh. The bleeding has stopped. The healing begins. Then you will be able to let a little of it come back in order to practice thinking about it as you would about other things in your life.

"At last it will fade until you can deal with it as a whole. But for at least the first several hours or days or months or years, this is the way to deal with it. You rest and think about the good things, about being comfortable and safe in bed in the darkness with nothing coming at you, nothing making any demands on you. You've earned your peace. You relax as if you had forever. And eventually you will sleep."

She had indeed "taken the pain", simply by distracting him.

"How long are you going to sit there?"

"Until you're asleep. Unless . . . unless you send me away.

"I hope you won't do that. You need someone here, making sure you stay free of any pain."

"This is highly irregular!"

"If you really can't accept it, at least you will have had a good laugh."

"That's true."

He thought the situation over. "Did you dream all this up just for me?"

There was a hesitation. "My father was sick for a long time. I tried to do this for him, though not in such a ludicrous fashion."

"What was his trouble?"

"He'd had a stroke. He—he would cry, sometimes. He couldn't talk, I mean he couldn't frame words, but he understood what I said to him. Sometimes he would cry because he hated being helpless; but mostly he cried over my mother's death. He'd cry and cry . . . and there was nothing I could do . . . . *Nothing*. Except love him.

"I just say, 'He was sick' to avoid having to explain. When you say, 'He'd had a stroke', people always want you to add, 'but he made a good recovery.' You cut too much beneath the surface of things when you admit that sometimes people don't get well. They only get worse.

"I think the old song refers to 'Death's bright angel' because death can be such a blessed release. We are given these bodies to enjoy and work with for such a short time before they start to break down!"

"Um-hm."

"But it comes to us all. We have to accept our condition, and be thankful that we've had whatever time we've had."

"Not punish ourselves for not being God."

"Exactly."

After another silence she asked, "Are you holding on okay?"

"I need convincing that you can hold my pain."

"I know some poetry about the pain. Would it help if I proved that?

> 'Charge once more, then, and be dumb.
> Let the heroes, when they come,
> When the forts of folly fall,
> Find thy body by the wall.'

> 'If you can bear to see the work you gave your life to broken,/And stoop to build it up with worn-out tools;/If you can bear to hear the truth you've spoken/Twisted by knaves to make a trap for fools, ...'

> 'That man is a success who has lived well, laughed often, and loved much; who has gained the respect of intelligent men and the love of children; who never lacked appreciation for Earth's beauty nor failed to express it; who has filled his niche and accomplished his task; and who leaves the world better than he found it, whether by an improved poppy, a perfect poem, or a rescued soul.'

"You're *already* a success. I can attest to that on the basis of my personal knowledge! If you don't believe it, you have adopted too many of the *world's* criteria for measuring your achievement of those tasks you had set yourself according to *your* criteria. That defines the situation so as to assure you that you're a failure no matter *what* you do. Doesn't that make sense?"

"Yes!"

"So watch your definitions."

After a few moments she added, "Sing to me not of lasting deeds, of a great, victorious life./Tell me a simple tale more true, a tale of endless strife./Few are the men who wear the wreath and sing the victor's song./I shall be one who clears the ground for others to build upon."

She paused, then confided in her little-girl voice, "That last part isn't as good as the others 'cause I just made it up on the spot. Actually, I sort of stole it from Matthew Arnold, but I didn't mean to."

He laughed. He actually laughed quietly, though tears came into his eyes.

She added, "And I probably scrambled some of the others 'cause I learned them when I was little."

After a few moments she said, "Words can't convey the truth, anyway. They only walk around it, talking about it, trying to stimulate appropriate memories of direct experience. To truly show you that I have your pain I would have to communicate through terrible visual images, like the ugliest nightmare.

"There is a proof much stronger. Its efficacy has been proven over and over for thousands of years, in the theatre: I can weep for you, as your surrogate. It gives catharsis to see another person release anguish which has built up but stayed buried in yourself. You can release it, all the pain, through me.

"That's why the theatre has always been a religious experience: the actual presence of living people can have such impact when they act out the passions we hardly dare admit even to our inner darkness, in our everlasting, unbreachable solitude."

She still spoke gently, but as always when she got started she had been unable to resist a digression, a substantiating and edifying footnote, and then gotten carried away by her own extemporaneous eloquence, her husky-sweet voice softly weaving together words which wept or thundered so they flowed into music.

"For the greatest bitterness, only blood will suffice. They say that's why Christ suffered, and why every generation relives that crucifixion again and again. But for this tears are sufficient."

"You could actually do that?"

"Some actresses can weep on cue by thinking of something sad. I have that much imagination too. It will take time, but tonight we have time. Shall I?"

"You would do that for me?"

"Of course! A person should never be embarrassed to help another!"

The intimacy of such a service in these circumstances seemed to elude her. But it was far more likely that this was one of those petty-minded objections she chose to ignore.

"How, then?"

"If I told you what I would be thinking about, it would risk letting the pain get back into you. I know you, Mr. Bjorklund. —I mean, I've seen enough of how hard you work, and we're enough alike, so that I could do it only by thinking about . . . you. Your hopes and your frustrations. Your . . . love for the world, your longing to ease its sufferings.

"Let me begin, not explain so that the cure becomes worse than useless."

"No!" He started up. "No! Don't do it! I—I believe you!"

"I don't mind. Just be quiet for a little, and it will begin."

"No! Now that I know what you would be thinking, it couldn't work. Because I would grieve too, not for myself, but for *you*."

She gasped. "Oh, *dear*."

"Let's just be still and rest," he went on hurriedly, "and your caring will be enough. It would be wrong for us to merely exchange grief. We should instead cooperate to disperse or at least divide the despair of one of us."

"Very well, but you mustn't let it come back. You mustn't let what I've done here be in vain. Promise to be careful, to make the effort to control your thoughts."

"Help me. Tell me how."

"You need to become better at sustaining yourself instead of tearing at yourself because your love and imagination far exceed your power. Your passion for good can never be satisfied for long, and you mustn't let it destroy you."

"No."

"Do you know the hymn, 'Could my tears forever flow,/Could my zeal no respite know,/All for sin could not atone'?"

"Oh, *yes*," he exclaimed, chuckling sadly. "They really did a number on me, didn't they?"

"I'm deeply grateful for people who've been shaped by letting that infinite burden be laid on them—rather than leaving the job of saving the world up to Christ, as the next line says. For all the mischief caused by their good intentions, they've been a beneficent influence on modern history. You can't truly regret your compulsion to be one of them."

"No. But it's killing me."

"You must learn how to rest your will; to let it be healed of its abrasions and replenished."

"But how?"

"Do you pray, Mr. Bjorklund?"

"Yes. I pray when I have hope. Praying when I'm desperate only makes me feel worse."

"Because you're seeing yourself as alone and helpless and doing something futile—not as talking to someone who is listening?"

"I suppose that must be it."

"There's a state of mind that's the same as wordless prayer, or meditation. In it you don't have to try to communicate with anyone. It seems superfluous to try to give it any real intellectual content. It's just a celebration of sentience, a thanksgiving for your existence: all the good things, even if the only good thing you can still believe in is your own consciousness.

"—And I believe consciousness *is* good, of itself, even if other things make experiencing it intolerable.

"Anyway, whatever the healing thought is, you bathe in it, using it as a healing solution.

"It's like saying, 'I'm alive. Praise God Almighty, I'm alive.' Usually you have a lot more to celebrate than that, like being in a comfortable bed, or having someone to love, even if they don't love you and you'll never see them again. It can be something very simple, like having had something to put in your stomach.

"You concentrate on all the positive things, things so simple as your own physical security and comfort at the moment and the sound of birds or a fragrant breeze; whatever there is to build up your sense of well-being, your awareness of having the right to rest until you have the strength to return to the struggle for others."

She paused, but he made no response, enjoying the quiet, simple way she spoke to him.

"Sometimes," she went on more shyly, "this state is so lovely that you realize this is what it's like to live your life in God, with God. This is where the idea of Heaven comes from. You know that you could live like this all your life and not need anyone else if you could order your life and your thought so that you always had this inside you. Heaven within you: the clichés are true.

"There's peace in knowing that. There's strength. It lets you carry all your burdens and disappointments and heartaches more lightly, even though they're no less important to you or to the world."

"I think you move away from such feelings as you grow older, if you ever had them to begin with."

"It's true that lyrical poets and martyred saints have usually been young. Most older people probably become afraid to feel much, once they learn how not to.

"But if you're afraid to feel, doesn't that reduce your active caring about other creatures' suffering—and therefore smother the passion necessary to take risks and make serious efforts to prevent it?"

Again she paused. But that obviously was a rhetorical question, so he said nothing, content to let her talk herself out.

Her voice resumed its quiet wandering, comforting him like a slow, casual, but professional massage, relaxing him. "You need *some* armor in order to survive, but if you care about others you have to be able to afford to let yourself be pierced by the world's hurts. So you need to keep practicing to develop your ability to heal yourself. It's important that you remain able to recapture your sense of inner wholeness when it starts to slip away.

"That can happen over and over—especially when the whole world seems to keep clamoring at you, Do this, Do that, Be this, Be that. And most of it useless, utterly beside the point. Poor world."

Suddenly she was compassionate as though petting a cringing, emaciated dog. "Poor, poor world! So mad, so blind, so wide of its good."

She fell silent.

"I should expect," he observed, trying not to sound unkind, "that a girl who gives top priority to visiting a church when she has only one night in Paris might have something like this to suggest."

"It seemed healthier and more responsible than getting drunk," she observed reasonably. "But homilies are out of fashion."

"That wasn't a criticism.

"You have a lovely voice, by the way. And I can tell that it's been trained by a good speech coach."

"Yes. Public speaking and I go 'way back."

"But most of all, you have a gift for words, Philanth."

"I didn't come here to display it. I didn't even take any extra time to rehearse. I just gobbled down my supper and came."

"Now, that was a real sacrifice," he teased, "if it was a good supper."

She tilted back her head to let the soft laughter flow free. "Yes. I try to make every calorie give a good account of itself on the way in.

"Speaking of which, Mr. Bjorklund, you need to keep up your strength; and a good, solid meal will make you feel better. Could you eat, now?"

"I don't want anything that could possibly . . . aggravate my physical problems. Do you have any bananas and nuts?" Those were staples in the food supply she kept in her room.

She went and got all of her current cache, then turned on his bedside lamp and nibbled some nuts to keep him company while he went through most of her supplies.

While she carried off the remains he went to repeat his bedtime ritual in the bathroom. He discovered that the water had been turned on and promptly went down the hall and took a shower.

He returned to his room to find Philanth back in her place on the floor at the foot of his bed, sitting on her pillow. She was resting her back against the side of his bed, leaning her arms on her knees and looking quite peaceful and comfortable. He threw off his wrapper and got back into bed.

She climbed to her feet and looked serenely down at him as she reached to switch off his bedside lamp.

"Go to sleep, now," she directed in a motherly tone.

"Yes, ma'am."

She chuckled and turned off the lamp.

By the soft light from her bedroom he watched as she resumed her position beside the foot of his bed. He lay watching her, concentrating on her relaxed presence, as she had instructed earlier. He found that by carefully keeping part of his mind walled off he could feel surprisingly peaceful.

Philanth knew how to live with an unbearable grief. So while she was there he could keep the monster shut up behind the wall; and as he learned to block his thoughts, the wall grew stronger. He felt as if a nursing sister were keeping vigil by his sickbed, lending him strength while his wound stabilized. It was enormously comforting to have someone there, caring for him.

"Thank you, Philanth."

Her voice was distant, contented, refusing any obligation: "I'm glad it worked out."

He closed his eyes, and the darkness which enveloped the two of them was quiet and friendly. He relaxed, feeling his refreshed skin and the settling of his guts, and emptied his mind.

Then it was morning, and there was no trace of her visit. It was as though he had fallen asleep and dreamed the entire episode.

Except that even in his dreams he could not have conceived of what Philanth had done.

In a way Carole had not imagined either, Philanth had comforted him with a gift of herself. Demanding his suspension of disbelief to make it possible, she had healed him "by the laying on of hands".

Whatever needed to be done, she was not too proud to do. Whatever needed to be given, she was prepared to give—shaped with only limited regard for the constraints imposed by other people's fears and prejudices.

She had taken care of her father, the "wise, kind" theologian, even when he cried . . . and cried. She must have been a wonderful nurse.

Chapter 15

A MERRY WIDOW

When he was dressed for his first official evening in Nairobi he tapped on the connecting door, and Philanth opened it to beam at him.

She was covered in black from neck to ankle, and her hair was drawn back into a round bun at the nape of her neck. "You look like an old maid," he protested.

Far from being insulted, she looked pleased. "That was part of the deal: you said I could come to this if I kept a very low profile."

"So I did."

She also had been included in the invitation, as the American ambassador had made quite plain to him that morning. Despite his precautions it looked like some discussion of her might have been traveling the diplomatic circuit within Africa. It also seemed that no matter what he did about her, he had a problem.

He hurried on, "I thought you looked terrific in that outfit in Morocco. What's different?"

"Only my hairstyle and makeup. I'm *supposed* to look uninteresting, isn't that the idea?"

"Oh. Well, actually I assumed you'd look the same . . . ."

He handed her a small parcel. "Here, put this on, it'll help."

"Something for me?"

"Yes. Just a bit of impulse buying."

"How very nice of you, Mr. Bjorklund!" She sat down on a corner of her bed and picked away the newspaper wrappings.

Carefully she disentangled the long, polished black beads. "Oooh, it's beau-ti-ful." She arranged the necklace in a circlet on the burgundy bedspread. "It's a Masai collar?"

"I don't know. Tony assures me it's ebony."

She gave him a glowing look. "What a wonderful present."

"It wasn't much. I spotted it when Tony guided me through the open-air market. I knew right away it'd look great with your dress. A compliment to the culture of our hosts tonight."

"Yes. . . . But . . . I can't wear it tonight, Mr. Bjorklund." She was all apology. "I'm sorry. You're right to be proud of it. Perhaps another time." She stole a look at his face. "It couldn't go with this lace cape. And I could never wear this dress without the cape."

"Why not?" The lace cape revealed that her soft knit dress under it was strapless. He began to smile. "Is it that you're too modest, Philanth?"

"Very definitely. And you did say I mustn't call attention to myself."

"I see the problem." Her walking into that party full of diplomats wearing a snug, strapless black dress with that sensuous necklace at her throat would "send the wrong message", since she would be accompanying him and he was to be the guest of honor. He was the American Peace Corps director, not an African politician who flaunted his women as proof of being "a Big Man".

But he was disappointed. "At least let me see how it looks on you. Someday you'll have a chance to strut your stuff, and I won't be there."

She forced a laugh. "Just use your imagination."

She checked her watch. "It's a little early. We're not due to be picked up for ten minutes or so. I was going to go back to my reading, rather than rushing to dress at the last minute."

"Same here.

"So let me see you in the necklace."

"I'm sorry, it wouldn't be proper."

"For cryin' out loud, woman!" He checked his impatience. "You're not being very gracious. It's all the thank-you I want, just to see you wearing it with that dress."

He began to get an uneasy feeling that he had crossed the boundary of acceptable conduct toward a female subordinate. But he thought his intentions had been entirely honorable, and he was not sure why he had just gone too far.

"Is that unreasonable?"

"N-no, I suppose not . . . except for the fact that the dress is too low-cut."

"Oh, come on! I can see how low-cut it is. And I wasn't just hatched, I see women in evening dress all the time. Considering what women wear nowadays, not to mention all the ads and movies, you're being ridiculous."

"But the lace is pretty dense, and the way the front hems of the cape lap over at the opening, it—it conceals . . . ."

His tone gentled. "You're not used to wearing anything that leaves your shoulders bare, are you, Philanth."

She shook her head vigorously, her eyes lowered.

"Well," he said after a moment, "I really don't think I'm asking very much."

"I beg you, Mr. Bjorklund, to take my word for it. I do know something about how I should look. I wasn't born yesterday."

"Not quite. But you should learn to accept being a woman, Philanth. It's part of growing up."

Her mouth tightened.

"Please," he said, stepping over to her and reaching for the top of the cape. She drew back.

"Philanth, it's for your own good. This is not that big a deal!"

"Don't do it, Mr. Bjorklund. It's—it's you I'm concerned about."

"Nonsense. I've made up my mind."

He found the top hook; she stiffened, lifting her chin away from his hands. As he began unlatching the little black hooks she quickly placed her fingertips along her collarbone, and she held the cape in place as he worked his way down to her waist.

"Please," she said. He ignored her.

As he started to tug the cape free of her fingers she let her arms drop to her sides, and the cape came away in his hand.

"Jesus," he whispered, hastily stepping back.

She took a deep, unsteady breath, and that only accentuated the effect. Whatever she was wearing inside that pliant knit embraced her so tightly that her soft flesh seemed about to overflow its stiff confines. The result was a sensuous ivory cameo, a tempting white marble bust. He was reminded of an unforgettable Italian Renaissance statue of a shy, young Psyche, innocent but exquisitely ripe, receiving the intense admiration of a beautiful youth, Cupid.

She was watching his face, her expression bleak and fearful.

Without the stiff cape covering her to her waist, the entire design of the dress drew the eye to her breasts from every direction. Though the knit fabric was snugly wrapped around her, along both sides of the seam which ran down the full length of the center front it was gathered up into folds of varying size. On each side of that central line in the skirt, below the waist these folds descended and spread out to vanish as their fullness became filled by the curves of her hips and legs. Above her waist the top-most gathers were filled by her breasts.

He cleared his throat, trying to find his voice. "That's . . . quite a dress."

"It cost me ten bucks," she snapped, turning her back to him. "Plus the zipper. I didn't even need to buy a pattern. I just adapted the design concept from a cigarette ad I'd seen, and pinned the fabric in place around . . . me. Wrong side out, of course."

"I meant . . . it's rather dramatic."

She bent and picked up the collar, turning so he couldn't watch what leaning over did to the exposed portions of her breasts. "Oh, come, Mr. Bjorklund!" she replied, probably the more scathingly to cover her embarrassment. "'Considering what women wear nowadays'!"

"It just startled me," he managed. "In all fairness, you don't expect a nun in traditional habit to take off her cape and suddenly look like one of the Playboy Bunnies."

"No," she admitted, tucking in her chin as though to look down to see what all the fuss was about, "you don't.

"However, I must say, I think it's unjust to compare this effect with what they did to Bunnies. The aesthetic principles are quite different: restraint versus padding."

Still keeping her back to him, she arranged the collar around her neck and hooked it.

Even without the contrast of the ebony and the dress, she had the whitest shoulders he had ever seen. During their long, hot excursion upcountry she had worn her office-quality, long-sleeved blouse and a floppy-brimmed hat she had insisted on going out

411

to buy in Morocco. During their stops outside the Jeep she had also added the shade of her folding umbrella, using it as though she literally feared the sun. Now he saw why. Nature had given her no protection.

There wasn't an excess ounce of fat, either.

"Have I got it right?" she murmured, and went to her dresser mirror. "Oh, it *is* a nice effect!"

An effect indeed, he thought, collapsing onto the chair. That dress made him want to throw her on her bed and . . . free her from it . . . and devour her. That dress was a dry hull being split open from within by the burgeoning ripeness of a succulent piece of fruit.

She turned to him, and he scrambled to adopt a formal sitting position. With slow dignity she sat down on the corner of her bed. Knees together, arms straight, and hands clasped on her knees, she reproachfully invited him to look his fill. He could see that she was steeling herself.

The glistening black collar sprayed out from her neck for at least three inches in all directions, mocking him with its fierce, nocturnal gaiety. It only sharpened the impact of her total image.

"You're lovely." That word was far too weak. She presented a study in contrasts, and he thought the mixture of austere simplicity and voluptuousness, innocence and maturity, made her the sexiest woman he had ever seen. But this he must not say.

"I don't want to hear that! Least of all from you."

"Your hair should be in keeping with the rest. If it were flowing freely, ..."

"Mr. Bjorklund, beneath that Nordic asceticism there lurks a sybarite. Heaven forbid you should ever succumb to the temptations offered to power."

"Amen," he answered, looking at her.

She threw him a glance of dismay.

Then she lifted her chin to return his gaze, and her voice deepened. "Especially when power is joined to a man's grace and beauty, and dressed as you are tonight."

She had turned the tables, and the unwelcomeness of her grave admiration made him face the fact that his attention to her physical attributes was equally unwanted. In response to Tony's attempts to flirt she had made clear her sentiments on the subject: complimenting her looks ignored her more significant attributes as a person, and her body was something she had to put up with: a utilitarian instrument with superfluous features she wished everyone would have the good sense to ignore.

Nevertheless, knowing a time like this might never come again, he could not turn his eyes away. He had no doubt that the rest of her was like the part he could see.

And the dress invited discovery. Indeed, it was a taunt: How long can you bear to look before you touch; how long before you grab?

"That dress is a gift for a woman to give her husband during a quiet evening at home. Forgive me for not respecting your usual good sense and agreeing to forgo any recompense for my little present."

"May I have my cape now?"

The cape. He looked in the direction she indicated and found that it was still between his fingers. He leaned forward and returned it to her.

She started to put it on, then laid it on the bed and began feeling around the back of her neck for the hook of the African collar.

"Allow me." He jumped up, nearly overturning the chair, and hurried over to stand behind her. Then he had to pause with his hands in mid-air to gather the

413

concentration necessary to make them do only what was strictly appropriate. A moment later he handed her the ebony collar.

She met his eyes as she accepted it. Slowly she rose, then walked over and laid it on the dresser. With her palms resting on the dresser top she said quietly, "Mr. Bjorklund, there's something . . . very unpleasant . . . you should know. So this misunderstanding won't . . . cause problems for you."

She looked at her watch. "Perhaps later tonight. When we come back. But I wish I could get it over and done with right away. If I think about it I probably won't be able to go through with it."

He consulted his own watch. "There's time."

"Very well!" She turned to face him, but then she leaned back and grasped the front of the dresser with both hands.

"I think Carole deliberately gift-wrapped me this way. Not to tempt you, because you were never to see me like this, but to bring more pressure on me: to make me feel . . . more desirable. It was an invitation to carnal self-indulgence. "

She walked over to the bed, then dropped onto it as if her legs suddenly refused to support her.

"I have to make you understand." She indicated the chair. "It will take a few minutes." He sat down.

"I . . ." She cleared her throat and began again. "I was already aware of the problem to some extent: I had always thought you were remarkably attractive, back when I worked for the Foundation. When I found myself planning my clothes for this trip I realized that everything had to be plain and wholesome and modest even if it wouldn't be as cool and comfortable as I'd prefer, so I wouldn't be yielding to the instinct to try to be attractive to you." Her words slowed to a full stop.

When she resumed, her voice shook. "And yet when Carole tempted me regarding this—" She moved her hands down the smooth curves of her sides. "I succumbed."

She seemed to try to shrink into herself, tightly closing her eyes. "I have to explain. It's ugly, I warn you. But you need the ugliness to balance what you see, to keep you realistic."

She paused again, taking a deep breath and letting it out fast. "This makes me sick," she muttered. "Let me back up a little and work into it."

She went on in a forced voice, "I had come to terms with the fact that I had to make an evening dress for this trip because I needed something that would travel well in very little space. And I wanted to be able to wear it without anything under it, again to minimize luggage, but also because it would be more comfortable in warm rooms. So I planned to make the top of the dress just a simple band of material with a couple of drawstrings and wear the cape over it."

She picked up the piece of black lace and fingered it as she went on, "This cape is ideal for travel, and it's very fine quality, something my mother bought for herself in Spain, on her honeymoon.

"Unfortunately that meant it would be best if the dress were black too. I didn't mind, since if I were in black I wouldn't be competing with all the other ladies in their beautiful, bright colors.

"I would've gone for something light and floaty, but I didn't have time to work with a delicate fabric, and it couldn't be at all fragile, something I'd have to fuss over to make it look flawless every time I wore it. I didn't know, of course, that I'd be barred from attending most of these affairs.

"Anyway, when Carole and I went shopping together I asked her to stop at a fabric store for just a few minutes. I had to explain what I was planning, and she was surprised because I was going to just 'throw something together' the afternoon before we left.

"She said I'd feel more self-confident and therefore function better intellectually if I had proper support, and a full foundation garment would actually be more comfortable than half-way measures. Time was running out because I'd put off dealing with the problem, hoping to think of a better solution, so I didn't try to argue with that.

"But I didn't have any black . . . underwear. And I'd been wanting a merry widow for a long time.

"That's the type of foundation garment I'm wearing. You'd probably call it a 'corset'. I'd never buy one for myself because it's so expensive, but I confessed to my hankering.

"Carole promptly insisted on my finding one that suited me and letting her pay for it. That took up most of the afternoon before the bon voyage party. She kept complaining because 'L.A. would have been a much better place to shop', but she insisted I had to be perfectly happy with it . . . ."

She took another unsteady breath and looked at her watch. "I'm running out of nerve again. But this won't take much longer. Shall I go on?"

He looked at his watch while he thought. Whatever she was leading up to with so much effort, it wouldn't be the same if she had time to think about it during the whole reception and dinner. The way she was behaving, she probably would back out altogether. He didn't want to be left wondering.

"We're on African time," he said, "and the later I arrive, the more important I look. Tony will keep the driver amused; and he'll think I got a phone call at the last minute. Go on, finish."

"I'm so ashamed. I loathe myself."

"Go on! By now I'm prepared to hear that you've engaged in human sacrifice and cannibalism."

"No. I wanted it for . . . my own pleasure. When a woman has sexual tension she craves stimulation all over. I wanted the merry widow as a sex aid—to gain some of the sensations I could have"—she gulped— "from being with a lover.

"Do you—can you understand? It must be quite different for a man.

"But a woman is conditioned to regard herself primarily as the receiver, the object acted upon. There's no need for her to focus her fantasies on an imaginary partner. To achieve climax she need only focus on her own body.

"So in female autoeroticism, self image is especially important." Her mouth kept twisting with effort. "Wearing an image-enhancer provides an aphrodisiac for her just as it would for an actual partner. In addition to the physical sensations. And I—I plan to go through life . . . without any sex partner. This was . . . a long-term investment.

"But you see, now, how disgusting I am. Please forgive me for telling you this."

He bit his lips, realizing that he was called on to respond. He started to speak several times but each time thought better of what he had been about to say.

Finally he adopted the relaxed tone of a reassuring parent. "You're not disgusting, Philanth. Relieving sexual tension alone is normal, especially for a responsible and conscientious person. You know that."

"But being constrained to *hear about* another person's most private physical problems *must* disgust you."

"Not when you already know that person inside out and really care about their distress."

This time her glance was worried.

Then she hung her head. "I'm also flawed in just the way Carole was trying to prove: owning an expensive self-indulgence like this violates everything I say I believe in."

"I understand your guilt at accepting something expensive that's strictly for your own pleasure. But you're always trying to see how little you can get by on. If you're getting something exactly right most of the time you can surely expect to stray a little over the line once in a while.

"And you do need something for yourself, sometimes. Because out of your sense of abundance you give. Your sense of the value of life comes not just from your convictions, it comes from the viscera: your physical enjoyment of existence—and having a body.

"It doesn't hurt you to be aware of your own frailties, either. You work at it, I know, and that, uh, merry widow will remind you, even as you enjoy it."

"Yes." She looked amazed at his insight. "But . . . you must be aware of . . . my dark side too."

"You don't really want me to feel you're ugly or disgusting. What you truly want is my ability to see you exactly as you are: loving, in a way that's both passionate and idealistic. I do see you that way."

She looked at him doubtfully.

It *was* getting late, and he had to get this contretemps sealed off satisfactorily in terms of her feelings about their relationship. Casting caution aside, he brought up his reserves.

"In fact, I love you, Philanth, *because* of your way of loving: because you struggle against the part of you that wants to be sexy . . . because you want to minimize your own needs, and sexuality creates a personal need. Because you would do anything to be beautiful in the ways which have nothing to do with appearances.

"In fact, you'd probably rather be pure spirit. For all your capacity for sensory pleasures, you'd rather be without . . . without a body at all. Wouldn't you?"

She nodded, brightening a little and stealing an admiring glance at him. Again she must think he was wonderfully insightful, marvelously in tune with her, when he only had the reasonably retentive memory for oral briefing which anyone in his line of work needed to develop.

He went on, "And knowing you as I do now, I'd love you no matter what you looked like, because however you looked, I'd love that because it was you inside that appearance. So this 'revelation' of how you look . . . *now* . . . can't matter."

"Thank you, Mr. Bjorklund. You're generous and compassionate, as usual."

It was obvious that she still felt humiliated. She had always wanted to protect the propriety of their relationship.

He went on, "Like you, I have times when I'm not the wholly rational person one expects to deal with in the ordinary course of doing business. You know that because you've helped me through one of them. —Maybe more than one.

"Ordinarily, as moral people we have to do our best to deal with these problems alone. You know better than most that we can't always expect to be so fortunate as to have someone special enough to empathize with us and help us back to peace and rationality. I was very lucky to have you with me and able to help.

"So now we're 'even', so to speak, and you mustn't feel bad or worry about this any more. After all, it wasn't your fault."

"Yes, it was. I should have had nothing to hide in the first place."

"Set too high a standard for yourself and it breaks you. Take it from one who knows. So just accept that you're human and you've done your best. Promise?"

She managed a tremulous smile. "No. But I promise to try to be rational."

He looked into her eyes. "You and me both."

A flare of what might have been hope briefly lighted her face in reaction to that.

Again he consulted his watch without seeing what time it showed. "Let's go," he said, getting to his feet, and she too stood up.

"Your orders are to relax and enjoy the party," he told her. "Watch how charming I can be when I've just had an enriching conversation with an assistant I cherish.

"And . . . " he raised a finger in admonition. "Keep your mouth shut."

She tried to smile, but her lips trembled, and her look was fearful. "I wish we were going home tomorrow."

Seeing that she was still afraid for *him*, he realized she had sensed his growing attraction to her. Instead of responding with more reassurance, he could only look at her.

Even her tightly bound hair and apparent absence of makeup made her look more defenseless, more purified by self-denial and willingly vulnerable. God, she was appealing! Make the tight black dress loose and white, unbind her hair and chain her to a rock, and she would look perfectly natural as the original subject of some old engraving. Made for martyrdom and sainthood, she had been born many centuries too late. He could imagine her praying herself into an ecstasy as she waited for the devouring monster to claim her as its virgin sacrifice from her people.

He shook his head, wishing he had a picture, and turned away, straightening his white dinner jacket and taking a deep breath.

But in fact he did have a picture, and it would be far safer to have it burned into his mind than as a physical object. Besides, as the sacrificial maiden on the rock she was religious pornography. Without wanting to, indeed *because* she did not want to, she drew a man to her at all levels of his being.

Or *did* she want to, unconsciously or even consciously? Beneath all her concern about the world's problems there had to be a desire for power to deal with them.

There also was her "project" of trying to improve Tony—and her efforts to "improve the character" of the boss she had driven to violence; also her peculiar relationship with Ken. Quite aside from her feelings about power, she was a frighteningly loving woman.

Without another word she put on her cape and they went downstairs.

His fine words about how virtuously he loved her had of course been whistling in the dark. They had been true, but they had not truly addressed the cause of her concern.

With reason. Even as the three of them were driven through the night to the reception at the U.S. embassy he knew he was in for some difficult times during those hours alone he had been talking about so glibly. Almost every night for the next three weeks he would have to go to sleep knowing she was in bed within sound of his voice, that lovely young body and passionately loving and selfless mind resting on the other side of an unlocked door; a woman who had dealt with her attraction to him by deciding to wear plain, dark-colored, loose-fitting denim, denim, and more denim while she was with him for five weeks, traveling through the tropics.

He was equally worried about all the hours with her, doing paperwork and traveling, since at those times he had to do a lot more than refrain from action. As if he didn't already have enough to worry about, he would have to watch his every move. She would be sensitized to notice the most subtle slips: eye contact that lasted a second too long, a glance that fell where it had no business, a smile or tone of voice a degree warmer than neutral, a momentary physical contact that could have been avoided but only at the cost of unnatural caution . . . .

Her fears would be confirmed. For her deliberately humiliating herself in her impulsive effort to inspire his repugnance and thus restore his sexual indifference to her had only hastened the culmination of a process that had been steadily accelerating anyway: it had caused him to look directly at her spiritual incandescence and realize that he was in love with Philanth, body and soul.

Her quiet, steadfast admiration made him play the role of the wise, kind statesman she thought him. When he had made his pretty little speech to reassure her, he had felt like the finest fellow alive, wise and strong and practically ten feet tall.

But it was sheer luck that he had been more rested and cheerful than usual this evening, and temporarily on the winning side in the trench warfare being waged to defend his colon from foreign domination. Basically he was mentally exhausted. Yet he was not even at the end of his hopscotching around in Africa, and when he did leave he would still have nearly two weeks for stops in Asia and for getting back to Washington.

He was married to a wonderful woman; and if ever there was a political job that called for a Mr. Clean, he was in it.

That was why his predecessor had been moved out so quickly when his improprieties toward women at the headquarters had exploded into a feature article in the *Post*. Almost overnight he had been gone, out of Washington altogether. Incompetence and peculation were tolerated, but getting female columnists up in arms was another matter. The man must have been an ignoramus not to have understood that, considering the precedents. It helped, of course, to have been reading the Washington newspapers for years.

The only thing he had going for him in this situation was that he was too scared of messing up on any of his innumerable job-related problems to get depressed again. But if he only got exhausted and stressed-out and down-hearted, knowing Philanth was in bed in the next room, ...

Dear God, how was he ever going to make it?

—Especially when in addition to all the other things he had to do or keep from happening, he had to make it back to Washington without blowing any chance of keeping Philanth working for him. Obviously he had to keep good control over himself, because Philanth would run like hell if she realized how he felt now.

Just wait, he thought, just wait until I get my hands on Carole. None of this would ever have happened if she hadn't . . . .

As he had implicitly promised Philanth, he outdid himself at the party. He had a substantive discussion with the Japanese ambassador and two of his staff, and as a result he developed great hope of wedding the human resources available to Peace

Corps with the large financial resources which Japan was committing in Africa but lacked the qualified personnel to administer. That alone made his stop in Kenya well worthwhile, and he was inwardly exultant when they concluded.

Moreover, he had a very promising conversation about the work of the UN Population Fund and the World Bank in addressing the deeply embedded cultural obstacles to smaller family sizes in the exploding slums of African cities. The governments were now sufficiently afraid of the destabilizing power of these huge, unemployed urban populations to cooperate with serious efforts from outside, provided that the approaches were not counter-productive due to cultural insensitivity. He laid the groundwork for important new efforts in this work which would be built into the Peace Corps' existing programming in maternal and child health education and services.

Thanks to his long interest in the problems involved and his study during the year he had been with Peace Corps, he was able to engender as much hope and commitment among the other officials as he felt himself. He used his pocket dictaphone to prepare for a confirming memorandum to each of the other principals. Philanth would be delighted when she transcribed and polished it for him.

When he climbed into his bed he felt rather pleased with himself.

But then he lay and thought about Philanth. It was exactly the activity which Carole had admonished Philanth not to make him waste his invaluable time on.

He wondered whether all the years of Carole's acting as if he were a future president and therefore should have anything he wanted/needed had impaired his ability to talk himself out of wanting something which he knew was absolutely out of the question—but which she had told Philanth he was entitled to have. He certainly was not capable of talking sense to himself about this and being sure he could make himself obey.

He had said something to Philanth about the obligation of a moral person to work through his irrational times alone if he had no one to help him. Certainly no one on earth could help him in this. Yet he needed help. His feelings toward her were no more amenable to reason than a fire-breathing monster roaring for his sacrificial maiden: he felt he had to have her, and that was all there was to it.

At the same time he was also the mythical hero, naked except for loincloth and sword and shield yet ready to be torn to pieces rather than let anyone or anything touch her.

Except maybe himself. When he helped her out of her chains.

He had never realized what desire could be aroused by the spectacle of humble righteousness—not by innocence in the sense of unawakened ignorance, but by painstaking striving for perfection under a full burden of carnal need.

His desire was augmented because in her constant effort to be perfect he saw mirrored his own striving to keep fighting the good fight: these were inward and outward battles with the same ultimate object, differentiated primarily because of their differences in age and gender and opportunity. In the central purposes of their lives they were the same, and he could not help feeling that rather than struggling separately they should have the strength of being joined. Not just by working together, but completely.

They both needed that. As Carole had expected. The Peace Corps organization temporarily sheltered them with a collection of sympathetic people, but outside it the world could be very lonely. Take those with serious commitment and scatter them around the globe, and they were spread very thin indeed.

And altruistic young people get into the work place, they get married, they get children, they get preoccupied with money and careers and security. And people get tired: tired of struggling against the intractability of a world full of other people with different priorities and even wildly different conceptions of reality. And people get discouraged. And lazy and self-indulgent; resigned to changing nothing, to just getting by, consoling themselves with their private comforts. They give up on the visions of all that this world and its people could be, all the grandeur and joy that could displace the squalor and despair.

The Bjorklunds hadn't. They had had each other, and no children, with all the pressures to conform to the prevailing patterns of family-sized me-ism brought on by the responsibilities of having children in a consumption-oriented society.

Philanth would not be deflected into the comforts and preoccupations and priorities of materialism, either. Her husband would never harass her for neglecting her duties to him; as for her children, her children were those laughing, little brown youngsters they had met in the thatched-roof, open-air schoolroom in The Gambia.

—Who will grow up to find themselves trapped by underemployment, the size of their generation, and their own impulses to acquire wives and children. Thus the misery and frustration continue and thus the discouragement is always there, waiting to absorb each new generation as the youthful illusions about "making a difference" are abraded away by experience.

Midway in the journey of his life, even he was finding it an ordeal to keep to the path, keep faith in the possibility that one ordinary individual could make the world a better place. Philanth had built her life and even her character around that goal, and she had taught him that needing to *see* that one made an impact was not only potentially fatal, but unnecessary: the commitment for its own sake was enough to live on.

But how hard, how very hard to live alone on commitment alone, when the obstacles are so tangible and ubiquitous! Hard enough when you are not quite alone, but the world seems bent on denigrating you and silencing you for the sin of telling truths it does not want to hear!

Philanth, can you have any idea what it means when I know that I must love you? To make love to you would be an act of affirmation of faith, a recreation of reality like the divine recreation which constantly sustains the world, according to the ancient religious idea you told me about.

It is my destiny, as each of us creates his destiny in the millions of minute decisions by which he determines from moment to moment the nature of the person he will be.

Let me show you how it must be with us.

Imagine a statue of the spirit which moves the Peace Corps. Make it physically perfect, as any young goddess is perfect. But give it the face and manner of Philanth: thoughtful expression, cheerful tilt of head; one hand offering, the other holding invisible abundance in its repose.

Put beside her, in the next alcove over in your pantheon, the tragicomic figure of a would-be statesman of that brash young republic which for one moment in the march of ages reared its head to declare in its

427

turn that it would be the bringer of light to all humanity. Since one image is as good as another, give it the form and features of Philanth's latest would-be lover, one son of Sweden and Beverly Hills named Charles Bjorklund.

Yes. Make our souls, our central intentions, deities because in our abysmal ignorance we dreamed—like Buddha and Jesus—of fulfilling the myth of Prometheus; of lifting all humanity out of the mud from which we all, at one time or another, came.

Then let us become one, for I who seek to govern and she who seeks to serve are the archetypal male and female yin and yang of that grand and ageless impulse of cultural evangelism which has often been foolish and self-righteously cruel but also has helped humankind stumble from huts to space stations, liberating the imaginations which build magnificent cultures and great empires to soar beyond the bounds of Earth and wander among the stars.

Yes, Philanth, you would understand if I put it in those terms: as though there were somehow a historical inevitability, something in the nature of the universe itself, which requires that I must love you; must long to wrap my soul around yours and envelope it, blend with it, completely.

Must press you down on your tidy bed and peel you, open you, devour you like a rich fruit created specifically for the engorgement and sustenance of my starving soul.

You would understand that it is what each of us is in our essence and our totality which makes our joining seem inevitable.

It's that sense of inevitability which is so frightening. People have let their lives be wrecked by that feeling of inevitability, of kismet, which is in fact only desire taking command of the entire organism with its biological prime directive. This, the body cries, is what I'm for, my destiny, my one fulfillment: coupling.

Again he fiercely reminded himself that he was, of all the world's secular government officials he knew about, the one who most had the obligation to appear to do no wrong. The Peace Corps was in its day-to-day focus bound to realism: it dealt with corrupt governments and wood-burning stoves and fish tanks. Nevertheless its genesis and ultimate objective was not only far-sighted foreign policy but also the institutionalization of altruism, a vision of the world which transcended races, religions, and nationalities with the realization that in the global village, poverty and ignorance anywhere are threats to the commonweal everywhere. And *he* was cast for a time as its representative to the world, the figure on whom all the spotlights fell.

So come off it, Bjorklund. You want to believe that because the rain must fall and the Earth must go 'round the Sun, your cute little temporary-special-assistant a.k.a. secretary must go to bed with you?

But she was beyond question what he needed.

He knew that because he knew her. Carole had warned him that his ability to know Philanth would depend upon his own capacities. That was axiomatic. Carole had meant it as a warning to open his eyes, to take the blinders from his middle-aged heart.

Then it had been easy . . . so easy. Seeing her ability to assist him threatened by his residual hostility, Philanth had let him know her as simply and unaffectedly as if she left the front door of her home wide open, allowing anyone who was interested to look inside. While she did not try to drag anyone in, she paused in her earnest comings and goings to offer whatever she had.

429

Inside he found warm, fragrant homemade bread, cool, whole milk, and pure honey. A beeswax candle stood half-consumed above a dog-eared collection of Wordsworth's poetry. A cheerful fire burned on a clean hearth, or a window with crisp, white curtains was open to a summer meadow.

Going all the way into her small, white cottage, he saw her narrow bed and framed pictures of her dead parents and husband and a bouquet of fresh flowers.

He was exhausted, and she offered a safe, peaceful place to rest. His clothes were caked with mud and stained with blood from his battles, but she would run him a hot bath and loan him her father's robe and bring him hot vegetable soup to eat. Before she washed his clothes she would even try to remove the ugly stains unless he told her not to, and she would mend a failing seam if allowed enough time.

Despite Philanth's complexity, her embrace of global concepts and her weakness for elegant phrases, this was how she lived; this was her character.

Who would fail to find it appealing? Only those whose regard for their own imperfect selves was threatened by the banality of goodness. Only those who refused to believe that the world still held or had ever actually held anyone who so defied their own experiences with life and people. They would refuse to believe that people could shape themselves around such selfless passion.

This quality of her love made Philanth's wish to escape her own sexuality all the more pitiful. She was still too young and too inexperienced to realize what potential for happiness she had as a woman and what fundamental human satisfactions she had dismissed in rejecting marriage. Her renunciation of what she could not know made him long to wrap himself around her, to protect her from herself.

And then, yes, to teach her what a wonderful thing it could be to be a woman. That had been Carole's phrase, engraved on his mind in a time when his world had still blossomed with promises. Carole knew, because with him she had learned. And with Carole he had learned what a splendid thing it could be to be a man.

Poor Philanth, who was so eager to learn and so grateful to anyone who would teach her! In the hope of making enrichment possible for others she felt obligated to forgo the most enriching lesson of all.

How far was the bottom of the well of pity from the underground stream of desire?

Well, that did not matter, because his feelings had a sound, objective basis in the way she had put herself together. Her sexuality was only part of a passion which was of far greater scope: beneath her fresh enthusiasms and cool intellectuality and crisp dedication to her duties he glimpsed the burning coal of a fire which could engulf the world and cleanse it. Even camouflaged as it was in the midst of her practicality because of her fear of being needlessly persecuted for being different, in his eyes it gave her a faint glow of potential greatness.

If only she could find a way of life which would allow her to let it grow stronger until it could blaze forth and spread to others, rather than continually requiring apology and concealment! He knew, he could not say how he knew except in her own philosophical terms, that she would never allow it to be completely quenched or smothered. She would still have that spark if she lived to be ninety.

Yet he feared for her. Being willing to give all, she was, as Carole had warned, in danger of being wasted if not destroyed. The humble cottage of her emotional simplicity seemed fated to be ravaged by one or another of the dead souls who marched through the modern world under the dark banner of knowledge that man was not only an animal among animals, but supreme in capacity for willful viciousness and depravity. She had not been toughened by normal exposure: the world had looked at her pale hair and modest mien and shy, dimpled smile, and it had been kind always. One rapist could destroy her.

The more he understood her, the more he feared for her and wanted to protect her. He wanted to protect from destruction by some other man the treasure he could not himself possess.

He needed to stop thinking about her and go to sleep! He set himself to concentrate on the horrible effects of a four percent population growth rate on Kenya. Nairobi was still a mecca for hundreds of thousands of tourists every year and a convention city for the world, but parts of its ever-growing slums must make Hell look clean and orderly and humane by comparison. A rich and beautiful country was in serious trouble.

In a sense, Philanth was correct: a person had no right to be absorbed in his or her own emotional life nor to live for pleasure or beauty, considering the millions of human beings who were being forced through no fault of their own to endure hunger, disease, filth, and hopelessness all their lives while their own officials terrorized and exploited them. Nevertheless, considering the extent to which the real battles for a better world had to be fought not against mere physical obstacles but against self-centeredness and pride and greed and the other forms of nastiness ineradicable from the human heart, didn't the struggle for everyday things like better schools require the solid nourishment of an everyday sort of love?

Chapter 16

THE TWO HALVES OF AN ARCH

He was just finishing typing his notes into his current briefing diskette when he felt the plane bank. He looked down, and sure enough, there was a city below them now. Another round of presentations, speeches, meetings. He sighed quietly. Just one full day in bed!

He put away his computer and took that day's "crib sheet" from his pocket for a quick review of the people and arrangements awaiting him below. Just before he left the plane he would fold it back into the jacket pocket beneath his right hand, its key facts having been placed uppermost in his mind. This was his routine now.

His schedule held him like the mechanism moving a metal duck back and forth in a shooting gallery. Regardless of anything, his body had to follow the pre-set routines from moment to moment. His mind came stumbling after.

He did not try to shirk any of it. He emphasized his openness to information and suggestions, his desire to hear what everyone wanted to tell him. He encouraged them all to talk, within the carefully calibrated time limits set by his schedule. Here it would be no different.

Seen from the air, this was a beautiful city. Its high-rise buildings were proud and clean-looking, their angular shapes neatly arrayed along wide, graceful boulevards. The overall impression of bristling modernity was softened by the greenery. From the air he could see no residue of the decades of exploitation, suffering and bitterness which had turned a thriving colony into a strife-torn nation and then into an economic disaster area.

While the politicians had argued and the true believers had fought, an awful lot of decent, ordinary people had died, quite unnecessarily and under appalling circumstances. Now that order was restored, there were plenty of things the Peace Corps could do here. Though interwoven with endless local variations, the basic needs were much the same everywhere: food supply, health care, education, income generation, and protecting or restoring the environment upon which food and income depended.

But Peace Corps did not have the necessary funds. As part of its basic mandate it always had to work out ways to get resources from other agencies for its people to use. The scale of Peace Corps involvement in a country was tied to the commitments from AID, and the U.S. ambassador had a lot to say about that. He needed to talk to the ambassador here.

Unfortunately this ambassador seemed to know a lot more about political fund-raising in the U.S. than he did about using funds for economic development programs in Africa. So what else was new? It was no wonder senators seemed to fall all over themselves in their eagerness to confirm any ambassadorial nominee who was actually qualified for a post by a relevant career in the Foreign Service.

At his meeting with Peace Corps country staff late that afternoon he was given his latest batch of mail from the State Department pouch. It included a lengthy cable keeping him "advised" regarding the rising level of unrest in Peru, but nothing else seemed especially significant except that Philanth had received a letter from Carole.

Carole had said she would not write to him via Marjorie's use of the pouch unless she had something particular to say, so it was not remarkable that she had not written to him as well. They also had agreed that since they had the use of the embassy pouch system, it would be foolish to have her letters trying to rendezvous with him by regular international mail while he was making such short stops at so many hotels in different countries where mail service was poor. Thus he had had only one letter from her, the one with the bad news about his legislation.

His occasional efforts to telephone her had only enabled him to leave "just called to say hello" messages on their answering machine. That was expensive; and considering the difficulty of being able to reach him in his hotel rooms without waking him up, she apparently had decided that no response was necessary. She would know from Marjorie that his trip was going according to plan.

However, he was anxious for Philanth to open her letter. He was afraid to imagine what she could be writing to Philanth about. Anyway, it might contain messages or a separate letter for him.

Philanth had not come to this staff meeting, for it involved personnel problems so sensitive that the country director had asked that the Peace Corps director himself be the only "outsider" present. Therefore as soon as the meeting ended he asked his driver for a quick return to his hotel before they went on to a "tea" with several government ministers.

In the corridor outside their rooms Tony went on to his own door, saying, "Knock when you're ready." Tony was in his bodyguard modality because of the country's residue of bitter factionalism.

He went to Philanth's door which opened onto the corridor and rapped sharply.

435

Philanth opened the door enough to peep around it, stared at him in astonishment, and started to shut the door in his face. "Excuse me," she said, and did shut it.

A few moments later she opened it again, apologizing because she had had to "put something on". She was still working her way down the buttons of her blue dress/housecoat. As usual when she was in her room for an extended period, she was barefooted.

"I thought you were going on to the embassy," she said, flustered.

"I thought *you* must be more hotel cleaning people wanting to tear up my room. I'm always having to fend them off. I've tried everything, even made my own copy of the international sign that means, 'No maid service wanted', but it seems the poor things feel they—"

"A letter for you," he interrupted, presenting it. She took it, saw the return address, and looked at him with trepidation. Carole had not written to her since they had left Washington.

She let him in then, apologizing profusely for the state of her room and fastening her housecoat.

The room was sultry and mildewy-smelling. Apparently the hotel was economizing on its mid-day air conditioning.

Her usually well-made bed looked as though she had just gotten out of it.

"Are you sick?"

"I'll be all right. Sit down." She sat on the bed and tore open the letter. He eased himself into the one chair, still looking her over while she skimmed through the pages.

Her mussed-up hair was in a loose braid down her back, and her incompletely fastened dress gaped above her knees, exposing thin, turquoise cotton.

She glanced up. "This is all just a lot of 'girl talk'. Nothing for us to be concerned about."

"What sort of 'girl talk'?"

"Carole's discussing plans she has for the house and entertaining a lot of people. It's nothing urgent."

"Entertaining?" he exclaimed. "'A lot of people'? Carole's idea of heavy entertaining is a couple of people stopping by for cake and coffee! She hates everything about it!"

"She's planning a formal dinner for the first Monday you're back in Washington."

"No!" He started to snatch the letter out of her hand but restrained himself. "Carole would rather be horsewhipped than give a formal dinner party!"

Philanth laughed, plainly not realizing how serious he was. "She and some of the headquarters staff have cooked up a black-tie event for two embassy couples, with the blessing of the proper officials at the State Department. She contacted the deputy director about her proposal first, and he thinks it's a fine idea. He 'took the ball and ran with it'."

"*What?*" He put out his hand for the letter.

She handed it over, looking nonplussed by his reaction. "Carole probably assumed you wouldn't mind, so there was no point in letting you know until the idea proved viable. People plan events around you all the time, assuming you'll have no objection."

He didn't answer, just started reading.

Dear Philanth,
    Please pass this on to Charles when you've finished with it. He will understand my writing to him through you just this once.

Considering Carole's game with her previous letter to Philanth, that had a possible layer of meaning that

Philanth might not pick up on; he did not like the sound of it.

"Take it with you," Philanth invited, "and get on to your meeting. You're going to be late."

"Thank you," he muttered, and left, mumbling, "See you later."

As soon as he and Tony were back in the car he resumed reading.

> I have been mulling over your comments about our home and have decided to begin implementing your ideas for its full utilization. If we are to use the house for overnight guests and dinner parties to promote Charles' influence regarding issues we will indeed require a freezer and an additional guest room. We may as well make it a suite while we're at it, since it would have to be in the basement and would need its own bathroom.

He managed to avoid exclaiming under his breath. What was Carole talking about?

First, who were all these people who would be overnight guests? It was he who might be the overnight guest, and on the other side of the country, if he decided to go for the senate race. It made no sense to spend large sums to prepare for his entertaining as a senator until he had won the election, especially considering how badly they would need the money in the meantime—not to mention if he ran and lost!

Besides, creating a "suite" which could please well-heeled and discriminating guests in that smelly, old basement would be no small matter: aside from the stunning costs of labor and materials in the Washington area, people tore their hair over the regulations

438

governing alterations to property in "historic old Georgetown". They would need a very good contractor, and that would not be cheap either. He felt his hair turning gray, just thinking about all the expense they could get into, not to mention the hassle.

He had always been happy to feel confident that he could leave such decisions to his wife. Carole had made no secret of her dislike of the aged kitchen and bathroom, but because she was not given to extravagance and had "better things to do than worship a house", she had kept the old town home much as they had found it.

Maybe she really was going off the deep end, from a brain tumor or something.

> I have gotten a contractor started on the project while Charles is away and won't be disturbed by the mess. He will enjoy using it as his study when we don't have company.

Carole went on to say what Philanth had told him about the dinner party she was planning. She added,

> I'm not much on playing hostess myself, but since you've said you enjoy it, perhaps you'd be interested in assisting me. I would see that you were well compensated in the currency you like best: information. I think you could make good use of my years of experience in Washington, so I've started making notes on topics I'd like to discuss with you.

He frowned. What would Carole have to teach Philanth that Philanth wanted to know? —Especially after Philanth had arranged for much larger formal affairs for the Hunger Foundation? Surely even

Philanth didn't really want to know *everything*. Moreover, what use could Philanth conceivably have for this suppositious new knowledge in her chosen career in poor countries' programs in birth control?

So what was Carole playing at?

Carole seemed to be suggesting to him a way of trying to get Philanth interested in his career so she would be more inclined to stick around. Philanth might be interested in *attending* his dinner parties because she wanted to talk with knowledgeable people, but Carole's offer seemed an unpromising ploy.

This effort to get Philanth involved in the *hostessing* sector suggested to him that Carole herself might be getting ready to pull out. Considering the state of the sexual aspect of their relationship and her previous letter to Philanth, that view of this new development gave him apprehension.

Maybe Carole had found someone else. Maybe she had gotten herself a ready-made senator, somebody with more drive and far better prospects of making it into the White House than he could hope to have. During the past eight years she had had opportunities to meet some of the senators who had been involved in his space/foreign aid initiative and/or his role with Peace Corps, and she might have come across someone reasonably attractive and at least unofficially available who recognized what a great asset she was.

The political costs of a divorce had declined a lot, and Carole would be worth getting divorced for. She was cold-blooded and extremely ambitious. And she had never enjoyed playing second fiddle to his career; she had only shared his conviction that it was a practical necessity because of the nature of the competition. She could have found a quicker route to the satisfactions of having a powerful husband.

This theory could explain her avoiding intimacy with him and pushing Philanth at him: with her usual thoroughness and cunning in getting whatever she wanted, Carole was angling for a painless divorce. So she had written this latest letter to encourage him to imagine Philanth fulfilling an important wifely function for him.

And since he had started this trip she had never been at home when he called, even in the middle of the night, Washington time. It all fit.

But was Carole capable of this? Perhaps she was only giving him a chance to acquire a useful mistress. Philanth would be a good second wife, as it were, to supplement the head wife in the areas where Carole did not measure up to her own exacting standards.

Horrible and risky as that notion was, could it be more than his own wishful thinking? Carole had always been tough-minded, especially where her ambition for him was involved. It *would* be nice, wouldn't it, to have two wives? It would be a lucky politician indeed who had two complementary wives harmoniously working together for him, one aggressive and experienced and the other warm and charming and pliant; able to stay out of each other's way in the office, the boudoir, and the kitchen.

Well, I believe I know how Philanth would react to that! I even think I know how she would begin: "*Theoretically* . . . ." Then, after having shown how sound the idea might be if only they were citizens of some other culture not remotely like their own, she would proceed to demolish it in terms of practical realities.

He arrived at the embassy and had to get back to business, but later, as he bathed and shaved to be fresh for the evening's activities, he toyed with imaginary openings for a letter home:

441

Dear Carole,

Sorry about your scheme to 'fix me up' with a mistress young enough to be my daughter so we could exploit her labor, but there's one wee snag: she's too committed to founding her own religious order, The Little Sisters of Responsible Parenthood, to have any aspirations for a decent sex life for *herself*.

Dear Carole,

Have you gone crazy? Have you never considered that I might already be having problems with this girl? If you want to reduce a grown man to a rigid state of suffering indignity, send him on an interminable journey with a sweet, young thing who also has brains, education, and character; say that you have no objection to his taking her into his bed; and tell her it's her duty to make love with him whenever he crooks his finger. Be sure that she's gentle, affectionate, and idealistic enough to be susceptible to your argument. And—dammit—also lock him into an iron maiden of guilt and ambiguity and her into a chastity belt of high principles to which not even angels have the keys.

He found his perspective improved by the exercise. Carole was *not* mentally ill or trying to trap him through some sinister plot. It must be wishful thinking to imagine that his wife really wanted him to be unfaithful to her. She had always trusted him implicitly, as he had trusted her. They were *married*: deeply committed to each other because they were linked in their feelings about their shared goals even when they were thousands of miles apart.

442

Besides, Carole was not only a good wife but also an astute judge of personnel. It was reasonable to suppose that she had merely found him a fledgling assistant who could be counted on not to complicate his life.

That supposition suggested that this latest letter of Carole's was only another way of testing Philanth to see whether she would respond as he and Carole would wish.

Serious entertaining of people interested in issues was what residents of Georgetown were noted for. If Philanth found merit in Carole's proposal, by all means, get her involved. Let her begin to take on chores which he *would* need done—he hoped—in future years and which Carole would be only too happy to delegate. If Philanth got more and more drawn into such roles and stayed on and on, before long all Carole would have to do when he needed to host a dinner party would be to take her place at the table and bear her part in the conversation.

As Carole had pointed out, because Philanth was young, she might learn and grow ever more useful to him for the rest of his working life. He could hope for another thirty-five active years: a period worth thinking about, especially considering the level he already had reached. The person Philanth could become would be invaluable behind the scenes at the White House; he could envision arguing with Carole over which of them was to have Philanth's services. In office or out, whatever he was trying to do, she could be a great help.

If she were interested.

He had to get her interested enough to stay with him as he undertook a perilous transition to another job. Aside from all her objections to his general proposal, changing her mind under the circumstances he faced at this juncture in his career seemed extremely improbable.

He probably would have to try for just enough of a short-term commitment from her to get him over that next hurdle. No, that wouldn't fly: working in a California senate campaign was hardly a step toward the overseas family-planning career Philanth wanted, so even that would require some sort of longer-term commitment.

However, he certainly was motivated—now. She was everything he could ask for.

Except that he would not be able to go to bed with her. That exception could give him some difficult moments.

But if she became willing he would have to send her away, because he was absolutely not that kind of politician: even in his own view of himself he had to be as immaculate as he knew how. Considering the constant walking of ethical tightropes inherent in the political jobs he wanted, he had to hang onto his clear conscience with great care. Besides, it was safer not to cut any unnecessary corners. And taking the larger view of his life, he would never risk letting sexual gossip tarnish his political objectives as it had tarnished the Kennedy legacy.

He turned the letter over to Philanth when he let her know that he and Tony were leaving. Seeing that she was still getting out of bed to deal with him, he told her not to wait up for his return.

It was not until the next day, while he and Philanth were waiting to board yet another plane and Tony was off "rassling with red tape", that he asked Philanth what she thought of Carole's letter.

"It was interesting," she responded politely.

She was acting unusually subdued. Suddenly gentle, he asked, "Are you under the weather?"

"I'll be okay," she answered listlessly. As though realizing how she had just sounded, she added, "Really!" But she was sitting quietly, not doing anything, and that was all the answer he needed.

He accessed his briefing materials for the next stop and began studying them. He couldn't do much with them, however, for he began having another attack of cramps and chills. This probably was what was bothering Philanth too, though they seldom ate the same things. Finally he excused himself and went to the men's room so he could at least get a little rest from the fear of losing control.

As he walked, torn between caution and haste and trying not to betray his discomfort in his gait, he thought, *God, what a miserable trip.* He was fed up with being hot. He never seemed to get caught up on his rest; he just kept getting tireder and tireder. His brain felt frazzled from information overload. He was tired of sun and dust and wretched little air terminals without air conditioning and delayed planes and anxious hosts. And all that was easy to cope with, compared with *always* having to be at his best with an endless succession of eager strangers. There was so much to try to do, there were so many things to try to remember.

These attacks of cramps and chills and the fears and nastiness which came with them were wearing him out, too.

And the food. He had almost forgotten to care about food. He had given up on the subject, partly because when it did turn out to be good he was always too distracted by the people around him, talking at him and demanding intelligent responses, to notice. Having to wonder whether everything that went into his mouth was going to make him sick made it too much trouble altogether. He had taken his belt up a notch, then another, and he had always been so thin that he avoided being seen in a swimsuit.

Poor me, he thought, glimpsing the funny side of being so completely miserable.

But some of the toilets he visited weren't so great, either. When he got to this one he couldn't do anything except be wretched. Driven out by the stink of the place, he left quickly. As he walked he ached. He ached . . . where did he *not* ache?

As he approached their waiting area he lifted his head to locate Philanth as the marker of his destination. He stopped in his tracks. Tony had joined her—in fact, Tony had practically wrapped himself around her, within the limits imposed by the intervening arms of their chairs. She was nestled in the black man's shoulder, and the brown hand which was not holding her shoulder was lifting a strand of her hair away from her face.

As he got closer he saw that Tony seemed to be checking her forehead for fever. Tony's expression was solicitous. But the black man was also smiling a little as he squeezed Philanth's arm with the hand that wasn't lingering by her face, and there could be no doubt that Tony was enjoying himself.

When they saw him they edged apart; Tony backed off completely. She said something, and Tony nodded.

As he came within earshot she was saying, "I wouldn't want Mr. Bjorklund to think I was weak."

"That's a big priority with you?" Tony asked with a hint of his habitual teasing.

As he eased into his seat on the other side of her she answered, "My Peace Corps medical officer decided I was weak, and I got sent home. It ended my tour of duty very prematurely. Not an auspicious beginning for a long career in Third World countries!"

446

"Oh, yeah," Tony said, leaning back and crossing his legs, his bare forearm finding a resting place behind Philanth's shoulders. "Somebody said you'd been given a medical separation."

Her tone became bitter. "That's a cover story. It would've been *administrative* separation, for non-compliance with medical policies under Manual Section two-eight-five, subsection seven-point-one. I had to cop a plea."

"*What?*" Tony quickly uncrossed his legs. "Wha' d'ya mean?"

"I couldn't even have afforded to pay my own way home! —Not to mention having to protect my record. So I had to 'make nice' regardless of whatever preposterous things anybody said. Never get on the wrong side of The Powers That Be."

Tony was all attention. "I don't get it. You mean you weren't really sick?"

"Oh, I was hospitalized at G.W.U., but only for a week or so. The point is, I wasn't allowed to go back. I was 'washed out'."

"You're saying they canned you, but they gave you an 'out' because what they were doing wasn't legit?"

"Well, what would *you* think? I was given a medical separation for dysentery."

"Come on! There had to be more to it than that!"

Wishing she were not sharing this story with someone who might put it into general circulation back at headquarters, he pointed out, "Lots of people have died of dysentery."

"Sure!" Tony scoffed. "If they didn't have any medical care!"

Philanth went on, "But they were making all sorts of allegations about how my being sick was related to my mental health. That would have bombed my record."

447

"'They'? Who's 'they'?" Tony asked sharply.

"The Peace Corps medical officer, who was also my supervisor." She added weakly, "Also my friend."

"Ha! Some friend!"

"I'm sure, now I've had time to think about it, that someone told him to 'fire' me. I don't blame him, but they felt they couldn't tell me the real reason. I think maybe I'd offended someone, maybe someone with a connection in the government or the diplomatic community."

"How?!"

"Well, birth control is a sensitive subject, for some people. Though goodness knows I *try* to be considerate of everybody's feelings! But when you get clinical experience you naturally become so used to dealing with it that if you're tired and in a hurry—"

"Baby, you never offended anybody in your life. You're incapable of it."

"You're sweet." She wasn't in the least consoled.

"You didn't appeal the decision?"

"No. I studied the relevant sections of the Peace Corps Manual, all forty-some pages of them. That was enough to scare *any*body in my situation. I saw that I had no clear-cut, objective criteria by which to defend myself before a Medical Review Board."

Her voice became hard. "When you're in a position you can't defend to your judges with incontestable proof, and somebody with more push than you've got wants you out, you're out. You're lucky if they let you slip away without a blotch on your record that makes it hard to get another job you want. After you're gone everyone can say whatever they please."

She went back to lamenting. "And I *had* to protect my record, so I could get another job overseas! So it was much safer to go quietly, as my—my friends advised me to."

448

Tony nodded, looking deep in thought.

She laid her hand on Tony's shoulder. "I think you understand, Tony!"

There it was again, that combination of unassuming but irrepressibly affectionate voice and gesture which could reach straight into people's hearts.

Tony pressed her hand but didn't look up from his reverie. "I know there's no justice," the black man replied dispassionately.

"Well, I don't care about justice for myself, but I had my life all planned." She was disconsolate. "All planned, for the third time! And it fell apart *again!*"

The crowd around them suddenly became active, and he realized that the boarding procedure was beginning. Apparently Tony had forgotten for watch and listen for the signal, and they relied on him for such matters.

"Le's go," Tony exclaimed, quickly shrugging into the strap of Philanth's shoulder bag and then gathering up their two suitcases.

"Here, Mr. Bjorklund." Philanth was helping him take up their two fully loaded briefcases and the computer. That done, she heaved onto her shoulder the strap of Tony's duffel bag and then picked up the flight bag containing Tony's on-board reading and their medicines.

They were very glad that Carole had offered Tony's dinner jacket a ride with his own in a hard-shelled case designed for men's suits, allowing Tony to stick with the "traveling-athlete" style of packing to which he was accustomed. For this flight was first-come-first-served in the sense that those who made it out to the plane first got to go, and the duffel bag had figured in their strategy. Their first thought had been that Tony should run ahead without more than two carry-ons. But while Tony was sure he could be among the first even at a dead run in the heat and carrying a fair amount of weight, he was not sure he would be allowed to hold off the other passengers to save two seats besides his own. Their present division of the bulk and weight of their possessions seemed their safest strategy, with Tony running interference for him and Philanth coming last with Tony's soft nylon bag to give her some protection from being jostled. He had wanted her in front of him, but Philanth had wanted him running interference for her and had promised to keep up.

Sure enough, the moment Philanth joined them Tony had to scramble to find himself a seat somewhere else in the plane. They watched to see that Tony had been successful. His bulk and color and football experience were helpful, but a number of the African men were even bigger and just as aggressive. When Tony gave them a wave and sank out of sight he and Philanth arranged their baggage as best they could and settled to rest, catch their breaths, and cool off.

But the air in the cabin was warm, smelly, and short on oxygen. His heart was pounding from his exertion in the hot sun, and his clothes were damp. Moreover, his bowels were shifting, and his chills gave him a cold sweat. He also felt a touch of nausea, but that could be from fatigue. There was a tape of hard rock music playing somewhere, not loudly but insistently, and in the metal shell of the plane the percussion played on his nerves like a dentist's drill. Some of the passengers were in ferociously bad tempers as they sought space for all manner of parcels.

He couldn't imagine anyone like Philanth traveling alone on an airline run like this one. She was panting too, and her eyes and mouth suggested she might be about to faint, though she certainly wasn't pale.

Seeing him scowling at her, she wiped her forehead with trembling fingers and dried them on her skirt, took a deep breath, and gave him a shaky smile which tried to say, "There, that wasn't so bad."

He wanted to kiss her for being a gallant little girl.

Had her gallantry been her undoing? She had attempted to make sense of getting kicked out of Peace Corps by trying to imagine how she could have offended someone. She *would*. Tony had her pegged: it wasn't likely.

It certainly was a strange case. *Surely* Philanth, even sick and refusing to stay in bed because there had been so much urgent human need around her, had not really seemed so neurotic that her Peace Corps doctor had truly believed she should be *terminated*. It wasn't neurotic to minimize your own problems because you saw them in the broader perspective of other people's needs.

Though if anybody ever really took that view of the world, it would soon kill them. One had to strike a balance between principles and survival.

"Philanth, about that termination Peace Corps gave you—sorry, 'separation': are you convinced it had to be a dirty trick to force you to 'leave quietly'?"

"I don't know *what* to think. I'd rather not speculate about it. There's nothing *you* should do, except give me a good recommendation, if I've earned it, when I try for my next job."

"That's quite a reasonable request."

If she really believed she had been railroaded and manipulated merely because some silly patient had complained to some local politician who had in turn put a bug in the ear of someone associated with Peace Corps, it was generous, considering what *he* might do about re-opening the case. Whatever the host-country politicians demanded, Volunteers were

451

at least entitled to honest record-keeping within Peace Corps.

Sensing another opening for pursuing his assignment from Carole, he shook his head. "Considering how much you want to go back overseas, you should use your influence with me to get your record cleared. But you don't want me to get involved in something messy. Isn't that right?"

"That's right."

"You know something, Philanth? You'll never amount to a hill of beans."

Her blue eyes widened. "Oh?"

"Because you won't even fight for yourself when it's a matter of protecting your power to do the good things you want to do."

Her dimples deepened with consternation. "I'm afraid you may be right."

"What're you going to do about it?"

"I—I guess I just have to put all my energy into whatever I'm allowed to do," she said weakly, "even though I keep noticing I seem to be smarter than some of the people who're running things."

"That's not smart. That's stupid. It's like the nun in the movie you were talking about, who let herself be wasted."

"Actually, she finally rejected the otherworldly priorities of the nuns and left the religious life. I believe the book the movie was based on was called, *I Climb Over the Wall*."

"Well, you know what *you* need to do to keep from being wasted, don't you? You need to get a mentor, someone who'll give you tasks that'll help you learn the things you need to know in order to move up."

Recognition sparked, but this time she looked impatient as well as amused. As though challenging him, she asked, "Someone who'll share his power to accomplish things I could never hope to do alone? —And who has the same policy priorities I do?"

He was a bit taken aback, but he had to accept that construction of his proposal. "Yes."

"We've already discussed this, Mr. Bjorklund. Are you implying you might become that mentor?"

"I am."

"It sounds wonderful." She smirked as she added, "Especially the way *I* put it.

"But the more attractive you make the position sound, the more certain I am that you don't need me to fill it. I'm very honored by your offer. But as I've said before, while lots of other bright young people are eagerly working for you—and your fine goals—I can be off doing many badly needed things that otherwise would go undone."

"Won't you at least think about it?"

"I'm not a closed-minded person. I did think about it, after we discussed it the last time. I could imagine myself having a wonderful time helping you write books and speeches and organize and influence people and all the other things you have to do to become more powerful and use your power well. But I could imagine other people doing the same things for you, if not in the same ways I would."

"But a lot of the people who'd want the position I'm offering you would be more susceptible to petty impulses. For example, they would be too concerned about protecting their careers to do what ought to be done about an important matter. Their very *perceptions*, their ability to ask the right questions, would be affected by their personal priorities. I'm talking about moral and psychological qualities, because those affect intellectual tasks. I'm sure that difference in you would eventually more than outweigh what you could have done in some job overseas."

"I'm sorry, but I don't agree. There can't be that much practical difference between myself and a lot of people my age who know and care much more than I do about the nitty-gritty of politics. You'd be the one to call the shots on their priorities, so—"

"Then you don't have adequate awareness of the significance of your personal qualities."

"Or it could be that your perspective has become distorted."

The fact that she even dared to mention that possibility was embarrassing.

However, his personal feelings about keeping her were a result of the qualities which she discounted.

He was striking out every time he came to bat, so he had better put shame aside and explain the crucial factor in his "perspective".

"It's not bias on my part," he began carefully. "But there's something about me you need to know in order to understand why I want to recruit you."

He leaned closer and lowered his voice, though the background noise of the engines and that awful music provided adequate security even if someone sitting nearest to them understood English.

"This is a hard confession for me to make to anyone. If you value me or our relationship at all, you must never even hint at what I'm about to tell you. Can I rely on you . . . completely?"

"I promise."

"Then here goes.

"I'm burned out, Philanth. That's the truth about me you've sensed but couldn't identify, probably because you didn't want to believe it. I need you working for me primarily because you're not, and I believe you never will be.

"Your integrity and idealism will reflect well on me if you're representing me to others, sure—but they're only part of this larger whole that I need to keep with me.

"You have a basis for renewing your strength that I don't have, one that I *can't* have. In my work I *must* be concerned about the visible effects of what I do, not content myself with 'right action' that'll be vindicated in the long run. Otherwise I look quixotic to people who don't look beyond their own lives and wants and prejudices. Being willing to be thought quixotic has aborted the careers of more brave and honest men than I can stand to think about."

"And you're not sustained by your own ego," she said, already grasping his developing theme. "That's why *you're* so needed; why you *mustn't* burn out."

"This will be my last job in Washington, my last job of any real impact, unless I can find a new source of . . . will to continue. I just don't believe in myself or in what I can do, anymore—not enough to make other people believe, and that's essential."

"Of course. But couldn't this be a temporary problem, from overwork?"

"No. It's been building for a long time.

"I could see that about myself past all doubt, the night I got the news about the fate of my pet legislation.

"And anybody who falls apart the way I did that night is in trouble. If he can't pull himself together fast he'd better get out before he disgraces himself publicly and demoralizes people who believed in him.

"Nobody but you could have helped me the way you did that night, and you were able to help me because I knew what you are; what you believe; what you're really committed to. Without you I'm . . . . I need you with me, Philanth. Because you know—not just philosophically, but in the center of your soul—that a life of honest service can't be made contingent on

455

the question of 'success'. To put it simply, I need your moral support."

"I see. I'd be like a battery charger. I must say, nobody's ever offered me a job like *that* before."

Seeing that she at least didn't seem hostile to the idea, he did not object to the ludicrous metaphor.

She went into a brown study.

Finally she said, "Let me balance your confidence with one of my own. You've told me that *I* had better get a reliable source of emotional nourishment. I said I would, but I wouldn't have to get married to do it. I had the thought then that just knowing you were somewhere in the world would be enough, if you went on demonstrating that you were the sort of person you seemed to be. Since then you've met that test.

"The two halves of an arch support each other. I think it could be that way for us, even though I go my way and you go yours."

He took a few moments to absorb the implications of that. "My God, Philanth! Do you realize what you're implying?"

"I hope you're able to . . . live with that information. I hope it solves your problem for you."

"Oh, I can live with it!

"But I couldn't live *on that basis*, even if you could. My argument stands: I need you *with* me."

She avoided his eyes. "Let's see how things go."

"That sounds like stalling until you can find an 'out' because you see that I won't accept 'no' for an answer."

She did not deny it. "I'd like a little more clarification of Carole's position before I try to deal with . . . yours. That's surely reasonable."

Chapter 17

COMFORT AT THE END OF THE DAY

When he returned to his room from the dinner and the protracted reception which had followed it he realized that he had been so distracted by the discussions he had not eaten enough to keep his stomach pacified until breakfast.

Philanth had learned to expect to work with him at this point in his schedule, so he tapped on her door. While she was getting him fresh cassette tapes in exchange for the ones to be transcribed he invited her to join him for a snack.

"I'd love to! Shall I order for us?"

"Yes, whatever sandwiches you can get."

"Stay in your black tie, would you? You look so smashing in formal wear."

Smiling at her tactful way of discouraging him from entertaining her in his pajamas, he said, "Okay."

He shut her door and got busy on putting his notes into his briefing diskette. This was one of their "easy" nights because the pouch-related paperwork could be dealt with during their flight the next day.

When the snack arrived she appeared in her black dress and cape. She had put up her hair in its usual swirl on top of her head.

"You didn't have to dress on my account."

"It took only a few minutes, and I thought it would be more fun, as if we'd both just come from a party— like in one of those slick magazine ads."

She had gone to considerable effort to act out such a modest fantasy. She must still be suffering from being excluded from his soirées.

457

She saw his expression change and hurried to the table to exclaim over the "elegant little sandwiches". She had been around Washington enough for this level of food service to be no novelty, but she knew when to be her childlike, delighted-with-everything self and when to act as if she ate canapés every day.

She ate with appetite, making appreciative little comments about various ingredients of the sandwiches and then falling silent in her effort to savor everything. He soon was satisfied, but with appropriate apologies and excuses she continued to work her way across the platter, switching to bottled water when her orange soda ran out. Considering how tight her corset was, he began to wonder where she was putting it all.

Finally he exclaimed, "What have you had to eat today?"

"Fruits and nuts. I was going to have more before I went to bed."

From her private cache, as usual. During his return trips to their hotels in various cities Tony occasionally asked that they stop because he saw a street vendor or market where he could quickly buy "hotel-room" food for Philanth. Tony knew her "wish list" and said she conscientiously paid her account.

All those fruits and nuts Tony had been buying must be getting tiresome by now.

"Don't you order from room service, when you can't leave the phone?"

"That seems extravagant when I'm eating alone. I can't enjoy something if it's expensive."

"Yes, but—Are you trying to save money on a government per diem?" It was an accusation that she lacked good sense.

"I know to be careful so I don't get anemic, because that happened once, when I was in college. I take vitamins plus minerals along with my anti-malaria drugs, et cetera. I asked Tony to find some supplements for me after I . . . stopped going . . . most of the places you do."

He grimaced. About every other night he was being feasted at an American embassy, where he could hope to escape the taboos on raw foods, dairy products, and anything which might have contacted unboiled water.

Meanwhile Philanth was living on fruits and nuts. He felt like accusing her of trying to make him feel guilty. But she had merely been stating facts in response to his questions.

Her spending so much time in her room had led to a similar problem when she had submitted her first bi-weekly time sheet for his signature: she had not claimed more than the minimal 80 hours, just as if her evenings and weekends had been free. They had had quite an argument about it, for besides assaulting the whole concept of reckoning a Returned Volunteer's work by the quarter-hour, she had insisted that she had a right to donate to Peace Corps the extra hours of credit she was entitled to accumulate while traveling on Peace Corps business. Such extra hours would mean time off with pay after she returned to Washington: a valuable commodity for anyone on a meager salary who planned to go job hunting. Yet he had had to simply order her to make out a new time sheet which reflected the standard allocation of hours of overtime credit.

Also there was the item of lingerie which Carole had insisted on buying her because she would not buy it for herself. It had become clear that this girl would not take anything unnecessary for herself unless someone else insisted. As Carole had said. She had taught herself to be comfortable doing without.

459

Yet she got so much out of such little things! They were on the top floor of the hotel, and the table and chairs in his room were arranged so that they could look out at the city as they ate. Philanth had just gotten up to dim the simple chandelier above the table so it did not compete with the lights outside. This was not a large metropolis with street lighting and neon signs all over the place, but it was pretty, seen from their window at night; and now that Philanth was no longer ravenous, she seemed mesmerized by the view.

"No matter how many planes we take, you always act as if staring at city lights were a rare treat."

"City lights represent all that civilization is able to achieve by making possible ordered lives; all the things I want people to have, now and forever."

She had ceased to apologize to him for her visions, and her pensive tone told him that she had slipped into the gentle, poetic frame of mind she was inclined to enjoy when there was time for reverie, especially in the evening. He would call it her "Sacré-Coeur" mood.

To protect it he spoke quietly. "What do you want for yourself, Philanth?"

"To help make it possible."

"For others." She nodded. "But for yourself? There must be *something* you want for *yourself*. Let your imagination roam free, tell me what you'd really *enjoy*: a mink coat, a cabin in the mountains, . . . ?"

"No, my conscience wouldn't let me enjoy anything like that."

She leaned back, folding her arms under her cape and looking dreamy. "Oh, there is one book I would enjoy having: a big, heavy pictorial atlas of the world. I found it at The Map Store when I was buying guide books for this trip. It's like holding the whole,

460

beautiful world in my hands. Going through it makes me feel like God contemplating his handiwork and loving it. Such a wonderful book.

"But it was costly, and I don't really need it. So it's just a fantasy, in the same class with the mink coat or the cabin in the mountains, even though I could buy it if I wanted to."

"Well, what sort of things do you *really want*, that wouldn't be too 'sinful' for you to let yourself have?"

"You mean, what things do I actually hope or wish for?"

He shrugged. She seemed to use even the most commonplace words as precision instruments.

"A safe, quiet, comfortable place to sleep is wonderful." Her tone was wistful. "And I enjoy wholesome food and practical clothes."

"Food, clothing, and shelter: 'safe' and 'practical'. What else?"

"A reasonable level of medical care, if possible."

"Come on! *Nothing* more?"

"Good health, and work I enjoy and believe in." Now her voice carried finality.

"D'you think you'd enjoy working for me?"

"Perhaps." Her flat tone and averted eyes said she didn't care to pursue that.

She added, "But enjoyment isn't relevant."

"No?"

"What matters is that I be properly utilized. The one thing I can't stand is feeling my time is being wasted! I wouldn't mind if my job made me miserable in other ways, as long as I believed in it."

"Well, I'm relieved to hear you say that," he admitted. "Working in a large organization like the U.S. Government isn't much fun for *any*body."

"If there's good management I think even working for the U.S. Government could be pleasant.

"—At least, if workers have realistic expectations. It's not so great if they suffer from hubris. People cut out to be St. George or St. Joan could still find it miserable.

"But as I say, I'm *prepared* to be miserable—as long as I don't feel I'm wasting my time."

He wondered whether she realized she had just compared herself to St. Joan, who would indeed have been a misfit in any large organization.

"Don't look like that!" she protested.

"Like what?"

"Like a spider watching a fly walk into its web."

"What're you afraid of?"

"That we'll have a fight if you still want me to go on working for you when I have to leave. I'd feel terrible if we didn't part on good terms."

"I guess it *would* be unpleasant to be on bad terms with someone you'd chosen as your source of inspiration for your life of lonely toil in the North African deserts."

Such a jab below the belt made her wince. "It would hurt me because *you* hurt."

"You're such a caring person," he burst out, "sometimes I can hardly stand it."

"What d'you mean?" she exclaimed.

"Don't you realize it humiliates other people when you put yourself on a pedestal by caring only about what's best for them?"

She nodded. "I'm sorry," she replied, her soft voice sounding genuinely penitent. "I had no intention of doing that. I must give the problem some thought."

He slammed his hand on the table, making her jump. "*Dam*mit, you've thought too much about such things already! So now how does anybody find you, inside that plaster saint you live in? Where's the real human being, the actual flesh-and-blood woman, when you take away all the five-dollar diction and righteous sentiments?"

She acted hurt but humble. "Suppose there's nobody left there? Suppose the center you're looking for, that you can relate to in terms of your own littleness, is only a commitment which requires denying the primacy of anyone's littleness and celebrating your greatness? Must you resent me in that case?"

He hunched forward, realizing that she must have explained herself in similar terms to Carole or to the journal Carole had read. "No. I suppose not."

Finally he complained, "It's just that your apparent perfection gets in the way of what *I* want. If you have your price, I certainly can't figure out what it is."

She straightened. "I'm not perfect at all! I'm just a good actress!" She had sidestepped his point.

She added quietly, "'Perfection' is not a useful concept where behavior is concerned—only control. As T.S. Eliot put it, 'Give; sympathize; control.'"

She tried to lighten her tone. "The last time I remember throwing aside control was when I was . . ." She paused to count on her figures. "Thirteen. I clipped a boy under the chin with a chessboard.

"I was playing a game on it with another boy, at the time. We never did find one of the white rooks; I always had to use a piece of eraser in its place, after that."

She could amuse him even in his blackest moods. "What made you so mad?"

"He was harassing us because he was an anti-intellectual male chauvinist."

"That bad!"

She gave him a warm almost-smile.

Then she leaned forward and spoke seriously, her voice husky-soft. "I'm not perfect any more than you are, Mr. Bjorklund. I'm simply no longer able to act spontaneously when I'm with another person. Every normal adult has learned to censor his behavior to some extent, but I've learned never to let my guard down. Everything is automatically weighed to calculate its effect. If I show anger there's a tactical reason to show it. That way I don't regret it afterwards unless it proves to have been a tactical blunder because of another person's reaction. To the extent that I'm caught unprepared, I'm unreadable. I've even been told so."

"You never struck *me* as the poker-faced type!"

"My apparent spontaneity is part of the act. In reality I'm following various prioritized, pre-set programs: I have one to edify a man who regards me in sexual terms, another to appease a woman who's jealous of me. If someone tries to insult me I pretend not to understand or attempt to mollify them; if they try to bully me to prove they're superior, I pretend to let them if there's nothing significant at stake. In an unfamiliar situation I behave within the range of what's conventionally acceptable to my audience without excessively compromising my own principles. When pressed, I 'waffle'; I'm very good at that. Later I decide what my real reaction is and consider how to deal with that type of situation in the future.

"It's not a likable trait. Anyone who begins to discover the extent to which I keep my true feelings and opinions to myself is going to feel I can't be trusted, that I'm two-faced and calculating."

"I wouldn't expect such a giving person as you are to have that way of dealing with people."

"I developed it because when I was in school the other students were always delighted to jump on me whenever I made the slightest mistake. An ill-chosen word, a tactless comment, any tiny thing they could find some silly excuse to take exception to, and even my closest friends would be all over me; *anything* to cut the 'genius' down to their own size, in revenge for making them feel stupid just by comparison.

"Isaac Asimov describes similar peer-group hostility in his autobiography. Like him, I refused to play dumb. I believed knowledge deserved respect, and having made an effort was nothing to be ashamed of.

"The abuse and ridicule I received because of other people's shaky self-esteem taught me never to trust, never to act naturally except as a calculated risk. Because the values I stood for would be denigrated along with me."

"I can see that it still hurts."

"Well," she replied, shrugging, "I'd never go back to a high school reunion. And I'm sorry girls still aren't allowed to be 'smart' as readily as boys are.

"That's partly because of the pervasive insecurity bred by the openness of our class structure and our competitive yet egalitarian culture. I'm very sorry that so many people don't base their self-esteem on goodness, because by that standard *every*body can be 'a winner'.

"But I'm not sorry to have learned my lesson: whatever I do or say, even if I'm acting annoyed or frivolous or flustered, I never say anything I don't mean—*in some sense*, possibly sarcastic or ironic. That way I never hurt anybody or make them angry unless there's a failure of communication. Also, I seldom have to eat my words, and if I do it's easy.

"I suppose being with such a careful person so much of the time *is* hard on you because you're under so much pressure anyway. I wouldn't make a good drinking companion, since I never morally relax.

"Maybe that's why I've proven to be unseducible. But it's more precise to attribute that to having more regard for my ability to reason than for my feelings, in this case my instinctive impulses. Not to mention being unable to commit myself sexually outside of the spiritual and formal public commitment of marriage."

He was wryly amused by her gambit. She was only underscoring an aspect of her character which he had not doubted, but it was helpful to have it spelled out.

Apparently she would have liked for him to make the next move, for there was a long silence.

At last she spoke. "You need something from me you think you can't have. But you *can* have it. It's just something you need to hear and believe and feel: I love you, Mr. Bjorklund."

"You say that the way you might say it to someone you don't even like. You seem determined to love *everybody*. Otherwise I couldn't have said it to you, at a moment when it seemed *you* needed to hear it."

"But you feel it's not enough."

"Was it enough for any of the other men you've hoped it would satisfy? Maybe for a casual acquaintance like 'Petya'!"

"Well, you need something like that, as comfort at the end of the day. At least, I do."

"You do? You admit it?"

"Of course. A person trying to do more than is possible needs a consolation for the intimation of mortality which comes when he has to admit that a day is over and he must take time to rest."

"Yes!"

"He holds himself to such impossible standards that he often feels like a failure. So he needs some consolation in his personal life."

"Yes again." He was glum.

"Also, you're giving others so much that you need something to replenish yourself. That must be why you've been so concerned about my having that problem."

"That may be one reason."

"Well, I do have something to give you that you need; and it's not the love I'd give just anybody. But can you receive it in the strength you need, yet in a way that leaves you unburdened? Can you receive it without tainting it by wanting more than is both necessary and right?"

Philanth could have loved him anyway. He could imagine how they could have enjoyed a deep mutual affection which was happy, undemanding, legitimate.

But Carole had taken their innocence in advance. He was obliged to evade her question.

"What do *you* turn to, Philanth? You admit you need *some*thing. Aren't there times when *you* need someone you can take in your arms, or something as completely satisfying as that can be, when you need to salve the abrasions of a hard day and make a truce with the world so you can let go of everything and sleep? It can be especially hard to 'let go' when it's the brain that's worked hard all day, not the body. You surely must have something that gives you more comfort at the end of the day than the sort of happy-think therapy you showed me that night I was so depressed."

"I do think about someone I love."

"Really?" He couldn't ask, Who?

"No one in particular. Someone I've never met or heard of. Just somebody who's alive somewhere right now. He's doctoring the sick for a hundred dollars a month, or he's teaching people things that will really help them. He's someone who wants only to spend his life making other people's lives better. Knowing he's doing that gives him happiness.

"He needs the comfort of knowing he's enriched the lives of the people he's trying to help, because due to his altruism he's been punished by all the pressures to be a 'success' in material terms. His parents will never stop letting him know that they expected more from him. They always thought that if they helped him through college and graduate work he'd be driving a fine car and living in a fine house by now, so they could feel vicariously successful in thinking and talking about how 'well' he was doing. 'Doing good well' leaves them cold. They know 'it's a dog-eat-dog world', and they feel 'nobody ever gave *them* much of anything', and they've 'paid their dues', and so forth. They never come right out and say that, of course. They just keep needling him, trying to get him to change. They're his parents and he's doing his best, so that disapproval hurts.

"And his wife has left him. Living in a developing country got to be too much for her, and she finally felt she had 'done enough'. And she's taken their two little children, to 'do what was best for them'. He loves them all, so he's very sad.

"Of course, their friends think he's in the wrong, sacrificing his marriage and family to his effort to help 'a bunch of ignorant peasants' who 'keep multiplying like rabbits'. As though some people chose to be ignorant and without opportunities for bettering their lot and therefore aren't worth helping!

"But he'll just go on working, because work in which you can lose yourself is a lifelong friend. And after a while he'll meet a nurse or teacher or aid worker who can see how fine he is and not want him to be different, ..." Philanth smiled at him.

"And they'll live happily ever after."

"No. They'll live happily until one of them dies. Then the other will try to go on living for both of them and try to make do with memories when the lost source of comfort is needed." As she spoke, her mouth began to go out of control. She folded her hands, then bowed her head and started twisting her wedding ring.

"That's not much of a bedtime story."

"I was just making that up as I went along, to give you an idea of the general type of person I wrap myself up in the thought of when I'm ready to go to sleep, if I can't be content with simple and immediate things because I'm too miserable."

"Why does that comfort you? It's such a sad story!"

"The capacity for love and intelligence is all I have to believe in. 'God is love,' and 'The glory of God is intelligence.' But people disappoint me so often and deeply that I have to remind myself that there *are* some outstandingly wise and loving ones around."

"But you fall back on some imagined ideal because a real person isn't ever good enough to outweigh all the bad and confirm your faith."

"That's not true! I didn't explain it well enough.

"I do know of an example: my dad had a sensitive, gentle, young doctor who had a very lucrative practice; and he frequently went to Africa to help the people *there*. He said he 'wanted to give something back because he had received so much'. Oh, he was such a kind man! And he was so handsome that his first-time women patients would exclaim about it.

"Yet once he said something which inadvertently implied that he was lonely late at night, and I thought it was terrible if such a lovely, good man hadn't found someone special enough to marry.

"He was even a gifted musician, like Albert Schweitzer; and surprisingly well-read.

"But *of course you and I* don't know of a *lot* of people like that! How can you expect to meet a certain kind of person if you don't go where they *are*? If I wanted to go looking for Jesus Christ today I'd go where the suffering is most widespread, in Africa."

469

"But I want to know how a truly good person who doesn't depend on God and has no close supportive human relationship manages to sustain herself."

"Yes. You've been asking what to do about needing stronger spiritual fare than our culture provides. Even a dirty, atheistic humanist like me may need that—may have an emotional life which needs powerful nourishment.

"I find it *can* be had by contemplating someone who represents an ideal. My imaginary doctor is such an example. He fills the need to find nourishment through worship. It's easy to do that by idolizing someone who's doing something heroic. Whether he's a particular person or an abstraction *doesn't matter*, since I know such good people do exist. As I believe I've said before, having known just one is all the proof that's necessary."

"If you were in love with someone, would you still need the comfort of this ideal?"

"Yes, if I were only in love—not happily married." Again she seemed to be reading his mind.

"And you're resolved not to marry."

"Yes."

"If you find someone who can serve as your comforting ideal, might you not fall in love with him? And if you did, could just thinking about him still give the comfort you need? Wouldn't being in love create an additional need?"

Their eyes were busy now, challenging or defending and then sliding away.

"I'd try not to think about him sexually. That's not too hard, because there's so much more to him that's so much more worth paying attention to."

"Would it really be that easy for you?"

"Of course I'd try to keep him at a distance, because simply idolizing takes less out of a person

470

than being in love, when there's no hope of . . . being happily married."

"But would you object to working with your idol?"

"Oh, no! That would be . . . the most wonderful thing I can imagine."

"If you're forced to be in close contact with your idol so you can't help seeing his frailties you might fall in love with him in spite of your effort not to."

"That doesn't follow: his frailties would have no bearing on the reasons I idolized him.

"If you mean his *faults*, you imply his faults would prove sufficient to make me stop idolizing, so I might begin to desire him as a man."

She began to smile. "But then he wouldn't be much good to me when I want to get to sleep, would he! The thought of him might give me comfort when I need something good to think about, but if I did think about him with desire I'd better be willing to spare the time to put aside the thought of him and . . . ." She took a deep breath and hardened her voice to a quietly matter-of-fact timbre. "And work up to a climax and enjoy myself."

Softening back to her normal tone, she added, "That's not such a bad idea, because if I do that I feel refreshed, and I have another reason to celebrate the goodness of life."

Her voice became feathery soft as she went on, "But I don't think about anyone in particular while I'm relieving a need for sexual release. That would be degrading. The only exception is when I'm deliberately working through my illicit feelings about someone I have no inclination to idolize."

She had led him away from his line of cross examination, but he could hardly object.

"You never think of anyone in particular, except for that purpose?" he dared to ask.

471

"No. I have far too much respect for people I love. Ordinarily when I reach the point of admitting I need a climax in order to stop wasting time struggling with sexual impulses I switch to fantasies which never involve real people—often not even myself.

"I find it hard to imagine having a climax while filled with awareness of the goodness of the person I'm making love with, because orgasm is an essentially self-centered experience even if it's enhanced by the excitement of one's partner. Does anyone really have orgasm because of some astral experience of perfect love and union? I at least am not that spiritual.

"The joy of loving a particular person for what they are is what makes the lovemaking beautiful in the first place: it's the foundation for enjoying complete intimacy and mutual giving without any unpleasantness at all, only happiness and tenderness and appreciation.

"But the joy of being with a certain person is not what one is likely to be focussing on during the moments when his wiring system takes on a life of its own. Then surely he or she is entitled to focus on getting the most out of that incredible, transitory pleasure so the body will cease its importunities at least for a while.

"I know I'm not supposed to talk about this with you. But I want you to be realistic in thinking about me, what I am as a woman: the human being inside the plaster saint. That realism is part of the friendship I want from you."

Knowing her motives were more complicated than that but appreciating both the stated motives and the reasons they were hypocritical, he nodded. "I appreciate your trust. And you're being clinical enough to do no harm. It's . . . informative."

472

"I'm not sure how much of what I've said is generalizable to other people," she demurred. "As I used to tell my sex-education students in Ohio, the scientific studies always have methodological problems which impair their reliability and validity; and a lot of fiction doesn't attempt to reflect reality."

"Are you asking for *my* impressions?"

"If you wish."

He knew she was asking for his past experience, but he took the time to imagine himself with her in his arms, entering her with the joy of complete possession. Avoiding looking at her, he began, "It's potentially more than you describe. I believe it's possible for the joy of closeness to a particular person and the joy of giving and taking pleasure to remain completely blended even in the moments of . . . sexual fulfillment. And that can be ecstasy indeed! When it occurs, love and pleasure do indeed enhance each other and increase together. That joining of erotic and spiritual love extends into all aspects of a shared life and blesses them so that all sufferings are made bearable."

The room was quiet for a moment.

"I defer to your greater experience. I'm glad to know that's possible."

"Your marriage probably didn't last long enough to mature into that much depth and completeness."

She dropped her eyes and sat very still. "No."

He had known he was taking a liberty in alluding to her personal circumstances so directly. He explained uneasily, "I only meant to point out that you're in no way at fault for not having completely united love and desire even when you were married."

She gave him a reassuring smile. "I understood."

To break the tension he sighed and leaned back, observing lightly, "Even a former sex-education

473

teacher can't know everything. I suppose there's always more to learn."

"Especially considering the vastness of the religious traditions which have incorporated sexual love."

She shook her head. "I never cease to marvel that Nature gave us such a gift. People who turn to drugs obviously have never experienced the power and wonder of love which combines strong sexual and spiritual passions; the enduring joy of people whose sexual love is vastly enriched by blending with a shared religious commitment to something far greater than themselves."

"But I thought you hadn't, either!"

"I'm speaking now of loving, including vicarious experience—not the moments at the height of orgasm."

"Excuse me."

Self-mockingly he added, "This is so edifying I could sit up talking with you all night."

She reverted to quiet compassion. "You welcome it because you're starved for a personal life from being away from home so long. Talking about sex is a high-potency surrogate for more diffuse forms of normal emotional intimacy. *That* is all you needed from me."

He heard the stillness of the night, felt their closeness, and knew she only wanted that to be true.

"Don't be afraid you've created a dependency in me, Philanth. A man crawling in off the desert doesn't become an addict because he sees a pool of clean, cool water right in front of him."

"Let's try to get you back on your feet with a little non-addictive recreation. Not taking enough 'water' into your system as you went along has to be what got you into this state."

"Any suggestions?" He was quietly sarcastic.

Resting her chin on her folded hands, she regarded him thoughtfully. "Wanna talk dirty some more?"

He smiled reluctantly. As usual, her instinct for personal interaction was sound: keep it light. She couldn't make him whole in an hour no matter what she did. So the best she could do for him right now was patch him up enough so he could relax and get the good night's rest he also needed. Physical and emotional exhaustion reinforced each other, so he was stronger psychologically when he was rested.

If he didn't have such a compulsion to keep pushing himself to the limit he wouldn't be such a cripple when deprived of emotional support from Carole. But he mustn't think about that now.

He slipped into an easy conversational tone. "You have a healthy attitude toward sex, for a dedicated celibate." He was taking her up on her offer.

"It was an eighteenth-birthday present from my mother. She was dead then, but she'd been putting together a box of things to give me: a 'curriculum'.

"It contained a long letter from her to me, plus photocopies of articles and selections from books, plus video tapes: medical advice, religious material, sex studies and manuals, fiction. She knew I'd had no 'experience', since I was too 'smart' to have a boyfriend. She cautioned that there was no way she could give me information only as I was ready for it, but she listed the items in order of increasing sophistication and noted risks of being exposed to the advanced levels too soon.

"She said I probably wouldn't find a suitable marriage partner before I got into graduate school, and she felt I'd be better prepared for life in general if I became well-informed and sexually self-sufficient when I needed to relieve the discomforts that begin during adolescence. So learning to give myself an orgasm was a bit like teaching myself to parallel park."

That made him smile again. He stretched comfortably, thinking that she didn't need to be apologetic for not being a good drinking companion.

"I'm talking too much," she said. "And it's getting very late."

"We could change the subject," he urged. "I . . . I'm not ready to end this."

She frowned thoughtfully. "What shall we talk about next? We haven't time for anything not to the purpose."

Desperation made him candid: "I want to use the little time we have alone together to ensure a lifetime supply. That would give me peace, tonight and every night."

"The memory can be enough."

"No. It would be too laden with regret because you'd gone."

"Not if you accept the rightness of my going.

"I don't want to be hard, Mr. Bjorklund, but this is nothing new to me. *Everybody* wants more of me, unless they're jealous or feel threatened.

"I go through life scattering flowers. I do it because others have scattered flowers for me. I don't claim any merit for it. But I do get discouraged by being made to feel like *la belle dame sans merci* just because I won't settle down and . . . ."

"Become a politician's mistress?" As soon as he blurted it out he could have bitten off his tongue.

"—cultivate somebody else's garden," she finished.

She added as though unsurprised, "That too. Both may be wrong and foolish, even if being a wife—or hired help—is more respectable."

She nodded toward the window. "There's such a crying need out there! It makes a comfortable life in

476

the States look immoral: each one of us consuming many times our per capita share of the world's resources, and so on. It's no wonder our people avoid seriously dealing with the question of what they should do about the inequities."

"They don't think much about it, do they."

"They use the excuse that they need all they can get for *their own* children. It's felt that 'A pretty dress for *my* little girl is more important than life-saving medicine for some other little girl I'll never see.' And of course everybody claims the right to have all the children they want, so they also claim the right to unlimited resources. Always grief comes from wanting too much."

"I've only asked for something I need in order to combat the same errors you want to fight."

"We disagree about what you need from me, then."

They sat in silence for a few moments.

She got up. "Good night, Mr. Bjorklund." She waited until he slowly rose to his feet. Then she stepped around the table toward him. She slipped her arms around his waist and gave him a warm hug, her cheek against his dinner jacket.

"You're a good man; a *wonderful* man." She looked up into his face as she stepped back. "Don't worry. This is only a transition as we come to—" She looked embarrassed, for she had been about to use Carole's phrase. "—As we become better acquainted. Things will settle down."

"I hope you're right." But he looked at her hungrily.

She stood firm. "You may rely on it."

He replied with a shrug which said he didn't have much choice. But he stood there feeling middle-aged and rejected.

"Are you humoring me, Philanth?"

She looked disconcerted. "I don't know what you mean! I really don't!"

"I mean, are you actually just playing along with me for as long as you have to put up with me?"

"You amaze me! One doesn't expect a handsome man to hold such a poor opinion of himself."

Slowly she moved up to him again. But this time she took his hand between both of hers.

"I truly think you must be the finest person in the world. I have no doubt at all that you're the most beautiful thing I'll ever turn away from." As she did literally turn away, releasing his hand, she added, "Without plenty of practice I couldn't manage it."

"You manage it all too gracefully. It's obvious that it costs you very little. But you don't want me, do you."

She paused in transit to the connecting door, but she did not turn to look at him. "*No.*"

"Well, your practiced compassion is cold charity. It only makes me more alone."

She turned to face him, pressing her hands and back against the door.

"I would rather die than make it harder for you to do what is right, what is necessary, if you're to fulfill your long-term obligations. So *I'm completely focussed on doing what I believe is best for you.*

"Don't be angry with me because that gives me more control and thus seems to set me above you . . . or because I reserve the right to follow my own judgment as to what I should do for you. In fact my love makes me more perfectly your servant, because *I forget my own needs when I'm with you.*"

His mouth twisted unhappily, but he nodded. She had squared the circle: given him an assurance he needed without compromising her integrity or undermining his already tottering control by admitting that she desired him. He sighed and turned away; they exchanged good-nights.

As she slipped out he registered the fact that she also had managed to make some suggestions about the acceptability of masturbation. For all her idealism, she was a practical lady.

To think that he had ever worried about her common sense.

The next morning he awoke very early. He was wanting her before he was properly awake. All he could think about was her body—he kept focussing on particular parts of it—and her smiles at him and her gentle, loving ways that in the aggregate seemed sexy just because they were so womanly. Her sense of humor. Her relaxed, down-to-earth comfortableness. The sometimes astonishing quickness with which her mind met his like an image in a mirror, confirming his right to be as he was—more deeply and perfectly than Carole ever could.

Her delight in even the most commonplace sensual pleasures. Nobody had to write him a letter to tell him that *she* could be wonderful in bed. As wife or mistress.

He tried to think of other things: the activities scheduled for that day, how the trip was shaping up, what he was learning.

But his body kept thinking about Philanth, about how supremely gratifying she would be to make love to. With that controlled yet boundless generosity and that acceptance of her natural responses, how completely suited she was to be a joyful, lusty, laughing playmate; a spiritual completion; a man's other self. Even her reluctance to accept anything for herself intensified his desire, for as her lover he would induce her to let herself move wholly into the ecstasies he would know how to give her. And he would have a great gift for giving to her because of the profound way she was capable of loving him.

He could not put into words the dimensions of his longing. As she had observed, words were artifacts of the mind. Not being able to compose great music, he was mute; for it was with his body and his soul that he needed to feed upon the pure happiness which this woman was capable of finding in his arms.

No matter how things worked out with Carole, he would need Philanth—sometimes desperately—for the rest of his life.

It was crazy to be lying there longing to touch her when she was on the other side of that door. Crazy. No wonder men of other cultures laughed at the sexual scruples of the white man.

His awareness of the invaluable time he was wasting on these thoughts combined with the prospect of continuing in this situation for the indefinite future drove him into action. It was either go next door and try to make love to Philanth—and be rejected—or start another day's work. He was still tired, and he had time to get more of the sleep he badly needed. But he ordered breakfast and set up his computer with the current briefing diskette and buckled down to do his homework.

But he was much too tired to do any good with all those dry summaries, and when room service tapped on the door to deliver breakfast he found he had been dozing at the table with his head on his arms.

This situation was ridiculous, he decided rather groggily as he began to eat. Something had to be done about it. But he hadn't a clue.

Oh, he had one, all right, but he wasn't too far gone to know that it was dangerous to take a course of action directly involving Philanth when he was too tired to think clearly. That night at Sacré Coeur she had spelled out the only acceptable way for a man to approach her sexually: verbally, as mind to mind. The fear of driving her away by following his impulses was all that was keeping him from doing something he surely would come to deeply regret.

On a flash of inspiration he left his coffee untouched; finished the warm, heavy breakfast; and went back to bed determined to finish his night's sleep.

Pleasure awakened him. A lingering shred of the dream snatched away by the intensity of his pleasure made him cover his head with his pillow. For it was Philanth his bed had become—not Carole. Philanth, his assistant for a month—not Carole, his wife for fourteen years! Not Carole, who had devoted her whole life to him, to his career, to his ambitions; who had helped him to achievements which otherwise would have remained wistful fantasies.

He was becoming infatuated with a girl almost young enough to be his daughter. He was despicable.

But his mind found pleasure in her mind. So his body wanted her body.

And though her body was dedicated to self denial, in his dream it had not denied itself to him. For her pleasure was in giving, a woman's sexual pleasure was in giving her body, and Carole had spelled out the mutually reinforcing dynamic between this young woman's vigorous sex drive and her ideology of selfless giving. Now he was dreaming of taking the fullest possible advantage of that synergism, strictly for himself.

Had it truly been a dream of enjoying *Philanth*? What made him think so?

He had felt her presence. When he tried to analyze what qualities, ideas, and emotions had blended to create her presence in his sleeping mind he had to

481

admit all over again that he loved her. Moreover, it would be impossible for him to persuade himself that he was wrong to love her.

So easy and natural a thing it is to love, after all. One can love an idea, a place, an object, without guilt. A man should likewise be able to love a woman.

Or so Philanth believed.

However, his poor, stupid body loved her in the only way it knew how. The body has its own laws. It is an animal innocent of knowledge of any law but the laws of nature. Ergo: it is without guilt.

The waking mind, if controlled by critical intelligence and self-scrutiny, may also be able to love without guilt. But in a very different way.

For the mind is responsible for dealing with the laws of society. These involve commitments and responsibilities, loyalties and expectations, fears and insecurities. Carried far enough, the concerns of society can cause a mind to let its body be destroyed with the full knowledge that it too will perish. For the laws of society are only tangential to the laws of nature.

However, the mind with all its responsibilities is influenced by the body and the body's amoral wants.

And so it is in the union of mind and body that we *homo sapiens* walk a treacherous path. For the body seduces the mind, especially when the mind is at its weakest, and they conspire together to ignore the laws of the waking world, the morals of a society, the commandments Philanth regards as derived from the laws of cause and effect which operate even in human nature and thus in the orderly societies it must struggle to create.

Philanth understands this conflict in our dual nature. She believes that no code for reconciling our conflicting needs as social animals is absolute. Thus she deals in consequences, in ethical relativism. Yet her conclusions are as demanding as those of earlier relativists have been permissive. For she not only looks at the possible consequences for society of self-centered actions, she also considers the inner consequences.

She even takes into account the fact that there are consequences imposed by social codes themselves, to the extent that they have been built into people's thoughts and feelings to make meeting society's needs easier and punishment more certain. That is why her immediate reaction to Carole's letter of instructions was that following it would be counter-productive because when passion cooled, having indulged it would impair my self-respect and thereby weaken my morale at its core.

Philanth's primary concern all along has been for my well-being. That has been her job, but it is also her nature. This provided the motivation for Carole's stratagem: Philanth would serve me well. It is also the reason why Carole's request of Philanth seems to acquire more and more plausibility as Philanth and I do come to know each other more fully.

Could she truly do it? He focussed on the Philanth who had been submitting to him in his dream.

Focussing on her face, suddenly he transposed onto it the face he had seen praying in Sacré Coeur. The ideas and emotions were the same. Could prayer and sexual intercourse be performed in the same spirit, whether or not the prayer was to a God no longer regarded as any more than an ideal and whether or not the intercourse was illicit?

For Philanth it could. She fused worship and sexual love; indeed, by comparing herself to a flower enjoying the sun she was using the first to control the second in her feelings toward him. Thus she would never give herself lightly, and even if somehow induced to give herself to him, she could do it . . . he could find no other word but "chastely".

It was a strange use of the word. But she could even be unchaste chastely: denying any right to pleasure for herself but accepting his physical need for her, she would accede to his illicit lust as she acceded to every other demand, with selfless love: unchaste in the deed but chaste in the spirit of the deed, with no desire but her ceaseless desire to do whatever he needed her to do. Moreover, when a soul was on its knees and she could give what it required, she had explicitly said that she believed in bending the rules.

He despised himself for even conceiving of such a self-serving, exploitative resolution of the dilemma posed by his desire for her. But he could find no flaw in it.

—Except for the one nagging little drawback: he could not imagine any way of persuading Philanth to agree to it. Unlike some women, she was not so stupid as to be unable to separate Bjorklund the champion of the poor from Bjorklund the man, Bjorklund the husband, and Bjorklund the public official.

Asking Philanth to give herself to him might have seemed to Carole a beguiling invitation to the sort of spiritual discipline through self-abnegation which Philanth courted. But in actuality Philanth was not vulnerable to that type of appeal: because as an atheist she believed that any form of self-abnegation had to serve a rational objective, she had consciously moved beyond the concept of any virtue, including self-abnegation or chastity, as a value in itself. Instead virtues were like sharpness or precision or other qualities of a tool: attributes of the mind which made it a better instrument for control and direction of the body toward constructive ends. Carole must have overlooked this point in her analysis.

Moreover, chastity, like many other virtues, was a type of self-control; and as a general rule Philanth had moved beyond the problems of controlling the wants of the self. Thus her chastity, like her other virtues, was only a by-product of her central commitment: it existed in reference to the spiritual core she had begun to reveal to him at Sacré Coeur. As she had implied the previous evening, this core was not a self to be controlled, but a commitment.

To put it another way, her reason and will arranged her character not around the center of a self which was in turn committed to her principle of service, but rather around the principle itself—the mediating self having been dissolved by ruthless self-discipline into pure altruistic intent.

Thus she did not have to make an effort to be "good", only to be rational in terms of her objectives. Convenient as that was for her in her daily life, it was also the reason why she was at risk of fruitless annihilation: she would not, she *could* not, make herself try to defend or protect what did not any longer effectively exist if some pressure to serve her exacting principles threatened it.

What interested him about that dissolving of the self and its temptations in some transcendent value was that Philanth's commitment was secular: to "the human family". She had knowingly chosen an abstraction in which to immerse herself in a way which made it psychologically equivalent to God. It had the advantage—and the distinct disadvantages!—of concrete referents. Yet she could subordinate her sense of personal identity to the concept of global humanity and its posterity just as a soul could lose itself in God.

Because of this immersion of her will in this idea, the demands of the body (and the unlimited possibilities for catering to the body's needs and pleasures) became incidental. Within the framework shaped by that central commitment, some subjectively imperative needs and experiences looked very small indeed. Thus her preference for chastity was only one corollary to the basic premise of her life.

Having such an overarching and unified framework of purpose, one would act modestly, selflessly, fearlessly, and with moral and intellectual deliberation, yet without making any conscious effort to do so. That perfectly described the way Philanth did everything, whether choosing her possessions, allocating her time, or interacting with others.

It was that constant, quiet absorption in doing her best, as well as her gentleness of speech and manner, which had deposited in his mind a cumulative impression of disciplined love and engendered his dream of a chaste girl accepting his wild and guilty desire out of devotion to the god she served by serving him. What made that idea irresistible was his feeling that she would accept him not in impersonal self-sacrifice but with pleasure because she loved him for the passion for service which she shared and which drew him to her.

Her design for her life was an integrated whole, shaping but also following the pattern of her character. It was a web of syllogisms which flowed out from a few simple premises about the value-neutrality of reality and the human need for values to culminate in a crushing combination of conclusions; and none of these could she refuse to bear, since each was inseparable from the whole. Any little pleasure she allowed herself had to make some defensible contribution to her ability to serve others, or like her new corset it became a cause for mental self-flagellation. If like the pictorial atlas it seemed more valuable in monetary terms than in terms of her needs, she would oblige herself to do without it. A visit to a church, on the other hand, cost only a few coins for the subway and met the same need.

Now he understood what Carole had meant when she had implied that any man of sensibility who came to know Philanth would be drawn to her. More than any of the practical benefits of her generosity and intelligence and the pleasures of her company and her body, a man would want to be loved in the unassuming and unqualified way in which she loved.

For while trying in every waking moment to follow the Christian injunction to be perfect, she, out of her bounty fed by order and beauty, imputes potential divinity to *us—to us all*! She even does it with compassion, understanding the burdens imposed by such a high calling. But by giving us the vision to see what good we do and what greater good we are capable of, she sets the heart on fire. In its humility, generosity, and inspiration this is a love which could enslave any man who could comprehend it.

How can I believe that the totality of all I think I have learned about her is valid? I'm a sensible man; I've seen the world; I can't believe it.

But I do. All of the tiny bits and pieces of her quiet, day-to-day self-revelation form a mosaic more integrated and coherent than the interior of Sacré Coeur. I know for a fact that she believes in *being* Christ in a way I never imagined. Ordinarily she tries to act ordinary, but in the times when she is most herself she reveals glimpses of that inner fire which illuminates her with its consuming purpose.

She makes me willing to die if dying could bring me close to her, to share the beauty of her vision.

But my life cannot be lived on that plane. I can't live on Communion wafers when I spend all my days chopping wood. I need the most concentrated and substantial nourishment I can get. And that is this woman in my arms!

To cover and embrace a woman who feels and thinks as she does, to move inside her, to touch her and feel her respond—oh, God, and to feel her understand me completely and yet worship because to her I embody what she adores! While I worship her for making me whole and strong.

Oh, Lord, please help me. I love her. I truly want to love her in a way that lets me not be ashamed with Carole. Philanth has mastered it, though she's fifteen years younger; but I don't see how that's possible for me. I need her too badly, just to go on living my life, to be capable of her kind of self-denial.

I have to love her, Lord. You must see that. She dreaded the thought of my having the "problem" of learning to desire her when I saw her sexually displayed because she already sensed that I was predisposed to love her. How can I not adore someone who understands so much, yet loves so purely?

When I saw her suffer at being exposed in that dress, I saw what this woman is as a living soul. The body, and sex by implication, really aren't important to a mind which embraces the world.

So she wants me to be as free of wanting, as free to give myself to everyone else, as she is. She truly wouldn't mind if we both were pure will.

If we *were* pure wills, I'd surely hate to met her head-on and moving under full power.

I thank you, Lord, for the fact that we're moving in the same direction. Please, God, let there be a way for us to somehow make the rest of the journey together! I've gotten awfully lonely.

Chapter 18

KATMANDU

His mind was getting so ground down by the constant travel and so worn out in the parts that did all the thinking and talking about economic development problems and programs and projects that when left to itself all it wanted to do was fantasize about lovely things he and Philanth could experience together. He could visualize her lying in bed with him, relaxed and playful; speaking softly, her seductive words making her dimples come and go like . . .

Like fireflies in the twilight of a spring evening in Virginia: in the background, fairyland vistas of blossoms and greenery, softened by a mist; in the foreground, savory food smells, the tinkle of ice in glasses, laughter, and the endless talk of politics.

Philanth would be happy and at ease with so many people she "could talk to on that level". She would approach his friends in Washington as though they were tempting dishes at a lavish feast. They would enjoy her, too. They would bring their weariness to bask in her admiration, willing to renew their enthusiasm for their labors from the amplitude of hers.

His friends would envy him. "Bjorklund's new wife certainly is young and lively," they would mutter to each other. "The sly dog. There's fire under all that lordly composure after all."

"New wife"? Now, where had *that* come from?

Probably from his fear that he was losing Carole. —Plus his fear that anywhere he took Philanth, it would be impossible to keep the cynics who populated politics from suspecting that he slept with her. He would be watched as he interacted with her, and if only because she was so skilled at eliciting people's warmer emotions, he could not constantly dissemble his feelings toward her. To the extent that she displayed her charm and intelligence she would remove any lingering shreds of doubt.

Word would get around. Staffers working for a prominent man they believed was fooling around just couldn't keep it to themselves, especially if he was reasonably young and attractive.

So he would be inviting gossip if he began moving around in his usual circles accompanied by Philanth.

But if Carole were preparing to dump him . . . .

He refocussed on the work in front of him.

He kept at it until Philanth absolutely insisted that he lean in front of her to look down at the snow-covered peaks rising through the clouds. She was awed at passing over the Himalayas and her chance to visit Katmandu. She promised him he could get all the good American food he wanted there, and it was cheap, and the restaurants had signs promising the safety of their food for travelers. It would be a wonderful city to look around in, she told him, "a marvelous mix of centuries and cultures", and she hoped he would get the chance.

"Even if you don't," he supplemented, looking at her, and she nodded, dimpling unhappily.

But she assured him softly, "I don't mind."

"Did you have things you wanted to see and do in every city on our itinerary?"

She hung her head. "Yes, sir. I imagined I'd have *bits* of free time, for quick forays. I didn't plan to buy much, just look around and try to get postcards.

"I didn't realize you needed 'round-the-clock phone accessibility to prevent the deputy director from taking over in some emergency involving the safety of

Volunteers anywhere in the world. Marjorie finally told me when I mentioned that you'd confined me to quarters after Banjul.

"If somebody had only pointed out that simple fact to me much sooner I'd have had more time for *productive* activities in the weeks before we left Washington. Of course I can understand why you preferred not to tell me, but you can always rely on my discretion, Mr. Bjorklund, if I'm adequately briefed."

"I'm sorry, Philanth."

"Don't be. I just wanted you to know that I'm not frivolous, I just didn't fully understand what was expected of me. I love beautiful things; and an unfamiliar country is full of things we'd never conceive of, created by the efforts of billions who came before us."

There seemed to be no capacity for personal resentment in her nature. She was always objective and analytical about what anyone did to her. She reserved her anger for abuses of power which caused waste of resources or hardships for others.

As soon as they were installed at their hotel he went downstairs and got a haircut. When he returned to his room the connecting door to Philanth's room was still open. Tony was standing beside her at her window, and they seemed in a sober mood.

At once he was afraid they had received some disturbing news during his absence. "Anything wrong?" he asked, quickly going in. Both speaking at once, they assured him that there had been no news and everything seemed to be "on track". He decided they must have been working on their personal agendas for each other.

He smoothed his hair, turning his head for their inspection. "Wha'd'you think?"

They came closer, scrutinizing, and Tony broke into a smile. Philanth frowned in consternation.

"That bad, huh! I thought so, too—when it was too late to do anything. It was a mistake for me to try to read while he worked. He must have a relative already in the shop or something."

"It's just that I like your usual barber much better," Philanth said too kindly. "The problem is, you wear it so short there's no margin for error."

"When I was campaigning I got a trim once a week. I'm subject to the same intensive scrutiny now."

He went to peer at himself in her dresser mirror.

"It'll do, Chuck. You know how women are."

"Sit down so I can see better," Philanth suggested.

He complied; Philanth started walking around him.

"Here's a place it would be easy to even up a little. Here's another. I could fix them if you like. I used to cut my dad's hair."

"Did you really?"

"He was too sick to go out."

"Well, if you think you can improve on it, I'll let you try. I don't want to risk a scene, maybe get a poor man thrown out of his livelihood and start unpleasant rumors, by going back downstairs. And I can't spare the time now to find another barbershop I'd be willing to turn myself over to.

"Do it right away. I have another full evening coming up." The country director had invited key Nepalese officials to a black-tie dinner for him.

"Yes, sir," she said, stepping toward the dresser.

"I'll run along," Tony said, moving toward the door.

"No, stay," she said quickly. "You and Mr. Bjorklund can, uh, confer." She opened her briefcase and took out her scissors.

"I think she wants a chaperon during this operation," Tony said with a chuckle, throwing himself onto her bed. "Well,—" Tony sprawled out, making himself comfortable. "Anything to oblige a lady."

Ignoring Tony, Philanth briskly set about her pre-parations: bath towel tucked around his neck just so, chair at the best angle to the combined artificial and natural light, his comb and her scissors at the ready.

"Now, tilt your head this way." She beckoned. "No, a little more like this." She pointed.

Tony snickered. "Go ahead, touch him, Philanth."

With one fingertip she pressed his chin at an angle. "Now I can see what I'm doing. Sorry my barber's chair is out of order."

"I know I need a shave too, but I'll do it before I go out." Even her light touch must have noticed.

Feeling her fingers in his hair made him stiffen. Had he not wanted to foresee the possibilities of this situation? No, fatigue had made him stupid. He had come unwound on the plane, and in the barber shop he had actually gotten drowsy and closed his eyes, welcoming the chance to relax before his busy evening. If he hadn't been so mentally disengaged he would have known better than to leave a foreigner he had never seen before to do whatever he—

"Hold still, please," Philanth murmured politely.

He looked up at her face. Her clear, blue eyes seemed to be focussed on the part in his hair.

She moved closer, and he found himself staring at the front of her blouse at close range. The soft, pale-blue cotton fabric was firmly pulled down into the waistband of her blue denim skirt.

He vividly remembered how her breasts looked when pressed upward and displayed to best advantage by the tight embrace of the merry widow. Rather than go on staring, he shut his eyes. But with his eyes shut tight he could still see her breasts, brimming the edge of her tight, black "merry widow".

She moved around and began arranging bits of his hair at the back of his part. The scissors snipped.

Everything she did was done carefully.

He could smell spicy warmth exhaled from her blouse. She had been rushing around that morning, doing Tony's usual sort of errands while Tony went shopping for underwear to replace what had been lost by the hotel laundry. In addition to handling calls to and from Nepal and Washington she had had to find out where to get photocopies made and send cables and so forth before she prepared their packet for the daily pouch to Washington and they were taken to the U.S. embassy for Thanksgiving dinner. Later, after the traditional "turkey 'n' fixin's", they had had that long, long wait in a warm, stuffy plane before taking off. Now he could not separate the scents of soap and spice and flesh; they were one with her body warmth. She smelled like a woman.

He had never noticed any other scent on her except shampoo and soap and that spicy deodorant smell. Not even when she was in evening dress.

"Don't you ever wear perfume, Philanth?"

"Such things have no place in my life," she replied swiftly but kindly, like a nun telling a little girl why she didn't wear jewelry.

"Haven't people given you some, from time to time?"

"When I was a child I kept it until it spoiled. Now if I'm given any I give it away.

"I like to keep my life simple: no jewelry, that sort of thing. —Unless of course I keep it as . . . a memento, a souvenir." She was recovering her tactless slip by promoting the necklace he had given her to the higher status of keepsake, then demoting it a little lest he think she valued it too much.

She concluded, "I like to fit the Peace Corps image."

"Living in a mud hut?"

"Sticking to the basics."

A butterfly seemed to be exploring his hair. Somehow that was an arousing image.

Philanth was a butterfly. A butterfly in a mud hut.

His brain was fuzzy from a too-short night's rest and the vibration of the plane—of too many planes. And too much talk. Even the short bursts of exposure to the sun and foreign smells outside the air-conditioned planes and cars and embassies and hotels were bothering him now. His fatigue was cumulative. He needed a nap. Hell, he needed a day or two in a dark, silent hole in the ground.

He needed a cool, dark mud hut. He needed Philanth there with him, to relax him and give him peace. Together with Philanth in a mud hut. Now, there was a concept. Alone together in cool darkness on a big pile of fresh, soft, fragrant straw. Sticking to the basics.

Philanth did not need any perfume.

"Chuck," Tony said, "isn't it true blondes are supposed to have more fun?"

"Yes."

"What about Philanth? You keep her chained to the phone *all the time*, and she'd just love to get—"

"He's speaking for himself, Tony," she cut in. "Tall, handsome, blond men have fun simply being what they are."

"Why doesn't he speak for a cute, little blonde like you?"

She paused in her combing and snipping. He felt like a statue receiving its final polish.

"How would *you* like to go through life being asked to sing, 'Old Man River'? That's how it feels to be a 'cute little blonde', except you're supposed to be capable of only the same kinds of 'fun' as a lap dog."

"Aw, Philanth, don't be a . . . ."

"Traditional nuns had the right idea about their hair," she said, ruffling his hair and making him want to grab her breasts just to get even. "Chop it off and hide the rest, to get on with the things that matter. Here people shave their heads—"

"*Looks* matter," Tony objected.

"Not to the eyes of the wise, my dear. And the sort of people to whom looks matter have certainly done a lot for *me*. —Starting when I was four or five years old and went to the movies and had to figure out what to do while Daddy was gone to buy popcorn and a strange man in the next seat kept putting his hand between my thighs."

That gave even Tony a moment's pause. "Well, you're not five years old *now*. And you should be having fun. This trip was the chance of a lifetime, and *you're blowing it*!"

"I like just *being* in Nepal. It's a very special place. Buddhist culture is a pleasure normally unavailable to me."

She started combing his hair.

The black man's tone became genuinely agonized. "But the temples you wanted to see, and all the fascinating shopping and scenery! The exotic foods! We were goin' to have so much fun together!"

"I'm having fun in my own way." She went on fiddling with his comb in his hair.

"Cuttin' the boss's hair?" Tony's voice became hostile. "That why you're takin' so long about it?"

"It's so silky it's hard to get hold of. And like I said, he wears it so short there's no margin for error. And these scissors have gotten much duller than I'd realized."

There was a pause. She again fluffed his hair this way and that with her fingertips. "I try to make the results unmussible," she confided to him, "so you always look tidy—even when it's windy and they're taking pictures of you for the newspapers. So I have to get it just so around the part and in front."

"Okay," he said, giving her a smile.

Tony grunted. "Watchin' you two together is like lookin' at a Norman Rockwell painting of puppy love. I don't know why you don't put each other out of your misery. I think if either one of you got smart, the other'd go along."

He tried not to react outwardly. The big black man was not all brawn and no brains, as he sometimes pretended. Like Philanth, Tony must have developed a non-threatening camouflage personality to fit stereotypes about his appearance, and that behavior had become second nature to him.

"Tony," Philanth replied in her sticky-sweet voice, "if you can't behave like a gentleman, you may leave the room."

"And miss this show? No way,—*ma'am*."

Now she gave his hair a thorough ruffling, and at once he forgot Tony. Oh, that felt wonderful. He'd never been sorry that she was a perfectionist when she was doing her usual paperwork for him, but now he was practically levitating.

It was over in a few seconds. She combed it all over, meticulously.

"Just a sec." She flitted into her bathroom, and when she came back she began dusting his face and neck with a damp washcloth. "*There*," she breathed in a maternal tone.

"You're like Carole. I've even known her to check my fingernails."

"Sorry if I made too big a deal out of this. I felt so ill-equipped and untrained! I'd never done anyone but my father before, and his hair was . . . quite

497

different. But you've got to look perfect for all these important people."

Now she even sounded like Carole. There really must be something about him that women found . . . impressive. They wanted him to be always picture-perfect, never human and ordinary.

Philanth gathered the white hotel towel away from his neck carefully so that nothing could escape from it. In his effort to cooperate he miscalculated her movements and turned his head the wrong way, causing his slightly stubbly cheek to brush against her arm. He was distracted from his pleasure by seeing goose bumps rise all over the back of her forearm. The room was not particularly cool.

"Thanks, Philanth." He had to clear his throat. "I'm sorry this guy is such a pest."

"He's much nicer when he's not playing his macho role, believe me."

She gave his forelock a final touch, smiling at it a little, and returned his comb to him.

"You'd make a wonderful mother," he murmured, looking into her eyes as he got to his feet.

"That's what the poor little world needs, after all the abuse and neglect it's suffered."

He was standing close to her and looking down into her eyes, enjoying it so much he didn't care that Tony was watching. "We need women who want to mother the world in politics."

"We need them everywhere." She turned away. "'Pity they seldom realize the seriousness of the need until after they've undertaken motherhood of the biological type, where we're so appallingly oversupplied.

"Better go and get ready for your dinner. I hope you can lie down and catch forty winks before you have to start your evening. No telling *what* opportunities may be awaiting you this evening."

She was right.

When at last he and Tony returned he took off his jacket and went to Philanth's room to deal with the day-to-day office paperwork.

First she wanted his cassettes. Reluctantly he informed her that the recorder had jammed on its first tape. Horrified, she took it from him and set it aside to try to fix it. She promised a new recorder would be in the next pouch, whether she succeeded in fixing this one or not.

"Would you like to dictate to me now, while everything's fresh?" Somehow she made it more of a command than a question.

He sat down and gave her a summary of what he had accomplished. It was brief.

"Anything more?"

"Well, . . . odds and ends. Sorry, I was pretty tired, I didn't put myself out as much as I might have."

"Oh, how I *wish* . . . . Nothing."

"That you could've been there," he finished.

"D'you really believe my just quietly *being* there would've interfered with your discussions?"

"Yes. It was an all-male working dinner."

He spoke apologetically, but she looked so grim that her biggest dimples appeared at maximum magnitude. After a moment she said, "Mr. Bjorklund, I'm very glad to be able to help you on this trip. I'm also glad you're sufficiently satisfied with my work to be interested in my staying on with you.

"But I'll be *damned* if I'll spend the rest of my life on details any literate, conscientious teenager could take care of while all the substantive things are done somewhere else.

"If I do have to go on doing this kind of work for the rest of my life, at least let it be in jobs I'm free to leave in search of something more challenging. Don't hem me in with a personal obligation that prevents me from ever having any *chance* to make some contribution which otherwise wouldn't be made. That would be like walling me up alive!"

"I would help you learn and advance."

"But that has to mean letting me go where people are sharing information and discussing problems." That sounded like an ultimatum.

"I can't change the fact that the world is run by all-male meetings."

"Not even a little?

"Anyway, let's just take an intermediate problem. I'd love to get some graduate degrees so I would have credentials to work my way into significant discussions on my own; and so I wouldn't always be at the mercy of somebody else's good will and good fortune.

"But let's be realistic: you'd never really leave me that much free time, because you'd always have something for me to do that was more pressing. *I* know how men in Washington use women: from early until late, subsistence wages, and with little or no real chance of advancement. You always 'need' us so badly for your slave work that you trap us and squander us. Meanwhile the men move up and move on, if only because *their* family obligations are respected by other men. You get away with that because there are always more women needing a job.

"Your motives are the best, but the effect is the same, because men's agendas are always more important than women's. That assumption continues to permeate the culture." Her voice held no trace of personal rancor. Again she was simply stating facts.

"The sort of work I'm doing now, tucked away with a telephone and routine chores while the negotiations and decisions are always transpiring somewhere else, is really all you offer me. You may claim it's going to be broccoli eventually if you can manage it, but 'I say it's spinach, and I say to hell with it.'"

He could not deny her contention. "I'm sorry. I don't want to be unfair to you."

"I understand. You'd undermine your own standing with other men if you stuck your neck out by being different. Because despite all the fine lip service to the contrary, the idea that men concerned about status or power would accept a woman as their equal any time they weren't pressured into it is a joke."

When he offered no rebuttal she went to bring a stack of folders from the dresser. Resuming her usual polite-little-secretary tone, she said, "I've dealt with today's pouch. Want to finish with the older stuff before we get into it?"

What else could he say? "Sure."

When everything was in order for the next day he got to his feet. "Good night, Philanth. I appreciate everything you do for me."

"Good night, Mr. Bjorklund." Her manner was loving, yet she might have been speaking to anybody.

As he returned to his room he felt sick. The one truly lovely thing in his life, and he could find no way to keep it!

Remembering Philanth's view of the matter, he made himself concentrate on the far larger broken dreams of Africa in order to forget his fears about his own. He visualized rusting machinery and rotting warehouses and gully-slashed roads until he fell asleep.

But his sleep was restless and sporadic, and he dreamed about questing for a place to sleep peacefully. Then he was contemplating the worthless souvenirs of a thwarted, wasted life. He was an old man, wandering in his mind; he lay down on the soft grass beside a country road and fell asleep in a patch of sunlight and died in his sleep. It was altogether a strange night. When at dawn he gave up on trying to rest any more he wondered whether Philanth, in her cultivation of wisdom through deprivation, had known many nights equally troubled.

It would have been altogether different if she had been with him. Being in the next room was not enough. Being in his heart and mind wasn't enough. Either a woman was able to share a man's bed and so get close enough to exorcise his demons or she wasn't.

To get his mind on track for the day ahead he shaved and dressed, then switched on his computer and began skimming through his notes regarding Peace Corps operations in Nepal.

Once again he felt that he was in a different world. Officially Nepal was a Hindu nation, though in fact the country had its own blend of Hinduism and Buddhism. Tradition permeated everyday life, and change was blocked at every turn: different castes would not even cooperate on a common water-supply project. The Nepalese hoped to improve their lot through reincarnation, but this must be achieved in part by accepting one's present suffering.

The spectacularly rough terrain for which the country was famous made medical care inaccessible to most of the population: roads were few, rivers mostly unnavigable. Moreover, there was one doctor for every twenty-six thousand people in Nepal, in contrast to one doctor per five thousand in India and one per eight hundred in the United States. Workers in the medical field became inured to dirt, disease, suffering, and death.

Yet there were changes even beyond the cities where tourists shopped and patronized restaurants offering the foods of their homelands. Despite the scarcity of roads among the steep mountains and the large areas where there was no refrigeration, heroic efforts were being made to reach rural districts with life-saving vaccines which must be kept cold. Understanding of the causes of infection also was being disseminated.

Providing safe water would make it easy to control half of the diseases. At present much of the population suffered from parasites, and their lowered resistance made them vulnerable to infection and disease. Every year diarrhea claimed forty-five thousand lives. Almost all of these victims were children.

The innovations in hygiene and health care were bringing to an end the time when a woman might bear nineteen children only to bury eighteen of them. Thus Nepal was moving into the same demographic-ecological crisis as most other underdeveloped countries, trying to support an exploding population on agricultural and forestry resources which continued to shrink due to overuse. Already two-thirds of the children under five suffered from malnutrition.

Peace Corps had been in Nepal since the 1960's. Indeed, the American community in Katmandu could trace its roots back to that time of young Americans' expanding horizons and spiritual questing. Some Peace Corpsmen had taken additional tours of duty, become field officers overseeing projects, moved on to jobs with AID, married, and returned to the States. Then, repelled by the superficiality of their own culture, they had come back to live in Nepal even though they had to find ways of circumventing the Nepali government's policy of making it difficult for foreigners to become permanent residents.

His review was ended by his usual morning phone call from Marjorie. Marjorie warned that she had a lot to go over with him, so he sent Philanth and Tony to the coffee shop to start their breakfasts.

When Marjorie had run through the usual sort of business she relayed the information that the chief of staff had met one-on-one with the deputy director and each of the associate directors and then reported to her that all had agreed that the ombudsman should be terminated for "stirring up trouble". Les was making his move, he thought. He told Marjorie to let it be known that he would defer dealing with that until he could meet with everyone after his return. He asked her to try to reassure Sandy.

He rushed through breakfast, after which he and Tony were picked up outside the hotel and taken to the first meeting of the day.

This was at the Peace Corps office. He hoped to do something about the staff's morale, which had been strained by conditions indirectly caused by the economic pressures imposed upon Nepal by India.

He began by mentioning his concerns, which had to do with complaints about adequacy of training and equipment. But he had been on the road for weeks, and the staff now treated these problems as old news. Therefore he soon opened the discussion to whatever other matters the staff wanted to bring to his attention. After some housekeeping matters the talk centered on whether and how Peace Corps could and should try to influence the host government's allocation of resources for building roads.

When the scheduled time ran out he realized that it had not been a worthwhile meeting. In fact, he was afraid it would only leave the P.C./Nepal staff feeling more discouraged: the dilemmas presented for their programs were all too clear, and now they saw that there was nothing that even he could do about them. They had to continue to simply slog along.

504

His diarrhea was back. He had another miserable session in a rest-room, then ate lunch in a restaurant with the country director. His host waited until he was full of steak flambé and grilled tomatoes and apples *en croute* and lemon meringue pie before he presented his "wish list". Of course there were no funds for any of the items on it, and he had to say so.

When his driver delivered the country director back to the Peace Corps offices one of the young men who had been at the staff meeting rushed out to give him his latest packet from the embassy pouch. "We couldn't give it to you earlier," the young man confessed, "because the embassy limo broke down, so the ambassador commandeered the courier's car."

There was a letter from Carole, addressed to him in care of Marjorie. At last! There was no return address on the envelope, but he recognized Carole's handwriting. The letter was thoroughly sealed in two hand-addressed envelopes, each of them marked "PERSONAL-EYES ONLY"; there was no sign of its having passed through the mail to reach Marjorie.

The letter itself had been printed by a computer. It bore no date, no salutation, and no signature.

He read it while being driven on to the first of his afternoon meetings with Nepalese officials.

I have a confession. I should not have let you hope that I might have good news for you when you return. It was sophistry on my part, resorted to because you wanted so much to hear it. I sincerely apologize.

The truth is that it can never again be as it was. I am not one of those cowards or fools who let their experience of what has been good be spoiled by consuming its dregs. I accept the fact that everything must end. I am sorry that this is over, but our memories have been kept beautiful.

I also must confess that the letter we discussed on the night of the bon voyage party intended more than I explained to you at the time. In the light of what I have just told you, you should feel free to ask for compliance with its instructions. If you hadn't found the carbon paper I would now be arranging for you to be shown the original. I am confident that this can provide compensation for the news I have just given you.

However, I cannot pretend that this is not a delicate matter. So I caution you as follows. She knows she can be destroyed in ways peculiar to her complete way of loving, so her duty of self preservation for the service of others may make her feel obliged to refuse to comply. On the other hand, she has unusual staying power once committed to a "cause", and for an adequate cause (*adequate to her*) she will risk allowing herself to be deeply and lastingly damaged.

You would never truly want to harm her, but you might fail to understand what certain actions mean to her as a woman. Unless you are certain that you are willing to dispense with her altogether and also willing to risk great injury to her, be very careful not to "use her up" in the way you gain her compliance in this matter.

Also—and this is *quite* important—if you are approaching the conclusion that she *is* dispensable, *please* phone me at once so we can discuss it! Because I can't recruit another like her, not even for you.

He had afternoon meetings with officials of the government who could not have concluded their business with him at the previous night's dinner; then with directors of the local offices of international development agencies; and finally with a group of journalists. These were much more difficult than the earlier meetings of the day had been. The encounter with the British, Japanese, German, Danish, and Dutch "country directors" was particularly harrowing: the other participants wanted to treat it as a discussion of substantive concerns rather than the get-acquainted session he had been led to expect, so it was far too brief and left everyone frustrated. It should have been set up in the two-hour morning time slot which had been wasted by the fruitless staff meeting.

He was dismayed by that mistake and the series of communication failures which must have led up to it. If he had been more deeply involved in all of the discussions and correspondence which had gone into the day-by-day schedules of his trip he might have prevented this from happening.

The final meeting, with the eager journalists, began more than half an hour late because of the difficulty in ending the previous meeting. Once it started it was so lively that he felt he had to let it run right up to the ambassador's reception and dinner, leaving him scarcely any time to freshen up and change.

Just as he was about to rush off to the reception Philanth stopped him and gave him his text for the after-dinner speech he was to deliver. She reminded him that it was the major item on the program, and he realized he had to muster his forces to present it. He looked it over in the car which had been sent for him by the embassy; he was relieved to find that Philanth had done a good job on it. Passion for one's causes was a great asset in one's speechwriter.

By the time the evening's scheduled events ended he was developing such good rapport with the U.S. ambassador that instead of saying good night they got involved in a really useful discussion. Their conversation moved from topic to topic, and it was proving so productive that finally the ambassador offered to accompany him back to his hotel.

When they arrived the driver found a place to park, and they went on talking for almost another hour. Tony felt it his duty to stand outside the car to give them his protective surveillance, and Tony was joined by the embassy driver, so he and the ambassador had a feeling of complete confidentiality even though he was taping their discussion for future reference.

When he said good night to Tony and let himself into his room he found lights already on. A moment later Philanth popped into view in the open connecting door, swaying with both hands on the sides of the door frame like a butterfly momentarily teetering forward in the act of perching.

He began removing his tie. Before he could come up with some routine pleasantry she burst out, "Mr. Bjorklund, the time! Where—pardon the cliché, but—*where have you been?*"

"Why," he said, startled but amused, "no place special. I was a bit delayed in getting here, ..."

"I called the embassy an *hour* ago, and they said you'd left the dinner with the ambassador and Tony and the driver! I was beginning to be afraid—for all of you."

He yanked off his tie and tossed it on the dresser. "Now, Philanth," he remonstrated, smiling. He began taking off his white dinner jacket, unhappily realizing that he was going to have to pack it and everything else again for the plane the next morning.

Philanth took a deep breath. "I'm glad to see you're all right."

"Of course I'm all right," he said, a bit irritated because he was going to have to get rid of her before he could continue to undress.

She put her hands on her hips and was momentarily snappish. "I'm not just a hysterical female who's gotten herself worked up over . . . someone she cares about."

Then the words came tumbling out: "The Nepalese are not *all* just a lot of peaceful, friendly peasants happy to stare at the tourists. They're undergoing a political awakening: they've been reached by the waves of democratic reform.

"But the vast majority are still illiterate. And a lot of people harbor bitter feelings toward members of the elite, especially if they knew any of the hundreds who were wantonly killed for demonstrating against the government's corruption."

He was tired; he didn't need this. He sat down on the bed and barely stopped himself from taking off his shoes in her presence.

"You're talking about before the reforms began. Things are reasonably quiet now."

"Yes, but the transition from divine monarch with corrupt, exploitative family and brutal repression of dissidents hasn't been easy."

"Well, that still has nothing to do with me," he said, looking around for the hanger for his jacket. "*I'm* not a member of the elite,—" He threw her a smile. "—Corrupt or otherwise."

"Aren't you? This is one of the poorest countries in the world. People who wear dinner jackets and ride around in limousines certainly *look* like exploiters of the oppressed, even if they're Peace Corps directors."

"I admit there's been turbulence from time to time, but the Nepalese are still regarded as a gentle people."

"Just the same, I read about one former Communist leader who had joined the puppets of the monarch, and his new cronies must've given him access to the goodies, because he was distributing earthquake relief to his supporters when a mob grabbed him. They . . . did ghastly things to him . . . before they beat him to death."

She clasped her hands at her waist. "I know this sounds silly. But it takes only one man with a grudge. I could imagine the ambassador's big, fancy car breaking down again and you being dragged out of it by some little group who felt they'd had all they could stand and figured you were some foreigner who was here to cook up some deal to line your own pockets, maybe use people as props for a movie, . . ."

"Oh, my dear girl. This is Katmandu—not New York City. As you can see, I'm completely intact."

He added wearily, "I've had a very busy evening."

"Was it successful?

"And what *took* you so long to get here from the embassy?"

"We were just sitting in the car talking after we got here. It was quite a useful discussion. The ambassador broke it off only when I ran out of cassette tape."

With a muffled exclamation of dismay she ran to her room. She returned with another cassette and hastily reloaded his dictaphone for him.

"I'll leave you to it," she said as she finished. "But please call me *the moment you're through*."

"Not necessary. I think I've gotten it all down."

"Oh, good," she began, but he was continuing.

"Thanks for the speech. I'd've had to speak off the cuff and done a miserable job, I was so . . . .

"You've improved in quantum leaps, Philanth."

"Why, uh, thank you, sir." She seemed distracted. "I needed a while to get my bearings."

"Well, . ." he said, waiting for her to go.

"Yes?" she asked quickly, shifting her weight from one foot to the other and back again.

"Well, I—I thought I'd go to bed," he stammered, puzzled by her hovering. "Unless there're any urgent messages . . . ."

"Well, there are! Things are falling apart in Peru."

"That's been expected. You've seen the advisories I've been getting. How bad is it?"

"A woman at the State Department said you may have to close the country post and withdraw the hundred-plus personnel Peace Corps has there. But it'll be a diplomatic disaster if you do it without its being clearly justified."

"Also a disaster if I leave them in place and one of them gets hurt."

"Exactly.

"It's already early afternoon in Washington and Lima. I've had to cover for you all evening when the calls came in from State and Peace Corps headquarters and the country director." Now he understood the strain in her voice. "They all want to talk to you. I expected you no later than an hour or two ago, and they agreed that the situation in Lima seemed too uncertain to require my pulling you away from your negotiations here. But they wanted you to call the moment you came in."

He rubbed his face, then his eyes. One full day's work done, and now it looked like he had to turn right around and do another.

Compassion crept into her voice. "I'll help you all I can."

"Mmph. Order some coffee."

As she moved off she said in her most motherly tone, "You get comfortable. I think we may need some sandwiches too, before we're finished. And the kitchen will be shut down." She stepped lightly through the connecting door to her room, closing it behind her.

Deciding he was not up to a shower, he started locating his one set of sports clothes.

Through the closed door he heard a silver peal of Philanth's merriest laughter; then Tony's deep, humorous voice. Their mood was so incongruous with the situation that he stepped over to look in on them.

Tony, still in black tie, was standing over Philanth, who was sitting on her bed with several crudely wrapped parcels beside her.

"Jus' deliverin' Philanth's groceries, Boss," Tony said, but he noted that Tony shifted position to speak to him so as to block his view of Philanth. "I was too rushed to do it before the reception, and I thought she might be gettin' hungry."

He came on into the room as he tactfully reminded her, "You'll still want to order the sandwiches . . . ."

He forgot what he had been talking about when he saw the shining brass figurine lying on the wrappings in Philanth's lap. He whistled. It was a dual form, male and female intertwined in what could only be highly enjoyable positions for both parties. The style was Hindu temple art. Those gods really knew how to have a good time!

"Philanth," he teased softly, "I had no idea."

She smiled but hastily pulled the wrappings over the embracing figures. "It's lovely, Tony," she said in her sugary voice. "I'm so thrilled to have such a souvenir. I'm going to keep it always. It was so very thoughtful of you. I'll always remember that you got it for me."

She turned from Tony to address him as she added, "Tony must've remembered I was hoping to see a little of the famous statuary while I was here."

She turned back to Tony as she added, "I suppose this is one of the copies from the local temple that was described in our tourist literature."

"That's right." Tony acted pleased with himself.

She turned from smiling at Tony to inform him, "There's a great deal of it around. It's thought to date from a period when lovemaking was revered as a celebration of life. Nowadays the guides get so embarrassed by some of it that they can't continue their spiels until they've moved on."

"Postcards." Tony handed her a small packet. "Also dirty."

Taking her lower lip between her teeth, she directed a wide-eyed smile at Tony, then hastily undid the packet. She flipped through them. "Oh, Tony, you're an angel! If you thought I'd get huffy and give them back, I hope you're not disappointed." She turned one over and began studying the white, printed side, then hastily put it back with the others, muttering, "I can read about them another time."

He reached for them, but she chuckled and held them to her breast. "I'm sorry, Mr. Bjorklund, you can't see them. You're too young."

Tony laughed. "I'll get you some too, Boss." Tony's expression fell a little. "At the airport, if I can."

"If he does, I'll explain them to you," Philanth promised demurely. "I learned about these when I was typing my dad's lectures for his History of Religion course. They usually depict Krishna and his favorite."

She put the statuary into a drawer without letting him or Tony see it again, then put the postcards with it as she went on, "The university library kept the illustrated books about Hinduism locked up in a special section of the stacks with the pornography. It's a shame."

She shut the drawer and went to her phone. "Want to share our order from room service, Tony? I'm afraid it'll be quite a while before Mr. Bjorklund and I can call it a day."

"No, thanks, I ate all that's good for me. I'm for bed. Sorry about Peru, Boss. Good luck."

Tony started to leave, but she hurried after him and reached up to throw her arms around his neck and give him a big hug.

"Tony, I'll always remember how kind you were to get these for me. Thank you, thank you!"

"It was nothing!" Tony threw him a glance and sounded almost angry. "If the boss would ever let you outa your room, you coulda gotten 'em yourself."

"Oh, and thank you for the food, too. I'm really looking forward to it. Don't forget to put it on my bill."

She hesitated, then added shyly, "And anything else you buy me, if you don't mind. Your getting them for me is quite present enough."

Tony threw him an embarrassed look and left, mumbling, "Sure. Sure."

"Now I'll order some proper food for you, and then to work," Philanth said in a school-teacherish tone.

"Let me change, first." He went back and changed into his sports clothes, pausing to splash cold water over his face and neck. He returned to her room.

Philanth put in a call to the State Department, and she had not finished briefing him for this first call when it was put through. From then on there was no let-up. Philanth stayed by him, feeding him papers and keeping him organized, prompting him in a whisper when he fumbled for a word, and making notes—

except when she had to go handle the phone in his room while he stayed on the one in hers.

It dawned on him as the hours crawled by that she was not only very good at this, she was enjoying it, in her graciously businesslike way. She understood the seriousness of the situation, yet she was thriving on it. That told him more than any words how bored she must have become by the normal pace and scope of her duties in this long succession of hotel rooms.

During one of his miserable trips to the toilet he saw that he looked glassy-eyed and haggard. The kitchen had sent up only drinks and pastries, so to keep him going Philanth made him eat some of the bread, cheese, and fruit that Tony had brought her.

"We're going onto a round-the-clock schedule," he warned her. "And your judgment will be crucial if developments justify calling me out of meetings."

She nodded, looking grave but not intimidated.

The calls tapered off at around three in the morning. He finally got to brief her about the rest of his day's activities, including the mistakes regarding the morning and afternoon meetings.

She was as upset about the misspent opportunities as he was. He had to order her not to try to find out how the misunderstandings had come about. These things happened, just because of communication failures and pertinent questions slipping into the cracks between various people's responsibilities, and it was far too late to make anybody back at headquarters unhappy after they had done their best. It was not as if the same thing might happen again.

"I'm going to bed," he mumbled, hauling himself to his feet. His brain swam, and he felt the nausea of acute fatigue. His back and lower body were masses of aches. "Call me if there's anything I should . . . you know, do, somebody who'll expect to talk to me."

515

"Yes, sir." He heard her sympathy.

He was stumbling toward the door to his room, but she surprised him by coming around in front of him and going up on tiptoes to throw her arms around his neck and hug him as she had Tony, her temple against his jaw. "You're a dear, good man, and I'm proud to be working for you," she said, suddenly sounding emotional. "You're a hero already, in my opinion."

He thought she might be regretting having been hard on him the previous night because he wouldn't take her to meetings. As though he didn't want her with him all the time!

He was feeling rocky, and as he grasped her shoulders he blinked tears from his eyes. "Philanth," he mumbled, "I got a letter from Carole today. It told me that her first letter to you meant exactly what it said. She says my relations with her are over. She tells me to . . . take good care of you." His voice was breaking. "I'll show you the letter, if you don't believe me."

Philanth drew back slightly and gave her head a quick shake as though to clear her ears. After a moment's hesitation she put her arms around his middle and pressed her cheek against his chest. "You poor man, you really have had a rotten day!"

She straightened. "Let's talk about that when we're fresh. Don't worry, now, you're doing your best."

Standing on tiptoes and pressing his shoulders to make him stoop, she gave him a peck on the cheek. He fumblingly tried to kiss her properly, but she dodged by giving him a final hug and moved away.

"Brush your teeth and say your prayers," she admonished, "and God will be good to you."

He chuckled numbly and went to fall onto his bed.

516

Chapter 19

A GOOD WAKING-UP PARTNER

As the reports came in he began to feel a cautious sense of elation. The last three days had been a nightmare, but it looked like he had done it! He had been in Shangri-La when he needed to be in the White House Situation Room, yet despite the terrific problems in getting adequate information, he had avoided making decisions which would have cost him and Peace Corps and countless other people for years to come. At the same time he had protected morale and Peace Corps' reputation by keeping his people informed and psychologically together.

Of course, he had ridden out events, not controlled them, and there had been a great deal of luck involved: the government had succeeded in curbing the insurrection, which had basically been useless protests against unavoidable economic hardships; in the meantime no Volunteers had been withdrawn, yet nobody in Peace Corps had been hurt. Just the same, he would get no critical headlines on this one.

Sitting at a table most of every night after keeping a heavy schedule of travel and meetings each day required more energy than he had, and Philanth had imparted to him her conviction that work done on a bed seemed much less like work and could be done longer with less fatigue unless a table or desk was clearly required. Because it was impractical to keep changing the arrangements by which calls were routed to her phone unless it was busy, he had by degrees taken over her bed, spreading his papers over it while he nursed the phone.

Now Philanth was lying across the foot of the bed. It had become routine for him to bring pillows from his room to ease his back, and Philanth had taken one of her own hotel pillows for herself. When her duties of taking notes and feeding him papers or folders had petered out and his final conversations had become a mopping-up operation she had clamped her personal pillow over her head and settled herself to rest, curled up with her back to him. As he concluded the final phone conversation she did not stir.

Slowly he got up and in his stockinged feet moved around to the foot of the bed. The little pillow had fallen to the floor, and she was sound asleep.

He hesitated over whether to disturb her by getting her properly into bed. When working with him in her room she tended to slump, and she always wore her smock minidress or smock wrapper; he had deduced that she wanted to keep him from noticing that she did not also wear her bra. Tonight had been no exception. Moreover, her hair was loosely gathered at the back of her neck, and she was barefoot. So she would be comfortable left as she was.

He gathered up his folders and loose papers and set them on the dresser for her to organize later. He drew down the sheet and bedspread and arranged them over her, tucking the covers up to her chin the way she liked her blanket on the plane.

He picked up his shoes and slipped into his own room and turned on a light. His phone had been moved as far toward hers as its line would reach, so he returned it to his bedside table. Finally he returned and switched off the lights in Philanth's room. He gathered his pillows and started to leave.

But it would feel so empty to merely return to his room. Philanth had been so patient and tirelessly helpful through all of this, and after everything they had just been through together, he needed a quiet little celebration with her.

A candlelight dinner, a visit to a cathedral, and then leisurely, quiet lovemaking: that would be ideal.

He needed to simply be with her. Close to her. He sat down beside her head and shoulders, watching her, hoping she would awaken.

Still she did not stir. There was only one thing he could do. Moving cautiously, he stretched out across the bed beside her, using his pillows to prop up one shoulder so he could relax but look down at the side of her face in the dim light from the other bedroom. This might be the only chance he would ever have to be alone with her and close to her like this.

He drank the sight of her. She seemed completely peaceful, yet intent; absolutely secure in her relationship to the big, dangerous, and sometimes very ugly world. In one of their airport-waiting-area conversations she had remarked that once one knew that if things got too bad one could always leave, there was nothing to really fear. Yet she also understood, better than anyone else he had ever known, the art of enjoying life; of getting the most out of each moment.

Wherever they were, she could create a pleasant, private little world for the two of them merely by being cheerful and loving, practical yet dedicated. Forgetting to be frightened by the guilt attached to some of his feelings toward her, he longed to cherish her. For the first time he understood why men had given fortunes for the favors of their mistresses.

He sank down onto his pillows, shutting his eyes, trying to absorb her nearness while he could.

The closeness he had enjoyed with Carole had been all he had ever imagined was possible. It had been constant, unintrusive, sustaining. He realized now that the woman a man chose to live in the compound adjoining his own could bring meat and drink fresh daily to his gate, could determine the sweetness or harshness of the stream running through his land, could determine the very weather and climate of his life. How she did all these things was of course greatly affected by how he did all of these things for her.

Because they had cared deeply for each other, their marriage had grown richer through the years. It was only during recent months that it had deteriorated.

Yet now that he thought about it, it seemed that he and Carole had seldom truly entered each other's compound for more than a few minutes at a time.

It had never occurred to him to make an effort in that regard, since up until this relatively recent breach in their relations Carole's feelings had never seemed a mystery. On the contrary, her single-mindedness had always made them seem obvious.

Carole had always seemed absorbed by her work. She lived quietly inside herself, concentrating on the imponderable external events and people upon which his interests and objectives depended. For these were the stuff of which her work had been made—until his move to Peace Corps had shut her out.

Yet she had shifted the focus of her activities without complaint. Her continued preoccupation with her work had been a major reason why he had been slow to notice when she had begun to deliberately shut him out.

Her letting the private passage through their common wall become filled *and stay filled* with everyday concerns also had for a long time seemed part of their normal cycle. Since he led a life filled with comings and goings and there never was enough of him to do everything that was wanted of him and that he wanted to do, their private interludes had always been slipped into the pauses in their lives. He had kept expecting a renewal of their intimacy without taking the initiative.

He could not deny that sometimes he had felt a need to share more with Carole than he did. But intimate communication required time and privacy, and meanwhile the world was always tapping on the door or calling on the phone or requiring him to rush off to catch a plane or attend a meeting.

Usually only in bed could they enjoy for as much as half an hour the preciousness of sharing which renewed the self by mutual affirmation. Thus those happy, stolen hours when need triumphed over duty were more than the only recreation he allowed himself. They had become his essential sustenance.

Surely they could be resumed. Surely Carole would allow it soon. Her latest letter must be only a ploy to promote her testing of Philanth so he would make the effort necessary to recruit her.

It *would* be a more severe test if he acted as if he believed both of the letters. Even if Philanth gave in to this pressure, he could of course decline to accept. His relationship with Carole was protected as long as he did not actually make love to Philanth.

But that was squalid calculation. Surely Carole had not thought things through that far.

However, if she had not, could the letter really mean what it said?

Well, even if Carole *had* meant it, she still could be won back. Even if things had not been so complete in recent months, the bond he and Carole had built up between them through the years was still there, underlying everything else.

That was the kind of marriage a man needed in order to face the merciless world bravely and patiently because he felt strong and content within himself.

521

He had felt confident that his private sides were constantly taken care of and nourished by that steadfast and dependable neighbor, his wife. With their hands clasped across the wall of their separate selves he had never felt completely alone. They had always been joined, complete together, free of awareness of any deeper need unmet.

But suppose there could be even more than this.

Only a man already so fortunate in his marriage as he had been could conceive of such a possibility. Most people did not even seem to realize what they were missing in not being as happily married as he and Carole were. (Had been?)

But *just suppose* there came a woman so open and so giving of herself that she seemed to have no compound of her own at all, nor any desire for such protection, because she had no fears for herself and no pride that seeks to conceal the blemishes of humanness; indeed, no personal agenda at all, except to give—but wisely. She was grateful merely to serve wherever someone would offer her minimal food and shelter and allow her to do what was most needed.

Yet suppose too that this woman's soul was in its essential characteristics the quintessence, uncomplicated by experience, of your own, so that when your two minds met there seemed no separateness, only a greater strength as when two flames fed by the same material move toward each other and become one. Suppose she lived only to lift the same burdens that you had bound to yourself so firmly for so long that the bindings had long since dissolved into the tissue of your mind. For she too had taken these burdens upon herself without reservation, even while recognizing that doing so would set her apart all her life from the many who lived only for themselves and those who were immediate extensions of themselves.

You need not open the gates of yourself to such a woman. You would simply awaken one morning with your high, strong gates of privacy still bolted and your guards still vigilant at their posts and see her standing there inside. She would be hesitantly smiling at you, her hands folded at her waist to say that she was ready to do your bidding, but she did not know how she came to be there any more than you did, and she was ready to leave at once if you wished it and could only show her the way.

But there would be no way for her to leave.

She might go out through the gate in her own will, and you might never see her body again. Her physical separation from you would impart an ache to your longing to join with what no man or woman alone can ever grow to become part of, that perfect, larger beauty which is visible in the starry heavens and palpable in the air around you when the world is still and cool and filled with moonlight.

Yet always she would remain as part of you: she would haunt the longings of your forever solitary consciousness, and the memory of her would shape you at your core. For having known her once for only one moment you will never, *can* never be separate from her again. Because what she is is what you are, and now you have seen it and known it, and also glimpsed, reflected in her, what you may be capable of becoming. You know that she exists or once existed, and by that knowledge you are forever changed: made larger by the understanding of what she is and made stronger by the joy of knowing that there is or can be such goodness, such wisdom merged with capacity for love, in a human mind.

You are made glorious by this knowing. For you can live in the certainty that her will is strong and beautiful yet of the same substance, even the same texture as your own, which seems so weak and flawed.

But if you are already married and faithful to your marriage, recognizing the good fortune of having already achieved that amount of union in something larger and better which is allowed to you, you will try—once you realize what has happened—never to look too closely at that silent apparition who stands in a corner with hands clasped, trying to stay out of your way because she can do so little for you of what she would wish to do and yet can never, never leave.

This was what Philanth tried to tell the men who wanted her: if a man was able to love her as she was, she could always be with him.

But her belief that this should be enough assumed that the blows a man felt because of his caring for the world would be no heavier than the ones which fell on her and that his mind would be as prepared as hers to bear them. Since it was in the nature of this woman to hope for less and prepare herself to endure more than a man would, neither of her assumptions could be true. Since she had yet to meet a man for whom both assumptions were true, allowing herself be known deeply enough to be loved risked giving more hurt than help.

She was finding that out. Moreover, she was learning that she could not go around admonishing those whose pain she had caused that the pain was from a rebirth, much less that suffering could be a great teacher. Most people would have preferred numbness, since that was at least safe and comfortable. Being reborn was frightening: one might not survive.

A sudden, loud clatter outside made Philanth start. He opened his eyes. The room looked onto an air shaft because she had wanted his party to be spared street noise, but light from the window was beginning to bathe the room in cool luminescence. Dawn was coming. Someone began shouting angrily in a tongue he could not identify.

Taking a deeper breath, Philanth turned onto her back, opening her eyes. She looked up at him, her eyes widening with the discovery of his closeness. Then she smiled warmly, her gaze unwavering. It was his own personal sunrise.

Her trust was a gift to him. He moved to kiss her.

At once she started to back away. Discovering that she was about to go off the end of the bed, she stopped. Her lips parted.

Seeing that she was trying to frame a jest to create distance between them, he pre-empted her: with a hint of reproach he told her, "People should wake up like this every morning."

"Oh, I agree," she answered quickly, though her husky voice was sleepy. She yawned a little as she went on, "It's worth any effort it takes to be good, just so you can find somebody good to be good with. But obviously too few people have figured that out. How else can you explain the fact that there's a terrific shortage of good waking-up partners in proportion to the supply of good sleeping partners?"

"Philanth, Philanth! What'm I going to do, carry your aphorisms around in my pockets on slips of paper? I'm being tempted beyond endurance."

He turned onto his stomach, too embarrassed by his inadvertent admission to go on facing her. Yet he forced himself to continue his confession. "Relieving the sexual tension wouldn't help. Because it's only an expression of the most enduring religious feeling I've ever had in my life."

She said nothing. He turned his head and saw her unblinking gaze at him. "Do you understand?"

She nodded, then looked away.

"Well, what do you recommend I do?"

"Put it in perspective. You must see that it's over-whelming you because of the many hours we've been together. Being pleasantly impersonal and efficient only eight hours a day is enough to take the starch out of anybody. Work sixteen hours and one inevitably starts getting personal. Work day and night for four weeks and then non-stop for three days and nights, and you wind up . . . ." Her eyes took in their present situation, and she dimpled meaningfully.

"'S no laughing matter," he said, "for me. All the hours you're holed up in your room, I'm on display to a crowd of people who study me like a rare specimen while I'm trying to be brilliant and exuberant and compassionate—or I'm studying my briefings, and soon I'm wanting you in my arms so badly I lose any ability to concentrate.

"And I can't tell you what I go through, thinking about how I want you when I'm supposed to be grab-bing a little desperately needed sleep. I don't know how much longer I can go on like this before I stumble into some stupid misstatement that makes headlines! A man like me can be haunted by something like that for the rest of his life!"

"I want you to know, Mr. Bjorklund, that when I see what you're going through I feel I must have a stone instead of a heart if I won't give you anything you want from me. But then I also think of how I would loathe to have you cheapened by any gossip that had some basis in fact. The jealous speculations I don't worry about: we do the best we can while not crippling our ability to get our jobs done."

"Philanth, I'm not the god everyone seems to ex-pect. I have only the strength of a normal human being. I certainly haven't got the stamina to be a model boss under these conditions. I'm sorry.

"But I want you. And I keep having the feeling I might as well be hanged for a sheep as a goat. I'm ashamed to be saying all this, but I don't know what else I can do . . . ."

"Try to be Christ to me. I know you're human, but I worship you; and the more perfect you are, the less there is to explain away."

"Hmph! Why should I encourage you to 'worship' me? That seems to be the way you propose to be able to go off to where I'll never see you again!"

"It was never *my* idea that you should want me to stay with you! Usually people *are* glad to have me around, and ordinarily they try to persuade me to stay longer. But they always come to accept the fact that I have to go on to do other things."

"That's because they have only their personal agendas. There's no one more selfish than a man with a great cause. I'm not prepared to do without you. I'm going to make you an offer you can't refuse . . . just as soon as I figure out what it is."

"It often takes us a long time to admit that something we want very much is not going to come to us. Especially when we're used to getting what we want."

"You've no right to condescend to me!"

"I was referring to myself too."

"What do *you* know about it! What do *you* know about life—about *anything*, except what you've read in some book!"

She meekly stood her ground. "I do know about letting go of a dream. I had everything *I* wanted, until my mother died. During the next several years I sat by my father's bed and watched my plans for grad school recede so far into the distance that it was easier to stop hoping than to think about what I'd have to do in order to fulfill them. I promised myself I'd take up the dream again as soon as I could.

527

"But by the time night-school graduate courses finally became possible for me I knew I'd die inside if I didn't go start a new life somewhere else. Since then I've had only jobs that left no time or energy to spare and haven't paid more than just enough to live on, because they were do-good work and didn't require any sort of graduate degree.

"I really wasn't feeding you easy platitudes from a book."

He had followed her general drift but was preoccupied. "I keep thinking, There's got to be a way!"

"To give me the same entrées into government and politics you'd be willing to give a young man you wanted to mentor, *and* to stop wanting to make love to me? Both are necessary. Surely you don't contend that those conditions are unreasonable."

"No. The conditions are not unreasonable. Just impossible."

"Well, give it some time."

"'Time will take care of it'? That's what Carole said—and started telling me how wonderful you are."

Before he could decide how to elaborate on that point so as to bring up Carole's most recent letter Philanth said, "I must've fallen asleep. But it's almost over in Peru, isn't it?"

He sighed, giving up for the moment as he remembered how tired he was, especially for the start of a whole new day. "Looks like it. When we get to Manila I hope we'll know for sure."

"Please go to your own room now, Mr. Bjorklund, and get some rest. We've done just fine since the crisis started, by sticking to our work. All we have to do is keep it on that basis for what time remains."

"More likely I'll have a nervous breakdown first." He moved closer, suddenly drawn by a desire that was quickly becoming overwhelming.

She forcefully turned away, hunching up her shoulders. "It's Monday morning! D'you realize that? On Thursday morning you'll leave Los Angeles for Washington."

"You think I won't want you in Washington?"

"You'll see this for what it was, and you'll be so embarrassed you'll *want* to get rid of me. Don't make what we've had together end *that* way!"

He hauled himself up to sit leaning on one hand. Again he felt a surge of nausea from fatigue. But he had to get one thing on record. "Believe me, Philanth: if it hadn't been for Carole's instructions I would have behaved a lot differently. I might not have felt differently, but at least this situation wouldn't have developed so quickly, and . . . ."

"I know," she answered, turning back to face him, quick to try to comfort him, as usual. "Carole put you into a difficult situation when you were quite innocent."

"She beat me over the head to get me to take you on this trip instead of anybody else! Now, why d'you suppose she would do *that*?"

She looked perplexed. "To get you to . . . ."

"To get me to recruit you for the long term because I'd notice how valuable you were . . ." His voice dropped. "Particularly to me."

They were quiet for a moment.

He added morosely, "Making this all one long trip to 'demonstrate my commitment to economizing': one guess whose bright idea *that* was."

"Carole's?"

"Bingo. She put us together like two rats in a maze. The male rat finds the female rat:" He smacked his palms together. "'Wow! One of my own kind! Fantastic!' And proceeds to act like a rat."

Philanth chuckled unhappily.

After a moment he added, "But I can't really blame Carole for what's happened to me—not when I'm so guilty now."

"If you start that, Charles," she warned—surely it was the first time she had ever called him that, but it slipped out naturally—"our relationship is doomed. Separate instinct from action, and hold yourself responsible only for the action."

"There's also the desire to act." Again he moved closer. "Philanth, I—want—you—so much, I think I'd deliberately make you pregnant, if . . . ."

She quickly put distance between them by easing off the end of the bed, kneeling to lean her forearms on it as she resumed facing him.

"Charles, Carole's set you up. Don't be angry until you understand all her reasons. But don't be too hard on yourself, either . . . just because I didn't leave it up to you . . . and thus force you into the realization that you wouldn't do what you want even if I were willing."

"Philanth, you can't know that about me! I don't know it about myself!"

"I do, though. You remind me of the famous poem by Stephen Spender: in essence it says, 'I think continually of those who were truly great. Born of the sun, they traveled a short distance toward the sun and left the vivid air signed with their honor.'"

"You scare me sometimes, Philanth. You don't live in the real world. You never have. You don't even know what it looks like. All you really see is projections of your own . . . beauty. Your basic way of reacting to everything is a 'kill-me-if-you-must-but-let-me-love-you' code of honor. You spent so much time with your mind among the saints and poets when you were growing up, it'll take you years, maybe decades, to separate all your self-delusions from the useful principles and valid inspirations you try to derive from them."

She looked thoughtful. "You may be right. I've told you what a hard time I have trying to understand some people—especially most rich people.

"Yet in the long run I think I'm building on stronger foundations than those who've never gotten below the surfaces of things. I see people messing up their marriages and careers and financial arrangements merely because they never got their own heads straight in the first place."

"Is it an underlying reality you've grabbed hold of, or only a set of pretty icons from the old culture that's slipping away?"

"There's no necessary distinction between the two. We each create our own underlying reality, and we selectively perceive the surfaces of things to fit it. I try to maintain an inner world with order and peace and love so the outer world can't destroy my will to affirm that there's reason to hope . . . and to act so as to alter reality for others."

That summed it up, apparently. As usual, she made it all sound simple.

"I might be able to understand how you go about that, but I can't undertake to do it too. My duty keeps me immersed in a reality where people want and hate. That's why I need you, Philanth: you're my retreat into solace and renewal."

"You think that now because of the way Carole has manipulated this situation."

"It's not an infatuation born of the conditions of this trip! I'll never 'get over' wanting you."

"I think you can get control over it," she said carefully, "when you understand what giving myself to you would cost me."

Carole's latest letter! "What d'you mean?"

531

"I . . . might also . . . inflame you if I spelled it out, however."

She acted as if she had expected him to understand enough to require no explanation. Perhaps his feelings were making him dense.

She went on softly, "Let's just say I love you more than anybody would want to hear about. We should leave it at that, if we possibly can."

He could see the sense of her deferring an explanation which might "inflame" him, under the current circumstances.

After a few moments he saw no alternative but to try a different approach.

"We haven't really talked about my latest letter from Carole. I think she truly wants me to have you because she wants to be able to divorce me with a clear conscience."

"How preposterous! For what possible reason?"

"Another man. Someone with more to offer."

"You're . . . You're very tired, and you've been away from familiar surroundings for a month. Your perspective has become distorted."

"Philanth, how can I trust her? In that letter to you she told you I'm a wonderful lover. She—hasn't—let me—touch her—since last spring!"

He turned away, shutting his eyes. Telling Philanth completed the humiliation. His wife, whom he had always admired because she was so self-possessed, hadn't found him, with his self-doubts and depressions, enough of a man. The sense of irony which finally had loosened his tongue had gone, leaving only despair.

"Charles," Philanth said hopefully, "couldn't it be medical? Cystitis can be excruciatingly painful. It's a urinary infection that can be cleared out by medication, but after a series of attacks the fear of bringing on another attack by having sex could make a woman feel like giving up having sexual relations altogether."

"If it were something like that, why wouldn't she have told me? It's not like her to refuse to discuss things! That's why our marriage used to be so good: we could talk about anything! And we told each other everything. No, Philanth. If it were just medical she would have explained it."

Philanth sank into a sitting position on the floor and was silent for a little.

"What*ever* Carole's intentions, she must have some rational reason for denying you your conjugal rights." Just as he was about to object to her terminology she dimpled at him and added, "Certainly no woman in her right mind would deny you those rights if she *didn't* have a compelling motive."

She went on reasonably, "So what could it be?"

"It makes sense if she's found somebody better."

"Not possible."

"Maybe not from *your* perspective."

"How could anybody be better from *her* perspective?"

"More able to fulfill her vicarious ambition. Somebody who got a faster start from his family background, somebody who makes his own advancement the top priority; someone so convinced he's God's gift to America that he's never even thought about it."

"Charles, you envy and despise a man who probably doesn't even exist. Carole isn't like that. She's tough-minded, sure. But she's ambitious for *you*: she wants to promote a man she can expect to pursue sound goals, not matter how hard it is. You take on that burden, and that's why she cares about you."

"There's more, Philanth. Carole has resigned from the state-level offices she held in big public-interest organizations in California: work she'd done all these years to help gain connections and name recognition for me among the core of committed activists—and thus build my resources for seeking statewide office.

"Instead of continuing to do that sort of work, you know what she's been spending her time on? She's been researching a series of personal-adjustment articles for women, maybe even a book: a study of values and motivation in relation to fulfillment.

"Meanwhile the political contacts grow cold and obsolete, and the name recognition fades." He sounded as desolate as he felt. "All that investment of her time and effort and talent is being abandoned just when it should be most strenuously cultivated to build toward a possible bid for the Senate in less than two years!"

"Why would she do a thing like that?"

"It's a return to her own field, psychological journalism. She wants recognition in her own right, after fourteen years of letting me get credit for practically everything she did. I can understand that, but only if she's really given up on me and doesn't want to tell me.

"And several times while we've been traveling I've tried to call her. She was never there, even in the middle of the night, Washington time! All I've ever gotten was the answering machine. But she's never told me she was going anywhere, and that's violating a principle of fourteen years of marriage: we're supposed to know how to reach each other.

"You can see there's a pattern. It all adds up."

"There might be some less sinister explanation than her planning to divorce you after you'd . . . become involved with me."

"Yeah." Again his voice became bitter. "Maybe she thinks we can form a *ménage à trois*, with you living in that new suite in the basement—giving me all the advantages of two wives."

"Ridiculous."

"Maybe not. There's no competitive advantage she's been willing for me to do without if she could find a way to get it for me. She's been pestering me for years about whether I couldn't benefit from having a car phone. I insisted I didn't need it, and I didn't want the car broken into by somebody wanting to steal it. If she's not promoting a car phone she's asking whether I couldn't use my own fax machine.

"So if she *isn't* getting ready to leave me, then I think this is her idea of a way to give me an edge over other politicians whose wives have more parochial attitudes toward their husbands' having second wives . . . or mistresses."

"Now, that's—" Philanth stopped. And looked intensely thoughtful.

"No, maybe it's not so ridiculous after all, is it! I believe both you and Carole could and would do it, if you made up your minds. You and she are a lot alike in some ways. I guess that's why she's always singing your praises: a tough-minded woman who isn't a self-promoting bitch is her kind of people."

Philanth sighed emphatically. "I begin to see why this situation is driving you crazy. It *had* to be more than my fantastic body."

He laughed wildly.

After a moment he said, "You don't even know about the warfare that's developing back at headquarters, among the top staff. They're going to be running to the press and the White House as soon as I don't make them happy. I'd be surprised if they haven't already started laying the groundwork.

"I'm in a delicate position anyway, because of the circumstances under which I got my appointment.

And the press attention that knocked out my predecessor will now be quick to turn the spotlight on *me*. By the time I get back the floor of my private office is going to be a solid expanse of trap doors."

"Oh, good Lord." Philanth turned away to sit hugging her knees.

"At least it makes letting me make love to you look like such a little thing, doesn't it."

"Mmm. I presume that's how these things always get started: poor, overworked, mistreated politician and devoted female subordinate who sees what he's going through. The next thing he knows he's holding a press conference on the Capitol steps with his wife beside him, apologizing to his constituents.

"But the Peace Corps director has such an awful lot of constituents, and just one or two of them can cut his political throat."

He collapsed onto his pillow. It was just his luck to fall in love with a woman who must have been reading the newspapers since she was twelve. Philanth found nothing romantic in being doomed to repeat history by not having learned from it. Between Philanth and Carole he wasn't being *allowed* to make mistakes, in the areas where they held veto power.

If only Carole were holding forth in her old role as his office manager she could certainly be keeping Les and Courtney and the deputy director in line and stopping them from trying to have Sandy garroted. She knew how to use fear to create credibility.

Marjorie was a good secretary partly because she had no desire to be *more* than a secretary. Thus while she had no problem with keeping minor pests from bothering him, she would not have enough assertiveness or taste for combat to try to help maintain discipline at the top level. Mark was reluctant to make enemies. In fact, nobody but a seasoned, qualified, official lieutenant who was known to have the full backing of the head man could lay down the law as his surrogate and make it stick.

However, people who did not respect his own authority because they had independent power bases would pay no attention to his stand-in anyway. This five-week trip had been a mistake when his chief of staff and deputy both owed their loyalties to somebody else. He should have had the guts and taken the time to clean house before he left town.

But if he had he would have been marked for life by the hard-nosed party regulars. And they would still control the resources of the White House and the National Committee if he became the party nominee for the senate seat in California in two years.

So as nearly as he could tell, he hadn't done anything really wrong. He had just unwittingly jumped in over his head by taking on the Peace Corps post with complete resolve to clean up the mess.

"Mr. Bjorklund." The voice was very gentle.

"Yes, Philanth."

"You do understand, don't you, that I'm trying to look out for your long-term political interests as well as your psychological interests by not making things easier for you in the short term?"

"Uh, you said in Paris it was a matter of my not taking on long-term guilt for a short-term benefit."

"It's more than that. Intimacy can be addictive under some circumstances, even though it's curative in others." She too realized that what they would find together could be virtually impossible to give up.

"Yes. That's the political aspect of the problem, all right. You know your sleeping with me during this trip isn't what would be likely to compromise me. It's what I might do afterwards."

"Just wanted to be sure you were clear about that."

They were talking about the consequences of becoming 'one flesh' as though deciding how to deal with a personnel problem.

"How did you become so dispassionate?"

"You mean, 'capable of disinterested judgment'.

"It was an effect of choosing to build my motives around something other than myself."

"Yes. But what *prompted* the decision?"

"It was just one of those decisions one makes in the process of growing up. One day I realized that taking on the burdens of others was a freer and more satisfying life than carrying the petty burdens of one's self. I saw how aimless or discontented people become when they live only to please themselves.

"Haven't I explained this before?"

"I'm trying to get at the moment when the whole realization began: the catalytic influence."

"Oh. It was the moment when I decided I'd already had enough happiness for one lifetime. Any more of the pleasures of life after that would be bonus, extra, not to be expected or hoped for. It was a consciousness of great spiritual wealth."

"Surely some event prompted such a decision."

Suddenly she was intensely shy. "It's very personal. I'd just received a little book of religious meditations from a boy I idolized. He wanted to become a priest. It was a token of our friendship: a good-by before he left for seminary."

"Do you remember how old you were?"

"I must've been sixteen."

There was a knock on Philanth's hall door. They looked at their watches. Climbing to her feet, she said, "That'll be Tony, ready for his breakfast."

PART THREE:

THE JOURNEY HOME

Chapter 20

MANILA

Tony thrust a section torn from a book between him and the briefing on Philippine agriculture he was supposed to be absorbing during the flight.

The title of the dismembered little book was not in evidence, but the section Tony had presented to him was entitled, "Night Life." He started reading the introductory section and at once was assaulted by its raunchy, leering, "You-ain't-seen-nothin'-yet" tone.

He skipped to the section on Japan, for he had considerable respect for that country, and learned that there one could find "evening entertainment" at a Turkish bath with an airport theme: the girls came dressed as stewardesses. At a Turkish bath based on a temple theme they came dressed as nuns.

The capital of another country had a first-class hotel that offered a package deal which included a female companion with the room.

The section was an explicit discussion of "where to get what for how much": a country-by-country guide to commercial sex in Asia. He could see what Asian women protesting organized sex tours had been complaining about for so many years.

"Where'd you *get* this?"

Tony flashed his toothpaste-ad grin. "Philanth bought it."

"You mean she bought the book and tore this out of it!"

"Naw, we'd *agreed* I'd have to tear them up. Obviously I couldn't bring whole books that contained sections on countries we didn't plan to visit.

541

"But they all were books she'd paid for. She freaked out when I told her I'd thrown away the sections I didn't need. She'd thought I understood she wanted them back."

Tony pointed. "Bet she liked this section, huh! Makes Western women sound like somethin' you wouldn't have as a gift."

"It says Asian women are 'unspoiled'. So is a dog that's never known anything but being starved and beaten."

"Aw, don't get mad. I was just suggesting we think in terms of the night life in general. Manila has a lot to offer. We've all been working hard, and we should get out for a little fun and relaxation. We could take Philanth to some classy nightclub where we can dance. I'll bet she'd like that." Tony leaned across him to add, "Huh, Philanth."

She looked up from the travel vouchers she was completing. "It's sweet of you, Tony, but . . . Mr. Bjorklund and I have lots of work to do."

She went on with her paperwork. Tony sat back, sighing and looking discouraged.

He handed the torn-out section back to Tony. "That's a piece of crap. Whoever wrote that must have serious problems in his relations with women. I'm amazed it was in a standard guide book."

"You didn't even *look* at the Manila nightclub section. *Surely* just before we start home we could spare *a few hours* . . . !"

"Forget it."

After a few moments Philanth began, "As long as you're interrupted anyway, Mr. Bjorklund . . . . I wanted to ask you about this big meeting tomorrow: it's going to involve only people who are American or quite Americanized, and I was . . . sort of hoping that if I sat in the back and took notes, I might be useful."

She waited, her mouth a thin line of anxiousness. She added, "Surely I won't still be needed to stay by the phone, all things considered."

His first thought was that it was a test of his willingness to be enough of a mentor to bring her into all-male meetings.

However, he tried to weigh the situation objectively and decided that she was right: the crisis in Peru seemed over, Manila was the last stop in his official itinerary, and the hotel switchboard operators' command of English should no longer be a reason for concern even if urgent messages came in. Besides, Manila was one of those cities where the phone system couldn't be relied on anyway.

So he couldn't duck the original reason why he had kept her in her rooms so relentlessly. Was he being paranoid in refusing to take her to meetings with him, or not? If he didn't loosen up she would decide his mentoring offer wasn't worth two cents; and once they were back in Washington his chance to recruit her soon would be over, for she would be gone.

He hesitated. "Try to make the right sort of impression. In the way you wear your hair, and so on." He lacked the nerve to tell her that he wanted something severe and unbecoming.

"Yes, sir. Thank you, sir." He saw her effort to conceal the degree of her happiness and regretted that she had to beg to be allowed to be fully utilized.

She looked through some folders and handed him one. "Your revised speech for tonight."

He looked it over. "Good job."

"Thank you." She always acted as if he should expect nothing less than excellence from her; it was his criticisms that really got her attention.

However, it really was a good speech—that he never got to give.

He was the guest of honor at a cocktail reception at the American embassy early that evening. When he was ready to leave it for his next engagement Tony guided him to the entrance where his car was waiting. But as he started to cross the few yards between the building and the curb he was tackled and flattened under a heavy body. The pavement had come up at him before he was aware that he was falling. Even as he realized that he had heard a series of sharp cracks and Tony shouting, "Get down!" before he landed, Tony was hauling him up and hustling him back into the building.

Tony made sure that he had not been hit, ordered him to "stay put", and rushed out again, collecting assurances from someone that the police were being called. He was joined at once by his Peace Corps driver, who had seen the whole thing and was mildly hysterical. A Filipino embassy employee began hovering around trying to find something to do for him. However, there did not seem to be much that could be done except send for an ice pack for a badly bruised elbow. Deciding that he should be guarding the car, his driver left as quickly as he had come.

After a little more hubbub he was escorted into a private office which afforded a large, comfortable couch, and there he was invited to rest. Someone brought the ice and carried away his suit jacket to attempt to clean the scuffed elbow. He was thankful that this event had been "business attire", for his white dinner jacket would not have come through this well, and he needed it that evening.

A young American he did not think he had met bustled around, straightening small objects and telling him how sorry he was and insisting that he accept a glass of sherry. He hadn't had time to be scared, and it seemed pointless to react now, so he behaved pretty calmly considering he was dealing with so many people who were more or less shaken.

He took the opportunity to rest while various official people ran in and out to assure him that they were doing everything they were supposed to do. One of the top embassy staff, apparently in a misguided attempt to make him feel better, told him that the intent probably had been to create a sensation for the news media, since "if they'd really been serious about killing you, they could have done it."

He was left to speculate about what sort of message could be intended by someone who deliberately aimed and missed. Killing token Americans to protest the government's continued failure to fulfill the hopes of the revolution had become a regular tactic. However, he decided that actually killing someone in his position could have backfired in terms of "public relations", whereas shooting at someone of his prominence had still managed to make a loud statement about how some Filipinos felt about all Americans.

Feeling that he was being too grim, he began wondering whether they hadn't all gotten excited because some jeepney had backfired. But when he asked Tony about that, Tony said the police were digging out the bullets.

Feeling a little ashamed of his previous contempt and more than a little in awe of the former football player's lightening-quick reflexes, he tried to thank Tony for what he had done. But Tony replied, "Just doing my job" in a way which said that saving his life had been nothing personal.

When they finally decided it was all right for him to leave he was able to step from another entrance straight into his car.

His next engagement was a larger reception followed by a dinner for two hundred people at a hotel which was much grander than the one where he was staying. But first he still had to go back to his hotel to shower, put on black tie, and pick up his speech and fresh cassette tapes. The heavy traffic on Roxas Boulevard offset the advantage of having been close to both hotels when he was delayed, so he had no choice but to be late. However, when he reached the Manila Hotel he was told that everything was running late because of extra security measures which had been thrown into place.

Realizing how upset people in the crowd at the reception were by the rumors about his narrow escape, he had someone line up an unused room and slipped off to it to revise his speech as quickly as possible. He found he had to junk some parts of it, devise diplomatic changes in the courtesy passages, and somehow tie what he had wanted to say into a context which put the most graceful face on what had happened: the cost to us all of allowing poverty and ignorance to continue.

It was a good thing he could handle his own speech-writing, because the news of the "attempt on his life" had spread across Manila so quickly that by the time he got up to give the main address of the evening the international press had turned out in force with all its light and sound equipment. He had not had such an eager audience in many years, and never one so important to his reputation.

Yet here he was speaking from a handful of scraps and scribbles. That was in the nature of political life: one had to rely on what sort of person he was, on the totality of his character and knowledge and experience, to see him through those times when careful preparation would have been most appropriate.

He thought the breast-beating because his country had not made more effective efforts to relieve the conditions which fostered terrorism went over well. After all, his audience included members of the country's small elite; having controlled most of the wealth of the Philippines for centuries, they were not conscience-stricken at being a large part of the problem. Like their counterparts anywhere, they saw their perpetuation of their privileges as the natural and proper order of things.

When he concluded his remarks he received a standing ovation, but this was surely an expression of regret rather than a reaction to his stiff delivery and soothing sentiments. At least it could be said that the speech had been well received, and that was helpful because certain chauvinists in Washington would find fault with him as usual. He had been given the Peace Corps post to keep him from continuing to make public statements critical of his party's and government's priorities, and here he was doing it anyway, and to a foreign audience at that.

The psychological situation created by the assassination attempt made it seem inappropriate for him to pursue any of the regular business he had planned for the period after the formalities ended. He decided that the best use of his time would be a prompt return to his hotel for a good night's rest.

That return was an exercise in heavy security: choreographed passages between car and service entrances, glimpses of hotel kitchens, service elevators—the works. If it hadn't been so much trouble for so many people it would have been rather fun, for it felt like being president.

Once in his room he had to accept more personal assurances of official regret and increased attention to his safety. Tony left with the last visitor, a Filipino government specialist on kidnapping and terrorism. To him it all seemed quite after-the-fact.

As soon as he was alone he took off his dinner jacket and tie and shoes and lay down, bunching one of the pillows under his head and then stretching out spread-eagled in complete relaxation.

He found ironic pleasure in realizing that he could count it a good day merely because he was still alive. It was an odd variation on the principle Philanth applied when she needed something to cheer herself up.

For a minute or so there was silence, blessed silence. Then came the predictable tap on his connecting door. He could almost see Philanth on the other side of it, on tiptoe with readiness to rush in.

"Come."

She peeped in, saw him flat on his back on the bed, and flitted over to alight in a sitting position beside him. She timidly touched the front of his dress shirt as though to make sure he was real.

Feeling that he would be a fool not to take advantage of this situation but knowing better than to go too far with her, he turned to face her, grasping her outstretched hand to keep it against his chest. He didn't bother with wise cracks or small talk or any sort of comforting noises, he simply held onto her.

"Oh, Mr. Bjorklund, I'm so glad you're . . . ."

He could see that his stillness confused her.

She had not indulged in any reaction to the incident during his flying stopover for his speech and exchange of cassettes because he had been running seriously behind schedule. But now they had time.

Making up his mind to enjoy it, he closed his eyes. If anything could make a man not mind getting shot at, ... Because of his acute awareness of the potential of this sort of situation the beginnings of arousal were surging through him.

But it would be folly to try to draw her to him. Indeed, the slightest miscalculation now could ruin everything. He must not resemble the sort of men who insulted her intelligence by trying to gain her favors through physical contact. *Slowly, go slowly,* he warned himself. *"Soflee soflee catchee monkey."*

Her hand became restive in his. "Mr. Bjorklund," she said as though gently attempting to awaken him.

He opened his eyes and studied her.

"Please," she said, tugging, and he let go.

She bent forward, tilting her head as though trying to read his thoughts in his eyes.

She was putting up her hair in braids a lot these days, whereas at the office she had ordinarily worn it drawn up and back, then swirled around the crown of her head. She looked better with that softer, more sophisticated style. But now she presumably was trying to look rigidly "moral", as a constant subliminal reminder to him that she was off-limits.

Braids were just as authentic an expression of her personality. This evening she had them woven across the top of her head, everything soft and molded to fill the top of a neat oval. She looked like a milkmaid on a dairy carton, except more earnest.

"You make me think of spring evenings on a farm," he murmured. "You're so fresh and wholesome and natural."

Smiling a little, she supplemented, "Like a supper of bread and milk, eaten sitting on a stone stoop, with cattle lowing in the background?"

"Like the Lost Eden, a past I never experienced but long to return to. What a pity, Philanth, that we have to forgo a simple life so others will have a better chance at it! We'd know so much better how to appreciate it than they could."

She looked as if she would have been glad to go on looking at him forever. She had to be as aware as he was that their journey was coming to an end.

He added,"I know I'm not the first to tell you this, but I can't stand the thought of the emptiness of going on with my life without having you with me."

"I understand," she admitted. "If you had died today I would have found the rest of my life more barren.

"But that's not the same: you'll be able to assume I'm still alive and working somewhere."

Shutting his eyes against that thought, he turned to lie face down beside her.

How good it felt to know Philanth was there beside him! He needed that solace always.

He sat up. "Philanth, let me try once more to explain something."

She nodded.

"Somehow in growing up I started feeling the same thing you do, that to have a full life you must accept a duty to do whatever you can for the future of the world."

"Not just all the people," she agreed, "but everything that's part of our uniquely precious planet."

"Yes." Her eyes were so open to him that it was easier to maintain his sense of separate identity and pursue his thought when he looked away. "But I have to be near the center of the decision-making, in government, because anything other than politics seems so much simpler that it feels like selling out."

"I understand. Anything that doesn't demand your best effort *is* selling out, becoming partly dead."

"But I'm not altogether cut out for it," he tried to warn her.

"No," she replied with a little smile, "we wouldn't have become this close if you were."

"And I wouldn't need you even closer. Whether it's my failings or my virtues we're talking about, it all comes to the same thing: I can't do a good day's work day in and day out for extended periods if I don't have a very special woman to feel close to, especially at the end of the day."

"Sometimes the pain of all those helpless people inside you becomes unbearable," she said in her low, husky voice. "And all those beautiful, innocent, dying animals, all those polluted streams and beaches. All the waste and unnecessary misery. Like the night you got the news that your legislation was dead: sometimes the pain of caring is terrible. I know.

"And always there's the pressure to do something about it, and over and over the feeling of being too weak, of falling short. Over and over I need someone to cuddle up to, someone whose goodness softens the harshness of the world's lovelessness.

"I know *you* need a special woman indeed. And Carole *is* special."

"I had her as my strength from the beginning of my public life. I can't live any other way. But now Carole has resigned from that post. And she's insisted that you're to take her place."

"She hasn't even hinted that she might divorce you, has she?" He shook his head. "Then why not wait until she does, before you start worrying about a replacement?"

"Because I *know* she wants me to have you. Whatever her intentions beyond that, I want that too."

He laid his hand on hers. "I want you for the rest of my life; and every night before that begins is another night wasted." He finished in an urgent half-whisper, "I want to make love to you!"

She withdrew from his touch.

"Tell me why not, Philanth," he begged. "Talk it out of me!"

She did not conceal her tension. "Making love is sacred to me, Mr. Bjorklund. I can't do it outside the

551

sanctity of marriage, not without doing psychological violence to myself in the process.

"I know all the arguments about that, but sex is like life, whatever one makes of it; and I'm content with cheap material things because in the intangibles I keep only the best. You'd be asking me to violate central elements in the pattern of all my thoughts and feelings; to betray myself."

Moving closer, he murmured, "You could transcend that pattern, Philanth. You could alter it because you love me—and you must know how I love you."

Their physical closeness was working on both of them. The space between them felt combustible, as if a spark could cause an explosion which would fuse them. He could see the effort to break free of a strong magnetic field as she half turned away.

"Sacred things require secular protection from the rottenness of human nature. Belief in the institutionalization of the sacred—belief in the sacredness of marriage—is needed to protect a fragile beauty from desecration. That's important, because 'Lilies, that fester, smell far worse than weeds.'"

She was starting to dissipate the mood without violating it. As though she were about to escape, he shifted position to grasp her shoulders.

"But I need you, Philanth! It would be wrong to keep apart when the life we've chosen demands more than we have. It's bad enough that someday one of us *is* going to die, and the other will have to go on living!"

She was looking into his eyes and her childlike face was starting to register her grief and sympathy when Tony breezed in from the hall. Startled, they turned their heads to look at him.

Tony stopped just inside the doorway, his mouth open as if he had been about to speak, staring at both of them from a short distance beyond the foot of the bed.

He expected Tony to laugh, seeing his earliest hints vindicated. But the black man quickly lowered his eyes and stepped back out and closed the door—leaving them as he had found them, Philanth sitting in the middle of the double bed with her boss, the middle-aged, married Peace Corps director, grasping her shoulders in the heat of urgently propositioning her. Despite all the control Philanth had been exerting, she had looked open to persuasion.

He and Philanth turned away from each other to sit on opposite sides of the bed.

"I'm very sorry, Mr. Bjorklund. It was my fault." Philanth got up and moved toward the connecting door. She paused with her hand on the knob. "Whatever happens, I'll accept full blame. If necessary you must say that *I* had approached *you*, and you were talking me out of it. I'll confirm that, to whatever extent is necessary." Then she was gone.

He looked at his watch. Her evening duties would bring her back before long. It was thirteen hours later in Manila than it was in Washington and Peru, and he was expected to be back from his evening engagements before long, so the overseas calls would soon begin to come in.

There was no point in getting ready for bed. He changed into his slacks and sports shirt and decided to get what rest he could while he waited.

Then he realized that Tony had intended to report something to him. The sooner he faced his young aide, the better. He went next door. Tony let him in, avoiding meeting his eyes.

"You wanted to see me?"

"Just wanted to let you know all the higher-ups had left, but we'll have extra people watching out for us until we fly out of Manila. They promised to keep a low profile.

"Sorry I barged in. I'd gotten out of the usual habits . . . ."

"What you saw wasn't what it may have looked like . . . ."

Tony still would not look at him. "Don't worry about it." Tony sat down on his bed and did look him in the eye as he added, "You could've caught me in the same position with her, if I'd been the one they were shooting at."

His first impulse was to hit Tony, but his aide's tone indicated that he was trying to be conciliatory, even magnanimous.

"You think so?"

Tony faced him squarely. "Now, I didn't mean nothin' against her! She'd hug the Devil if she found an excuse. She thinks affection is good for people's character."

The young man obviously was doing his best to avoid a fight. Tony had something on his boss and his boss knew it, and Tony had a healthy instinct for self-preservation.

He backed down. "I understand. She's just got an affectionate nature."

Tony began taking off his shoes. "Right."

He returned to his room and lay down.

By the end of the previous business day in Washington everyone had been assured that the Volunteers would remain in place in Peru unless there were significant new developments. The only exceptions were a few who were leaving for rest and reassignment because their project had been burned out. Now he needed only the latest reports from the field, which were continuing to be collected at the country headquarters in the still-unsettled capital.

The sound of knocking came from the connecting door. The pattern of raps told him that he had a phone call waiting. When he opened the door she half-whispered, "The country director for Peru." He nodded and went to pick up the receiver.

"Danny? How's it goin', pal?" They had never met, but after the last several days they were practically bosom buddies. Philanth drew a chair over and sat down facing him, poised to make notes.

The country director greeted him in his deep, pleasant voice and reported that things were returning to normal. They said the necessary words of earnest appreciation, they joked about how they both were going to catch up on their rest, and they exchanged courtesies regarding the prospects for his visit to Peru the following summer.

After hanging up he told Philanth to call Marjorie for a general check-in and to be sure she relayed his news from Peru to all the appropriate parties.

She nodded and got up.

As she displaced him beside her phone he remarked, "We ought to celebrate."

"I'm not one for celebrations," she apologized. "I savor an achievement by lying down to read a book or tackling a major chore. You could say I stay on a plateau rather than scaling peaks to escape valleys."

"We're mighty dull people." He took her hand. "But there are mountains on that plateau that are . . . wonderful—beyond the dreams of those who . . ."

"Mr. Bjorklund, I know you're very tired and a little shaken, but in view of what happened with Tony just a little while ago, ..."

"I'm sorry, you're quite right." He stepped back, shoving his hands into his pockets.

"I admire the way you've handled this," he said. "—The crisis, I mean."

"The same to you."

"I'll want you with me during all the crises I hope are ahead of me."

Philanth's reaction—half sigh, half shrug—made him realize that his compulsion to harp on that theme could only build her resistance. If he had any sense he would defer any more discussion or even thinking about the matter until he could talk with Carole face-to-face. Somehow, on the basis of that,—

The phone rang. He reached over and answered it himself, still watching Philanth as she again traded places with him.

It was Marjorie. She took the good news from Peru in stride and promised to be sure that everybody was informed of it according to the protocol.

"I have more bad news," she added. "Sandy's submitted his resignation." He uttered such a heart-felt groan that Philanth reacted with alarm. "He says he can't do any more in his present position, and he's going to look for a job in the private sector. He says he hopes you'll serve as a reference."

"What's been happening there?" he exclaimed.

"Most of it's been behind closed doors, and I'm sorry to say that the participants don't confide in me. All I get is official messages.

"Except that Les said to tell you if this thing goes any further somebody's going to leak 'the whole business' to the press."

"Is that a warning that I have to accept Sandy's resignation, or a blatant threat?"

"No comment."

"Okay." Sitting up straighter in the effort to marshal his forces, he thought quickly. "Tell Sandy I accept his resignation." He caught a glimpse of Philanth's instant dismay and quickly swung away to tune it out. "Then let Les know that I've done so.

"But that's just to buy some time. You tell Sandy privately that I said for him to take a vacation, and by no means is he to commit himself to anyone else. Tell him *very confidentially* that there'll be major changes and a much better job for him around the first of the year. But he mustn't breathe a word of that to *anyone*. He's going to tell anyone who presses him that he's going to relax during the holidays, then decide what to do next."

There was silence at the other end, and he knew that Marjorie was writing everything down, not using the recording system in her phone. "Got it."

"Make Sandy believe I can deliver! It's very important, Marge!"

"Yes, sir. I understand."

"There's more involved than the problems at Peace Corps. If I go far, so can he. I want to take him with me. All I'm asking is that he rely on me for a few weeks. It's not a time of year for job hunting anyway."

"I should tell him *all* of that?"

"Yes. But no hint of what job I have in mind for him. If *that* leaked, it could . . . ."

"I understand. I'll invite myself to his home, to talk to him and his wife. They've got three little children to think about, and it's a pinch to live here on just his government salary. His wife is starting to take in other people's children for day-care, as it is."

"Do that. Thanks, Marge." They went on to discuss the assassination attempt, and he dictated a memo about it for the headquarters staff.

When that was taken care of his mind reverted to his troublesome lieutenants. "Any news regarding the deputy director?"

"He seems to be just tending to his business. He and Eric have been doing so many luncheons and banquets that Eric's being teased about putting on weight."

"What? Archie's using my driver as a flack?"

At once he realized how that sounded. That sort of short-hand thinking was hard not to slip into, but it was corrupting. Eric was the Peace Corps director's driver, not Charles Bjorklund's.

"What's the harm. Les—"

"Marge, it's irregular. Eric has no part in negotiations with businessmen, he's not entitled to their free meals, and he's supposed to be spending his waiting time tending to his own responsibilities."

"I'll speak to him."

Marjorie's tone became confidential. "But it seemed like a low-cost sop to the deputy director's ego. Les reportedly has discouraged him from trying to get involved in budget meetings."

"Thank God."

"No, it's not that good. Sandy told me it seems to be all over the headquarters that the chief of staff cut the deputy director to pieces in front of the budget specialists. I'd love to have seen their faces."

"Lord." Poor Archie. "A meeting of the budget analysts from *all the offices*, or just the specialists from Management?"

"Just from Management. But apparently it got around anyway."

He could understand why. Every office had its link to Budget through its budget analyst; through the resulting network of personal contacts, gossip about infighting between the top people would get a fast start.

Behind his perpetual tight smile the plump, balding little deputy director must be ready to turn on Les. Maybe he could use that. Archie might be feeling he could use a friend about now; if the problem people were becoming bold enough to attack *each other*, ...

He sighed. "Just keep things from blowing apart until I get there on Friday, Marjorie," he pleaded. "It's all I ask."

"Whatever happens, you'll fix it. Meanwhile you'd better get some rest. You're going to need it."

Things really must be going sour. "Understood. Call me before I start tomorrow's schedule, unless something truly urgent comes up in the meantime. I'm going to bed."

"Right. 'Bye."

He put down the phone. While Lima had been a battleground and Volunteers' lives had been in danger, the bickering among those creatures at Peace Corps headquarters who owed their jobs to political connections had worsened. The time he had been given to celebrate the end of the Lima crisis had been measurable in seconds.

The time for merely fretting about Les and his allies outside of Peace Corps was past. To keep things from continuing to deteriorate he was going to have to take some firm positions and make the President's loyalists accept them. That meant he had to be prepared for political bloodshed. He could not gauge how much of the blood would have to be his own, since at the moment he had no inkling of what he was going to do.

The only thing he was sure of was that he would have to be a lot more clear-headed and self-possessed than he was now. His hopes for the future depended upon minimizing bad press and trying to keep the White House at least neutral in a major senate race. If the press made hot copy out of him as just a pretty face too weak to keep order even in one small federal agency and the White House turned on him before he could win the primary which was well over a year away, he was on the skids and he definitely should have stayed in the House.

Maybe if he had just taken a long, intensive vacation instead of trying for the Peace Corps opening because he was burned out in the House he would now be a reasonably contented congressman with a safe seat representing a wealthy, highly educated district, continuing to produce brave speeches and articles while building seniority for committee assignments. Carole would still be his cheerful, vigorous assistant. All his problems were his own fault because he had gotten too big for his britches.

He felt disheartened and drained. His next order of business was to sleep like the dead. But he was a long way from being able to relax. He was all too likely to spend the rest of the night chewing things over and becoming more and more tired and frustrated and depressed.

Drugs and alcohol were unreliable at best, and he could not risk being caught groggy at the wrong moment.

There was one source of tranquillity he could always trust, and it was within arms' reach.

"Philanth. Stay with me until I can go to sleep. Let me hold you, just for a little while! That's all I ask—I swear—just to feel you close to me for a little while."

She looked at him and wrung her hands.

"Think about it," he urged. "What harm can it do that hasn't already been done? I'm on my way home, I just need a little peace to get me the last bit of the way without flying apart and making some big, stupid mistake tomorrow. I know I'm taking advantage, but I—at least I'm sorry!"

She bit her lips, and tendons stood out in her neck.

"As a—a friend. You go around hugging everybody in sight. Do I have to be the exception? And does it matter so much if you're that close for just a little longer?

"Don't make me beg, Philanth! But I must have your help! I *know* what I *need* right now! I'm strung out, and I need to relax and sleep. You're the only safe tranquillizer around. *I need you.*"

She hesitated a little more, then spoke quietly. "Go get ready for bed."

He complied. When he opened the connecting door and told her he was ready she told him to get in bed and then came in, leaving the light from her room to guide her after she turned off his bedside lamp.

She lay down outside the covers, her back to him and her arms folded, and let him slip his arms under her neck and across her middle. She tugged down the sleeve of his pajama before resting her arm over his where it lay along her midriff. Then she stretched slightly from head to toe as though willing herself to relax. He too settled into position and closed his eyes; and suddenly the bickerings in the offices in Washington shrank into the far distance and vanished like a film trick in a movie.

He caught the floral scent from her hair and longed to nuzzle it. After a few minutes he felt serenity begin to flow into him as though by osmosis.

"You have a unique ability to make me happy," he murmured. "Even when you're doing nothing at all. Just being with me."

"That's my middle name: Beatrice: 'she who makes happy'."

The childish response carried a hidden reminder: she felt it was her duty to give happiness, but not only to him. It followed that he could not keep her, at least not on any terms he had been able to devise.

"May I ask you something?" she inquired. He could feel in her body how difficult it was for her to relax. She was trying to distract herself from their intimate position.

"Ask." Anything to keep her close like this.

"There's one thing about you that puzzles me: How can you be so fundamentally unassuming? I've never seen any sign of vanity or pride. When you're not acting officially you have no pretensions at all. Yet you have such obvious reasons to stand on your dignity, and you're dignified by nature. Your humility doesn't fit with any of your other characteristics except your being a Christian."

He sucked in his breath, chagrinned that she had recognized a source of frequent subliminal humiliation for him.

"Actually my mother is Jewish, though not so Jewish that she minded Dad's seeing that I received a Christian education. What counts is that my parents are down-to-earth people, and they made a sustained effort to prevent me from becoming 'too full of myself'. They repeatedly warned me that they were 'not going to tolerate having a son who's just another handsome, arrogant rich kid'. Going through Beverly Hills High School, I could see what they were talking about: arrogance isn't attractive."

"It indicates shallowness."

"Just the same, I think they overdid it a little."

"It makes your inner life harder, but it makes you a better person, and therefore more effective."

"It makes me look like a wimp."

He could feel her instant indignation, though she took a moment to put it into words. "Only if that label is used as a put-down of caring and sensible people by the sort of cretinous, bullying mentalities that really would enjoy kicking sand in the faces of

'90-pound weaklings' on the beach! Their conception of masculinity is atavistic! They—"

"Spare me this discussion right now, please."

She caressed his arm. "Of course. But just let anyone try to hang that aspersion on you, and I'll tear them to pieces."

"Obviously just the solution."

She laughed and pressed his arm. "You're such a dear."

"*Everybody's* 'a dear', by the time *you* get through with them. Spare me any meaningless compliments."

"You're clever even when you're worn out. I'll let you go to sleep now."

"Thank you."

But used up as he was, he found he could not go to sleep. He disdained to worry any more about Les, and there was nothing more he could do at the moment about whatever suffering Sandy was enduring. Thanks to that confidential talk he had had with Marjorie about the problem of replacing Les, she could be counted on to relay his message without distortion. If his young ombudsman proved mature enough to stick with him through this and emerge with his standards of conduct intact, wonderful. But for the present that had to be up to Sandy.

Having been shot at wasn't bothering him either, because that still was not real to him. And he was too mentally used up to be concerned about the final day of Peace Corps conferences immediately ahead.

His mind kept circling the pit he must not allow it to plunge into, the thought that he would feel like crying out with an ecstasy that really had nothing to do with sex if he could ever feel that Philanth was truly his, that she had every right to lie in his arms as she was doing now, whenever they found an opportunity. He would have no peace until then.

But if that time ever came he would have everything he needed.

"Philanth." He hesitated for a while, but she didn't speak, only pressed his arm. "Philanth, I want you to seriously consider becoming my mistress."

Withdrawing from his arms, without haste she sat up.

After a few moments she said in a neutral tone, "Because with me you'd find the will to do such great things that every disadvantage would be justified?"

"That's the assumption. Now, what were you saying a minute ago about my remarkable humility?"

"It wouldn't work out that way, because it's dishonorable."

"You're usually rational, Philanth. You take pride in it. And when a woman really loves a man who likewise loves her but can't marry her, it's not dishonorable."

"Yes, it is. Even aside from what it does regarding women's status, it sets a bad example for those who aren't able to avoid the consequences if they too ignore the rules: fatherless infants, children hurt by divorce, sexually transmitted diseases,—the list is interminable."

"How can it set an example if nobody knows about it?"

"The Bible points out that deeds done in darkness will be shouted from the housetops. That's hardly less true in our society than it was thousands of years ago."

"There are ways to provide a cover . . . ." That new suite Carole was having built in the basement.

"There must be no town in the world where people love gossip more than they do in Washington. They *come* there, and put up with the costs and the climate and the crime rate, out of the desire to feel like 'insiders'. And if jealous or spiteful people can't dig up any dirt about people they envy, they manufacture it."

"That could be taken as all the more reason to ignore what such people think. A lot of people in Washington seem to come to that conclusion. As long as a relationship is discreet it needn't have negative consequences. Even the voters generally allow their officials the right to private lives."

"It's still wrong to set a bad example, and people *will* find out."

"Is the 'example' set by a clandestine relationship that hurts nobody really important?"

"Society begins to disintegrate when too many people ignore the rules. That's a real danger for ours because we cherish the freedom of the individual at the expense of everything else. Therefore respecting the rules is part of the code I've set for myself."

"You don't have that big a share of the collective obligation to keep society going!"

"Responsibility is proportionate to knowledge. Why else d'you suppose there are so many who 'don't want to know'? We who are able to understand the value of rules developed over thousands of years of human experience have a particular obligation to honor those rules. We're the backbone of society, the ones who keep things going and make things work as well as they do. Therefore honoring the principles that preserve civilization is part of my duty to the society that produced and nurtured me."

Oh, wow, he thought, *excuse me!*

However, he could tell just from her didactic tone that this idea was not going anywhere as an abstract proposition. She was not going to be reasoned into it any time soon if only because it was too contrary to her nature. Maybe when she was older and had become less religious in her emotions . . . .

Meanwhile the more she enunciated her positions to him, the harder it would be for her to abandon them later. Besides, he did not enjoy arguing his side of this issue.

"All right, forget the idea of being my mistress. I really just want you *with* me. I—I guess I felt the personal bond would ensure more permanence for the working relationship. Don't you want to take advantage of everything I could do for you as a mentor?"

"Why pursue this, when I've already realized that you wouldn't drag me around as your aide, getting me into all the interesting things you get into, because of my gender and age?"

"And because of your appearance, Philanth—be fair, even if my position still seems indefensible to you. A girl who looks like you has a terrible time convincing men she has brains."

"I've never even tried. I just need opportunity for them to discover it for themselves."

"I'm prepared to provide that."

"Wha—Really?"

"Yes. You've made me feel like a coward. I'm now prepared to put up with what people might suspect."

She was cautious. "That will work out—in the long run—provided what they suspect isn't true."

He couldn't accept any aspect of that proposition: the prediction, its proviso, or the premise implicit in the proviso. "One can only hope so."

"Look, do I have to keep convincing *you* that I'm not a 'dizzy blonde'? Just give me the same opportunities you'd give a man of the same age and abilities, and I'll do the rest. I can cut it, if you can only accept the fact that I'm off-limits and treat me accordingly. I'm sure you can if you really want to."

"Okay, okay."

She got to her feet. Gently she said, "We'd better start getting back in practice right now, Mr. Bjorklund. Knowing I'll stay with you is surely a more than adequate substitute for holding me."

"It will have to be."

She ignored his note of despair. "Wonderful. Good night, then."

She paused. "Thank you for reopening the negotiations. I was feeling pretty bad at having to pass up such a once-in-a-lifetime opportunity. I'll love working with you."

"Now, don't start getting mushy on me, I can't take it. Keep your hugs and pecks on the cheek . . . ."

"For everyone else? I think my excluding you might look suspect. It might make things a bit harder, but when even Tony noticed, ..."

"Hell, Philanth! I'm an ordinary man, not a saint like you!"

They both hesitated.

"This will pass, Mr. Bjorklund. You'll get used to me."

"I know what you mean, and I'll do it or die trying. I've got to keep you with me."

"It won't be so bad. People can enjoy being together without becoming erotic. They do it all the time. They just don't take five-week trips together with connecting bedrooms."

He had no response which would help his cause, so he said nothing.

As she went to her door she remarked, "Just wait til I get a chance to have a word with Carole. There *must* have been a way for her to get you to achieve this agreement without making you feel she'd deserted you! This 'test' has gotten completely out of hand."

Philanth went out, closing the connecting door.

Philanth was mistaken. Only Carole's at least pretending to desert him could have made him feel he needed Philanth so desperately that he would agree to this compact. Whether or not Carole had in fact been pretending, now at least he had Philanth if he could master himself sufficiently to keep her on her own quite reasonable terms.

He went to sleep on that thought.

The next morning he was awakened for Marjorie's call, which was followed by a difficult conference with Les. The chief of staff wanted to "clear" certain "decisions" with him, and it was apparent that Les was in fact pressuring him to back up several actions which indicated that Les was throwing his considerable weight around. Having succeeded in getting rid of Sandy, Les seemed to be pressing to consolidate his position and reduce his nominal superior to a ceremonial figurehead.

To Les's open displeasure, he mildly declined to say that he agreed to these rulings, taking the position that they could wait for a few days and he would "look into them" when he got back. Les started warning him that it was "necessary for the chief of staff to have the director's full support in order to function", but he sidestepped that by urging Les to "focus on more routine matters for just a few more days." Les growled that he would "see about that" and hung up.

His relationship with Les was coming to resemble a game of "chicken" because Les believed he was not prepared to risk a head-on collision. A man in his position needed to make himself feared by his opponents; he had failed to do that. Beginning to wonder whether Les was trying to push him into a confrontation in which he, not Les, would be replaced, he slouched over to the connecting door and knocked.

There was no answer, so he opened it and looked in. Philanth's room was neat and empty.

He found Tony lounging in the corridor outside his door. "Philanth went on down to get a table and order for us, thinking you must be almost through. Two more minutes and I woulda knocked to tell you what time it was."

He consulted his watch. He would have to hurry to get the breakfast he definitely needed if his stomach was not to become a participant in the formal discussions. He went to grab his briefcase and they took off.

Inside the entrance to the coffee shop they paused, looking for Philanth. The hostess came up to them, smiling. "Two?"

"Three. We're looking for a little blonde girl," Tony informed her.

She smiled some more as she turned. "Thees way, please." She led them to a table where a girl in pigtails was eating alone. He was about to protest the mistake when the girl looked up from her almost-empty plate and he finally recognized her.

"Philanth!" She had braided her hair into pigtails and looped them up behind each ear with a bow of blue gift-wrapping ribbon. She wore her office-style blouse and had her denim suit jacket over the back of her chair, but from the neck up she looked about ten years old.

Oh, God. Now he was in for it.

She pretended to ignore the look he gave her as he sat down. "I ordered for you, finally, and I hope they haven't ruined it trying to keep it warm. If you can't stand it I'll have them package it so I can have it for lunch, and you can order fresh. The same goes for you, Tony."

"I'll eat anything," Tony replied, sitting down.

"You look cute, kid." Tony was beaming at her. "I like the way everything matches your eyes."

"Well, thank you, kind sir," she said with mock flirtatiousness. "You're pretty cute, y'self."

He glared at her, but the waitress was setting a stack of pancakes in front of him. He studied his watch and decided that there was no point in commenting, since there would be no time for her to change the style. He was unwilling to risk being late for the meeting. In fact, considering his role there he ought to arrive early in order to be available for informal contacts. He ate quickly.

Philanth laid before him a small newspaper clipping. "This was in this morning's major local paper, and it may be referred to in your discussions today. You'll see that it was occasioned by the conference we're involved in."

"Thank you."

He glanced over it without slackening his pace with his fork. It was a wire service item with a Manila dateline and carried a series of unequivocal statements regarding facts of life about which he had been publicly expressing concern for many years:

Population was growing faster than ever before, and the coming decade would add the equivalent of an extra China to the world's population.

The increased numbers threatened to erase the gains which many countries had struggled to achieve, and the biggest increases would occur in the poorest countries. The executive director of the U.N. Population Fund had pointed out that there should be "a strong and determined family planning program as part of a package which includes other types of investments in human resources".

The article noted that Indonesia, South Korea and Thailand had made great economic strides while at the same time controlling population growth. This was "the first step toward a sustainable development."

He thanked Philanth again and returned the article, knowing she would want it for her collection.

He got annoyed all over again every time he looked in her direction and saw her childish braids. Despite his resolve, he couldn't help muttering as he followed her out of the coffee shop, "You do nothing by halves."

"My parents noticed that too."

His brain finally slipped into gear. "Go up and fix your hair properly! You can slip in late."

"With all respect, sir," she replied sharply, "this was not a flippant gesture. I know one thing about men that they don't know about themselves: few of them ever see past the surface of a woman, especially one they don't know well. Most of them will actually see me as closer to fifteen than twenty-five.

"On the other hand, if I try to slip in late I might have trouble getting past the door, and you'd be subjected to having to vouch for me.

"We've had a difference of opinion before regarding which of us knew more about how a woman should appear in a given set of circumstances, and I believe you finally recognized the adequacy of my judgment in that area."

Her clipped speech and five-dollar phrasing, not to mention her reference to their argument over her removing her cape, constituted a warning: she had arranged her hair as she judged would best address his concerns, and if he was on the verge of revoking her chance to attend the last of his important meetings, that could be the last straw for his recruitment efforts. She had pined in her room through all of the other interesting meetings he had been engaged in

571

since they had left The Gambia, and she had had enough.

"Okay, okay," he muttered. Their new agreement was off to a rocky start. Her eyes flashed as she turned to lead the way to the section of the ground floor devoted to ballrooms which divided into meeting rooms.

Tony escorted them to the appointed conference room but stayed outside to talk to a hotel security officer who was stationed there.

As he entered he strode ahead, avoiding any inter-action with Philanth. At once he was recognized and taken around and introduced, while the girl who had come in behind him took a chair behind the place reserved for him at the conference table.

But as he took his seat she showed him the papers he had told her he wanted reproduced for distribution at this meeting and promptly forgotten about. She asked with a gesture, "Shall I . . . ?" He nodded, tight-lipped at being unable to avoid being linked with her.

As she moved around the rectangle of narrow tables, laying a handout by each water glass, keeping her eyes downcast like the well-bred teenager she was pretending to be, he saw that the other men at the table seemed unable to look at her round, snub-nosed, rosy-cheeked young face and neat, blonde pig-tails without a paternal smile. She had not been able to do anything about her breasts except wear her usual loose-fitting blouse and her suit jacket, but he reminded himself that he had been only thirteen when he had been fascinated by a very well-developed classmate.

However, so had everybody else in the class.

The man next to him said, "That your daughter, Mr. Bjorklund? She's a cutie."

That word again. It was the Cupie-doll dimple that really did it. He nodded, tried to look pleased, and said nothing.

He prayed the man would not suspect that he had his daughter traveling with him at government expense. There was just no way to win, in this type of situation. It was amazing what a man would agree to undertake in order to keep a certain woman with him.

She worked around to him. As she laid his copy before him their eyes met, and her expression of repressed amusement almost cost him his composure.

She had no idea why he was so upset, or if she did she probably felt it was time for a courageous crusader like him to take a "meaningful" stand in the Equal Opportunity department. As far as Philanth was concerned, the boys in the political clubhouse would just have to learn to keep their minds on their business when there were attractive women sitting around the table, just like the boys in other clubhouses were now required to do.

He saw with relief that the meeting was being called to order. The moderator was remarking that they had enough to discuss that day to keep them busy for a week, and some of the men settling into their seats made noises of agreement.

There was only one woman at the tables, a stout, middle-aged Filipina.

Because this was an international and interagency conference to work on funding allocation and integration of programs in the Philippines, he was backed up by program development and administration specialists from Peace Corps/Washington as well as the top PC/Philippines staff. His own role required considerable political and diplomatic skills because they were dealing face-to-face with officials representing a government which continued to have serious built-in obstacles to sound policy-making and honest, effective implementation. Moreover, he had his own agenda of Peace Corps concerns, and he particularly wanted to watch the country director, who sat beside him, in action.

He had been fascinated by his study of the situation in the Philippines. The scope and complexity of the problems were daunting, yet the prospects were more hopeful than in a lot of other places. The political situation gave development programs more elbow room than did more corrupt and repressive governments in some of the other Peace Corps countries, and aside from the Catholic Church's opposition to artificial birth control, the culture presented few obstacles to addressing the poverty of most of the population. Many of the people had informed views on what they needed, and solutions were available.

As the meeting progressed he could see that he was dealing with people who had both extensive training and many years of field experience in this sphere, whereas he was still a neophyte. Lots of other directors might not have even made the effort he was making to understand Peace Corps' tasks in such detail. But he already had a strong background in the kinds of activities which might have absorbed most of their energies, especially congressional relations and public speaking; and however much one knew going into a new position, he believed that a political position should always be treated as on-the-job training.

And he was there because he was a politician: not the most popular kind of animal with the specialists who were trying to actually get things done, but a necessary intermediary for the work of tapping, organizing, and channeling collective resources. He was there partly to serve as a conduit of informed opinion back to the congressional appropriations process.

The morning passed quickly. As they left the conference room for the lunch break Philanth began telling him how interesting she had found the session. He had never seen her so happy unless it had been after the reception in Morocco—before he had clipped her wings as they left Banjul.

"I'm glad you're learning so much," he replied absently, cutting off what promised to be a lengthy analysis. They were entering the wide corridor between the various meeting rooms. Delicious food smells were in the air, and through an open door they could glimpse round tables prettily set with linens and crystal.

Her steps slowed; she was looking at him questioningly.

"See you around dinner-time, I guess," he said.

"Uh . . . What about this afternoon?" She knew the afternoon meetings would involve smaller groups, more personal interaction—and more sensitive, therefore even more interesting, discussions.

"You have things you can do in your room."

Aware that he sounded callous, he tried to justify himself by giving her a project on the computer which could well consume the entire afternoon. It would have to be done when they got back to Washington anyway, and they would need to hit the ground running.

Philanth probably knew what he was up to, but she gave no sign. She opened her notebook and made sure that she understood what he wanted.

She started to outline a better procedure for getting the desired results, and he perfunctorily left that up to her. He was looking around, thinking that he should visit the men's room before going to lunch; but if he didn't get moving he was going to be late and have to rush through his meal instead of carrying his proper role in the conversation. Tony was watchfully loitering nearby, waiting for him to set a course.

"Mr. Bjorklund," Philanth was asking in her meek and plaintive tone, "is there any way I could have fixed my hair that would have made it okay for me to go to the meetings this afternoon?" Now he saw pain in her eyes, though she was trying to make it sound like a routine question.

"I think not. Sorry."

"Could Marjorie have gone with you? Or Carole?"

"No."

"Yet a presentable male aide could have gone with you as easily as a briefcase."

"I don't know. Perhaps you're right."

"Well, have a good afternoon." She spoke without rancor, but as she turned toward the elevators he saw how she drooped.

Belatedly he remembered that on the way in from the airport Tony had tried to persuade Philanth to go sightseeing with him this afternoon. She had declined with obvious regret. As usual, she had seemed to know all about the "marvelous" local attractions and to have written off all hopes of seeing any of them. Feeling like a heartless monster, he took a step after her as he considered inviting her to take the afternoon off.

He was distracted by noticing Harleigh Adams watching him. As their eyes met, Harleigh started forward from his position lounging against a slick, pink pseudo-marble wall in the off-hand way he always had. Harleigh had been in the meeting, representing the U.S. ambassador to the Philippines.

Harleigh had once expressed an interest in a position at the White House if he should ever become president. Harleigh seemed to be a reasonably sharp individual, and he certainly had polish, but it was difficult to imagine being so hard up for help as to choose someone whose manner suggested that he expected life to amuse him. Nevertheless it had been a clever way for Harleigh to assure himself a niche in another man's memory.

They had known each other for years because of Harleigh's penchant for showing up at the more significant Washington parties; and as Harleigh came toward him he realized that Harleigh had often seemed to have his eye on him. It was almost as if he were a race horse on which Harleigh thought he might decide to place a bet.

—Correction, "a small wager". Harleigh had a lazy charm which made others slow to notice that he occasionally enjoyed putting poison on the tip of his rapier. But there was a cold glint in Harleigh's manner on the rare occasions when he smilingly took the trouble to make sure you saw his point.

He found the other man's attention unpleasant also because Harleigh always seemed to be looking at things from an angle: it was in the way he tucked in his chin, as though he were keeping something to himself that you'd like to know about and he might condescend to tell you.

They murmured renewed greetings as they shook hands.

Harleigh had softly curling, golden-brown hair and a tall, lean build somewhat like his own. But he was a prep-school, polo-playing, pseudo-English type that could have been repellent had he not been so well-mannered. Harleigh gave the impression of always being flawlessly dressed and almost excessively composed, as though he had just stepped out of his family mansion in the pages of *Architectural Digest*.

After they had exchanged pleasantries about the morning meeting Harleigh observed that they both had full schedules all day, but they "ought to get together for a drink after dinner this evening".

He had no objection, for besides a style which always made him an interesting study, Harleigh had the political magnetism of inherited wealth and interesting connections. He consulted the printed schedule he carried in his pocket, and they set a time; and as Harleigh put it, he wrote him in on his "dance card".

When the afternoon meetings ended he was scheduled for an early, informal dinner with the Peace Corps staff who worked out of Manila. There were nearly a dozen Americans, plus three times that many Foreign-Service Nationals, to support several hundred Volunteers in the field. Before they all sat down to eat there was a reception in the private dining room they had reserved, and during this time he made a point of being introduced to all the Filipinos and having Tony take his picture with whatever groupings would include all of the guests. Tony had quickly become adept at stage-managing this type of activity, so instead of being stiff and embarrassing as it might have been, it served as an entertaining ice-breaker.

He sat down to dinner at a table with the staff people he had particularly wanted to get to know, and as they ate he drew them out. Once they got warmed up he started taping them, and they responded well to the implicit compliment.

However, in deference to his very long flight the next day, they took their leave promptly at half-past seven. In parting he got their solemn promises that they would feel free to write to him at any time about any of their concerns.

This routine had become easy to him by now. He had sown many good seeds during this trip; when he was back in Washington he would have a full-time job just in nurturing them.

Inevitably, unless he could gain much better co-operation from his top staff, most of these precious seedlings would wither and die from lack of tending.

His last official appearance of this five-week journey was over. That realization struck him as he went up in the elevator. And it had gone well. That made him mildly happy.

Tony left him at his door, and when he had freshened up he dictated some notes about the afternoon meetings and the staff with whom he had talked. Then he went next door to see how Philanth was doing.

She was doing just fine, thank you, barefooted and wearing her denim trousers and matching smock-like minidress. She had eliminated the offending pigtails by adopting the hair style he liked best, the swirl around the crown of her head. She had piles of papers and folders ranged along the top of her dresser with Post-It labels on them: "H.Q.", "plane", "Mr. B.".

He gave her the cassette he had just recorded, accepted a fresh one "just in case" he had "any random inspirations about anything", and took the chair to rest for a minute. Once reassured as to his intention, she ensconced herself against the pillows at the head of her bed and resumed her work on her clipboard.

"I hope your afternoon wasn't too dull," he said.

"If a job's dull I make it interesting by finding ways to do it faster or better. The problem is that if you're functioning as part of a bureaucracy you've got to do your best to do everything exactly the way everybody else is supposed to. That's the sort of work that can drive a high-pressure person nuts."

He could not blame Philanth for not wanting a lifetime of such tasks. St. Joans and St. Georges generally left "public service" after a season or two, embracing the openness and flexibility of the private sector with relief.

Philanth could have been St. Joan, and she knew it. She knew she was not cut out to be the docile, life-long slave Carole intended. How had Carole made such a mistake?

Probably the same way everybody did, by seeing what she had wanted to see: the fact that Philanth was dedicated, idealistic, and docile in manner.

It had become apparent to him that in this case all of these traits were misleading. Nevertheless Carole wanted him to override Philanth's ambition to break new ground because Carole was trying to wean him from the dependency which she herself had created by being so capable and so determined to help him advance.

He could hardly fail to understand Carole's motivation. He hoped to get into the Senate. She wanted that for him. However, once he was there he could be counted on to pressure Carole to fall back into the all-purpose role she had filled when he had been in the House. Before that happened and caused serious conflict between them Carole intended to make Philanth a new reliable dumping ground for otherwise undelegatable "details".

Carole had often complained that she too needed a wife, since she didn't find handling all the details of food, clothing, and shelter much more interesting than he did. Of course she was talking about needing a high-quality servant, and the money she saved by doing all those countless little things as she found time was needed to help bear the costs of his political career, especially of all those plane tickets. Besides, *somebody* reliable had to keep track of their affairs, or he could easily be ruined.

Just the same, he could understand the strength of Carole's determination to get some cheap and reliable help, in view of the countless arguments and discussions he and Carole had had over the years because of his insistence that nobody he could hire could properly do those things which he wanted Carole to do for him.

The girl in front of him was supposed to be Carole's proof that he was wrong. But Philanth's misgivings about what sort of work he was recruiting her for only supported his own position in his long struggle with Carole: truly good help *was* scarce and expensive, and training for a responsible position took time even if a person's character and abilities proved suited to the work.

Thus it was far faster, easier, safer, and cheaper to turn things over to one's wife. Even when a situation was impossible you and she could simply yell at each other until the problems were worked out or at least the goal was re-defined into something do-able . . . or acknowledged to be impossible.

Carole, however, had come to hate spending her life serving as his adjunct. She liked the high-level staff work, but she had served him so well in all other respects that he kept asking her to do things he could have done himself while he was asking; he asked her to drop whatever she was doing and get him things when he knew perfectly well where they were kept; he laid on her any chore he did not want to bother with even when he could have done it more easily. Though she had always kept in mind the fact that he was preoccupied and under pressure, that had not always been an adequate excuse. Over the years those lapses into self-indulgence had aggravated Carole's growing anxiety over having "nothing to show" for her life. The swiftness with which her temper could flare over such trivial questions as which of them was to hang up his shirt indicated that she had become fed up.

But exactly how fed up with her role in their marriage *was* Carole? Enough to sexually withhold herself while throwing Philanth at him, just as a way of trying to ensure that she would be free to make better use of her skills as the top assistant of a senator?

Yes. Just look at Carole's education and natural abilities and some of the tasks on which he had induced her to waste them. She could be so disenchanted with him that she was willing to have another woman sleep with him, if that would help motivate the woman to take her place as his personal slave. He no longer doubted that it was possible.

The only hitch seemed to be that Philanth exceeded even Carole's specifications.

He looked at his watch. It was time to go meet Harleigh in the lounge. Tony followed him down.

Harleigh was watching for him. "Sorry to keep you waiting," he said automatically.

"Oh, I was prepared to wait longer." Harleigh's breath indicated that he had in fact started without him. "I suppose you'll be going right back to her. Don't blame you a bit." Harleigh winked, raising his glass to him in a mock toast, then had another sip.

Before he could decide how to deal with that, the bartender came for his order. It had been a full day, and now this. Determined to keep a clear head, he ordered tonic water with a dash of lime. Harleigh asked for more of whatever he already had.

"You insisted they ring my room instead of my assistant's, there was no answer so you got switched over to my aide, and he told you I was with her?"

Harleigh winked again and pointed at him to say, "You've got it."

The man's uncharacteristic jocularity was a bit hard to take, especially considering the sort of ambition he suspected it masked. He had a feeling he was deliberately being set up for future pressure. His reaction was dismay, for Harleigh was bound to be more rested for this game than he was.

He needed a character witness. He managed a sly smile. "Like to meet her?" He realized that Harleigh must have been following him while Philanth chattered to him as he made his way out of the conference room. "You hardly got an adequate picture this morning!"

Harleigh's eyebrows went up; his smile broadened.

Their drinks were delivered, and he let Harleigh pay. They relocated to a deeply upholstered red booth.

As they set down their drinks Harleigh signaled to a waiter and gestured that he wanted a phone. It was quickly brought and plugged in, and Harleigh tossed out a tip. Harleigh presented the receiver to him with a flourishing wrist action which surely had been perfected by practice. The man must be a demon at squash and tennis.

He dialed and summoned Philanth.

As he was about to hang up he decided to set Harleigh up, in turn. "Wait, hold on a second. Put on your sexiest dress, I want to show you off to someone." He hoped Harleigh did not overhear her scandalized, "*Mis*ter *Bjork*lund!"

Harleigh laughed as he hung up. "Why doesn't she wear her hair in a little bagel-shaped bun at the nape of her neck, when she goes to meetings like this morning?"

"Didn't you find the braids convincing?"

Harleigh laughed again. "Is she good? —I mean, at what she does?"

"In a class by herself."

"Been with you long?"

"She was recruited specifically for this five-week trip. For qualities you can scarcely begin to appreciate, without personal contact."

"You don't say.

"Clean?" Harleigh *would* think of that right away.

"Would I have her with me if she weren't?"

He added, "She's a widow with very high standards. But loaded with natural talent."

The light in Harleigh's eyes almost made him burst out laughing. Harleigh saw the impulse and, having interpreted it, looked even more impressed. He seemed to be playing right into Harleigh's hands, being so proud of his acquisition that he could not resist a rare chance to boast—and thereby reveal a side he had always kept hidden, a very vulnerable side. He realized that Harleigh was proving gullible because Harleigh would assume that any intelligent person who went around in an aura of rectitude must be a practiced hypocrite.

He could not anticipate how long Philanth would take to get downstairs to them, considering what he had said to her, so he half-heartedly tried to turn the conversation to the embassy's view of Peace Corps programs in the Philippines. As he expected, Harleigh didn't bother with that and kept watching the entrance through which Philanth would come.

The only outward sign of tension was that Harleigh lit a long, filter-tipped cigarette, adding to the stink of stale smoke and booze. The piped music was an old Frank Sinatra number, full of *ennui*. Philanth was going to go right up the wall when she realized she had been summoned for a game as sleazy as its setting.

Harleigh kept trying to come up with safe ways of asking more questions. To each awkward query he responded, "Maybe you'll see for yourself."

The third time he said that, Harleigh asked, "Could I? See for myself?"

"I thought that's why you wanted to meet her. I figure you'll return the favor sometime.

"Just be frank with her. She's remarkably sophisticated, for her age.

"She's also very partial to tall blonds. And she's the affectionate type. If she likes you, believe me, she'll let you know.

"One word of warning." He leaned forward and looked serious, so Harleigh did too. "Don't waste her time. That's the one thing she doesn't tolerate."

He inhaled with his mouth open to avoid laughing in Harleigh's astonished face.

"Just the thing for a man with a heavy schedule," Harleigh commented.

He winked and pointed to his companion to say, "You got it."

Harleigh shook his head, warning, "You're giving her quite a buildup."

"After you've met her you must go over everything I've said and see whether any of it was untrue." Now he was openly preening.

After a long pause Harleigh asked, "What sort of pay does she get for . . . overtime?"

"I insist on being generous with her."

Harleigh would not have dreamed of pulling out his wallet at that juncture, but he could almost see this bon vivant contemplating his supply of cash and trying to decide whether it would be adequate; and speculating regretfully about whether hotel jewelry stores could be open in late evening, since cash was so crass a token of appreciation.

Philanth appeared within fifteen minutes of his summons, looking as pretty as she had at the reception in Morocco. Knowing she was to be "shown off", she moved with an upright, regally graceful bearing, and as she came toward them past the crowded bar, heads turned.

Harleigh rose and grasped her elbow through the cape as he urged her to sit between them with her back to the wall.

Her acknowledgment of the introduction was polite but guarded. She looked at him for some cue, but he only gave her a small smile with an intent stare.

At once she began to find out everything she could about Harleigh, and Harleigh set out to floor her with his background in the most modest tones possible. He aided their effort by throwing in credentials even Harleigh would find it difficult to present in person, especially the Rhodes Scholarship. At once she began politely quizzing Harleigh about exactly how he had won it and what he had done with it.

Harleigh soon cut that discussion short and asked whether she would be interested in his showing her a few night spots, her last night in Manila. "We can continue talking as we go," Harleigh added.

She looked disconcerted. "Mr.—Mr. Bjorklund and I are here to work," she managed. "I'm sorry, and I thank you for the invitation."

This was the first false note, and Harleigh looked at her closely. She could not have failed to notice that Harleigh was male-model handsome, and she had assurances from both of them that Harleigh had a lot of other things going for him as well.

"Your boss says he has no objection to your going with me."

She looked at him rather than at Harleigh, and she clearly didn't believe it. "Really?"

"I said . . ." Harleigh had moved too fast for him. "I, uh, said Harleigh could see for himself whether you, uh, were 'the sort who's always ready for a good time.'" That had been one of Harleigh's questions to which he had answered that Harleigh might see for himself.

Philanth reacted with an alarmed frown. "Well, strictly speaking, that's true, of course, . . . though it has connotations I'm not altogether comfortable with, ..."

There was a pause.

Turning her head to look at each of them in turn, Philanth asked, "What's the agenda, here?"

He smirked and gestured to Harleigh, It's your show.

Harleigh hesitated, then took her hand. "I'd like to spend the rest of the evening with you, Philanth. I'm told you're a lady I'd like to know better."

"Well, that's very nice," she said, withdrawing her hand. "Perhaps another time."

She added, "I don't believe I've mentioned it to Mr. Bjorklund, but I don't date."

"I beg your pardon?"

"I don't go anywhere with a man for social purposes," she clarified as though Harleigh were from some other planet. "I swore off some time ago." She watched Harleigh as she added, "I got tired of men trying to get me into bed."

"Don't you like to go to bed?" Harleigh teased.

"I'm strictly opposed to sexual relations outside of marriage, Mr. . . . Mr. Adams. No exceptions, not where I'm concerned."

Harleigh looked disconcerted by her bluntness and threw him a glance. "That's not what I was led to believe."

Philanth acted shocked. "I'm sure you misunderstood! Mr. Bjorklund would have no reason to suggest that I'm ever sexually available to *anyone*! I find it impossible to imagine otherwise! He's a very moral gentleman!

"And he knows I have no intention of remarrying. We've discussed it in relation to my plans regarding my career development. I've told him I plan to devote my life to work overseas, and he's refused to believe that that precludes marriage to some young man similarly dedicated."

Harleigh raised his eyebrows and shot a suspicious glance across the table at him before refocussing on Philanth. "You don't say."

"I do indeed. My present job is only an interim arrangement between my being medevacked from my Peace Corps post and my returning to my life's work for the poor in North Africa." Philanth was warmed up now, starting to throw in grace notes, laying it on just thick enough to be having her own fun with Harleigh.

Harleigh slouched back on the seat and threw him an accusing scowl. "I see."

"Isn't she something special, just as I told you?" he asked innocently. "She's excellent in her work, and she's completely dedicated to the Peace Corps ideals. Right, Philanth?"

"I do my best."

"I see," Harleigh said again, more grimly.

"Thanks for coming down to meet Mr. Adams," he said to Philanth. "He wanted to see for himself what an excellent assistant I have."

Philanth began to smile at him. "Not at all."

She turned to Harleigh. Practically oozing syrup, she said, "If you'll excuse me, Mr. Adams, I have such a lot of work to do! I've got to have it under control before we get back to Washington."

Harleigh got up to let her out. They exchanged very polite "Good nights." Philanth glided away like a princess with a book on her head.

Harleigh slid back into the booth. "Now, that wasn't nice."

"You'd never have taken my word for it, would you?"

Harleigh shook his head. "She's 'straight arrow' even if she's not built for it. What a waste."

"And she thinks I'm the greatest thing since Jesus Christ."

"Wonderful." Harleigh had the grace to laugh and offer to shake hands. "My condolences, Bjorklund."

As their hands parted he added, "My wife selected her for the trip, knowing exactly what she was like."

"That so.

"Now," Harleigh said, starting to get up again, "since you've wasted enough of *my* time tonight, I think you'll understand if I say good-by. 'Til we meet again."

They shook hands a second time. "Good-by. 'Happy trails'."

Harleigh laughed again and took off with a waving salute which conceded that the match had been fairly won. Harleigh would hold him in higher regard in future, though not for his morals.

He tossed off the rest of his tonic water, then hurried upstairs so fast that Tony had to make a dash in order to join him in the elevator.

They found Philanth pulling pins out of her hair. "Thank you, Philanth. That guy was trying to set me up for a little political blackmail. I knew I could count on you to set him straight. You handled it very well."

"Hmph." She pulled her hair down. "Some people have the morals of dogs and cats. Is he married?"

"Divorced."

"Figures. Who'd have him?"

He had met two of Harleigh's lovely wives but did not feel it necessary to tell her that some people must enter marriage with the same attitude they would bring to roulette. "We'll leave, I can see we're interfering with your efforts to 'get comfortable'."

"No, wait," Tony said. "Philanth, you have certain rights. You're 'way overdue for some time off. This evening, while we're both still dressed, I'm taking you dancing. You're gonna have a great time.

"Now, you gonna come quietly, or are we gonna have a big hassle?"

Philanth looked at him. He looked at her.

"I'm coming too," he informed Tony. "I hope there's no objection."

"Hell, no!

"Brush your hair, Philanth, and le's get outa here. The ev'nin's passin'!"

Tony hurried out into the corridor, calling, "Be right back, gotta get my guide book."

"Hold him down for a few minutes, will you, Mr. Bjorklund? I have to alter the design of my dress."

"Omigosh, I didn't think. That cape would be ridiculous, on a dance floor—unless maybe you did nothing but slow waltzes. What're you going to do?"

"Just tack my scarf around the top edge for extra coverage. I figured it out as a fantasy, after Tony suggested the idea on the plane coming in. It won't take long. I'll let you know when I'm ready."

He nodded and went to have a talk with Tony about their choice of night clubs.

Chapter 21

FORBIDDEN PLEASURE

Watching Philanth dance was an experience he expected to look back on in old age as a reminder of the simple joy of existence. As though she flicked an internal switch, she could relax and enjoy everything so completely that her pleasure was contagious.

She was literally dancing circles around her partner, and she kept encouraging Tony to master her variations. When Tony began to clown in the melodramatic style of a Valentino she responded with a flair she must have picked up from watching televised dance competitions. Tony was delighted.

Ken the ex-family man and self-absorbed professor must have taken considerable time and attention from his research and devoted it to her lessons. He found it impossible not to pity Ken for having taught her so much and then destroyed their relationship by asking for more than she was prepared to give.

Only a fool would refuse to learn from someone else's mistake.

At least Ken had not had to watch her slinking and bouncing around in this "basic little black dress".

Then the orchestra shifted into a quieter number, and Tony, resplendent in his white jacket and black tie, gave her a tender smile and drew her close. She did not resist. They looked extremely content, and in the simplest of dance steps they moved well together. Very close together.

Oh, this was more than he could endure! He was identifying with Tony and feeling it in his gut. He got up and went to claim a right to every other dance.

"Groups of three," Tony counter-offered, and he agreed. Tony, who had already had about that number, left them to "scout around".

At last. He put his arm around her waist, moving closer. It was so special a moment that he was very careful.

But as he began to lead she did not move with him. Her fluid movements had become wooden. He paused and frowned at her in perplexity, and she looked up at him as if suddenly anxious. He tried to draw her close; that caused her to break from him altogether.

More firmly, this time, he restored the position from which he could lead. "Relax, Ms. Devon," he murmured. "I won't fire you if you step on my foot."

He felt her try to obey. But she was following his lead skittishly, throwing them out of synch. He realized that she was trying to avoid some mistake which would momentarily cause body contact. Imagining how he wanted them to move together, he growled with impatience. But he couldn't force her.

He stopped trying and drew back from her in exasperation. "Please, Philanth!"

"I—I'm sorry." She looked anguished. "This was a mistake. I . . . I can't go through with it."

"What're you afraid of?"

"Of touching you!"

"But Philanth," he breathed, trying to take her hand and not succeeding, "this was the whole point! If you wanted a disco you should've said something. I knew how much you liked ballroom dancing, so I fought Tony to bring you here."

"I'm grateful. But I didn't think it through. I was too busy fussing with my dress." She turned away disconsolately, repeating, "I'm sorry."

He put his arm across her shoulders and guided her off the floor, murmuring, "Let's talk about it."

He found a dark corner near a fire door where Tony would have a hard time spotting them.

"It's too smokey and noisy in here," she complained. "Does that door go out?"

He found that it did and that it was not locked from the outside. It opened onto a fire escape. The light bulb beside the door was barely adequate for lowering the stairs and finding one's way down into the alley: an allocation of electricity which was considerate as well as economical.

He turned to her. "What's the problem?"

"I told you," she said agitatedly. Each word came out with difficulty: "I can't . . . dare to touch you."

He tried to be jocular. "Have I been promoted to Almighty God, and you'll be burned . . . ."

She was painfully serious. "*Yes*. Don't you see, I've *conditioned* myself to feel that way."

"Philanth—" He lifted both hands in dismay and turned to stare vacantly into the darkness. "It must take great mental control to preserve such an attitude when my behavior has deteriorated so much."

"But I *need* to preserve that feeling!"

"So I can serve as some nourishing ideal, like some idol you've never met?"

Her "Yes" was barely audible.

He turned to her. "I think there's more to it than that. Aren't you afraid to come to terms with the fact that I'm just another man, someone you can feel ordinary emotions about, . . . including desire?"

At first she didn't answer, only stood with head bowed, hugging herself.

"Besides that, there's another reason. Dancing is like making love, only less." The anguish returned. "So it would hurt all the more when I know I'll never see you again: the more I feel close, the more bereft I'll feel afterward. The pleasure of having could be

593

only for an hour, if I could accept it at all; but the pain of having lost would go on and on."

"You believe you're going to have to go away, after everything that's been said?"

"It's become obvious to me today that you were right: my terms for staying *are* unrealistic."

"Look, Philanth, about today, I owe you an apology—"

"No, I understand. That evil Mr. Adams proves you were right . . . about me . . . after all. It isn't necessary for everybody to be like him. Only one or two, in strategic positions, could get you pilloried in the media. As you've said, there's always somebody wanting to cut you down when you're doing anything significant. And apparently to a certain kind of mentality I do look like I couldn't possibly be with you because of my brain."

He wondered how early she had understood the true nature of his fears. Probably she had faced the truth only as she spoke it, for in her last sentence he could hear that she was even now hurting as only the young can hurt when someone spurns their claims to the dignity of being dealt with according to their merits.

Head bowed, she went on. "I've always refused to accept how some men think: how they refuse to believe in innocence and so feel a compulsion to destroy it to vindicate their disbelief; how they reduce a woman to her body parts to prove she's inferior to themselves.

"But degrading someone is usually a manifestation of insecurity or lack of regard for oneself. When I'm treated like a floozie by somebody so obviously pleased with himself as Harleigh Adams, it's time to stop fooling myself: I've got no hope of being 'equal'. Because it's not just sexism we're up against: it's

594

sexism reinforced by cynicism fed by other women's descent from the pedestal."

"I'm so sorry, I didn't realize . . . ." At first she had put up a front of tough-minded contempt for Harleigh. But as her anger at the insult had subsided, the humiliation had begun to sink in.

She replied firmly, "You did what you had to do under the circumstances. It was right for me to get you out of something I'd gotten you into."

With despair she concluded, "But I'm right to want to get away from this culture as soon and completely as I can. I can hardly wait!"

"You can't run away from something so pervasive as sexism. You have to fight it from the strongest professional position you can find."

"I choose more basic battles: women can't try to achieve equality anywhere until they can control their child-bearing.

"Besides, you're trying to suggest a position with you. But I've caused you problems today from morning until night. Know-it-alls like Harleigh infest the circles you move in. So there's no way I can work for you on a basis that wouldn't waste me or compromise you."

"There's one way out of the problem. What if Carole has decided to divorce me?"

"That's unthinkable. If you weren't angry with her because of all the trouble her 'testing' me has caused, you wouldn't consider such a notion."

"I've been married for fourteen years, and I surely know more about my marriage than you do. Carole's thoroughly sick of my exploiting her. She knows I have to do it because I'm competing with other men who do the same thing. But she's got more graduate degrees than I have, and she wants more to show for her life than I can promise her."

"She hasn't told you that."

"Not 'the bottom line', not yet. But soon, I think. Because everything fits—once I have the courage to believe it because she's lined you up to take her place—because she wants out of the servitude of marriage to a politician who may not be going anywhere."

"*I* wouldn't believe that even if she jumped up and down and screamed it at me!"

"Why not? *You* don't want that servitude, either."

"No. A mind is a particularly terrible thing to waste if it's your own.

"But you can change. You can re-design your ways of doing things so your wife finds fulfillment while also helping you."

"I do intend to change. It's just that today's meetings were a bad place to begin."

"Carole will be glad."

"That won't be enough for *her*. At her age, she wants much more. And she's earned the right."

"I don't believe she's as disloyal as you fear."

"She's pushed you into my arms in order to get free of me."

"You have non-logical reasons for being willing to believe that. Don't trust your judgment."

She was again implying that he was thinking with his groin. He took a moment to curb his annoyance.

"Speaking of non-logical motivations, I brought you out here to persuade you to dance with me."

"You don't really mind not dancing with me. You just want me in your arms." She moved forward to snuggle against his chest, and he automatically put his arms around her as she added, "Like this."

"You prefer this because it's easier for you to keep it platonic. Okay." He held her close.

But it was not a platonic experience for him. Because of the way she was dressed, his hands found no place where they could rest easy except clasping each other behind her waist.

After a minute he began murmuring to her. "Stay with me, Philanth. Let me be close to you forever. Because I have the same creed that you do: if there is this one loving, decent, dedicated person in all the world, there is adequate cause for hope, adequate reason to go on keeping the faith, serving the principle that humanity is worth whatever we can do for it, whatever the evidence to the contrary."

He grasped her shoulders and looked down into her eyes as he continued, "The only difference between us is that I'm experienced enough to know I need to be with that person in order to do the best I'm capable of. Abstractions alone can't sustain a whole-hearted effort when the enemies are all too concrete."

"I understand. That's why I would stay with you if the necessary terms could be adhered to."

"That is, if I can treat you as I would a man, personally and professionally."

She nodded.

Chuckling at the hopelessness of that, he released her and took a few paces away from her and back again. The metal treads of the old platform clanked under his feet.

A light breeze stirred, carrying a fetid stench. Garbage, raw sewage and industrial waste carried by the river were being washed up against squatters' shacks along its banks. This city was among the worst in the world in air pollution. The litany was familiar: grinding poverty, unregulated industry, and poor planning. With government services swamped by demand from the burgeoning population, the city was out of control.

It was places like Manila, with their massive deprivation which would always make his own needs look ridiculous, that were wooing Philanth away from him. He had to make a better pitch than he had managed so far, or he would lose her and all the strength that she could give him. He thought hard . . . and cursed Harleigh Adams for his wanton blow to Philanth's already feeble hope.

He was supposed to have a silver tongue, and he had better start using it in his own behalf. "You mustn't let a destroyer like Harleigh Adams defeat you before you start. He's the enemy: he's one of the many kinds of enemy we have to fight, in the center of power where they congregate like vultures and can do the most harm.

"There are dead souls in Washington, moving around in living bodies. You can find them in all the nicest places. Most of them seem never to have been alive, never to have understood that principles are all that lifts us above the other animals and makes us wonderful as well as preposterous. The 'clever' ones cite Machiavelli as though they had never heard of Washington or Jefferson or Lincoln.

"But if you and I have each other they won't get us down; their poison won't infect us; the rot that per-vades them won't be able to take hold. It will go on spreading, the cynicism and despair, killing other souls by stages. But it won't even weaken us, be-cause when we feel it we can turn to each other."

"All right, Mr. Bjorklund. If you're willing to stand up to the Harleighs and all the other sorts of dirty minds that want to bring you down to their own lev-el, I can work with you. But it'll take the two of us, with nothing personal to hide, to out-face them."

"The other horn of the dilemma: 'nothing personal to hide'. I must have a motivated blindness about this the way you do about unintended double enten-dres. I feel I shouldn't *be* obliged to hide what I feel toward you. I've told you what you are to me. When I turn to you for the qualities that make my life worth living, how can I not want union with you?"

"Let's talk to Carole. I find it impossible to believe that *she* truly doesn't want to share that with you. Then . . ." Her voice faltered. "You won't need it from someone else."

He swung away. "Oh, God."

"I'm sorry," she added hastily, "you may be right about her, but I just don't see it. And I can't take your word for it, when . . . ."

"That's not what I was swearing about. Don't you know you make me want you more even with the reasons you keep pushing me away?"

"You must mean you keep loving me more. That's not the problem. Love isn't the same as desire."

"I don't separate loving you from wanting you! If that takes me off your pedestal, that's okay by me!"

"It's *not* okay," she exclaimed, "it's not *a bit* okay, when you need to be on a lot of *other* people's pedestals in order to lead them where they need to go!

"And I'm fed up with your romanticism! *Surely* my whole future doesn't have to be shaped by the unfortunate fact that you've developed an instinctive attraction to my *body*! *That's primitive.*"

There was a pause. Then he remarked, "Don't you find the very idea of 'lovers' quarrels' boring?"

She nodded vigorously. "But why must you keep wanting more from me than I do from you? That's what's causing the whole problem."

"I've tried to explain, Philanth, that if you truly share my work you'll need all the comfort you can get, as I do."

"I could share your work, but I wouldn't truly share your burdens, because we're not equals."

"I *want* you to share—"

"I see what it is! 'We're not equals'! It's the fact that we're not equals that makes you need more from me than I do from you! I'm satisfied to worship you. I don't even know how to truly want more, even though I love you deeply. But all I have can scarcely be enough for you.

"So it's *not* a matter of your not being selfless in the way I try to be. I believe you're so close to that as makes no difference.

"*This* is why we can't agree on the terms of our relationship: what we need from each other is irreconcilable because of the distance between our levels. You want me so much *especially because*—being so far beneath you—I've disciplined myself to be too 'pure' to want you.

"That dynamic will go on reinforcing itself until your reputation is tarnished, if I don't leave you soon after we get home."

"No, wait, you're going too fast on the wrong assumptions! In the first place, we're not necessarily unequal. You would feel we were on the same level if I were as young as you are."

"Why debate that? The fact remains that you can't tell me you don't look down on me when you're among your own kind, as you were this morning."

He winced. "Just the same, that *is* because of the difference in our ages. It's what a person is *at* his age that matters. You're special to me because of your potential. Think what you could be at my age!"

"A woman." Her voice was flat. "Whether young or old or middle-aged doesn't matter; the word 'woman' is like the word 'politician': there's no adjective you can put in front of it that credibly redeems it. Mel Lazarus pointed that out about politicians long ago."

"Don't be so negative. You keep a lot of your fire and drive and brilliance under wraps almost all the time. You've told me you're tired of 'hiding your light under a bushel' because it's so 'crippling'. That's because you're too young and traditionally feminine to feel you're *allowed* to display—and use—your genius the way a flaming egoist would.

"But in, say, thirty years you could be . . . as magnificent as you believe I could be at the same age—*if you let me give you a start!*"

"I don't see how I could be spared from helping you in order to achieve that," she said as though thinking aloud, "unless you'd consciously shaped my work to develop me so that when you retired I could move into your position."

Something flashed in him: she was offering to become the inheritor of his career. After a lifetime of struggle to build a personal power base for doing good he might have someone to take over whatever he had built so it need not dissolve into nothing. His bridge to power was made of intangibles, personal contacts and organization and reputation; but an elective office *could* be passed on if the forces which had won it could be used to persuade enough people that the chosen successor was the best candidate.

"Excellent idea. I agree without reservation."

It took her only a moment to catch her breath. "And our personal relationship?"

He spoke stiffly. "I'm ashamed of having pressured you. I'll play it your way."

Philanth nodded. Her expression was impassive, and he sensed that she did not believe he could do it over the long haul any more than he did.

Nevertheless he had achieved a breakthrough: he not only had found something Philanth couldn't help wanting, he had found an adequate personal motivation for putting up with the political inconveniences for himself: *he* was going to have an heir!

Philanth added reassuringly, "I'm sure Carole has been withholding herself to turn your attention onto me—even though she may claim that that was only part of her test for me. I still say, when you resume normal relations with your wife you'll have no problem in maintaining propriety with me."

It took him a moment to decide which aspect of that statement he could allow to make him mad. "*Assuming* all that, I may be able to behave myself; but don't dare to suggest I'll have 'no problem'! What's between you and me goes far beyond—"

He broke off as he realized that asserting the dignity of his passion for her by promising it would endure even if he resumed a normal sex life had not been one of his smartest moves.

Philanth chewed her lip. "Perhaps," she ventured, "if you understood what intimacy with you would cost *me*, you could use that to motivate yourself to reject the idea each time it surfaces."

"Then *tell* me! Help me to understand!"

"It takes several pages in my current journal. I'd give them to you, but I sent that section home in my final package of clippings."

"Go ahead. Tony's probably just hitting his stride with some 'foxy lady' about now."

Philanth took a deep breath and released it, then began slowly. "Just let yourself think about what we're talking about, in the complete way we're ordinarily afraid to think about it when we're together. I couldn't give less than all of myself to you. The complete sharing you must have is what I must also give. It's in the nature of what we are—and know each other to be."

"I understand that."

"But you apparently haven't thought enough about the psychological after-effects. From then on I would feel that most of what I am was yours, in a sense some people never understand. My sense of identity and my body would feel bound to you for the rest of my life."

"I want that oneness with you more than words can express."

"But I would feel that I had betrayed something sacred, and that feeling would always remain with me. I'd feel degraded even with you."

"My dear, don't you know that's wrong?"

"Not right or wrong, just out of fashion. It's a matter of assigning values; and that is ultimately beyond questions of validation, apart from what's functional in terms of achieving some other value."

"Then surely you can let it go."

"I hold to that strict old code of confining sexual expressions of love to marriage because it can beautifully shape the lives of those who have the depth and strength to see and protect its beauty. The order it imposes on life provides for the needs of children."

"But you needn't let yourself be enslaved by a principle just because it's useful."

"But I *honor* the principle: it's *part* of me, and I wouldn't want to be otherwise.

"Serving a principle has costs, yet that service gives you pleasure if that principle has become integrated into your nature: you feel good when you adhere to a standard you're conditioned to feel is good. People have died rather than violate that integrity, because the inner harmony it gives is so precious.

"You're like me in this, so you can understand. We couldn't love each other as we do if we weren't both built around socially functional commitments which cause us all sorts of personal pain and hardship. 'I could not love thee, dear, so much, loved I not honor more.'"

"But why define honor in these particular terms?"

"When I was little the Christian church taught me that the body is a temple of the spirit. As I went through adolescence I fitted my feelings about sexual intimacy into that increasingly self-fulfilling premise. Nothing in physiological psychology challenges it, since science only describes the 'hardware' which supports the 'software', the mind: the character and personality and will; and their conceptual and emotional essence, the soul and spirit.

"I grew up very religious, and becoming an atheist doesn't require you to turn your back on the human capacity for setting some things apart as sacred. It's an emotional enhancement of the definitions of ultimate reality and meaning and purpose we need to supplement the individual's survival instincts; it helps to make culture and society possible by drawing people together in self-transcending objectives.

"Like conceptions of ultimate meaning and purpose, sacredness is a fruit of our ability to create abstractions. We become willing to live and die for abstractions because we understand that they can transcend our short lives and finite selves, making us far more than our animal nature.

"Am I losing you?"

"You've explained how an atheist can keep things sacred."

"Yes, but calling me a 'secular humanist' is more precise." She added humorously, "Besides, it sounds better."

"Ha! Only a little. Ask any politician.

"Okay. So you're telling me intimacy can be sacred." He could not keep from touching her as he spoke. "Tell me something I don't know."

"You must have seen that keeping it sacred is problematical, especially outside the commitments made in marriage. The increasing degradation of sex in our culture strengthens my conviction that I could never settle for anything less than the protection of lovemaking provided by the sacrament of marriage; that intimacy should be reserved as the ultimate expression of the commitment of marriage.

"All this may be regarded as either painfully obvious or hopelessly old-fashioned. But even casting aside its great value to society and regarding it as only my personal way of valuing things makes it no less deeply valid and binding *for me*."

"I understand."

"And that's why I would feel I dishonored myself in giving myself to anyone except my husband.

"Apparently that's one of the things that's meant by the old expressions, 'an honest woman', or 'a good woman'. It's been said that 'nice girls do'. But 'a good woman' doesn't. 'A good woman' makes a complete commitment to 'a good man' and then is faithful to it. She has too much of a spiritual self and too close an interweaving between that core of her personality and her sexuality to be willing to give herself conditionally. A woman's sexual feelings can be so inseparable from her sense of identity that letting a man into the center of her body means letting him into the core of her psychological self as well.

"That's one basis for the tradition by which women who've lost their 'honor' through rape or seduction want to commit suicide. I believe that suicide of any potentially socially productive person is immoral. Yet I understand their feelings. When a woman's self-esteem is made fragile by her cultural environment it's easily damaged."

That helped explain the warning in Carole's last letter. He had assumed that this referred only to such limited matters as scars from being persecuted for her intelligence by adolescent boys. But because of the high standards she had set and the depths to which she knew her culture had sunk, her vulnerability went far beyond that.

"I see," he murmured.

"Let me be sure you do. Suppose we were so monumentally discreet that you felt safe in seeking the presidency. Then I'd certainly have to go away. Having left with you all of myself that mattered to me personally, I'd wish to die. But I'd have no right to. I'd have to go on working for others. You know what I'm trying to do with my life. Surely you wouldn't ask me to carry a burden like that too!"

"I wish you could have spelled this out for me earlier. I can see that you couldn't because it admits too much about your own feelings. But it makes your argument that gossip might damage my career look trivial by comparison."

"But this is only a matter of my personal happiness. That would have a bearing on my productivity, but your power for good is so far greater than mine that I would nevertheless do what you've wanted, if I believed it to be the best course of action in terms of your basic objectives."

"Your generosity shames me."

"I tell you this only in the hope that it'll help you stop getting carried away into believing it's necessary. Because by *believing* it's necessary, you *make* it 'necessary'; and acting accordingly *would* jeopardize your personal reputation.

"You may think it's silly to harp on that. But I care about your potential value as an ideal, a symbol. I believe that if John Kennedy had foreseen the effects of womanizing on his legacy as a cultural icon he would have made it impossible for anyone to credit the stories that surfaced after his death, in spite

606

of his inclinations and his attractiveness to women.

"Have I helped you?"

"Philanth, if I ever ask that of you again, please remind me of what you've just said!"

"I didn't want to play on your emotions, but I had to give you facts strong enough to help you manage your own feelings so as to protect your own interests."

"I understood that.

"Well," he said after a moment, "I think we've got both aspects of our future together worked out now."

"I know you've promised a lot." He was transparently unhappy, so she was commiserating.

Trying to justify his tone, he added, "But all I really have to offer beyond the next couple of years is a share of power I may never attain . . . and work experience in the meantime that you may find of little value afterwards."

"I'm aware of that." She laid her hand on his sleeve above the wrist as she added gently, "It still doesn't look very good for us, Mr. Bjorklund."

"No. But I think we can make it work. Maybe it's a close decision for you, but the world's a hard place, so I *am* your best bet. It's a gamble of two years that could have a worthwhile payoff for the rest of your life."

"We can make it work," she agreed, but it was plain that she was not any more optimistic than he was. They had confronted reality, and it wasn't nearly as appealing as their fantasies.

"Come on," he said, gesturing toward the fire door, "let's do what we came here to do."

As soon as they reached the dance floor he started leading her. He held her at a distance, and she accepted this degree of closeness in the manner of someone tolerating a temporary discomfort.

He gazed unseeingly over her shoulder, intensely aware of her. For the first time she was accepting contact with him without the defense of acting like a little girl.

He would be patient and tender, and eventually she could surely outgrow her inhibitions enough to accept him as if he were her husband. If his worst fears about Carole proved true there would be no "as if", but the reality. Whatever the obstacles, it was inevitable that in time they could rest in shared fulfillment in each other's arms.

His falling in love with her had been as inevitable as water flying earthward over Victoria Falls. Flashing in the sunlight as it hurtled down, joyfully yielding to the inescapable order of things: that was how he felt about loving her.

We are but flecks of spray that catch the light for a moment before sinking back into the ocean from which we came, she once said, leaning back in her seat on a plane and easing the weariness of evening with one of her pleasant reveries. We flare with precious sentience and are gone. And yet, she insisted, if there is order in our minds, our emotions, and our lives, if we understand how we are parts of the cosmos which enable it to contemplate itself, we can only give thanks for this one shining moment when *we*—as precious specks of sentient dust—are able to think about eternity.

All that multi-hued complexity, painstakingly put together like some mosque filled with Islamic mosaic; one lovely, orderly design for a life of unstinting loving—in a world where those who give freely will be left with nothing! What a treasure. What a pearl of great price. If only—

There was a tap on his shoulder. "So *there* you are!" Tony's deep voice said. "I was starting to worry that you'd grabbed the girl and given your poor old bodyguard the slip."

Philanth's hand had eased off his shoulder, for Tony's arm was already sliding around her waist.

Tony added with a touch of his old cheekiness, "I'll let you know when it's your turn again, Mr. Director. I was timing you, wherever you were, so don't hold your breath." He bowed and returned to their table.

He had been sitting there for maybe fifteen seconds when a smooth male voice from behind him said, "Good evening, Mr. Bjorklund. May I join you?"

It was the State Department area specialist for the Philippines who had been working with them before the trip. He had been at the meeting that morning. Now he was darkly handsome in black tie.

"Kingsley Merriweather," he exclaimed as though nothing could have given him greater pleasure. He gestured toward the closest chair. "By all means."

He had unfinished business with this man. "You were with Philanth at the Peace Corps' bon voyage party." He added meaningfully, "You and 'Petya', the Russian from the CIA."

Merriweather positioned himself with his fist beside his mouth and began speaking into his ear. "I can't discuss this, Mr. Bjorklund—not here, and not now. Except to tell you that he is *not* CIA."

"Philanth certainly believed he was."

"That may have been a lie to protect her."

"Look, don't play with *me*. Wasn't a Peace Corps party used as a cover for intelligence activity?"

"In a sense you can hardly disapprove. He had information relevant to efforts to protect you from incidents like the one yesterday."

"I suppose I have to accept that at face value. But you can tell your superiors I'm very displeased."

"I can see that Philanth has told you nothing."

"I was not happy with her evasiveness."

"She was told to say nothing to anyone, but she might have made you the exception."

"She made my wife the exception, apparently as a way of getting my 'consent' after the fact to having CIA personnel at a Peace Corps gathering."

"Well, I'm sorry she was worried enough by his supposed position with CIA to get that passed along to you. If I'd been aware that he had told her—"

"What does a man *really* do, if he uses a job at CIA as a cover story? Petya doesn't look the sort to need to impress girls—"

"Really, Bjorklund, you overestimate this as a sufficiently secure setting for this discussion." Merriweather was still muttering behind his fist. "He insisted on using the party as a cover because it was Philanth who put us in contact with him. It was a worthwhile opportunity Philanth gave us. Since he isn't CIA, Peace Corps has not been compromised. Now, can't we let it go at that?"

As he spoke, Philanth came hurrying over to greet Merriweather. He watched the man from State smile as he stood up and took in her shapely dress and general appearance at close range. Dancing had brought high color to her cheeks, her unbound hair had developed the fashionable look of just-having-gotten-out-of-bed, and the sinuous beads of her African collar looked like a couple of dozen glittering black snakes gathered to admire her white neck.

Tony came sauntering up and shook hands rather reluctantly with Merriweather.

Merriweather lost no time in politely requesting his and Tony's permission to ask Philanth to dance.

He and Tony seemed to be of one mind: give the man a dance and get rid of him. Merriweather soon had Philanth swaying in his arms.

610

"Look how he operates," Tony said. "Swings her around so he can feast his eyes, then reels her in tight: having his cake and eating it too. Bastard prob'ly started dancing school when he was a pimply kid o' thirteen."

He didn't bother to answer.

"He'll spend his life shufflin papers and kissin ass and attendin parties," Tony went on. "I wonder about guys like that, I really do."

"Not about *him*, surely. Look at the way he looks at her."

"Awright, he's straight enough. *Now* lookit. She's playin up to him, givin him the big blue eyes like Miss Debutante. Bet the dimples alone are givin him a hardon."

"Watch your language."

"'Scuse *me*." Tony took a slug of his drink.

They went on watching. Merriweather apparently was too preoccupied with their conversation to think of moving her around the floor. Philanth was wholly attentive.

"He's the only one of the three mousketeers who's had the brains to stay single," Tony remarked. "Wonder if she'd get serious about a gink like that. They're actin' like they got chemistry."

"Oh, she's got chemistry with a number of men!" He realized that Tony seemed to have become one of them. "That's *their* problem.

"No. She might find the idea interesting, but that life would be too easy for her."

"Plottin the overthrow of inconvenient gover'ments? I think she'd go for that, if she thought Kingsley was up to doin *his* part. My bet is he hasn't got the balls to work up to that level."

"They'd have to be married quite a while for her to find out; and she has so little regard for men's good sense to begin with that she sometimes underestimates it."

Tony nodded consideringly.

When the music began to move toward a closing phrase Tony sprang up to claim the next dance.

As soon as they were alone again Kingsley said, "I'd like to ask you a favor, Mr. Bjorklund." Again leaning forward on his elbow to speak into his ear and arranging his fist so that he casually shielded his lips from anyone else's view, Merriweather went on, "Please don't look around or otherwise react openly to what I'm about to tell you.

"First, we'd appreciate it if you'd wind up your appearance here very soon. There's only so much we can do for you in a setting like this, and you're at risk."

"Another assassin has followed me here from the hotel?" he asked acidly.

"Aside from that, there's risk of another kind when you and Ms. Devon are together. You'd be wise not to dance with her any more. Your face is well known here now, thanks to the front-page stories about yesterday's shooting. Nifty things can be done with cameras these days, even in a bad light. Photographers make their living at places like this. And you're not always the one they offer to sell your picture to."

He *did* react, turning to scowl at the man. "Who's *this* concerned about 'protecting' me?"

"You weren't to know about our involvement unless you began making things difficult."

"By whose orders was I assigned this sort of protection?" he persisted.

"It comes from the very top."

"No!"

"That's my information."

"Since when?"

Merriweather shook his head.

"Well, why? D'you know that?"

"Put yourself in his place," was the evasive reply. "Somebody lets him know you're traveling around the world for five weeks with a baby-faced, hourglass-shaped little blonde and your own press agent. He may even hear about all those connecting bedrooms your people reserved for you. The President doesn't need to have two Peace Corps directors in a row departing his service under unpleasant circumstances, especially with women involved in both cases. The stand-up comics would pretend it reflected on his whole administration: if even Peace Corps has such a problem, what must his appointees in the other agencies be up to? You know that kind of 'reasoning'.

"And of course we all want the Peace Corps to retain the best public image possible.

"That brings me to my second message. You've done about as well as can be expected, but you're starting to slip."

"What are you talking about?"

"Bringing her to that big meeting this morning was not a good move. She looked about as authentic as Judy Garland in *The Wizard of Oz*, if you know what I'm referring to, and . . . .

"Anyway, this night-club fling crosses the line. I would've come forward earlier, but I was off duty, and I'm afraid my colleague reacted too slowly. You weren't at much risk on the fire escape, so we covered the alley and left you alone.

"But this has got to stop, Bjorklund. I'm sorry, I can see the attraction. She's fun to be with, and Petya assures me that inside that cuddly personality is a mind like a straight-edged razor. Unfortunately the good sense you showed earlier by confining her to quarters has gone the way of all good judgment when a woman starts getting under a man's skin. And once the gossip begins, ..." The flick of a finger indicated a beheading.

"I think a man really does have to keep his wife with him every minute and smiling like a puppet if he wants to stay in politics."

"One more thing I'd like to mention before you get mad. You're not the only one with a soft spot for Ms. Devon. We want to protect her as well as you from public attention. We're interested in her services as soon as she's ready to go back overseas. She's indicated she's likewise interested in working for us. After she's been separated from Peace Corps again, *of course.*"

"*Philanth?*"

"She's quite gifted, and not just in languages."

"How d'you mean?"

"Face like a child, that nice body, a whole wardrobe of personalities, multidimensional mentality—even for a woman—and excellent control." Merriweather chuckled. "Or as a layman might put it, a cold-blooded liar who could sleep with a man and pick his brains, then blow them out and not let it spoil her breakfast.

"Now," Merriweather concluded as he got up, "I must move along. I'm sorry to have cut short your evening." A slap on the back, and the dinner jacket vanished into the crowd.

He sat staring after it. Kingsley Merriweather had deliberately made him out to be a fool. He had stayed out of trouble through a lot of busy years.

At least up to now.

He dismissed Merriweather's characterization of Philanth. He had seen for himself that Philanth could be different things to different people.

But he wondered whether Harleigh, who had been chumming around with Merriweather that morning, also had known about the protective surveillance and thus had welcomed an excuse to check out the nature of his relationship with Philanth for himself.

Maybe also to check out Philanth for a report to Kingsley. If so, they must have had a thigh-slapping time over the "Miss Prude" number she had laid on for Harleigh.

Twice in one day Philanth had overplayed her role as virtuous maiden merely out of excess drive. He was going to have to get her into work challenging enough to keep her fully occupied. Unfortunately that would involve more of the close contact which already was getting him into trouble.

He turned to look into the dark crowd of dancers, searching for Tony and Philanth.

When he saw them he had trouble believing his eyes.

Philanth looked as though she had gone to sleep snuggled up against Tony. Tony had his arms locked around her as though to hold her up and had his head bent so his jaw could rest against Philanth's temple. They were shifting their weight to sway to the music, but it could hardly be called dancing. Philanth had one hand on Tony's chest, and her other arm rested around the side of her partner's waist under his jacket. That reflected the fact that they were about as close as they could get while remaining fully clothed.

They looked like teenagers in love, and if he had been a chaperon he would have waded in and told them then and there what a proper position for dancing was. Tony's expression, as nearly as he could make it out in the dim light, was as pensive and tender as the music. He and Tony both had set out to steal some forbidden pleasure this evening. Tony had succeeded.

The number ended, and Tony half released his partner. They seemed to confer. Then, ignoring the new number which was beginning and to which Tony was most certainly entitled under their equal-time rule, they came back to their table.

"You look tired, Mr. Bjorklund," Philanth said by way of greeting. "We should call it a night so we'll be able to do some work during our flights tomorrow."

"And somebody might be getting ready to take another shot at you," Tony added.

He looked at them. They were younger and more rested than he was. They could have danced all night.

"Merriweather is a kill-joy."

They nodded as one, confirming his suspicion that Merriweather had also warned Philanth during their dance together.

"The hell with it," he said. "'You're only young once.' I'm going back to the hotel by myself. I'm confident I'll be tailed by someone vitally interested in my welfare. You two stay as long as you want."

They thanked him with alacrity. Philanth started to kiss his cheek but arrested the movement and wished him good night. Tony vigorously shook his hand, thanking him again.

As he rode back to the hotel he realized that he had not gotten even one complete dance with Philanth.

616

Chapter 22

THE BET

The next morning he and Philanth were filling the time until their flight was called by putting their heads together over their routine paperwork when he noticed that she was not wearing her wedding ring.

"You spilled something that made your ring finger inflamed again?" He took her hand, glad of the excuse, and gently turned it to examine the finger on all sides. "Looks all right now."

Withdrawing her hand, she said, "I don't have a ring anymore." She kept her eyes on the papers before her as she added, "It's Tony's now."

"Philanth!" He searched her face; she remained expressionless. "You don't mean—you *can't* mean . . . ."

"He's going to collect the other guy's five hundred dollars. Wonder who it is."

"I'm sorry, this doesn't make any sense! I can't believe what you're implying!"

"It was just something I wanted to do. I see no reason to discuss it.

"Now, which way should I change this to say what you meant more clearly?"

He slammed his hand down on the papers lying on the lid of her briefcase. "Talk to me, Philanth!"

"I think you should figure it out for yourself."

She gathered all the papers into their folder. "We'll deal with these when you're ready to concentrate." She slipped the folder into her briefcase, shut and fastened it, and went off toward the rest-rooms.

A few minutes later Tony showed up with several new paperback books in one hand. "Where's our gal?"

"She'll be right back.

"Why doesn't she have her wedding ring? She says you have it."

Tony's immediate uneasiness was palpable. "Sure, I bought it from her. She said last night she'd decided to sell it, and she wondered if I'd like to have it. I paid her for it, and that's all there is to it."

"I don't believe you! She wouldn't sell it!"

"Well, she did. No use causin a fuss and makin her feel bad, now, is there."

Tony looked around. "Oh, good, here she comes." Tony bent to collect their carry-ons. "Le's go." Tony started off. He and Philanth followed.

Tony and Philanth were tight-lipped as the flight got underway. Tony read, then dozed; Philanth kept rustling papers of one sort or another.

He pretended to look through the airline magazine he had found in the seat pocket in front of him. It was impossible to believe Philanth, and Tony's story wasn't plausible either. What was going on?

He remembered something Merriweather had said the previous evening: Philanth could sleep with a man, then blow his brains out and not let it spoil her breakfast. Carole had said that Philanth could reason herself into *anything*. With a philosopher's thoroughness, she truly believed in nothing; and with that core of total nihilism came a complete moral freedom whenever she chose to exercise it. In fact, that absolute freedom upon which all her principles rested was at the center of Carole's test.

Philanth voluntarily circumscribed her own absolute freedom only by pragmatic calculations of what ends and means blended to serve her arbitrarily chosen ethic. Being arbitrarily chosen, that ethic might prove exceedingly elastic in interpretation and application; indeed, it could arbitrarily shelve itself.

618

So perhaps she had simply given all her intellectualizing an hour or two off. She had said it was just something she wanted to do. To a sensuous and fundamentally amoral woman a roll in the hay with a no-strings type like Tony could simply be "an experience". As a calculating, super-controlled type she might kick over the traces all the more coolly.

Of course Philanth, ever rational and organized, would have insisted upon precautions. Tony was the type who always carried a condom as a badge of his virility, but Philanth the sex-education teacher and family-planning professional would have had no qualms about purchasing her own supply.

There was also the caring and mutual loyalty which had developed between Philanth and Tony. The three of them had been virtually living together for a month, and the way Philanth and Tony had been dancing together the previous night showed that Philanth did not find physical contact with Tony unpleasant. They had stayed on when he left.

If she had decided to exercise her freedom, she could have found excuses enough. To the extent that she felt any need for moral rationalization she could have presented the fling to herself as a logical culmination of her "project" of improving Tony.

It might also be rationalized as a demonstration for his own benefit: it showed him that no illicit action could ever be truly without hurtful consequences for some innocent bystander.

Tony's explanation could be only the cover story they had agreed upon. Philanth had not employed it because she wanted to stop his overtures to her.

Philanth seemed different this morning: not cheerful, as usual, but not unhappy, either; just completely neutral. It was as though she had shut all her emotions up in a vault, and there was no telling when she would decide to let them out again. Tony seemed different this morning, too: more serious and reserved. His usual happy-go-lucky manner was gone. Indeed, he seemed to have a bad conscience and a determination not to let it get him down.

Philanth's outrageous claim had begun to take on a shadow of credibility. He still could not accept it, but finding no alternative, he would have to suspend judgment.

If he tried to look at the matter as a disinterested observer it seemed logical that Philanth had let Tony take the ring and had let him know why Tony had it because he had given her so much trouble about "the nature of their relationship". Doing this had been a chance to show him that he had no claim on her.

In the cool light of this morning the nature of their relationship seemed a dead issue. He had believed that he knew her to her core, but maybe he did not truly grasp her essential nature much better than Merriweather did.

She was capable of *anything*, Carole had said. Now he had to absorb an implication of that proposition. When someone got as sophisticated in one's ethical reasoning as Philanth was, there was no telling where she might wind up.

Then *why not with him*? That was what Carole had believed was possible, and Carole's knowledge of Philanth had proven phenomenal. So had he had a chance to take her to bed after all and simply failed to hit upon the right arguments?

And why *Tony*? That made him sick. If only it had been Stuart, the black man Tony had replaced, or some other man of complexity and sensibility, he could have understood a little better.

He tried not to react. It was her life, her body, after all. He had no claim to it.

620

Another part of his mind still kept insisting, It can't be so. She's lying, it can't be so.

But that was only what he wanted to believe, and one thing he was learning from this was that he might be as inclined to believe what he wanted to believe as the next person. Right now he did not know whether he was trying to overcompensate for his tendency to avoid unpleasant truths or not.

Philanth's suggestions that he had lost his perspective because of their constant association for so many weeks must have been valid. She was after all just another imperfect human being.

But he felt robbed. She had been important to him as proof of the validity of ideals, just as she claimed he was to her. Now that was gone.

No! Philanth—even with all her regard for "the feminine virtues"—was a feminist; she would not give herself to any man knowing he had been involved in a bet like Tony's.

But like Carole, she regarded herself as tough-minded. Her pride in her intellect made her capable of suicide, not to mention personally repugnant decisions. The very fact that an action would be a violation of her deepest personal feelings could spur her to override them. She would simply refuse to think about the game of exploiting and degrading women in which she was letting herself be used.

Because she had hoped to shame Tony. A woman who regarded femininity a practically synonymous with self-sacrifice, she was prepared to humble herself whenever someone else's moral well-being was at stake. Indeed, she was clear about that objective in allowing herself to be "walked over".

Besides, there was also in her makeup a recurring despair regarding her prospects for significant accomplishment. Carole had suggested that this could make

Philanth willing to throw herself away. So she had been willing to let herself be used even to that degree in the hope that Tony would as a result develop new depth.

Perhaps she had even succeeded. How could *any* man be intimate with such a woman who was already on affectionate terms with him and not be changed?

But when was a willing sacrificial maiden not a fool?

Even the idea that she *might* have done it made him gag, the possibility was so humiliating. He felt vastly superior to Tony, and if Philanth could have accepted Tony's casual lust while rejecting his own idealistic passion it was possible that all his supposed virtues were nothing because of the one attribute he did not have: he wasn't macho.

Maybe the good guys did always finish last. Not just in politics, but even in this, even with a woman for whom the highest principles supposedly mattered above all. Maybe even here, idealism and compassion and "honor" were what kept the decent people hanging back and debating while the unscrupulous, self-seeking, shallow, simple-minded, or vulgar went ahead with serene indifference to any consideration but their own wants. Moral indifference was more than a match for moral courage.

He stole a look at Tony. What had she seen in him? Oh, all right: Tony was good-looking, warm, genial, tall and powerful and quick, easy-going yet reliable on a day-to-day basis, and surprisingly good in a crisis. Not so stupid as he had seemed at first, either—just psychologically hung together differently from himself.

Moreover, Tony had been consistently considerate of Philanth's needs, even concerned and compassionate when he learned that she had a problem.

The cabin intercom came to life, and the stewardesses came though checking to see that everything was tucked away, belted down, or "in the upright position". It was amazing how short the two-and-a-half-hour flight to Tokyo had seemed.

When they debarked there was no need to rush to the boarding gate for their flight to Los Angeles, so Tony invited Philanth to come with him to browse in the duty-free shops. Airport shops held no appeal for him, so they left him in their boarding area to guard their carry-ons and read some newspapers.

When they returned he asked Philanth to stay with the carry-ons while he went to freshen up in the men's room. "You come with me, Bodyguard."

As they began washing their hands he asked, "How much did you pay for Philanth's ring?"

"That's none of your business. Sir."

"Yes, it is. I'll buy it back from you."

"Sorry, no sale. I got personal reasons for wanting to keep it."

"A memento of your association with her?"

"Sort of."

"How much, Tony!"

"Sorry. I said, 'No sale.'"

His brain finally started turning over. He needed the name of Tony's co-conspirator in the bet.

"You know what I think? I think you raped our little Philanth last night: just a garden-variety 'date rape'. She feels there's nothing to be accomplished by making an issue of it, since you would claim she led you on. And she knows you're aggressive in your own interests, but she's not."

Tony registered growing alarm. Just then a couple of other men came in, so he motioned for Tony to follow him out. They found an inactive departure lounge and sat down.

623

"You took her ring while she was helpless, and you're holding onto it as a form of blackmail to encourage her to keep quiet in the hope you'll return it when you leave your job at headquarters.

"Obviously I'm going to have to take some harsh measures. You'll be explaining things or trying to conceal things in your precious record for the rest of your life. I don't want to do that, but you'd have to convince me that there's a more innocent explanation. But your story that Philanth sold you her ring is just impossible for me to believe, because I know how she cherishes it."

"She cherishes it, all right," Tony growled. "She wouldn't take a penny less 'n five hundred bucks for it. Said somebody'd told her it was worth that, so that had to be her price."

He bit his lips to keep from laughing. Tony was planning to save face with his pal at the office and break even. Philanth's knowledge of basic economic concepts and her precision in choosing her words had given her a slick argument, and Tony the economics major had not been in a position to call her on it: what a thing was "worth" was whatever somebody was willing to pay for it.

He had been a fool. He wanted to laugh with delight. Making his repressed laughter pass for ridicule, he said, "That's preposterous. You're aiming for the top business school that'll take you, yet you claim you paid five hundred for a cheap little ring like that? She told *me* it cost seventeen dollars. Now I *know* you're lying!"

"Philanth will show you the traveler's checks I signed over to her las' night."

He paused as though to consider that, then admitted some bewilderment into his voice. "Philanth works extra hours for nothing. She isn't interested

in money. Why would she part with the most precious object she owns, even for that much?"

"She's goin to give it all to Zero Population Growth
and Planned Parenthood. Said she didn't have the
right even to her husband's ring when it had so
much market value."

His hilarity went dead. "You stinking bastard."

"Now what've I done to deserve that?" Tony was
astonished and hurt.

"Taking that one thing that means so much to her,
something you could duplicate for twenty dollars, just
because she cared about how much human suffering
could be prevented with your filthy five hundred
dollars. Every woman has her price, all right, and
you sure found hers! You *swine*."

"Hey, now, Chuck, that's uncalled-for."

"It sure as hell *is* called for! I happen to know you
want that ring because of a sick little bet you made
with another man at headquarters, that you'd 'score'
with Philanth. What's his name?"

Blind-sided, Tony hesitated.

"It's your neck or his, buddy! His name!"

"It's your chief of staff: Les. He's a nasty guy, and
he'll deny it. But I'll die sayin he's the one.

"He invited me to have a few drinks with him after
the goin-away party; said he had 'an offer I couldn't
refuse'. You don't turn down an invite from the chief
of staff, and naturally I was curious. I followed him
to a swank little watering-hole he knew about in
McLean. Then he bought me drinks and fed me a
line until he'd backed me into it."

"What sort of line?"

"He told me Philanth was 'one of those broads that
need taking down a peg or two'. I didn't buy that,
but I . . . had my own reasons for liking the idea.
And he'd gotten me sloshed enough so I couldn't see

any way out of it. He made out like it was important to him, that I'd be doin him a favor to make his job easier. And he made fun of me for not jumpin at the idea: wasn't five weeks enough for me to work with, was I afraid of a sure thing, didn't I like women, and so on. He's quite an arm-twister."

It all fit. Les was the type who would consider personal sabotage clever, and Tony's protective camouflage as an empty-headed jock had encouraged Les to consider the bet a low-risk as well as low-cost venture.

Motive was not hard to figure out. Les would regard anybody sharper than he was as one of his natural enemies, since such a person undermined him by comparison. Philanth had elicited that reaction from her classmates in high school. Les had anticipated that her applying her abilities during the trip was likely to lead to her being offered new responsibilities instead of being let go; but if Tony came back claiming he had won the bet, Les could use that—against a girl as uninterested in defending herself as Philanth—to make her decide not to stay.

"Tell Les the bet's off."

"How can I do that? Besides, I'd still be out five hundred dollars! —Unless Philanth would let me change my mind and . . . ."

"Don't embarrass yourself by asking, 'cause she'll do as I say. The money goes to her causes."

Tony's shoulders sagged.

He held out his hand. "Let's have it. That ring is not going to be used to cause trouble for Philanth."

"Five hundred dollars!" Tony lamented, sticking his hand into his left trouser pocket. "I can't afford to be out that kinda bread. And if Les puts on the screws insisting I pay up, I'll be out a thousand!"

"You deserve to be hung out to dry."

He gestured impatiently with his extended hand. Tony began transferring coins and chewing gum from his trouser pocket to the empty seat beside him.

"As for Les's giving you any trouble, . . ." The lawyer in him began to stir. "Don't let him talk to you until you've got somebody overhearing. The offices are full of sound-proofing materials, so you'll probably have to corner him in the men's room. Then get him to incriminate himself by offering to pay him your forfeit. Pull out your checkbook and offer to write him your personal check. If he accepts it, wonderful: we'll wait for him to deposit it.

"Don't worry about incriminating yourself along with him. Your key witness will have to know the score in advance anyway, in order to retain what he hears as well as possible.

"Les may try to give you the brush-off and slip away, so have his line of retreat covered by other witnesses. Then do your best to get him to raise his voice by raising yours, and hassle him for pushing you into the bet. Anything he says can be analyzed afterwards for implicit self-incrimination. Get it on tape if you can. Borrow a dictaphone from Marjorie."

Tony's eyes lit up. "A sting operation!"

"The moment it's over, let me know exactly what's transpired. If I'm not immediately available, write down every detail you can remember, immediately, and work at it until you have the most accurate blow-by-blow account you can manage. Have your accomplice primed to do the same, independently.

"Are you with me?"

Tony nodded, looking fascinated.

"I may call on you to repeat in front of him and some other people what you've just told me, plus whatever else you've been able to get. You may find yourself being closely cross-examined by some tough customers, men who are used to playing hardball of a kind you're not familiar with. Will you be prepared to stick to your story like an honest man and keep your cool, no matter what anybody says?"

"Hell, yes!"

"Getting Les to incriminate himself within earshot of third parties who'll make highly respectable witnesses has top priority. But with that in mind, avoid doing or saying *anything* that could let Les suspect *I* know something—right up until I send for you and tell you to spill the whole story. Make it seem to him like pure personal hostility on your part.

"You do this right, and Les is out of the picture altogether. But he has influence and he's unscrupulous about using it, so we have to play rough."

"You can count on me."

"I hope so! This guy has a pipeline into the White House Office of Presidential Personnel. That's why he is where he is. And he's blackmailing the Peace Corps with a threat of bad press in order to hang onto his fat little political appointment. I answer to those White House people. I also answer to the United States Senate, including former Peace Corps Volunteers who really care about what the Peace Corps stands for and tries to do. *If you blow this for me, . . .*"

Tony's dark eyes glittered at this new perspective on the situation. *I won't!*" Nothing like giving a man a high moral purpose and far-reaching practical stakes to vindicate his efforts at personal revenge.

Having emptied his left trouser pocket, Tony pulled it inside out. He began working on it with both hands, saying, "Philanth gave me a safety pin and we discussed where to pin it so it wouldn't get lost."

If he had had any remaining doubt about Tony's version of the previous night's events this would have

628

wiped it away: he could see his assistant in her motherly mode now, providing Tony with a safety pin so he wouldn't lose his new toy.

Tony handed him the small, plain gold band. Noticing that it was lined with tiny lettering, he began trying to read what the engraving said.

"'For thou art with me,'" Tony told him as he worked his pocket back inside his trousers. "She was in the valley of the shadow of death, but her husband was with her."

He nodded, seeing that Tony was correct. "Seems like that should have been her message to her husband on *his* ring."

"They got by with just the one. She wanted it on hers as a reminder to herself."

He slipped it into his pocket, making sure the pocket was sound before he let go.

"One more thing," he said, caressing the little ring. "Philanth led me to believe she'd gone to bed with you last night. Why would she do that?"

Tony registered surprise, then anger. "Oh, hell, you can't figure *that* out? I *swear*, you're . . . ."

"Obviously it made you feel she's not good enough for you anymore, since she was good enough for *me*. Idn't that the way you honkies think? She wanted you to feel like I'd polluted her with my black body.

"I figured that's how it was with the creep who suggested the bet: *he* can't get anything out of her except her standard politeness, 'cause he looks like a toad and acts like a toad. So he tried to get me to drag her in the mud for him.

"You're not much better. And since you'd be cured o' lookin at her the way a starvin slave would look at his master's sizzlin steak, she could hope to go home and pal around with your wife jus like before, without worryin there'll be trouble. All very tidy.

"Well, you and your ol' lady can work *that* out any way you want to, it's nothin to do with me!"

Tony got up, then bent to scoop up his coins and miscellany and shove them back into his pocket.

"Spare me your righteous anger. You were willing to screw Philanth's reputation at the office, even just now—when she's been as nice to you as she could, even though *she's known all along* about your stupid, disgusting, high-school-locker-room bet!"

Tony's face appeared to stretch several ways at once. "Philanth's . . . known . . . ?"

"That's why she set such a ridiculous price: she didn't want you to make a profit, just preserve your precious reputation as a stud—at her expense."

"You—you self-righteous sonuvabitch!" Tony left.

He hurried back to their boarding area. Glancing up as he approached, Philanth saw his expression and looked frightened.

Without ceremony he seized her left hand and worked the ring onto the proper finger. "I won't have you distracted by such garbage at the office!"

"How—how did you . . . ."

"Never mind that!" He sat down beside her. "You've behaved very irresponsibly, and you should be thoroughly ashamed of yourself."

"I'm sorry you're angry, sir." She worked the ring over the second joint of her finger and settled it in its proper place.

"Had you thought at all about what might happen to your effectiveness at the office after Tony collected his five hundred dollars? Did you imagine for one moment that both of those louts would keep their mouths shut?"

"People at headquarters are a decent, sophisticated, live-and-let-live sort. They'd despise anybody who started a story like that."

630

She added in the same tight voice, "They'd also be less likely to speculate about whether there'd been any attraction between you and me."

That last reason hurt. He pretended to get madder. "You're so naïve! It *would* have impaired your effectiveness as my assistant, even if nobody were so crass as to give you some overt sign of it. Jealous people love nasty gossip, and your position excites jealousy."

"Tony won't be around long, he's temporary same as I am. I was prepared to try to live it down—if I stayed on."

"You have considerable faith in your charm."

"Yes." She spoke absently. "Though that's not what I'd be using in this case. The weapon of choice would be what they used to call 'Christian fortitude'."

She stared at the ring, and he could almost see wheels turning inside her head. She probably had expected him to be too crushed by her sexual adventurism to feel like confronting her supposed partner.

"Poor Tony," she said. "He's going to have to pay somebody at the office five hundred dollars."

She wanted to know how much of the truth he had gotten from Tony. Since she had not made Tony a party to her attempt to deceive him because she had been manipulating Tony in a different way, she would hesitate to question either of them directly.

"Ohhh, 'poor Tony'! Really, Philanth!" He almost added, You're too forgiving. But he likewise had to be careful. He did not want her to guess that her attempt to dismantle his passion had miscarried.

He picked up a section of the newspaper she had scavenged from his leavings.

He leaned over to mutter to her, "Next time you want a fuck, pick somebody with a little more class."

That told her what she wanted to know, and from her point of view it was very good news, but she reacted as if he had smacked her across the face. Her mouth went out of control, and as usual when that happened, she put up her fist to steady it.

"My God," he muttered, not daring to look at her as he remembered the first time he had seen her mouth and fist do that, at the little café in Paris. He surmised that she was moving to soften him up just enough to restore their working relationship. But regardless of how real this grief was, she was a complicated woman. It was complexity, not control alone, which made her an accomplished actress.

So Merriweather wanted her for intelligence work, and she was "interested". It didn't bear thinking about. He began pretending to read the newspaper.

Then, wanting to see how far she could be pushed, he whispered, "You needed it so bad as all that?"

Her eyes flashed. "I did it for Tony."

"Huh! *I* saw how you'd been dancing with him! You can't tell *me* you didn't enjoy it!"

She retorted for his ears alone, "I didn't need it 'so bad as all that'! I'll bet I could teach *you* a few things about how to satisfy a woman."

Keeping his face averted so she couldn't see how he was beginning to enjoy this, he growled, "Oh, yeah?"

"Yeah!" she whispered. "There's a lot of heavily wired terrain to work with, and the possibilities are endless.

"Why, 'needing a man' has become a joke! In a gift shop down the street from the office I saw a booklet entitled, 'Why a Cucumber is Better Than a Man'. It pointed out that cucumbers don't snore, et cetera. Et cetera, et cetera, et cetera.

"Women in all times and places have found substitutes that make realistic dildos look like the juvenile concept they are. In the East Indies they developed special little balls—"

"Bitch," he muttered, still feigning revulsion. But she was pouring scorn on his suggestion because she was genuinely proud of her ability to satisfy herself. He couldn't help admiring her even as he pitied her.

"They can leave them in as long as they want, I suppose," she continued, "because the balls adapt to the shape of the vagina.

"The main concern regarding such measures—well, there are several because people invariably have impaired judgment when they're aroused—but obviously there's the risk of infection due to inadequate sanitary precautions."

Lord, now she had slipped into a lecturer's cadence which meant that she could go on indefinitely. He could bet that she had never dealt with this topic in her high school classes or her Peace Corps work, but Philanth always wanted plenty of background before she could feel comfortable with a subject.

"The natural defenses can be bolstered by a douche of vinegar and water, the recommended proportions being two tablespoons to a quart. However, when used for itching caused by dryness or yeast this measure can be very counterproductive, and gynecologists now view the practice with disfavor—"

"Now, that's enough!" he thundered, and several Japanese in nearby seats jumped but then pretended not to have noticed. A very small boy began staring, fingers in his mouth, but his mother corrected him.

Shouldering her bag and picking up her briefcase as though both were even heavier for her than usual, Philanth hauled herself to her feet. Walking a little lopsided and trying to ease the strap of the bag on her shoulder with her free hand, she made her way over to the wide expanse of windows overlooking a stretch of tarmac. She stopped close to the window and set down her burdens at her feet.

He bent his head as though reading but watched her over the edge of his newspaper. She stood very still, shoulders slightly hunched, facing out. After a few minutes she stooped to take something from her bag, and then he thought she might be crying into a tissue. That was confirmed when he saw a white man near her do a double-take, hesitate, and then move away. Airports were a special venue for open expression of emotions, and tears in a boarding area must ordinarily concern matters beyond the help of strangers.

Such a lonely figure, there against the window. It made *such* a difference, having just one person who truly knew all about you and yet believed in you!

He longed to go over and tell her that he knew what she had actually done. But she had decided that she preferred his contempt to his desire. Protecting his peace of mind must have become second nature to her, and she calculated that his contempt might fade into sexual indifference. It should be his problem alone if the respite from desire which Philanth had won for him had been brief.

A public address system began delivering an announcement in which he could pick out the lovely words, "Los Angeles", and other waiting passengers began to form a line even as the announcement was repeated in English. He checked his seat number. His section of the plane had not been called yet. Nevertheless he got to his feet, eased his carry-ons up onto his chair, and looked around.

Tony appeared in the crowd, weaving across its flow in a struggle toward Philanth. As the big black man reached her he started talking. Suddenly she threw her arms around Tony's neck and burst into tears, trying to answer, but Tony went on talking at the same time, shaking his head for emphasis and rubbing her back. They went into a clinch and kissed each other's cheek. The other passengers edging past them were making a well-bred effort not to notice.

It took them a minute or so to pull themselves together. Tony took her tissue from her and gently wiped her face, then kissed her again. They were still billing and cooing when their section was called.

Tony urged Philanth on toward the gate, and she took up her carry-ons and started off. Tony threaded his way back to him and picked up his carry-ons.

"Our deal stands, Mr. Director," Tony said coldly. "I'm gonna help you nail the guy that started this."

"Fine."

He and Tony had nothing else to say to each other. As he and Philanth stowed their carry-ons in the overhead compartment Philanth said, "Mr. Bjorklund, would you take my window seat?" She was changing their normal seating to separate him from Tony.

Before he could reply Tony said, "Philanth, no! I'm gonna ask the stewardess to let me change my seat." As soon as most of the other passengers were settled, Tony spoke to a stewardess. Tony took his personal belongings and left for another part of the cabin.

At once Philanth turned on him like a wife who had scarcely been able to wait until she could get her husband alone. "Do you realize what you've done? Can't you imagine at all how you made him feel, telling him I knew *that* about him all along?"

He was nonplussed. "He had it coming."

"You don't understand!" She lowered her voice. "Put yourself into a black man's skin. You come to care for a white woman; you'd like for her to be the mother of your children. You feel you can't marry her because you're black, so it would hurt your chances in life and hers, for the rest of your lives."

"Oh, surely you didn't swallow a line like that!"

"Who can say how real a feeling is or isn't, even when someone becomes committed to action? Hear me out.

"You've still got to play the macho male, pretend a white woman is only a potential trophy and a relationship with her is only a way of getting back a little for all the generations of abuse.

"But if you come to want to take care of her, and then the mature, generous image of yourself you've built up with her—and thereby discovered in yourself—is destroyed, *how must you feel?* What you did to Tony was like *murder!* You wrecked something beautiful he and I had carefully built together, in the times we had alone. You smashed something that was new and precious and vulnerable: a better Tony Hall. *Dam*mit! You must be really proud of yourself.

"I can only hope he's able to pick up some of the pieces of what you shattered, and put them together for himself. You've certainly ended *my* chances of doing anything more for him: he can hardly bear to look at me."

Rather than make her madder, he said nothing.

"Can't you believe he's found the ability to care about a woman as a person, to care deeply about what she thinks of him, to love in a mature way?"

He turned to look into her eyes. "I can believe anything of you. Forget charming birds out of the trees—I can believe a rhinoceros in the jungle would eat out of your hand and nudge you to be petted."

He needed to know whether Tony had let her know that her implicit lie had been discredited. "Wha'd Tony say to you, there at the window?"

"He said he was so sorry about that dumb bet."

"And what did you say to that?"

"I told him I understood, that I knew he was really a wonderful person, and I wanted never to lose touch

636

with him, I was sure he was going to have a beautiful life."

"Yechh, that's so sweet, I think I'm going to be sick."

She set about opening her briefcase on her lap. "Barf bags are usually in the seat pocket," she flung over her shoulder. "Too bad your taste is so corrupt. One man's saccharine is another man's passion."

He was glad Philanth was pretending to concentrate on the contents of her briefcase so she couldn't see any glimmer of his carefully concealed satisfaction. Surely a woman who could get so heartily mad at a man was no longer keeping him on a pedestal.

He would have liked to ask one final question to be sure Tony had not said any more to her, but he could nott find any way to do it that would not arouse her suspicions.

Tony was now on his own and headed for a connecting flight to Chicago; he would not be in Washington until the following Monday. Then he would be at headquarters only to wind up his business and go through the sign-out procedures with various offices. Tony would hardly want to go out of his way to talk about this episode any more with Philanth. So he would take his chances that Tony would never tell Philanth that her attempt to cool his desire had been aborted.

"I have a first draft of your speech for next Tuesday," Philanth said, "and if you'll look it over before you get too tired, I can—"

"Next *Tuesday?*"

"The one for Town Hall. Marjorie mentioned it, and the title refers to everything this trip has been about, so it seemed to be in my bailiwick. I took a chance that I might give you something that could be used in the final version if I just started from that."

If she had guessed that he had not thought once about the event since it had been scheduled, she had been right.

He accepted the folder. She selected another for herself, then stowed the briefcase for take-off.

He read the speech through once, quickly.

Without his noticing, she had gradually taken over more and more of the adapting, organizing, and revising of prepared components involved in producing his formal speeches for the trip. This time, starting from scratch on her own initiative, she had reached a new level of proficiency in saying what he was willing to say in ways he might say it.

The speech addressed Peace Corps' major problems squarely yet in a way that was inspiring. It would make him look good with a wide cross-section of Washingtonians interested in Peace Corps or in him.

He went over the speech again, looking for weaknesses, and found additional reasons to be impressed. His countless tapes of dictation regarding his discussions and findings had borne unexpected fruit. It made him realize how much she might have achieved if she had not been shut out of the meetings and conversations themselves.

She might also have captured his preferred style more closely if she could have watched him present more of the speeches she had helped to prepare. Just the same, she had done well in that regard too.

He started through a third time, determined to find flaws. A clear, logical outline had been carefully implemented, yet she had woven all the elements into a beautifully integrated, conceptually simple and emotionally satisfying rhetorical whole. Her attending his previous Town Hall speech, before either of them had become associated with Peace Corps, must have given her a sense of the appropriate tone.

But this speech didn't just flow nicely. It built until at the end it quietly reverberated. It would require his best delivery to do it justice. He would find the time to prepare for that somewhere in all the time she had saved him because he did not have to supervise the writing and revisions.

He handed it back. "Fine. Very good."

"But is it in line with what you had in mind?"

"No problem. The conceptual level is just right."

"Overall structure acceptable?"

"Of course."

"Transitions okay? Appropriate distribution of emphasis?"

"Yes, yes."

"No changes *at all*?"

He shook his head.

"Well, when you go over it to prepare to deliver it I'm sure there'll be some refinements."

"My taking the time to tinker with it wouldn't be justified."

She put it back into her briefcase, looking timidly pleased.

After a moment she began talking about the importance of organizing as quickly as possible the long-term, large-scale program re-evaluation which would draw heavily from the masses of information and ideas collected by or for the trip.

". . . So would it be all right if I stay late at the office on Friday and come in on Saturday too?"

"My gosh, Philanth, why don't you just come over to our house for the weekend? Our computer is compatible with this one, and we'll keep them both busy on the essentials. Schedule blocks of time for other staff to join us, and with Carole working too we'll get the whole mess reorganized and parceled out for converting into rough drafts by Monday morning."

639

"I think you're joking."

"I most certainly am!"

"But it's a fine idea. Doing it along with the regular flow of the office will drag it out for *weeks*."

"Months."

"Whereas if we could provide all the key staff with preliminary drafts to work from right away, the whole thing would proceed swiftly and in a rational, coordinated fashion."

"Maybe if you and I were thoroughly rested instead of needing several days in a darkened room, I would admit it could be tempting."

"I suppose civil service regulations make it out of the question. That wouldn't be a problem, you realize, if Peace Corps personnel who don't perform tasks analogous to those of assembly-line workers were allowed to function as professionals: working whatever hours a time-bound job requires, and taking projects out of the distractions and interruptions of the work place when they need concentration or creativity."

"Yes, and you want to remind me about that proposal for taking Peace Corps out of the bureaucratic quagmire that's inapplicable to its nature and making it a foundation."

She smiled, probably for the first time that day. "Just wanted to get in a plug. The present system is so incredibly inappropriate, and everything could be done so much better, considering the calibre and dedication of the people who come in excited to be working at Peace Corps headquarters. Before long they may hate their jobs worse than the career bureaucrats who didn't particularly care which agency they worked for to begin with."

"Okay. Okay." His voice was patient.

She bent over the papers lying on her briefcase. "In the meantime, I've started developing ideas for structuring the work on this project. I'm sure there's a way to do it so that we get a better report done faster in spite of the restrictions on working conditions. It's basically a matter of following the same organizational approach I used to coordinate preparation of your materials for the trip."

He grabbed her wrist. "*You* were the one who masterminded all that organization and polishing?"

"Well, yes," she said, pulling out of his grip, "I thought that was part of my job. I started editing pieces as they were developed because some people can't spell, et cetera, and I had to keep working toward consistency of format so you'd find it easy to use. One thing led to another."

"You designed the overall structure, the sections within each country on the itinerary?"

"Well, yes. But I had help on the implementation. This computerized five-week 'briefing book' was an ad hoc project, and Les said the regular people for this type of job were tied up, so pulling together the briefing and scheduling documents was up to me. But I had so much traffic from the regular aspects of my job, and there was such a mass of raw material, I couldn't possibly have done it all myself. It was all *I* could do to deal with the politics of coordinating and revisions. Fortunately the nice Office Support man in IRM happened to know of several people who are good at fast revision of formats, and one of them turned out to be a whiz at whipping the masses of material into consistent form with optimum layout. She worked unpaid overtime to execute my requests.

"She's got powerful desktop-publishing software at home for her husband's use, and she also promised to help with any quantitative profiles or projections or other graphics we want developed, for appendices or whatever, if the regular production process—"

"Never mind spreading the bouquets around." Les obviously had tried to overload Philanth in the first place by denying her the regular channels for getting such materials prepared. "I had the impression Les was responsible for that beautiful set of diskettes and folders. He certainly let me think so."

"Oh, well, of course since I was working under him, he naturally would be the one *responsible*, . . ."

He chuckled dismissively. By playing her role of innocent victim Philanth had avoided conflict with a powerful oppressor and developed her own strengths in overcoming the obstacles Les had set for her.

He leaned back and relaxed, conceiving an additional maneuver for his campaign to drive out Les. After a few minutes he took out his dictaphone and began talking into it while addressing her. "Go ahead and develop a procedure for the whole project of organizing and disseminating my findings. Give a write-up of your proposal to those who're going to be most deeply involved, asking for their suggestions. You and Jeff get together with the others and come up with a game plan that won't take until summer to get the key documents to the printers."

"You need the information accessible ASAP, for preparing the Annual Report to Congress and your budget testimony. Not to mention all the possible immediate applications. We need it broken down, restructured in various forms, and ready for dissemination no later than Friday of next week."

"That's the spirit. Seriously: schedule backwards from the preparation of appropriations testimony, with slack for schedule overruns. Tell Jeff to coordinate the necessary resources in IO."

"What about Les?"

He switched off the recorder. "You tell Les he's out of the loop, this is your baby and from now on you're reporting only to me. Have some preliminary drafts on hand to wave around, but refuse to let him look at them."

"Les won't like that!"

"He's not supposed to."

"But *he runs Peace Corps!*"

"Funny, I thought I did."

"But everything goes through the chief of staff! It's the first thing a new employee is told."

"Well, that may change. I intend to get rid of him."

"Oh? Well, . . . fine. But why make him mad? There's never a rational justification for making any enemy unnecessarily, even if you're firing him."

"Rationality has limited usefulness in human relations. Sometimes an enemy is so intractable and dangerous he has to led into the open and then broken. Your role is to make him mad; mine is to cause him to accept his defeat as the result of his own actions, rather than bearing a grudge against us when it's over. So be firm and impersonal. Act like his equal."

"All the floors occupied by Peace Corps will ripple all the way from K Street to I Street!"

"Les would like you to think so. But you *must* give it to him straight. Understand?"

"No, sir. But I'll do exactly as you say."

"Good girl.

"Not even Marjorie is to know about these instructions. If she's present, give her no sign that you know anything about anything. She already knows I want Les out and I have to be careful how I do it."

"Whatever you say, Mr. Bjorklund. I understand that if whatever you're planning doesn't work out, you may have to . . . hang me. As though I'd gotten on a power trip completely on my own."

"Don't worry about it."

"I won't."

643

He chuckled. Her carefully crafted independence could make her useful when one wanted to make changes which were bound to upset somebody. Most people in Washington seemed to be so afraid for their jobs that they couldn't think straight.

He flipped his dictaphone back on. "And start setting aside a copy of anything you think I might like to look at before the appropriations testimony on the Hill that isn't getting into the reports you're working on. Duplications can be eliminated later . . . when you move on to helping me get ready to testify." He watched her eyes light up.

"Yes, sir. I'll track down past testimony to get a feeling for what's expected."

"Okay." He went on dictating, expanding upon the authorizations he had been giving her.

Then he added, "Oh, and you're going to need your own office. Get started on that as soon as you get back, and come to me when you want implementation so we can explore the options."

It was good to be able to rely on her to handle something that delicate without giving her detailed instructions and without worry that she couldn't manage it without ruffling somebody's feathers.

She tried not to look pleased. "Yes, sir."

"That's all for now."

She "borrowed" his computer and went to work, developing an elaborate outline and what appeared to be a work-flow/organization chart and a list of queries for various people at headquarters. She asked him not to watch because it was a creative job.

"And because you're embarrassed at having such fun?"

"Ye-es." She began bubbling, "I can't remember when I've enjoyed anything more. I love organizing a big, new, complicated project; and this is the best one I've ever worked on. Setting up the Hunger Foundation wasn't nearly this satisfying."

"'Setting up'?"

"Uh, yes." She seemed embarrassed by the slip, but his reaction made it impossible for her to back-track. "When Rod hired me all he had to start the operation was the estate, the chef, and the gardener. I found myself supervising carpenters, doing interior decorating, putting together brochures, and what-have-you."

He grunted, closed his eyes, and began imagining her working on other such projects in other years.

Inside the plane hours and days of the week were academic. They were flying into the past, reaching the end of the fifteen-hour journey from Manila only about an hour after they had begun it.

He welcomed the coming of darkness. It might be his last night with Philanth beside him. When they agreed to try to get some sleep Philanth snuggled down, pulling her airline blanket up to her chin.

"It's going to be good," she remarked as she strove to persuade the skimpy brown blanket to also cover her feet. "You know what to expect of me now, and you won't be bothered by my idiosyncracies. No wasted time, no interpersonal problems or distractions. This was like a shake-down cruise."

"Yes."

"I'm so lucky."

"I think I am too." He wished he were a lot more sure. Apparently she intended to will him to behave as she wanted him to, for the indefinite future.

After a while he asked, "Philanth, are you still worried about whether I might let you down on a matter of principle, if I went on in politics and you worked for me as a career commitment? You mentioned that objection early on, but you've never pressed it."

"Do you want to give me some assurance?" Straight to the point, as usual.

"Yes. I'll put this in writing: if I ever start losing your respect by too much compromising, I'll give priority to keeping your support. I can only pray that that policy won't leave me so powerless that I won't be worth your help so I'll lose you anyway."

"You mean, because I'd 'move on'?"

"If you didn't I'd have to send you away. Because I *promise* not to waste you, if I can possibly avoid it."

"What can I have to offer you, that's worth such promises?"

"You know the answer to that, Philanth: Let us be Christ to each other."

"I . . . can't do that for you, now. I've . . . already let *you* down."

She struggled for words. "That was all you truly needed from me all along, and now I've negated the very reason I was really useful to you. I'm sorry, Mr. Bjorklund! How can you still need me, in the personal way you've said?"

"I understand, now, why you did what you did, and I appreciate your motives."

"Why, thank you, Mr. Bjorklund!"

Chapter 23

LOS ANGELES

Philanth was opening her suitcase on the dinette table. The apartment was so small that the table was the only separation between kitchen and living room. He opened the sofa and showed her where the bedding was kept. She thanked him and began to make her bed. He shut himself in the bedroom.

He and Philanth had begun avoiding meeting each other's eyes. He wondered what Carole could have been thinking of, setting this situation up. Scratch that. He knew very well what Carole must have been thinking of. She really must be getting ready to dump him. Well, he couldn't stop her.

He called Kate to confirm the plans for the evening, and they chatted about his trip for a few minutes. She said Carole had told her that he and Philanth would have their hours turned around by their flight across the Pacific and would need to get some work done after the party, so she and Ned planned to retire for the night to the "maid's apartment"; that would leave him and Philanth the run of the house. He protested, but she assured him it was no problem. He thanked her and Ned very warmly, for now he had something to look forward to after the hours of wooing his supporters.

Next he called Marty Sheinbaum, his long-time campaign manager, at Marty's office outside San Francisco. Marty told him that since the November elections he had made laying groundwork for the senate contest his "pet project". Like Carole, Marty considered the race essential to his future. Marty refused to believe that there were any other potential candidates whose entry should discourage him from making an all-out effort.

Marty's preoccupation at the moment was the party that night. His mailing list for the invitations had been designed to identify big-money Bjorklund supporters in Ventura and Los Angeles Counties, and he was proud of the collection of "high-powered people" on the list of those who had sent in their "donations" in anticipation of attending.

Marty was not happy because Carole had refused to fly out to join him for the evening. That would be perceived as a slight, and Marty urged him not to forget to say in his speech that Carole was taking care of essential business for him in Washington but sent her warmest regards. He promised, but Marty went on complaining about Carole's absence and how much she could have contributed to the contact-building at the party because of her years of organizational work in California—especially among groups where issue-oriented activists might cross party lines.

Considering the nature of the gathering, Marty was right. He wondered why Carole had not planned to come with him. It could not be just because of the long flight across the country and back. It must be because her game with him was coming to a close, so she did not need to pretend any more.

When Marty finished he replied, "It would've cost too much."

"'It takes money to make money.'"

"'Politics is a rich man's game,'" he quoted in turn. "I'm hoarding my pennies to become a pauper member of 'the millionaires' club'."

"Now, no poor-mouthing, you'll get nailed on it. Besides, convince the average citizen you're not rich, and rich contributors won't take you seriously."

He snorted.

"Avoid the question," Marty continued, "and we'll show 'em it can still be done without being independently wealthy. You were made for the Age of Television.

"You just be sure to give 'em some of your old thousand-watt charm tonight. Carole says you've forgotten how to smile, but I know a good, warm crowd will be just the tonic you need.

"By the way, be sure you have a real winner of a speech ready tonight. Make 'em laugh, remember? You're too sober, it's your biggest liability. And let 'em know how much clout you got: the scope of the operations you're in charge of, size of your budget, number of people involved—just a few fat, round statistics that carry some punch."

He realized he was flat out of the energy required to do justice to the occasion. He needed a week at the mountain cabin his parents had once owned, letting his nerves uncoil and rest. But they had given him the sale price of their cabin as seed money to help start his first congressional campaign.

"I'll try," he replied, thinking that it was too early for him to have to put up with this sort of thing.

"Don't just 'try', *do* it, boy," Marty snapped, and hung up.

"Yes, Coach," he murmured as he set down the receiver. Marty had been with him since his first campaign and was like a parent who would never admit that his "child" was now an adult. But at least Marty still thought of him as thrilled to be in politics.

He escaped from the claustrophobic bedroom only to be confronted by Philanth's neatly made bed as he stood in the tiny hall, for opened out, the sofa filled most of the available floor space between the front door and the dinette table.

The place had gotten shabby over the years, and it smelled musty. It had been purchased because rents were exorbitant and he had needed a voting address in his district; and it had been kept when he left Congress because he would continue to need California residency if he wanted to sit in the Senate. It was a legitimate residence paid for out of his own pocket, so it was unlikely to draw political flak. It was also a private place in which to shower, change, and sleep when he was in town, though considering the condo fees and taxes, it could not be called cheap.

He deduced that Philanth was in the bathroom. He went back into the bedroom and changed into a robe.

He had to take care to leave town with the same clothes he had brought, for otherwise he would disrupt Carole's Washington/Los Angeles inventories. Just the same, it was a relief to be "home".

But where was "home", really? The town house in Georgetown had too many slick, hard, sharp-cornered surfaces and a rotting, antiquey quality he could never feel comfortable with. Moreover, since they could not afford to renovate, they lived with steam radiators, high ceilings, and ancient wiring and plumbing which continually caused problems. The new rooms in the basement probably would make the rest of the house look even tackier by contrast.

Yet Georgetown was "home". "Home." Interesting concept.

He noticed again that the apartment smelled musty. Carole had been smart to put cedar lining even in the chest of drawers and dresser. He found some pajamas and came out into the hall again.

Philanth was already bedded down. She must have hurried her shower with that objective. She was lying with her back toward him; she didn't move, though she must know that he was standing there.

Well, bless her for always doing her best to try to make things as easy for him as possible. "Bless the beasts and children": that should cover both of them.

As he shut himself in the bathroom and turned on the shower he mused about how different from our anticipatory fantasies the reality usually turns out to be. During the past week or so he had had some passionate visions of being alone with Philanth in his tiny apartment all this afternoon. But now they were playing it strictly by the book.

It wasn't hard. They were tired, especially from the hours and hours of vibration on the plane from Tokyo. They also felt more self-sufficient now that they were back on American soil, having left behind all the possibilities of minor mishaps which could have been disasters while they were overseas. Now they both could be models of practical rationality.

Of course, there was the long night ahead in that huge house looking out over a sea of little cities, while the sage-scented wind swept across the northern faces of the mountains and a few coyotes sang in the brush-covered slopes behind the house, up toward the crest beyond which the moon would rise.

Kate and Ned had had the house built on an "unbuildable" lot, using their own architect. Whenever there was a respectable earthquake the whole house swayed on pillars containing enough steel and concrete to support an office building. It was a marvel of thickly carpeted spaciousness with a long, long wall of windows looking out from each of its three floors at the populous San Fernando Valley below and at the San Gabriel Mountains in the far distance.

The top floor and carport were at "street level", the "street" being a small plateau at the top of a steep and winding private road: as on a roller coaster, one always seemed about to drive off into thin air just before the car nosed down to level off. The bottom floor opened onto a grassy yard overhung by a sheer drop which was traversed by wooden steps leading up to the level of the top floor. At the edge of the grass the side of the mountain dropped steeply past the storage caves under the house to the winding public road, then down again to an expensive private school. A "hillside home", indeed.

To him the best part was the big, flat, railing-enclosed roof open to the vastness of the Valley and the sky. Lounging in a deck chair, one was elevated to a solitude visited only by the desert wind which rushed across the flatlands to throw itself against the mountains. He loved that windy openness: it gave him feelings of freedom and dominion.

As he entered the bedroom after his shower he laid out a coat for himself and a coat of Carole's for Philanth so they would be comfortable in the mild chilliness of an early December night in Southern California. He also identified the black-tie ensemble he would wear, thinking how grand it was to be able to put on something he had not worn until he was tired of the sight of it.

He climbed into bed with a grateful sigh. Then he set the alarm so they would not be late for dinner. He had felt like he couldn't unwind under the present circumstances, but thinking about that night on the roof when everyone had gone, relaxed by his shower and by knowing he was in his own bed at last, he drifted into sleep.

He awoke from a nonsensical dream about trying to get Philanth alone so he could make love to her while they were at a crowded, chaotic party where everyone wanted to shake his hand. She had been wearing something soft and white and fluffy and had been regarding him with a particularly appealing

652

expression. He let himself look at the clock, then wished he hadn't. He groaned softly. The alarm would not go off for almost an hour.

His body had waked him up because it was aware that Philanth was in bed in the next room and there was no longer any chance of being interrupted by Tony. He needed to be as fresh as possible in order to do what he had to do that evening, and his body knew very well how that could best be accomplished.

He wanted very badly to make proper use of this last hour of complete privacy with Philanth. If he only went in there and sat down by her, at least he could look at her. If she woke up they could talk.

But the more he thought about what he wanted to do, the clearer it became that he could not go in there without wiping out the last vestige of benefit they had derived from her attempt to deceive him about Tony. Within five minutes he would be begging her to let him make love to her then and there. And she would be making up her mind that she must leave him soon after they got back to headquarters.

She must be in love with him, because otherwise by now she would surely have become repelled by his overtures. Instead she had tried to "cure" his desire so she could go on working for him. Yes, she loved him, in just the way Carole had said she would.

Was she awake now, feeling safe and rested and therefore having the same struggle to keep from going through that bedroom door? She must know he would not hesitate to welcome her into his bed.

This bed, where so many times he had made love to Carole. If Carole was going to divorce him he and Philanth would be married among his old friends in his old congressional district, and this would be the bed to which he would bring Philanth on their wedding night.

So many politicians' wives called it quits any more, and Californians were so tolerant of even a messy political divorce, that he might still have a future worth offering Philanth.

Knowing how unequivocally Philanth held married love sacred and regarded any other lovemaking as qualitatively inferior and an irrevocable blemish on a relationship, he found the strength he needed: he would reserve this bed for Philanth to use for the first time as his bride. Therefore he would stay in here and thus manage to keep from disgracing himself until the buzzer on the clock went off.

He was a heel anyway, even fantasizing about the benefits of being divorced by Carole. Carole was a wonderful wife and a splendid woman. He never forgot that he was extraordinarily lucky to have a wife he could respect and admire as he did Carole. She was so hard-boiled about his career probably in order to compensate for his squeamishness about the vulgarity of self-promotion and the appalling power of money to determine the outcome of elections.

But he needed Philanth for his soul as he needed Carole for his career. An intelligent society would have allowed him a second wife.

But a president's having more than one woman was no longer tolerated. And that was what this was all about, really: he would give up his manhood far more readily than his aspiration to the presidency.

He and Philanth always came back to the same common-sense answer: he must stop letting his love for her take the form of desire. It didn't sound that difficult.

But considering how often he went through periods when practically every thought he had got displaced by a vision of taking Philanth in his arms, . . . as Philanth had observed, it didn't look good for them.

654

Well, it was now less than twenty-four hours until he could be with Carole. Until then he would hope his tough-minded wife could give him some answers both he and Philanth could accept.

He and Philanth were quiet as they prepared to leave for the party and their overnight visit, quiet as he drove through the heavy early-evening street traffic to get onto the San Diego Freeway. The freeway going north into the mountains also was slow-moving, so once he had gotten into a middle lane he felt little need to attend to his driving.

He began to relax and savor the evening. He enjoyed the drive through these mountains, which had been kept relatively undeveloped by many years of political struggles. They provided a glimpse of rugged Nature, a break in the urban sprawl.

The traffic continued to move slowly, and he wished he already had the material Kate was holding for him so Philanth could feed him the information about the guests while he drove. In lieu of that, he started collecting his thoughts for his speech.

Before he realized it they were descending from the mountain pass. He began signaling and working his car over into the lane which became the first off-ramp into the Valley. The brake lights of the homeward-bound vehicles slowing in front of them down the long slope flared like a merry string of red Christmas lights in the December twilight.

God, it was good to be back home again, where the traffic signs were easy to find and easy to read and said what they meant and didn't wait to point to a turnoff until you were passing it! The only route he knew of in all of Greater Washington that had good signage was the one to Dulles Airport. V.I.P.'s leaving Washington in a hurry must have raised hell to get such results, because countless letters to the *Post* saying in effect, "Why can't we do it like they do in California?" seemed to have made no impression.

It was good to be back in the city where he had grown up, good to be back where dress was casual, people seldom stood on ceremony, and the weather was usually perfect. Here snobbery was the exception, and everybody was free to do his own thing. Best of all, there were rational cause-and-effect relationships between working hard and having something to show for it: a blessing of private life missing in the political arena. No wonder he kept feeling tempted to come back here and spend the rest of his life trying to forget that Washington existed.

Philanth was secretly more Californian than she knew, even if this was her first day in the state. Secure in her self-esteem as "a good person" and glad to be free to go her own way independent of other people's skewed values, she felt no need to appear prosperous or "tasteful" or important.

At dinner Philanth played the mousy ingenue, and he chatted with Ned and Kate. After dinner Philanth began working with Kate and Ned and their caterer on the top floor, and he shut himself into his bedroom on the ground floor. There he drafted and then memorized the main points of his "few remarks" which would highlight the guests' expensive evening.

When he felt that was under control he began studying the computerized data on the guests which Marty had had sent to Kate for this purpose. It was the usual mix: mostly lawyers, media and business people, and miscellaneous activists. He was hard-pressed to absorb as much information as he could before he had to make his appearance.

Realizing what enormous amounts of work had been invested in this list and how much these people might do for a senate campaign, he felt ashamed of his priorities. He should have put more time and effort into studying the people he had to meet here. He could have had a copy of the print-out on the guests sent to his hotel in Manila so he would have it for the long flight from Tokyo. He also should have had his speech for tonight down cold when he got off the plane in Los Angeles that morning. Instead he seemed to have spent much of that travel time mooning over his relationship with Philanth.

He could see more clearly than ever why Carole had said he should simply take Philanth to bed when he felt like it: a man with the scale of responsibilities he had was wrong to waste time on anything so trivial as his sex life. Following traditional conventions could be agony, given the working conditions of his profession; and the importance of deviations from those conventions was wildly overblown by journalists who pandered to the prurient interests of their audience. No matter what else he did, collecting money or endorsements or staff, he would not truly be ready to run for the Senate until he had gotten this matter straightened out and onto a sound, long-term footing. Philanth was too logical to be a real problem. He simply had not yet approached her with the right argument.

At least his present cramming session was eased by the fact that he had known some of these people for years and by the fact that all of the guests would be given printed name tags when they arrived at the church parking lot to be shuttled up the hill.

When Kate came down in a flowing, scarlet-and-gold hostess gown to tell him that the first guests were about to arrive he put on his jacket and went upstairs with her.

He put on the proper mind-set along with his tuxedo jacket: these were all his good friends who had come to party with him and wish him well and bask in the glory of his present and prospective status. Putting forth a natural, genial warmth, he must make sure every hand was firmly shaken and its owner made to feel that he or she was—or was then becoming—a significant individual in his mind.

Ned was greeting three men and a woman at the door, peering at their name tags as he did so.

Immediately he fell into a familiar routine which would carry him until the last guest had departed, with a break only during the brief period when he spoke to the group as a whole. No one entered without his excusing himself from whomever he had been talking to and welcoming the latest arrivals individually, eliciting and reacting to each guest's occupation and previous association with himself or other distinctive service or achievement.

If this started to cause a pile-up at the door the process took on the character of a receiving line, but he controlled the pace by eliciting promises to talk to him later from those experienced enough to appreciate his plea and follow through. By the end of the evening everybody must feel that they had gotten adequate contact with him and that he was a rising star whom they must continue to support because of his obvious diligence, charm, intelligence, vitality, enthusiasm for public service, grasp of the issues, active concern about others' opinions, and inside understanding of the current processes of politics and the workings of government.

Several ladies from the local party central committees or his own congressional district kissed him, and some women he didn't know also took the liberty. He dispensed warm hugs and special greetings to those who had labored to get him into Congress, and although some of these ladies were getting along in years, they were thrilled. He loved them for that, for being made happy by so little.

The hard work was with the men, each of whom had resources which could be helpful to him and each of whom expected certain considerations in return. They wanted to know what he knew and what he believed about every issue under the sun. He had to demonstrate his ability to meet their demands without compromising himself.

Once as he worked the crowd he caught a glimpse of Philanth watching him. The little blonde serving-maid's wide, serious eyes indicated that she realized how much quick memory, concentration and political skill were required to do what he was doing.

But he was managing pretty well, and he realized he was more impressive because his weariness relaxed his voice despite the effort he was making.

And people were responding nicely. In fact, the emotional contagion of a crowd had begun working in his favor even among these experienced critics.

The spotlight of adulation in which he was now moving would have brought other women to his bed as soon as the party was over.

He pushed that thought out of his mind. He was on his home turf now, and these were "his" people, so the standards he set for himself were higher than when he had to deal with strangers about whom he was expected to know little or nothing.

A professional photographer was on hand when the ceremonial decorated sheet cake was brought out at around ten o'clock, and he posed for pictures with everybody who wanted one. Ned and two other long-time supporters organized the groupings so this didn't take forever, and meanwhile Ned helped him keep a patter going so that toasts were drunk and jokes exchanged. Meanwhile Kate got the names of

659

those in each shot so they could receive prints. Additional sets of those prints, labeled with these names, would find their way into campaign files.

To many of the guests those signed color photos would be worth the cost of their evening. The pictures would be framed and displayed in offices, especially if he won this next election. Dollar for dollar, they could be more valuable to his long-term career than paid advertisements. In this world everything was quid pro quo.

Finally the last group was dispatched. He began the cutting of the cake, formally presenting the first pieces to Kate and Ned. After that he let the caterer take over the job, with Philanth's assistance.

Then—surprise, surprise!—there were cries of, "Speech, speech!" plus various ad libs which never got too hackneyed to be used again: "Let's hear from our very own celebrated orator." "Hey, Congressman, how about it?" and so forth. The chemistry of the crowd was so good that there had to be some experienced and politically savvy "facilitators" scattered among the less demonstrative guests.

But these particular calls were also the simple-hearted *politicus Americanus* crying for its favorite meat. What candidate could be so boorish as to refuse? For those who loved the drama of the American political process, memories were made of this.

"Okay, okay," he said, raising his hands in surrender after just the right amount of pressure.

The crowd quieted, pressing closer, and he raised his voice to project it conversationally as far as it could travel. "Thank you. Thank you for your enthusiasm. And thank you for coming to share this happy evening of homecoming with me."

He paused because Kate, looking smug, had thrust a tiny microphone at him. With a tap he verified that it was live and slipped its clip onto his lapel. He gave Kate a smiling nod.

He briefly repeated his thanks, then began his "I've just returned from . . ." formula. He moved smoothly from point to point, spending several minutes covering all the bases as Peace Corps director and husband of a well-known California activist and cause-conscious former congressman and budding statesman.

As he waxed eloquent, beginning to tie together all his themes so that "the human agenda" was somehow extended to include "the hopes that bring us all together here tonight", he swung his head to his extreme right in order to make eye contact with more of his listeners and caught sight of Philanth watching from the kitchen doorway. As his gaze lingered she snapped out of her sober intentness to give him a smile with a shake of her head for emphasis as if to say, "Thank you for letting me be here for this!"

He gave no sign and moved on.

"It's a great privilege," he concluded, "to be part of this effort. There were times during the trip when I was tired or sick or discouraged by the obstacles I was studying during my tour. But I never for an instant doubted that this was worth doing.

"And my experience as director of this outstanding organization is invaluable training which I hope to use well later, in other sectors of this generations-long battle against poverty and ignorance and disease. We will never give up our dream of making a more peaceful and prosperous and humane world for ourselves, our children, and future generations. This aspiration is our greatest heritage as a nation.

"Thank you so much for doing your part tonight. Thank you for this wonderful welcome back to the land, the state, and the city that I love so much. Thank you, Kate and Ned, for bringing us together

661

in your beautiful house. You have no idea how good it is to be home!"

The applause was warm and prolonged. Philanth was clapping as madly as anyone, and he saw her fighting back tears as she gazed at him.

That and the emotion of the rest of the crowd almost brought tears to his own eyes, for he realized how much it would mean to him if he could have her with him at each of the endless stream of events like this which should lie ahead of him. If he could love her only across a crowded room then let him have that, let him have moments like this. The thought that even this could be made impossible by his own inexcusable failings was hard to bear.

The applause faded and he resumed his circulating, seeking out people who had not yet taken much of his time. At first this was difficult. But it became easier as soon as guests began to leave: he could stay near the door and reverse the procedure he had used as people arrived.

Gradually the crowd thinned, leaving some people brousing along the buffet tables or lingering beside the windows, looking out at the lights and talking among themselves.

Some political junkies would have stayed all night just to watch him and listen to him and question him or promote their own agendas for him. However, it was in the nature of his role as a busy and important politician that even the grace with which he got rid of people should impress them with his professional competence. Therefore when the time for the event to end stated on the invitations was less than twenty minutes away he began to use still another routine: when a group had had their share of him and one guest in a cluster had just had his latest question answered, he would pause before moving off to some other conversational cluster to thank the lingerers for coming, implicitly demonstrating that they now had a secure place in his mind. Thus they had little choice except to leave soon afterward, but they could not take offense; and the fact that he was concluding the event on time by saying good-by helped them feel they would not be shorting themselves by leaving.

When the last guest had finally departed and Kate was satisfied that "everything else" could be left for the next day Kate and Ned gave him and Philanth their farewells, accepted his heartfelt thanks, and started downstairs to cross the lower yard to the apartment.

He accompanied them downstairs to the outside door. He suspected that Kate was uncomfortable about leaving him and Philanth alone in the house, so as they descended the two flights to the ground floor he good-naturedly complained about having had to confine Philanth to her room during the trip so she would not keep climbing on her soap box with diplomats and foreign officials to tell them how the world should be set straight. He seemed to achieve the desired effect.

"Just the same," Kate warned when they reached the first-floor entrance, "you be careful with her. It's plain she thinks the sun shines out of you."

"Thanks, but don't worry," he replied off-handedly. "She's a good girl, or Carole wouldn't have insisted on my taking her with me. She wouldn't dream of causing any trouble."

They half-heartedly assured him they were confident he could "handle anything of that kind". No one could, of course; and "that kind" of trouble could leave a lasting scar on any politician's reputation.

He watched to see Ned and Kate go inside the apartment. The moment their door closed he hurried back to his assigned bedroom and snatched up his coat. He climbed the stairs two-at-a-time back to the third floor. He plucked Carole's coat from the bed assigned to Philanth and went to the kitchen.

The dishwasher was churning along, and Philanth was swabbing the now empty expanses of counter while munching from a plate of leftovers. He watched idly for a moment, enjoying seeing her usual efficiency transferred to domestic endeavors. Then he said, "Come up to the roof with me for a while."

"Could I change into my pants-dress first? This uniform is too tight."

He tried not to smile. He could see that it must be uncomfortable around her chest, though the problem was not noticeable from the front because of her frilly white apron.

"By all means, get comfortable."

He turned away, saying, "I'll be waiting." He put on his coat and took the woman's coat with him up to the deck on the roof.

When he arrived he paused to inhale the cool, damp, pungent air. Then he walked over to the side overlooking the valley. He chose a well-cushioned chaise longue and stretched out, closing his eyes and willing himself to relax. He started to undo his tie, then decided he would need any advantage he could get in the situation which was about to develop.

After a few minutes she came up the wooden steps, treading quietly as though to avoid disturbing the great silence which enveloped the huge house and the mountainside on which it stood. She was leaving the lighted stairway for the darkness on the roof, so he stood up to help her locate him against the glowing haze of the valley floor.

As Philanth slowly approached he had the feeling that he should have been looking out over the city lights and smoking, like the hero in the climactic scene of some Humphrey Bogart movie. He had never smoked, but somehow smoking suggested a despair about life that she made him feel at this moment. He wondered how many years it would take for her youthful idealism to catch up and merge with his adult realism, regarding sex or politics.

Well, tonight questions like that didn't matter. It was going to take him quite a while to unwind anyhow, after his full evening of "calibrating every nuance" with scores of watchful and potentially invaluable people, and they had the whole night ahead of them. Tomorrow they had only the four-hour flight to Washington, since the time zones would shorten their day by several hours. For the present they had this enormous house and its view of the glittering valley all to themselves.

And nothing was going to happen.

He would have liked to bring a huge bed up here on the roof and give her a floating, white Grecian gown to wear. With a balmy Santa Ana wind blowing over them and the stars twinkling in the vast dome of black sky overhead . . . . It was such a waste.

She stopped beside him and stood motionless for several minutes, watching the cities of the plain shimmering in the mist.

"Thank you for this, Charles. Thank you for this moment, with you, in this wonderful place." Her voice told him that she had slipped into her Sacré-Coeur mood. "I know Carole arranged this whole thing, but you let her plan be carried out; and you knew it was partly because of this."

"I don't deserve any thanks. I wanted it for both of us."

He helped her into Carole's coat, and she thanked him for that too. She had let her hair out of its bun.

He helped her pull it free of Carole's coat, and his fingers lingered. As he expected, she shied from the contact, this time by turning to look at the lights.

"I won't spoil this, the way I did our night in Manila," she promised. "And I'll treasure it always."

Leaning both arms along the rail beside her, he remarked cautiously, "That sounds like the beginning of 'Good-by'. I thought we'd finally gotten everything worked out."

She did not answer at once.

Finally she murmured, "Sorry. I try to preserve a beautiful memory the moment I have the experience, because you never know when everything else will be taken away."

He considered that.

She had sensed that his feelings were unchanged. She was too smart to suppose that he had truly believed what she had implied about letting Tony have her ring. Like her "taking the pain" from his chest with her hands, the maneuver had been only a gesture which might put him through a salutary emotional exercise. The truth between them was a rock on which they stood, and such diversions were drifts of sand which blew back and forth over its surface. Therefore she had not mistaken his restraint during the past twenty-four hours for indifference.

He had been foolish to hope he could get by on that basis. He could not help speaking kindly to her, and—as Tony had implied—there were times when he no longer knew how to look or not look at her.

"You don't have much faith in me," he observed. "That must be because I haven't deserved it."

"It's no disgrace to have difficulty with a problem you've never dealt with before."

When he didn't answer at once she went on, "Mr. Bjorklund, I know I'm much younger than you, but

I've been along this road before, and apparently you haven't. I was named after the Beatrice who was Dante's guide. If you let me help you we might make a journey together that would make the one we've just finished look prosaic by comparison."

"An internal journey."

"Yes. Imagine what it could be like. A quarter of a century from now we could be standing here again, an important public official and his assistant, after some other fund-raising party for you. What must it be like to have loved someone for that long as I love you, and been with them all that time, and never . . . even properly kissed them? I'm willing to find out."

"I daresay you are." He turned his back on the lights, resting his elbows on the rough-hewn rails behind him, and looked down at her pale, upturned face. "I daresay you'd have welcomed crucifixion, too, as a great opportunity for personal growth."

"Please be serious. I'm not interested in personal growth I can't hope will be useful later.

"But suffering is a price we have to pay for doing what we must until we learn how to get what we need without it. So you might as well try to learn something while doing what you have to.

"That's all I meant. You can't have me on any basis that's in your interest, so rather than leave you to carry your burdens without me, I was offering the only kind of comfort for your dilemma that I could."

"You call that 'comfort'?"

"I know. To you it would be Purgatory because you can't spare the effort to struggle against wanting.

"But if you can't deal with your instinctive attraction so that it's no longer a problem for you, the only other comfort I can offer is knowing that you're loved by someone who isn't physically present any more."

"It's not enough."

"Again, it's not enough only because you haven't learned how to make it enough. It's like the comfort of the presence of God, if you believe in that."

"You're talking about self-delusion. I need something real to counter the realities I have to deal in."

"Is my love for you not real just because it's limited in its permissible forms of expression? Is the knowledge that you're loved not real? An idea in the mind is not unreal just because it's not material!"

"No. But an idea in the mind *is* fragile and likely to be transitory."

"Not if it's clearly defined. If I were only a character in a story you'd read, couldn't I still be real as a model or an image of a principle? Is such an idea or attitude necessarily less durable than a marble monument or a Grecian urn? Shakespeare and Keats proved otherwise even in the ways they said so!"

"In my chosen life, Philanth, poetry is something you quote in a speech; not something you can hold onto in order to stay afloat during a disaster."

"My point is that you can *make* a fact like loving or being loved transcend the material world. It's what *you* make the center of your emotional life that matters, and you control that—not anybody else, not ever. The love can grow and mature as you do even when the object is an abstraction."

There was no use in arguing the reality of ideals with an idealist. She would remain convinced that she could fill his need for her from the other side of the world. Or from the next.

"Philanth, if you do give up on me and go back overseas, at least promise me one thing: don't get involved with the CIA. It may sound exciting, but it's not your kind of outfit."

"Oh, Kingsley mentioned that to you?" She sounded pleased by this proof of serious interest.

"Somehow he's realized how fearless you are."

"Well, I didn't envision myself doing anything dramatic, just passing along information on local conditions," she argued. "Our country will always need intelligence! Otherwise it'll go on blundering around wasting billions of dollars and causing disasters from bad judgment."

"Just by being an American in the wrong place at the wrong time you could get yourself killed, and tortured first. They especially enjoy working on women."

"I've read the horror stories put out by Amnesty International; I know what some human beings are capable of doing to others.

"But one can die only once, even if it takes a long time. I don't have an egoistic preoccupation with the sanctity of my body, and my ideology helps me keep my natural feelings in check. All that really matters is that civilization continue."

"Philanth, if something horrible happened to you I'd feel it was my fault because I hadn't used my brains well enough to keep you with me. I'd never forgive myself."

"Your instincts are getting in the way again. You want to protect me."

"Damned right."

"Well, we can't very often do what our instincts tell us—not on the major things. If I could, I'd buy a big house in the country and fill it with other people's children who are being neglected or mistreated."

"Philanth, I beg you, promise you won't ever go into intelligence work."

"The Peace Corps treated me like I was a fragile orchid. Kingsley said the CIA might help me find the sort of regular job I really want, as a cover. I'd love to have opportunities to move around and see things."

"Go to work for AID, or any of the other development organizations."

"Let me tell you something, Mr. Bjorklund. I've learned a lot during this trip. I didn't get the sort of experiences I'd hoped for, but at least while you kept me stuck in my room I had time to read and think and discover some connections I hadn't seen before.

"I'd known for some time that it's not politically feasible to bring population and environment into sustainable balance when the non-renewable resources are already depleted and the ecosystem is already damaged, yet the expectations of the world's poor are being raised by the global network of mass media just when their populations are exploding and doctors are being murdered over the issue of abortion because of lingering prescientific conceptions of what constitutes a human being. If we're to have a tolerable future we also need to cooperate on the international level so we can mobilize the resources needed for reaching out to the unlimited resources of space.

"It's unspeakably urgent that we clear the major hurdle of beginning a major flow of resources from space *before* we've finished using up the surpluses generated by the Age of Fossil Fuels. That means *now—yesterday*. If we do we can go onward and upward. If we don't we're doomed to extinction, and the cosmos will never see our like again."

"You figured that out all by yourself?"

"Of course not—I just pieced it together. You know I've been studying such matters for years.

"But recently I saw an article about the fabulous mineral wealth out in the asteroid belt and how we could rig asteroids to propel themselves by their own fuel into earth orbit for mining. The refined ore could be dropped into uninhabited areas of the earth's surface like meteorites, like the huge, rich lode of iron ore that landed in Canada. An expert at the Johnson Space Center said the only real obstacle would be getting nations to trust each other enough to let it be done, instead of being afraid that the same technology could be used as weapons.

"I want to work where my talents can be useful to help bring that kind of cooperation about. Right now the international scene is dominated by a lot of little groups that want to cut their neighbors' throats; and with all the economic interests still pushing high-tech weapons sales, it's a mess. We're wasting assets we'll badly need in order to survive as a species.

"There are things a woman can do toward fostering a just and stable world order that a man can't, and Kingsley says I'd be particularly good at them. I think he's right. Might as well make my physical attributes an asset instead of a liability."

"Oh, Jesus, Philanth!"

As Carole had said, it was only a question of who got to her first, to give her altruistic ambition a specific course of action by fitting her agenda into their own.

"I'm not a good, virtuous person, Mr. Bjorklund— not in the sense of being hobbled by ethical or moral premises I regard as absolute. It's only my personality which gives that illusion. In fact I'm prepared to do whatever a situation and my own priorities seem to call for. So my personality and appearance and pristine background will serve as part of my 'cover'."

"You may be used and tossed away. You may be treated as a dirty joke by the men you're risking your life to help!"

She was silent for a moment. "I can accept that."

He gripped her wrists. "I'd keep you bound hand and foot, to prevent anything from happening to you! You're precious just in the fact of your existence."

"As to that, my having existed should be enough."

"Well, it isn't! Even if it were for me, what about the rest of the world?"

"The rest of the world will always have to look to whoever's left, whoever still remembers and believes that duty to the best we know is all we have."

He released her. "Sometimes dying is easier."

"Yes." Despite the darkness, she peered up into his eyes. "But dying for the wrong reasons demoralizes people, too."

"So neither of us is allowed to throw ourselves away, Philanth."

"No. But each of us must accept certain risks."

They were quiet for a while, watching the floor of the valley appear to shimmer as its many square miles of lights shone through the rising warmer air and the haze.

"Why has she done this to us?" Philanth asked. "You wouldn't have been this distressed five weeks ago. I wouldn't have hurt like this because you hurt. I answered to no one, then. Why has she forced us to know each other in this way? What possessed her to offer you something which was so obviously forbidden to you as the public man you must be? Even if she planned to divorce you, why do it this way?"

"'Why has she done this to us?'" he repeated. "It's all perfectly straightforward. Is it too ugly to be real to you, is that why you don't believe it even now?"

"What, that she merely wants a painless divorce?"

"Don't forget the other possible explanation: she wants you to stay with us; to live with us, to share our goals, our work, our lives. She wants you with us enough to be willing to share me with you."

Philanth's voice was incredulous. "But we couldn't be sure of hiding that forever, even if I lived with you and Carole continued to be so friendly to me!"

"We could get by, if we were careful when other people were around. You could date other men, to avert suspicion."

"But that—that would be thought monstrous!"

"It *would* be monstrous. Because all we could do is provide for you, including rewriting our will. It's to be assumed that Carole will outlive me, ..."

"You make it sound like a real plan!"

"Philanth, Carole and I discussed her scheme to let you fall in love with me and then enlist your help for the rest of your life *before you and I took this trip*. That plan was the reason she wrote that letter."

She stared at him, open-mouthed. He quickly looked away, down the steep, dark mountainside, and hunched over the rail.

"I'm sorry, Philanth!

"Carole will be furious if she finds out I've driven you away by telling you just how ugly it is.

"But I couldn't lead you into it blind: couldn't take advantage of your inexperience and illusions and lack of concern about your own interests. The guilt I'd have felt when I looked at you ten or twenty years down the road . . . . I'd have told you everything, sooner or later."

Once he had started he could not stop the flow of self-loathing. "You'd have had grounds for murder, if you'd let us steal your life and exploit you that much for the sake of our own ambitions."

Philanth remained speechless, and he began to absorb the import of what he had just done. This knowledge could be all she needed to free herself from him altogether. He had been a fool to risk that when it looked like Carole might be preparing to leave his "wife" position legally vacant.

Remaining draped over the railing rather than face her, he went on. "Of course, Carole believes we'd be justified in co-opting you because we'd be saving you from wasting yourself. This hare-brained scheme you got from Merriweather—try to understand why it got me so upset: you're just too willing to put yourself at risk! And *God knows* what it would take to teach you to defend yourself! They'd do better to have <u>you</u> teaching classes on how to endure torture.

"But you're basically sensible. In a few years—if you make it that far—you probably *will* find a niche in some reasonably stable country, marry somebody as wise and good as you think I am, and have a contented life of solid if unheralded achievements.

"And I hope you do, Philanth." His voice broke. "Because I'll love you always, and I've got no real prospects. I mustn't pressure you to accept being exploited in even the *standard* ways men exploit women, much less bind you to me just because you've insisted on believing I'm better than I am, just because Nature blessed me with this face and well-educated parents and—" He pressed his fist to his mouth much as he had seen Philanth do.

"Oh, Mr. Bjorklund, ..." She pressed against his back, trying to hug his shoulders. Her compassion for his remorse seemed quite unadulterated by any resentment. She had treated Tony the same way after he had blasted Tony for buying her wedding ring to break even on his bet. It was the compassion she would have given a grieving child.

Such generosity made him turn and take her in his arms.

She huddled against his chest, and his own emotion subsided as he realized that she was holding her breath. He stroked her hair.

"I know this hurts you," he said, "for a lot of reasons. But you'll get over it. Because it's only the evil we see prevailing that's so hard to live with. At least you'll know I really did love you more than my personal ambitions."

"But that's not being fair to yourself—"

"Don't make excuses for me. People who believe their cause has top priority can justify anything.

"It's impossible to be sure how to set a proper level for our aspirations. But at some points we have to accept the world's limitations on what we're allowed . . . no matter how that cripples us in trying to help others, and no matter how irrational and mean-spirited the world may be in giving things out.

"I'm not allowed to have you. I don't deserve you. I have little to offer you.

"I can't even offer you marriage, not as long as Carole wants me as her husband. I still love her very much, in spite of what she's done to us. And I owe her everything.

"If she doesn't leave me I can't have you on any terms, except as I allot you a government salary. And that half-a-loaf may not be better than none: it's probably not bearable for me or fair to you for more than a matter of a few more weeks."

He held her from him. "I finally understand that I've got to let you go, Philanth. I finally grasp the truth you've shaped yourself around: we hurt unbearably and we hurt others when we let ourselves want too much *for ourselves*.

"Let's try to at least part friends. Can we, in spite of everything?"

"Of course," she managed. "Thank you, Mr. Bjorklund." She turned and walked to the stairs.

Her plaster saint had been shattered. Philanth was free to choose some more realistic talisman against the despair of caring for the world.

But what could keep despair at bay except having someone to love—and believe in? Philanth was right:

675

the traditional sources of inspiration were obsolete: contrary to the words of the song, he did not believe in angels. Only in Philanth.

There must be a man somewhere who was pure-hearted yet sensible enough to be worthy of her, one who could offer her what she wanted in marriage. It wasn't so much as she thought: a childless partner-ship in worthwhile service, a way of life which did not waste resources, and sustaining mutual love. Such a man should be able to convince her that he would not be satisfied with anyone else and that there was no point in their both accepting the practical and emotional impediments of being alone.

Anyway, their own future was settled. Again. Every time they tried to "settle" it, it got worse.

And Philanth was downstairs grieving over her broken idol. She could no longer close her eyes to the fact that he was not merely subject to the normal human shortcomings: he was morally mediocre at best, if only because he had not spelled everything out for her long before now.

If he went down and helped her recover she could begin to get the rest she needed. He couldn't bear to leave her down there alone and unhappy.

Besides, now that she no longer could persuade herself that she "worshipped" him, he might be allowed to get as close to her as other men did.

Hell, it was worth a try. His own struggle to survive would make him go on trying any way of being close to her that he could talk himself into, as long as she stayed with him. He couldn't leave her alone.

That was why she had to leave soon. Aside from what other people might see and report and remember ever afterwards, he did not want to look back after she had gone and shudder at having done some of the things he already found himself considering.

Philanth still wanted and needed to go on respecting him as much as she could.

Kate had given Philanth the master bedroom on the third floor, thereby maximizing the distance from his much more secluded bedroom with its own private garden entrance on the ground floor. He had only to descend the steps from the roof, enter the house, and turn left in order to be in her bedroom. Indeed, there was no door to pass through, for this bedroom was separated from the long, wide living room only by a folding partition which remained open so that those in the bedroom could look out through the wall of windows toward the distant mountains. A couple of table lamps on mahogany cabinets softly showed him his way.

Philanth had laid Carole's coat on a chair and flung herself diagonally across the king-sized bed, burying her face in the pillow she was hugging.

He tossed his coat on top of hers, then his tuxedo jacket, and crawled over to sit beside her.

"Philanth. I'm sorry I had to destroy your idol. They must be hard to come by, for someone as discriminating as you are. You must have really needed one, or you wouldn't have chosen me."

"Oh, don't!" Rising onto her elbows, she half turned to him. "Please don't keep saying things like that about yourself. Can't you see what you are?"

"I'm aware of all that other people want me to be. But I see only too well what I actually am."

"You reject the mirror of other people's admiration because you assume they're reacting only to your physical attributes. You despise yourself because you can do so little."

"Yes."

"Well, the distance between the god they want to admire and the man you're inclined to despise can destroy you, if that's all you let yourself look at."

He was leaning on one hand. She moved to momentarily lay her cheek against it, kissed it, and quickly moved back. He started, then kept his position as he realized that she was counting on his self-control to keep him from reacting.

She sat up. "But look what you've done to *me*: planted a whole new meadow of love and commitment, out of the bounty of your goodness. It'll go on ripening and reseeding itself to feed the hungry world for as long as I live.

"This shows your future: what you've done for me you can do for countless others. You're one of those who can change the course of people's lives with a speech, a smile, a handshake and a few words . . . simply by being what you are, intending nothing but to do your duty."

Suddenly she seemed about to weep, but she overcame the impulse and went on. "If you hadn't stooped to desire me I could have stayed and watched it happen . . . and felt as close to God as I could ever come."

"My child, my child." He had failed to free her.

"I tried to be that," she reproached him. "I've tried and tried."

"I know," he apologized.

She returned to her original position, nursing the pillow, and they were quiet for a while.

"I can't believe what you told me was as bad as you made it out to be," she told him. "So telling me was only an effort to make me stop loving you. You can't kick me hard enough to free yourself of the burden of being loved. I'm like 'The Hound of Heaven', I don't stop no matter what."

"How can you still love me? I'm so thoroughly flawed! And now you must see that."

She rolled over to look at him again, her expression bewildered. Finally she answered, "I didn't love you because I thought you were perfect. I didn't love you because I idolized an illusion about you. I loved you, and I still do, because what you *are* includes the potential to be magnificent. I idolize your ability—the essence of what you are right now—to be wise, to be good, to be great."

"You look at people in terms of their potential: not what they *are or are likely to become*, but what they *are capable of.*"

She nodded.

He sank onto his forearms to look into her face. "You can't expect people who've been damaged by the disdain of others not to adore you for that. If they learn to respect your judgment you give them the gift of faith: in what they can be, or become, and do. It's small wonder some men want to hold onto you for dear life."

"What I do wouldn't seem remarkable if it were supported by our culture. In a way it's what the Gospel was all about."

"Please leave religion out of this. It takes *a living person* to make all those abstractions part of people's lives."

"Yes." She began to smile. "I know of a religion which makes much of that fact."

"Well, if you want to be Christ to me, you can't just go off and leave me to fend for myself. You've got to be with me always, 'even to the end of the world'. *Physically* present. Why d'you suppose people have always gone crazy when they found someone who not only offers them glory but also walks among them?"

She raised her eyebrows, approving his point. "It's hard, I know. But if we just have some object to cling to in the darkness: a cross, a ring, a book . . . ."

"Not good enough! With you, Philanth, I might be all that you imagine. Without you I won't have all that abundance to give away. I'll have only the self-doubt and despair which make 'goodness' seem synonymous with weakness, and altruism the same as naïveté."

"Because you're too honest to let *anyone* idolize you? You must overcome that excessive honesty. It's a sacrifice of integrity that charismatic leaders make in order to inspire great collective actions."

At first she seemed to be evading his argument. Then he saw that she was more interested in making her own, assuming that his following her suggestion would bring him enough success to eliminate his morale problem.

That course was not for him, but she was capable of making herself one of those charismatic individuals who understand how to apply behavioral science in appealing to religious impulses. She would have no scruple against that because she was convinced that there was no intrinsic distinction or legitimate boundary between religion and science. That was scary even given her principles.

However, at present she was scarcely more than a school girl, applying bravely amoral theories about expedient acceptance of folly in others and impurity in oneself in order to "do good" in an imperfect world. If he wanted to help shape her future he had to deal with her on her own terms.

"I understand that some people need that idolatry," he mused. "They hate those who deny it to them. They *want* the 'image' of perfection to believe in.

"But I couldn't tolerate it in someone close to me, because that would imply that I believed in it myself.

"And I couldn't allow it in you because I very much want you to become all that *you're* capable of. It's time for you to join the real world, where mature men and women, with all their weaknesses and blindnesses, have only each other to sustain them even when the world seems very, very dark."

"Oh, I'm sorry, Mr. Bjorklund! Is idolatry necessarily unfair to a human object? Have I used you?"

"Oh, Philanth. Are you going to keep showing me new reasons to love you up to the moment you board a plane back to Africa?"

Her mouth went out of control, and she crumpled back down onto the pillow. She cringed but did not move away when he laid his arm across her.

The house was so silent and the floors and stairs creaked so much that no one could surprise them together like this. So he too sank down; he relaxed, closing his eyes, grasping her shoulder.

A man who could call such a woman his wife would never cease to be conscious of his good fortune. On the other hand, a man who had loved and lost her would never cease to feel bereft.

She turned onto her side facing him again. "I don't need to be comforted," she apologized shyly.

The transparency of this excuse made him chuckle sadly. "Even now you won't let another woman's husband stay close to you."

"I wouldn't even let him *get* close, if seeing me vulnerable might arouse a tenderness. With you it's too late.

"I have only myself to blame. I wanted to end your hostility so I could help you during the trip. So I let you know me. I would never have believed how susceptible you were. I feel I deserve a heavy penance for letting you come to love me."

"'To seek to love, rather than to be loved': that's one of the objectives set for you by the Prayer of St. Francis you carry in your wallet."

"You didn't know that just from the glance that let you identify it among my photos."

"No. My father's mother had it, illuminated on a parchment scroll like yours, in a picture frame on her bedroom wall, when I was a boy. It was the cover of a Sunday church bulletin. She saw me studying it and gave it to me. I still have it."

She studied him. It was the expression of someone contemplating a favorite work of art.

He got up. "Good night, Philanth."

"Good night, Mr. Bjorklund."

He picked up his jacket and coat and left quickly.

He would never respect himself again if he didn't make her either his wife or his mistress. If he was such a spineless fool as to let her walk out of his life now, he deserved forty years of living death in political oblivion.

He wakened her at first light. They left the usual morning rituals to be done when they reached his apartment, for he wanted to avoid wasting time in the rush-hour traffic south through the mountains. They slipped out of the house as quietly as possible.

A low-lying mist made everything look otherworldly, and huge, exotic clumps of tasseled broom plants and scrubby thorn trees were frosted and glittering with dew in the perfumed hush of the verdant, half-wild mountainside. California was holding out her arms, making it difficult for him to leave. He dared not think about all her loveliness now, her mountains and shores and redwood forests.

They got into his car and set off. Philanth remarked that it was a lovely morning and then was silent. Even he felt that there was nothing more to be said; no more need to speculate or argue. For by nightfall he would be with Carole.

682

Chapter 24

VIRGINIA

As he entered the boarding area for the flight to Washington he heard a woman exclaim, "Charles Bjorklund! How lovely to see you!" He turned to see a beautiful, smartly dressed lady come up beside him.

It took him a moment because she was out of context, but he covered his hesitation by setting down his briefcase. "Virginia," he managed as he took her hand between both of his, "how kind of you to come all this way to meet me."

She lifted her chin, causing some of her unnaturally perfect blonde hair to shift back as she laughed at his sally. Her delicately molded face was artfully made up. Carole had relayed the story that she had even had bones broken in order to attain her perfect features. He had the impression that she was somewhere around his own age, though only her eyes suggested that she was over thirty.

"It seems we're on the same flight," she observed. "May we sit together?" At his momentary hesitation she added enticingly, "I'll fill you in on the latest."

He turned to find Philanth, who was all eyes but immediately tried to shrink a little. "May I introduce my assistant, Philanth Devon.

"Philanth, may I present Virginia, who goes to my church in Georgetown." He was not sure what last name Virginia was using since she had traded up from a wealthy plastic surgeon to a very wealthy mass-mailing entrepreneur.

The two women nodded, each moving her hand in case the other offered to shake it. Completing her movement by shifting the shoulder strap of her impeccable designer-label beige-tapestry carry-on, Virginia smiled a little as she looked Philanth up and down. Philanth's shoulder bag was blue vinyl and was showing signs of wear. Philanth completed her hand movement by smoothing her blue denim skirt. Virginia was wearing a matched set of gold jewelry which complemented her brown cashmere ensemble and brown eyes while emphasizing her shining hair. Makeup, jewelry, shoes . . . even in the fingernail competition Virginia would have been a perfect 10 and Philanth a 2, since in the world to which they were returning no points were awarded for using only what one had been given by Nature.

"I'm pleased to meet you," Philanth said politely.

At once she added, "I have lots to get done during the flight, Mr. Bjorklund, but I know what to do, so you should arrange the seating to suit yourselves."

He nodded, licking his lips in displeasure. Virginia had offered him a dubious *quid pro quo* to get what she wanted, while Philanth was being her usual self-effacing self. It bothered him that Philanth would be content to keep this servile role with him indefinitely if he would only make full use of her mind.

The clerk at the check-in desk was happy to be able to put Virginia into the seat which had been left empty between him and Philanth. Virginia hastily asked whether she could have the aisle seat, so she could "get up and run around without disturbing" him. He disliked being scrunched between other people with no place for his elbows as much as anybody, and he had just finished five weeks with long flights spent trying to work while scrunched between Philanth and Tony, but naturally he nodded and said politely, "Of course." His mother had brought him up to be a gentleman.

"I assumed you'd be in the first-class section," Virginia commented as they left the check-in desk. She had inadvertently given up her seat there and was not happy about it. "Of course you're on government business, but considering your position, ..."

"The regulations let me travel first class, but not any of my subordinates," he explained. Philanth did not look up to see Virginia's deprecating glance at her.

During the long flight, across deserts and mountains and plains, Philanth shuffled papers and Virginia talked. Virginia was so obviously putting her charm through its paces that he wondered whether her new marriage was proving less than satisfactory and she was already on the look-out for her next social and economic parachute. One never knew when an attractive man might turn out to be marriageable, allowing for a transition period.

He did not suppose for a moment that Philanth was getting much work done, though she occasionally made a notation or rearranged some pages. Virginia would be something out of a different moral universe, and he knew Philanth must be fascinated; he could almost see her ears expand whenever Virginia dropped her voice. Fortunately Virginia was using a high, cocktail-party voice which carried well.

He and Philanth learned about the terrible problems of Virginia's new neighbor, a lobbyist who had been delayed for months in moving into the multi-million-dollar town house he had had built because of problems involving the contractor, the architect, and the restrictions imposed under Georgetown's Historic Preservation regulations. The hapless man had complained that it was costing him a thousand dollars a week to be delayed in moving in.

Though he had an active social life, the lobbyist lived alone; and due to a problem with his dog-sitting service, he had left his two thoroughbred dogs, "large, noisy, stupid, clumsy animals", alone in the big house during the day. The dogs had set off the security system, and the police had been "all over the place" but unable to get in to shut off the horrible noise and flashing red lights. Virginia had gone out and did not know what had finally been done.

They learned about the First Lady's recent visit to the home of a retired senator, another neighbor, who had recently lost his wife. Now in his eighties, the senator had let the First Lady persuade him to have the entire back of his four-story house restored, now that her good friend his late wife would not be disturbed. The appearance of the house would be the same as was preserved in the photographs in the books and articles in which the history of the house was discussed.

"Have you been through it?" Virginia rattled on. "Busts of Kennedy and Churchill, inscribed photos of him with all the Presidents for the last thirty years. And of course all those wonderful porcelains he and Ellie brought back from his time in the Far East."

"Y-Yes," he said, stammering a little because he felt it polite to finally try to get a word in edgewise. "A very interesting house. Carole loves the colors in the drawing room and sitting room, and the extraordinary treasures in the dining room." He was playing the odds to see whether he could get away with it, for his comment could have applied to any number of Georgetown mansions.

"I was in a house last week that has nine ovens, two kitchens, and two dining rooms all on one floor," Virginia said as though she had not heard him, "and the latest style in swimming pools.

"Oh, by the by, Lord and Lady Grey have bought a lovely house that's perfect for their dogs: it has little doors that open onto the garden below the back windows . . . ."

Philanth was nudging him. She had arranged her steno pad so that it was angled toward him, and on the exposed page she had printed, "I was on the same Georgetown House Tour she's getting some of her material from. And *I* saw a huge inflated alligator in a swimming pool on that tour, *so there*." He quickly covered his mouth and coughed.

Virginia was saying, "And they have 'his and hers' studies on the floor with the master bedroom and dressing rooms. His is spacious, with displays promoting his novel.

"Have you read it?" He shook his head. He did not even know who Lord Grey was. "Well, I do wish he'd used more of his government experience. I find drawing-room comedies a bit tedious, especially period pieces. I prefer more serious literature."

This made him hope that Virginia was about to take the risk of embarking on some interesting "bull-crit" by discussing books she had not actually read. He would have liked to at least hear what she had read or heard about them.

But instead of elaborating on that provocative remark or giving him any chance to ask her to do so, Virginia quickly went on to discuss in equally general terms that season's offerings at the Kennedy Center, where she had season tickets, and the programs the previous summer at Wolf Trap. Then she talked about other Washington-area theatres, explaining why she liked or disliked their physical characteristics. The vibration of the plane played on his fatigue, and he was having trouble keeping from nodding off when they touched down in Cleveland.

During the brief stop he and Virginia visited the rest-rooms and stretched their legs in the terminal while Philanth stayed with their belongings in the plane. He knew from experience that there would be no lunch served on this flight because of this stop-over, so he also picked up a few sandwiches, a coffee, and a carton of milk for himself and Philanth. Virginia claimed she didn't need anything.

Even in Cleveland Virginia apparently felt she had to entertain him. Her stamina wasn't remarkable: he was used to people who laid on the charm for him, since so many people wanted something from him. Virginia would at least be able to add the fact that she had made this flight with him to her conversational repertoire.

By Cleveland he had realized that Virginia never tried to say anything serious, so it wasn't necessary to listen closely, and he could nod and smile and let his mind take a much-needed rest. She was not unpleasant to be around: her skinniness distressed him because it conjured up images of exercise machines and celery sticks, but she had a cheerful liveliness that could get through a dinner party well, she wore a very good perfume, and she had the high cheekbones and flawless skin he had seen all over Georgetown among women of all ages.

In fact, he saw them all over Washington, including Peace Corps headquarters, especially among the women in the political-appointee positions.

Carole had once "explained" that if a woman in Washington wasn't beautiful it meant that she was neither rich enough to get herself made over nor the naturally beautiful daughter of a prominent man's beautiful wife; and if she wasn't rich or well-connected she wasn't important. So not being beautiful and therefore presumably important would affect where she was given a seat in a restaurant when she came to lunch with a friend, how much effort store clerks and service workers expended to satisfy her wishes, and how she got treated in general.

He imagined all the women who wanted to be treated as if they were important in Washington marching out of some plastic surgeon's office like Barbie Dolls coming off of an assembly line. Female political appointees tended to be good-looking and come from the financial elites of their respective states, but they might well get themselves fixed up further when they came and saw the competition in Washington.

But like many ambitious people he knew, Virginia took herself very seriously and displayed little sense of humor or mental flexibility. Moreover, Virginia didn't seem to have paid much attention to the sort of "think pieces" that Philanth was always cutting out in a manner which said, "Oh, yum" even when she didn't actually vocalize the words.

At Dulles they said good-by to Virginia before descending to the baggage-claim area, for Virginia had not checked any luggage.

As they reached the baggage carousel he saw his driver coming toward them, waving his arms and beaming. Eric gave him a two-handed shake like a long-lost friend, and Philanth and Eric hugged each other. Eric made no secret of his relief that they had returned intact.

Eric helped them get their luggage across the street to the parking lot and into the car. As he turned the car toward the exit his driver asked, "Straight home, Mr. Bjorklund?"

Realizing that this was a unique opportunity for him to see how Philanth actually lived, he invented a need to stop by the Peace Corps headquarters. "I'll want you to wait in the car while I run up to my office, so take Philanth home first." After leaving her place he would cancel the stop downtown because he was too tired to work on the papers he had wanted.

He settled down to rest during the long ride in from Dulles. Now that he was so close to confronting Carole, he was dreading it. All he wanted to do was shuck his outer clothes, fall into bed, and sleep for a week.

After they swung around the Capitol, Pennsylvania Avenue took them into an old neighborhood packed with tiny town houses in varied states of repair. Soon there were no white faces on the streets. They passed a carry-out rib joint and several boarded-up houses and businesses; then a vacant lot littered with rubbish and a burned-out shell of a town house and a liquor store and a convenience store. This must be part of the area with homicide rates which made nationwide headlines.

"I'm on a one-way street," Philanth warned Eric, leaning forward. "Turn right here and circle the block." She continued to give directions until they stopped in front of one of the entrances in a long wooden building divided into little town houses.

"Stay with the car, Eric," he ordered, "and keep the windows up and the doors locked. If anybody approaches, take off; circle the block for a little."

"Yessir, if that's the way you want it." His normally sanguine driver seemed nervous. Eric tried to joke: "Too bad Tony stopped off in Chicago. You need a bodyguard even in your own town."

"Got your door key ready, Philanth?"

"Yes, sir."

"Then, let's go."

Eric jumped out to open the trunk for him, and when he had Philanth's suitcase and briefcase Eric slammed the trunk shut and hurried back to the driver's seat and slammed the door. He told Philanth to precede him and get her door open. They strode briskly up onto the wide, old wooden porch, and she unlocked the front door.

They stepped inside, and Philanth paused before the first door on her right to select another key. Raucous music was blaring from the basement, making the floor vibrate under his feet. Shutting the front door with his foot, he glanced around.

They were in a dark, stinking, narrow hall which ran straight back into a dingy little kitchen. A strip of threadbare carpet ran down the middle of the scarred, painted floor. To his left was a little, old telephone table with a rotary-dial phone on it and a hand-lettered sign on the wall above it. Behind that was a stairway to the upper floor. On his right were two painted wooden doors, Philanth's and a second on the way to the kitchen.

Hampered by the slipping strap of her shoulder bag, Philanth was struggling with the lock, which required both hands. Bending toward her so he didn't have to shout above the din, he remarked, "You'll have to ask your landlord to get that volume turned down."

She gave him a lopsided smile. "That *is* the landlord," she replied in an equally projected voice. "I did call downstairs at three a.m. one time, and since then it's been a little better. I think."

The building really stank. It was appalling. She opened the battered solid-wood door, and as he followed her through it with her cases he asked, "How do you keep your clothes from smelling?"

She heaved her bag off her shoulder onto the broken-down bed as she answered, "I do things to my closet and keep it shut as much as possible." Keys still in hand, she indicated the heavy gray tape around a door on her left. "It's been sealed while I was gone, so I may smell like a cedar chest tomorrow."

Because town houses of the area tended to be laid out much alike, he could see that her closet had been created by the half-painted wall which had been built to create two small rooms out of the original living-dining-room.

Philanth's room was virtually filled by the sagging double bed, a tall, ancient chest of drawers, and—the one new item in the shabby room—a five-foot set of flimsy-looking aluminum shelves.

In addition to an assortment of books, the shelves contained kitchen items: a jar of applesauce, a can of tomato juice, a few cans of soup; a stack of empty plastic cottage cheese cartons and a roll of paper towels; two saucepans, a peanut-butter jar of eating utensils, a few plastic plates and glasses. A small bedside table held a sorry-looking lamp, a tin can containing pencils and a pair of scissors, and an alarm clock atop a stack of reading materials.

The silky, dark-blue quilted bedspread had once been beautiful and expensive. Now frayed and mended, it harked back to better times and places now remote.

On the closet door below a calendar was hung a framed xerox of a newspaper clipping. He was surprised to recognize a column he had gotten published in the opinion section of the Los Angeles *Times*. It was several years old, and Philanth must have photocopied it from the original in Carole's scrapbook.

"Why is my essay given such honor?"

"It says everything that needs to be said about why our economic assistance to developing countries should focus on family planning services. And it was an act of courage for a congressman to publish such a piece, when so many people are so irrational as to feel personally threatened by any advocacy of birth control."

"Hah! It only repeats what Tony Beilenson said in a similar article, when *he* was the congressman from Beverly Hills, in 1990."

"Well, Malthus set forth the underlying principles two centuries ago. And well-informed people have been saying these things since the 1960's. The point is, *how many of them were elected officials?* It was brave, even for a congressman from Beverly Hills."

He shrugged and turned to survey the rest of the room more closely. He realized that Philanth by various circumlocutions habitually sought to conceal the fact that her "living quarters" were not an apartment, as everyone would assume. They would never conceive of the reality. He could imagine Virginia taking one look and putting a perfumed lace handkerchief over her nose, like a member of the nobility in a movie depicting life in the seventeenth century.

On the wall to his right the faded, moisture-stained wallpaper was partially concealed by a large two-page spread of a "Popline" newspaper. It carried a set of charts and the headline, "State of World Population Report" for the current year. A front page of the same publication carried a large photograph of a woman from the Bangladesh region standing beside her toddler just inside a huge section of pipeline. Behind her, smaller sections of pipeline were stacked inside the greater section; one held rolled bedding, another held cooking utensils. That was her home.

He looked around Philanth's home and saw that it met her moral standard: it was a great deal more than the woman in the picture had, but it constituted the minimum she required in order to try to help that woman or others like her. She had not planned to live there for more than four months, so she had not expended her resources to do more for it.

There was only one window. And no air conditioner!

"This place must have been *miserable* when you moved in around the end of August! Even if you'd just come back from North Africa a few weeks before, how could you endure it?"

"The same ways people do when they have no choice in such matters. It's much easier, remember, when it's voluntary."

She took her suitcase from his hand and heaved it onto her bed. The briefcase followed.

"It was kind of you to help me. Thank you. I'm sure you're quite anxious to get home. I'll see you tomorrow."

"Just answer one question, Philanth: Why? Why do you live like this?"

"I'm promoting racial integration and paying my rent to a good man who needs the money for his sickly wife and four little children." She rattled it off as if weary of having to explain. "He's a school teacher, and he appreciates having me here because I'm a good tenant: I never cause trouble, and I always do my share of the housecleaning.

"I'm sorry about the smell. It has nothing to do with the tenants' housekeeping duties. I'm sure new paint and wallpaper and carpeting would help a lot. But the landlord is saving so his little girls won't have to go on sleeping in the same room with his little boys when they get older."

She squeezed around him to get out the hall door, opened the door onto the front porch, and looked out. "Eric's waiting." He stepped into the hall, noting all the locks on the outer door.

694

She took his hand between both of hers and shook it. "Thank you for taking me, Mr. Bjorklund. I'm sorry if I was less help and more trouble than Marjorie would've been. It's been an experience I'll always treasure."

"Get some rest, Philanth."

As he spoke the second door in the hall burst open, and two little black boys, maybe nine and ten years old, ran out.

"Philant'," the smaller boy shouted, waving his arms. The older boy turned to yell back into the room from which he had just erupted, "Hey, Star, I *tol* ya I felt somethin! Philant's back!"

He deduced from the boy's gesture that he had felt Philanth's door bump against the half-painted wall as she had edged out past him to look for Eric. The boombox downstairs had covered the sounds of their entry.

A little girl of about seven came out, saw him, and stopped, putting her thumb in her mouth. The little boys also paused, looking up at him as if they had never seen a white man before. It was like being back in the village in The Gambia.

Philanth smiled wearily at him. "Sure." There was a note of sarcasm. But she added, "You get some rest too."

She turned to the children, stooping to their level, and her voice switched to warm endearment. "Lester, Star, Box, how *are* you? I'm so glad to see you!" She began hugging each one as she went on, "Were you good? Did you get my postcards? Where's Emmy?"

The children all started talking at once, and as he turned away he gathered that Philanth's popularity with them was founded upon her ability to help them generate unlimited quantities of fresh, warm oatmeal cookies. They were claiming to have been "good" and therefore entitled to a prodigious batch at the earliest possible moment. The oldest boy magnanimously proposed to let her wait until Saturday afternoon.

Philanth started asking whether they didn't need help with their homework first. She began quizzing the older boy to see whether he had completed his mastery of the multiplication tables. Something in her voice told him that she was happy and at home.

He left.

Chapter 25

A VERY MANAGING WOMAN

He was met at his door by a thin, middle-aged black woman. "Mrs. Bjorklund is upstairs," she told him. Extending her hand, she added, "I'm Barbara Michaels. Welcome home." Having shaken his hand, she went off to the kitchen.

His eyebrows rose in puzzlement. Maybe he had somehow failed to receive a letter mentioning that Carole had engaged a housekeeper.

He found Carole working on their double bed, wearing a long, green-and-orange muumuu she had kept in Los Angeles for many years to wear at summer parties. As he entered the room, setting down his cases, Carole smiled at him and closed a loose-leaf binder, laying it aside. He hurried forward.

But as he approached her his attention was diverted by a set of metal shelves which stood beside the bed. It was of the same ugly nuts-and-bolts construction as the shelves Philanth had in her bedroom; but its legs were black, its shelves were brown, it was only half as tall though it had about the same number of shelves, and it had been assembled so that its shelves were clustered close together at heights where Carole could reach them without abandoning her position on the bed. It was filled with the contents of a busy person's desk.

"What's *that* monstrosity?" he asked, flicking a thumb at it. He moved one end of it aside so he could get at his wife.

"Is that how you greet me after five weeks?" she teased, extending her arms to him.

Heartened by her response, he started to sit down to get into her embrace, but she cried, "Watchit!" and snatched some papers out of the way. He finished sitting down and managed to grasp her upper arms, but since she did not move forward to meet him he awkwardly kissed her cheek.

"Now," she said, "tell me how things went. Did—"

"Where did that *come* from?" Again he indicated the shelving. It would not have been allowed near any of the bedrooms Virginia had been in, that was certain.

"Philanth put it together for me and left it in the basement. I'd happened to mention that I wished I had a little table with lots of shelves to hold my things when I wanted to work on the bed, so she asked me to take her to a Hechinger's. Then she borrowed one of our screwdrivers, and there it is. She was happy to be able to do something for me."

"Is this a permanent addition to the decor?"

"She said any time we want, she'll use its other four posts and reassemble it as utility shelving. It can go into the attic."

"Good, it's ugly as sin, it *should* be hidden."

"It wasn't expensive, though. That girl knows how to get value for a dollar.

"Now, tell me: how's she working out?"

"Pretty well."

"Care to elaborate?"

"Her common sense is in good working order."

Carole gave him what he thought of as her Mona Lisa smile. That smile had been the first thing he had ever noticed about her: the subtle smile of a woman who had added things up, accepted the total, and understood more about it than she would ever say. Despite his misgivings about her intentions toward him, he began to be embarrassed.

"Do you think she might decide to stay with you indefinitely?"

"That's very uncertain."

"Why?"

"She's not sure I can be relied on to meet her terms."

"*What* terms?"

"That—that I act as her mentor. Take her with me into meetings. Give her challenging things to do."

"Oh." Carole seemed amused.

"Carole," he exploded, "how do you *expect* her to react, after—after those two letters you wrote saying I should . . . go to bed with her?"

"You discussed both letters with her?"

He nodded. "Wasn't I supposed to?" His tone became acid. "Wasn't that part of the 'test'?"

"Of course! And now that you've had time to study her, how do you feel about her?"

"She's a good person. She'd make a good Sunday-school teacher for children. I can also see her teaching other nuns in a convent. They'd like and respect her—the nuns *or* the children. She can be all things to all people as well as anyone can, yet be true to herself. A remarkable young lady, just as you said."

"That's good." Carole frowned. "What's the problem about mentoring?"

"She wants certain assurances from you regarding those letters. The second letter made the first one a lot more convincing."

Carole faced him squarely. "My letters to you and Philanth meant what they said. I'd be happy for you to start going to bed with her."

"I want to go to bed with my wife!"

"You won't have a wife much longer, Charles." Seeing him react, she added hastily, "I'm sorry, I don't know how to be less abrupt . . . ."

Quickly he edged toward her, babbling, "No, wait, look, whatever the problem is, surely we can work it out to your satisfaction! I'm prepared to do anything you—"

"No," she tried to cut him off, "I'm afraid not."

"I can't just let you walk away! If it's—"

"You can't do anything about it—"

"But what's the problem? Tell me!"

She hesitated for a moment. "No. I'm afraid first we have to go back and talk about Philanth."

His worst fears were confirmed. Something heavy seemed to topple over inside him. Leaning on his knees, he held his head and swore under his breath.

"How do you feel about her?" Carole asked. "I want the bald truth. I've just told you you're going to lose me anyway, and I want you to take her to bed if you ever want to. Tell the truth. If you ever tell the truth in your life, let it be now. Do you think you could you ever feel like making love to her?"

"I refuse *to discuss* such a question!"

"Charles, you must. Could you consider her an acceptable wife?"

"She won't consider your leaving me. She wouldn't accept it."

"Oh, we're getting nowhere!"

Carole paused, took a deep breath, and then said quietly, "I've been deceiving you, Charles."

He was glad he had been preparing himself for this. He took her hand, for he felt no anger, now, only regret at having failed to keep such an unlimited treasure. It was his own fault. He had been too ambitious, too self-absorbed, too inconsiderate.

And he had not measured up as a man.

"Who is he?"

"I beg your pardon?" Carole seemed affronted by his approaching the matter on such a primitive level.

"I know I've exploited you. Just because I was the man and you were the woman, and you let me do it. I know it was wrong, even though I don't know how to go about things any differently, being in a field where that's part of the rules of competition.

"But I want to know how badly I've been out-classed." His lips could hardly frame the words: "Who is it that's taking you away from me?"

"*No one* could take me away from you, Charles Bjorklund, but the Grim Reaper himself!"

He almost laughed. "Then how have you been deceiving me? What in the world are you involved in? Unethical research? Some CIA operation?"

"No, no. Cancer."

He stared at her, hoping he had misunderstood. "*Cancer?*"

"At first I didn't want to worry you: you already had too many problems. Then I got so involved in making things easy for you that I found I was escaping my own feelings by planning how to provide for you.

"I don't apologize. If things work out as I hope, I believe I'm even entitled to be a little proud of myself.

"But the time has come when you have to know." Her tone became heavier. "I'm going to be leaving you, possibly before very long. If you can love Philanth, if you can accept her, I mean, I can—"

"It's hopeless? There's nothing they can do?"

"They've been doing it. I stopped sleeping with you because I'd had surgery while you were in Puerto Rico. When we made love before you left for that trip I knew it had to be for the last time, because otherwise you would've found out far too soon."

Her voice was not completely steady as she went on, "But it was out of control before they even found it. It's a fast, lethal kind. I'm running out of time. I want to start training Philanth . . . ."

Something in him began gathering, something so vast and awful that he did not want her to see what it did when it crashed over him. Quickly he kissed her cheek as he got up, and as he hurried sideways to the door he was muttering, "Excuse me, Carole, excuse me . . ."

He turned and ran down the two flights of stairs, hurtling past the strange woman in the kitchen, and when he reached the basement he found it transformed into a study and bedroom. Even in his present state, one glance told him it was a first-class job.

It looked ready for occupancy. Carole must have had the materials selected and the contractor ready to start work the day he left Washington.

He threw himself down on the shimmering expanse of velvety dark-blue bedspread and began releasing labored, voiceless sobs.

He knew he was being selfish, leaving Carole upstairs alone as though her death were wholly his own hurt, his own loss, but he couldn't let her see him like this. By degrees he began to cry. He hadn't cried for thirty years, and he wasn't very good at it. It was even harder because he didn't want to be overheard by the woman at the top of the basement stairs. Then he heard the door close quietly, and grief engulfed him, and he didn't care whether she still heard him or not.

As soon as he could put aside his feelings for a while he took off his jacket and tie and cleaned himself up in the beautiful new bathroom—its perfection made him cry a little more—and returned to Carole.

He cleared the bed of her papers and closed the bedroom door against the stranger downstairs and yanked off his shoes and lay down beside her. Carole did not want to change position, so he put his arm across her waist and held onto her.

She stroked his hair, playing with it a little. She had always been fascinated by his hair.

"I know it's a shock for you. But this coming year will be crucial for your career, so you mustn't let what's happening to me interfere with your time-table. Not when we've worked so hard to come this far! I was hoping Philanth might please you. She could step in and—"

He turned his face into the pillow she lay on, gagging on his guilt.

"Please don't take it this way," she begged. "If you give way to it, it'll drain you; and there's *so* much to do!"

"You should have told me—from the beginning, the way we've always shared everything."

"You've had all the burdens you could bear, and it's no reflection on you that that's true. You're so conscientious, Charles."

"Is *everyone* sure it's *hopeless*? How can it be that bad? You've seemed all right."

"I made a special effort to be out of the hospital when you came home today. I was very lucky to be able to manage everything that was going on so I could keep things seeming normal to you right up until you left. Sometimes I really got worried, but you were so preoccupied that my little slips and problems apparently went unnoticed.

"Once you did ask whether I felt all right. Hell, I felt like I was dying *then*! I said I was short of sleep, I'd been working too hard. And you accepted that, just said to take it easier."

He tried to imagine all she had been doing without his guessing.

"Marjorie was a big help," she went on, "without knowing what I was up to. Your out-of-town trips were always set up well in advance, and they were nice and frequent! She got used to giving me the current state of your day-to-day scheduling whenever I needed to make plans. I used the excuse that I was no longer working with you, but I still needed to anticipate your needs regarding packing or whatever, without pestering you. I was glad you can always be counted on to be where you're supposed to be."

"Still, how could I not have noticed *any*thing?"

"You were so wrapped up in your work! Flying in and out and tearing your hair over the mess you'd inherited when that sleazeball was forced to resign .
. . ."

"And you're a very indulgent and trusting husband, while I'm a very managing woman."

"But I'm gone five weeks and all of a sudden you've just gotten out of the hospital!"

"You may vaguely recall that I made several trips to California during the summer, to wind up my organizational work?"

"Yes."

"Well, I never left Washington."

He leaned back to stare at her in astonishment. Then he looked at her closely for the first time since arriving home, trying to decide whether she looked different.

As though reading the movements of his eyes she touched her hair and asked, "Notice anything different about my hair?"

"Uh, ..."

"It's a wig. I got it the last time I was in Los Angeles, because they don't know much about them here.

"You see how easy to deceive you are?

"In all fairness, though, it's mostly a matter of letting me handle all of the mail, the money, and the incoming phone calls; and not being around much except to sleep. I could have had three affairs and gotten an abortion in the past year, and you'd never have had a clue—as long as breakfast was always properly served and the house was neat and your closet and dresser drawers never ran empty."

"I'm sorry, Carole! I've been a lousy husband."

"You're exactly what I ordered. A little too lordly at times, but it's the pressure, and you couldn't do what you do without putting yourself under pressure.

"I wanted to help you by keeping it from you until you had someone you knew was still going to be there for you.

"*Please* tell me whether Philanth might be acceptable! I don't want to force her on you, I have other candidates, I honestly just want to know as soon as possible!"

When he didn't speak she went on, "Getting times when I feel up to reasonably normal work has been an ongoing problem, the treatments have such side effects. I timed my medication around your arrival today, and called to be sure your flight was on time." She looked at the bedside clock. "What kept you? Problems with the luggage?"

"I had Eric take Philanth home first so I could see where she lives. I've gone on following your instructions to 'get to know' her."

"Did you see her bedroom?" He nodded. "Good. Then you can see she's real: she lives her principles."

"How-how much time do they give you, Carole?" His voice was thick.

"Oh, cancer is so unpredictable that nobody can be sure."

"They must have given you *some* prognosis."

"Assuming things continue their present course, my doctors are talking now in terms of months. When they get more exact than that it probably doesn't mean anything anyway, because doctors are so locked into their medical-ethics mindset that they can't stop using their usual way of reckoning life: mere mechanical ability to keep certain organs working, rather than the normal functioning of the mind. I don't care what the pious say—you're a human being only to the degree that you've got a mind.

"I've been doing all I can to record what I know and want to pass along to your next wife; even whipping our favorite recipes into order.

"But I don't owe it to God or anybody else to stay alive when I'm no longer of use. And the financial costs of that sort of spinelessness can be horrendous! I'm trying to keep down our expenses while buying all the quality time I can."

He tightened his arm around her. "I can't let you go."

"Philanth will be able to help you even if you don't want to marry her. She's come to terms with death.

"When you do that it makes you fundamentally different from practically everybody else. There's only one thing harder, coming to terms with the fact that your're likely to die much earlier than you'd hoped."

She caressed his shoulder. "Don't you think she's a lovely person? I realize she's a little young for you, but there are advantages to that, especially her being more malleable."

"Oh, Carole! Please! Don't!"

She stopped caressing and grasped his arm. "Charles, I know these weeks must have been hard for you, but try to imagine what they've been like for *me*. And the worst part was not having any way to find out what was going on between you and her.

"Just tell me yes or no, deliberately ignoring our normal feelings in the matter: could you accept having Philanth as your wife *if I were gone?*"

He made the effort. "Yes."

"Good!" Carole stirred as though she wanted to get up. "I've put some things into another letter for you. It's time for you to read it."

"I don't want it." Even to himself he sounded like a stubborn child refusing to eat a vegetable.

"Charles, *we must keep moving.* Take my word for it, it's the only way. I avoid looking directly at my situation; I focus on what I want to get done."

He took her hand. "How could you be so brave all by yourself?"

"The investment of all of my life was at stake, and I had so much to do!

"Figuring out how to find someone you'd care about was difficult—not because I'm so wonderful, but because you'd be so discriminating, at your age and with your aspirations.

"And there'd be unsuitable women all too willing to solve the problem for you. While your impulse would be to crawl into a hole and pull it in after you. It would have been more than you could handle and keep working as effectively as you must. You *know* how crucial these next two years will be for you!"

"For my bloody marvelous career."

"Yes, for your very promising career. Don't get sentimental on me, you know how important it is.

"I knew darn well you'd take much too long to get around to finding someone else to help you. Loyalty to my memory, and all that."

Loyalty! He had been loyal, all right. He had been complaining to himself like a spoiled child. He had begun making excuses, persuading himself that he might be "entitled" to enjoy Philanth. If Philanth had only yielded to the pressure he had put on her, he would now feel ready to slit his throat. And his wife was praising his loyalty to her! No, she was complaining that it was excessive.

"But we have to be rational," Carole was saying, "because there are too many people affected by what we do. We have to think of all the good people who have sacrificed to help you get this far, especially in California. So we have to face facts and get busy.

"Please get the letter. You've moved the shelving so far I can't reach it. Third shelf down on the right."

He got up and went around the bed to the shelving. The envelope bore his name. It was sealed. It contained two pages neatly typed by their computer's printer. Carole had wanted to get the wording just right without wasting time. A dying farewell written with a computer. How like her that was.

The wintry afternoon was giving way to evening. He switched on the bedside lamp.

He flattened the folds in the papers and settled himself in the middle of the bed to read, his right arm behind Carole's neck and her right hand holding onto his. She gave his hand a little kiss just as she had used to do. It was so good to be close to her again.

> My dearest Charles,
>
> I am writing this because I want to say it exactly and completely and because I want you to have it to read again at other times.
>
> When I first began to contemplate dying within a few years I immediately remembered an old film in which a woman with terminal cancer asks her husband's co-worker to prepare to become his second wife and mother to

their child. I began devising methods for making a thorough search to find someone suitable for a very special man with very special needs in a wife: a woman who could become a First Lady truly worthy of the position.

I settled on an approach which would use my background: recruiting subjects and conducting interviews for a series of articles about unmarried women's ways of defining their goals in life and which kinds of goals structures provide the greatest satisfaction and best all-around adjustment. Knowing how much you need the best help you can get in order to continue the work we have begun, I gave the project my best effort.

I hope you will forgive me for keeping all this from you when you consider how much more bearable our final separation will be for me if I know I will not be leaving you the task of going on with our life's work alone.

Since we cannot continue to be everything to each other we must begin to include someone else. Please remember that what we have together now did not begin when we married. It has developed through all our years together. You can be loyal to our relationship only by forming another loving partnership which carries on our work and gives you happiness, and you can do that only by accepting the necessity of building it from equally slender beginnings.

Easing the transition for you has enabled me to live with what was happening to me. That externalizing of my concern was helped, incidentally, by studying Philanth's journal, where she explains how centering one's values

709

outside one's self can lay the foundation for a satisfying life.

—And for enduring the prospect of death. It is not so dreadful to die even when the best years should still be ahead if you know that what you have done will continue to benefit the lives of others, since you can invest your sense of identity in them.

In itself death is only the ultimate analgesic. In death I will be closer to you than if I lay sleeping beside you because my will must become merged into yours.

But you also must free yourself of pain, the pain of wanting more than has been granted to us. It is natural to forever want more of what has been good. And our life, our sharing, has been so very good! Now we must try to avoid increasing each other's pain by wanting so much for the other that we impair what we can have. We are finite beings, so we must learn to accept limits, including endings, even though our aspirations can be infinite.

I can hear Philanth saying such things now. Philanth would be my choice as someone with whom to confront death. With her passion for self-transcendence she would defy the limits it sets.

She is also my choice as someone with whom you might want to share the rest of your life. I cannot tell you what to feel, but I expect you will discover that I have chosen carefully.

Philanth has taught herself how to live generously even while she manages to get by on the emotional rags she forages for herself. Meanwhile you and I have been dressing each other in silks and velvets. She thinks of love and purpose as the fabric and tailoring of the identity in which she clothes herself, and she is well qualified to inherit my portion of the rich wardrobe of love and purpose which you and I have created. I want her to have it because I believe she will faithfully continue to develop it. To the extent that she does, my life will continue also.

But bear in mind that Philanth deserves to be cherished not as my surrogate, but for herself and what she can become. Your reward for encouraging her to become all that she can with you will be your becoming all that you can with her. It gives me joy to imagine what you and she can become together.

As for me, I have so much for which to be thankful. I could have lived twice as long and not experienced one twentieth as much happiness as I have had with you, my love. Thank you for being what you are and for uniting your life, your mind and heart, with mine. How can I complain that I have not received my share of the goodness of life?

I love you so much.

He put the letter under the bedside lamp. That was where Carole had kept Philanth's letters from North Africa in order to read them at bedtime and share passages with him. He hadn't been interested: one Volunteer among thousands, that had been his feeling at the end of a typically difficult day.

He switched off the lamp and slipped down beside her, taking her in his arms—carefully, for already he was learning how to avoid hurting her. He kissed her.

"Carole, you're wonderful."

"Senator Bayh's wife Marvella showed how it could be done. He wrote a book about her. It's in the collection of books on cancer I found at UCLA."

"Always the careful researcher, that's my wife. Like Philanth: she always wants solid background."

"*Do* you care for her, Charles? *Please* tell me, I've been *so* anxious!"

"I couldn't help it."

"You weren't supposed to be able to help it. If someone goes to as much trouble as I have to choose a gift and get it exactly right, and when the gift is so important to both the giver and the receiver, it would be sickening to find that it had been an unsatisfactory choice."

She added severely, "Anyway, I'd think there was something wrong with you if you *didn't* . . . love her. You'd have some fancy explaining to do."

He said nothing.

She went on gently, "And I hope you can feel toward her the way a mature man can feel toward a lovable woman of a suitable age. I did what I could to make you look at her in that light because otherwise you both would have been so proper that you would never have realized what she can be to you."

He lay perfectly still.

After a few moments she caressed his cheek and murmured, "Bonnie Prince Charlie."

It was her private pet name for him. It bound together intimate moments scattered through all their years, extending back to that moment in their courtship when she had touched his fair hair and spoken it for the first time. It meant all that they had been to each other, all that they had shared: the flirting and the laughter, the passion and the commitment.

All that was going to be taken away from them.

Because she was going to die. She had done so much, endured so much, accepted so much, for him. Yet she was going to die, having received nothing in return.

The world was unbearable.

"My Bonnie Prince Charlie," she repeated, trying to comfort him, urging him to respond in the old ways, from the days when they had been happy. After all, she seemed to be trying to say, Would their love not continue?

No. She would not be touching him, speaking to him, when she was dead. To pretend that that didn't matter was Philanth's type of consolation, and he had called it self-delusion.

In fact this wonderful mind and will soon would be extinguished—just as a candle is blown out, but with no hope of ever relighting it. A world which permitted that was really too terrible to be endured.

He broke down and wept again, turning away from her. At once she began caressing his shoulder and murmuring how much she loved him.

But in her voice he could hear her own tears beginning. "Oh, Charles, it's not the dying, it's the leaving you I mind! Don't make me feel how much! I couldn't bear it!"

To distract both of them he asked, "How could you do this for me? How could you?"

"I can't hang onto you anyway. It's necessary to know when to let go, if you don't want to drag anyone else down with you."

"You're wonderful. Really wonderful."

Suddenly he was bitterly mocking his political persona: "'My wonderful wife Carole'."

"No," she replied, her voice becoming steadier, "only just as beady-eyed ambitious for you as ever. What

713

would be wonderful would be a woman prepared to give the man she loved to another woman when she *wasn't* preparing to let go of life itself. Philanth might be that type, but I'm not."

Her voice forced, she became mockingly ferocious: "I'd scratch out the eyes of any bitch who so much as smiled at you." This was so far from her utterly secure attitude that it had the effect she obviously intended: he grunted with unwilling amusement and wiped his eyes with his fingers.

When he had gotten a tissue and blown his nose he commented, "Ken must have gotten your share of the self pity."

"You may have something there. Growing up with a little brother who was always dramatizing his suffering gave me a hearty aversion to it. Just as my mother's irrationality made me prize rationality."

"Well, however you came to be, you've done something splendid."

"You love her?" Carole needed to hear it again.

He nodded. But the room was quite dark now. "Yes." That still wouldn't be enough for her. He chose his next words carefully: "As much as you could wish."

Easing herself into a more relaxed position, Carole sighed. "What a relief." She actually chuckled.

"You imagine I'm going to be able to simply explain the situation and 'pop the question', and she'll accept me?"

"That would be nice."

"Well, it may not be that simple. She's complicated, and she's got an awful lot of reasons for not wanting to remarry. She's had occasion to consider it thoroughly."

"She likes men, all right, and they like her—if they're not afraid of women with brains."

"She keeps a man interested the way a psychologist might train a rat: if he presses the wrong bar to get the wrong kind of affection he gets the right kind, accompanied by an electric shock of moral edification. That's unpleasant, but not enough to make him quit. And the whole set-up is uncalculated: it's in the structure of her moralistic personality."

"Maybe I shouldn't have done that series of rat articles when we were first married. You were too young and impressionable."

"I'm not young any more. I'm too old for Philanth."

"Nonsense! A busy politician needs an adoring wife at his beck and call, and far better if she accepts that role because she's much younger than because she's stupid."

"Ouch."

"I can't wait for you to propose to her. Let's have her over for dinner tomorrow night and talk about it. Or you can take her out to dinner and court her. Barbara and I will send out for Chinese, and I'll have Barbara rent me a funny movie, to celebrate the fact that my relief pitcher has arrived."

"For pity's sake, Carole!" He paused to avoid attacking her for her gallantry. "My office tomorrow will be a madhouse! Marjorie's got me booked solid with long-overdue meetings, besides which I'm heading into a show-down over staff conflicts that's going to end with blood all over the floor, some of it probably mine. If I get through it I'm going to come straight home and go to bed!

"Oh, and incidentally, I just got back from a killer of a trip."

"Wellll, Saturday night, then," she conceded reluctantly.

"Have a heart, Carole! I can't just . . . . I need time to get my head together. This is all so . . . .

"Besides, I have a huge amount of paperwork to deal with. Philanth and I were half joking about her bringing some staff over here for the weekend."

"Oh, that's ridiculous, Charles. You're a hopeless workaholic. Stop hiding in your everlasting paperwork and listen: you can do all that while I'm starting to tutor Philanth.

"Think of *my* agenda. There are all sorts of places I'd like to *take* Philanth, and it's already too late for that. I'm losing ground, and it drives me crazy to realize that whatever things I won't think to tell her, she may not figure out for herself until years after she needed to know them."

"I need some time. I mustn't be tired when I talk to her."

"I don't see what the big deal is. You're each just what the other needs. I'm proud of the psychological fit.

"Doesn't she love you? —I mean, in a special way?"

"Even if she can still stand me after the past five weeks, you told me yourself how little she lets emotions affect her decisions."

"Yes, but you can offer her so much opportunity for 'good works'."

"Not as a sure thing. And she's really committed to the work to be done overseas. I certainly can't pretend I don't understand why. I've had to do considerable work on the terms of my offer to get her to even *consider* a staff position as an alternative.

"Besides, I am not the answer to every woman's prayer. She does *not* have a high opinion of the life of a politician's wife!"

"Of course not. But you can give her exciting things to do."

"Maybe. Maybe not. But even if I can, she much prefers to organize and run things herself. Besides

716

that, she's reluctant to trust anyone in my sort of position not to let her down or misapply her. Believe me, we've discussed this."

"Now you've got me all the more anxious to get the matter settled. Sunday dinner. That's absolutely my final offer."

When he said nothing she went on, "I must insist it be taken care of before the weekend's over. I'm going to have to go back into the hospital soon—I only hope I can make it through Sunday—and even if Philanth gives us no trouble, other things have to be arranged before I go. *Heaven help us* if this falls through and I have to go back to square one!"

"I don't understand! Why do you have to go into the hospital?"

"They don't usually give the principal pain medication I'm taking except in the hospital. I had to talk Dr. Stein into letting me have it. And it's becoming less adequate. That scares me: I'm afraid the next things they give me may make me drowsy and unable to express myself properly, and from then on it'll be a continuous battle between grogginess and pain. I dread that! And side effects and complications . . . . Ugh!

"I may not be able to come home again until I have to insist on it in order to be able to die in my own way."

Feeling how tense he had become, she paused to pet him, saying, "I'm sorry, sweetheart. I know all this would strain anybody's ability to adjust.

"But I can't tell how much time I've got before I'm unable to think and express myself well enough to be worth talking with. During the last month one can go down quite fast. And judging from what I've seen and heard in the oncology unit and from what I've read, ..."

There was a silence. "Sometimes you just sleep, toward the end. How meaningless and expensive! But there may be such an ordeal of suffering that you keep screaming and begging for death. But nobody will help you. Some nurses in oncology become so case-hardened that they're too monstrously indifferent to patients' suffering to even consult their doctors about pain control. So the road to death has to be a journey through hell. When you've already been used up by many months of suffering, you lose your last shreds of human dignity pretty fast.

"I'm sorry to lay it out for you like this. But when life can be so good, even the idea of leaving it that way offends me deeply. I see no otherworldly justification for suffering that serves no earthly purpose.

"And I'm already tired of enduring, of hoping for a little time that brings the joy of feeling like a normal person. So it may not be long before it'll be time for me to say, 'I'm used up, the price of doing more is more than I should have to pay; I want the relief of dreamless sleep even if it has to last forever.'"

As though sensing his reactions she added, "I'm lucky, in a way: some people are dead a few months after their diagnosis."

To speak was too hazardous. Silence was the best he could manage.

Finally she went on, "I need to be able to talk to Philanth after she's had time to go through the materials I've prepared for her. There's so much experience I'd like to just transplant into her! In the next few years you'll have no time for coaching a political neophyte. And she can learn more quickly from her own observations if she's at least been sensitized by the best summaries of the fruits of my years in California and Washington politics that I can give her."

718

"Relax, it can't be this urgent."

"I'm sorry, Charles," she said, suddenly angry, "but I don't want to waste this precious dose of lucidity arguing with you because you're not ready to accept this! I don't want to be cruel to you, but I've got to talk you through it so we can get on with what needs to be done, as quickly as possible!

"I've talked candidly with one of the nurses. She says she sees patients come in lively and looking normal, and a few weeks later they've become . . . suffering vegetables. The doctor said that once it gets into the lungs it goes quickly. As I said, I've got a fast-spreading form; and the problems aggravate each other as the whole system starts breaking down."

Carole's voice became unsteady. "I'm not . . . too afraid . . . to die, Charles, even at my age, but I'm terrified of doing it badly.

"You think of me as a strong, steady person, but now I've been through such a lot that you don't know about, and my reserves are dangerously low. I have times when . . . I'm not at all proud of myself. I'm talking about fear and anger and desperation and despair. I don't want to spill any of that in front of anybody else. And that's just the matter of attitude, of dignity and self-respect.

"There's the other aspect of not dying on my own terms that I'm much *more* afraid of: the useless suffering of those who've abandoned themselves to others who have no right to end their misery. I mustn't take a chance of losing control so the medical system takes over. It's a monstrous robot—the law can't control it humanely because this society hasn't even worked out *how to deal with* the basic question of what gives life value.

"Besides, that could bankrupt you—at a time in your life when you need money so badly!

"To put it all in simplest terms, I abhor the thought of hanging around making a horrible spectacle of myself. Vomiting my feces, as my grandmother did; wanting nothing but for it to be over. I mustn't go that way! I *mustn't!*"

Her voice quieted. "But dying well also means I mustn't leave unfinished a life's work I believe is important, fearing it won't be finished without me." Carole's voice again tightened as she continued, "And if somehow the, the prospects with Philanth should fall through, it would be . . . so hard for me to pick up the pieces: to try to offer you someone else when I was already . . . failing . . . ." Her agitation was on the edge of giving way to tears, and he saw how afraid she really was that her plan for Philanth would be rejected—afraid just because so much depended on it.

He embraced her carefully. "I'm sorry, love. I don't want to make things harder."

Again she made her voice become steady. "And I won't be able to wait around. I've got to go while I still have the ability to manage it on my own."

"Oh, love, ..."

"You know it's true. California *still* hasn't passed a 'death with dignity' law I can trust myself to. How I wish it had! I could simply fall asleep in our dear little bedroom in Los Angeles, with you holding me in your arms. I'd have nothing to fear. But it would cost too much for us to go to Europe so I could get some Dutch doctor to help me die that way.

"And here euthanasia is such a horribly squishy legal area. Besides, a legal hassle would have the media vultures crawling all over you.

"So I have to orchestrate things carefully: I'm desperate to do all I can to train Philanth for you, but I'll have to give up some quality time in order to be sure I don't try for too much and lose the ability to act effectively when it's most important.

"It follows that you must enlist Philanth *right away*. You can't allow yourself the luxury of mulling it over, given the fact that you do . . . agree that she would make you a very good wife."

He made himself frame the response she was asking for: "I do agree to that."

"May I phone and invite her for Sunday dinner?"

"Have I got a choice?"

"Please turn on the light. And hand me the phone. And my rotary file."

Carole spoke into the phone with her usual businesslike pleasantness, the same tone in which she had ruled the ordered little kingdom of his congressional offices. The exchange was very brief. Carole hung up and handed him the phone.

"She accepts."

He did not comment. He didn't want to think about what Philanth was trying to deduce from that invitation. He could hardly call and give her a blanket reassurance, and he was going to seem crass enough, telling her that his wife was dying so he wanted to marry her, without doing it over the phone. He would talk to her in the morning. In the meantime he would have to let her worry. She was a sensible, strong-minded girl, so she would set even this aside to focus on more immediate matters.

He switched off the light and held Carole for a while, telling her over and over that he loved her.

Footsteps came up the stairs. He jumped up and turned the light back on.

It was the gray-haired woman. Barbara. Carole introduced her as a licensed vocational nurse. She announced that she had thought it was time to come up and "help Carole take care of some things".

Carole replied that she had finished her agenda for the day and was looking forward to her bedtime sedative. So he kissed her and wished her a good night. Barbara told him his dinner was waiting for him.

He took his suitcase and briefcase back downstairs and ate his dinner, then took all his luggage on down to the basement.

The new bedroom/study really was beautiful, now that he took time to look at it. After unpacking—he found the clothes and toiletries he had left behind already neatly arranged in the drawers and closet and bathroom of the new suite—he went to bed in his new "study".

At last he understood why a house which for years had served quite well with one functioning bedroom now needed three.

Chapter 26

A LONG-TERM PROJECT

The next morning he arrived at the office at a quarter to eight. Philanth was already at her desk. Marjorie was already at her desk too, nursing a cup of coffee.

He and Marjorie greeted each other with a hug and expressions of the greatest satisfaction over his return. Philanth only looked at him soberly.

Marjorie waved in Philanth's direction. "Take a look at what Tony sent Philanth before you people left Nairobi. It came as a complete surprise. And he shipped it air mail!"

He stepped over to Philanth's desk, and Philanth opened the well-traveled carton box before her. After removing some crumpled newsprint she pulled out a polished mahogany carving and set it on her desk.

It was an African madonna with child. It was quite lovely, stylized in a fashion at once both African and modernistic.

"Philanth loves it."

He looked at Philanth's unhappy face. "Do you, now."

"It reminds me of the elongated style El Greco gave his figures, to suggest spirituality."

He looked at her skeptically. "Uh-huh."

She threw him a defiant glance. "I love El Greco."

"Mm." Africa was full of wood carvings of all sorts. Tony's choice was interesting. Maybe Philanth had been right about Tony's potential after all.

Well, Philanth's misery over having left a crack in a man's heart by trying to improve his character was getting to be old news; such diversions were inconsequential compared to the future he had in mind for her. He had to harden himself to maintain rational priorities.

"Marjorie, before the war starts around here I need to talk to Philanth. Hold everybody off until I'm through."

Marjorie nodded. "Les is out with flu, and the deputy director has been delayed in getting back from taking care of personal business in Indiana. They both expect to be in on Monday. But you have a packed schedule today anyhow."

"Fine."

He unlocked his door and motioned Philanth into his office, then closed the door behind her.

"You'd better sit down." He hung his overcoat and his neck scarf on his personal coat tree. She sat.

He took his place behind his desk. Seeing her tense expression and stiff position, he spoke gently.

"Philanth, there was one motive for Carole's mischief we didn't think of. She does want us to love each other. She does want us to become intimate." If he poured this out too fast he would only have to repeat it. He paused.

"Because she's going to die soon. She's known since last spring that she had terminal cancer, and . . ." Rather than watch Philanth react, he looked down at his long, pale hands. "She's *very anxious* for you to be my next wife." He paused again.

"I debated how much of that I should tell you this morning. I knew you had a full day ahead too. But I also knew how much this situation has worried you.

"I hope you can, um, keep your head up so the rest of the staff can be kept unaware of my wife's condition for as long as possible. It would be a distraction and distort normal behavior."

She nodded and half-whispered, "I understand."

"Can you do it?"

"I can do it.

"Oh, Mr. Bjorklund, *I'm so sorry!*"

"Thank you, Philanth.

"I—I wouldn't—I don't know how to—I wouldn't be saying this now, but Carole is *terribly* anxious about it. *She* needs to know as soon as possible. I know this is terrible timing, but . . . please, won't you agree to marry me . . . when it becomes appropriate for me to ask you?"

His question seemed to intensify her silent grief.

Realizing he should not be proposing from behind a desk and looking for any way to put pressure on her, he got up and came around to take her hands. Impulsively he dropped onto one knee.

"Just say, 'Yes', Philanth. I know I'm too old for you, but I'm better than nothing, and—and I'm going to need you in this, this role more than you can begin to imagine. I know it's too soon to say 'yes', but just nod, and I can call Carole and tell her, and we can go on about our business . . . ."

She violently shook her head.

"No? *Why?*"

"I've told you all along, Mr. Bjorklund." Her voice cracked. "I'm not good enough for you."

He got to his feet. "Now, that's preposterous."

"And I'm not what you need in a wife.

"Now, please, let's not get embroiled in a personal argument. People are converging from all directions, anxious to talk to you."

She dropped her eyes. "I'll stay with you and help you for as long as you need me."

"Will you help me in all the ways *I* think I need, or all the ways *you* think I need . . . for the next twenty-five or fifty years?"

"I'll help you in all the ways I can without risking compromising you."

"I'm afraid that means the same thing it always has."

"I'm sure it does."

He hesitated, then went back to his position of authority in his big chair behind his big desk.

"This situation is new to you, Philanth, and you haven't considered that without . . . that when . . . I'm left alone, I . . . ."

Before he could find a way to put his fear into words which wouldn't repel her by making him appear to grovel as Ken had, she said, "A man who's lost his legs and never needed crutches would find it hard to take on the monsters which lay siege to the future of humanity. Unless he had a suitable . . . helper."

"*Yes.* Alone and without his legs, he'd be as good as dead. Probably better off. But . . . like a woman who's . . . given herself and then been obliged to tear herself away from the man she feels she belongs to, a man would have to go on living, doing small things with pain rather than great things with pleasure."

She nodded.

He waited.

"It wouldn't be that way for a man as good as you. Because you're not only beautiful inside, you're also beautiful outside. I saw how women at that party in Los Angeles reacted to you. You're so—so perfect and yet so unassuming, they can hardly help . . . showing how you attract them. When it's known you're eligible you'll have to fight them off.

"There must be a great many sensible, altruistic women who'd adore you as I do, for the same reasons. One day, perhaps as soon as you could remarry anyway, you'll find someone appropriate to be your

726

wife—and help you toward a glorious future—as I could not."

"Why not?"

"You need somebody like Virginia, that lady on the plane, but with brains and social conscience. And I don't know how much money Virginia has, but plenty of that too, and the ability to help you get lots more. You'll need everything that an upper-class wife can do for you, to help you get into the White House."

"I'm a has-been congressman, and you're planning for my winning the presidency? Hadn't I better at least make it into the Senate first?"

"Let's not waste time. I will stay with you, I can see that I'm needed . . . until your next wife wants me to go."

Her mouth worked nervously. "But if I ever made love with you I'm afraid you wouldn't want to remarry. And you'd get careless, so I'd have to leave you, to protect your image."

"Oh, *why* are you so stubborn?"

"You saw how Virginia and I went together: like queen and peasant. She's from the circles you move in. The only way I can 'fit in' with that sort of people is by putting on a maid's uniform."

"You can hold your own with ministers of state from another culture, like Thomas Ablebody. And with sharks like Harleigh Adams, and weasels like Kingsley Merriweather."

"I'm talking about acceptance as an equal, not as an oddity. Those who set the standards for social acceptability among the power elite—and those who aspire to it—want nothing to do with anyone who undercuts their claims to their privileges or threatens their self-esteem.

"But my values do exactly that. I test everything by the question, 'Does the poorest person in the

727

world have this, and if he doesn't, why must I, in my effort to help him toward a more fulfilling life?' All my habits are developed from that question."

"That's why you'd be *more comfortable* among people whose poverty forces them to have similar habits. But being where you're most comfortable hardly guarantees that you're where you're most needed."

"That's true. But I can't be of any use when I'm consistently rejected.

"My mere existence impugns every aspect of the affluent life-style; and because of our traditions that value thrift and self-denial and charity, it impugns what people who practice conspicuous consumption are as persons. I'm a walking indictment of conspicuous waste—and glad of it. So even middle-income materialists have little tolerance for me, to the extent that they see me as I really am. Many people will go out of their way to put me down in direct proportion to their aspirations to be like the wealthy.

"You must see that bringing me into your circles as your wife could hurt you. I could never qualify for their acceptance except as your assistant because I would never accept their practices or assumptions regarding values: regarding appropriate dress or home furnishings or styles of entertaining or recreation or anything else involving the uses of time or money or other resources.

"Don't undertake to gratuitously challenge something so fundamental in human nature as the wrath of swine kicked in the snout on the way to the trough. I won't let you do that to yourself; and you would, in marrying me."

"You've told me you're skilled in controlling how you behave because people persecuted you in high school—for the same thing, making them feel inferior. You can be agreeable with people you despise, and you have no compunction about deceiving people you think would misuse the truth."

"Yes," she admitted. "But you're asking for duplicity on such a scale that I would be forced into major compromises, as I was at the Hunger Foundation."

"You can always do what you believe is right in inconspicuous ways until you don't have to be inconspicuous any more. And meanwhile you could still do a great deal of good."

"That's a compromise which has the same effect as total capitulation except that it soothes the conscience. This country must be full of people who go on contributing to our decline and fall while invoking rationales as flimsy as that."

"I can see that this is going to take negotiation.

"We're expecting you at one o'clock on Sunday. In the meanwhile find the time to ask yourself very seriously whether you're not reacting to my offer with irrational self-denial and self-abasement."

"My 'self-denial and self-abasement' are thoroughly grounded in a rational philosophy which governs my existence and keeps me peaceful and happy. So I won't be talked out of them. Surely you can understand that you'd be foolish to try.

"But I'll come to dinner hoping the three of us can constructively discuss what I can do for you and Carole." She rose. "Since Carole's sick, may I bring something? I'll bring the whole meal if you'd like."

"You're going to whip it up in your gourmet kitchen and carry it all the way from Foggy Bottom Metro Station to P Street?"

"No, if I were bringing a load I would take a bus up from the Metro station."

"You've got too many other things to do. Just bring yourself."

"I'm so very sorry about Carole, Mr. Bjorklund!"

"Then don't give us any trouble. Just be a good girl and say you'll marry me."

She shook her head and left.

On Sunday afternoon he and Philanth ate high-quality and expensive carry-out food from trays in the master bedroom while Carole entertained them. Carole had no appetite, but she sipped milk while she described her "adventures" during their long absence.

Carole's account had nothing to do with doctors or medical procedures, however: she used the problems caused by her decreased mobility as a pretext for a rambling discourse which served to acquaint her guest with her arrangements with her cleaning woman, her choice of local stores and services, and her philosophy of meal planning. Philanth showed polite interest, acting a little stiff but giving no overt sign of understanding that Carole was using the time to brief a prospective replacement.

Since Carole did all the talking, he and Philanth soon finished dinner. Philanth took everything downstairs to see that the leftovers were safely put away.

She returned and sat down again. Without preamble she asked, "Carole, did you persuade Dr. Sabagh to get me forced out of Peace Corps because you wanted me here?"

"What are you talking about!" he began, but Carole silenced him with an emphatic gesture.

"You suspected that at the time, didn't you?" Carole's tone seemed to admit it was true.

"I would have accused you openly if I'd had any idea you might have a motive."

"Aren't you furious with me, even though you know, now, what my motive was?"

"Weren't you afraid that I'd figure it out, considering the 'coincidences' and how important it was to me, and be angry?"

730

"What is it that makes you angry? That I abused your trust? That I deliberately exaggerated confidential information and used improper influence?"

"No. What I can't accept is that you damaged my ability to get work overseas. You 'filched from me my good name, and made me poor indeed'."

"We must protect your record," Carole acknowledged. "Leave it to Charles and me: if anyone ever raises the issue, you'll be prepared to knock the question out of the park. You must have an impeccable past when your husband seeks high office."

"But I'm not going to marry him."

"Well, consider it an act of penance on my part, and forgive me, Philanth." Carole sounded sincere.

Philanth made a palms-up gesture. "Suppose we deal with more immediate concerns."

"Let's do. I want to ask for your help."

"Before I started to get mad at you I wanted to ask what I can do for you. You're my dearest friend, and any allocation of my time that's acceptable to you and Mr. Bjorklund is acceptable to me."

"I'm glad you made the offer open-ended." Carole drew a folded sheet of paper from under her pillow and opened it. "As a matter of fact, I've got quite a list of matters I've been looking forward to discussing with you."

"Fine."

"I'm going back into the hospital tomorrow morning. I may not be able to leave until I come home to die. That's just the worst-case scenario. But I want to teach you everything I know, to the extent my remaining time permits. Will you let me do that?"

"I'm willing to help Mr. Bjorklund for as long as he needs me. Your instruction would be most helpful, and I'd be honored to receive it."

"Lovely. Now, in order to spend the necessary time with me you're to phase yourself out of your work at Peace Corps as soon as you can, just as it was expected you would do when you were hired as a temporary. Charles has told me he's authorized you to manage the organization of findings from the trip. But surely you can set that up without building yourself into the middle of it."

"Of course."

The three of them spent several minutes working out plans for Philanth's orderly transition between jobs so that the needs of all the parties involved would be met. Her instructions regarding Les were unaffected.

"Now, as to your housing," Carole resumed, again consulting her list. "Charles and I both will have far more that must be done than can be done, and you'll be trying to help both of us. So all three of us will have to conserve our time carefully. That means you and Charles will need to coordinate your movements between here and the office and the hospital as efficiently as possible. Therefore this floor is to be rented to you and another tenant.

"That means you can whiz back and forth in our car during the bad weather, stop for groceries or other errands, and so on." Carole looked up from her notes. "Doesn't that make sense to you?"

Philanth nodded uncertainly. "But who would be the other tenant?"

"Your fellow tenant will be Barbara Michaels. She's a nurse who works with cancer patients. Her living here should reduce the time I have to be hospitalized."

"It also makes it permissible for you to be living here when Carole's in the hospital, which is the main point," he murmured. Philanth looked embarrassed.

Carole acted as if she had not heard. "You'll be paying all of your rent in the form of miscellaneous household activities and secretarial work that you'd be doing anyway in the course of spending time with me. No cleaning or laundry, that's taken care of, as you know." Philanth knew because of Carole's briefing during dinner. "Just groceries and serving simple meals and generally seeing that the household runs on time so Charles doesn't have to take on any more than he has to think about at present.

"I want you to give your landlord notice immediately, forfeit whatever rent you have to, and move in on Tuesday evening. Any problem there?"

Philanth's eyes widened, but after a moment she shook her head.

"Lovely. Charles will come for you and your boxes. I tried to persuade him to pay Eric to take care of that, but he insists he wants to do it himself."

Carole gestured toward the hall door. "You'll have the small second bedroom and any additional space you need throughout the rest of the house. Barbara will have only this room. She'll move in on Tuesday afternoon. Right now she's camping in the room that'll be yours because she needed to move in last week, on the first of the month.

"Any problem with any of that?"

Philanth's glance at him attempted to address the question to both of them: "Are *you* planning to live in the new suite in the basement?"

"Charles already has his study and bedroom down there. He's quite comfortable."

"But what about you, Carole? With Barbara and me living here it might be feasible for you to be cared for in your own home most of the time. A hospital is no place for anybody who can manage to be somewhere else."

733

"The next time I come home the dining table and some of the chairs will go to the basement. During the periods I'm here I'll be behind that oriental screen we have shutting off the dining-room area. A bedridden patient is easier to deal with on the ground floor, and everyone else's arrangements will be kept as normal as possible whether I'm here or in the hospital or in a hospice.

"Our costs will also be kept down by the four of us being here under the same roof. Barbara has other private patients and flexible hours, so she needn't be underfoot.

"It's all a matter of consultation and coordination. I've upgraded our phone service and had an intercom installed, and my parents are giving Charles a car phone for Christmas.

"Any other questions?"

"After I'm completely phased out of my work for Mr. Bjorklund at Peace Corps, will I be working full-time for you?" Philanth clearly realized the delicacy of the question.

Carole smiled wryly. "When I'm no longer able to train you we'd like for you to move directly from your apprenticeship with me into helping put together Charles's senate campaign in California. You'll be well-briefed by the time you leave Washington, and Charles's manager in California will give you the supervision you need in order to do statewide organizational work. There's no reason to be intimidated by that: you'll be fitted into a framework of experienced campaign professionals and activists.

"You'll learn a great deal that will make you invaluable to Charles when he becomes a senator. Would you like that?"

"It would let me work for him while staying out of his way until he's happily ensconced in the Senate."

Like Carole, Philanth did not seem to think it worthwhile to consider any alternative scenario in which he lost the election or decided not to run because he couldn't raise enough money.

"And by then you envision him finding somebody better qualified than you to become his wife." Carole conveyed her contempt for this idea.

"Yes."

"So what would you gain by having learned so much that a California senator's staff person needs to know, if you then went back to Africa?"

"I'd have helped someone over a difficult patch on his way to a splendid career of public service."

"You wouldn't want to stay, if you'd worked into a position that made full and effective use of all your abilities? —And even applied your interest in policy issues?"

"His wife wouldn't like me, so I'd have to go anyhow. And I wouldn't want to stay in Washington working for somebody else."

"Why not? Why waste all that useful knowledge when you could get a position with someone in California's big congressional delegation?"

"It would be better to start fresh in Africa."

"Because you're in love with Charles?"

Philanth glanced at him. If she said, "No," her acceding to their plans for her next two years made so little sense that her honesty would be open to question; if she said, "Yes," she would undermine their ability to work together on her terms.

"It would be better for everyone," she managed. "I have to go where I'm most needed."

Carole accepted in silence this failure of her line of questioning. "Well, at least we're agreed on the program for the next two years: *everything* is to be directed toward getting Charles into the Senate.

735

"Now, as for the immediate future. I'll need you to hostess the formal dinner here tomorrow evening. Marjorie will have the information on the guests for you tomorrow morning. The cleaning woman and the caterer have their instructions. All you'll have to do is show up dressed for a black-tie affair and assist Charles. Any problem with that?"

"N-no, ..."

"Great. I know you'll enjoy it."

There was another pause.

"Carole," he ventured, "I think you're being a little rough on Philanth. You're laying everything on her pretty fast."

Carole looked at Philanth. "Am I going too fast?"

"No. I admire you, Carole. I like to think that I would have thought things out as well as you have."

"Well, thank you. I *expected* you'd feel that way.

"Is there anything on your mind, Philanth, besides the fact that I'm prepared to dunk you in boiling oil until you agree to become the second Mrs. Charles Bjorklund?"

Philanth sat up straighter. "We can talk about that if you insist . . . ."

"It would enhance your ability to master all the material I want you to learn if you expected to have a long-term use for it. In my present circumstances I'm very reluctant to take on a poorly motivated student, especially when her reasons for not accepting the larger agenda are ridiculous."

"I'm sorry. But I believe the reasons I've given Mr. Bjorklund for not marrying him should be allowed to stand so we can move on to deal with other immediate concerns."

"Such as whether you'll make love with him until he finds a rich socialite with all of your personal, intellectual and psychological attributes?"

"Uh, I don't think we need discuss that."

"Philanth, I have no time to be delicate even about delicate matters."

"Well, I couldn't possibly!" Philanth's pliant tentativeness was gone. "It would—could—create a bond between us which would cause problems for him later."

"Suppose your *not* sleeping with him became a problem?"

There was a long silence. At last Philanth answered in a low voice, "If desire becomes a problem for two people who cannot or must not marry then they must regain their equilibrium apart from each other."

"Under the present circumstances that would be insane. Do you want to help Charles or don't you?"

"We disagree about the most important thing I can do for him: for me, it's facilitating his marrying someone more suitable."

"He needs you, Philanth."

"If I died he could still make a successful bid for the Senate. He just got into a habit of feeling otherwise during the special conditions he was experiencing during the trip.

"Don't force this, Carole, or it'll blow up in your face. Because I *will not* marry him or impair his willingness to marry someone else by sleeping with him."

Before Carole could respond he interjected, "I don't see how you can attack that position, Carole, except to the extent that you depict me as so flawed I hardly seem worth helping. She's agreed to your immediate plans, and we have to leave it at that."

"Well, I really had hoped for more!" Carole was pettish. "It's such a nuisance when major decisions are left hanging. And you're both very affectionate people."

"That doesn't mean we must love each other erotically," Philanth reproved Carole. "Please don't embarrass us further."

"My point was that your refusal to marry him envisions an unnatural situation. It's so unrealistic in this and other respects that it seems motivated by a compulsive self-denial."

"It may look that way to you, but my position is thoroughly supported by facts and logic."

"Philanth, do you maintain that if you found yourself trapped by facts and logic into admitting that marrying Charles is the right thing for you to do, you'd agree to accept a whole life you've consciously conditioned yourself not to want?"

"Ye-es, ... But the facts—"

"You're on!" Carole sounded triumphant.

Philanth frowned inquiringly.

"You're in trouble, Philanth! Because 'the facts' can never be completely determined, and logic is only a tool for organizing the facts you've got in hand. So each of us starts from something more basic. Even your system that's freestanding in the midst of the void of nihilism is founded on what you call an 'arbitrary choice'. But it's not a *random* choice—far from it! You admit it's made on the basis of psychological needs."

"So I do: the need to love, and—"

"Yet you've just agreed to let facts and logic override your fundamental compulsions."

"But they can't!

"The 'compulsion' on which my system and my resulting refusal to marry Charles is based isn't an unreasoning fear I can set aside if I identify it and find reason to: it's a basic commitment I need as the basis for my existence. So it *isn't* vulnerable to attack by facts and logic.

"It's simply the need to love, and the need to believe in something as worth loving and living for. Facts and logic can only point up worse and better ways of doing that under given circumstances. My circumstances in relation to Mr. Bjorklund have been thoroughly considered."

"But you've come to an irrational conclusion."

"No. We just weigh the facts differently: we place different emphasis on various aspects of complex realities involving the need for political money, possible alternative marriage partners, and the nature of Mr. Bjorklund's need for me. My evaluating these realities differently from you doesn't prove that I'm operating on the basis of any irrational compulsion."

"You are, I'm convinced of it."

"Accusing someone of unconscious motivation behind the way he weighs his facts is hitting below the belt. You can accuse anyone who doesn't see things your way of having a 'hang-up', and there's no way he can rebut you."

"Just the same, we've agreed that people make their basic decisions for psychological reasons, then call in logic and maybe ideology to justify them. Surely it's obvious you're doing that here, because Charles and I are offering everything you could possibly want, even including life in the city that hurts you because you love it so much and you thought you had no right to stay here."

Philanth was adamant: "Not proceeding from conventional self-interest or desire for personal gratification to reach a conclusion is not inherently irrational. I don't care if all the psychologists and psychiatrists in the world say otherwise. If they do they must've been combatting the psychological perversions of Christianity so long that they can't see that there's validity in altruism. The ethnologists—"

"Oh, dear, all right," Carole interrupted, chuckling. "Leave that point aside—please don't tell us just now how other species do things.

"You refuse because you say you don't have enough to offer Charles. That's nonsense."

"It's not nonsense! People buy elections in this country, because the incumbents aren't about to enact new campaign procedures that would let people challenge them just by fighting out the issues with free time on radio and television! And the voters are too lazy to force the change, so they hide behind self-serving cynicism. Since that order of things is so entrenched, and since Mr. Bjorklund's going to get married again anyway, he's got to marry a rich wife! Once he's in the Senate he'll be very sought-after, and there *can be* nice women who are rich, especially since it's usually men who make the money."

Carole shrugged and sighed as though giving up all at once. "Well, at least we can start tutoring sessions." She took a deep breath and helped herself to a drink of water. "I'll call you about that as soon as I get into the hospital on a new regimen that lets me feel better than I do right now. But I'm *very disappointed.*

"Thank you for coming. I don't like to cut this short, but I'm beginning to feel rather bad, and I'd like to be alone."

"Surely there's something I can say that'll make you feel better."

"But you won't say it. So run along."

Philanth rose and gave her friend a token embrace. "I love you, Carole. I'll do everything I can for Charles."

"Then go home and get your head straight."

Philanth took up her purse and disconsolately walked toward the door.

Getting up, he told Carole, "I'll see her out." Carole flicked her hand in acknowledgment.

When they were downstairs he turned and grasped Philanth's shoulders. "Didn't you see what your refusal did to her?" he asked under his breath.

Philanth nodded, hunching as though prepared for another onslaught.

"Then why can't you pretend to go along? She knows you're not going to be 'brought 'round' by my 'friendly persuasion' over time. She knows that if I press you, you'll leave me. And she knows you're a wonderful choice for me. You can't let her die knowing she's failed in this!"

When Philanth only hung her head he went on, "Her bringing you to me without my knowing what was going on has been a long-term project that's made it possible for her to accept dying. She's used it to cope alone with the fact that in spite of everything she's got to live for, she may not see her fortieth birthday.

"Can't you even begin to imagine how important this is to her? It's her life's work that's on the line. Because all the work she's done has gone into promoting me, and she wants you to act for her to ensure that it won't be wasted. You're her solution to the biggest personal problem most people ever have to face, their own mortality. Her life can continue through you even while you're fulfilling yourself.

"She knows how much I need her. She knows what she and I have had together. She knows what my losing her is going to mean to me. I'm losing half of *myself* at a time when I need her most. The entire remainder of my life may be determined by how effectively I can function during the next two years. The senatorial election in California a year from next November is the sort of opportunity that comes once

741

in a generation, and I'm already under serious pressure to start knocking myself out raising money for it."

He gave her a little shake. "Philanth, please at least pretend to change your mind! At least let Carole die in peace! I'm begging you!"

"How could you and I fool Carole on something so close to the heart as this?" Philanth whispered. "And suppose she pretended to believe us? How much more alone and wretched would she be then?"

He let go of her with a sigh. "I'll run you home."

"I'd rather do it the hard way. I need the fresh air and exercise. Thank you anyway. And thank you for the delicious dinner." She hesitated, gave him her standard peck on the cheek, and turned to get her coat and scarf from the closet.

Leaving her to let herself out, he returned to Carole.

As he walked in, Carole growled, "I'm going to die before I can take that medication again—and then it won't do the job. I think I'll go to Britain, where they're not such fools as to guard against addiction in people who are terminal. Or maybe you could go out and get heroin from a pusher on the corner."

He sat down and took her hand. "Shall I call Dr. Stein?"

"I called him while you were out buying the dinner. He says I can have Barbara give me a shot at bedtime. Until then I'm to continue my 'present course of treatment.' Damned ghouls."

"What can I do for you, Carole? I'll do anything."

"There *is* something you can do. Give me the rest of today. I have an assignment for you.

"Go up into the attic.

"Be prepared for a mess: I had them put up there all the stuff that was stored in the basement.

"Find a big box labeled 'Christmas' in *purple* marking pen. Remove the strapping tape. One side will fall open, giving you access to a two-drawer filing cabinet in its factory carton. Open the top drawer. The names are alphabetical. Find the manila folders in the hanging folder labeled 'Devon, Philanth'. Bring them down to me. All of them."

Carole relaxed, closing her eyes.

He went out into the hall and pulled down on the cord which began the lowering of the stairs from the attic. He clambered up them and rummaged around until he unearthed the box she had described. He opened it and found a brown filing cabinet.

The top drawer was labeled, "Finalists". The lower drawer was labeled, "Others".

He opened the first drawer. It was packed with folders. They all bore names. *Women's* names.

He opened the second drawer. It too was packed with folders bearing women's names.

He sat down on a box labeled "Law Books" and buried his face in his hands. He sat there for several minutes, rubbing his face.

Carole was waiting. She was going to go on feeling worse and worse until her next pain medication. He checked his watch. About forty-five minutes longer to wait. Then she had to make it to bedtime.

He found the hanging folder bearing Philanth's name and took out the three manila folders. A thin one was labeled, "Notes". A thicker one was labeled, "Letters". A very thick one was labeled, "Journal".

He shut the filing cabinet and took the folders downstairs and pushed the stairway back up into the ceiling.

He sat down beside Carole's bed. "Read the letters first," she instructed. "Then the journal."

"What am I reading them for?"

743

"The key to Philanth. A way to turn her around. It's in there, it has to be. Maybe it's a lot of little pieces, and we'll have to fit them together."

He examined the journal. "Lot of reading here. And it's not even typed copy. I thought I'd have time for my 'In' box this afternoon."

"Which is more important right now: laying the foundation for your future, or your everlasting routine paperwork? I've seen some of those memos you bring home to read, and they're ridiculous. Your staff structure really must be a mess."

"What's in the folder you're keeping?"

"My interview notes, though she never suspected she was being 'interviewed'. Also my notes on behavior and other observations. I'm going to review and condense them for you . . . as soon as I take my next pills. When you're finished reading, come back to me."

Getting up, he replied, "All right, sweetheart."

"Wait. Sit down." He complied.

"Charles, Philanth doesn't know I made a copy of her journal. Don't ever let her suspect that this existed, much less that I let you see it. Promise me solemnly."

"I swear this never existed."

"I'd rather you didn't mention that I gave you all her letters, either."

"Is all this material that personal?"

Carole nodded.

"I hate to abuse her trust."

"I've already decided that the circumstances justify it. Only bear in mind that if you ever reveal your familiarity with something that's in it, she mustn't suspect that this was your source. That could damage your relationship with her."

Carole hesitated. "D'you think I'm unscrupulous?"

"You have the courage to make hard choices in other people's behalf, and for that no one will ever thank you." He kissed her hand. "Except me."

"Thank me when we've got her.

"Go down and inaugurate your new study by finding the ammunition you're going to need. At least work on getting your bearings inside her mind so you can spot weaknesses in her defenses. And learn where any land mines are buried."

Carole was launching them on a massive fishing expedition. That was hardly encouraging.

He got up again. "Anything *you're* going to need, before I go?"

"Yeah, come to think of it: the damn' bedpan."

During this business Carole tried to distract both of them by talking. "I'm excited about her, Charles, because her idealism is going to be tempered, not adulterated, by the tough nastiness in this town. And with all her other qualities and abilities she could become the most effective politician this country has seen since Franklin Roosevelt."

When the bedpan was rinsed and put away he picked up the two folders. "Well, if there's nothing else, ..."

She shook her head and blew him a kiss.

"Call me if you need something," he said, indicating her new intercom box.

"Go, Charles."

He went down to the "study", stretched out on the double bed, and switched on the lamp.

He began skimming through the letters, expecting to get through them fast and go on to the more valuable material.

At first they seemed to be merely typical letters from a Peace Corps Volunteer to her chum. Gradually, though, he was caught up in them. He saw that

745

Philanth had fallen in love with a part of the world which he had always regarded with aversion and dismay. As in their travels together, Philanth had looked at dirt and destitution and alien ways of life and seen the marvels of the human spirit.

As the novelty of her environment faded and she felt closer to Carole the letters became more personal. In one letter, after going on about how much her Peace Corps work meant to her, she commented,

> It's obvious why I plan to ask for an extension. I'd like to stay here for the rest of my life! I've got the rest of my Peace Corps service in which to work out my next step as well as prove myself worth hiring by some other agency, and I could become one of those people who evade the five-year rule by moving in and out of Peace Corps over many years.
>
> You asked about my health. Dr. Sabagh is unhappy about my weight loss and wants me to rest more, but my main problem is not letting the problem interfere with my efficiency.
>
> I keep in mind that my efficacy would have been greater if Larry and I could have made a loving home for scores of unhappy children over the course of our lives, as we had planned to do when we married. That realization helps me keep pushing on, trying to do as much as I can alone.
>
> I apologize if this letter seems maudlin. Perhaps I shouldn't send it, but you'll understand when I say I don't have the energy to write another. I know I'm not quite "myself".
>
> But it's so wonderful to have a real person to "talk" to rather than just a journal. You're very kind to let me write to you. Don't feel

you have to reply.
Love,
Philanth

Dear Carole,

I was bewildered by your latest letter, the one in which you tell me I must "come home". In the first place, my health is not that bad. I've just had a series of problems because my resistance was lowered by inadequate rest and the antibiotics had troublesome side effects.

While I concede that sick people may be the worst judges of their own needs because illness can distort their thinking, I do not agree that my illness is prompting me to take serious risks. Dr. Sabagh's care of me seems quite adequate.

Furthermore, if worse came to worst and I were forced to stop work until I had fully recuperated I could do it here. I understand that my problems may have been started by the local bacteria, but I hope to adapt to the local conditions in time. If on the other hand I were returned to the States I might risk being reassigned after I was pronounced well enough to return to work, and you know how happy I am here. As I've said, I feel that this is where I belong; this is my home now.

Please accept my promise that I'll try to take better care of myself. I'll even accede to your insistence that I ask for a week off—or a few days at least. . . .

Dear Carole,

I feel as though I'd walked into a trap. I started fulfilling my promise to you that I'd take better care of myself by suggesting to Dr. Sabagh that the work schedules be shifted to give me a few days in bed. He reacted as if this meant I must be in serious condition. He ordered more tests, and he also kept asking questions—but as if he didn't want to believe my answers. He kept talking about the importance of not keeping "a stiff upper lip".

Then he said he was going to have me medevacked. Naturally I started to protest. Next he told me he would want a thorough psychiatric evaluation before he would agree to my returning to duty. I told him not-too-politely what I thought of that idea, and he said, "Well, then I'm going to recommend that you be terminated." And he wouldn't discuss it any more!

I guess he expected me to back down at once and agree to try to get "a clean bill of health" from a psychiatrist. But a psychiatrist would ask me all about my past, and then he'd probably decide I needed a few years of "analysis" to "work all these things out".

I don't know what to do! I seem to have no choice but to appeal, if Sabagh tries to get me terminated. This is crazy, and I understand enough about bureaucracy to be scared: "Kafkaesque" became a cliché for a reason.

You'll say I should be glad I'm going to be forced to take time to get well before my health is permanently impaired. But I still think I would have adapted, and in the meantime I'd rather have gone on feeling physically lousy and had something worth getting out of bed for in the morning. How can I get physically well when I'm this unhappy?

I'm being sent to George Washington University Medical Center, and I'd be very, very grateful if you could come and see me there—just once, 'cause I know you must be very busy with important things. I'd like to feel there's one person who cares whether I live or die. It'll help me get back on my feet and come out fighting for the right to go back to my post.

Don't worry, I am going to come back here. I'm going to appeal and fight this all the way. I have no choice: my whole future is at stake.

But I'm more discouraged about my life right now than you can imagine. There's only one thing I'm sure of. I got it years ago, from an editorial in an old *Psychology Today*. When there is nothing else left to keep us going, there is duty. Think of the caterpillar: beauty, pleasure, reason, hope, even the future are nothing to him. Perhaps he may become a butterfly, perhaps not; he neither knows nor cares. All he knows is his duty. Right now my only duty is to go "home" and get well.

If you don't have time to see me while I'm in Washington I'll let you know where I go next, as I treasure your friendship.

Love,
Philanth

Dear Carole,

I was surprised and honored by the swiftness of your response to my last letter. It's very kind of you to be so anxious to keep me from doing something which would have messed up my record even worse.

Your urging that I accept a medical separation is persuasive. Your analogy of how the

F.B.I. has seriously and repeatedly damaged innocent people's careers by uncritical record-keeping from their investigations was a compelling illustration of how little control I would have over my official record if I required an appeals board to venture into the swamp of "proofs" of "mental health". They could indeed be expected to take the safe course, the defense of their colleagues' judgments. The fact that it's my friend Dr. Sabagh who suggests I might be neurotic *is* pretty devastating to my case.

You're probably right also in suggesting that my illness puts me in a poor position from which to defend myself. As you described my position it does seem hopeless, given the way human nature operates within a bureaucracy.

So rather than completely lose control over my official record, I'll take your advice and follow the "safe" course. But it really goes down hard! I never would have believed I'd even consider letting myself be forced out of Peace Corps by such a thoroughly ridiculous, apparently trumped-up allegation. However, you're right: when one is up against a bureaucracy trying to protect itself and the issue involves allegations about one's mental health, one has no attractive options. I can only try to minimize my losses.

Thank you for your help at this crucial time.

There were no more letters in the folder.

He turned to the journal. He had to look twice to make sure he was in the correct folder because the opening page had a date followed by the salutation, "Dear Larry". But the date was from the previous

January, and the letter proved to be Philanth's effort to work herself out of a depression by addressing her dead husband about her obligation to live for both of them.

He began flipping through the journal. Some of the entries were letters to Larry. They were tender and reminiscent and grateful. She often wrote to Larry when she needed to cheer herself up.

Other entries were simply Philanth talking to herself, reasoning things out to be sure she was on the right track. She could go on for pages, developing a point of moral philosophy in psychological terms. She tested herself for faults and figured out ways to remember to "do better". At one point she went on for two pages about the importance of wanting nothing. She speculated about the roots of her fantasies, analyzed her dreams when she could remember them, left no part of her consciousness unexamined when analysis could help her perfect her motives or her conduct. Perfection, to her, included the ability to live a life of constant giving without ever being loved.

Carole had been right: there were bits and pieces of Philanth's psychology here, and together they could form a pattern, the mosaic of Philanth's mind. If he could only see the whole pattern or at least the details of that self-abnegation which was at the center of the pattern of her motives he might find a route through which to get at her so that she could not push him away.

To freshen his brain he went upstairs, got a glass of cold juice, and stuck his head out of the front door for some fresh air.

The day had grown darker, and a cold wind carried the scent of rain. The street of old brick town houses and brick sidewalks was without any sign of life. He heard a distant hum of traffic, but everything he saw looked dead.

He returned to his study.

The next entry began,

> I've been thinking about my disastrous relationship with Rod, and I believe now that I failed in what one of Father's books on ethics called "the duty to think". I was so much in need of someone to love and Rod was so unlovable and I was so isolated there on the World Hunger Foundation estate that I lost perspective in my desire to improve him and thereby perfect our relationship. In my solitude I felt such passion for him that I would have wanted to stay with him until he killed me if I hadn't felt the obligation to carry on the lives of those who had done so much for me—and the life of Larry, who had been given so little and then had even that little snatched away from him.
>
> I had much to learn. However, an unhappy love is like breaking a bone: after a while one seems to be going on as before, but something has been permanently changed, and it can be a strengthening experience.
>
> Therefore I am stronger and richer for having loved Rod, even though he brought me to my knees as nothing else ever had, even—to my shame—the needless deaths of those three I honored above all others.

"Philanth," he whispered, touching the page as if it were a window and she were on the other side of it, writing and musing and unable to see him watching.

But this was only a photocopy, and the Philanth who had touched paper with those words had lived nearly a year before, in a dusty town on the edge of the Sahara.

In the next entry she discussed the relationships between sexuality and religion, analyzing one of her favorite passages from T.S. Eliot: "Waking alone/At the hour when we are trembling with tenderness/Lips that would kiss/Form prayers to broken stone." As he read he felt ready to weep with frustrated desire.

Again and again she hauled herself out of despondency over the burden of her vast and hopeless longings by clinging to the belief that in spite of her total insignificance, in spite of "the degradations of wearing a woman's body", she was needed. Over and over again she talked herself out of despair over her powerlessness in a world crying with need, inspiring herself to persevere. Her hope of service, of becoming part of something larger and better than her individual self, was the central value of her existence and the floating log to which she clung in the midst of the flood.

> This day, this week will pass. Pain and the humiliations of the body and their poisons in the mind cannot last forever. What can endure is the affirmation: "Faith, hope, and love endure, these three. But the greatest of these is love." Duty is love in action. "To work is to pray"; and to pray is to be in God, who is love.

In a fantasy-meditation Philanth moved out beyond the farthest planets of the solar system, moved at last to the edge of the universe, moved beyond time and matter.

> You are in an infinite void: no color, no sound, no movement—only total emptiness

and complete darkness.

But in that void is a speck of light, a spark so tiny that it is visible only to your own mind. It *is* your own mind: your consciousness in the midst of nothingness.

You stand at ease on the vast emptiness. You look into the depths of nothingness. What are you looking for?

You have come to look for God. You wish a word with him, only a moment of his eternity. Whether or not he is, you have left time and space and matter behind and come into the depths of darkness to seek him.

Speak to him. Speak to him, for perhaps speaking to him will bring him into existence out of the deaf eternal silence.

Oh God, whether or not you are, I am. I will, I know, I choose. In this moment of peace I am utterly and completely only myself. And what I am now I shall continue to be amidst the buzzing confusion of the world, the only world I can know.

If you are, give me strength, only the strength to live one more day, one more hour, struggling to create the meaning of my life among the stars, not among the beetles in the dung.

If you are not, still I shall will, know, and choose, just the same. And I shall have this strength as often as I come to ask for it, because I ask it of whatever I can find to ask, I ask it in the depths of myself and find it where I ask, in the asking. For the question of divinity is to be answered only in the self-awareness of sentience, and *I* will, *I* choose. And in that depth where I can still choose, there is silence, peace, and a silent voice forever silently calling, calling: "God, give me strength."

So the mind turning in upon itself finds its own strength, its own "courage to be". You stoop to take the peace of the silence of nothingness itself, and as hemp is twisted into rope you condense it into strength. And as you begin to return with it to the world of things, you know that however many times you may be close to breaking, no matter how many forces keep pressing in on you until you're hanging onto your purpose and love of life by your fingernails, you will ask again and find again the will, the choice to get up and go on. And on.

This is the last strength we have, the determination to go on trying to make something good out of our shabby little lives when misery and despair unite to clothe death in the beauty of peace, rest, and oblivion.

As he read he began to feel that he rested beside a flicker of light at the center of that vast darkness. He was alone at the bottom of the silent house, at the end of a cold and dying afternoon, nearing the end of a miserable year. Only one light burned, the lamp above his bed. The dark, quiet chamber became Philanth's mind, the single burning light was her consciousness, and he was unknown to her resting beside that center of light, eavesdropping as her mind murmured to itself in its profound aloneness.

The dark, silent room was not only the inside of Philanth's mind. It was also a subterranean vault, a tomb, the inside of his own mind, and Carole was lying dead in his mind, and he was left alone, a single speck of light, his own solitary consciousness. The whole world was a tomb thick with death, darkness, and silence, every other reality having gone away to another universe where people still moved in the bright, tinsel dreams of hope and the evergreen illusions of happiness. He was left alone in the tomb of his solitary skull with Philanth's past mind shining through white pages to his present mind in the small circle of light. If that fragile bulb were smashed, all would be darkness and void.

But the words moved upon the face of the darkness, for Philanth kept saying, Let there be light; kept saying, There will be light; kept saying, There is light,

> for *I have willed it so*. I *choose* it to be so! And *therefore it is so*. While I live, while one human intelligence exists anywhere in the cosmos, GOD IS. Not merely expanding the universe and maintaining every atom in its orbit, but also God knowing and God loving and God thinking.

Carole could not have imagined what an ordeal this journal would become for him. She could not understand how or why he loved Philanth as he did—oh, she had glimpsed common traits in their natures and must have seen the possibility of emotional symbiosis, but she could not know how fully the anguish with which Philanth loved the world, her frustration at her impotence and her longing to do more than was possible, paralleled his own.

Philanth's anguish became his all the more because as he watched her suffering he felt he could reach out and end it if she would let him, and thereby ease his own. —As two halves of an arch reach out and keep each other from collapsing into rubble. For having the reassurance that they shared longings which the self-centered would fear and thus work to despise, they would feel that those impulses had a dignity, a right to exist. The comfort and strength which came with that reassurance would be more precious than any hope.

But he saw also that the roots of Philanth's rejection of him reached into a depth which even this journal did not fully illuminate. Moreover, he saw as Carole had how Philanth kept working at the net of principles she had woven to hold her life above the abyss of meaninglessness so that its cords grew stronger and firmer; and with the passage of time she grew more secure and comfortable in its confines. As he finished reading he had a new comprehension of the vast dimensions of his and Carole's failure to "change her mind".

He also felt sure that he was not equal to the task he saw before him. He had arrived home close to nervous exhaustion, only to receive a stunning blow which left him numb.

The trip seemed to have taken even more out of him than he had realized at the time. Sometimes when he had gone into a new, highly charged situation as the center of attention he had had a momentary feeling of unreality and exaltation, a fleeting thought along the general line of, "I can't actually be doing this. It's more than I had dreamed I was capable of. It's wonderful." He had been too preoccupied with the emotional and intellectual demands of each moment for that thought to register except as it set new standards of professional and personal excellence by which he would measure his performances in the future. He should have expected that this let-down would follow.

The time had come when he must sink back into himself and rest. But he could not, he must not.

Add to the mental labors and physical stresses of the trip the abrasion of worrying about Carole's machinations while wrestling with impulses which in the weary final weeks had kept him on the brink of disgracing himself with Philanth, and it was no wonder if he felt ready, after reading this journal which was a series of journeys into Hell and struggles out again, to consign himself to an indefinite sojourn in oblivion.

But upstairs his dying wife was waiting to try to help him plan a new strategy by which he could thrive and prosper—before her painkiller faded out again.

Sometimes it was too painful to move on to the next moment of life and too painful not to. But at those times the only retreat from pain was sideways into such hopelessness as made continued existence problematical. Oblivion beckoned.

At the fundamental level of his being the accelerating erosion of all his aspiration (*acedia*, spiritual aridity, Philanth's journal called it) had been in the making during all the years he had been struggling toward realization of the dream which was threatened now. For only the most soul-destroying perseverance, the most determined initiative and overwhelming conviction, could keep anyone pounding away against self-aggrandizing cynicism, apathy, and business-as-usual ritualism long enough and hard enough to make any impression whatsoever. Therefore without commitment to some formula, some hunger of the ego, or some powerful inspiration outside the self, an ordinary individual's political effort could soon exhaust itself in puny assaults against the vast world's intransigence.

He had no formula for the world's salvation. Not even a sophisticated elaboration of the neo-Malthusian principle had all the answers, though Philanth had been delighted to find him roundly endorsing it in his speech at Town Hall the day she had first approached Carole. Nor did he have that hunger for power itself which he had seen in some political people. The inspiration he had once had, altruistic commitment to substantive measures for addressing the world's most obvious needs, had been used up and not replenished.

It was so used up that right now he felt emotionally unfit to cope with even the most minor traffic accident; unfit to do anything above the level of moronic physical routine: walk, eat, dress himself, make automatic gestures of civility. And those no longer had any point.

Once he had hoped that a reasonably gifted man like himself could build a decent career in public service much as he could build one in the private sector: rising through proven ability, hard work, and the art of pleasing.

That was nonsense, he knew now. He was left trapped in his commitments, buried under the rubble of his aspirations and no longer with any adequate incentive to dig himself out. For Carole, the founder and sustainer of the vaulting aspirations by which he had come to measure himself, must die.

For a long time there had been nothing for him to live for on the personal level—now that his youthful illusions were gone—except his relationship with Carole. As he had tried to tell Philanth, that one relationship had been his compensation for everything. That source of comfort and strength was going to end within a matter of months. When Carole was gone there would be nothing left to live for.

Nothing at all.

Except to make love to Philanth.

Even now that thought could rouse him to desire continued life. The animal in him had fixated upon what he must have in order to survive, to grow anew the hopes which had been worn away; to find a new source of joy in a world subtly but pervasively altered by the knowledge of her presence in it: her nature, her consciousness, her nearness; but most of all, by his awareness of the fact of her existence and by the moral and philosophical implications of that fact.

He moved his fingertips over the plain manila folder which contained the copy of her journal. That journal had served as her confessor, her therapist, her lover. It had been an anvil upon which she had forged a more perfect self.

Then she had moved on. If someone showed it to her now she probably would say, "Burn it." It was not purified like romantic literature, like Carole's letter of instructions to her, but befouled with the unlovely processes of all organic life: a mother-of-pearl-lined shell formed, outgrown, and cast off by some innocent, striving sea creature, with sand and traces of dead connective tissue still clinging to it.

Without egoism or ideology, defenseless yet secure in the completeness of her commitment to others, she would continue to move on. With him or without him. He must not be left behind.

He got up from the bed, took up the folders and switched off the reading light, and went upstairs.

As he entered Carole's room he asked, "How're you feeling now?"

"Able to do what we can. Have you gotten any leads from your reading?"

He shook his head. "Some new insights, but nothing I can see how to work with. This missile has lots of payload, but it has a pretty tamper-proof guidance system." He hefted the two thick folders as he laid them on the bed, adding, "And this is not what I'd call a succinct instruction manual."

"Well, I've had more time to digest all that than you have, and I've gone over my notes. I think I have some things that may be useful."

"Let's have them."

Carole took a dictaphone from her bedside shelves. She replaced the cassette which was in it and switched it on. Then she turned her attention to the top page of the lined tablet on her clipboard.

"Philanth thinks of people as fragile. Why not, when all of those who were important to her have died? That may be one reason she's afraid of committing herself to any one person."

"That makes sense, but what can anyone *do* about something like that?"

"Make clear that she's committing herself not to you, but to political objectives you both believe in."

"Right. Romantic appeals are out, anyway."

"Here's something else—several things, in fact.

"Today I appealed to her practicality as well as her interest in doing things for you, and that was successful as far as it went.

"But I also invoked her personal desires the way you would with most people, and in dealing with her that's stupid.

"I also failed when I tried to challenge her basic way of defining her objectives for her life. That consists of self-denial and undertaking jobs nobody else wants, so it looks like we have to accept these as givens."

"She's used those principles to turn aside conventional offers. But now . . . I think I see how they might be used against her."

"But be careful, Charles. She can't help wanting you and everything you have to offer; and the more she wants something, the stronger her impulse to counteract the want."

"Wonderful."

"So if you go at this too straightforwardly she'll find excuses to shoot each point down as it comes through the pass single-file. She's—to switch metaphors—a bird poised for flight, so you'll have to throw a woven net of arguments over her."

"That sounds difficult."

"I admit it does to me too.

"Be sure to include whatever arguments you can find to show that you really need *her*. Also appeal to her sense of duty and her practical concern about how to serve *effectively*."

"Yes, yes," he said wearily. "I'd better take a nap before I try to figure out how to do *any* of this, much less how to do all of it at once."

Carole switched off the recorder and sorted her papers for a minute, then sighed wearily. "It's better if you skim these yourself, so you can be as sensitized as possible to the way she thinks. You're going to have to prime yourself as though you were preparing for a debate or going into court to argue a case.

"Because I think you'd better be ready to tackle her about this after the dinner here tomorrow night."

"Oh, *no!*"

"I'm afraid so. Because the longer you leave it unresolved, the more risk there is that she'll find what she considers an acceptable excuse for leaving you. She'll soon start re-examining the way our plans fit against her priorities, and she'll want to start working on some way to back out of her highly imprudent commitment to give you her next two years without building toward her own future ability to accomplish things overseas. It's dangerous to feel that time is now on our side, since she will use it to seek routes back to her own goals; she won't be worn down."

"But I thought you were moving her in here so I could gradually get her to relent."

"Actually it's to bring the matter to a head because of the risks of letting it drag on indefinitely. But it may backfire so that she finds the escape hatch she'll be looking for.

"You see, by bringing Philanth into this household we're laying a heavy emotional burden on her—and not just because she loves us. She also would feel she'd died and gone to heaven if she could live our sort of life, as one of those who truly belong in Washington."

"She never said anything like that to me!"

"Probably because your recruitment effort put her on the defensive. Sex and power are not drives one advertises when trying to keep a tight rein on them.

"Philanth won't take long to realize that the whole living arrangement I proposed today is a trap. However, she has confidence in her self-control, so she won't back out. Instead she'll begin teaching herself how to endure the powerful emotions of this situation.

"The time will come when I ask you to help me get out of the hospital for the last time. I already have the sleeping pills and champagne I plan to use. But you must know I don't want you making it harder for me by making it harder on yourself than it has to be.

"Add Philanth to that situation, with her empathy and compassion, and she's going to be in your arms whether you want her there or not. *What happens then?*

"Since it'll be emotional for you too, she can just wait until the pressure builds up to some sort of explosion, then manipulate and interpret it to serve her own need to extricate herself from the obligation posed by your need for her."

Carole avoided meeting his eyes, and she wasn't saying whose emotions would precipitate the blow-up.

"Considering what she's willing to do for you and that she's capable of anything, she might well succeed.

"But assume she finds no way to leave you without hurting you or otherwise violating her principles. Assume she then goes off to California and you start working on your campaign there. You'll be feeling very alone, and you'll be seeing her under conditions of considerable stress. Those stresses too might build to a blow-up that could push her away . . . unless she's already promised to marry you.

"I realize that's only a possibility, but I'm anxious to minimize the risks."

Carole was being tactful, but she was making a cogent point. People involved in the prolonged stress of a campaign got a little crazy. If in a fit of nervous exhaustion he became insistent Philanth would opt to gracefully let him have his way rather than let things get ugly. Then, having surrendered her "honor", she would want to slip out of sight like a used paper napkin going into the trash.

Besides, what she was likely to see and hear among his female supporters during the campaign would reinforce her feeling that he could find someone more "suitable" from among the many who would be willing.

Moreover, she would leave him quickly if it seemed possible to her that his indiscretions might provide his opponent with material for smear tactics in the final days before the election. As a recently bereaved widower with the millstone of "sex appeal" around his neck he could be especially vulnerable to the suggestion that he was already exploiting his charisma as a prospective major political figure.

He didn't feel like saying anything, so he nodded.

"Besides," Carole went on, "even without the risks posed by delay, it'll save a lot of time we can't spare if you insist on settling this tomorrow night."

"You're right. I have to get this thing nailed down as quickly as possible. I've got battles looming on too many other fronts."

"And nailed down thoroughly: you've got to deal with her from all angles, knock her completely out of the game. Because as long as she can find one weakness in your position, you've lost. When all her logical arguments are shot down she'll look for psychological ones. For example, she's so determined to make *herself* strong enough to go it alone that she might try to make *you* prove your manhood by letting her go."

"Yes," he said, remembering his attempted renunciation of her after the party in Sherman Oaks, "I might even be fool enough to let her get away with that, if it weren't for your involvement in this."

"She's a hard sell because she's afraid to hope to be happy in her own right," Carole summarized, pulling her notes together in a pile.

"Is it fear? Can't her selflessness arise from a sense of abundance and an aversion to pettiness, as she believes it does?"

"But why does she choose to interpret her life experience this way rather than some other way? That's always the psychological question underlying a person's ideology."

"Even if that can be known, does it matter?"

"It might have practical value if there's an element of fear reinforcing the self-denial. For example, maybe she feels she can't take any more personal hurt or loss or disappointment—though that suggests a deeper wound deeper than I'm aware of."

"Maybe not a wound, just a vulnerability."

"True. But implying it's intrinsic to someone's nature begs the question.

"Anyhow, I would like to know why she's *so* unworldly as to feel compelled to reject the opportunities we offer her. If we knew that, we might be able to get her to deal with it, because she's a 'coper', not an 'avoider'."

"Before the trip you mentioned her low level of aspiration because of the obstacles to her as a woman."

"Yes. But her scars and frustrations from finding herself with a brilliant mind trapped in a very womanly body in this society don't adequately account for what looks like self-hatred under all that determination to love. She's too sensible in other respects to be letting sexism make her renounce all hope for herself. That attitude seems out-of-date."

"It needn't be any particular things that have wounded her, just general experience of life," he said a bit impatiently. Psychologists always seemed to be trying to oversimplify cause and effect.

"All right," Carole said, giving up. "The fact remains that she's almost completely written herself off; and that poses a problem for us. How do we deal with it?

"For one thing, continue to be explicit about this issue of excessive self-denial, because she's aware that her vulnerability to the temptations of self-sacrifice threatens what she does value.

"Incidentally, that's why the strength of her emotions scares her: because under certain circumstances

766

it could threaten her judgment and her will to live, even her convictions about the duty to live.

"That's why she wants a quiet, solitary life; gradualism, not drama; intellectual work, not emotion. That's all she can hope to be able to manage, alone; and she doesn't understand what stability a close, healthy, loving relationship can give."

"And if she did she'd feel sure that it should go to somebody who needs it more."

"Right. So naturally what she must feel toward you frightens her."

"Oh, it does." He could still see Philanth's distress over dancing with him.

"So by all means, keep her on intellectual grounds, where she feels secure.

"She has scant regard for emotion as a reason for doing anything anyway, so you've really got no choice. You've *got* to defeat her *on her own ground*: logic, logic, logic."

"Understood. Fortunately I've had time to get used to that."

"And get her committed quickly; get complete capitulation that she can't find a way out of," Carole reiterated, closing the third folder on all of her notes. She presented it to him.

"Otherwise she'll think about all the people she can help with her own hands and all the other 'reasons' you should marry somebody else, and she'll harden her resistance. And sooner or later we'll lose her, because she's the one in a million who will always put duty 'to the world' above everything else."

"And choose the harder path."

"Yes. If a course of action requires giving up everything anyone could want, all the more reason it must be the right choice, since the path to greatness must always be hard.

"When someone consistently measures herself against the handful of men who've shaped the world, you mustn't expect her to do the ordinary."

"I've seen that clearly, in her journal. She *will* go on rejecting me, will leave me and not shed a tear. It's only a question of . . . how soon I drive her away."

Carole's face registered his hesitation, but she let the statement pass without comment.

She lay looking at him, her hands folded. "She's my insurance policy on all I've every worked for and cared about. And I don't want to see you looking at me with all your love in your face and without her to turn to, when I must tell you good-by. I don't see how I could bear to die that way."

Chapter 27

THE MARRIAGE COUNSELOR

He was relaxing on his new bed in his new study, going through Carole's notes on Philanth and developing his own, when the phone rang. Carole's brother Ken greeted him and asked to speak to Carole.

"She's gone to bed, with a sedative. She won't be available until tomorrow. Can I take a message?"

"Oh." There was a long silence.

"Anything I can do, Ken?" he prompted.

"Well, I'm in town for a couple of days, and I was, uh, trying to contact a friend of Carole's. Philanth Devon. I was hoping Carole might happen to know where Philanth is, this weekend."

"Philanth was here for dinner, early this afternoon. It's after ten o'clock now, and tomorrow's a working day, so I assume she's at home."

"I've been trying that number for two solid days. Several times the other roomer who shares the phone picked up, but Philanth must not've gotten the messages I left, 'cause she hasn't returned my calls."

"It's strange she's not answering now. Isn't the phone right outside her door?"

"Yeah, and she usually answers it." There was a pause.

"Is it possible," he asked tactfully, "that she doesn't want to talk to you?"

There was no response.

"Hello, Ken? Are you still there?"

"Yeah. Well, I came to Washington to see her. I guess I have no choice but to go over there unannounced."

"You'll have to knock on the outer door, and the boombox downstairs is probably going full blast, so nobody'll hear you unless you rouse the whole neighborhood."

"But her window opens onto the front porch. I can knock on that."

"The lady who's staying with us came in earlier, saying the rain was turning to ice and it's no night to be out. Why not just wait until tomorrow and call her at the office?"

"I've got to *see* her, Chuck. I should have gone over there and camped on the porch when I came in from the airport *yesterday morning*." A note of desperation crept into Ken's voice. "I've got to be giving exams in California starting tomorrow evening, so I got a non-refundable plane ticket, and . . . ."

"You came three thousand miles to see Philanth, without checking with her first?"

"I thought I'd surprise her, her first weekend back after her long trip."

"That was taking a bit of a chance, wasn't it?"

"It looks that way.

"Well, thanks, Chuck. Give my love to Carole. Sorry I missed seeing you people this trip, but I've been wasting my time sitting by the phone. I'll be in touch."

"No, wait. Uh, Ken, would you please come over? I think we should have a talk."

"I really can't spare the time right now, Chuck. You understand. I'm at a hotel in Arlington, and I've got to get a cab on a rainy night and go all the way across the District to Philanth's place, and I know she has to be at work early—"

"I believe you dated Philanth during September."

"Yes."

"And you asked her to marry you, but she refused."

"Well, now, that was a misunderstanding. I'm surprised she told you about that."

"She was very upset by your . . . uh, reaction to her refusal, when she'd specifically told you—"

"Well, I'm here to clear all that up. We had this little problem just before I had to go back to San Francisco for my fall quarter, and I haven't been able to patch things up by phoning or exchanging letters. I've just *got* to talk to her in person, tonight if at all possible!"

"Ken, you've got to come over here and talk to me, first."

"Why?"

"Philanth's going to marry somebody else."

"On my way." Ken hung up.

Ken arrived about forty-five minutes later. He shed his rain-spattered trench coat onto the knob of the bannister, shivering and exclaiming, "*Rotten* climate you've got here. Thought the winter had to be better than the summer, but now I'm not so sure." Ken bent his head to brush off his shaggy brown hair as he added, "It sure must cut down on the competition for jobs: no *intelligent* person would—"

"Sit down, Ken. I believe Scotch is your drink."

"Uh, no, thanks. I'm not fool enough to go see Philanth with whiskey on my breath. Besides, I've got the cab waiting."

"Tell him to go on his way. Trust me."

Ken glared at him but put his coat back on and went outside for half a minute.

He returned and again dripped and shed and brushed off his hair.

As he did so he was saying, "Now, what's all this about Philanth? Who's she planning to marry?"

"Me," he lied, regretting Ken's choice of words but too preoccupied to rephrase the question.

Ken swung around to face him, half laughing but ready to hit him. "You're already married! To *my sister*!"

"Ken, . . . please sit down."

Ken slowly walked over to the fireplace and sat down, wiping rain off his face.

He drew the matching chair closer to Ken's and sat down. Laying his hand on Ken's arm, he began, "Ken, get ready for some bad news. Are you . . . braced?"

"You've just told me Philanth's involved with a married man. Poor Philanth, of all the girls for it to happen to. What a mess. But I suppose that long trip with you—"

"Ken, your sister is suffering from a deadly cancer. She probably has only a few months left."

Ken registered the appropriate dismay. "That's horrible."

"Yes. Please promise you won't tell your parents. Let me break it to them when I'm in California a few weeks from now."

"I promise."

"And don't tell this to anyone either, but Carole has chosen Philanth to be my next wife."

"What! Carole *chose* Philanth . . . ."

"Yes. She'd made an exhaustive search."

"Oh, good Lord. 'A search'."

Visibly shaken, Ken got up and took a walk around the little sitting room. "All summer Carole gave me phone numbers of 'nice, eligible women'. Claimed she knew a lot of them because of some magazine article she was researching. Then she made me report on them—in detail— so she'd 'know which one to recommend next'."

Ken turned on him. "Carole was *using* me to *screen prospective wives* for *you*!" Ken added a string of heartfelt expletives, gesticulating.

"Well, that must be how you met Philanth," he pointed out.

"Yes," Ken conceded.

"However, you can see that your chasing on over to Philanth's rooming house tonight is not a good idea."

"Are you in love with Philanth?"

"I can recognize a no-win question when I hear one! It's been an awkward situation—I don't deny it."

"I'll just bet it has." There was a pause.

"Carole sure wasn't relying on my report when she picked Philanth for you! I said, 'Your crazy little friend should be locked up in a convent, for the protection of all men.' 'If you're going to play matchmaker,' I said, 'take her out of your file.' Carole promised she would. Probably actually enjoyed keeping a straight face while she did. *Damn!*"

"Please, try to take it easy. I know all this is a shock, ..."

"'Take it easy', the man says! How can I? Philanth is everything a man could ask for, unless he wants a mindless toy. She was born to serve; she wants only to make people happy. She'd . . . she'll make a wonderful wife."

"I noticed that too."

Ken did not seem to register his irony, only stared into space with his fist against his jaw.

Ken had to let off more steam before he could accept defeat and go away defused. "She really got to you, in the course of a month of Friday and Saturday evenings," he observed. "I don't see how you could become seriously involved in such a small amount of time." He made his tone half question, half persuasion to explain.

"Our relationship got off to a fast start because I practically forced my way into her 'apartment' at the end of our first date. We'd hit it off, and I wanted to get to know her more quickly by seeing how she'd fixed up her place.

"Have you seen it? The room where she sleeps, I mean?" He nodded. "Well, of course I was floored. So she sat down and gave me a whole list of reasons why she lived that way. By the time I left that night I was ready to ask her to marry me."

"I understand."

"But I went on finding more reasons after that. The very next night she was in my kitchen, unpacking the groceries we'd bought and starting dinner like she had never wanted to be anywhere else. She's so giving, even in the subtlest ways, it lends grace to everything she does.

"You notice she *never* says anything that takes even the tiniest piece out of another person's ego? She has to make an effort to say anything unpleasant at all."

"We could sit here all night if you just want to tell me why you like her," he complained. "I have a particularly long and heavy day tomorrow."

"Right.

"But I told Carole that Philanth was the nearest thing to a female Alyosha Karamazov I ever hoped to meet. Scholars have suggested that if Dostoevski had lived longer he would've gone on to show Alyosha destroyed because of his purity. No one can survive in this world and yet remain that holy.

"I wanted Philanth to have, well, a chance to remain as . . . as she is. An 'angel in the house': that was a Victorian ideal. I could've dedicated my life to that."

"I can understand."

"Can you? But *you* won't try to protect her. *You'll* throw her into the trenches like she was Carole!"

774

Ken's voice became drenched with disgust: "All that kissing up to people who spit in your face."

"She wouldn't want to be protected any more than is necessary in order to function. I expect I can comfort her when the realities are hard to bear."

"I expect you *can*. Carole gets a lot of fun out of how women groove to your 'manner'.

"But has Philanth agreed to marry you? Have you had the indecency to ask her already?" Again Ken was ready to attack him, either way.

"Carole was very anxious that it be settled." He prayed Ken wouldn't pursue this.

He added, "Carole's been preparing to start training her as quickly as possible, now that we're back and aware of her wishes."

"Yeah, she's young enough to be 'trainable'!

"You realize you're old enough to be her father? You're obviously exploiting a young woman's respect for an older man's accomplishments."

"I've discussed this very point with Philanth. I intend to see that she benefits from whatever opportunities for personal development I can give her."

"Well, I hope you can also straighten her out sexually." Ken sounded bitter.

"I have no idea what you're talking about."

"Considering your situation, I guess you wouldn't have gotten into that area. All I can say is, Good luck."

"Ken, you're implying something's wrong with her. If this isn't just sour grapes, if you really care about Philanth's happiness, surely you have an obligation to tell me what you think you know about her that I don't."

"How much time've you got?"

"For this, as much as it takes."

Ken sat rubbing his mouth.

775

"Philanth feels a great deal of remorse over having made you unhappy," he reminded Ken. "She believes she was selfish because her enjoyment of your company made her ignore the nature of your interest in her. She grieves whenever she thinks of you.

"You know how much of what happened between the two of you was really her fault. Is there anything you can tell me that might help me to make things easier for her?"

"I don't know if she's capable of normal sex," Ken began. "When I started getting passionate she reacted as if she thought I was going to rape her. She—she begged me not to make her relive her mother's rape."

"Her mother's?"

"Yeah, that threw me, too—which is what I think she intended.

"She was lying there on my bed ready to cry, not resisting at all."

"Fighting for herself isn't in her nature."

"I would imagine not.

"But she was like a little girl who had absolutely no reaction patterns for that sort of situation *at all*. I couldn't believe she was that inexperienced. She'd been married, for heaven's sake! But she didn't seem to know what to do except cringe away from me and try to hang onto her clothes. It was like I was trying to debauch my little Lisa. Only a monster could make love to a frightened child. It was awful."

"She has an intuitive sense of what behavior will have the most edifying effect on people while getting them to do what she wants."

"You don't mean that performance was calculated! I'm tellin' you, Chuck, she was scared witless!"

Philanth would never show that much fear except for the benefit of an audience. Ken would not want

776

to know about Philanth's ability to psych herself into crying like an actress, so he only raised his eyebrows, and Ken went on.

"Ever know a woman who could drive you crazy to have her at the same time she was makin' you hate yourself for wanting her? It's a gorgeous sensation. The only thing you can do is throw her clothes at her and stalk off."

"Did she *ever* respond normally to your overtures, in less threatening situations?"

"Well, she kept trying to keep things casual, that's why I became insistent. I wouldn't have done that if I hadn't been sure she really cared about me."

"Ken, I hate to tell you this, but her normal behavior tends to be what you'd expect from a woman who's in love with you."

"But we 'clicked' from the beginning!"

"She 'clicks' with a remarkable variety of people. It's a point of honor with her to get people to like her because that's a way of making them feel good. And she's got so much warmth and empathy that it often makes for 'chemistry' where men are involved."

"But . . . really, it's not just my ego, it's hard to believe that's all there was to it." He could hear genuine hurt, now.

"That's not all there was to it," he soothed. "She wouldn't consider marriage because she had broader commitments. She's going to marry me *because* of her broader commitments: as you said, I'm going to 'throw her into the trenches'."

"It never would've occurred to me that she might want the sort of life you would offer her! She's not at all the type."

"Precisely the reason she should be involved. And she's willing to put up with the externalities because of her dedication to certain objectives."

"Well, I really must've been on the wrong track. Maybe if I'd seen how she was with other people instead of always being alone with me I wouldn't have . . . gotten off into my own reality."

"And when it didn't fit with hers you 'threw her clothes at her and stalked off'?"

"I went back into the living room. When she'd gotten herself together she joined me. She told me that if I'd call her a cab she'd tell me something she'd never told anyone else. Then I'd understand why she'd gotten so upset. I called the cab. She started talking, and she held my attention until it honked outside.

"I realize now that she wanted to distract me until it came so I would let her walk out without any discussion of what had just happened. She knew I had to vacate my apartment and leave town the next day, and she didn't want to see me again. But of course she wouldn't want to say that.

"Instead she began by telling me about her mother's rape. It was one of those horrible cases you occasionally read about: the autopsy showed that her mother had been sexually abused and tortured and mutilated until she died. It apparently had taken several days."

He tried to imagine Philanth's reaction to that at the time. She had been no more than seventeen and without sexual experience.

Finally he managed, "Do you know whether they found the man?"

"For all the good it did. He had a long record of sex offenses, but he's intelligent—and mentally ill more or less when he wants to be. You can't keep that type locked up for very long.

"He'd been drawn to Philanth's mother because of her public lecture at the university: 'Sadomasochism in Contemporary Women's Fiction'. This guy practically had a library on the subject of sadism toward women—but not the sort her mother had worked with.

"Her mother had just received an award for a book-length scholarly study of the subject: all about popular forms of entertainment that appeal to women's self-hatred arising from generations of cultural subjugation—like the self-hatred among blacks. She placed it within the dual context of the movement for sexual equality and sexploitation in advertising. She also took a look at contemporary films and television, to show parallels. From what Philanth told me, her mother must have been a brave, socially concerned woman as well as a hard-working and disciplined scholar.

"But both of her parents died slowly and—in her words—'fully aware of their degradation in having bodies.'"

"How did her father die? I just knew he had a stroke."

"He had the first one when his wife's mutilated body was found. He lingered for about five years, unable to do much except fret or cry the way a baby does. Knowing how he would have hated an institution, Philanth took care of him herself at home.

"Besides, there was the problem of money. She was only a sheltered teenager, so she had to find some things out the hard way. She started using up the equity in the house through a reverse mortgage, but she had to try to ensure that her father would never be forced out of his home, and with one thing and another, her finances were a juggling act.

"Then she met a scholarship student not much older than she was, and he started helping her take care of her father—first in exchange for meals, and then—as a matter of practical necessity, since they both were working and going to school at all hours—for room and board.

"But gossip started in the neighborhood. She was aiming toward a teaching position in that same little college town in order to keep her father in his own home, and she was afraid the ugly talk might keep her from getting it.

"She's got a phobia about 'what people will say', even now. She'd expected everyone to assume her innocence, under the circumstances, so those nasty-minded busybodies made a lasting impression.

"They both were having trouble making ends meet, so to continue to pool their resources and yet kill the gossip the boy offered to marry her. She persuaded herself that she loved him. She still believes it.

"After her husband was killed she took in other female students as roomers, and eventually she finished teacher training and was able to get her finances in order."

"So she did marry for economic reasons."

"Yes. But it wasn't worth it, even though Philanth should have been able to make it work if anybody could. Unfortunately she and Larry had a misunderstanding on their wedding night, and from then on her darling Larry abused her.

"—Until he was killed in a highway accident nine months later. It was lucky *he* didn't linger, because another burden like that would've . . . ."

"Philanth's husband abused her?"

"Yes. All this begins to sound like *The Perils of Pauline*, doesn't it? But sometimes one disaster makes everything fall apart, and this all stemmed from her mother's death."

"Tell me about her . . . her problem with Larry. If you can."

"He was from a coal-mining family, and suddenly he found himself married to the daughter of two professors who were highly regarded in the community. He felt out of his league. He saw Philanth as 'just a sweet, old-fashioned girl', so of course he expected a timid virgin.

"But Philanth was well-read and affectionate and grateful to him for marrying her, and at the wrong moment she came on too strong for him. He jumped to the conclusion that she'd played him for a sucker.

"He accused her of having deceived him about her experience with men. She found that very funny and denied it.

"She still appreciates the irony. Boys had denigrated her *as a female* because she was smarter than they were. Since the culture's laced with misogynistic put-downs, they found weapons ready to hand. As a result she'll never be able to trust a man enough to have sex with him unless she's married to him: she's afraid he'll turn her own accessibility against her. So she's never been made love to.

"Yet her new husband became suspicious of the passion she was ready to offer him on their wedding night. She blames herself because she was too 'brazen'."

"Jesus." He was remembering how Philanth always wanted to revert to early childhood in his arms; and how such a minor incident as Harleigh's treating her like a "floozie" had wounded her.

"He started slapping her around," Ken went on, "insisting she confess she'd tricked him into marrying her. She became hysterical; and of course she was not about to confess to what she hadn't done.

"He wasn't mollified when he found her a virgin. He said she could've had herself repaired, considering the money her parents had had until her father's medical expenses wiped it out. His class hatred came out: it was as though being a daughter of privilege, she had to pay for the sufferings of *his* people.

"He treated her like dirt when they had sex, and much of the rest of the time too.

"Philanth persisted in trying to prove her virtue by being submissive. It wasn't rape, she told me, because she wouldn't let it be: she kept it from being rape by doing her best to cooperate and by saying over and over, 'I love you.'

"That only gained her more insults. Even her everyday gentleness made him mad. He insisted on misinterpreting everything she did.

"She was sure that eventually he would relent and then things would be all right. So the more he abused her, the more submissive she became. She also cried a lot.

"But apparently he couldn't stop punishing her for disappointing his idealism about what a 'pure' young girl she had been."

"That sounds like her interpretation at the time."

"Well, you can see why she thought so: after all, she was only nineteen, and she was . . . well, you know what she's like even now, a meek little heroine out of one of her favorite stories. It's not remarkable that Larry didn't expect an eager partner on his wedding night."

"Just the same, it sounds like she had touched off something perverse in his nature. She later worked for a man who sexually tormented her, and eventually he almost raped her. She has a kind of innocence some people would want to destroy to preserve their own version of reality—not to mention their self-esteem."

"Well, they may find they have to destroy *her* to destroy her innocence."

"Yes. It's conscious; informed; based on choice."

"So she accepted the role of whipping-boy for whatever was really eating him. She insists she wasn't being masochistic, because it wasn't seriously hurting her—Larry's cruelty was psychological, he wasn't a batterer—and she felt he needed to work things out of his system before he could recognize how wrong his behavior was and renounce it.

"But as it went on she began to realize that he was going too far to be able to turn back. She once asked whether he wanted a divorce, and he said no. She had no time or money for legal hassles even if she'd wanted to fight free of him. She couldn't leave her father, so she was trapped.

"Then Larry went out making liquor deliveries on a snowy Christmas Eve and met a drunk who didn't know—or maybe didn't care—which lane of the highway he was supposed to be in."

Ken paused. "I guess that's all of it."

"So she's *never* been made love to?"

Ken shook his head, closing his eyes.

"She must've done an extraordinary job of walling off her marital experiences, to have the positive attitudes she does."

"No. She's just kept that personal experience from affecting the glorification of sex within marriage she'd built up as a student of various religions.

"The question is whether her horrible experiences or her lovely ideals will come into play when she's put into a sexual situation—especially if she's under stress."

"I take your point."

"You could get lucky," Ken conceded, but his tone made it clear that he did not consider this likely. "She has a beautiful nature."

"Yes. She does. Her parents were special people."

783

"I sure hope you know what you're doing when you take her into a life of politics." Ken's scathing tone said he was *certain that* wasn't so.

His brother-in-law finally had gone too far. His reaction to this crack was partly a release of tension from what he had just heard, but he also had been waiting for years for a chance to get some things off his chest. Now he had justification for seizing the opportunity when it was offered.

"Ken," he began, "I understand your prejudice against people who've given the pursuit of power top priority in their lives. They're likely to be self-centered, exploitative, and various other definitely unattractive traits—and hypocritical besides, if they seek power at the ballot box.

"I don't claim to be free of any of those traits. But please bear in mind that at least I didn't go into politics on the strength of my own ego. Carole convinced me that I could aspire to a seat in Congress because I'd been getting good press for my work on a *cause célèbre*.

"And the work she did to clear that path for me wasn't demeaning to anybody. Either it was done in another person's behalf, or it was organizational effort for worthy causes.

"So maybe I have a little more honesty and altruism and independence than you've had occasion to notice, in our brief and superficial contacts over the years."

He thought Ken looked at him with a bit more respect. "Yes. I concede that."

Ken's gaze slid away. "But I'm still concerned about Philanth."

"I'm aware of what I'm doing when I recruit her. The last thing I'd do is let her feel wasted. I want her to be happy so she can keep *me* happy. So I plan to initiate her gradually and systematically."

Ken answered simply, "I appreciate your reassurance."

Ken hesitated, then added shyly, "Speaking of initiating her, Chuck, I, uh, want to be sure you understand: she does have a problem regarding her body. When I put her on my bed she started citing St. Augustine's doctrine about the spiritual virginity of young Christian women raped by barbarians. It goes deep."

"The problem, or the spiritual virginity?"

"Oh, touché." Ken appeared to chew that over with appreciation.

"I'll be careful. I'd do anything to keep her happy; and whether you regard her attitude toward sex as a 'problem' or not, she also has a willingness to be hurt: not a martyr complex or masochism, simply a generosity based on the profound personal security of wanting nothing for herself; of *needing* nothing, in order to have happiness to share."

"That's what makes her so appealing."

Again they sat in silence for a moment.

"Well!" Ken got up. "May I use your phone to call a cab?"

"Phone is in the kitchen. Cab phone numbers on the cork board above it."

Ken thanked him and walked back through the dining area. A minute later he was back, sitting down in the love seat under the front window and adjusting the shutters so he could watch the street.

During the lull he had been musing. "Would you say Philanth is someone who's afraid to relinquish control?"

"I hadn't thought about it," Ken said. "But maybe you're onto something. A woman has to do that if she's going to be made love to."

"Yes. She's spoken of having been conditioned to never let down her guard. She won't even accept a social drink."

"Well, I can believe that she's not prepared to give herself to anybody she doesn't trust *completely*," Ken agreed. "And from all we've said, her trust at that level must be very hard to come by. I'd say you'd better be very careful about the circumstances when you ask her to trust *you* that much. She's got to feel totally secure, or all sorts of things could go wrong."

Ken was telling him that he must forget about making love to Philanth before they were married. To let Ken know he understood he said, "Otherwise she'll feel violated even though she loves me."

Ken nodded. They were quiet for a little.

"Another point," Ken said, "since I'm turning marriage counselor. How does she feel about the age difference and all that comes with it?"

"She puts me on a pedestal."

"That must be nice."

"It's demanding. She couldn't want to marry anybody who wasn't a priest of her religion. People are her religion, and to her I'm their consecrated servant. I know that sounds stange, ..."

"No, it figures. Religion and sex are all bound up together, for her. As just one example, she cites Castiglione—who wrote four hundred years ago, by the way—and who said that for a woman, passion is only a step on the ladder up to the same burning passion for God. Now that God has moved away and left no forwarding address, *she* says, the passion once directed toward Him flows out in all directions.

"That was in a wonderful letter she wrote me while she was with you in Africa."

"I can see why the religious dimension of sex is important to her. It's not just her parents' influence.

It keeps her safe from all the ugliness that sex can get into.

"She has a lot of things pretty well figured out."

"Yes."

Ken hit his forehead with the heel of his hand. "God, what I wouldn't give to have known at the beginning of September all that I know now! I'd have done a lot of things differently."

"It wasn't there, Ken. She couldn't have accepted anything so easy as loving and living with you."

Ken smiled painfully. "Thanks for putting it that way. It's pretty much what she told me in that letter."

Ken turned to peer out the window for a few moments before he spoke again.

"Anyway, so you let her keep sex sacred as much as she wants to; and after a while you might get her to relax back to where she was when she got married, with her natural enjoyment of everything reunited with her sexuality, able to accept simple, animal pleasure as well as the tenderness she specializes in."

"I appreciate your attitude and your advice," he replied sincerely.

He had never known Ken to be so magnanimous or so perceptive. That was in fact part of what he was appreciating. He suspected that Philanth had started the subtle changes he saw in this man, though the ensuing months had been required for the germination of seeds she had planted. She would say that she had only brought out potential already there.

"I know when I'm out of the running. Sorry I took it so badly at first."

"Forget it."

After a few moments, remembering Ken's abject appeal to Philanth in his letter, he asked, "What will you do, Ken?"

"Philanth's answer to everything was to lose yourself in good works."

"It still is."

"That's not so easy for someone in my line. I'm thinking of leaving my profession."

"What? Really? After you've worked so hard, sacrificed so much, to get to where you can relax?"

"There are more exciting things to do with fluency in Russian these days than extolling the genius of long-dead writers to youngsters busy mentally undressing the coeds in front of them. And I can see now that my cozy stagnation was a major reason Philanth had no interest in marrying me.

"Whatever I do, I'd like to keep in touch when we're no longer related. I hope that'll be all right."

"You know it will."

Chapter 28

TO LIVE IN THAT WHITE MANSION

It was Monday morning, and he had to get up and go to work. Like a small boy he hid his head under his pillow. He felt more like an aging man, too old and tired and used up to get out of bed in this strange, new room. Today was going to go on forever, one ordeal after another, into the night.

The end of it would be the most demanding part of all; and losing this bout with Philanth would make it even more difficult to win another one later.

Besides, Carole didn't have time to wait around. For Carole's peace, he must not fail. Death at thirty-nine was bad enough, but if he let her die without assurance that she had safeguarded the hopes which had sustained her he would regret it for the rest of his life, even if Philanth married him later.

Therefore he set himself in motion like an automated machine which would run smoothly through the appropriate routines once it was switched on. He showered and got half-dressed.

Barbara called down a good-by, and he heard the front door close. She knew he and Carole preferred to be on their own this morning.

He put on a robe and went upstairs to see Carole. Barbara had prepared a tempting breakfast for her, and he did his best to coax Carole to eat some of it. She was too miserable to be interested, so he carried it downstairs and finished it himself.

A few minutes after eight he called his office and told Marjorie he would not be in until around half-past ten. She agreed that he could be spared until then, "but just barely". He did not give Marjorie any reason for his tardiness on his second day back at the office. He was in no mood for concocting some half-truth, and he could hardly explain about the hospital's check-in time or why he wanted to be with his wife until then.

When he had made Carole as comfortable as he could he lay down beside her. He held her, and she stroked his hair. She didn't feel like talking, but she told him how wonderful she thought he was. He felt like making his mind a blank, but he realized that would be selfish. He tried to make a joke about how thrilled the voters were going to be when he campaigned all over California, since he was so wonderful.

At nine o'clock he helped Carole dress and finished getting himself ready for his day's work.

To simplify the maneuver of getting Carole into the car he went out the back door and brought the car around to the front of the house, then left the front door open as he came in.

She gasped with pain when he helped her into her coat. As he positioned himself to lift her he realized he was taking her away from her home, her life, in an almost final way.

He was still struggling to convince himself that all this was real, that the changes she had told him to expect were actually going to occur. He kept expecting to wake up and discover that all this had been a very bad dream. But it kept going on and on.

He wrapped his arms around her shoulders and knees, and she put her arms around his neck to ease her weight for him.

She was lighter than he had expected, but the memory of the times in their first years together when he had carried her laughing into his "cave" made him stagger as he swung around to head for the door.

They had had fourteen years together. Fourteen years were a long time in a young man's life. Such good years. He was carrying his dying youth in his arms.

And he was taking Carole away from himself, from their sustaining physical closeness. He was taking away the source of his strength of purpose, his courage, the wellspring of his happiness and zest for life.

With a little cry and a pointing finger she reminded him of the small suitcase the nurse had packed for her. "I'll come back for it, honey," he assured her. "When I lock up."

Very slowly and carefully, lest he slip on the steep, polished stairs, he carried her down to the front door. She probably would never go up those stairs again, never go into their bedroom again. The thought was heavy in him.

This was the beginning of the end, the end when the glory of a disciplined mind must succumb to the destruction of its sustaining body. It was only the end itself she planned to control with courage and intelligence.

He stepped out onto the little stone stoop above their tiny, ivy-covered yard bisected by its red-brick sidewalk. His front entrance had always reminded him that he lived in Georgetown, among the rich and famous. That meant less than ever now.

The sky was heavy, thick and gray, and all the trees were bare.

It was only a few steps to the car. He helped her in, and though her breathing was shallow from renewed pain, she shakily caught at him and drew him close for a kiss before he closed her door. He went to get her little case and lock up the house.

791

As he pulled away from the curb he glanced over at her, and a wisp of memory troubled the peripheral vision of his mind.

He looked at her more closely. She was leaning against the car door with her eyes closed. He put on the brake, reached across her, and made sure he had shut her door securely and locked it.

As he released the brake and drove on he decided he had been troubled by the memory of a dream: driving, driving, trying to get somewhere, with Carole beside him, unheeding.

Carole—of all people!—was letting go.

The wave of devastation started to rise in him again, the strongest in several days. He wanted to throw himself against her, wailing, Carole, you *must* not leave me! How can I let you go, even into release from suffering and weariness and the ugliness of this stupid, selfish world?

Better I should go with you than remain without you in a world that must let you die so young! —So full of the strength that is beyond mere goodness, the strength of a self lost in its will to act, to do whatever needs to be done no matter how vast and dull and complicated, no matter how thankless and exhausting and impossible.

Let the child Philanth go on her lonely way into sainthood, since that's what she wants. Service and contemplation in the midst of the world: it's a good life, serene if not happy, and it's a life she's suited to if anyone is. And *let me go with you.*

Oh, Carole, my love! I'm going to bleed to death anyhow, in all the thousand parts of me where you are torn away.

But he had no right to such thoughts. He had a duty to Carole. He was a grown man, and he was on his own now. She had done all she could.

Yet her present helplessness was deceptive. Her mind had reached out from its doomed body, still plotting to shape the rest of his life, and picked a woman to fill his heart and share his bed as if trying to select the ripest melon at the grocery store.

And she had done everything she could to commandeer Philanth's life, without serious regard for anything except her own program. She had taken some big chances.

Yet how carefully she had proceeded to question him about his feelings toward Philanth when he had arrived home! Lord, this woman was cold.

No, not cold, she had never been cold in his arms until she began this last campaign. But hard in her purpose like a Toledo blade. Carole had taken the trouble to know her own mind, had settled on what she truly wanted to do with her life, and had done all she could in the time she had.

Now she rested beside him, immobilized by pain but armed against her mortality with books and notes and cassette tapes, the gleanings from her mind with which she could move on to the final stage of her plan, teaching an unworldly young woman to become the invaluable wife of a rising politician. She would have been glad to be going to the hospital today to have her brain grafted onto Philanth's, the diseased body to be discarded afterwards.

Only confrontation with death could have made even Carole this strong. What she had done was inhuman. It was as though she already had died to herself, died in her emotions, and only her reasoning mind remained: calculating, planning, manipulating, working to perpetuate its will beyond its own end.

Carole made him feel weak. Even in trying to help him to go on being strong without her she had made him more vulnerable, for now he needed Philanth.

Moreover, the longer he had studied all of his information on Philanth, the more clearly he had seen that he could not shake her loose from the wide and deep and carefully designed foundations of commitment which obliged her to reject him. Her selflessness was too complete and profound, her indifference to her personal happiness was too relentless.

Yet he had to persuade her to marry him, for it was clear now that—for a combination of reasons, none of which was necessarily compelling in itself— he could not keep her long on any other terms. As she had realized in Manila, the qualities which made her withhold herself from him were the same qualities that made him obsessed with desire for her.

So he had to do his best to break that impasse, if only out of duty to Carole.

Looking only at the day ahead, he seemed to have a lot of duty to perform. Duty was like a body cast, holding you together when nothing else did. Duty was the blind drive of necessity to go on acting which kept caterpillars struggling on, even writhing to keep moving when partially crushed. Duty.

The present duty was to get Carole to George Washington Hospital through all this morning traffic. Just then he was following Pennsylvania Avenue.

It seemed strange that the hospital was so close to the Watergate complex and the Kennedy Center, with their liveliness and elegance. To him George Washington University now meant only its hospital.

But as Philanth had remarked, in Washington everything was just around the corner from something else. Layers upon layers of time, of lives, of meanings, were piled upon each other like snow in an alpine winter. Sometimes an avalanche.

As Pennsylvania went around Washington Circle he turned right onto Twenty-third, then left into the half-circle driveway serving the entrance. Thick, damp air hovered, absorbing auto exhausts below the heavy sky. Through it the lights of the hospital were beacons of futility.

The Christmas lights were a mockery; the regular lights and the building were hollow boasts of man's power, the power of technology to save. The proud building and the lights were there to promise life and hope, but they could not keep their promise. They were all there in this cold, gray December morning to put the best possible face on Death.

He helped Carole into the building, telling the guard he would be only a few minutes. They bore left from the entrance. The t.v. above the waiting area for the admissions section was chattering about a stabbing in the District. It seemed like somebody was *always* getting stabbed or shot, in the District. He found an admissions clerk who would start "processing" Carole.

Carole was barely coherent, but the computer had been informed of her early arrival. Good doctor. He went to park the car.

When he returned he sat beside Carole and signed hospital forms in one of the bright-green upholstered chairs that matched the carpeting.

Green, the color of life and hope and spring. Might Carole see another spring, another weeks-long riot of blossoms, cherry and apple and dogwood and forsythia and tulips and irises and hyacinths and jonquils and all the others which made the area an endless array of flower shows? Today it seemed not.

Once the clerk was satisfied and said Carole would be called as soon as a room was ready he kissed his wife good-by and left. He knew she would rather be alone, even with that horrible, idiotic television, than make him any later in arriving at the office.

As he moved through the well-groomed lobby and quiet voices and ordered routines he was suddenly filled with a towering rage at its neatness, its civility, its calm perfection. What pretense! What hypocrisy! He would have felt instantly at home among the fire and chaos and screaming wretches of Hell. He was furious with the world, outraged like a child deprived of his heart's desire by some unreachably high power. Like a child, he wanted to screech and beat his fists against the iron gates of the world, insisting that it lose its polite indifference to his anguish.

Yet only a slight impatience had crept into his voice in dealing with the compassionate admissions clerk's routine questions, and to Carole he had been his usual calm, patient, tirelessly amiable self. He had a strong shell of good manners and restraint, having been suckled on the commandment that there was no excuse for bad manners or lack of self-control. His mother had called it "breeding"; in Washington it was part of what they meant by "polish".

It was not something one decided to acquire the day one decided to seek public office. It was only a way of getting through the world without making life any more trying for others than it had to be and without compromising one's self-respect. With such self-respect one earned the respect of others.

But the shell of reserve did nothing to relieve the pressure of massive internal hemorrhaging. As he passed through the doors to the sidewalk a thick flood of heavy pain suddenly welled up from his chest, choking him and threatening his control of his mouth, even reaching into his eyes. The colored lights of the Christmas decorations around the hospital entrance became blurred.

He walked back to the parking structure in the next block. A light snow was falling.

Oh, God, let the snow fall thick and heavy. Thick and heavy. Let us all be buried under it and finally be at peace. Carole, lie beside me under the snow, and I will be at peace. I have learned to want even less than Philanth: only the end of pain.

But he was a well-programmed piece of machinery with a lot of years left in his motor. He found his car, unlocked it and got into it, and drove it out of the parking structure, one machine operating another.

But all we can do while we live is keep trying. Keep struggling, keep working, keep pushing. Fight the good fight, keep the faith, finish the race.

Yes, it's hard. As his father had admonished him, it "takes everything you've got". By definition, since only doing your best is ever good enough!

But once you've gotten used to living life on that basis, any other way of doing it is hollow; you see the bitterness, the emptiness and cynicism, of those who profess to be rich and happy in the pleasures of this world, and you wouldn't trade griefs with them for anything.

So you find yourself trapped in your own virtue, and when it gets hard you start grasping for whatever comfort is permitted, to make it bearable.

A car's bumper sticker called out, "One Planet—One People . . . Please."

He was only a few blocks from his office, but the one-way streets got him fouled up and he found himself passing the National Law Center. He had given a speech there, at one time or another.

That all seemed so meaningless, now. All those years of hurrying around, spreading the light of his own little fragment of The Truth: choking down his supper in some barren fast-food joint while he missed Carole's sensible meals and leisurely sense of humor;

changing his shirt in some men's room, looking for a parking space, shaking hands, trying to look and sound brilliant and inspired when all he felt was distracted and exhausted and wound up too tight to think after a full day of running from one group of expectant, critical, demanding people to another. Always the man on the make, the campaigning politician trying to make his mark, make a good impression. For what?

In God's name, for what? If one man couldn't make any difference anyway, why not at least a peaceful, sensible life, plodding through a nice, safe little career out among decent, ordinary people in God's Country? Why must he feel obligated to take on so much more?

On the right a beautiful little stone church said, "All Are Invited." He wished he could stop and pray, soak up the calm of a sanctuary.

Like Philanth in her journal, he wanted a few words with the Almighty: just a few moments of His eternity.

God, the injustice! How can you take her before she's had time to do more than lay the foundations of her life's work? This woman would have deserved to live in the White House!

And how can you leave me without her when I'm only forty? You know how weak I am, Lord. I was the fair-haired boy, born with looks and brains and money, and everything was handed to me. I was never much of a fighter: the school debate team was just *my* speed.

No wonder the world's in such a mess, Lord, if this is how you manage things. No wonder Philanth gave up on you. No wonder she mocked her own impulse to pray by going out into the emptiness between stars to look for you.

But there had to be someone, somewhere, who could give comfort at a time like this. *Had* to be.

There was Philanth. Philanth, the devout atheist for whom the best place to spend one's life was a great cathedral. ". . . And I shall dwell in the house of the Lord forever."

And who could prove to be as stubbornly determined to martyr herself as any of the martyrs of past centuries, as soon as she could persuade herself that he would be better off without her.

He got honked at and realized he had not been paying much attention to his driving. Downtown Washington on a weekday morning was no place for behind-the-wheel philosophizing. He forced himself to get oriented and keep his concentration until he reached his destination, another parking structure.

Soon he was striding down the wide, polished hallway lined with little shops and eating places, heading toward the elevators in the center of the building. The handsome white-haired, one-handed security guard nodded and smiled to him as he passed.

En route to his suite of offices he stopped by the small one-desk office Philanth had arranged to borrow while someone about to retire from Finance was taking extended leave. "Now," she had said on Friday as she reported her move to him, "I'll really be able to get some good work done without getting worn out by the effort to concentrate. And all the people I have to meet with about my project can drop in on me, singly or together." She flung out her arms, exclaiming, "It's *wonderful!*" He had chuckled to himself after she left his office as if walking on air.

After exchanging "Good morning"s with her he asked, "What time do you need to leave this afternoon to get ready for that dinner?"

"Three o'clock, if you want me to pull out all the stops for these people."

"Yes. Black tie; quality but restraint. That's the best we can do because it's such a modest house."

"I've been worrying about that: Do you prefer too much restraint or too little? My black dress and Spanish cape, or a white brocade intended to make me an impressive hostess to a large crowd?"

"If those are the only choices, wear white."

"I'll have to borrow a car to get it from my little public storage unit. The last time I went to it Carole drove me. I called her about the problem after I got home yesterday, and she said if I decided I had to go there I should ask you to have Eric drive you home today so I could take your car from the office."

"Okay." He gave her his car keys and claim stub and told her how to find his car.

He added, "I'll leave around then, too. I need a nap to get me through the evening. I did a lot of reading last night." He absently traced the circle on the front of the demoralized-looking dictionary she still kept on her desk. "And I have a very important meeting even after the dinner guests leave."

"Really?" She looked impressed.

"Uh-huh. I have to talk a stubborn young woman into getting married."

Philanth made those inverted "U"'s at the corners of her mouth which he found so endearingly comical. But she spoke gravely. "I want you to live in that white mansion on Pennsylvania Avenue as much as Carole does. This country can no longer afford the luxury of electing incompetents. The world needs your combination of compassion and intelligence and leadership.

"I know there must be lots of women who say the same sort of thing to the men they admire, but none of them is willing to do more to help make it possible than I am. That includes refusing to marry you so you'll be free to marry someone else."

As she spoke her phone had started trilling, and while she was quickly promising to call someone back and making a note he found himself once again staring at Philanth's crippled dictionary.

When she hung up he said, "Get rid of that dictionary."

"Now, we've been over this before. I'll take it home only if you insist."

"*Throw. It. Away.* It's not healthy to keep a memento of that man."

"When you're alone, Mr. Bjorklund, you give personalities to broken pitchers."

"When you're engaged to me you won't need such a sorry emotional attachment as that."

"I'll never be engaged to you. So I'll never part with it."

"You'll part with it the day I give you something better." That gave him an idea which made him add, "And that will be soon."

She folded her arms and looked stubborn.

It was a shame, in a way, that doing this made her look like eight-year-old Shirley Temple in one of her cute little pouts. It was no wonder Philanth favored serious hair styles instead of the looks which were in fashion.

"Why do you cling to it?"

"I love him. He's a loathsome man, but in his meanness and exploitativeness he's pathetic. He's a human being, and I was close to him."

"You're really hard up for someone to love, aren't you."

"Please don't hound me. You know what I think and I know what you think, and needling me about

801

our differences will only make even a working relationship impossible. If you must press this, do it tonight, as you said you intend to. But if you can't let it rest even after that then I'll have to leave."

"Fair enough." Carole would have been pleased. But he was the one who had to win the showdown.

When he reached his own office Marjorie raised her hand to stop him from marching past her desk.

"Yes, Marjorie?"

"Tony's reported in, and he wants to know whether you'll need Eric today or tomorrow. He'd like to get Eric's help on a special project you gave him. Tony says if he lines up Eric to help him and then you need a driver it could create 'big problems'." Marjorie tried to imitate Tony's deep voice.

"Tell Tony and Eric I want Eric to be available to Tony full-time for as long as he's needed. Neither of them is to be called away to do anything for anybody else. That includes the deputy director. My orders."

"Yes, sir."

"But tell Eric I'll need the official car from three o'clock today until I get here tomorrow at eight. It's necessary because of the Peace Corps dinner tonight."

"Yes, sir. I'll get the keys for you."

He took the deputy director to lunch at a quiet restaurant down the street. During the salad course he remarked, "I'm planning to replace some top personnel and shift some of the workload out of Les's position. I'm hoping you'll be willing to take on some of that responsibility, now that you've had time to get a better understanding of how the federal bureaucracy differs from the private sector. Would you like that?"

Archie's normally cheerful face had suddenly become saggily sober. After a moment he asked, "You're getting rid of Les?"

"He's going to resign."

Archie waited for him to explain, but he calmly went on enjoying his crisp, cold salad and well-spiced yogurt dressing. Ah, American food. He was a new fan of the simple salad.

"Oh." Archie hesitated, and he knew Archie was not anxious to take sides in any power struggle. "What sort of additional duties did you have in mind for me?"

"What sort would you like?"

There was a cautious brightening of the other man's expression. "Bjorklund, after four years of being treated like a useless nincompoop, I'm willing to let you take the lead on that."

"Good man," he said, nodding. "I'm sorry your initiation into government service has been frustrating. It is for a lot of people. But I'll try to see that the situation improves markedly for you."

Archie didn't respond. Their meat course arrived, and they ate and drank in silence for several minutes.

The waiter dropped by to see that everything was satisfactory and to pour some more water. Ah, water too. Lovely, safe water, with safe ice in it.

He resumed in the same casual voice, "One delicate project I'd particularly like your help on is working with the White House on selecting a replacement for the associate director for Management."

He thought Archie might choke, but a little water did the trick.

"Is she resigning too?"

He almost smiled. "Yes. But don't let her know it yet. It might upset her."

He leaned back, arching his fingers. He had a poor hand, but if he played it with enough skill it could suffice to bluff a tableful of opponents into letting

803

him take the pot. Together they could clean him out, but if they didn't get sufficiently suspicious to compare notes, each might even help him against the others.

Fortunately these people weren't the sort to compare notes: they were too preoccupied with Looking Out For Number One. When one of them was whispered to be "in trouble" no one else would rush to champion that person's frailties, much less his or her gross shortcomings and offenses. If he succeeded in getting Les out of the way without weakening himself he could use the *Common Cause* article to justify getting rid of the Public Affairs director and the associate director for Management: he could tell the White House they had to be sacrificed to prove his responsiveness, as part of his mandate to rehabilitate the image of Peace Corps.

Once the three worst appointees were quietly cleared out and Archie was loyally pulling his weight he should have gained enough clout to at least curb the smaller fry who were also damaging or exploiting Peace Corps. It all might work even if Tony didn't actually deliver the one modest high card he was playing from, and it would be cleaner if he never had to show his hand at all.

Meanwhile he laid out solid meat for his hungry deputy. "We need a seasoned federal bureaucrat with flexibility and courage to act decisively but without an oversized ego; a skillful administrator with strong interpersonal skills, genuine commitment to the interests of Peace Corps, and a strong background in federal fiscal procedures.

"I know that's a tall order. But perhaps you could induce the White House to let us have somebody they want to displace anyway to open a slot for a more political type in some other agency. We need to start soundings on that idea right away so it can perculate through the network while the post-election rewards are still being sorted out."

Archie nodded.

"This is an important position, Arch, and it would be worth a great deal of your time to do a good job of filling it for us. It'd give you a chance to develop your own contacts for future reference. Think you could put out feelers and wangle some good cooperation on that? I want everybody in town to know it's open, but it would have to be handled with great care because—God forgive whoever was originally responsible!—it's supposed to be a patronage position."

Archie's brows went up. "I got to hand it to you, Bjorklund . . . ." Archie managed not to finish that.

Good. The poor man had been learning discretion the hard way, but that seemed unavoidable at the introductory levels.

He took time to devise a discreet reply which wouldn't chill Archie's growing regard and trust. "I know it's asking a lot. But I know you'll put your heart into doing whatever can be done. When you get discouraged, remember that all you really have to do is improve on the situation we've got. *That* shouldn't be too hard!"

Archie gave his head a shake which could mean anything. He gave Archie a confidential smile, and Archie managed a worried little smile in return.

When they got back from lunch Marjorie told him that Les had left a message asking to see him "right away". He was on a roll, and self-confidence was the key to keeping it going. Tony's operation needn't be jeopardized by changing the order of events, if he didn't tip his hand to Les in the meantime. By the time Tony started yelling in front of witnesses Les could be ready to panic and say something that

805

would help to finish him. He told Marjorie to let Les come in.

Les walked in and began without preamble, "I started talking to Philanth about the paperwork I wanted from your trip, and she told me *she's* in charge of it, and I'm 'out of the loop'. I said, 'Since when?' She showed me her preliminary plan for organizing and reporting your findings, but when I started to take it she said it 'wasn't ready for circulation'. She flatly refused to let me even look at it! And she's taken over somebody else's office. What have you got to say about that?"

"Sit down, Les."

Les deposited his portly frame in a chair, but his scowl deepened, accentuating the toad-like shape of his mouth to which Tony had alluded.

He thought Les must have looked at himself in the mirror when he was little and decided that trying to be liked was something he might as well dispense with. But he could defer pitying Les until the man was no longer dangerous.

"Certain facts have come to my attention which make it unlikely that you'll be with us beyond the end of this year. So I've started assigning your responsibilities to other people, to ease the transition to a new chief of staff."

"What're you talking about?"

"No need to raise your voice.

"You'll want to offer your resignation rather than be the subject of a scandal which would be very annoying to the White House and extremely damaging to your career. The damage to Peace Corps would be minimal, fortunately, due to the circumstances."

"I still don't know what you're talking about."

"Then you certainly should!

"That's all I have to say to you at present."

He leaned to look at his schedule. He pressed the intercom. "Marjorie."

"Yes."

"Do I have any submissions to look over before my 1:30 appointment?"

"Just a moment."

Les got up and left.

Even if Les felt sure he hadn't done anything that bad, he now knew that his superior was prepared to make things look as though he had; and in politics, where standards for conduct and even the meanings of words had unlimited flexibility thanks to moral pragmatism combined with sloppy thinking, reality could be defined by whoever spoke with the most authority and determination—always assuming he got the media on his side. Les also had to know that well-chosen allegations alone could make a position like his untenable.

—Unless the charges could be completely discredited, and in practice that depended largely upon the credibility of the accuser. There too he would have the upper hand with Les *if* the White House were implicitly involved as potential judge of a malefactor rather than allowing itself to be dragged in prematurely as the defender of the spoils system. A couple of pre-emptive phone calls over to the White House seemed to be in order right away. He knew just the men to call.

As for allegations which were true, these could not even be vitiated when the witnesses were beyond the reach of the accused. The house of fear from which Les ruled was made of cards, and a light breeze had come up.

He must make time his friend. He must continue to joggle the ground under people's feet, not raise any flag which would put them on the defensive: that

807

could inspire them to mass together and attack.

But things were finally starting to break his way; and by the end of the week, he hoped, he would be able to call Sandy with a good job offer. To throw just one more metaphor into his silent celebration, he was beginning delicate surgery to remove dangerous parasites, and when the incisions were cleaned and closed, the healing of a healthier Peace Corps could begin.

He picked up the phone.

Chapter 29

A REASONED PRESENTATION

He left the office a little after three. Marjorie said Philanth already had signed out.

However, before going to the parking structure for the Peace Corps vehicle he walked on over to The Map Store. He found what he wanted: the pictorial world atlas. This Philanth would be able to enjoy without much guilt even though it was so "costly". Actually it cost no more than dinner for two at a medium-priced restaurant.

The Map Store did not gift wrap, so he carried the large, heavy book to a nearby drugstore. There he got a packet of wrapping paper and matching ribbon. He could not remember the last time he had gift-wrapped something himself, but since the bow was ready to stick on he didn't think it would be too difficult. He knew Philanth would make allowances.

As he drove over to Georgetown he worked out what he would inscribe in the book before he wrapped it. The task was remarkably easy.

How strange it was to be in love and courting when his wife was suffering in the hospital! Carole would consider it healthy, and Philanth would imagine flowers springing out of ashes. Everybody else would think it was awful. But people consumed by a cause were known for being indifferent to such secondary considerations as what other people thought.

When he got home he saw that the cleaning woman had come and left everything immaculate. But the house felt empty, soulless, without center or purpose, like a house between owners. Because of the dinner party Barbara was spending the evening with relatives, and no one would sleep there that night except himself.

The next day would be moving day: Barbara would shift Carole's belongings downstairs and her own personal effects into Carole's room.

His task would be to destroy all of their notes and files on Philanth. The other secret files would stay in the attic, for if the doctors managed to give Carole enough time she hoped to publish at least one article about how unmarried women's values affected their satisfaction with their lives.

In the evening Philanth would move in. Then the little old house would again know who its mistress was and come back to life.

He inscribed and wrapped the book, then showered and shaved and climbed into his new bed in his underwear. As he prepared to go over his notes for his presentation to Philanth it occurred to him that he had never reviewed the tape of his discussion with Carole. He went upstairs, found it, and did so.

His spirits lifted when one fleeting thought he had mentioned on the tape began pulling other items together in an outline. He went on to develop it.

When the key phrases of his main points were etched on his mind in a framework which would help him avoid leaving one out of his oral argument even under cross-examination he set aside his new crib sheet and relaxed. He was careful not to look at the clock to see how much time was left before he had to get up. He filled his mind with peaceful, affirmative thoughts.

When the alarm went off he had been dreaming about gobs and gobs of soft pink and white ice cream. He assured himself that he felt clear-headed and reasonably fresh. He resolved to pace himself through the evening, reserving energy for the crucial period at the end. He told himself that he had weapons now, and as he got out of bed he closed his mind to grief.

In the beautiful new blue-and-gold bathroom he combed his hair and splashed his face with the aftershave lotion Carole had always said it was "immoral" for him to put on when he was going *out*.

As he slowly finished dressing he kept referring to his outline and mumbling its phrases over and over, strengthening his hold on his arguments. He had to be able to fling them out in a body like an African hunter casting a net over a priceless, fragile wild thing which must be captured without being injured.

Philanth must be partial to blue since she had obliged herself to wear it for five solid weeks, so from the closet he selected his fancy light-blue dress shirt and dark-blue satin-trimmed tuxedo.

Purists frowned on so much color and decoration in a man's evening wear, and he had thought the ensemble effeminate when Carole bought it to make him "stand out in a crowd". However, Californians were known for colorful dress, and now all of the frills, especially on the shirt, gave him an odd sensation of being gay and dashing.

Yet as he put them on he also felt solemn, like a knight donning his armor for a holy battle. He was committed to a terribly important jousting match tonight, and a fair lady was the prize. It was an adolescent fantasy, but rising adrenalin had wiped away his fatigue, and he felt almost young this evening. Call the image "archetypal" rather than "adolescent". Yes. Philanth would like that.

From the front door above came the rap of the brass knocker. It was six twenty-seven, so that had to be Philanth. He put on his jacket, checked himself in the mirror on the closet door, and went up to let her in.

"Where's my car?" he greeted her, looking up and down the car-lined street.

Her quick response sounded anxious. "I've heard that Georgetown has scarcely enough parking for its residents, and I knew your garage would hold only the Peace Corps vehicle. So rather than park several blocks away and expose your car to thieves and vandals I picked up Eric at his place, and he let me out here. This way you and he only have to exchange car keys tomorrow at the office.

"I'll get a cab home—no charge to Peace Corps, since it was my fault I needed a car to go retrieve my dress."

He nodded, not bothering to quibble since he planned to drive her home, and let her in.

Only then did he really look at her. She had done something to her hair which made it shine like gold. It was in a style he had never seen her wear before: piled in soft waves on top, but smoothly swept back on the sides to a pair of long, white combs at the back of her head, where it became bouncy ringlets cascading down onto her shoulders.

As he shut the door she turned to allow him to remove her blue wool coat, which she was holding closed from inside with both hands as though it were a cape.

"No purse?" he asked as he took it from her shoulders.

"I have essentials in my coat pockets."

He stood holding her coat and staring.

She was dressed for marriage to a lord. That thought popped into his mind because of the Chaucer story she had told him during the trip. Her long, white brocade gown had fine gold threads in it, and its high, soft collar made gold jewelry unnecessary. Each sleeve suggested a shaft of light, tapering outward from a pair of pleats at the top.

She was staring back at him, but she was looking askance. "What's the matter?"

"I wasn't prepared for how splendid you look in dark-blue evening attire."

"I'm glad you like it."

"I like *every*thing you wear," she replied with the artlessness which could make her candor so disarming. "When I saw you in the white dinner jacket you wore during the trip I thought that if an angel came to me he'd look just like you, except his tie and trousers wouldn't be black, and he might not have real raw silk on his lapels. You looked stunning, so tall and slim and blond and with that gentle, strong face you have. Quiet, shining power is greater than the dark, roaring kind."

"But what I'm wearing now must not be so stunning," he teased, moving closer as he enjoyed the effect he was having on her, "since it's surely not angelic."

"A simple black-and-white ensemble for the tropics compared to all that dark-blue satin and embroidered ruffling? You've moved from stunning to overpowering."

He moved still closer, deliberately invading her personal space. "It's not bad that you think so."

"No." She stepped back. "It does seem unfair, though, for Carole to keep setting us up the way she has . . . over and over and over! She's like a fairy godmother, pairing me off with a handsome prince."

"She had to use every stratagem she could think of to assure me that I was supposed to be determined to get another woman even while she was still alive.

"She knows *you're* not going to let your judgment be influenced by your feelings."

He touched her shoulder. "Just the same, wouldn't you at least find it interesting to be kissed by a man who looks so 'overpowering'?"

She shrank from him. "I wouldn't want that at all. I'd risk fainting, silly as that sounds. It might take me a while to react again to simple things: the sound of rain, the smells of warm earth and fresh-cut grass . . . ."

He touched her chin just enough to make her look him in the eye. "Is that *so* important, Philanth?"

"It's essential." She evaded his touch. "All pleasure is relative to what one is accustomed to. One can enjoy life just as much with very little—it's only a matter of paying attention to the right things. And the less you need to be happy, the more independent you are; so the less you have to compromise in order to survive and get things done." She edged away from him.

"And when you realize how little others in the world have," he supplemented, "your conscience can be clear because you're taking for yourself only what you truly need in order to try to make their lives better."

She gave him a shy smile. "Yes."

"So you don't really approve of the way I look when I'm dressed up, nor of yourself."

"Dresses like this cost very little, since I shop carefully for fabrics and then do the sewing as recreation. I *had* to acquire *this*, for my job with the Hunger Foundation. My boss insisted—he even went to the store with me and helped me pick out the material and the pattern.

"As for you, I understand that you're obliged to live by the same set of rules as the governing elites."

"A tactful evasion. Could you accept living as I do, Philanth, or are the compromises involved a significant part of your reluctance to marry me?"

"It *would* be difficult. I'm a sensual person by nature, Mr. Bjorklund. I'm an ascetic only by conviction. Being as indifferent to the finer points of decisionmaking about consumption choices as most other men of secure position, you can't imagine what a constant inner battle I would have if I were your wife."

She touched her hair and smoothed her dress. "Do you imagine I could get myself up like this so quickly if part of me didn't love the pleasures of the senses? I know there are lots of people who genuinely don't care about beautiful clothes and lovely homes, but I'm not one of them. Often I wake up to discover I've been dreaming about trying to get a nice bedroom or fixing up a house for myself; or I've been dreaming about a store full of wonderful fabrics or china and linens. I can lose myself in a great crystal chandelier. I *love* beautiful things!

"And I love comfort, even luxury. The only thing that saves me is always asking whether I have any more right to these pleasures than the poorest person on earth.

"I've wanted to get away from the practical conflicts of my present way of life for a long time. The constant weighing of alternatives—practicality versus beauty, how to meet expectations which I thoroughly understand yet morally reject—it takes time I could be expending on things that matter."

"Think what you could do about that duality in your nature if you became First Lady."

"I suppose I could get snobbish rich people to hate me even more than they did President Carter's wife."

"Oh, much more."

815

"You'd like to pretend that wouldn't hurt you. But remember that Carter was a one-term president. A reponsible attitude toward waste was denigrated as though turning down the thermostat were really the same as wearing a hair shirt."

"Much has changed since then.

"And leaving that aside, think how superlatively suited you are to the balancing act required of a senator's wife: frugal, and able to share the feelings of the poorest of his constituents, yet able to find common ground with his wealthy supporters."

"But that's common ground I don't want!

"Let's not get into all this, please, until we've taken care of the evening's official business."

"You're right," he conceded reluctantly.

She made her voice more businesslike. "I've studied the briefings on tonight's guests. Do you have some instructions for me regarding the conversation?"

"Just a minute." He went to hang up her coat and take the atlas from the hall closet.

He presented it to her, saying, "Call it a gesture of appreciation for your services as my hostess this evening."

"That wasn't necessary, Mr. Bjorklund," she said in her dulcet voice. "But it was very thoughtful." She smiled into his eyes as she spoke. It was the same warm look she might have given anybody, yet it left a glow.

As he let the weight of the package shift into her hands she hefted it and chuckled. "Goodness, it's heavy! I'll bet it's a book. I do hope you bought it second-hand, whatever it is!"

"No. You'll soon see why."

She was charming him silly. In the few minutes since she had appeared at his door she had kept shifting her personae before his eyes, and yet all of the sets of voices and mannerisms were legitimately hers. Now she was the unspoiled but well-bred young lady.

She sat down in one of the blue satin chairs by the white fireplace to undo the wrappings. The package was all in soft blues, the room was all blue and white and golden brown, and she was all gold and white and pink. He wished Renoir or Degas could have painted her as she picked open the package without damaging its wrappings.

Above her on the mantel was a large color photograph of his swearing-in as Peace Corps director by the vice president, compliments of the White House. He had curbed the practice of decorating the halls and meeting rooms in the vicinity of his office with framed photos of the current director in various ceremonial poses, but Carole had insisted on displaying his copy of this one in his home. Since they had few visitors, he had not fought her on the point.

He stepped over and removed it in preparation for the evening's entertaining.

His nearness made Philanth look up, and she saw it in his hand. She turned to look at the empty space on the mantle, then smiled at him. "You're going to hide it?" She could see only the brown backing, but apparently she knew what it was.

He nodded and slipped into the dining-room area to put it into a lower drawer of the buffet.

He returned quickly.

Watching Philanth open his gift, knowing how she felt about him and how much of that feeling must derive from his official position, he felt that the psychological rewards of such jobs as his justified not only modest salaries but many of the miseries of public life. Considering the temptations, it was quite reasonable to be strict with men guilty of taking advantage of young women equally susceptible and less conscientious than Philanth.

Her sober expression gave way to wonder and delight when she pulled the heavy book from its tightly fitted carton. "Oh, Charles!—I mean, Mr. Bjorklund!" For a few moments, as if she couldn't help herself, she peeked into the colorful pages. The stiff, new book might have been breakable at a touch, she handled it so carefully.

She looked at him reproachfully, shaking her head a little. "You're very thoughtful. And generous! Thank you so much."

"I've inscribed it so you can't give it away."

She peeped inside the cover. He had written,

> Philanth,
>   We give each other the world
> when we give our love.
> Charles

Her mouth became tremulous.

Studying her, he realized that she had skill with hair and makeup which she seldom used. He had never seen her look so pretty as she did now. She moved ably between the world she believed in and the one imposed upon her by the values of her country as they assumed extreme forms in its capital.

Ably but not willingly. His hold on her was weak, and if she slipped from it she would gladly leave the world of hair rinses and brocade dresses and eye makeup behind her forever.

That must not happen.

She carefully folded the wrapping paper into its original creases, placed paper and ribbon inside the book so that only the flower-like bow protruded at the top like an extravagant bookmark, and slid the book into its box.

"I'll keep it all my life," she vowed. "If I go away and never see you again I'll still be as rich a woman as ever lived." She went to place the box in the closet with her coat. As she returned she added, "And I swear I won't ever tell anyone who 'Charles' was."

The knocker sounded before he could frame a reply. The house had no service entrance, so the caterer and her assistant were on the front steps. The evening's work began.

Philanth had the African couple at her end of the table because they were more comfortable in French than in English, and they clearly enjoyed her company. They were so impressed by the fact that she was a recently returned Peace Corps Volunteer that their support of Peace Corps in coming negotiations with their governments seemed assured. Several times she had them in stitches over her clinical experiences in family planning, and she blushed prettily.

The evening went smoothly, and Philanth gracefully assisted him with the substantive discussion of possible new Peace Corps involvement in their guests' countries. At ten-thirty the African couple left, saying they had heartily enjoyed themselves but pleading their commitment to their baby-sitter. The Asian couple followed their example.

As soon as the front door closed behind them Philanth said, "You've had a busy day. It's not fair to you for us to try to have this out now. We can talk another time."

"*No*," he replied, going to sit down with his back to the fireplace. "We can talk *now*. This has got to be settled, for all sorts of reasons. We're very busy people!"

"All right." She settled on the sofa facing him. "But if you go forward with this it means you promise not to keep trying for a re-match when the outcome doesn't suit you. As you say, it's got to be settled."

He nodded reluctantly.

She added, "As far as I'm concerned, it already *is* settled. If you still can't leave it alone, I won't keep wasting time and energy arguing with you. I'll leave even before the next two years are up. Because I'm not that valuable to you in your effort to get elected, and I don't want all the positive emotions to be leached out of our relationship by conflict. I'd rather remember you with regret than with irritation."

She shook her head. "You're in a weak position to pursue this, Mr. Bjorklund! I've offered you a lovely working relationship—for the indefinite future, if you can keep your next wife from becoming jealous of me—and you're risking it all."

"Understood." Now he was grim.

Smoothing her dress under her as she shifted so as to avoid pressing creases into the brocade, she leaned sideways onto the small cushion inside the arm of the sofa, rested her cheek upon her fist, and tucked up her nylon-stockinged feet inside her long skirt. In the husky-deep voice which sounded seductive even though it only reflected the fact that she was relaxed, she announced, "It's your serve."

"Pretty sure I'm going to fail, aren't you," he remarked. She nodded unsmilingly. "Why, when your position is so quixotic?"

"Because there's no way you can demonstrate that there's no one better suited to become your wife than I am. It's always hard to prove a negative, and in this case it's obviously impossible." Forewarned, she too had come armed.

820

"Then if you really are prepared to be fair," he rejoined, "the burden of proof must rest on you."

"Okay. I'm willing to do all I can to locate candidates and bring them to your attention. Since I'm to be one of the first to join your campaign staff I could get into a good position to do that. I'd put a lot of effort into sifting your supporters even as I sought to recruit more. It would challenge my ingenuity, but having studied the sociology of mate selection would give me a strong theoretical foundation for designing appropriate methods. And who would know better what personal qualities to look for than I do?"

"But Carole already has expended an enormous amount of intelligent effort and selected you. We could go on conducting systematic wife-hunts until we're withered with age, but that would waste far more than time. Carole's main objective was to help me get on with my life during a crucial period."

Her expression of contemptuous amusement made him add hastily, "I'm not saying I have to rush into marriage and therefore I have to marry you, of course. But Carole certainly found a woman I need to keep with me, and especially during a difficult period. Marriage is by far the most practical way."

"You make marriage sound like a mere convenience, if not a temporary expedient."

"You're lying there waiting for me to make a reasoned presentation. If I don't make marriage sound like a matter of practicality you'll accuse me again of emotionalism. I'm condemned either way."

She smiled at his scoring that point. But she answered firmly, "I don't accept your arguments of *either* type."

He rubbed his forehead and sighed. "Let's call a 'time out'. I didn't get enough to eat, I was so busy talking. Let's go see what the caterer left us."

Philanth followed him out into the kitchen, and they assembled two plates of goodies. Her glass of milk in one hand and her plate in the other, she preceded him to the dining table.

As he sat down opposite her she remarked, "I've read that there are so many charming, intelligent and politically sophisticated widows in Washington that they create a problem in making out guest lists for dinner parties because they so far outnumber the eligible men."

"But if they're not frail with advanced age, why are they here? Why aren't they in places like the ones where you want to be? I'll tell you: values. And it's far easier to teach you what they know than to give them the values I've found in you."

"Their values would make them much more suited to the role of wife of an important politician."

"But they wouldn't be in fundamental conflict with what's wrong with this culture, so how could I care for them, having known you?"

"All you've ever said is special about me is my value system. The Peace Corps *collects* such people." She began her shrimp salad.

He took a fork to his slice of the chicken roulade. "There are a lot of fine people associated with Peace Corps. But I've got exactly what I want in a prospective wife, so I have no incentive to look further."

Philanth speared a prawn. "But *I* lack incentive to learn any more than I already know about how to conform to the outrageously corrupt standards of Washington."

She held up the little shrimp on her fork. "Even eating shellfish is corrupting: d'you realize how much these succulent little goodies cost per pound at Safeway?" She popped it into her mouth and mimed an ecstasy: "Oh, please, Mr. Satan, where do I sign?" She was so secure in her position that once again she was making a joke of it.

"Oh! Lord!" she shrieked in mock horror, "I just used the wrong fork!"

"I don't *care* about how you fit in with snobs when they're showing off how discriminatingly they fling money around! I'm looking at the fact that you're willing and able to do things that they think they're 'too good' or 'too smart' to undertake.

"And I don't want you wasted overseas."

"I admit it seems hopeless for some countries. They're looking to us to help them out of their demographic disasters with our surplus wealth, while we're teaching them to expect too much by exporting our profligate mass culture.

"But even though we're not interested in giving them enough of the right kinds of help to get them on their feet so they won't keep tearing themselves apart, shouldn't we at least plant a few seeds of the best that we have to offer? Don't we at least owe their youth the food of hope, considering all the suffering and bloodshed it will cost them to fight their own leaders and recreate their cultures so their societies work in terms of the twenty-first century?

"So why stay here to comfortably watch the decay of a society institutionally committed to short-term goals and rampant materialism?"

"You know the answer: because this is where the key resources are, the technologies and surpluses of educated talent that can be channeled to make the difference in our global future."

"But it won't *be* channeled! The wrong values are too deeply entrenched, and getting stronger in reaction to the increasing economic pressures as scarcity increases!"

823

"It's not an all-or-nothing situation."

"I tell you again," she declared, "this country is hopeless. It's already been ruined from within by a dynamic intrinsic to its basic nature as a classless society: money is made the measure of all things. Its whole culture has been shaped by the goal of stimulating consumption rather than the motivation to do and produce what is socially needed. 'The profit motive' has produced wonderful products, but it's become a rot which permeates the fiber of our culture.

"I saw what kinds of people we're producing by looking at the youngsters I taught. Even in 'middle America' they're being influenced by movies and television that don't even *nod in the direction* of sound values.

"The country's already sinking. The economic problems are only symptoms of the cultural disease: an inversion of values caused by the cultural obsolescence of the supernatural support for traditional values and the absence of a shared secular definition of reality which can functionally take its place."

They both had laid down their forks.

"Look, Philanth, in taking that position you're abandoning the most realistic hope we have: that our country can be persuaded to outgrow its parochial nationalism and save itself in the only way possible in the long term, by making the sacrifices necessary to serve as an effective leader of the global community.

"I know what really bothers you about trying to fight the basic battles here rather than putting out brush fires overseas: the battleground here feels too comfortable; you enjoy being here too much."

"I'm not a masochist. I do love it here. But I'm not needed here because so many other people feel the same about it!"

"Never mind all those others just now—there's no surplus of Philanths even in Washington, because God needs them to manage things in Heaven.

"Right now my point is that working where conditions are worse certainly doesn't guarantee that you're more needed in the sense that you can really accomplish more. Without adequate resources you can wind up completely wasting your time."

"So it's an uncertain venture. But *politics* must be the most uncertain enterprise there is. In a clinic you can say, 'We perform four vasectomies per hour, twenty hours a week, and if we just keep at it long enough we can reduce the average family size in this area from twelve children to five. This will result in improved nutrition, better health, reduced unemployment, and increased per capita income,'—and so forth, on into less and less direct benefits. That's a safer investment than anything politics has to offer."

"And the funding for your wretched little clinic gets cut off if the wrong man happens to get into the White House or appointed to an ambassadorship. What do you do then?"

She sighed emphatically. "We could argue about this indefinitely and not convince each other. We're still deadlocked; and I'm sorry, but that means . . . we both lose. You may as well make up your mind to tell me good-by as soon as I've—"

"No, wait. 'We have yet had but a small fight.'" He hesitated, gathering his forces yet refraining from saying anything which could put her on guard.

"Let's take a close look at the job of 'politician's wife' I'm offering you." At last he had found an opening for his prepared statement. "Let me explain what makes you specially qualified for it. Then you'll understand why I need you—you specifically—to stay with me and help me."

"This sounds like something we'd both better get comfortable for. Besides, it's always a waste to eat good food while thinking about something else. Let's finish here."

They ate what was left on their plates, then took them back to the kitchen and rinsed them off. Preoccupied as he was, he nevertheless found pleasure in brushing against her and had to fight down the impulse to take her in his arms.

They returned to the living room. She arranged herself at one end of the deeply soft sofa like a luxury-loving cat, and again he took one of the chairs facing her.

"Quiet words, in quiet rooms," he began, "plant the fields in which private lives must flourish or wither. They—"

"Oh, that's good, that's good," she murmured. At his look she held up a hand and murmured, "Sorry.

"—But have you used that before?" It was obvious she knew it wasn't likely.

"Shut up, dammit. Play fair."

She smirked and subsided.

"They decree the growth or death of cities. They raise or lower the cost of bread and meat. They cause homes to be built or bombs to be dropped. They inspire people to love and sacrifice or they arouse people to hate and destroy. They cause a nation to tackle its problems intelligently or indulge in self-delusion and folly. Words, only words, can do all those things. You know that.

"You have a gift with words. So politics, the essential substance of which is words and only words, is an area in which you can serve effectively."

She nodded appreciatively. Seeing at last that he too had come prepared, she was all pleased attention. More discomfited by this disinterested pleasure in discourse than by her warnings or teasing, he made an effort and went on.

"Dealing in words, you have little immediate experience of the effects of what you do. And you get little thanks that's meaningful. You only get the chance to keep on doing it, if you're lucky.

"So it's work that requires whole-hearted, unswerving commitment if it's to be done honestly and conscientiously and honorably, without arrogance or self-seeking, over a period of years. The years of constant opportunity corrupt; and people can quickly lose perspective on what they're doing. Yet many years are required to learn the political landscape of your time and develop the skills for working effectively in it.

"A person who essentially lives for himself or herself becomes drunk with self-importance when given power or responsibility. Washington collects people like that. At all levels politics also attracts people who regard pettiness, vengefulness, and all sorts of other egoism as virtues, as proofs of strength and self-confidence and moral rightness.

"You're not such a person, yet you have enough wisdom and forbearance and tact to be able to work with those who are.

"Also, with your generous mind and wider vision you can not only tolerate close association with these people, you can resist the seductions which attract them: adulation and 'glamour', luxury and privilege, power and prestige. Your perspective, combined with your commitment to others as a matter of reasoned principle rather than as a matter of ideology, also makes you especially needed.

"Most important, you understand that principles can be more practical for the common good, over the long run, than partisan or personal expedients. And you would never confuse your personal interests and practical or psychological vested interests with the broad, long-term interests of society. You're as incorruptible as anyone can be.

"Politics requires a great capacity for caring about people, including people who take pleasure in denigrating and insulting you—not to mention people who will gladly exploit you or manipulate you or even kill you if they get a chance. You need to be able to care for the stupid, the coarse, and the unscrupulous; the unbelievably self-centered and shallow and childish and the unashamedly greedy; to care about them all unfailingly, and under difficult circumstances.

"You must have compassion toward people you despise, because you have to work closely with some of them. You have to be able to get along with some really ugly characters, because politics attracts them as dung attracts flies.

"How many people who've been close to the centers of power for many years can still do all of this? Every week that you're involved you have to fight impulses to become like the people you deplore, because of the effort required to work with them and survive among them.

"You have these qualifications, Philanth.

"You also have the imagination to see the human beings behind the abstractions and visualize remote consequences; you have the compassion to avoid becoming inured to the seriousness of the impacts of what you do. You care enough to endure the costs of promoting realistic approaches to problems when many people refuse to accept the reality that's right in front of them.

"There are costs in addition to the abuse heaped upon you: constant frustration; the insider's awareness of the emptiness of fine appearances and lofty sentiments; the understanding of painful facts behind the comforting fictions so many people cling to.

"Yet in spite of such requirements and costs, the position I offer has no title or salary or security at all. The widows of many men who helped make history while giving their all for their country have been left penniless. I want to enlist you in behalf of sound, concrete goals we both understand. But hard work and abuse are all I have to offer you. And the nastiness will increase in proportion to whatever success I—*we*—achieve."

"How serious are you when you downgrade your political prospects? Is it your version of obligatory modesty, or do you really think you can't make it into the Senate and you'll have to go back to being just a crusading lawyer?"

Her tacit acceptance of all he had just said gave him hope. Her concern about his prospects made this hope tremble. But if he successfully misled her now she would realize it later and despise him. Knowing, now, how that would feel, he would not risk it.

"I'm afraid I'll aim too high in my methods. I'm well aware that everything I've been saying sounds terribly self-righteous. But I feel very strongly that the job of an elected official in a democracy is to lead, inspire, and educate—not just do what's expedient in terms of the mood of the public at any given time. That's why I've risked being branded as a maverick: I always *want* to press too hard when I'm sure I'm right, be too candid, ask voters and other officials to understand more than they're ready for. To stay in office one has to pick his battles wisely and not move too quickly.

"Also, a lot of luck—chance—is involved in a successful career in politics. Thus far I've been lucky. My luck can change."

"If you find yourself out of power I'm stuck in the political wilderness with you if I marry you."

He had no attractive answer to that, yet he had to be honest. "Yes. There's not even job security, beyond the advantages of incumbency."

"But you would go on working for the same causes, spending as little on yourself as you could get by with in order to have your time free for them?"

He met her searching gaze, pleased because she had formulated his answer for him in this critically weak sector. "I promise. We would never stop being crusaders for the policies we know are badly needed.

"I know you feel ordinary persuasion to do right is a waste of time and politics is empty gestures concealing fruitless compromise. I know there's a firebrand lurking inside you. I promise that her energies will be allowed effective release one way or another. I'll do my best to stay in public office, but if I'm wiped out at the polls I'll give up seeking election and support you by staying in the thick of good legal battles while you work for your central cause, Zero Population Growth. You could become to the ideal of childless altruism what Martin Luther King, Jr. was to the ideal of racial equality.

"And while you were getting started on that you'd have time to get some graduate degrees that would strengthen you in your causes or in battles where I could make proper use of your full-time support. However things go, we make the best use of our combined strengths, but not in a short-term way that cuts off your personal development to provide me with cheap help. Does that deal with your concern adequately?"

She nodded.

"But in any case, I promise you only hard work. Surely you can see that being the wife of a man with my concern about the long-term future isn't a position for which there would be many aspirants if the women who considered me eligible understood what was really involved."

"But suppose you succeed. Suppose you even make it into the White House. The role of a senator's or president's wife doesn't have much going for it as far as I'm concerned, except as it subsumes a regular staff position. I *don't* want to get stuck deciding what to do about table decorations while you're deciding what to do about energy policies! So I'd much rather go on working for you in roles that have more dignity because they're assigned on the basis of professional qualifications. You can get somebody else for a position that really requires only the ability to dodge questions, look nice, and smile a lot."

"I promise you over and over, I wouldn't let you be wasted. You've made your requirements clear. I'll put them into a detailed contract and sign it in my blood if you want me to."

"But I can't believe *you* believe you can keep that promise if you're president! People who don't like a politician's policies can be completely unreasonable about his delegating any of his authority to his wife. And functioning as his wife, she never has opportunity to develop her own formal political credentials. If she lacks them they carp over whether she has undue influence on his decisions."

"Well, they're fools who must be faced down if they think I'd be unduly influenced by my most trusted adviser because I sleep with her. I sleep with her because I respect her judgment, not the other way around."

"This is a crucial consideration, Mr. Bjorklund. I take Carole's view, not yours, regarding your prospects. So I expect you to have to face this problem."

831

"I take the promises I've made you very seriously; more than any of the conventional marriage vows, though I take those very seriously indeed. —In spite of unfortunate recent evidence to the contrary."

"I can't press you on this because I have no right to demand that you share what is officially given only to you."

"*I* think you do. A woman married to her boss or colleague has a moral right to be given work commensurate with her ability in spite of having accepted marriage instead of a salary. That should be as true in politics as anywhere else."

"But it's a quagmire because there are no objective criteria for qualifications for political work. And even your staff would want me relegated to the usual mindless 'wife' roles so *they* could do the interesting things."

"Philanth, if you ever think I'm being run by my staff rather than the other way around, let me know.

"In fact, I'll put you in charge of them the way Carole was, so a basic qualification for joining my staff must be an eagerness to take orders from you as my deputy."

She dimpled. "Now we're cookin'. But if I'm to inherit your position I should be involved in your issue support work, including speeches and writing for publication and directing research. Somebody else can be chief of staff—Sandy, for instance. You need to start thinking in terms of a much larger operation."

"Good point.

"I'm not easy to work with sometimes, Philanth. I'm a worrier. I'm inclined to squeeze the last ounce of energy out of the people who work for me, and I expect results, not excuses . . . ." She had begun laughing softly. "What's so funny?"

"I know better than you do what you're like to work with, Mr. Bjorklund. You're a very sweet man, and when you're trying to act mean you're not very convincing." She hunched her shoulders and added, "I only pretend to be scared."

"Oh, great. That does so much for my self-image.

"I was about to say, Philanth, that only you could be self-denying enough to really want to be married to me once it was clear what it would entail." He waited.

She tilted her head. "You've tried to make it sound as awful as possible, hoping I'd pretend to believe that being a senator's wife could be worse than working in a refugee camp—and accept marriage to you on that basis."

"That was the idea. I was rather proud of myself when I came up with it."

She hid her smile with her hand. "Well, you've presented a good case. I think I *would* like the refugee camp more than shelling out money for a caterer. And I'd never waste *my* time deboning and flattening a raw chicken breast so I could stuff it."

"So will you accept my offer?"

"May I think about it for a while?"

"For how long? Five minutes, or five years?"

She smiled. "Five years sounds reasonable."

"Oh, Philanth! Maybe *you* enjoy a good parlor debate, but *I'm* working toward the real thing: I need to raise millions of dollars *during the next year* before I can even decide whether I can risk becoming a candidate. To say I'm pressed for time is a world-class understatement.

"As for you, we need to have you functioning as a political professional as soon as possible, and there's a very great deal for you to learn. Carole and I have already agreed on a reading program for you, and you can start the moment you're ready: preferably tomorrow evening, after you've spent five minutes getting your vast possessions settled in upstairs."

"I'll be happy to accept a reading program, but I don't want to accept you, Mr. Bjorklund. Let's leave it open. We'll talk again soon."

"I think you just want time to find a way out."

"You're right."

"But as you warned me going in, re-matches aren't allowed."

"Oh, it didn't occur to me that applied to *me*. I'm not at all prepared to capitulate just because you made a good speech. I feel sure there are points we haven't covered."

"What objections have I not adequately dealt with?"

"You still haven't dealt with my basic feeling about the idea, and it's very strong. I simply can't accept anything so . . . exciting and luxurious as being your wife."

"It won't be that exciting and luxurious once you get into it: just lots of hard work, as I've said. Only believing in what you're trying to accomplish makes it satisfying.

"And *think*, Philanth! Ask yourself what you want to do with the rest of your life! Consider everything you could possibly want to accomplish, once you set aside the idea that you have to be weak and power-less and alone. Ask yourself what could be the best ways to try to accomplish those things, accepting the fact that here *or anywhere else*, you can spend years building something up only to see it destroyed in a few moments.

"Also ask yourself what you truly need for yourself as a human being in order to accomplish all your goals as well as possible. Ask what working condi-tions would make you the most effective and free you from emotional burdens so you could have maximum energy and concentration for serving others.

"Ask how you can do the most good in the world, and consider all the mighty things that are done by the leadership of this country because of what this country is in the world at this time. They may be stupid things, destructive things, but they are big things; and often they're set in motion by only a few poorly qualified people in a haphazard fashion. Then vast armies of 'little people' work hard and maybe give up their lives to make the big people's decisions work out as well as possible. These big things will be done one way or the other, with or without your trying to affect them. How can you walk away from a challenge like that?

"Finally, ask yourself how you can possibly turn your back on a chance to use all of your abilities fully and develop them further, rather than drifting around and very probably being more or less wasted for most of the rest of your life.

"The answers to all those questions will give you one conclusion: the best way you can love the world and serve it is by joining forces with me. I need all that you are. I can direct your efforts to where they can do a great deal that needs to be done and that other people lack the wit or the will to undertake. You need to give love, and I can show you the best ways. They're not easy, but they're important.

"Our strengths and weaknesses will mesh to become one great strength. I know that we can be and do much more together than we ever could, working apart from each other."

There was another pause. "I won't marry any other woman, Philanth. This is my ultimatum, and I won't change it later: It's you or no one. I can be just as stubborn as you can.

"Think what far-reaching effects our impasse on this matter would have! Could you do that much damage to Carole and to me, considering how much our life's work depends on what I have to do in the next two years?"

"That's a pretty strong argument."

"I should hope so. Is there *anything* I haven't covered to your satisfaction?"

"No." After a moment she slowly turned away.

She simply lay there on the sofa for a while with her back to him. The bird was under the net and saw that there was no escape. It was too smart to struggle, but if he reached for it, it would only edge away. If he made a grab it would find a hole and slip through it and take flight, for it would be telling itself he was not worthy to cage it.

He got up and went over to sit in the space left by the crook of her knees. He was careful to keep his hands in his lap. "I didn't expect you to jump up and down with joy, but is it so bad to lose this one, fair and square?"

She turned to him. "Ah, there is another problem I've never mentioned."

Carole had warned him. Any excuse would serve, if he did not demolish it. "What's that?"

"I'm a member of the American Humanist Association." She waited.

He laughed. "You make it sound like being a Communist in the 1950's."

"Well, isn't it? Most Americans tell the pollsters they believe in God. And here as everywhere, fundamentalists are fighting a bitter rear-guard action.

"But secular humanists don't accept supernatural explanations for *anything*. If my religious views became a campaign issue I'd have to deny my beliefs, or it could sink you politically.

"That's just as serious a matter to me as it would be to any other deeply religious person—more so, since I want to free people to be spiritual and good without being hobbled in their thinking about life-and-death issues affecting the general welfare."

"I see," he said. He was still struggling to keep a straight face, even though she was quite serious.

"I'm sorry, Charles." There was a pause. "My membership would be bound to come out. Some reporter would get a tip and study their recruitment literature, and then I'd be on the spot."

"You couldn't claim you'd joined just to read their publications?"

"It's true that they've got some very smart and famous people in their membership; they always have had. And they do some fine work on ethics education and crucial social issues. But I know a lot of people would get up in arms at the idea of an atheist living in the White House."

He saw her point. Since she had made it clear even to Ablebody that she found religious orthodoxies a reactionary influence which crippled rational problem-solving in contemporary society, denying her humanism would make her feel like Peter denying that he knew Jesus. She had a compulsion for bearing witness to the truth, since the truth would make people free.

"Couldn't you simply let your membership lapse and trust that all records of it will have been obliterated by the time I'm a national candidate, if that ever happens?"

"And go to services pretending to be a believer when your position 'requires' it?"

"Philanth, you sought out Sacré Coeur and told me you 'worship everywhere'."

"So I did.

"Well," she said with a sigh, "I guess at the prospect of becoming a 'mole' for humanism I could deprive the Association of my dues."

"It's necessary, Philanth. Politicians are despised because they're willing to do what is necessary."

She nodded. Since her journal made it clear that she was willing to accept "damnation" to save others, he had touched the right key.

There was a pause. "What else've you got?" he asked bluntly. There was a long silence.

"I didn't expect to lose," she confessed. "I still can't quite believe it's happened."

"Well, you have. 'Hoisted on your own petard'."

"But I'm *still* not good enough for you, Charles! I'm not! You've won on logic, but *please* let me go! I'm afraid . . . ."

Slowly he put his hand on her arm. "Are you afraid I might hurt you, Philanth?"

"Oh, no, no." Far be it from Philanth to confess to being afraid of anything. "I really just want to keep my distance from you. I really don't want to . . . spread my leaves so close to the sun. I can't give any reason, except that I'm not good enough for you."

"It's the age difference. It makes me feel I'm not good enough for you, either. When you're forty-five I'll be sixty. That bothers me, probably more than you can understand."

"'Love is not love, that alters when it alteration finds,'. . . ."

"And I wish you could have children, even though you don't want them at the intellectual level. I had a vasectomy just before I got married, and my reasons for doing that haven't changed: I don't feel we have the resources to be good parents and also do the things I see crying to be done. I know you agree with that. But I feel I'm depriving you anyway."

"There are other ways of fulfilling that need."

There was a pause. "Charles, there is one thing I want you to change your mind about."

"And that is?"

"Stop fighting the fact that you're handsome. Accept women's admiration with benign patience. Let your hair grow so it can be more attractively styled.

"If women squeal and moan when you show up to give a speech, that's in the natural order of things: they're like the females of other species. Simple people respond openly to primitive appeals. It's not necessarily wrong to exploit that fact in order to unite them in great collective endeavors. Accept power even when it's not based on rationality."

"I don't want to corrupt the process any further than it has been already."

"It won't undermine the Constitution if you adopt a different hairstyle."

"You can be very persuasive."

She shyly returned his smile. After a moment she asked, "It's definite that I'll be going to work in California as soon as—as Carole . . . dies?"

"Unless something develops that causes me to refrain from entering the senate race. And if that's what you want."

"It makes excellent sense to me. And you'd better be able to make a *very* good case against running for the Senate!"

"Very well," he said, sighing. "You'll go to California. It's the only way I can imagine keeping my hands off you until we're married."

He sought a change of subject.

"Are we going to fight about money, Philanth? Sometimes, like tonight, I'm required to look prosperous in order to be accepted by people I have to deal with. But I'm always pinched by the expense of doing that, like other politicians who aren't wealthy. And keeping official expenses separate from personal ones can be impossible unless you impoverish yourself to err on the side of being careful, as I do.

"But I'm used to a thrifty wife who does all my shopping for me, and I won't question your judgments as long as I'm convinced that you're keeping a sound balance between the value of your time and the availability of money, and provided you don't make me look poor at the wrong times."

"That's quite reasonable. And you're giving me some fine resources to work with: this house, the apartment in Los Angeles . . . . I'll cope with my abundance by trying to make the most of everything."

"I'm sure you will."

He kept wanting to kiss her and caress her, but he maintained his quiet, fatherly manner. "And I'll expect you to study hard whenever there's nothing more pressing to be done. We're going to go as far as we can, and I'm going to be proud of you.

"Assuming you outlive me by, say, a few decades, you'll carry on my—our—work. For Carole and for me. Does that bother you?"

"I was going to do that anyway, Charles. Only without ever seeing you again."

"And I hope you'll marry again."

"Wha-at?"

"Remember that passage you wrote for one of my speeches, when I was trying to find reasons not to take you on the trip? I threw it out, and you conceded it was 'too motherly' to come from me.

"It said that we have to stay connected to the network of people who care about the things we do, and draw comfort from each other. It's pretty idealistic to imagine a global network of people who have that much in common, because everybody is normally so wedded to his own perspectives and personal agenda.

"But I don't want you to be lonely in the most influential years of your life because you got browbeaten into marrying a man so much older than you were. So when you're alone again, make the effort to find a good man who can completely commit himself to the same objectives you're committed to.

"You won't have any problem finding men who want you. You'll only have to pick one who's worthy of you in the way that he loves you.

"Promise me you'll be open to a new love when I'm gone, Philanth. Even if it's forty years from now, remember that I asked you to promise."

"Charles, please don't . . . ." She was moving into his arms, and he settled her against him.

"Believe me, my love, I know how this hurts. I'm on the other end of it these days. Make me easy as I'm making Carole easy."

"All right," she said, the words dragged out of her. "If you should leave me before I'm very old, and if there's a man fit to help me carry whatever legacy you leave me, I'll give him a chance to prove it."

"Good enough. With your help, I hope to leave you more than a wedding ring."

She rubbed her cheek against his shoulder. "Oh, Charles. I don't know how I could stand . . . ."

"None of that," he soothed her. "This is a happy time. I only wanted to protect the future before we get on with the present. Just imagine how glad Carole is going to be when I give her the news in the morning."

She nodded, and he caressed her hair. "I love you so much, Philanth. I never imagined it was possible to love someone so completely as I love you. I go out of my mind when I think about spending the rest of my life with you. Carole has given me a great deal to try to be worthy of, but you're her greatest gift."

She pressed against him, he held her close, and they were quiet for a while.

The way she cuddled against his chest, her arms bent in front of her, gave him the feeling that she was about to fall asleep expecting him to carry her off to be tucked into her trundle bed. "Philanth."

He hesitated so long that she asked, "Is there still something you're afraid to say to me?"

"Every time I've touched you, when you weren't calibrating your movements to be merely 'affectionate', you've shied away."

"Yes."

"If I were to try to make love to you now, would you still react that way, or . . . respond?"

"I'd still withdraw."

"You're not a little girl, Philanth. You're not supposed to play the role of sexual innocent when you're in the arms of the man who loves you."

"Naturally there's an adjustment when what was forbidden becomes not only permitted, but proper."

"That's all there is to it? You're sure you can respond as a woman who's in love with her husband?"

"When you are my husband, yes."

"But you can't respond to me that way now?"

"You're not my husband now. You won't be, for a long time."

"I'm asking whether you *can*. I know better than to try to experiment with you, but I worry." She looked up at him, puzzled. "You've conditioned yourself to separate love from sexuality. That ability serves you well. However, ..."

She spoke just above a whisper. "You want me to prove I can kiss you as if I were your wife? Now?"

He answered with a look.

"Mr. Bjorklund, I continue to want to make everything as easy for you as I can. Yesterday, by appealing to my vanity about being tough-minded, Carole talked me into sharing rides with you, managing your home, buying and preparing food for you, sharing meals with you, living under your roof . . . for the coming months. It's a practical arrangement, and I look forward to it for many reasons. But surely you don't want me to inflame your imagination with a demonstration that would put your fears about my 'maidenly reserve' to rest! Do you propose to bed me while your wife is . . . ?"

"Point taken.

"But I've talked with Ken, Philanth. He was here last night, and he told me things that made me believe it would be a miracle if you weren't conditioned to regard the act of love with repugnance."

"I see," she breathed, and he could feel her tension.

After a moment she said, "I did my best to keep my husband from hurting me in any lasting way. I cried when I needed to, but I always tried to be objective about the situation. And I never stopped loving him and pitying the reasons he degraded me. I believe I've succeeded in keeping myself able to enjoy sexual contact.

"To the extent I haven't, what's done by conditioning can be undone by re-conditioning. I'm not old and 'set in my ways'.

"I have a vivid imagination, Charles. And I have no fear of intimacy with you.

"I know you'll be considerate; and if you did something that aroused feelings of aversion, I'd tell you— just like the magazine articles tell people to. If there were a problem or difference in preference I'm sure we'd work it out. Educated people are equipped to

843

communicate effectively about such things, so they have higher marital satisfaction."

"But that's all your usual second-hand information and theorizing. You've done such a thorough job of locking away your normal feelings that I'm afraid of what it may take to let them out. Are we looking at twenty years of therapy? Will I have to go on letting you take care of your orgasms by yourself, knowing I could give you so much more? You have reasons for not even letting me kiss you now, and how can I know that intercourse won't repel you when all your theories about psychology and society say it's finally permissible?"

"It's only my control that worries you."

"How do I know that? From things you've said, I gather that you're in love with me; yet if actions mean anything, you don't want me. You've even said so."

"That's true."

"How is it possible?"

"You don't understand because you haven't mastered loving without wanting."

"No, I haven't. But can you tell me how I can make love to you if you don't want me?"

"I think we have a semantics problem."

"Oh, do we?" he snapped.

"Well, partly. To me, 'wanting' means both desire and aspiration. I don't *aspire* to make love to *anyone*."

"Have you ever?"

"For a short while, years ago. But if you keep batting down the fire of hope, eventually it goes out. And I'm not interested in rekindling an aspiration *for myself.*"

She had desired to see the splendid sights of Africa and Asia, but in Paris he had instructed her not to

844

hope to do any more sightseeing. Ablebody had elicited a momentary spark of interest, but she had obeyed her boss: she had stopped aspiring.

"By refusing to 'aspire' you make yourself an emotional eunuch."

"No. I do desire you, in the sense of . . . desiring to be . . . touched. And . . . possessed. To make love, and to be made love to.

"A woman doesn't always 'want' a man in the way a man wants a woman. She's the object that desires to be taken. Night after night I would've died to be in your arms, giving myself to you, and I . . . ." She saw his look and stopped, leaving her mouth open.

"You don't say."

"Yes, I *do*."

"Philanth," he began in a dangerous tone, "you had it all worked out just now, the distinction between 'aspire' and 'desire', and between 'wanting' and 'desiring to be possessed'. You couldn't have explained any of that earlier, could you?"

"Of course not! Heavens, didn't we have enough trouble as it was?"

"Yes! A little more than we could've handled would have been very much to the purpose."

"I have no regrets."

"And no wonder you ruined our one chance to be in each other's arms by dancing in a night club. You really couldn't handle it!"

She didn't disagree. "I tried not to be a hypocrite, but there were times when you put me through some verbal tightrope walking."

"I imagine so! I asked you flat out, 'Do you want me?' and you said, 'No.' I suppose by your reasoning you were telling the truth."

"Sort of." She dimpled, quite unrepentant.

"You little vixen! You should've been a lawyer!"

845

"Now, don't cuss me out like that, Mr. Bjorklund. If you don't appreciate my hair-splitting, why not call me a fallen amateur theologian?"

"Never mind. Just assure me now that you're . . . really normal."

"Do you want to hear it now? Are you *shu-ure*? Knowing I won't *want* to make love with you until you're my husband?"

"Yes," he said through clenched jaws. "I want to hear it now. I want to know I'm not getting a wife who really does feel that 'a cucumber is better than a man'. I don't propose to live in sexual Never-Never Land with a mystic."

"All right, I *do desire* to belong to you. I desire *to exist only for your pleasure*." Though still cradled against his chest, she waved her hands to match her extravagant tone. "I have marvelous fantasies, incredible impulses. My instincts are in excellent working order."

She crossed her arms on her midriff, and by degrees her voice returned to normal. "I may have succeeded in not wanting you, but sometimes my need to give myself to you is all but overwhelming. I become possessed by the desire to be possessed." Her cheeks began to redden.

"Only not when I'm with you, because then I'm focussed on doing what's best for you. Tough luck.

"But if it's any comfort to you, yes, I will have some distractions to deal with, in working with you during the next year or so. I'm already starting to want you, now that I have the right. But I enjoy being with you and working with you far too much to mind the inconvenience.

"One reason I don't believe it would've been wrong for me to go on working with someone I was in love with for the indefinite future is that the problem you've found unacceptable is for me business-as-usual. Once I even resorted to masturbating in a women's rest-room, and it had nothing to do with thinking about any man: I was just having a fertile period. You wouldn't believe how little it can take to make me horny, or amorous."

She paused. "I think I've said enough."

"So instead of worrying, I'm to think about initiating you into idyllic lovemaking."

The corners of her mouth quirked. "That sounds much more constructive."

After a moment she added softly, "But it shouldn't be something you have to prepare for, when I always imagine that all the angels will hold their breaths 'the first time ever I touch your face'."

He looked at her so intently, trying to will her to do it then, that she turned shy.

She snuggled against him. "You know how to love, Charles—not just 'make love'. Don't worry another second. We're going to have a wonderful marriage."

"I'm through worrying." He relaxed also, and again they were close and quiet for a while. Again he felt all that was good come seeping into him from their physical contact.

"I'm sorry I had to press the point. I just didn't want any heartache later on. If there was a problem I wanted to start working on it.

"Because this marriage is going to be the one beautiful certainty that sustains us through all the hard work and nastiness."

"I understand."

"There's so much more to understand than you've experienced, yet! When I'm totally drained by a long period of work and feel completely bogged down, just hearing one special voice on the phone is all it takes to turn me around, to make me happy and put me back on track with what I ought to be doing. That's why as time goes on, Philanth, you'll see that deciding on this marriage was the essential first step toward everything else.

"It's the way Carole and I have lived. That's why she cared so much about making it possible for me to have it again."

"It's going to be so good to be close to you, Charles. I know how Cinderella felt when the prince tracked her down and the whole world opened up for her. But her hopes were paultry, compared to all I want to do for you."

Easing out of his arms, she sat up, looking at her watch. "But speaking of Cinderella, it's close to midnight, and I've got to wash all this stuff out of my hair and get some sleep. Otherwise I won't be worth a nickel tomorrow. And—as the crooked politician says on his death bed in that old movie—'there's so much I gotta do.'"

As he drove Philanth back across the District from Georgetown to Southeast their route took them past the White House and around the Capitol. The two illuminated white landmarks which dominated his life seemed to glow with his own renewal.

Decked out for Christmas and white with fresh snow, the "city of many dreams" had never looked lovelier to him. The cold and darkness outside his car sharpened his pleasure in the glittering night.

Looking into the night and into the future with this woman beside him, he realized he was immortal in the ways she believed were possible: understanding how powerfully one person could transcend death, his mind found the courage to embrace all times simultaneously.

Thus in thought he would always be close to those two people whom he loved and was loved by with thorough knowledge yet without reservation. And they would be with him always, for they also understood this way of loving beyond time. He was assured of the one aspect of his future which mattered to him personally, assured beyond reach of the deaths of all of them.

People who lived only for themselves and those closest to them could never have this kind of love, and whatever desire or respect or tenderness they felt for each other, he was sorry for them. A caring which embraced the world and its future was a burden, certainly—but from those with similar breadth of dedication this commitment could elicit a depth of devotion beyond any love born only of more limited emotional horizons. He could count on this most powerful mutual loyalty and the shared passion it engendered, knowing these would sustain him.

There was more. He not only had been made immortal and given the courage to endure, he felt no older at forty than Philanth was at twenty-five. He was more filled with both-feet-on-the-ground hopes than he had ever been before. During the weeks while he had been coming to know Philanth his spirit had begun molting its dead skin and growing younger, and now he had been reborn: for once again—miraculously, considering all its pain and power to hurt,—the world lay all before him, full of promises.

THE BJORKLUND LEGACY,

like any vision which shapes many lives,
must be built upon the dreams of ordinary people.
Once caught up in it,
what will they *not* sacrifice?

The Bjorklund charisma

is seductive. How can those who have it *not* use it
for their cause?

The Seduction of the Chief of Staff

can provide an invaluable pillar for the building of
the Legacy: an unassuming model of honesty, loyalty,
and managerial skills is needed to oversee a rising
politician's expanding enterprises. He is also needed
as a surrogate father to the Bjorklunds' daughter.

But "our faithful Sandy" finds that everything else
he has ever cherished must be relinquished:
home, family, even hope of heaven.

For how could he resist Bjorklund's charm,
Bjorklund's altruism, or Bjorklund's wife?

THE BJORKLUND LEGACY

The Seduction of the Chief of Staff

Library of Congress Catalog Card Number: 94-92336
ISBN: 1-886087-04-0 (hardcover) $12.50
1-886087-05-9 (softcover) $4.95
First publication date: September 1995.

THE BJORKLUND LEGACY

becomes the responsibility of

Bjorklund's Daughter.

In her early teens Hypatia was rescued from her abusive family by the Bjorklunds. Growing up in the top levels of Washington, she has come to feel that fulfilling her adoptive parents' aspirations for her precludes marrying any of the men in her circles: she fears they might tarnish her politically or insist on co-opting her for their own ambitions.

Dr. Steven Wysislavsky, a NASA executive based in Houston, offers to help Congresswoman Bjorklund's senate campaign in California. She decides he is the man she has been waiting for, but their work keeps them rooted in different parts of the country.

Steve would do anything for the space program; Hypatia would do anything for her political agenda, which includes the space program; and President William Farber will do anything to possess

Bjorklund's Daughter.

THE BJORKLUND LEGACY
Bjorklund's Daughter

Library of Congress Catalog Card Number: 94-92334
ISBN: 1-886087-06-7 (hardcover) $19.95
       1-886087-07-5 (softcover) $11.95
First publication date: May 1995.

THE BJORKLUND LEGACY

has helped William Farber win the White House because Hypatia Bjorklund became his running mate in order to get a chance at the presidency. But a vice president's political prospects are subject to the president's whim; and where Hypatia is concerned, Farber has a whim of iron. The resulting crisis is resolved quite differently than it seemed to have been at the end of *Bjorklund's Daughter*. An outspoken black maid is entrusted with the secret.

So begins a four-year struggle over political priorities, conflicting ambitions, and different kinds of love; a stretching of the definitions of marriage and parenthood; and an excursion into shifting definitions of reality to explore the complexities of loyalty, integrity, and desire.

It could have been an eight-year campaign, but

The Farber-Bjorklund Presidency Ended Strangely.

THE BJORKLUND LEGACY

The Farber-Bjorklund Presidency Ended Strangely.

Library of Congress Catalog Card Number: 94-92335
ISBN: 1-886087-08-3 (hardcover) $27.50
1-886087-09-1 (softcover) $17.50
First publication date: May 1995.